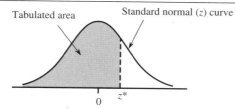

Tabulated area

Standard normal (z) curve

z^*	.00	.01	.02	.03	.04	.05	.06	.07	.08	.09
0.0	.5000	.5040	.5080	.5120	.5160	.5199	.5239	.5279	.5319	.5359
0.1	.5398	.5438	.5478	.5517	.5557	.5596	.5636	.5675	.5714	.5753
0.2	.5793	.5832	.5871	.5910	.5948	.5987	.6026	.6064	.6103	.6141
0.3	.6179	.6217	.6255	.6293	.6331	.6368	.6406	.6443	.6480	.6517
0.4	.6554	.6591	.6628	.6664	.6700	.6736	.6772	.6808	.6844	.6879
0.5	.6915	.6950	.6985	.7019	.7054	.7088	.7123	.7157	.7190	.7224
0.6	.7257	.7291	.7324	.7357	.7389	.7422	.7454	.7486	.7517	.7549
0.7	.7580	.7611	.7642	.7673	.7704	.7734	.7764	.7794	.7823	.7852
0.8	.7881	.7910	.7939	.7967	.7995	.8023	.8051	.8078	.8106	.8133
0.9	.8159	.8186	.8212	.8238	.8264	.8289	.8315	.8340	.8365	.8389
1.0	.8413	.8438	.8461	.8485	.8508	.8531	.8554	.8577	.8599	.8621
1.1	.8643	.8665	.8686	.8708	.8729	.8749	.8770	.8790	.8810	.8830
1.2	.8849	.8869	.8888	.8907	.8925	.8944	.8962	.8980	.8997	.9015
1.3	.9032	.9049	.9066	.9082	.9099	.9115	.9131	.9147	.9162	.9177
1.4	.9192	.9207	.9222	.9236	.9251	.9265	.9279	.9292	.9306	.9319
1.5	.9332	.9345	.9357	.9370	.9382	.9394	.9406	.9418	.9429	.9441
1.6	.9452	.9463	.9474	.9484	.9495	.9505	.9515	.9525	.9535	.9545
1.7	.9554	.9564	.9573	.9582	.9591	.9599	.9608	.9616	.9625	.9633
1.8	.9641	.9649	.9656	.9664	.9671	.9678	.9686	.9693	.9699	.9706
1.9	.9713	.9719	.9726	.9732	.9738	.9744	.9750	.9756	.9761	.9767
2.0	.9772	.9778	.9783	.9788	.9793	.9798	.9803	.9808	.9812	.9817
2.1	.9821	.9826	.9830	.9834	.9838	.9842	.9846	.9850	.9854	.9857
2.2	.9861	.9864	.9868	.9871	.9875	.9878	.9881	.9884	.9887	.9890
2.3	.9893	.9896	.9898	.9901	.9904	.9906	.9909	.9911	.9913	.9916
2.4	.9918	.9920	.9922	.9925	.9927	.9929	.9931	.9932	.9934	.9936
2.5	.9938	.9940	.9941	.9943	.9945	.9946	.9948	.9949	.9951	.9952
2.6	.9953	.9955	.9956	.9957	.9959	.9960	.9961	.9962	.9963	.9964
2.7	.9965	.9966	.9967	.9968	.9969	.9970	.9971	.9972	.9973	.9974
2.8	.9974	.9975	.9976	.9977	.9977	.9978	.9979	.9979	.9980	.9981
2.9	.9981	.9982	.9982	.9983	.9984	.9984	.9985	.9985	.9986	.9986
3.0	.9987	.9987	.9987	.9988	.9988	.9989	.9989	.9989	.9990	.9990
3.1	.9990	.9991	.9991	.9991	.9992	.9992	.9992	.9992	.9993	.9993
3.2	.9993	.9993	.9994	.9994	.9994	.9994	.9994	.9995	.9995	.9995
3.3	.9995	.9995	.9995	.9996	.9996	.9996	.9996	.9996	.9996	.9997
3.4	.9997	.9997	.9997	.9997	.9997	.9997	.9997	.9997	.9997	.9998
3.5	.9998	.9998	.9998	.9998	.9998	.9998	.9998	.9998	.9998	.9998
3.6	.9998	.9998	.9999	.9999	.9999	.9999	.9999	.9999	.9999	.9999
3.7	.9999	.9999	.9999	.9999	.9999	.9999	.9999	.9999	.9999	.9999
3.8	.9999	.9999	.9999	.9999	.9999	.9999	.9999	.9999	.9999	1.0000

Applied Statistics
for Engineers
and Scientists

THIRD EDITION

Applied Statistics
for Engineers
and Scientists

Jay Devore
California Polytechnic State University, San Luis Obispo

Nicholas Farnum
California State University, Fullerton

Jimmy Doi
California Polytechnic State University, San Luis Obispo

CENGAGE
Learning·

Australia • Brazil • Mexico • Singapore • United Kingdom • United States

CENGAGE
Learning·

Applied Statistics for Engineers and Scientists, Third Edition
Jay Devore, Nicholas Farnum, Jimmy Doi

Publisher: Richard Stratton

Senior Sponsoring Editor: Molly Taylor

Development Editor: Laura Wheel

Editorial Assistant: Danielle Hallock

Associate Media Editor: Andrew Coppola

Brand Manager: Gordon Lee

Content Project Manager: Jill Quinn

Senior Art Director: Linda May

Manufacturing Planner: Sandee Milewski

Rights Acquisition Specialist: Shalice Shah-Caldwell

Production Service: Prashant Kumar Das, MPS Limited

Text and Cover Designer: Jenny Willingham

Cover Image: Female Scientist: wavebreakmedia/Shutterstock.com; Solar Panels: portumen/Shutterstock.com; Nanotubes: PASIEKA/SPL/Getty images

Compositor: MPS Limited

For product information and technology assistance, contact us at **Cengage Learning Customer & Sales Support, 1-800-354-9706**

For permission to use material from this text or product, submit all requests online at **www.cengage.com/permissions** Further permissions questions can be emailed to **permissionrequest@cengage.com**

Library of Congress Control Number: 2013944181

ISBN-13: 978-1-133-11136-8

ISBN-10: 1-133-11136-X

Cengage Learning
200 First Stamford Place, 4th Floor
Stamford, CT 06902
USA

Cengage Learning is a leading provider of customized learning solutions with office locations around the globe, including Singapore, the United Kingdom, Australia, Mexico, Brazil and Japan. Locate your local office at **international.cengage.com/region**

Cengage Learning products are represented in Canada by Nelson Education, Ltd.

For your course and learning solutions, visit **www.cengage.com**

Purchase any of our products at your local college store or at our preferred online store **www.cengagebrain.com**

Instructors: Please visit **login.cengage.com** and log in to access instructor-specific resources.

Printed in the United States of America
1 2 3 4 5 6 7 17 16 15 14 13

This book is dedicated to

My grandsons, Philip and Elliot
J.L.D.

My grandchildren, Ava and Leo
N.R.F.

My wife and daughter, Midori and Alicia
J.A.D.

Contents

Preface

PURPOSE

The use of statistical models and methods for describing and analyzing data has become common practice in virtually all scientific disciplines. This book provides a comprehensive introduction to those models and methods most likely to be encountered and used by students in their careers in engineering and the natural sciences. It is appropriate for courses of one term (semester or quarter) in duration.

APPROACH

Students in a statistics course designed to serve other majors are too often initially skeptical of the value and relevance of the subject matter. Our experience, however, is that students *can* be turned on to the subject by the use of good examples and exercises that blend their everyday experiences with their scientific interests. We have worked hard to find examples involving real, rather than artificial, data—data that someone thought was worth collecting and analyzing. Many of the methods presented throughout the book are illustrated by analyzing data taken from a published source.

The exercises form a very important component of the book. A really good lecturer can deceive students into thinking they have an excellent mastery of the subject, only to discover otherwise when they start working problems. We have therefore provided a rich assortment of exercises designed to reinforce understanding of the material. A substantial majority of these are based on real data, and we have tried as much as possible to avoid mathematical manipulation for its own sake. Someone who attempts a good portion of the exercises will gain a greater appreciation of the scope and applicability of the subject than would be gleaned simply by reading the text.

Sometimes the reader may be unfamiliar with the context of a particular problem situation (as indeed we often were), but we believe that students will find scenarios, such as the one below, more appealing than they would in patently artificial situations dealing with widgets or brand A versus brand B.

---

64. The use of microorganisms to dissolve metals from ores has offered an ecologically friendly and less expensive alternative to traditional methods. The dissolution of metals by this method can be done in a two-stage bioleaching process: (1) microorganisms are grown in culture to produce metabolites (e.g. organic acids) and (2) ore is added to the culture medium to initiate leaching. The article "Two-Stage Fungal Leaching of Vanadium from Uranium Ore Residue of the Leaching Stage using Statistical Experimental Design" (*Annals of Nuclear Energy*, 2013: 48–52) reported on a two-stage bioleaching process of vanadium by using the fungus *Aspergillus niger*. In one study, the authors examined the impact of the variables $x_1 = $ pH, $x_2 = $ sucrose concentration (g/L), and $x_3 = $ spore population (10^6 cells/ml) on $y = $ oxalic acid production (mg/L). The accompanying SAS output resulted from a request to fit the model with predictors x_1, x_2, and x_3 only.

Source	DF	Sum of Squares	Mean Square	F Value	Pr > F
Model	3	5861301	1953767	7.53	0.0052
Error	11	2855951	259632		
Corrected Total	14	8717252			

Fitting the complete second-order model resulted in SSResid = 541,632. Carry out a test at significance level .01 to decide whether at least one of the second-order predictors provides useful information about oxalic acid production.

MATHEMATICAL AND COMPUTING LEVEL

The exposition is relatively modest in terms of mathematical development. Limited use of univariate calculus is made in the first two chapters, and a bit of univariate and multivariate calculus is employed later on. Matrix algebra appears nowhere in the book. Thus virtually all of the exposition should be accessible to those whose mathematical background includes one semester or two quarters of differential and integral calculus.

The computer is an indispensable tool these days for organizing, displaying, and analyzing data. We have included many examples, as illustrated on the next page, of output from the most widely used statistical computer packages, including Minitab, SAS, R, and JMP, both to convince students that the statistical methods discussed herein are available in these packages and to expose them to format and contents of typical output. Because availability of packages and nature of platforms vary widely from institution to institution, we decided not to include instructions for obtaining output from any particular package. Based on our experience, it should be straightforward to supplement the text by independently introducing students to any one of the aforementioned packages. They can then be asked to use the computer in working the many problems that contain raw data.

Example 10.2   Over the past decade researchers and consumers have shown increased interest in renewable fuels such as biodiesel, a form of diesel fuel derived from vegetable oils and animal fats. According to www.fueleconomy.gov, compared to petroleum diesel, the advantages of using biodiesel include its nontoxicity, biodegradability, and lower greenhouse gas emissions. One popular biodiesel fuel is fatty acid ethyl ester (FAEE). The authors of "Application of the Full Factorial Design to Optimization of Base-Catalyzed Sunflower Oil Ethanolysis" (*Fuel*, 2013: 433–442) performed an experiment to determine optimal process conditions for producing FAEE from the ethanolysis of sunflower oils. In one study, the effects of three process factors on FAEE purity (%) were investigated.

Factor	Factor name	Factor levels
A	Reaction Temperature	25°C, 50°C, 75°C
B	Ethanol-to-oil molar ratio	6:1, 9:1, 12:1
C	Catalyst loading	.75 wt.%, 1.00 wt.%, 1.25 wt.%

(See Page 467 for the complete data)

Plots of all two-factor interactions are shown in Figure 10.18, along with the main effects Plots for the three factors. Suppose we are interested in maximizing the value of the response variable, FAEE purity. Looking at the interaction plots, the combination of factor levels that best accomplishes this objective is $A = 75°C$, $B = 12:1$, and $C = 1.25\%$. In this example, the conclusions from the interaction plots agree with the conclusions that we would have drawn from inspecting the main effects plots.

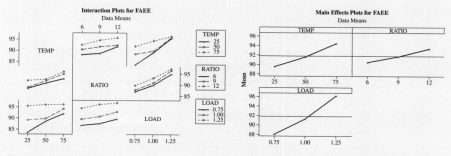

Figure 10.18 Two-factor interaction plots and main effects plots for Example 10.2

FOCUS AND CONTENT

We have written this book for an audience whose primary interest is in statistical methodology and the analysis of data. The ordering of topics herein is rather different from what is found in virtually all competing texts. The usual approach is to inject a heavy dose of probability at the outset, then develop probability distributions and use these as a basis for inferential methods (drawing conclusions from data). Unfortunately, an introductory one-term course rarely allows sufficient time for comprehensive treatments of both probability and statistical inference. If probability is emphasized, statistics gets short shrift. An additional problem is that many students find probability to be a difficult and intimidating subject, so starting out in this way creates an aura of mathematical formalism that makes it all too easy to lose sight of the applied and practical aspects of statistics.

Certainly descriptive statistical methods can be developed in detail with virtually no probability background, and even an understanding of the most commonly used inferential techniques requires familiarity with only the most basic of probability properties. So we decided to proceed along a path first blazed by David Moore and George McCabe in their book *Introduction to the Practice of Statistics*, written for a non-science audience. In their Chapter 1, the normal distribution is introduced and employed to address many interesting questions, whereas probability does not surface until much later in the book. Our Chapter 1 first presents some basic concepts and terminology, continues with an introduction to some descriptive techniques, and then extends the notion of a histogram for sample data to a distribution of values for an entire population or process. This allows us to develop and use not only the family of normal distributions but also other continuous and discrete distributions such as the lognormal, Weibull, Poisson, and binomial. Chapter 2 covers numerical summary measures for sample data (e.g., the sample mean $\bar{x}$ and sample standard deviation s) in tandem with analogous measures for populations and processes (e.g., the population or process mean μ and standard deviation σ).

The focus of the first two chapters is on univariate data (observations on or values of a single variable, such as tensile strength). In the third chapter we consider descriptive methods for bivariate data (e.g., measuring both thickness and strength for wire specimens) and then multivariate data, emphasizing in particular correlation and regression. This chapter should be especially useful for courses in which there is insufficient time to cover regression models from a probabilistic viewpoint (such models and inferences based on them are the subject of Chapter 11).

Most other books intended for our target audience say rather little about how data is obtained. Yet statistics has much to say not only about how to analyze data once it is available but also about sensible and efficient techniques for collecting data. Several lower-level texts, notably the one by Moore and McCabe cited earlier, successfully and entertainingly covered this territory prior to probability and inference, and we follow their lead with our Chapter 4. Sampling and experimental design are discussed, and the last section contains an introduction to various aspects of measurement.

At last probability makes its appearance in Chapter 5. Our minimalist treatment of this subject is intended to move readers expeditiously into the inferential part of the book. Since only the notion of probability as limiting or long-run relative frequency is needed to understand the basis for most of the usual inferential procedures, little time is spent on topics such as addition and multiplication rules and conditional probability, and no material on counting techniques is included here (combinations enter briefly in Chapter 1 in connection with the binomial distribution). The concept of a random variable and its probability distribution is then introduced and related to the distributional material in Chapter 1. Finally, the notion of a statistic and its sampling distribution is discussed and illustrated.

The remaining six chapters focus on the most widely used methods from statistical inference. Descriptive techniques from earlier chapters, such as boxplots and quantile plots, are employed in many of our examples. Chapter 6 covers topics from quality control and reliability. Estimation and various statistical intervals—confidence, prediction, and tolerance—are introduced in Chapter 7. Hypothesis testing is discussed in Chapter 8. Chapter 9 covers the analysis of variance for comparing more than two populations or treatments, and these ideas are extended in Chapter 10 to the analysis of data from designed multifactor experiments. Finally, regression models and associated inferential procedures are covered in Chapter 11.

SOME SUGGESTIONS CONCERNING COVERAGE

It should be possible to cover virtually all the material in the book in a semester-long course that meets four hours per week. For a course of this duration that meets only three times per week or for a one-quarter course, some pruning will have to be done (perhaps combined with reading assignments on topics not discussed in lecture). The first four sections of Chapter 1 are essential, but Section 5 on other (than the normal) continuous distributions and Section 6 on the binomial and Poisson distributions can be covered very lightly or even omitted altogether. The first two sections of Chapter 2, on measures of center and spread, are also required. The material on more detailed summary measures (e.g., boxplots) in Section 3 can be just touched on or skipped, and quantile plots from Section 4 can be presented very quickly.

When time does not allow for coverage of inferences in regression, we strongly recommend that at least a bit of bivariate descriptive methods from Chapter 3 be covered. At minimum, this could consume just two or three one-hour lectures in which scatterplots, correlation, and fitting a line by least squares are discussed. More time would provide the opportunity to introduce r^2 as an assessment of fit, nonlinear relationships, and even multiple regression. If inference in regression is to be covered, this chapter can be skipped over for the moment and then combined with Chapter 11 at the end of the course.

Chapter 4, on obtaining data, can be covered next or postponed until later. There is no mathematics here, only some definitions and examples, so this is one place where a minimal amount of lecture time can be expended along with a request that students read on their own. Most of Chapter 5 is crucial; inferential methods cannot be understood without a modest exposure to probability and sampling distributions of various statistics. The quality control and reliability techniques of Chapter 6 are attractive applications of sampling distribution and probability properties. When time is limited, as few as two lectures might be devoted to some general concepts and a single type of control chart. Another possibility is to postpone this material until after hypothesis testing has been introduced.

From this point on, it is local option as to what is covered and in how much detail. We certainly believe that students deserve at least minimal exposure to point estimation, confidence intervals, and hypothesis testing. Time may permit presentation of just some selected one-sample procedures (Sections 7.1, 7.2, 8.1, and perhaps a bit of Sections 7.4 and 8.2). A longer course would accommodate topics from among prediction and tolerance intervals, two-sample situations, chi-squared tests, testing the plausibility of some particular type of distribution (e.g., testing the assumption that the data came from a normal distribution), analysis of variance and experimental design, and more on regression.

Changes for the Third Edition

- There are nearly 200 new exercises and 40 new examples, most of which include real data or other information from published sources.
- Chapter 1 contains a new subsection on "The Scope of Modern Statistics" to illustrate how statisticians continue to develop new methodology while working on problems in a wide spectrum of disciplines.
- Section 8.3, on hypothesis testing based on categorical data, now contains a subsection on Fisher's Exact Test that is a useful alternative when assumptions for the standard chi-squared test fail.
- Section 11.6, on regression, now contains a subsection on the multiple logistic regression model that accommodates multiple predictor variables for a dichotomous response.
- In general, the exposition has been polished, tightened, and improved.

ACKNOWLEDGMENTS

We greatly appreciate the feedback and useful advice from the many individuals who reviewed various parts of our manuscript: Christine Anderson-Cook, Virginia Tech; Olcay Arslan, St. Cloud State; Peyton Cook, The University of Tulsa; Jean-Yves "Pip" Courbois, University of Washington; Charles Donaghey, University of Houston; Dale O. Everson, University of Idaho; William P. Fox, United States Military Academy; William Fulkerson, Deere & Company; Roger Hoerl, General Electric Company; Marianne Huebner, Michigan State University; Alan M. Johnson, University of Arkansas, Little Rock; Steven L. Johnson, University of Arkansas; Janusz Kawczak, University of North Carolina, Charlotte; Mohammed Kazemi, University of North Carolina, Charlotte; David P. Kessler, Purdue University; Barbara McKinney, Western Michigan University; Jang W. Ra, University of Alaska, Anchorage; John Ramberg, University of Arizona; Stephen E. Rigdon, Southern Illinois University at Edwardsville; Amy L. Rocha, San Jose State University; Joe Romano, Stanford University; Lewis H. Shoemaker, Millersville University; and Paul Wilson, Rochester Institute of Technology.

The editorial and production services provided by numerous people from Cengage Learning are greatly appreciated, especially the support of Shaylin Walsh, Laura Wheel, and Jill Quinn. It was indeed a great pleasure to have Prashant Kumar Das overseeing production of the book; his attention to detail, timely feedback, and willingness to tolerate the authors' idiosyncrasies made our work during production much more tolerable than would otherwise have been the case. A special thanks goes to Soma Roy for her accuracy checking and work on the solutions manuals. Finally, the continuing support of family, colleagues, and friends has helped smooth out the bumps in the road. We are truly grateful to all of you.

Crepesoles/Shutterstock.com

Data and Distributions

INTRODUCTION

Statistical concepts and methods are not only useful but indeed often indispensable in understanding the world around us. They provide ways of gaining new insights into the behavior of many phenomena that you will encounter in your chosen field of specialization in engineering or science.

The discipline of statistics teaches us how to make intelligent judgments and informed decisions in the presence of uncertainty and variation. Without uncertainty or variation, there would be little need for statistical methods or statisticians. If every component of a particular type had exactly the same lifetime, if all resistors produced by a certain manufacturer had the same resistance value, if pH determinations for soil specimens from a particular locale gave identical results, and so on, then a single observation would reveal all desired information.

An interesting manifestation of variation appeared in connection with an effort to determine the "greenest" way to travel. The article titled "Carbon Conundrum" (*Consumer Reports*, 2008: 9) described websites that help consumers calculate carbon output. The results for carbon output for a flight from New York to Los Angeles appear in the accompanying table.

Carbon Calculator	CO_2 (lb)
Terra Pass	1924
Conservation International	3000
Cool It	3049
World Resources Institute/Safe Climate	3163
National Wildlife Federation	3465
Sustainable Travel International	3577
Native Energy	3960
Environmental Defense	4000
Carbonfund.org	4820
The Climate Trust/CarbonCounter.org	5860
Bonneville Environmental Foundation	6732

Substantial disagreement clearly exists among these online calculators as to exactly how much carbon is emitted, characterized in the article as "from a ballerina's to Bigfoot's." A website also was provided where readers could learn more about how the various calculators work.

How can statistical techniques be used to gather information and draw conclusions? Suppose, for example, that a materials engineer has developed a coating for retarding corrosion in metal pipe under specified circumstances. If this coating is applied to different segments of pipe, variation in environmental conditions and in the segments themselves will result in more substantial corrosion on some segments than on others. Methods of statistical analysis could be used on data from such an experiment to decide whether the *average* amount of corrosion exceeds an upper specification limit of some sort or to predict how much corrosion will occur on a single piece of pipe.

Alternatively, suppose the engineer has developed the coating in the belief that it will be superior to the currently used coating. A comparative experiment could be carried out to investigate this issue by applying the current coating to some segments of pipe and the new coating to other segments. This must be done with care lest the wrong conclusion emerge. For example, perhaps the average amount of corrosion is identical for the two coatings. However, the new coating may be applied to segments that have superior ability to resist corrosion and under less stressful environmental conditions compared to the segments and conditions for the current coating. The investigator would then likely observe a difference between the two coatings attributable not to the coatings themselves but just to extraneous variation. Statistics offers not only methods for analyzing the results of experiments once they have been carried out but also suggestions for how experiments can be performed in an efficient manner to mitigate the effects of variation and have a better chance of producing correct conclusions.

In Chapters 1–3, we concentrate on describing and summarizing statistical information obtained from populations or processes under investigation. Chapter 4 discusses how information can be collected either by the mechanism of sampling or by designing and carrying out an experiment. Chapter 5 formalizes the notion of randomness and uncertainty by introducing the language of probability. The remainder of the book focuses on the development of inferential methods for drawing interesting conclusions from data in a wide variety of situations. We hope you will find the subject matter and our presentation to be as interesting, relevant, and exciting as we do.

1.1 POPULATIONS, SAMPLES, AND PROCESSES

Engineers and scientists are constantly exposed to collections of facts, or **data,** both in their professional capacities and in everyday activities. The discipline of statistics provides methods for organizing and summarizing data and for drawing conclusions based on information contained in the data.

An investigation will typically focus on a well-defined collection of objects constituting a **population** of interest. In one study, the population might consist of all gelatin capsules of a particular type produced during a specified period. Another investigation might involve the population consisting of all individuals who received a B.S. in engineering during the most recent academic year. When desired information is available for all objects in the population, we have what is called a **census.** Constraints on time, money, and other scarce resources usually make a census impractical or infeasible. Instead, a subset of the population—a **sample**—is selected in some prescribed manner. Thus we might obtain a sample of bearings from a particular production run as a basis for investigating whether bearings are conforming to manufacturing specifications, or we might select a sample of last year's engineering graduates to obtain feedback about the quality of the curricula.

We are usually interested only in certain characteristics of the objects in a population: the number of flaws on the surface of each casing, the thickness of each capsule wall, the gender of an engineering graduate, the age at which the individual graduated, and so on. A characteristic may be categorical, such as gender or type of malfunction, or it may be numerical in nature. In the former case, the *value* of the characteristic is a category (e.g., female or insufficient solder), whereas in the latter case, the value is a number (e.g., age = 23 years or diameter = .502 cm). A **variable** is any characteristic whose value may change from one object to another in the population. We shall generally denote variables by lowercase letters from the end of our alphabet. Examples include

x = gender of a graduating engineer

y = number of major defects on a newly manufactured automobile

z = braking distance of an automobile under specified conditions

Data results from making observations either on a single variable or simultaneously on two or more variables. A **univariate** data set consists of observations on a single variable. For example, we might determine the type of transmission, automatic (A) or manual (M), on each of ten automobiles recently purchased at a certain dealership, resulting in the categorical data set

M A A A M A A M A A

The following sample of lifetimes (hours) of brand X batteries put to a certain use is a numerical univariate data set:

5.6 5.1 6.2 6.0 5.8 6.5 5.8 5.5

We have **bivariate** data when observations are made on each of two variables. Our data set might consist of a (height, weight) pair for each basketball player on a team, with the first observation as (72, 168), the second as (75, 212), and so on. If an engineer determines the value of both x = component lifetime and y = reason for component failure, the resulting data set is bivariate with one variable numerical and the other categorical. **Multivariate** data arises when observations are made on more than two variables. For example, a research physician might determine the systolic blood pressure, diastolic blood pressure, and serum cholesterol level for each patient participating in a study. Each observation would be a triple of numbers, such as (120, 80, 146). In many multivariate data sets, some variables are numerical and others are categorical. Thus the annual automobile issue of *Consumer Reports* gives values of such variables as type of vehicle (small, sporty, compact, mid-size, large), city fuel efficiency (mpg), highway fuel efficiency (mpg), drivetrain type (rear wheel, front wheel, four wheel), and so on.

Branches of Statistics

An investigator who has collected data may wish simply to summarize and describe important features of the data. This entails using methods from **descriptive statistics.** Some of these methods are graphical in nature—the construction of histograms, boxplots, and scatterplots are primary examples. Other descriptive methods involve calculation of numerical summary measures, such as means, standard deviations, and correlation coefficients. The wide availability of statistical computer software packages has made these tasks much easier to carry out than they used to be. Computers are much more efficient than human beings at calculation and the creation of pictures (once they have received appropriate instructions from the user!). This means that the investigator doesn't have to expend much effort on "grunt work" and will have more time to study the data and extract important messages. Throughout this book, we will present output from various packages such as Minitab, SAS, and R. The R software can be downloaded without charge from www.r-project.org.

| Example 1.1 | Charity is a big business in the United States. The website charitynavigator.com gives information on approximately 5500 charitable organizations, and many smaller charities fly below the navigator's radar screen. Some charities operate very efficiently, with fund-raising and administrative expenses only a small percentage of total expenses, whereas others spend a high percentage of what they take in to perform the same activities. Here is data on fund-raising expenses as a percentage of total expenditures for a random sample of 60 charities: |

6.1	12.6	34.7	1.6	18.8	2.2	3.0	2.2	5.6	3.8
2.2	3.1	1.3	1.1	14.1	4.0	21.0	6.1	1.3	20.4
7.5	3.9	10.1	8.1	19.5	5.2	12.0	15.8	10.4	5.2
6.4	10.8	83.1	3.6	6.2	6.3	16.3	12.7	1.3	0.8
8.8	5.1	3.7	26.3	6.0	48.0	8.2	11.7	7.2	3.9
15.3	16.6	8.8	12.0	4.7	14.7	6.4	17.0	2.5	16.2

Without any organization, making sense of the data's most prominent features is difficult: What is a typical (i.e., representative) value? Are values highly concentrated about a typical value or are they quite dispersed? Are there any gaps in the data? What fraction of the values are less than 20%? Figure 1.1 shows what is called a *stem-and-leaf display* as well as a *histogram*. In Section 1.2, we will discuss construction and interpretation of these data summaries. For the moment, we hope you see how they begin to describe how the percentages are distributed over the range of possible values from 0 to 100. A substantial majority of the charities in the sample obviously spend less than 20% on fund-raising, and only a few percentages might be viewed as beyond the bounds of sensible practice.

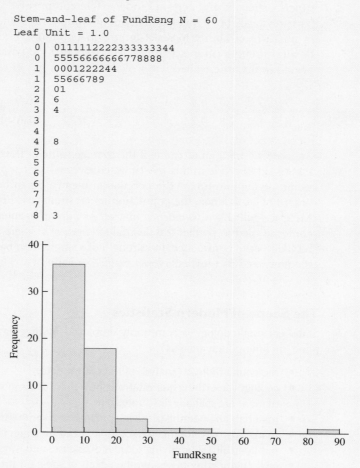

```
Stem-and-leaf of FundRsng N = 60
Leaf Unit = 1.0
    0 | 0111112222333333344
    0 | 555566666666778888
    1 | 0001222244
    1 | 55666789
    2 | 01
    2 | 6
    3 | 4
    3 |
    4 |
    4 | 8
    5 |
    5 |
    6 |
    6 |
    7 |
    7 |
    8 | 3
```

Figure 1.1 A Minitab stem-and-leaf display (10ths digit truncated) and histogram for the charity fund-raising percentage data

Having obtained a sample from a population, an investigator would frequently like to use sample information to draw some type of conclusion (make an inference of some sort) about the population. That is, the sample is a means to an end rather than an end in itself.

Techniques for generalizing from a sample to a population are gathered within the branch of our discipline called **inferential statistics.**

Example 1.2	Material strength investigations provide a rich area of application for statistical methods. The article "Effects of Aggregates and Microfillers on the Flexural Properties of Concrete" (*Magazine of Concrete Research*, 1997: 81–98) reported on a study of strength properties of high-performance concrete obtained by using superplasticizers and certain binders. The compressive strength of such concrete had previously been investigated, but not much was known about flexural strength (a measure of ability to resist failure in bending). The accompanying data on flexural strength (in megapascals, MPa, where 1 Pa (pascal) = 1.45×10^{-4} psi) appeared in the article cited:

$$5.9 \quad 7.2 \quad 7.3 \quad 6.3 \quad 8.1 \quad 6.8 \quad 7.0 \quad 7.6 \quad 6.8 \quad 6.5 \quad 7.0 \quad 6.3 \quad 7.9 \quad 9.0$$
$$8.2 \quad 8.7 \quad 7.8 \quad 9.7 \quad 7.4 \quad 7.7 \quad 9.7 \quad 7.8 \quad 7.7 \quad 11.6 \quad 11.3 \quad 11.8 \quad 10.7$$

Suppose we want an *estimate* of the average value of flexural strength for all beams that could be made in this way (if we conceptualize a population of all such beams, we are trying to estimate the population mean). It can be shown that, with a high degree of confidence, the population mean strength is between 7.48 MPa and 8.80 MPa; we call this a *confidence interval* or *interval estimate*. Alternatively, this data could be used to predict the flexural strength of a *single* beam of this type. With a high degree of confidence, the strength of a single such beam will exceed 7.35 MPa; the number 7.35 is called a *lower prediction bound.*

The Scope of Modern Statistics

Statistical methodology is commonly employed by investigators in virtually every discipline, including such areas as

- molecular biology (analysis of microarray data)
- ecology (describing quantitatively how individuals in various animal and plant populations are spatially distributed)
- materials engineering (studying properties of various treatments to retard corrosion)
- marketing (developing market surveys and strategies for marketing new products)
- public health (identifying sources of diseases and ways to treat them)
- civil engineering (assessing the effects of stress on structural elements and the impacts of traffic flows on communities)

As you progress through the book, you'll encounter a wide spectrum of different scenarios in the examples and exercises that illustrate the application of techniques from probability and statistics. Many of these scenarios involve data or other material extracted from articles in engineering and science journals. The methods presented here have become established and trusted tools in the arsenal of those who work with data. Meanwhile, statisticians continue to develop new models to describe randomness and uncertainty and new methodology to analyze data. As evidence of the continuing creative efforts in the statistical community, here are titles and capsule descriptions of some articles that have recently appeared in statistics journals (*Journal of the American Statistical Association* is abbreviated JASA, and APS is short for the *Annals of Applied Statistics*, just two of the many prominent journals in the discipline):

- "Application of Branching Models in the Study of Invasive Species" (*JASA*, 2012: 467–476): Seismologists often predict earthquake occurrences using what is known as *epidemic-type aftershock sequence* (ETAS) models. The name stems from the model feature that allows earthquakes to cause aftershocks, which in turn may induce subsequent aftershocks, and so on, thereby generating a cascading effect. The authors propose the use of ETAS models in studying invasive plant and animal species. In particular, the article considers the spread of an invasive species in Costa Rica (*Musa velutina*, or red banana). The authors determine the estimated spatial–temporal rate of spread of red banana plants using a space–time ETAS model.

- "Spatio-Spectral Mixed-Effects Model for Functional Magnetic Resonance Imaging Data" (*JASA*, 2012: 568–577): For many years, scientists have attempted to model cognitive control-related activation among specific regions of the human brain. Researchers measure this brain activity through *functional magnetic resonance imaging* (fMRI). fMRI data often exhibit spatial and temporal correlations (i.e., observations made at nearby locations or time points are often strongly related). Standard approaches to fMRI analysis, however, fail to incorporate these relationships. The article proposes a statistical model to study activation in specific regions in the prefrontal cortex while also incorporating the underlying spatio–temporal correlations. The authors provide a simulation study that shows that significant errors can occur by ignoring the correlation structure in the network.

- "Active Learning Through Sequential Design, with Applications to the Detection of Money Laundering" (*JASA*, 2009: 969–981): Money laundering involves concealing the origin of funds obtained through illegal activities. The huge number of transactions occurring daily at financial institutions makes detection of money laundering difficult. The standard approach has been to extract various summary quantities from the transaction history and conduct a time consuming investigation of suspicious activities. The article proposes a more efficient statistical method and illustrates its use in a case study.

- "Robust Internal Benchmarking and False Discovery Rates for Detecting Racial Bias in Police Stops" (*JASA*, 2009: 661–668): Allegations of police actions that are at least partly attributable to racial bias have become a contentious issue in many communities. This article proposes a new method that is designed to reduce the risk of flagging a substantial number of "false positives" (individuals falsely identified as manifesting bias). The method was applied to data on 500,000 pedestrian stops from New York City in 2006; 15 officers from the pool of 3000 regularly involved in pedestrian stops were identified as having stopped a substantially greater fraction of black and Hispanic people than what would be predicted if bias were absent.

- "Measuring the Vulnerability of the Uruguayan Population to Vector-Borne Diseases via Spatially Hierarchical Factor Models" (*APS*, 2012: 284–303): Vector-borne diseases are illnesses caused by infections transmitted to people by organisms such as insects and spiders. According to the World Health Organization, the most deadly vector-borne disease is malaria, which kills more than 1 million people annually, mostly African children under age five. The authors develop a statistical index to model the vulnerability of Uruguayans to vector-borne diseases by accounting for variation attributable to factors such as different census tracts within cities and different cities in the country.

- "Self-Exciting Hurdle Models for Terrorist Activity" (*APS*, 2012: 106–124): The authors develop a predictive model of terrorist activity by considering the daily number of terrorist attacks in Indonesia from 1994 through 2007. The model estimates the chance of future attacks as a function of the times since past attacks. One feature of the model considers the excess of nonattack days coupled with the presence of multiple coordinated attacks on the same day. The article provides an interpretation of various model characteristics and assesses its predictive performance.

- "The BARISTA: A Model for Bid Arrivals in Online Auctions" (*APS*, 2007: 412–441): Online auctions such as those on eBay and uBid often have characteristics that differentiate them from traditional auctions. One particularly important such property is that the number of bidders at the outset of many traditional auctions is fixed, whereas in online auctions this number and the number of resulting bids are not predetermined. The article proposes a new BARISTA (for Bid ARivals In STAges) model for describing the way in which bids arrive that allows for higher bidding intensity not only at the outset of the auction but also as the auction comes to a close. Various properties of the model are investigated and then validated using data from eBay.com on auctions for Palm M515 personal assistants, Microsoft Xbox games, and Cartier watches.

Statistical information now appears with increasing frequency in the popular media, and occasionally the spotlight is even turned on statisticians. For example, "Behind Cancer Guidelines, Quest for Data," a *New York Times* article from November 23, 2009, reported that the new science for cancer investigations and more sophisticated methods for data analysis spurred the U.S. Preventive Services task force to reexamine guidelines for how frequently middle-aged and older women should have mammograms. The panel commissioned six independent groups to do statistical modeling. The result was a new set of conclusions, in particular one that mammograms every two years give nearly the same benefit as annual ones and confer only half the risk of harm. Donald Berry, a prominent biostatistician, was quoted as saying he was pleasantly surprised that the task force took the new research to heart in making its recommendations. The task force's report has generated much controversy among cancer organizations, politicians, and women themselves.

We hope you will become increasingly convinced of the importance and relevance of the discipline of statistics as you dig more deeply into the book and subject. We also anticipate you'll be intrigued enough to want to continue your statistical education beyond your current course.

Enumerative Versus Analytic Studies

W. E. Deming was a very influential American statistician whose ideas concerning the use of statistical methods in industrial production found great favor with Japanese companies in the years after World War II. He used the phrase **enumerative study** to describe investigations involving a finite collection of identifiable, unchanging objects that make up a population. In such studies, a *sampling frame*—that is, a listing of the objects to be sampled—is available or can be created. One example of such a frame is the collection of all signatures on petitions to qualify an initiative for inclusion on the ballot for an upcoming election. A sample is usually selected to ascertain whether the number of *valid* signatures exceeds a specified value. The variable on which observations are made is dichotomous, the two possible values being *valid* (S, for success) and *not valid* (F, for failure). As another example, the frame may contain serial numbers of all ovens manufactured by a particular company during a particular period. A sample may be selected to infer something about the average actual temperature of these units when the temperature control is set to 400°F (an inference about the population mean temperature).

Many problem situations faced by engineers involve some sort of ongoing **process**—a group of interrelated activities undertaken to accomplish some objective—rather than a specified, unchanging population. An investigator wants to learn something about how the process is operating so that the process can then be modified to better achieve the desired goal. Deming described such scenarios as **analytic studies.**

Example 1.3

The process of making ignition keys for automobiles consists of trimming and pressing raw key blanks, cutting grooves and notches, and then plating the keys. Dimensions associated with groove and notch cutting are crucial to proper key functioning. There will always be "normal" variation in dimensions because of fluctuations in materials, worker behavior, and environmental conditions. It is important, though, to monitor production to ensure that there are no unusual sources of variation, such as incorrect machine settings or contaminated material, which might result in nonconforming units or substantial changes in product characteristics. For this purpose, a sample (subgroup) of five keys is selected every 20 minutes, and critical dimensions are measured. Here are a few of the resulting observations for one particular dimension (in thousandths of an inch):

Subgroup 1:	6.1	8.4	7.6	7.5	4.4
Subgroup 2:	8.8	8.3	5.9	7.4	7.6
Subgroup 3:	8.0	7.5	7.0	6.8	9.3

This is indeed sample data, which can be used as a basis for drawing conclusions. However, the conclusions are about production process behavior rather than about a particular population of keys.

Analytic studies sometimes involve figuring out what actions to take to improve the performance of a future product.

Example 1.4

Failure in fluorescent lamps occurs when their luminosity falls below a predetermined level. The article "Using Degradation Data to Improve Fluorescent Lamp Reliability" (*J. of Quality Technology*, 1995: 363–369) described a case study involving fluorescent lamps of a certain type. The project engineer suggested focusing on three factors thought to be crucial to reliability:

1. The amount of electric current in the exhaustive process
2. The concentration of the mercury dispenser in the coating process
3. The concentration of argon in the filling process

Two levels, low and high, of each factor were established, leading to eight combinations of factor levels (e.g., low current, high mercury concentration, and low argon concentration). Luminance levels were then monitored over time for certain factor-level combinations. (Because of limited resources, only four of the eight combinations were included in the experiment, with five lamps used at each one.) Here is data for one particular lamp for which all factor levels were low:

Time (hr):	100	500	1000	2000	3000	4000	5000	6000
Luminance (lumens):	2810	2490	2460	2370	2320	2160	2140	2080

Statistical methods were used on the resulting data to draw conclusions about how lamp reliability could be improved. In particular, it was recommended that high concentration levels should be used with a low current level.

1.2   VISUAL DISPLAYS FOR UNIVARIATE DATA

Some preliminary organization of a data set often reveals useful information and opens paths of inquiry. Pictures are particularly effective in this respect. In this section, we introduce several of the most frequently used pictorial techniques.

Stem-and-Leaf Displays

A stem-and-leaf display can be an effective way to organize numerical data without expending much effort. It is based on separating each observation into two parts: (1) a **stem**, consisting of one or more leading digits, and (2) a **leaf**, consisting of the remaining or trailing digit(s). Suppose, for example, that data on calibration times (sec) for certain test devices has been gathered and that the smallest and largest times are 11.3 and 18.8, respectively. Then we could use the tens and ones digits as the stem of an observation, leaving the tenths digit for the leaf. Thus 11.3 would have a stem of 11 and a leaf of 3, 16.0 would have a stem of 16 and a leaf of 0, and so on. Once stem values have been chosen, they should be listed in a single column. Then the leaf of each observation should be placed on the row of the corresponding stem.

Example 1.5   The use of alcohol by college students is of great concern not only to those in the academic community but also, because of potential health and safety consequences, to society at large. The article "Health and Behavioral Consequences of Binge Drinking in College" (*J. of the Amer. Med. Assoc.*, 1994: 1672–1677) reported on a comprehensive study of heavy drinking on campuses across the United States. A binge episode was defined as five or more drinks in a row for males and four or more for females. Figure 1.2 shows a stem-and-leaf display of 140 values of x = the percentage of undergraduate students who are binge drinkers. (These values were not given in the cited article, but our display agrees with a picture of the data that did appear.)

0	4
1	1345678889
2	1223456666777889999	Stem: tens digit
3	0112233344555666677777888899999	Leaf: ones digit
4	111222223344445566666677788888999
5	00111222233455666667777888899
6	01111244455666778

Figure 1.2 Stem-and-leaf display for percentage binge drinkers at each of 140 colleges

The first leaf on the stem 2 row is 1, which tells us that 21% of the students at one of the colleges in the sample were binge drinkers. Without the identification of stem digits and leaf digits on the display, we wouldn't know whether the stem 2, leaf 1 observation should be read as 21%, 2.1%, or .21%.

When creating a display by hand, ordering the leaves from smallest to largest on each line can be time-consuming, and this ordering usually contributes little if any extra information. Suppose the observations had been listed in alphabetical order by school name, as

16%   33%   64%   37%   31%   . . .

Then placing these values on the display in this order would result in the stem 1 row having 6 as its first leaf, and the beginning of the stem 3 row would be

3 | 371 . . .

The display suggests that a typical or representative value is in the stem 4 row, perhaps in the mid-40% range. The observations are not highly concentrated about this typical value, as would be the case if all values were between 20% and 49%. The display rises to a single peak as we move downward, and then declines; there are no gaps in the display. The shape of the display is not perfectly symmetric, but instead appears to stretch out a bit more in the direction of low leaves than in the direction of high leaves. Lastly, there are no observations that are unusually far from the bulk of the data (no **outliers**), as would be the case if one of the 26% values had instead been 86%. The most surprising feature of this data is that at most colleges in the sample, at least one-quarter of the students are binge drinkers. The problem of heavy drinking on campuses is much more pervasive than many had suspected.

A stem-and-leaf display conveys information about the following aspects of the data:

- Identification of a typical or representative value
- Extent of spread about the typical value
- Presence of any gaps in the data
- Extent of symmetry in the distribution of values
- Number and location of peaks
- Presence of any outlying values

Suppose in Example 1.5 that each observation had included a tenths digit as well as the tens and ones digits: 16.4%, 36.5%, and so on. We could use two-digit leaves, so that 16.4 would have a stem of 1 and a leaf of 64; in this case, the decimal point can be omitted, but commas are necessary between successive leaves. Because such a display can become very unwieldy, it is customary to use single-digit leaves obtained by *truncation* (not rounding). Thus 36.7 would have stem 3 and leaf 6, and information about the tenths digit would be suppressed.

Consider a data set consisting of exam scores all of which are in the 70s, 80s, and 90s (an instructor's dream!). A stem-and-leaf display with the tens digit as the stem would have only three rows. However, a more informative display can be created by repeating each stem value twice, once for the low leaves 0, 1, 2, 3, 4 and again for the high leaves 5, 6, 7, 8, 9. A display of the binge-drinking data with repeated stems is shown in Figure 1.3. (The 11 on the far left in the fourth row indicates that there are 11 observations on or above that row; the (14) row contains the middle data value.)

```
Stem-and-leaf of pct binge  N = 140
Leaf Unit = 1.0

    1   0   4
    1   0
    4   1   134
   11   1   5678889
   16   2   12234
   30   2   56666777889999
   40   3   0112233344
   61   3   5556666777778888899999
  (14)  4   11122222334444
   65   4   5566666677788888999
   46   5   001112222334
   34   5   55666667777888899
   17   6   011112444
    8   6   55666778
```

Figure 1.3 Minitab stem-and-leaf display using repeated stems

Suppose that a final exam in physics contained questions worth a total of 200 points and that the only student who scored in the 100s earned 186 points. Rather than include rows 10, 11, . . . , and 18 just to show the extreme outlier 186, it is better to stop the display with a stem 9 row and place the information HI: 186 in a prominent place to the right of the display. The same thing can be done with outliers on the low end.

Consider two different data sets, each consisting of observations on the same variable, for example, exam scores for two different classes or stopping distances for cars equipped with two different braking systems. An investigator would naturally want to know in what ways the two sets were similar and how they differed. This can be accomplished by using a **comparative stem-and-leaf display,** in which the leaves for one data set are listed to the right of the stems and the leaves for the other to the left. Figure 1.4 shows a small example; the two sides of the display are quite similar, except that the right side appears to be shifted up one row (about 10 points) from the other side.

	9	658618
9447	8	13754380
2208965655	7	5312267
2432875	6	45104
5882	5	9

Figure 1.4 A comparative stem-and-leaf display of exam scores

Dotplots

A dotplot is an attractive summary of numerical data when the data set is reasonably small or there are relatively few distinct data values. Each observation is represented by a dot above the corresponding location on a horizontal measurement scale. When a value occurs more than once, there is a dot for each occurrence, and these dots are stacked vertically. As with a stem-and-leaf display, a dotplot gives information about location, spread, extremes, and gaps.

Example 1.6　Here is data on state-by-state appropriations for higher education as a percentage of state and local tax revenue for fiscal year 2009–2010 (from the *Statistical Abstract of the United States*). Values are listed in order of state abbreviations (AL first, WY last):

14.0	3.1	8.6	9.6	7.4	4.0	4.5	6.5	6.1	8.8
8.2	8.6	6.4	6.7	8.0	8.5	9.4	9.5	4.6	6.8
3.9	6.9	6.3	11.9	5.8	5.8	9.9	5.9	2.7	4.2
14.9	4.0	12.1	8.0	5.2	9.2	6.8	4.3	3.9	9.6
8.0	8.6	8.6	8.7	3.1	5.8	6.2	8.7	6.8	8.9

Figure 1.5 shows a dotplot of the data. The most striking feature is the substantial state-to-state variability. The largest values (for New Mexico, Alabama, North Carolina, and Mississippi) are somewhat separated from the bulk of the data and may possibly qualify as outliers.

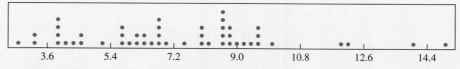

Figure 1.5 A dotplot of the data from Example 1.6

If the data set discussed in Example 1.6 had consisted of many more observations (e.g. average per pupil spending for each school district in the U.S.), it would be quite cumbersome to construct a corresponding dotplot. Our next technique is well suited to such situations.

Histograms

Some numerical data is obtained by counting to determine the value of a variable (the number of traffic citations a person received during the last year, the number of persons arriving for service during a particular period), whereas other data is obtained by taking measurements (weight of an individual, reaction time to a particular stimulus). The prescription for drawing a histogram is different for these two cases.

DEFINITIONS

> A variable is **discrete** if its set of possible values either is finite or else can be listed in an infinite sequence (one in which there is a first number, a second number, and so on). A variable is **continuous** if its possible values consist of an entire interval on the number line.

A discrete variable x almost always results from counting, in which case possible values are 0, 1, 2, 3, . . . or some subset of these integers. Continuous variables arise from making measurements. For example, if x is the pH of a chemical substance, then in theory x could be any number between 0 and 14: 7.0, 7.03, 7.032, and so on. Of course, in practice there are limitations on the degree of accuracy of any measuring instrument, so we may not be able to determine pH, reaction time, height, and concentration to an arbitrarily large number of decimal places. However, from the point of view of creating mathematical models for distributions of data, it is helpful to imagine an entire continuum of possible values.

Consider data consisting of observations on a discrete variable x. The **frequency** of any particular x value is the number of times that value occurs in the data set. The **relative frequency** of a value is the fraction or proportion of time the value occurs:

$$\text{relative frequency of a value} = \frac{\text{number of times the value occurs}}{\text{number of observations in the data set}}$$

Suppose, for example, that our data set consists of 200 observations on $x =$ the number of major defects on a new car of a certain type. If 70 of these x values are 1, then

frequency of the x value 1:   70

relative frequency of the x value 1: $\dfrac{70}{200} = .35$

Multiplying a relative frequency by 100 gives a percentage; in the defect example, 35% of the cars in the sample had just one major defect. The relative frequencies, or percentages, are usually of more interest than the frequencies themselves. In theory, the relative frequencies should sum to 1, but in practice the sum may differ slightly from 1 because of rounding.

Constructing a Histogram for Discrete Data

First, determine the frequency and relative frequency of each *x* value. Then mark possible *x* values on a horizontal scale. Above each value, draw a rectangle whose height is the relative frequency (or, alternatively, the frequency) of that value (all rectangles should have the same base width).

This construction ensures that the *area* of each rectangle is proportional to the relative frequency of the value. Thus if the relative frequencies of $x = 1$ and $x = 5$ are .35 and .07, respectively, then the area of the rectangle above 1 is five times the area of the rectangle above 5.

Example 1.7   Every corporation has a governing board of directors. The number of individuals on a board varies from one corporation to another. One of the authors of the article "Does Optimal Corporate Board Size Exist? An Empirical Analysis" (*Journal of Applied Finance*, 2010: 57–69) provided the accompanying data on the number of directors on the boards of a random sample of 204 corporations.

Board Size	Frequency	Relative Frequency	Board Size	Frequency	Relative Frequency
4	3	0.0147	19	0	0.0000
5	12	0.0588	20	0	0.0000
6	13	0.0637	21	1	0.0049
7	25	0.1225	22	0	0.0000
8	24	0.1176	23	0	0.0000
9	42	0.2059	24	1	0.0049
10	23	0.1127	25	0	0.0000
11	19	0.0931	26	0	0.0000
12	16	0.0784	27	0	0.0000
13	11	0.0539	28	0	0.0000
14	5	0.0245	29	0	0.0000
15	4	0.0196	30	0	0.0000
16	1	0.0049	31	0	0.0000
17	3	0.0147	32	1	0.0049
18	0	0.0000		204	0.9997

The corresponding histogram in Figure 1.6 rises to a peak and then declines. The histogram extends a bit more on the right (toward large values) than it does on the left—a slight *positive skew*.

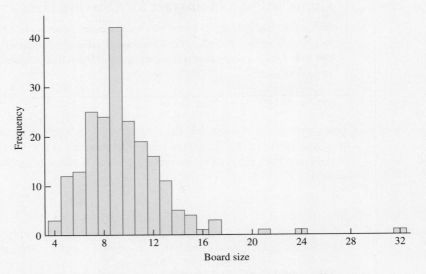

Figure 1.6 Histogram of number of corporate board members

From either the tabulated information or the histogram itself, we can determine the following:

Proportion of boards with at most 10 directors = (relative frequency for x = 4) + (relative frequency for x = 5) + ⋯ + (relative frequency for x = 10)

= 0.0147 + 0.0588 + 0.0637 + 0.1225 + 0.1176 + 0.2059 + 0.1127 = 0.6959

Similarly,

Proportion of boards with more than 15 directors = (relative frequency for x = 16) + (relative frequency for x = 17) + ⋯ + (relative frequency for x = 32)

= 0.0049 + 0.0147 + ⋯ + 0.0049 = 0.0343

Constructing a histogram for continuous data (measurements) entails subdividing the measurement axis into a suitable number of **class intervals** or **classes,** such that each observation is contained in exactly one class. Suppose, for example, that we have 50 observations on x = fuel efficiency of an automobile (mpg), the smallest of which is 27.8 and the largest of which is 31.4. Then we could use the class boundaries 27.5, 28.0, 28.5, . . . , and 31.5 as shown here:

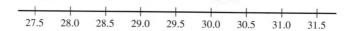

A potential difficulty is that an observation such as 29.0 lies on a class boundary so it doesn't lie in exactly one interval. One way to deal with this problem is to use boundaries like 27.55, 28.05, . . . , 31.55. Adding a hundredths digit to the class boundaries prevents observations from falling on the resulting boundaries. Another way to deal with this problem is to use the classes $27.5 - <28.0, 28.0 - <28.5, . . . , 31.0 - <31.5$. Then 29.0 falls in the class $29.0 - <29.5$ rather than in the class $28.5 - <29.0$. In other words, with this convention, an observation on a boundary is placed in the interval to the right of the boundary. This is how Minitab constructs a histogram.

Constructing a Histogram for Continuous Data: Equal Class Widths

Determine the frequency and relative frequency for each class. Mark the class boundaries on a horizontal measurement axis. Above each class interval, draw a rectangle whose height is the corresponding relative frequency (or frequency).

Example 1.8   Power companies need information about customer usage to obtain accurate forecasts of demand. Investigators from Wisconsin Power and Light determined energy consumption (BTUs) during a particular period for a sample of 90 gas-heated homes. An adjusted consumption value was calculated as follows:

$$\text{adjusted consumption} = \frac{\text{consumption}}{(\text{weather, in degree days})(\text{house area})}$$

This resulted in the accompanying data (part of the stored data set FURNACE. MTW available in Minitab, which we have ordered from smallest to largest):

2.97	4.00	5.20	5.56	5.94	5.98	6.35	6.62	6.72	6.78
6.80	6.85	6.94	7.15	7.16	7.23	7.29	7.62	7.62	7.69
7.73	7.87	7.93	8.00	8.26	8.29	8.37	8.47	8.54	8.58
8.61	8.67	8.69	8.81	9.07	9.27	9.37	9.43	9.52	9.58
9.60	9.76	9.82	9.83	9.83	9.84	9.96	10.04	10.21	10.28
10.28	10.30	10.35	10.36	10.40	10.49	10.50	10.64	10.95	11.09
11.12	11.21	11.29	11.43	11.62	11.70	11.70	12.16	12.19	12.28
12.31	12.62	12.69	12.71	12.91	12.92	13.11	13.38	13.42	13.43
13.47	13.60	13.96	14.24	14.35	15.12	15.24	16.06	16.90	18.26

We let Minitab select the class intervals. The most striking feature of the histogram in Figure 1.7 is its resemblance to a bell-shaped (and therefore symmetric) curve, with the point of symmetry at roughly 10.

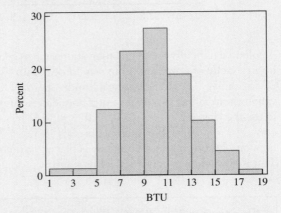

Figure 1.7 Histogram of the energy consumption data from Example 1.8

Class:	1−<3	3−<5	5−<7	7−<9	9−<11	11−<13	13−<15	15−<17	17−<19
Frequency:	1	1	11	21	25	17	9	4	1
Relative frequency:	.011	.011	.122	.233	.278	.189	.100	.044	.011

From the histogram,

$$\begin{array}{l}\text{proportion of observations} \\ \text{less than 9}\end{array} \approx .01 + .01 + .12 + .23 = .37$$

(exact value = 34/90 = .378)

The relative frequency for the 9 − <11 class is about .27, so roughly half of this, or .135, should be between 9 and 10. Thus

$$\begin{array}{l}\text{proportion of observations} \\ \text{less than 10}\end{array} \approx .37 + .135 = .505 \quad \text{(slightly more than 50\%)}$$

The exact value of this proportion is 47/90 = .522.

There are no hard-and-fast rules concerning either the number of classes or the choice of classes themselves. Between 5 and 20 classes will be satisfactory for most data sets. Generally, the larger the number of observations in a data set, the more classes should be used. A reasonable rule of thumb is

$$\text{number of classes} \approx \sqrt{\text{number of observations}}$$

Equal-width classes may not be a sensible choice if a data set has at least one "stretched-out tail." Figure 1.8 (page 19) shows a dotplot of such a data set. Using a small number of

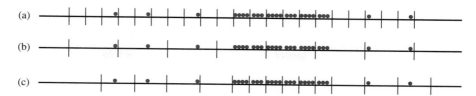

Figure 1.8 Selecting class intervals when there are outliers: (a) many short equal width intervals; (b) a few wide equal-width intervals; (c) unequal-width intervals

equal-width classes results in almost all observations falling in just one or two of the classes. If a large number of equal-width classes are used, many classes will have zero frequency. A sound choice is to use a few wider intervals near extreme observations and narrower intervals in the region of high concentration.

Constructing a Histogram for Continuous Data: Unequal Class Widths

After determining frequencies and relative frequencies, calculate the height of each rectangle using the formula

$$\text{rectangle height} = \frac{\text{relative frequency of the class}}{\text{class width}}$$

The resulting rectangle heights are usually called *densities*, and the vertical scale is the **density scale**. This prescription will also work when class widths are equal.

Example 1.9   Corrosion of reinforcing steel is a serious problem in concrete structures located in environments affected by severe weather conditions. For this reason, researchers have been investigating the use of reinforcing bars made of composite material. One study was carried out to develop guidelines for bonding glass-fiber-reinforced plastic rebars to concrete ("Design Recommendations for Bond of GFRP Rebars to Concrete," *J. of Structural Engr.*, 1996: 247–254). Consider the following 48 observations on measured bond strength:

11.5	12.1	9.9	9.3	7.8	6.2	6.6	7.0	13.4	17.1	9.3	5.6
5.7	5.4	5.2	5.1	4.9	10.7	15.2	8.5	4.2	4.0	3.9	3.8
3.6	3.4	20.6	25.5	13.8	12.6	13.1	8.9	8.2	10.7	14.2	7.6
5.2	5.5	5.1	5.0	5.2	4.8	4.1	3.8	3.7	3.6	3.6	3.6

Class:	2–<4	4–<6	6–<8	8–<12	12–<20	20–<30
Frequency:	9	15	5	9	8	2
Relative frequency:	.1875	.3125	.1042	.1875	.1667	.0417
Density:	.094	.156	.052	.047	.021	.004

The resulting histogram appears in Figure 1.9. The right or upper tail stretches out much farther than does the left or lower tail—a substantial departure from symmetry.

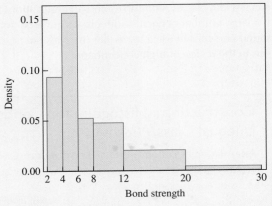

Figure 1.9 A Minitab density histogram for the bond strength data of Example 1.9

When class widths are unequal, not using a density scale will give a picture with distorted areas. For equal class widths, the divisor is the same in each density calculation, and the extra arithmetic simply results in a rescaling of the vertical axis (i.e., the histogram using relative frequency and the one using density will have exactly the same appearance). A density histogram does have one interesting property. Multiplying both sides of the formula for density by the class width gives

$$\text{relative frequency} = (\text{class width})(\text{density})$$

$$= (\text{rectangle width})(\text{rectangle height})$$

$$= \text{rectangle area}$$

That is, *the area of each rectangle is the relative frequency of the corresponding class.* Furthermore, since the sum of relative frequencies must be 1.0 (except for roundoff), *the total area of all rectangles in a density histogram is 1.* It is always possible to draw a histogram so that the area equals the relative frequency (this is true also for a histogram of discrete data—just use the density scale). This property will play an important role in creating models for distributions in Section 1.3.

Histogram Shapes

Histograms come in a variety of shapes. A **unimodal** histogram is one that rises to a single peak and then declines. A **bimodal** histogram has two different peaks. Bimodality occurs when the data set consists of observations on two quite different kinds of individuals or objects. For example, consider a large data set consisting of driving times for automobiles traveling between San Luis Obispo, California, and Monterey, California (exclusive of stopping time for sightseeing, eating, etc.). This histogram would show two peaks, one for those cars that took the inland route (roughly 2.5 hours) and another for those cars traveling up the coast (3.5–4 hours). However, bimodality does not automatically follow in such situations. Only if the two separate histograms are "far apart" relative to their spreads will bimodality occur in the histogram of combined data. Thus a large data set consisting of heights of college students should not result in a bimodal histogram because the typical male height of about 69 inches is not far enough above the typical female height of about 64–65 inches. A histogram with more than two peaks is said to be **multimodal.** Of course, the number of peaks may well depend on the choice of class intervals, particularly with a small number of observations. The larger the number of classes, the more likely it is that bimodality or multimodality will manifest itself.

Example 1.10   Figure 1.10(a) shows a Minitab histogram of the weights (lbs) of the 121 players listed on the rosters of the San Francisco 49ers and the New England Patriots as of November 28, 2012. Figure 1.10(b) is a smoothed histogram (actually what is called a *density estimate*) of the data from the R software package. Both the histogram and the smoothed histogram show three distinct peaks: The one on the right is for linemen, the middle peak corresponds to linebacker weights, and the peak on the left is for all other players (wide receivers, quarterbacks, etc.).

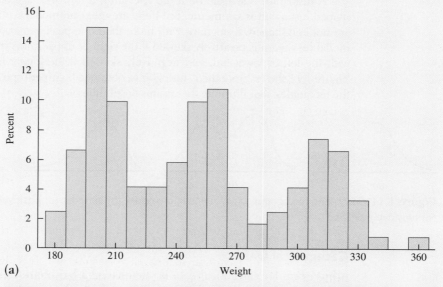

(a)

Figure 1.10 NFL player weights: (a) histogram, (b) smoothed histogram

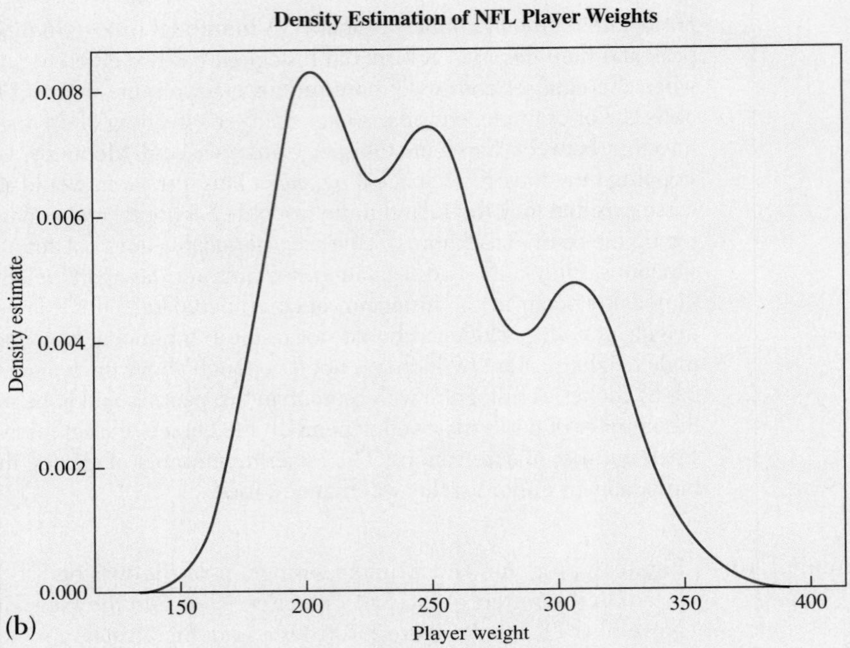

Density Estimation of NFL Player Weights
0.008
0.006
0.004
0.002
0.000
Density estimate
150
200
250
300
350
400
Player weight
(b)

Figure 1.10 *(continued)*

A histogram is **symmetric** if the left half is a mirror image of the right half. A bell-shaped histogram is symmetric, but there are other unimodal symmetric histograms that are not bell-shaped; histograms with more than one peak can also be symmetric. A unimodal histogram is **positively skewed** if the right or upper tail is stretched out compared with the left or lower tail, and **negatively skewed** if the longer tail extends to the left. Figure 1.11 shows "smoothed" histograms, obtained by superimposing a smooth curve on the rectangles, that illustrate the various possibilities.

(a)
(b)
(c)
(d)

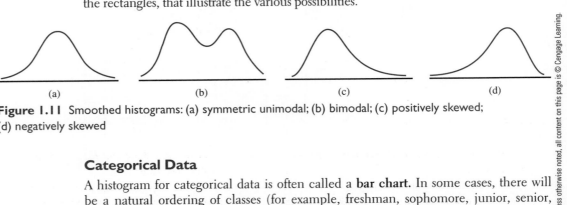

Figure 1.11 Smoothed histograms: (a) symmetric unimodal; (b) bimodal; (c) positively skewed; (d) negatively skewed

Categorical Data

A histogram for categorical data is often called a **bar chart.** In some cases, there will be a natural ordering of classes (for example, freshman, sophomore, junior, senior, graduate student), whereas in other cases, the order will be arbitrary (Honda, Yamaha,

Harley-Davidson, etc.). A **Pareto diagram** is a bar chart resulting from a quality control study in which each category represents a different type of product nonconformity or production problem. The categories appear in order of decreasing frequency (if a miscellaneous category is needed, it is the last one).

Example 1.11   In the manufacture of printed circuit boards, finished boards are subjected to a final inspection before they are shipped to customers. Here is data on the type of defect for each board rejected at final inspection during a particular time period:

Type of defect	Frequency	Relative frequency
Low copper plating	112	.615
Poor electroless coverage	35	.192
Lamination problems	10	.055
Plating separation	8	.044
Etching problems	5	.027
Miscellaneous	12	.066

Figure 1.12 is a Pareto diagram. Roughly 80% (.615 + .192) of the defects were of one of the first two types.

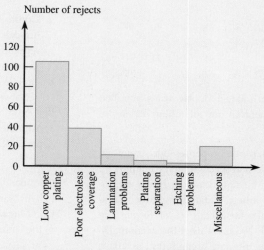

Figure 1.12 A Pareto diagram for Example 1.11

Section 1.2 Exercises

1. Consider the strength data for beams given in Example 1.2.
   a. Construct a stem-and-leaf display of the data. What appears to be a representative strength value? Do the observations appear to be highly concentrated about the representative value or rather spread out?
   b. Does the display appear to be reasonably symmetric about a representative value, or would you describe its shape in some other way?

c. Do there appear to be any outlying strength values?

d. What proportion of strength observations in this sample exceed 10 MPa?

2. The article cited in Example 1.2 also gave the accompanying strength observations for cylinders:

6.1 5.8 7.8 7.1 7.2 9.2 6.6 8.3   7.0   8.3
7.8 8.1 7.4 8.5 8.9 9.8 9.7 14.1 12.6 11.2

a. Construct a comparative stem-and-leaf display of the beam and cylinder data, then answer the questions in parts (b)–(d) of Exercise 1 for the observations on cylinders.

b. In what ways are the two sides of the display similar? Are there any obvious differences between the beam observations and the cylinder observations?

3. The accompanying specific gravity values for various wood types used in construction appeared in the article "Bolted Connection Design Values Based on European Yield Model" (*J. of Structural Engr.*, 1993: 2169–2186):

.31 .35 .36 .36 .37 .38 .40 .40 .40
.41 .41 .42 .42 .42 .42 .42 .43 .44
.45 .46 .46 .47 .48 .48 .48 .51 .54
.54 .55 .58 .62 .66 .66 .67 .68 .75

Construct a stem-and-leaf display using repeated stems, and comment on any interesting features of the display.

4. Allowable mechanical properties for structural design of metallic aerospace vehicles requires an approved method for statistically analyzing empirical test data. The article "Establishing Mechanical Property Allowables for Metals" (*J. of Testing and Evaluation*, 1998: 293–299) used the accompanying data on tensile ultimate strength (ksi) as a basis for addressing the difficulties in developing such a method:

122.2 124.2 124.3 125.6 126.3 126.5
126.5 127.2 127.3 127.5 127.9 128.6
128.8 129.0 129.2 129.4 129.6 130.2
130.4 130.8 131.3 131.4 131.4 131.5
131.6 131.6 131.8 131.8 132.3 132.4
132.4 132.5 132.5 132.5 132.5 132.6

132.7 132.9 133.0 133.1 133.1 133.1
133.1 133.2 133.2 133.2 133.3 133.3
133.5 133.5 133.5 133.8 133.9 134.0
134.0 134.0 134.0 134.1 134.2 134.3
134.4 134.4 134.6 134.7 134.7 134.7
134.8 134.8 134.8 134.9 134.9 135.2
135.2 135.2 135.3 135.3 135.4 135.5
135.5 135.6 135.6 135.7 135.8 135.8
135.8 135.8 135.8 135.9 135.9 135.9
135.9 136.0 136.0 136.1 136.2 136.2
136.3 136.4 136.4 136.6 136.8 136.9
136.9 137.0 137.1 137.2 137.6 137.6
137.8 137.8 137.8 137.9 137.9 138.2
138.2 138.3 138.3 138.4 138.4 138.4
138.5 138.5 138.6 138.7 138.7 139.0
139.1 139.5 139.6 139.8 139.8 140.0
140.0 140.7 140.7 140.9 140.9 141.2
141.4 141.5 141.6 142.9 143.4 143.5
143.6 143.8 143.8 143.9 144.1 144.5
144.5 147.7 147.7

a. Construct a stem-and-leaf display of the data by first deleting (truncating) the tenths digit and then repeating each stem value five times (once for leaves 0 and 1, a second time for leaves 2 and 3, etc.). Why is it relatively easy to identify a representative strength value?

b. Construct a histogram using equal-width classes with the first class having a lower limit of 122 and an upper limit of 124. Then comment on any interesting features of the histogram.

5. Consider the accompanying values of golf course lengths (yards) for a sample of courses designated by *Golf Magazine* as being among the most challenging in the United States:

6433 6435 6464 6470 6506 6526 6527
6583 6605 6614 6694 6700 6713 6745
6770 6770 6790 6798 6850 6870 6873
6890 6900 6904 6927 6936 7005 7011
7022 7040 7050 7051 7105 7113 7131
7165 7168 7169 7209 7280

a. Would it be best to use one-digit, two-digit, or three-digit stems as a basis for a stem-and-leaf display? Explain your reasoning.

b. Construct a stem-and-leaf display based on two-digit stems and two-digit leaves, with successive leaves separated by either a comma or a space.

c. Construct a stem-and-leaf display in which the leaf of each observation is its tens digit (so the ones digit is truncated). Does this display appear to be significantly less informative about course lengths than the display of part (b)? What advantage would this display have over the one in part (b) if there had been 200 courses in the sample?

6. Construct two stem-and-leaf displays for the accompanying set of exam scores, one in which each stem value appears just once and the other in which stem values are repeated:

74 89 80 93 64 67 72 70 66 85 89 81
81 71 74 82 85 63 72 81 81 95 84 81
80 70 69 66 60 83 85 98 84 68 90 82
69 72 87 88

What feature of the data is revealed by the display with repeated stems that is not so readily apparent in the first display?

7. Temperature transducers of a certain type are shipped in batches of 50. A sample of 60 batches was selected, and the number of transducers in each batch not conforming to design specifications was determined, resulting in the following data:

2 1 2 4 0 1 3 2 0 5 3 3 1 3 2 4 7 0 2 3
0 4 2 1 3 1 1 3 4 1 2 3 2 2 8 4 5 1 3 1
5 0 2 3 2 1 0 6 4 2 1 6 0 3 3 3 6 1 2 3

a. Determine frequencies and relative frequencies for the observed values of $x = $ number of nonconforming transducers in a batch.

b. What proportion of batches in the sample have at most five nonconforming transducers? What proportion have fewer than five? What proportion have at least five nonconforming units?

c. Draw a histogram of the data using relative frequency on the vertical scale, and comment on its features.

8. In a study of author productivity ("Lotka's Test," *Collection Mgmt.*, 1982: 111–118), a large number of authors were classified according to the number of articles they had published during a certain period. The results were presented in the accompanying frequency distribution:

Number of papers:	1	2	3	4	5	6	7	8
Frequency:	784	204	127	50	33	28	19	19

Number of papers:	9	10	11	12	13	14	15	16	17
Frequency:	6	7	6	7	4	4	5	3	3

a. Construct a histogram corresponding to this frequency distribution. What is the most interesting feature of the shape of the distribution?

b. What proportion of these authors published at least five papers? At least ten papers? More than ten papers?

c. Suppose the five 15s, three 16s, and three 17s had been lumped into a single category displayed as "≥15." Would you be able to draw a histogram? Explain.

d. Suppose that instead of the values 15, 16, and 17 being listed separately, they had been combined into a 15–17 category with frequency 11. Would you be able to draw a histogram? Explain.

9. The number of contaminating particles on a silicon wafer prior to a certain rinsing process was determined for each wafer in a sample of size 100, resulting in the following frequencies:

Number of particles:	0	1	2	3	4	5	6	7
Frequency:	1	2	3	12	11	15	18	10

Number of particles:	8	9	10	11	12	13	14
Frequency:	12	4	5	3	1	2	1

a. What proportion of the sampled wafers had at least one particle? At least five particles?

b. What proportion of the sampled wafers had between five and ten particles, inclusive? Strictly between five and ten particles?

c. Draw a histogram using relative frequency on the vertical axis. How would you describe the shape of the histogram?

10. The article "Knee Injuries in Women Collegiate Rugby Players" (*Amer. J. of Sports Medicine*, 1997: 360–362) gave the following data on type of injury (A = mensical tear, B = MCL tear, C = ACL tear, D = patella dislocation, E = PCL tear):

A B B A C A A D B A C E B
B A A C D C A C B C C C A
B B C A A B C C A C B B D
A B A C B A A C A B B E B
B B C C A C A A B D A A C
B C C A B B A D C A B

Construct a Pareto diagram for this data. The three most frequently occurring types of injuries account for what proportion of all injuries?

11. The article "Determination of Most Representative Subdivision" (*J. of Energy Engr.*, 1993: 43–55) gave data on various characteristics of subdivisions that could be used in deciding whether to provide electrical power using overhead lines or underground lines. Here are the values of the variable x = total length of streets within a subdivision:

1280 5320 4390 2100 1240 3060 4770
1050   360 3330 3380   340 1000   960
1320   530 3350   540 3870 1250 2400
 960 1120 2120   450 2250 2320 2400
3150 5700 5220   500 1850 2460 5850
2700 2730 1670   100 5770 3150 1890
 510   240   396 1419 2109

a. Construct a stem-and-leaf display using the thousands digit as the stem and the hundreds digit as the leaf, and comment on the various features of the display.

b. Construct a histogram using class boundaries 0, 1000, 2000, 3000, 4000, 5000, and 6000. What proportion of subdivisions have total length less than 2000? Between 2000 and 4000? How would you describe the shape of the histogram?

12. The article cited in Exercise 11 also gave the following values of the variables y = number of culs-de-sac and z = number of intersections:

y: 1 0 1 0 0 2 0 1 1 1 2 1 0 0 1 1
   0 1 1 1 0 0 0 1 1 2 0 1 2 2 1
   1 0 2 1 1 0 1 5 0 3 0 1 1 0 0

z: 1 8 6 1 1 5 3 0 0 4 4 0 0 1 2 1
   4 0 4 0 3 0 1 1 0 1 3 2 4 6 6 0
   1 1 8 3 3 5 0 5 2 3 1 0 0 0 3

a. Construct a histogram for the y data. What proportion of these subdivisions had no culs-de-sac? At least one cul-de-sac?

b. Construct a histogram for the z data. What proportion of these subdivisions had at most five intersections? Fewer than five intersections?

13. The article "Ecological Determinants of Herd Size in the Thorncraft's Giraffe of Zambia" (*Afric. J. Ecol.*, 2010: 962–971) gave the following data (read from a graph) on herd size for a sample of 1570 herds over a 34-year period.

Herd size:	1	2	3	4	5	6	7	8
Frequency:	589	190	176	157	115	89	57	55

Herd size:	9	10	11	12	13	14	15	17
Frequency:	33	31	22	10	4	10	11	5

Herd size:	18	19	20	22	23	24	26	32
Frequency:	2	4	2	2	2	2	1	1

a. What proportion of the sampled herds had just one giraffe?

b. What proportion of the sampled herds had six or more giraffes (characterized in the article as "large herds")?

c. What proportion of the sampled herds had between 5 and 10 giraffes inclusive?

d. Draw a histogram using relative frequency on the vertical axis. How would you describe the shape of this histogram?

14. The article "Statistical Modeling of the Time Course of Tantrum Anger" (*J. of Applied Stats*, 2009: 1013–1034) discussed how anger intensity in children's tantrums could be related to tantrum duration as well as behavioral indicators such as shouting, stamping, pushing, and pulling. The following frequency distribution was given (as well as the corresponding histogram):

$0-<2$:   136 $2-<4$: 92 $4-<11$: 71
$11-<20$: 26 $20-<30$: 7 $30-<40$: 3

Draw the histogram and then comment on any interesting features.

15. Automated electron backscattered diffraction is now being used in the study of fracture phenomena. The following information on misorientation angle (degrees) was extracted from the article "Observations on the Faceted Initiation Site in the Dwell-Fatigue Tested Ti-6242 Alloy: Crystallographic Orientation and Size Effects" (*Metallurgical and Materials Trans.*, 2006: 1507–1518)

Class:	0 – <5	5 – <10	10 – <15	15 – <20
Rel Freq:	.177	.166	.175	.136

Class:	20 – <30	30 – <40	40 – < 60	60 – <90
Rel Freq:	.194	.078	.044	.030

  a. Is it true that more than 50% of the sampled angles are smaller than 15°, as asserted in the paper?
  b. What proportion of the sampled angles are at least 30°?
  c. Roughly what proportion of angles are between 10° and 25°?
  d. Construct a histogram and comment on any interesting features.

16. A transformation of data values by means of some mathematical function, such as $\sqrt{x}$ or $1/x$, can often yield a set of numbers that has "nicer" statistical properties than the original data. In particular, it may be possible to find a function for which the histogram of transformed values is more symmetric (or even better, more like a bell-shaped curve) than the original data. For example, the article "Time Lapse Cinematographic Analysis of Beryllium–Lung Ibroblast Interactions" (*Envir. Research*, 1983: 34–43) reported the results of experiments designed to study the behavior of certain individual cells that had been exposed to beryllium. An important characteristic of such an individual cell is its interdivision time (IDT). IDTs were determined for a number of cells both in exposed (treatment) and in unexposed (control) conditions. The authors of the article used a logarithmic transformation. Consider the following representative IDT data:

28.1	31.2	13.7	46.0	25.8	16.8	34.8	62.3
28.0	17.9	19.5	21.1	31.9	28.9	60.1	23.7
18.6	21.4	26.6	26.2	32.0	43.5	17.4	38.8
30.6	55.6	25.5	52.1	21.0	22.3	15.5	36.3
19.1	38.4	72.8	48.9	21.4	20.7	57.3	40.9

Construct a histogram of this data based on classes with boundaries 10, 20, 30, ... Then calculate $\log_{10}(x)$ for each observation, and construct a histogram of the transformed data using class boundaries 1.1, 1.2, 1.3, ... What is the effect of the transformation?

17. The accompanying data set consists of observations on shear strength (lb) of ultrasonic spot welds made on a certain type of alclad sheet. Construct a relative frequency histogram based on ten equal-width classes with boundaries 4000, 4200, ... (The histogram will agree with the one in "Comparison of Properties of Joints Prepared by Ultrasonic Welding and Other Means," *J. of Aircraft*, 1983: 552–556.) Comment on its features.

5434	4948	4521	4570	4990	5702	5241
5112	5015	4659	4806	4637	5670	4381
4820	5043	4886	4599	5288	5299	4848
5378	5260	5055	5828	5218	4859	4780
5027	5008	4609	4772	5133	5095	4618
4848	5089	5518	5333	5164	5342	5069
4755	4925	5001	4803	4951	5679	5256
5207	5621	4918	5138	4786	4500	5461
5049	4974	4592	4173	5296	4965	5170
4740	5173	4568	5653	5078	4900	4968
5248	5245	4723	5275	5419	5205	4452
5227	5555	5388	5498	4681	5076	4774
4931	4493	5309	5582	4308	4823	4417
5364	5640	5069	5188	5764	5273	5042
5189	4986					

18. The paper "Study on the Life Distribution of Microdrills" (*J. of Engr. Manufacture*, 2002: 301–305) reported the following observations, listed in increasing order, on drill lifetimes (number of holes that a drill machines before it breaks) when holes were drilled in a certain brass alloy.

11	14	20	23	31	36	39	44	47	50
59	61	65	67	68	71	74	76	78	79
81	84	85	89	91	93	96	99	101	104
105	105	112	118	123	136	139	141	148	158
161	168	184	206	248	263	289	322	388	513

a. Why can a frequency distribution not be based on the class intervals 0–50, 50–100, 100–150, and so on?

b. Construct a frequency distribution and histogram of the data using class boundaries 0, 50, 100, . . . and then comment on interesting characteristics.

c. Construct a frequency distribution and histogram of the natural logarithms of the lifetime observations and comment on interesting characteristics.

d. What proportion of the lifetime observations in this sample are less than 100? What proportion of the observations are at least 200?

I.3   DESCRIBING DISTRIBUTIONS

In Section 1.2, we saw that a histogram could be used to describe how values of a variable x are distributed in a data set. In practice, a histogram is virtually always constructed from sample data. Consider the population or process from which a sample might be selected. It is often possible to give a concise mathematical description of how the possible values of x are distributed or dispersed along the number line or measurement scale. Suppose, for example, that x is the fuel efficiency (mpg) of a vehicle of a particular type (a continuous variable), so that the value of x varies from vehicle to vehicle. Knowing the distribution of x enables us to determine the proportion of vehicles for which x is less than 32, the proportion for which x exceeds 30.5, the proportion of vehicles having $31.5 < x < 32.5$, and so on. If x is the number of defects on an item produced by some process (a discrete variable), then the x distribution will describe what proportion of items produced will have $x = 0$, what proportion will have $x = 1$, and so on. We now describe the essential features of distributions for continuous variables and those for discrete variables.

Continuous Distributions

Let x be a continuous variable, one whose value is determined by making a measurement of some sort. Suppose we have a sample of x values from a population or ongoing process. For example, the sample might consist of fuel efficiencies of cars selected from a large rental fleet (a population) or waiting times for a succession of patients entering a large medical clinic (a patient arrival process). If the sample size is small, a histogram based on only a small number of relatively wide class intervals is appropriate. For a large sample size, many narrow classes should be used. Let's agree to draw our histograms using the density scale discussed in Section 1.2 so that

- For each rectangle, area = relative frequency of the class
- Total area of all rectangles = 1

With a large amount of data, a histogram based on any reasonable choice of classes should have roughly the same shape and can very frequently be well approximated by a smooth curve. This type of approximation is illustrated in Figure 1.13.

Many approximating curves that arise in practice can be obtained as graphs of reasonably simple mathematical functions. Such a mathematical function provides a very concise description of the x distribution.

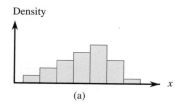 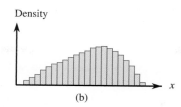 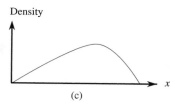

(a) (b) (c)

Figure 1.13 Histograms of continuous data: (a) small number of wide classes; (b) large number of narrow classes; (c) approximation by a smooth curve

DEFINITIONS

A **density function** $f(x)$ is used to describe (at least approximately) the population or process distribution of a continuous variable x. The graph of $f(x)$ is called the **density curve.** The following properties must be satisfied:

1. $f(x) \geq 0$
2. $\int_{-\infty}^{\infty} f(x)\, dx = 1$ (the total area under the density curve is 1.0)
3. For any two numbers a and b with $a < b$,

$$\text{proportion of } x \text{ values between } a \text{ and } b = \int_{a}^{b} f(x)\, dx$$

(This proportion is the area under the density curve and above the interval with endpoints a and b, as illustrated in Figure 1.14.)

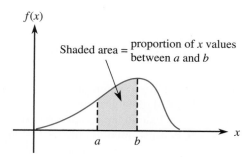

Figure 1.14 The area under the density curve is equal to the proportion of values in an interval

There is no area under the density curve and above a single value (e.g., above 2.50), which implies that

$$\text{proportion of } x \text{ values satisfying } a \leq x \leq b = \text{proportion of } x \text{ values satisfying } a < x < b$$

That is, the area under the curve between a and b does not depend on whether the two interval endpoints are included or excluded.

Example 1.12   A certain daily program on a public radio station lasts 1 hour. Let x denote the amount of time (hr) during which music is played. (There are no advertisements, but the host provides occasional commentary and makes announcements.) A potential program sponsor is interested in knowing how the value of x varies from program to program. Consider the density function

$$f(x) = \begin{cases} 90x^8(1 - x) & 0 \le x \le 1 \\ 0 & \text{otherwise} \end{cases}$$

This looks complicated, but the corresponding density curve in Figure 1.15 has a simple and appealing shape.

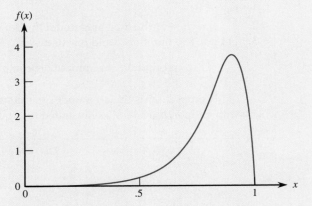

Figure 1.15 Density curve for Example 1.12

We see immediately that most x values are quite close to 1 and very few are smaller than .5 (almost all programs consist of at least a half hour of music). The constant 90 in $f(x)$ ensures that the total area under the density curve is 1.0 [$f(x) = kx^8(1 - x)$ is a legitimate density function only for $k = 90$]. Various proportions of interest can now be obtained by integration. For example,

$$\begin{aligned}\text{proportion of programs} \atop \text{with } x \text{ between } .7 \text{ and } .9 \quad &= \int_{.7}^{.9} 90x^8(1 - x)\, dx = 90\int_{.7}^{.9} x^8\, dx - 90\int_{.7}^{.9} x^9\, dx \\[2mm] &= 90\left(\frac{x^9}{9} - \frac{x^{10}}{10}\right)\Bigg|_{.7}^{.9} = .587\end{aligned}$$

$$\text{proportion of programs} \atop \text{for which } x \text{ is at least } .8 \quad = \int_{.8}^{1} 90x^8(1 - x)\, dx = .624$$

What duration value c separates the smallest 50% of all x values from the largest 50%? Figure 1.16 shows the location of c; the corresponding equation is

$$\int_0^c 90x^8(1-x)\,dx = .5$$

which becomes

$$90\left(\frac{c^9}{9} - \frac{c^{10}}{10}\right) = .5$$

Newton's method or some other numerical technique is used to obtain the solution: $c \approx .838$. That is, about 50% of all programs have music for more than .838 hr, and about 50% have music for less than .838 hr. The value .838 is called the *median* of the x distribution.

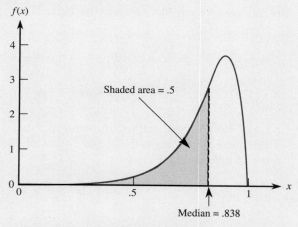

Figure 1.16 Determining the median of the distribution in Example 1.12

Example 1.13 Let x denote the response time (sec) at a certain on-line computer; that is, x is the time between the end of a user's inquiry and the beginning of the system's response to that inquiry. The value of x varies from inquiry to inquiry. Suppose the density function for the distribution of x is

$$f(x) = \begin{cases} .2e^{-.2x} & x \geq 0 \\ 0 & \text{otherwise} \end{cases}$$

where e represents the base of the natural logarithm system and approximately equals 2.71828. A graph of $f(x)$ is shown in Figure 1.17. By inspection, $f(x) \geq 0$, and

$$\int_{-\infty}^{\infty} f(x)\,dx = \int_0^{\infty} .2e^{-.2x}\,dx = -e^{-.2x}\Big|_0^{\infty} = 1$$

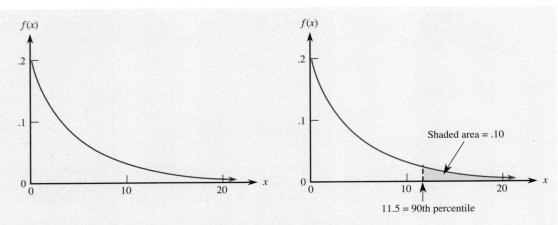

Figure 1.17 The density curve and 90th percentile for Example 1.13

The proportion of inquiries with a response time less than 5 sec is

$$\int_0^5 .2e^{-.2x}\, dx = 1 - e^{-.2(5)} = .632$$

So 63.2% of all response times are at most 5 sec, and 36.8% of all times exceed 5 sec. The value c that separates the largest 10% of all times from the smallest 90% (called the *90th percentile*) satisfies

$$.9 = \int_0^c .2e^{-.2x}\, dx = 1 - e^{-.2c}$$

from which $c = -[\ln(.1)]/.2 = 11.5$. Only about 10% of all inquiries will have response times exceeding 11.5 sec.

The density function in Example 1.13 is a particular case of a more general function.

DEFINITION

A variable x is said to have an **exponential distribution** with parameter $\lambda > 0$ if the density function for x is

$$f(x) = \begin{cases} \lambda e^{-\lambda x} & x \geq 0 \\ 0 & \text{otherwise} \end{cases}$$

Each different value of λ prescribes a different exponential distribution, so we have an entire *family* of distributions. The shape of each density curve is like the curve in Figure 1.17; the curve starts at height λ above $x = 0$ and decreases exponentially as x increases. The exponential distribution has been used to model many different phenomena, including time between successive arrivals at a service facility, the amount of time to complete a specified task, and the 1-hr concentration of carbon monoxide in an air sample. In Sections 1.4 and 1.5, we introduce several other important continuous distributions.

Discrete Distributions

Let's focus on a variable x whose possible values are nonnegative integers; usually the value of x results from counting something. A histogram of sample data will have rectangles centered at values 0, 1, 2, ... (or some subset of these) regardless of the sample size. However, as the sample size increases, the relative frequencies (sample proportions of various x values) tend to get closer and closer to their true population or process counterparts. We will use the following notation:

$$p(0) = \frac{\text{proportion of } x \text{ values in the population that equal 0, or the long run}}{\text{proportion of } x \text{ values in a process that equal 0}}$$

$$p(1) = \frac{\text{proportion of } x \text{ values in the population that equal 1, or the long-run}}{\text{proportion of } x \text{ values in a process that equal 1}}$$

and so on. None of these proportions can be negative, and their sum must be 1 (so that 100% of the x values are included).

DEFINITION

A population or process distribution for a discrete variable x is specified by a **mass function** $p(x)$ satisfying

$$p(x) \geq 0 \qquad \sum p(x) = 1$$

where the summation is over all possible x values. Other interesting proportions can be obtained by adding various $p(x)$ values. In particular, if a and b are integers with $a < b$, then

$$\text{proportion of } x \text{ values between } a \text{ and } b \text{ (inclusive)} = p(a) + p(a + 1) + \cdots + p(b)$$

Example 1.14

Consider a package of four batteries of a particular type, and let x denote the number of satisfactory (i.e., nondefective) batteries in the package. Possible values of x are 0, 1, 2, 3, and 4. One reasonable distribution for x is specified by the following mass function:

$$p(x) = \frac{24}{x!(4 - x)!} (.9)^x (.1)^{4-x} \qquad x = 0, 1, 2, 3, 4$$

where "!" is the factorial symbol (e.g., $4! = (4)(3)(2)(1) = 24$, $1! = 1$, and $0! = 1$). This looks a bit intimidating, but there is an intuitive argument leading to $p(x)$ that we will mention shortly. Substituting $x = 3$, we get

$$p(3) = \frac{24}{(6)(1)} (.9)^3 (.1)^1 = .2916$$

That is, roughly 29% of all packages will have three good batteries. Substituting the other x values gives us the following tabulation:

x:	0	1	2	3	4
$p(x)$:	.0001	.0036	.0486	.2916	.6561

The proportion of packages with at least two good batteries is

$$\begin{array}{l}\text{proportion of packages with } x \\ \text{values between 2 and 4 (inclusive)}\end{array} = p(2) + p(3) + p(4) = .9963$$

More than 99% of all packages have at least two good batteries.

In Section 1.6, we will generalize the distribution of Example 1.14 and introduce one additional important discrete distribution.

Section 1.3 Exercises

19. A continuous variable x is said to have a *uniform* distribution if the density function is given by

$$f(x) = \begin{cases} \dfrac{1}{b-a} & a < x < b \\ 0 & \text{otherwise} \end{cases}$$

The corresponding density "curve" has constant height over the interval from a to b. Suppose the time (min) taken by a clerk to process a certain application form has a uniform distribution with $a = 4$ and $b = 6$.
   a. Draw the density curve, and verify that the total area under the curve is indeed 1.
   b. In the long run, what proportion of forms will take between 4.5 min and 5.5 min to process? At least 4.5 min to process?
   c. What value separates the slowest 50% of all processing times from the fastest 50% (the median of the distribution)?
   d. What value separates the best 10% of all processing times from the remaining 90%?

20. Suppose that the reaction temperature x (°C) in a certain chemical process has a uniform distribution with $a = -5$ and $b = 5$ (refer to Exercise 19 for a description of a uniform distribution).
   a. In the long run, what proportion of these reactions will have a negative value of temperature?
   b. In the long run, what proportion of temperatures will be between -2 and 2? Between -2 and 3?

c. For any number k satisfying $-5 < k < k + 4 < 5$, what long-run proportion of temperatures will be between k and $k + 4$?

21. Suppose that your morning waiting time for a bus has a uniform distribution on the interval from 0 to 5 min, and your afternoon waiting time also has this distribution. Then if x denotes the total waiting time on any particular day, the density function of x can be shown to be

$$f(x) = \begin{cases} .04x & \text{for } 0 < x < 5 \\ .4 - .04x & \text{for } 5 \le x < 10 \\ 0 & \text{for other values of } x \end{cases}$$

   a. Draw the density curve, and verify that $f(x)$ specifies a legitimate distribution.
   b. In the long run, what proportion of your total daily waiting times will be at most 3 min? At least 7 min? At least 4 min? Between 4 min and 7 min?
   c. What value separates the longest 10% of your daily waiting times from the remaining 90%?

22. Data collected at Toronto Pearson International Airport suggests that an exponential distribution with $\lambda = .37$ is a good model for rainfall duration in hours (*Urban Stormwater Management Planning with Analytical Probabilistic Models*, 2000, p. 69).

a. What proportion of rainfall durations at this location are at least 2 hours? At most 3 hours? Between 2 and 3 hours?

b. What must the duration of a rainfall be to place it among the longest 5% of all times?

23. Extensive experience with fans of a certain type used in diesel engines has suggested that the exponential distribution with $\lambda = .00004$ provides a good model for time until failure (hr).

a. Sketch a graph of the density function.

b. What proportion of fans will last at least 20,000 hr? At most 30,000 hr? Between 20,000 and 30,000 hr?

c. What must the lifetime of a fan be to place it among the best 1% of all fans? Among the worst 1%?

24. The article "Probabilistic Fatigue Evaluation of Riveted Railway Bridges" (*J. of Bridge Engr.*, 2008: 237–244) suggested the exponential distribution with $\lambda = 1/6$ as a model for the distribution of stress range (MPa) in certain bridge connections.

a. What proportion of stress ranges are at least 2 MPa? At most 7 MPa? Between 5 and 10 MPa?

b. What value separates the highest 2% of the stress ranges from the remaining 98%?

25. The actual tracking weight of a stereo cartridge set to track at 3 g can be regarded as a continuous variable with density function $f(x) = c[1 - (x - 3)^2]$ for $2 < x < 4$ and $f(x) = 0$ otherwise.

a. Determine the value of c [you might find it helpful to graph $f(x)$].

b. What proportion of actual tracking weights exceed the target weight?

c. What proportion of actual tracking weights are within .25 g of the target weight?

26. Let x represent the number of underinflated tires on an automobile.

a. Which of the following $p(x)$ functions specifies a legitimate distribution for x, and why are the other two not legitimate?

   (i) $p(0) = .3,\ p(1) = .2,$
       $p(2) = .1,\ p(3) = .05,\ p(4) = .05$
   (ii) $p(0) = .4,$
       $p(1) = p(2) = p(3) = .1,\ p(4) = .3$
   (iii) $p(x) = .2(3 - x)$ for $x = 0, 1, 2, 3, 4$

b. For the legitimate distribution of part (a), determine the long-run proportion of cars having at most two underinflated tires, the proportion having fewer than two underinflated tires, and the proportion having at least one underinflated tire.

27. A mail-order computer business has six telephone lines. Let x denote the number of lines in use at a specified time. Suppose the mass function of x is given by

x:	0	1	2	3	4	5	6
$p(x)$:	.10	.15	.20	.25	.20	?	?

a. In the long run, what proportion of the time will at most three lines be in use? Fewer than three lines?

b. In the long run, what proportion of the time will at least five lines be in use?

c. In the long run, what proportion of the time will between two and four lines, inclusive, be in use?

d. In the long run, what proportion of the time will at least four lines *not* be in use?

28. A contractor is required by a county planning department to submit 1, 2, 3, 4, or 5 forms (depending on the nature of the project) when applying for a building permit. Let y denote the number of forms required for an application, and suppose the mass function is given by $p(y) = cy$ for $y = 1, 2, 3, 4,$ or 5. Determine the value of c, as well as the long-run proportion of applications that require at most three forms and the long-run proportion that require between two and four forms, inclusive.

29. Many manufacturers have quality control programs that include inspection of incoming materials for defects. Suppose a computer manufacturer receives computer boards in batches of five. Two boards are randomly selected from each batch for inspection. Consider batches for which exactly two of the boards are defective; for convenience, number the defective boards as 1 and 2, and the nondefective boards as 3, 4, and 5. Let x denote the number of defective boards among the two actually inspected, and determine the mass function of x. *Hint:* One possible sample of size 2 consists of boards 1 and 2, another of boards 1 and 3, and so on. How many such samples are there, and what is the value of x for each sample?

1.4 THE NORMAL DISTRIBUTION

The normal distribution is the most important distribution in statistics. A typical normal density curve is shown in Figure 1.18. Many population and process variables have distributions that can be very closely fit by an appropriate normal curve. Examples include heights, weights, and other physical characteristics of humans and animals, anthropometric measurements on fossils, measurement errors in scientific experiments, reaction times in psychological experiments, pollutant concentrations of various sorts, amounts dispensed into containers by machines, thicknesses of material specimens, and numerous economic measures and indicators. In addition, even when individual variables themselves are not normally distributed, sums and averages of the variables will, under suitable conditions, have approximately a normal distribution; this is the content of the Central Limit Theorem, discussed in Chapter 5.

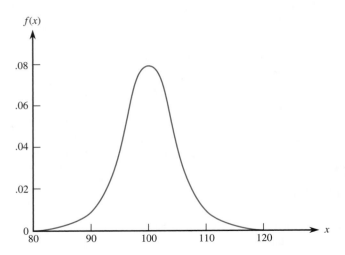

Figure 1.18 A typical normal density curve

DEFINITION

> A continuous variable x is said to have a **normal distribution with parameters μ and σ**, where $-\infty < \mu < \infty$ and $\sigma > 0$, if the density function of x is
>
> $$f(x) = \frac{1}{\sqrt{2\pi}\sigma}\, e^{-(x-\mu)^2/(2\sigma^2)} \qquad -\infty < x < \infty$$

Again, e denotes the base of the natural logarithm system and has an approximate value of 2.71828, whereas π represents the familiar mathematical constant approximately equal to 3.14159.

Clearly, $f(x) \geq 0$ for any number x, but techniques from multivariable calculus must be used to show that $\int_{-\infty}^{\infty} f(x)\, dx = 1$. The graph of $f(x)$—the density curve—is always a bell-shaped curve (and hence symmetric) centered at μ, so μ is the median of the distribution. If the value of σ is close to zero, the normal curve is highly concentrated about μ (little variability in the distribution), whereas a large value of σ corresponds to a curve that spreads out a great deal (a substantial amount of variability). Figure 1.19 displays several different normal density curves. Any normal curve has two inflection points—points at which the curve changes from being concave downward to concave upward—that are equidistant from μ. It can be shown that the value of σ is the distance from μ to each inflection point, as illustrated in Figure 1.20.

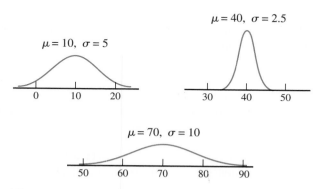

Figure 1.19 Several normal density curves

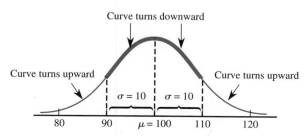

Figure 1.20 Visual identification of μ and σ

Suppose that capacitors of a certain type have resistances that vary according to a normal distribution, with $\mu = 800$ megohms and $\sigma = 200$ megohms. If a particular application requires a resistance between 775 megohms and 850 megohms, the proportion of capacitors with satisfactory values of resistance (x) is

$$\text{proportion of } x \text{ values between 775 and 85} = \int_{775}^{850} \frac{1}{\sqrt{2\pi}(200)} e^{-(x-800)^2/[2(40,000)]} \, dx$$

Unfortunately, none of the standard integration techniques can be used to evaluate this integral. To calculate proportions of this sort, a special normal reference distribution is needed.

The Standard Normal Distribution

DEFINITIONS

The normal distribution with parameter values $\mu = 0$ and $\sigma = 1$ is called the **standard normal distribution.** We shall use the letter z to denote a variable that has this distribution. The corresponding density function is

$$f(z) = \frac{1}{\sqrt{2\pi}} e^{-z^2/2} \qquad -\infty < z < \infty$$

The standard normal density curve, or **z curve,** is shown in Figure 1.21. It is centered at 0 and has inflection points at ± 1.

Appendix Table I, which also appears on the inside front cover of the book, is a tabulation of cumulative z curve areas; that is, the table gives areas under the z curve to the left of various values (to $-\infty$), as illustrated in Figure 1.21. Entries in this table were obtained by using numerical integration techniques, since the standard normal density function cannot be integrated in a straightforward way. Let's first use this table to obtain various z curve areas and other z curve information, and then see how the table applies to *any* normal curve.

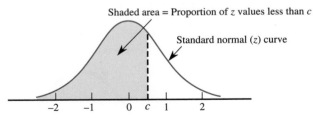

Shaded area = Proportion of z values less than c

Standard normal (z) curve

Figure 1.21 The standard normal (z) curve and a cumulative z curve area

Example 1.15   The proportion of values in a standard normal distribution that are less than 1.25 is

$$\text{proportion of } z \text{ values satisfying } z < 1.25 = \text{entry in Appendix Table I at the intersection of the 1.2 row and .05 column}$$

$$= .8944$$

It is also true that

$$\text{proportion of } z \text{ values satisfying } z \leq 1.25 = .8944$$

Similarly,

$$\text{proportion of } z \text{ values satisfying } z < -.38 = \text{entry in } -0.3 \text{ row and .08 column of Appendix Table I}$$

$$= .3520$$

Figure 1.22 illustrates the simple relationship between an upper-tail area and a cumulative area.

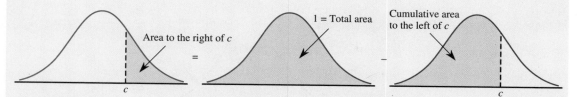

Figure 1.22 Obtaining an "area to the right" from a cumulative z curve area

In particular,

$$\begin{array}{l}\text{proportion of values} \\ \text{satisfying } z > 1.25\end{array} = \text{area under } z \text{ curve to the right of } 1.25$$

$$= 1 - \text{area to the left of } 1.25$$

$$= 1 - .8944$$

$$= .1056$$

What about the area under the z curve and above the interval between $-.38$ and 1.25? Figure 1.23 shows that this is a difference between two cumulative areas:

$$\begin{array}{l}\text{proportion of } z \text{ values} \\ \text{satisfying } -.38 < z < 1.25\end{array} = (\text{area to the left of } 1.25)$$

$$-(\text{area of the left of } -.38)$$

$$= .8944 - .3520 = .5424$$

The proportion of z values satisfying $-.38 \le z \le 1.25$ is also $.5424$.

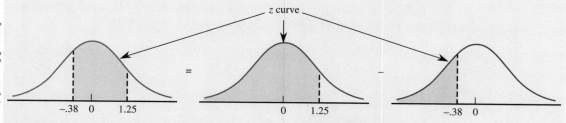

Figure 1.23 The area above an interval is the difference between two cumulative areas

In Example 1.15, a value on the horizontal z scale was specified and a curve area was determined. We now reverse this process by showing how to select a value or values to capture a specified curve area.

Example 1.16 What value c on the horizontal z axis is such that the area under the z curve to the left of c is .67? Figure 1.24 illustrates the situation.

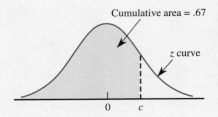

Figure 1.24 Determining c to capture a specified cumulative area

In Appendix Table I, we must look in the main body for .6700 (or the closest entry to it). The value .6700 does indeed appear; it is at the intersection of the 0.4 row and the .04 column. Thus $c = .44$. That is, 67% of the area under the z curve lies to the left of .44. Another way of expressing this is to say that .44 is the *67th percentile* of the standard normal distribution. If .6710 replaces .6700 in the question posed, the closest tabulated entry is .6700. Rather than use linear interpolation, we generally recommend simply using the closest entry to answer the question; our answer to the revised question would also be (approximately) .44.

What value c captures the upper-tail z curve area .05, as illustrated in Figure 1.25? The cumulative area to the left of c must be .9500. A search for this area in Appendix Table I reveals the following information about the two closest entries:

.9495 is in the 1.6 row and .04 column
.9505 is in the 1.6 row and .05 column

Because the desired area .9500 is halfway between the two closest entries, we use interpolation to find $c = 1.645$ (1.64 or 1.65 would also be acceptable answers).

Finally, what interval, symmetrically placed about zero, captures 95% of the area under the z curve? This situation is illustrated in Figure 1.26.

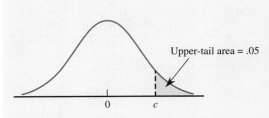

Figure 1.25 Finding the value c to capture a specified upper-tail area

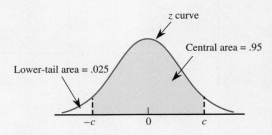

Figure 1.26 Determining c to capture a specified central z curve area

Since the lower-tail area to the left of $-c$ must be .025, the cumulative area to the left of c is $.9500 + .0250 = .9750$. This cumulative area is in the 1.9 row and .06 column of the z table, so $c = 1.96$. Alternatively, the desired lower-tail area .0250 lies in the -1.9 row and .06 column of the z table, so $-c = -1.96$ and again $c = 1.96$.

Nonstandard Normal Distributions

Any normal curve area can be obtained by first calculating a "standardized" limit or limits, and then determining the corresponding area under the z curve. The particulars are presented in the following proposition.

PROPOSITION   Let x have a normal distribution with parameters μ and σ. Then the **standardized variable**

$$z = \frac{x - \mu}{\sigma}$$

has a standard normal distribution. This implies that if we form the **standardized limits**

$$a^* = \frac{a - \mu}{\sigma} \qquad b^* = \frac{b - \mu}{\sigma}$$

then

proportion of x values satisfying $a < x < b$ = proportion of z values satisfying $a^* < z < b^*$

proportion of x values satisfying $x < a$ = proportion of z values satisfying $z < a^*$

proportion of x values satisfying $x > b$ = proportion of z values satisfying $z > b^*$ ■

Example 1.17   The time that it takes a driver to react to the brake light on a decelerating vehicle is critical in avoiding rear-end collisions. The article "Fast-Rise Brake Lamp as a Collision-Prevention Device" (*Ergonomics*, 1993: 391–395) suggests that reaction time for an in-traffic response to a brake signal from standard brake lights can be modeled with a normal distribution having parameters $\mu = 1.25$ sec and $\sigma = .46$ sec. In the long run, what proportion of reaction times will be between 1.00 sec and 1.75 sec? Let x denote reaction time. The standardized limits are

$$\frac{1.00 - 1.25}{.46} = -.54 \qquad \frac{1.75 - 1.25}{.46} = 1.09$$

Thus

$$\begin{array}{l} \text{proporiton of } x \text{ values} \\ \text{satisfying } 1.00 < x < 1.75 \end{array} = \begin{array}{l} \text{proportion of } z \text{ values satisfying} \\ -.54 < z < 1.09 \end{array}$$

$$= \begin{array}{l} \text{entry in } 1.0 \text{ row}, .09 \\ \text{column of } z \text{ table} \end{array} - \begin{array}{l} \text{entry in } -0.5 \text{ row,} \\ .04 \text{ column of } z \text{ tables} \end{array}$$

$$= .8621 - .2946$$

$$= .5675$$

This calculation is illustrated in Figure 1.27.

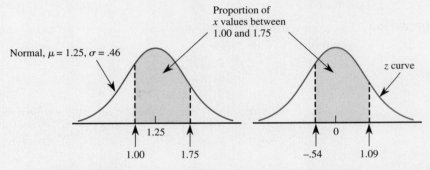

Figure 1.27 Standardizing to calculate the desired proportion in Example 1.17

Similarly, if 2 sec is viewed as a critically long reaction time, the proportion of reaction times that exceed this value is, since $(2 - 1.25)/.46 = 1.63$,

$$\begin{array}{l} \text{proportion of } x \text{ values} \\ \text{that exceed } 2.0 \end{array} = \text{proportion of } z \text{ values that exceed } 1.63$$

$$= 1 - \text{area under } z \text{ curve to the left of } 1.63$$

$$= 1 - .9484$$

$$= .0516$$

Only a bit more than 5% of all reaction times will exceed 2 sec.

Example 1.18   The amount of distilled water dispensed by a certain machine has a normal distribution with $\mu = 64$ oz and $\sigma = .78$ oz. What container size c will ensure that overflow occurs only .5% of the time? Let x denote the amount of water dispensed. The density curve for x is pictured in Figure 1.28, which shows that c captures a cumulative

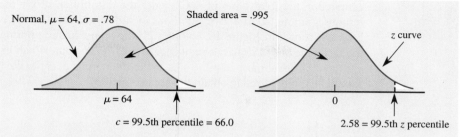

Figure 1.28 Distribution of amount dispensed and desired percentile for Example 1.18

area of .995 under this normal curve. That is, c is the 99.5th percentile of this normal distribution. Standardizing then tells us that

$$\begin{matrix}\text{proportion of } x \text{ values} \\ \text{satisfying } x < c\end{matrix} = \text{proportion of } z \text{ values satisfying } z < \frac{c - 64}{.78}$$

$$= .995$$

How can we capture cumulative area .9950 under the z curve? The 2.5 row of Appendix Table I has entries .9949 and .9951 in the .07 and .08 columns, respectively. Let's use the value 2.58 (a more detailed tabulation gives 2.576). This implies that

$$\frac{c - 64}{.78} = 2.58$$

giving

$$c = 64 + 2.58(.78) = 64 + 2.0 = 66 \text{ oz}$$

Notice that the general form of the expression for c in Example 1.18 is

$$c = \mu + (z \text{ critical value}) \, \sigma$$

where the z critical value captures the desired cumulative area under the z curve. Once we know how to capture a particular cumulative area under the z curve, it is easy to determine how to capture the same area under any other normal curve.

A histogram of sample data may suggest that a normal curve specifies a reasonable population or process distribution, but appropriate values of μ and σ still remain to be chosen. In Chapter 2, we begin to see how this can be done.

The Normal Distribution and Discrete Populations

The normal distribution is often used as an approximation to the distribution of values in a discrete population. For example, the distribution of $x = $ IQ in many populations is taken to be approximately normal with $\mu = 100$ and $\sigma = 15$, though IQ is an integer-valued variable. A picture of the population distribution consists of a histogram with rectangles centered at possible values of x. Consider the distribution of $x =$ the number of correct responses among 20 true–false questions included on a final exam. A picture of the distribution is shown in Figure 1.29 along with the approximating normal curve. Notice that the rectangle above 10 has its right edge at 10.5, so an approximation to the proportion of x values that are at most 10 is the area under the normal curve to the left of 10.5 (i.e., 10.5 should be standardized to obtain the approximation).

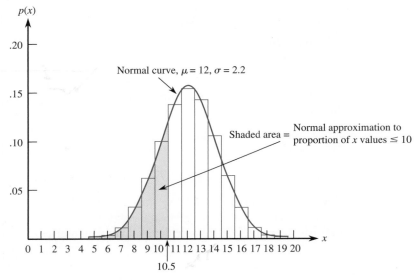

Figure 1.29 A normal approximation to the distribution of $x =$ number of correct responses on a 20-question true–false test

Section 1.4 Exercises

30. Suppose that values are repeatedly chosen from a standard normal distribution.
   a. In the long run, what proportion of values will be at most 2.15? Less than 2.15?
   b. What is the long-run proportion of selected values that will exceed 1.50? That will exceed -2.00?
   c. What is the long-run proportion of values that will be between -1.23 and 2.85?
   d. What is the long-run proportion of values that will exceed 5? That will exceed -5?
   e. In the long run, what proportion of selected values z will satisfy $|z| < 2.50$?

31. In the long run, what proportion of values selected from the standard normal distribution will satisfy each of the following conditions?

   a. Be at most 1.78   b. Exceed .55
   c. Exceed $-.80$ d. Be between .21 and 1.21
   e. Be either at most -2.00 or at least 2.00
   f. Be at most -4.2 g. Be at least 4.33

32. a. What value z^* is such that the area under the standard normal curve to the left of z^* is .9082?
   b. What value z^* is such that the area under the standard normal curve to the left of that value is .9080?
   c. What value z^* is such that the area under the standard normal curve to the right of z^* is .121?
   d. What value z^* is such that the area under the standard normal curve between $-z^*$ and z^* is .754?
   e. How far to the right of 0 would you have to go to capture an upper-tail z curve area of .002? How

far to the left of 0 would you have to go to capture this same lower-tail area?

33. Suppose that values are successively chosen from the standard normal distribution.
    a. How large must a value be to be among the largest 15% of all values selected?
    b. How small must a value be to be among the smallest 25% of all values selected?
    c. What values are among the 4% that are farthest from 0?

34. Determine the following percentiles for the standard normal distribution:
    a. 91st   b. 9th   c. 22nd   d. 99.9th

35. Suppose that the thicknesses of bolts (mm) manufactured by a certain process can be modeled with a normal distribution having $\mu = 10$ and $\sigma = 1$. *Note:* The density curve here is just the standard normal curve shifted to be centered at 10 rather than 0.
    a. What is the long-run proportion of bolts whose thicknesses are at most 11 mm? *Hint:* The corresponding normal curve area is identical to what z curve area?
    b. In the long run, what proportion of these bolts will have thickness values between 7.5 mm and 12.5 mm?
    c. In the long run, what proportion of these bolts will have thicknesses that exceed 11.5 mm?

36. Suppose the flow of current (milliamps) in wire strips of a certain type under specified conditions can be modeled with a normal distribution having $\mu = 20$ and $\sigma = 1$ (think about how the corresponding density curve relates to the standard normal curve).
    a. What proportion of strips will have a current flow of between 18.5 and 22 milliamps?
    b. What proportion of strips will have a current flow exceeding 15 milliamps?
    c. How large must a current flow be to be among the largest 5% of all flows?

37. Mopeds (small motorcycles with an engine capacity below 50 cm³) are popular in Europe because of their mobility, ease of operation, and low cost. The article "Procedure to Verify the Maximum Speed of Automatic Transmission Mopeds in Periodic Motor Vehi-

cle Inspections" (*J. of Automobile Engr.*, 2008: 1615–1623) described a rolling bench test for determining maximum vehicle speed. A normal distribution with $\mu = 46.8$ km/h and $\sigma = 1.75$ km/h is postulated.
    a. What proportion of mopeds have a maximum speed that is at most 50 km/h?
    b. What proportion of mopeds have a maximum speed that is at least 48 km/h?
    c. What speed separates the fastest 75% of all mopeds from the others?

38. Spray drift is a constant concern for pesticide applicators and agricultural producers. The inverse relationship between droplet size and drift potential is well known. The paper "Effects of 2,4-D Formulation and Quinclorac on Spray Droplet Size and Deposition" (*Weed Technology*, 2005: 1030–1036) investigated the effects of herbicide formulation on spray atomization. A figure in the paper suggested the normal distribution with $\mu = 1050$ μm and $\sigma = 150$ μm was a reasonable model for droplet size for water (the "control treatment") sprayed through a 760 ml/min nozzle.
    a. What proportion of all droplets have a size that is less than 1500 μm? At least 1000 μm?
    b. What proportion of all droplets have a size that is between 1000 and 1500 μm?
    c. How would you characterize the smallest 2% of all droplets?

39. The article "Reliability of Domestic-Waste Biofilm Reactors" (*J. of Envir. Engr.*, 1995: 785–790) suggests that substrate concentration (mg/cm³) of influent to a reactor is normally distributed with $\mu = .30$ and $\sigma = .06$.
    a. What proportion of concentration values exceed .25?
    b. What proportion of concentration values are at most .10?
    c. How would you characterize the largest 5% of all concentration values?

40. Consider babies born in the "normal range" of 37–43 weeks gestational age. Extensive data supports the assumption that for such babies born in the United States, birth weight is normally distributed with $\mu = 3432$ g and $\sigma = 482$ g. [The article "Are Babies Normal?" (*The American Statistician*, 1999: 298–302) analyzed

data from a particular year; for a sensible choice of class intervals, a histogram did not look normal but further investigation revealed that this was because some hospitals measured weight in grams and others measured to the nearest ounce and then converted the data to grams. A modified choice of class intervals that allowed for this gave a histogram that was well described by a normal distribution.]

a. For babies of this type, what proportion of all birth weights exceeds 4000 g?

b. For babies of this type, what proportion of all birth weights is between 3000 and 4000 g?

c. How would you characterize the highest .1% of all birth weights?

d. What value c is such that the interval $(3432 - c, 3432 + c)$ includes 98% of all birth weights?

41. Let x denote the number of flaws along a 100-m reel of magnetic tape (values of x are whole numbers). Suppose x has approximately a normal distribution with $\mu = 25$ and $\sigma = 5$.

a. What proportion of reels will have between 20 and 40 flaws, inclusive?

b. What proportion of reels will have at most 30 flaws? Fewer than 30 flaws?

42. Based on extensive data from an urban freeway near Toronto, Canada, "it is assumed that free speeds can best be represented by a normal distribution" ("Impact of Driver Compliance on the Safety and Operational Impacts of Freeway Variable Speed Limit Systems" (*J. of Transp. Engr.*, 2011: 260–268)). The values of μ and σ reported in the article were 119 km/h and 13.1 km/h, respectively.

a. What percentage of vehicles have speeds that are between 100 and 120 km/hr?

b. What speed characterizes the fastest 10% of all speeds?

c. The posted speed limit was 100 km/hr. What percentage of vehicles were traveling at speeds exceeding this posted limit?

d. What two values, symmetrically placed about 119, capture 90% of all vehicle speeds.

e. What values symmetrically placed about 119 separate .1% of the most extreme vehicle speeds from the rest?

1.5   OTHER CONTINUOUS DISTRIBUTIONS

Normal density curves are always bell-shaped and therefore symmetric. Exponential density curves are positively skewed but have their maximum at $x = 0$ and decrease as x increases. Many histograms of data encountered in applied work are skewed and unimodal, rising to a maximum and then declining. We now present several useful distributions that have this property. Our survey is not exhaustive. Consult the bibliography at the end of the chapter for information on the gamma, beta, and other distributions not discussed here.

The Lognormal Distribution

Lognormal distributions are related to normal distributions in exactly the way the name suggests.

DEFINITION

A nonnegative variable x is said to have a **lognormal distribution** if $\ln(x)$ has a normal distribution with parameters μ and σ. It can be shown that the density function of x is

$$f(x) = \begin{cases} \dfrac{1}{\sqrt{2\pi}\sigma x} e^{-[\ln(x)-\mu]^2/(2\sigma^2)} & x > 0 \\ 0 & \text{for } x \le 0 \end{cases}$$

Figure 1.30 illustrates density curves for several different combinations of μ and σ. Every lognormal distribution is positively skewed. The following example shows that by taking logarithms, calculation of any lognormal curve area reduces to a normal distribution computation.

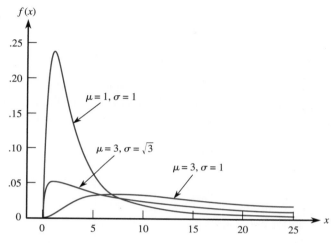

Figure 1.30 Lognormal density curves

Example 1.19 According to the article "Predictive Model for Pitting Corrosion in Buried Oil and Gas Pipelines" (*Corrosion*, 2009: 332–342), the lognormal distribution has been reported as the best option for describing the distribution of maximum pit depth data from cast iron pipes in soil. The authors suggest that a lognormal distribution with $\mu = .353$ and $\sigma = .754$ is appropriate for maximum pit depth (mm) of buried pipelines.

Since $x < 2$ is equivalent to $\ln(x) < \ln(2) = .693$,

$$\begin{aligned}
\text{proportion of pipelines} \atop \text{with } x < 2 &= \text{proportion of pipelines with } \ln(x) < .693 \\
&= \text{area under normal } (.353, .754) \text{ curve to the left of } .693 \\
&= \text{area under } z \text{ curve to the left of } (.693 - .353)/.754 \\
&= \text{area under } z \text{ curve to the left of } .45 \\
&= .6736
\end{aligned}$$

Similarly, since $\ln(1) = 0$ and $(0 - .353)/.754 = -0.47$,

$$\begin{aligned}
\text{proportion of pipelines} \atop \text{with } 1 < x < 2 &= \text{area under } z \text{ curve between } -0.47 \text{ and } 0.45 \\
&= .6736 - .3192 \\
&= .3544
\end{aligned}$$

The Weibull Distribution

This distribution was introduced in 1939 by a Swedish physicist who developed many applications over the course of the following two decades.

DEFINITION

A variable x has a **Weibull distribution** with parameters α and β if the density function of x is

$$f(x) = \begin{cases} \dfrac{\alpha}{\beta^\alpha} x^{\alpha-1} e^{-(x/\beta)^\alpha} & x > 0 \\[2mm] 0 & x \le 0 \end{cases}$$

When $\alpha = 1$, the Weibull density function reduces to the exponential density function (with $\lambda = 1/\beta$). Figure 1.31 shows several Weibull density curves. Some combinations of α and β result in a positive skew and others, a negative skew.

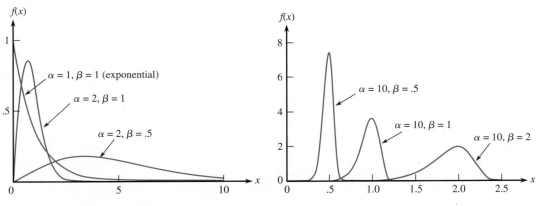

Figure 1.31 Weibull density curves

Let t represent some positive number. The proportion of x values satisfying $x < t$ is

$$\text{area under density curve to the left of } t = \int_0^t f(x)\, dx$$
$$= 1 - e^{-(t/\beta)^\alpha}$$

Thus, rather than needing a table of cumulative areas, such as the z table for normal distribution calculations, we use a simple mathematical function to get this information.

Example 1.20

In recent years the Weibull distribution has been used to model engine emissions of various pollutants. Let x denote the amount of NO_x emission (g/gal) from a certain type of four-stroke engine, and suppose that x has a Weibull distribution with $\alpha = 2$ and $\beta = 10$ (suggested by information in the article "Quantification of Variability and Uncertainty in Lawn and Garden Equipment NO_x and Total Hydrocarbon Emission Factors," *J. of the Air and Waste Management Assoc.*, 2002: 435–448). The corresponding density curve looks exactly like the one in Figure 1.31 for $\alpha = 2$,

$\beta = 1$ except that now the values 50 and 100 replace 5 and 10 on the horizontal axis (because β is a "scale parameter"). Then

$$\begin{array}{l}\text{proportion of engines emitting} \\ \text{less than 10 g/gal}\end{array} = 1 - e^{-(10/10)^2} = 1 - e^{-1} = .632$$

The proportion of engines emitting at most 25 g/gal is .998, so the distribution is almost entirely concentrated on values between 0 and 25. The value c which separates the 5% of all engines having the largest amounts of NO_x emissions from the remaining 95% satisfies

$$.95 = 1 - e^{-(c/10)^2}$$

Isolating the exponential term on one side, taking logarithms, and solving the resulting equation gives $c \approx 17.3$ as the 95th percentile of the emissions distribution.

Selecting an Appropriate Distribution

The choice of an appropriate distribution for a continuous variable x is usually based on sample data. An investigator must first decide whether a particular family, such as the Weibull family or the normal family, is reasonable. Then any parameters of the chosen family must be *estimated* to find a particular member of the family that in some sense best fits the data. These issues are considered in subsequent chapters.

Section 1.5 Exercises

43. A theoretical justification based on a certain material failure mechanism underlies the assumption that ductile strength of a material has a lognormal distribution. Suppose the values of the parameters are $\mu = 5$ and $\sigma = .1$.
    a. What proportion of material specimens have a ductile strength exceeding 120? What proportion have a ductile strength of at least 120?
    b. What proportion of material specimens have a ductile strength between 110 and 130?
    c. If the smallest 5% of strength values were unacceptable, what would be the minimum acceptable strength?

44. Nonpoint source loads are chemical masses that travel to the main stem of a river and its tributaries in flows that are distributed over relatively long stream reaches in contrast to those that enter at well-defined and regulated points. The article "Assessing Uncertainty in Mass Balance Calculation of River Nonpoint Source Loads" (*J. of Envir. Engr.*, 2008: 247–258) suggested that for a certain time period and location, x = nonpoint source load of total dissolved solids (in

kg/day/km) could be modeled with a lognormal distribution having $\mu = 9.164$ and $\sigma = .385$.
    a. What proportion of source loads are at most 15,000 kg/day/km?
    b. What interval (a, b) is such that 95% of all source loads have values in this interval, 2.5% have values less than a, and 2.5% have values exceeding b?

45. The article "Response of SiG_f/Si_3N_4 Composites Under Static and Cyclic Loading—An Experimental and Statistical Analysis" (*J. of Engr. Materials and Technology*, 1997: 186–193) suggests that tensile strength (MPa) of composites under specified conditions can be modeled by a Weibull distribution with $\alpha = 9$ and $\sigma = 180$.
    a. Sketch a graph of the density function.
    b. What proportion of specimens of this type have strength values exceeding 175?
    c. What proportion of specimens of this type have strength values between 150 and 175?
    d. What strength value separates the weakest 10% of all specimens from the remaining 90%?

46. Suppose that fracture strength (MPa) of silicon nitride braze joints under certain conditions has a Weibull distribution with $\alpha = 5$ and $\beta = 125$ (suggested by data in the article "Heat-Resistant Active Brazing of Silicon Nitride: Mechanical Evaluation of Braze Joints," *Welding J.*, August 1997: 300s–304s).
   a. What proportion of such joints have a fracture strength of at most 100? Between 100 and 150?
   b. What strength value separates the weakest 50% of all joints from the strongest 50%?
   c. What strength value characterizes the weakest 5% of all joints?

47. The Weibull distribution discussed in this section has a positive density function for all $x > 0$. In some situations, the smallest possible value of x will be some number γ that exceeds zero. A *shifted* Weibull distribution, appropriate in such situations, has a density function for $x > \gamma$ obtained by replacing x with $x - \gamma$ in the earlier density function formula. The article "Predictive Posterior Distributions from a Bayesian Version of a Slash Pine Yield Model" (*Forest Science*, 1996: 456–463) suggests that the values $\gamma = 1.3$ cm, $\alpha = 4$, and $\beta = 5.8$ specify an appropriate distribution for diameters of trees in a particular location.
   a. What proportion of trees have diameters between 2 and 4 cm?
   b. What proportion of trees have diameters that are at least 5 cm?
   c. What is the median diameter of trees, that is, the value separating the smallest 50% from the largest 50% of all diameters?

48. The paper "Study on the Life Distribution of Microdrills" (*J. of Engr. Manufacture*, 2002: 301–305) reported the following observations, listed in increasing order, on drill lifetime (number of holes that a drill machines before it breaks) when holes were drilled in a certain brass alloy.
   a. Construct a histogram of the data using class boundaries 0, 50, 100, . . . , and then comment on interesting characteristics.

b. Construct a histogram of the natural logarithms of the lifetime observations, and comment on interesting characteristics.

11	14	20	23	31	36	39	44
47	50	59	61	65	67	68	71
74	76	78	79	81	84	85	91
93	96	99	101	104	105	105	112
118	123	136	139	141	148	158	161
168	184	206	248	263	289	322	388
513							

49. The authors of the paper from which the data in the previous exercise was extracted suggested that a reasonable probability model for drill lifetime was a lognormal distribution with $\mu = 4.5$ and $\sigma = .8$.
   a. What proportion of lifetime values are at most 100?
   b. What proportion of lifetime values are at least 200? Greater than 200?

50. The article cited in Example 1.20 proposed the lognormal distribution with $\mu = 4.5$ and $\sigma = .625$ as a model for total hydrocarbon emissions (g/gal).
   a. What proportion of engines emit at least 50 g/gal? Between 50 and 150 g/gal?
   b. What value c separates the best 1% of engines with respect to THC emissions from the remaining 99%?

51. The article "On Assessing the Accuracy of Offshore Wind Turbine Reliability-Based Design Loads from the Environmental Contour Method" (*Intl. J. of Offshore and Polar Engr.*, 2005: 132–140) proposes the Weibull distribution with $\alpha = 1.817$ and $\beta = .863$ as a model for 1-hour significant wave height (m) at a certain site.
   a. What proportion of wave heights are at most 0.5 m?
   b. What proportion of wave heights are between 0.2 and 0.6 m?
   c. What is the 90th percentile of the wave height distribution? The 10th percentile?

1.6   SEVERAL USEFUL DISCRETE DISTRIBUTIONS _____

A distribution for a discrete variable x is specified by a mass function $p(x)$ satisfying $p(x) > 0$ for every possible value and $\Sigma p(x) = 1$. Here $p(0)$ is the population or long-run process proportion of x values that equal 0, $p(1)$ is the proportion of values that equal 1, and so on. We now introduce the two discrete distributions that appear most frequently in statistical applications: the binomial and the Poisson distributions.

The Binomial Distribution

Cartridges for a certain type of rollerball pen are sold two to a package. Suppose that 20% of all such cartridges leak, making them unsatisfactory, and the other 80% do not leak. Let's also assume that the condition of the second cartridge in a package—satisfactory or unsatisfactory—is *independent* of the first cartridge's condition. By this we mean that in packages with a satisfactory first cartridge, 80% of the second cartridges are satisfactory, and in packages with an unsatisfactory first cartridge, 80% of the second cartridges are satisfactory. In other words, the percentage of satisfactory second cartridges is not affected by the condition of the first cartridge. We will use SS to denote a package with two satisfactory cartridges, and SF to denote a package with a satisfactory first cartridge and an unsatisfactory second cartridge (S for success and F for failure). Then 80% of all packages will have a first S, and of these, a further 80% will have a second S, giving 80% of 80% or 64% SS's. Similarly, of the 80% of all packages that have a first S, 20% will have a second cartridge that is an F, so 20% of 80% or 16% of all packages will be SF's. This is also the percentage of all packages that are FS's: 80% of 20% or 16%. Finally, 20% of 20% or 4% of all packages are FF's. Notice that these percentages result from multiplying pairs of proportions:

SS: $(.8)(.8) = .64$ or 64%

SF: $(.8)(.2) = .16$ or 16%

FS: $(.2)(.8) = .16$ or 16%

FF: $(.2)(.2) = .04$ or 4%

Now let x be the number of S's in a package. Possible values of x are 0, 1, and 2. Our calculations imply that the proportion of all packages with $x = 0$ is .04 and the proportion of all packages with $x = 2$ is .64. Because 16% of all packages are SF's and 16% are FS's,

(proportion of all packages with $x = 1$) $= .16 + .16 = .32$

That is, in the long run, 32% of all packages will have $x = 1$ (this also comes from $1 - .04 - .64$).

Suppose instead that cartridges come in packages of four. Again let x be the number of S's in a package. One way to get a package with $x = 2$ is SSFF, and by independence, the percentage of all such packages is 20% of 20% of 80% of 80% or $100[(.8)(.8)(.2)(.2)] = 2.56\%$, or a proportion of .0256. But there are in fact five other ways, for a total of six possibilities:

Outcome for which $x = 2$	Proportion
SSFF (S in 1 and 2)	$(.8)(.8)(.2)(.2) = .0256$
SFSF (S in 1 and 3)	$(.8)(.2)(.8)(.2) = .0256$
SFFS (S in 1 and 4)	.0256
FSSF (S in 2 and 3)	.0256
FSFS (S in 2 and 4)	.0256
FFSS (S in 3 and 4)	.0256

The population or long-run process proportion of packages having $x = 2$ is then the sum of these six values of .0256, or $6(.0256) = .1536$. Similarly, there are four possibilities for $x = 1$—the single satisfactory cartridge could be the first, second, third, or fourth one in the package. The proportion of SFFF's is $(.8)(.2)(.2)(.2) = .0064$, which is also the proportion of FSFF's, FFSF's, and FFFS's. Adding .0064 four times gives

$$(\text{proportion of packages with } x = 1) = 4(.8)(.2)^3 = .0256$$

By the same reasoning,

$$(\text{proportion of packages with } x = 3) = 4(.8)^3(.2) = .4096$$

so roughly 41% of all packages will have three satisfactory cartridges.

What if packages have ten cartridges, and you want to know what proportion have six S's? It is extremely tedious to list all possibilities, but fortunately this is unnecessary. There is a straightforward counting technique to determine the number of possible outcomes having any particular x value.

The Binomial Distribution

Suppose that items or entities of some sort come in batches or groups of size n. Let π denote the proportion of all items in the population or process that are satisfactory (S, for success), so the proportion of all items that are unsatisfactory (F, for failure) is $1 - \pi$. Assume that the condition of any particular item (S or F) is independent of that of any other item. The **binomial variable** x is the number of S's in a batch or group. The mass function of x is given by the formula

$p(x) = $ proportion of batches with x S's

$$= \frac{n!}{x!(n-x)!} \times \pi^x(1 - \pi)^{n-x} \qquad x = 0, 1, \ldots, n$$

In the case of a population, the formula gives good approximations as long as the total number of items examined in all batches is at most 5% of the population size (answers are exact if the population size is infinite). For a process, it is required that the value of π remain constant over time (a *stable* process).

In the mass function formula, $\pi^x(1 - \pi)^{n-x}$ generalizes the multiplications $(.8)^3(.2)$ and $(.8)^2(.2)^2$ in the pen cartridge example. The factorial expression is the number of possible outcomes for a batch of size n that have x S's. For example, when $n = 4$ and $x = 2$,

$$\frac{n!}{x!(n-x)!} = \frac{4!}{(2!)(2!)} = \frac{(4)(3)(2)(1)}{(2)(1)(2)(1)} = \frac{(4)(3)}{(2)(1)} = 6$$

as we saw previously. You can find a derivation of this formula in several of the references listed in the bibliography.

Example 1.21   The binomial distribution is used extensively in genetic applications. An early genetics article ("The Progeny in Generation F_{12} to F_{17} of a Cross Between a Yellow-Wrinkled and a Green-Round Seeded Pea," *J. of Genetics*, 1923: 255–331) reported on an experiment in which four-seeded pea pods from a dihybrid cross were examined. The variable of interest was x = the number of YR (yellow *and* round) peas in a pod. Mendelian laws of inheritance imply that $\pi = 9/16 = .5625$ [from (3/4)(3/4)]. Now consider peas with eight-seeded pods. The proportion of all pods with five YR peas is

$$(\text{proportion with } x = 5) = \frac{8!}{(5!)(3!)}(.5625)^5(.4375)^3$$

$$= 56(.5625)^5(.4375)^3 = 2641$$

The proportion of all pods with at least five such peas is

$$(\text{proportion with } x \geq 5) = p(5) + p(6) + p(7) + p(8)$$

$$= .2641 + .1698 + .0624 + 0.100 = .5063$$

In the long run, slightly more than 50% of all pods will have five or more YR peas and slightly less than 50% will have four or fewer YR peas. The complete distribution of x is as follows:

x:	0	1	2	3	4	5	6	7	8
$p(x)$:	.0013	.0138	.0621	.1598	.2567	.2641	.1698	.0624	.0100

Figure 1.32 shows a picture of this distribution. The binomial histogram has a slight negative skew (it is symmetric only when $\pi = .5$).

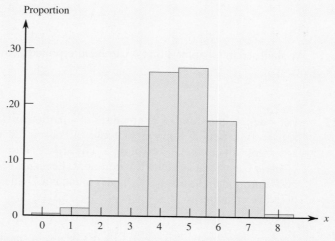

Figure 1.32 A binomial histogram when $n = 8$ and $\pi = .5625$

Use of the binomial distribution formula can be tedious when n is large. Appendix Table II gives a tabulation of $p(x)$ for a few selected values of n and π. This will allow you to practice binomial calculations without referring to the formula. Alternatively, values of $p(x)$ for any n and π can be obtained from Minitab and other statistical computer packages.

The Poisson Distribution

The Poisson distribution is usually used as a model for the number of times an "event" of some sort occurs during a specified time period or in a particular region of space. Examples include the number of accidents that occur on a segment of highway during a particular 24-hour period, the number of blemishes on the exterior of a new automobile, the number of customers in a grocery store's express line on Wednesday at 6 P.M., and the number of plants of a particular species that are found in a chosen geographic sampling region.

The Poisson Distribution

The **Poisson mass function** is

$$p(x) = \frac{e^{-\lambda}\lambda^x}{x!} \qquad x = 0, 1, 2, 3, \ldots$$

where the parameter λ must satisfy $\lambda > 0$.

The condition $p(x) \geq 0$ is clearly satisfied. The fact that $\sum_{x=0}^{\infty} p(x) = 1$ is a consequence of multiplying both sides of the following infinite series expansion by $e^{-\lambda}$:

$$e^{\lambda} = 1 + \lambda + \frac{\lambda^2}{2!} + \frac{\lambda^3}{3!} + \cdots$$

We shall see in Chapter 2 that λ can be interpreted as the average rate at which events occur.

Example 1.22   Let x denote the number of creatures of a particular type captured in a trap during a given time period. Suppose that x has a Poisson distribution with 4.5, so, on average, traps will contain 4.5 creatures. [The article "Dispersal Dynamics of the Bivalve *Gemma Gemma* in a Patchy Environment (*Ecological Monographs*, 1995: 1–20) suggests this model; the bivalve *Gemma gemma* is a small clam]. The proportion of traps with five creatures is

$$(\text{proportion with } x = 5) = \frac{e^{-4.5}(4.5)^5}{5!} = .1708$$

The proportion of traps having at most five creatures is

(proportion with $x \le 5$) = $p(0) + p(1) + \cdots + p(5)$ = .7029 (roughly 70%)

so the proportion of traps with at least six creatures is $1 - .7029 = .2971$. As x increases, $p(x)$ decreases but never quite reaches zero. The proportions for the first 13 x values follow; their sum is .9992. Figure 1.33 shows the corresponding Poisson histogram.

x:	0	1	2	3	4	5	6
$p(x)$:	.0111	.0500	.1125	.1687	.1898	.1708	.1281

x:	7	8	9	10	11	12
$p(x)$:	.0824	.0463	.0232	.0104	.0043	.0016

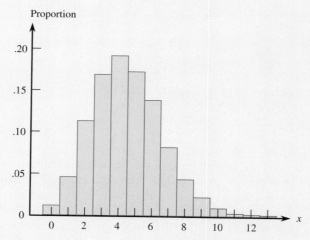

Figure 1.33 Poisson histogram when $\lambda = 4.5$

A small tabulation of the Poisson mass function for selected values of λ appears in Appendix Table III.

The Poisson Approximation to the Binomial Distribution

Often a binomial scenario involves a group size n that is quite large in combination with a success proportion π close to zero. Under such circumstances, the binomial mass function can be well approximated by the Poisson mass function with $\lambda = n\pi$. In particular, if $n \ge 100$, $\pi \le .01$, and $\lambda = n\pi \le 20$, then

$$\frac{n!}{x!(n-x)!} \pi^x (1-\pi)^{n-x} \approx \frac{e^{-\lambda}\lambda^x}{x!}$$

A more formal statement of this result is that the Poisson mass function on the right-hand side is the limit of the binomial function on the left as $n \to \infty$, $\pi \to 0$ in such a way that $n\pi \to \lambda$.

Example 1.23   Components of a certain type are shipped from a supplier to customers in lots of 5000. Because the purchaser cannot check the condition of each component, a sample of 25 is selected and tested. The entire lot will then be accepted only if the number of components x that *do not* conform to specification is at most three (so here S's are nonconforming units, not what we usually think of as a success). Suppose that .5% of all components are nonconforming, giving $\lambda = 100(.005) = .5$. Then the proportion of acceptable lots is

proportion of lots with $x \leq 3$

$$= p(0) + p(1) + p(2) + p(3)$$

$$= \frac{100!}{0!100!}(.005)^0(.995)^{100} + \cdots + \frac{100!}{3!97!}(.005)^3(.995)^{97}$$

$$\approx \frac{e^{-.5}(.5)^0}{0!} + \cdots + \frac{e^{-.5}(.5)^3}{3!}$$

$$= .6065 + .3033 + .0758 + .0126$$

$$= .9982$$

The exact proportion using the binomial mass function is $.6058 + .3044 + .0757 + .0124 = .9983$.

Many applications of the Poisson distribution are in fact based on an underlying binomial situation without the values of n and π being stated explicitly. For example, a very large number of vehicles may pass over a given stretch of highway during a particular time period, but the long-run proportion of vehicles receiving speeding tickets will be quite small, so the number of ticketed vehicles will have at least approximately a Poisson distribution.

Section 1.6 Exercises

52. When circuit boards used in the manufacture of compact disc players are tested, the long-run percentage of defectives is 5%. Let x denote the number of defective boards in a batch of 25 boards, so that x has a binomial distribution with $n = 25$ and $\pi = .05$.
    a. What proportion of batches have at most 2 defective boards?
    b. What proportion of batches have at least 5 defective boards?
    c. What proportion of batches will have all 25 boards free of defects?

53. A company packages its crystal goblets in boxes containing six goblets. Suppose that 12% of all its goblets have cosmetic flaws and that the condition of any particular goblet with respect to flaws is independent of the condition of any other goblet.
    a. What proportion of boxes will contain only one goblet with a cosmetic flaw?
    b. What proportion of boxes will contain at least two goblets with cosmetic flaws?
    c. What proportion of boxes will have between one and three goblets, inclusive, with cosmetic flaws?

54. On his way to work, a friend of ours must pass through ten traffic signals. Suppose that in the long run, she encounters a red light at 40% of these signals and that

whether any particular signal is red is independent of whether any other one is red.

  a. On what proportion of days will our friend encounter at most two red lights? At least five red lights?

  b. On what proportion of days will our friend encounter between two and five (inclusive) red lights?

55. Suppose that 10% of all bits transmitted through a digital communication channel are erroneously received and that whether any particular bit is erroneously received is independent of whether any other bit is erroneously received. Consider sending a very large number of messages, each consisting of 20 bits.

  a. What proportion of these messages will have at most 2 erroneously received bits?

  b. What proportion of these messages will have at least 5 erroneously received bits?

  c. For what proportion of these messages will more than half the bits be erroneously received?

56. Components arrive at a distributor in very large batches. A batch can be characterized as acceptable only if the fraction of defective components in the batch is at most .10. The distributor decides to randomly select ten components from the batch, test each one, and accept the batch only if the sample contains at most two defective components. Assume that the condition of any particular component is independent of any other.

  a. If the actual fraction of defectives in each batch is only $\pi = .01$, what proportion of batches will be accepted? Repeat this calculation for the following values of π: .05, .10, .20, and .25.

  b. A graph of the proportion of batches accepted versus the actual fraction of defectives π is called the *operating characteristic curve*. Use the results of part (a) to sketch this curve for $0 \leq \pi \leq 1$ (proportion of batches accepted is on the vertical axis and π is on the horizontal axis).

  c. Suppose the distributor decides to be more demanding by accepting a batch only if the sample contains at most one defective component. Repeat parts (a) and (b) with

this new acceptance sampling plan. Does this plan appear more satisfactory than the original plan?

57. Suppose that the number of drivers who travel between a particular origin and destination during a designated time period has a Poisson distribution with parameter $\lambda = 20$ (suggested in the article "Dynamic Ride Sharing: Theory and Practice," *J. of Transp. Engr.*, 1997: 308–312). In the long run, in what proportion of time periods will the number of drivers

  a. Be at most 10?

  b. Exceed 20?

  c. Be between 10 and 20, inclusive? Be strictly between 10 and 20?

58. Let x be the number of material anomalies occurring in a particular region of an aircraft gas-turbine disk. The article "Methodology for Probabilistic Life Prediction of Multiple-Anomaly Materials" (*Amer. Inst. of Aeronautics and Astronautics J.*, 2006: 787–793) proposes a Poisson distribution for x. Suppose that $\lambda = 4$.

  a. What proportion of gas-turbine disks have exactly one anomaly?

  b. What proportion of gas-turbine disks have at least three anomalies?

  c. What proportion of gas-turbine disks have between one and six anomalies inclusive?

59. Let x denote the number of trees in a quarter-acre plot within a certain forest. Suppose that x has a Poisson distribution with $\lambda = 20$ (corresponding to an average density of 80 trees per acre). In what proportion of such plots will there be at least 15 trees? At most 25 trees?

60. An article in the *Los Angeles Times* (Dec. 3, 1993) reports that 1 in 200 people carry the defective gene that causes colon cancer. Let x denote the number of people in a group of size 1000 who carry this defective gene. What is the approximate distribution of x? Use this approximate distribution to determine the proportion of all such groups having at least 8 people who carry the defective gene, as well as the proportion of all such groups for which between 5 and 10 people (inclusive) carry the defective gene.

Supplementary Exercises

61. The accompanying frequency distribution of fracture strength (MPa) observations for ceramic bars fired in a particular kiln appeared in the article "Evaluating Tunnel Kiln Performance" (*Amer. Ceramic Soc. Bull.*, August 1997: 59–63).

Class:	$81-$ <83	$83-$ <85	$85-$ <87	$87-$ <89	$89-$ <91
Frequency:	6	7	17	30	43

Class:	$91-$ <93	$93-$ <95	$95-$ <97	$97-$ <99
Frequency:	28	22	13	3

   a. Construct a histogram based on relative frequencies, and comment on any interesting features.
   b. What proportion of the strength observations are at least 85? Less than 95?
   c. Roughly what proportion of the observations are less than 90?

62. The article cited in Exercise 61 presented compelling evidence for assuming that fracture strength (MPa) of ceramic bars fired in a particular kiln is normally distributed (while commenting that the Weibull distribution is traditionally used as a model). Suppose that $\mu = 90$ and $\sigma = 3.75$, which is consistent with data given in the article.
   a. In the long run, what proportion of bars would have strength values less than 90? Less than 95? At least 95?
   b. In the long run, what proportion of bars would have strength values between 85 and 95? Between 80 and 100?
   c. What value is exceeded by 90% of the fracture strengths for all such bars?
   d. What interval centered at 90 includes 99% of all fracture strength values?

63. Once an individual has been infected with a certain disease, let x represent the time (days) that elapses before the individual becomes infectious. The article "The Probability of Containment for Multitype Branching Process Models for Emerging Epidemics" (*J. of Applied Probability*, 2011: 173–188) proposes a Weibull distribution with $\alpha = 2.2$ and $\beta = 1.1$ for $x - .5$ (i.e. the Weibull density curve is shifted to the right of 0 by .5; Minitab refers to .5 as the value of the *threshold parameter*).

   a. What proportion of elapsed times exceed 1.5 days?
   b. What is the 90th percentile of the elapsed time distribution?

64. Let x denote the distance (m) that an animal moves from its birth site to the first territorial vacancy it encounters. Suppose that for banner-tailed kangaroo rats, x has an exponential distribution with parameter $\lambda = .01386$ (as suggested in the article "Competition and Dispersal from Multiple Nests," *Ecology*, 1997: 873–883).
   a. What proportion of distances are at most 100 m? At most 200 m? Between 100 m and 200 m?
   b. What proportion of distances are at least 50 m?
   c. What is the median distance, that is, the value that separates the smallest 50% of all distances from the largest 50%?

65. Suppose the unloading time x (centiminutes) of a forwarder in a harvesting operation could be assumed to be lognormal with $\mu = 6.5$ and $\sigma = .75$, as suggested in the article "Simulating a Harvester-Forwarder Softwood Thinning" (*Forest Products J.*, May 1997: 36–41).
   a. What proportion of unloading times exceed 1000? 2000? 3000?
   b. What proportion of times are between 2500 and 5000?
   c. What value characterizes the fastest 10% of all times?
   d. Sketch a graph of the density function of x. Is the positive skewness quite pronounced?

66. In an experiment, 25 laminated glass units configured in a particular way are subjected to an impact test (cf. "Performance of Laminated Glass Units Under Simulated Windborne Debris Impacts," *J. of Architectural Engr.*, 1996: 95–99). We are interested in the number of units that sustain an inner glass ply fracture. Suppose that the long-run proportion of all such units that fracture is .20. In the long run, for what proportion of such experiments will the number of fractures be
   a. At least 10?
   b. At most 5?
   c. Between 5 and 10 inclusive?
   d. Strictly between 5 and 10?

67. Airlines frequently overbook flights. Suppose that for a plane with 100 seats, an airline takes 110 reservations. Let x represent the number of people with reservations who actually show up for a sold-out flight. From past experience, we know that the distribution of x is as follows:

x:	95	96	97	98	99	100	101	102	103
$p(x)$:	.05	.10	.12	.14	.24	.17	.06	.04	.03

x:	104	105	106	107	108	109	110
$p(x)$:	.02	.01	.005	.005	.005	.0037	.0013

   a. For what proportion of such flights is the airline able to accommodate everyone who shows up for the flight?
   b. For what proportion of all such flights is it not possible to accommodate all passengers?
   c. For someone who is trying to get a seat on such a flight and is number 1 on the standby list, what proportion of the time is such an individual able to take the flight? Answer the question for individuals who are number 3 on the standby list.

68. The accompanying data are observations on shower flow rate for a sample of 129 houses in Perth, Australia ("An Application of Bayes Methodology to the Analysis of Diary Records in a Water Use Study," *J. Amer. Stat. Assoc.*, 1987: 705–711):

4.6	12.3	7.1	7.0	4.0	9.2	6.7	6.9
11.5	5.1	3.8	11.2	10.5	14.3	8.0	8.8
6.4	5.1	5.6	9.6	7.5	7.5	6.2	5.8
2.3	3.4	10.4	9.8	6.6	3.7	6.4	6.0
8.3	6.5	7.6	9.3	9.2	7.3	5.0	6.3
13.8	6.2	5.4	4.8	7.5	6.0	6.9	10.8
7.5	6.6	5.0	3.3	7.6	3.9	11.9	2.2
15.0	7.2	6.1	15.3	18.9	7.2	5.4	5.5
4.3	9.0	12.7	11.3	7.4	5.0	3.5	8.2
8.4	7.3	10.3	11.9	6.0	5.6	9.5	9.3
10.4	9.7	5.1	6.7	10.2	6.2	8.4	7.0
4.8	5.6	10.5	14.6	10.8	15.5	7.5	6.4
3.4	5.5	6.6	5.9	15.0	9.6	7.8	7.0
6.9	4.1	3.6	11.9	3.7	5.7	6.8	11.3
9.3	9.6	10.4	9.3	6.9	9.8	9.1	10.6
4.5	6.2	8.3	3.2	4.9	5.0	6.0	8.2
6.3							

   a. Construct a stem-and-leaf display of the data.
   b. What is a typical or representative flow value? Does the data appear to be highly concentrated or quite spread out about this typical value?
   c. Does the distribution of values appear to be reasonably symmetric? If not, how would you describe the departure from symmetry?
   d. Does the data set appear to contain any outliers?
   e. Construct a histogram using class boundaries 2, 3, 4, 5, 6, 7, 8, 9, 10, 12, 14, 16, and 20. From your histogram, approximately what proportion of the observations are at most 11? Compare this with the exact proportion that are at most 11.

69. Let x denote the vibratory stress (psi) on a wind turbine blade at a particular wind speed in a wind tunnel. The article "Blade Fatigue Life Assessment with Applications to VAWTS" (*J. of Solar Energy Engr.*, 1982: 107–111) proposes the *Rayleigh distribution* as a model; the density function is

$$f(x) = \begin{cases} \dfrac{x}{\theta^2} \cdot e^{-x^2/(2\theta^2)} & x > 0 \\ 0 & \text{otherwise} \end{cases}$$

   a. Verify that $f(x)$ is a legitimate density function.
   b. Suppose that $\theta = 100$ (a value suggested by a graph in the cited article). What proportion of vibratory stress values will be at most 200? At least 200? Between 100 and 200?

70. The article "Error Distribution in Navigation" (*J. Institute of Navigation*, 1971: 429–442) suggests that the frequency distribution of positive errors (magnitudes of errors) is well approximated by an exponential distribution. Let x denote the lateral position error (nautical miles), which can be either positive or negative, and suppose the density function of x is $f(x) = (.1)e^{-.2|x|}$ for $-\infty < x < \infty$.
   a. Sketch the corresponding density curve, and verify that $f(x)$ is a legitimate density function.
   b. What proportion of errors are negative? At most 2? Between -1 and 2?

71. "Time headway" in traffic flow is the elapsed time between the time that one car finishes passing a fixed point and the instant that the next car begins to pass that point. Let x be the time headway (sec) for two consecutive cars on a freeway during a period of heavy flow. The following density function is essentially the

one suggested in "The Statistical Properties of Freeway Traffic" (*Transportation Research*, 1977: 221–228):

$$f(x) = \begin{cases} .15e^{-.15(x-.5)} & x > .5 \\ 0 & \text{otherwise} \end{cases}$$

a. Sketch the corresponding density curve, and verify that $f(x)$ is a legitimate density function.

b. What proportion of time headways are at most 5 sec? Between 5 and 10 sec?

c. What value separates the smallest 50% of all time headways from the largest 50%?

d. What value characterizes the largest 10% of all time headways?

72. A *k-out-of-n* system is one that will function if and only if at least k out of the n individual components in the system function. If individual components function independently of one another and the long-run proportion of components that function is .9, what is the long-run proportion of 3-out-of-5 systems that will function?

73. An insurance company offers its policyholders a number of different premium payment options. Let x denote the number of months between successive payments chosen by a policyholder. For any particular number k, the proportion of x values that are at most k (i.e., $\leq k$) is called a *cumulative* proportion. Consider the following cumulative proportions: 0 for $x < 1$, .30 for $1 \leq x < 3$, .40 for $3 \leq x < 4$, .45 for $4 \leq x < 6$, .60 for $6 \leq x < 12$, and 1 for $x \geq 12$.

a. Graph this cumulative proportion function, that is, graph (proportion of x values $\leq k$) versus k.

b. Determine the mass function of x. *Hint:* The cumulative proportion function jumps only at possible values of x.

c. Use the cumulative proportion function to determine the proportion of all policyholders for which $3 \leq x \leq 6$, and check to see that the mass function gives this same proportion.

74. Based on data from a dart-throwing experiment, the article "Shooting Darts" (*Chance*, Summer 1997, 16–19) proposed that the horizontal and vertical errors from aiming at a point target should be independent of one another, each with a normal distribution having parameters $\mu = 0$ and σ. It can then be shown that the density function of the distance from the target to the landing point is

$$f(v) = \frac{v}{\sigma^2} \cdot e^{-v^2/2\sigma^2} \qquad v > 0$$

a. This pdf is a member of what family introduced in this chapter?

b. If $\sigma = 20$ mm (close to the value suggested in the paper), what proportion of darts will land within 25 mm (roughly 1 in.) of the target?

75. The bursting strength of wine bottles of a certain type is normally distributed with parameters $\mu = 250$ psi and $\sigma = 30$ psi. If these bottles are shipped 12 to a carton, in what proportion of cartons will at least one of the bottles have a bursting strength exceeding 300 psi? *Hint:* Think of a bottle as a success S if its bursting strength exceeds 300 psi.

Bibliography

Chambers, John, William Cleveland, Beat Kleiner, and Paul Tukey, **Graphical Methods for Data Analysis,** Wadsworth, Belmont, CA, 1983. *A very readable source for information on constructing histograms, checking the plausibility of various distributions, and other visual techniques.*

Cleveland, William, **The Elements of Graphing Data** (2nd ed.), Hobart Press, Summit, NJ, 1994. *An informal and informative introduction to various aspects of graphical analysis.*

Johnson, Norman, Samuel Kotz, and Adrienne Kemp, **Univariate Discrete Distributions** (2nd ed.), Wiley, New York, 1992. *A veritable encyclopedia of information on discrete distributions.*

Johnson, Norman, Samuel Kotz, and N. Balakrishnan, **Continuous Univariate Distributions** (vol. 1, 2nd ed.), Wiley, New York, 1994. *An encyclopedic reference for continuous distributions.*

Olkin, Ingram, Cyrus Derman, and Leon Gleser, **Probability Models and Applications** (2nd ed.), Macmillan, New York, 1994. *Contains in-depth discussions of both general properties of discrete and continuous distributions and results for specific distributions.*

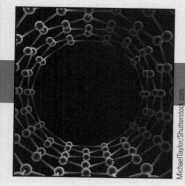

MichaelTaylor/Shutterstock.com

2

Numerical Summary Measures

INTRODUCTION

In Chapter 1, we learned how to describe sample data using either a stem-and-leaf display or a histogram. We then saw how a density function or mass function could be used to represent the distribution of a variable x in an entire population or process. Often an investigator will want to obtain or convey information about particular characteristics of data. In this chapter, we first introduce several numerical summary measures that describe where a sample or distribution is centered. Another important aspect of a sample or distribution is the extent of spread about the center. In Section 2.2, we develop the most useful measures of variability. In Section 2.3, we consider more detailed data summaries and how they can be combined to yield concise yet informative data descriptions. Once sample data has been obtained, it is often important to know whether it is plausible that the data came from a particular type of distribution, such as a normal distribution or a Weibull distribution. In Section 2.4, we show how to construct a picture from which the plausibility of any particular type of underlying distribution can be judged.

2.1   MEASURES OF CENTER

A preliminary sense of where a data set is centered can be gleaned from a stem-and-leaf display or a histogram. A precise quantitative assessment entails calculating a measure of center such as the mean or median; the resulting number can then be regarded as being representative or typical of the data. First, we consider measures of center for sample data, and then we turn our attention to analogous measures for distributions of a numerical variable x.

Measures of Center for Data

Suppose that the sample consists of observations on a numerical variable x. We shall use the letter n to represent the sample size (number of observations in the sample, e.g., $n = 10$). The individual observations will be denoted by $x_1, x_2, \ldots, x_n$. The subscripts typically refer to the time order in which the observations were obtained—the first observation is x_1, the second observation is x_2, and so on. In general, the subscripts are unrelated to the magnitudes of the observations: x_1 is not usually the smallest observation, nor is x_n the largest sample value.

The Sample Mean

The most frequently used measure of center is simply the arithmetic average of the n observations.

DEFINITION

> The **sample mean** of observations $x_1, \ldots, x_n$, denoted by $\bar{x}$, is given by
>
> $$\bar{x} = \frac{x_1 + x_2 + \cdots + x_n}{n} = \frac{\sum_{i=1}^{n} x_i}{n}$$
>
> The numerator of $\bar{x}$ can be written more informally as Σx_i, where the summation is over all sample observations.

For reporting $\bar{x}$, we recommend using decimal accuracy of one digit more than the accuracy of the x_i's. Thus if observations are stopping distances with $x_1 = 125$, $x_2 = 131$, and so on, we might have $\bar{x} = 127.3$ ft.

Example 2.1   In recent years there has been growing commercial interest in the use of what is known as *internally cured concrete*. This concrete contains porous inclusions most commonly in the form of lightweight aggregate (LWA). In the article "Characterizing Lightweight Aggregate Desorption at High Relative Humidities Using a Pressure Plate Apparatus" (*J. of Materials in Civil Engr.*, 2012: 961–969), researchers examined various physical properties of 14 LWA specimens. The following are the 24-hour water absorption percentages for the 14 specimens:

$$x_1 = 16.0 \quad x_2 = 30.5 \quad x_3 = 17.7 \quad x_4 = 17.5 \quad x_5 = 14.1$$
$$x_6 = 10.0 \quad x_7 = 15.6 \quad x_8 = 15.0 \quad x_9 = 19.1 \quad x_{10} = 17.9$$
$$x_{11} = 18.9 \quad x_{12} = 18.5 \quad x_{13} = 12.2 \quad x_{14} = 6.0$$

Figure 2.1 shows a stem-and-leaf display of the data (the tenths digit is truncated); a water absorption percentage in the midteens appears to be "typical." With $\Sigma x_i = 229.0$, the sample mean is $\bar{x} = \frac{229.0}{14} = 16.36$, a value consistent with information conveyed by the stem-and-leaf display.

```
0H   6
1L   024
1H   556777889
2L
2H
3L   0
```

Figure 2.1 A stem-and-leaf display of the water absorption data

The mean suffers from one deficiency that makes it an inappropriate measure of center under some circumstances: Its value can be greatly affected by the presence of even a single outlier (unusually large or small observation). In Example 2.1, the value $x_2 = 30.5$ is obviously an outlier. Without this observation, $\bar{x} = 15.27$; the outlier increases the mean by more than 1%. If the 30.5 observation were replaced by the relatively large value 90.0, a really extreme outlier, then $\bar{x} = 288.5/14 = 20.61$, which is larger than any of the other observations!

The Sample Median

An alternative measure of center that resists the effects of outliers is the median. The median strip of a roadway divides the roadway into two equal parts, and the sample median does the same for the sample. If, for example, $n = 5$ and the observations are ordered from smallest to largest, the third observation from either end is the median. When $n = 6$, though, there are two middle values in the ordered list; the median is the average of these two values.

DEFINITION

The **sample median,** denoted by $\tilde{x}$, is obtained by first ordering the sample observations from smallest to largest. Then

$$\tilde{x} = \begin{cases} \text{single middle value} = \left(\dfrac{n+1}{2}\right)\text{th value on ordered list} & n \text{ odd} \\ \begin{array}{l} \text{average of two} \\ \text{middle values} \end{array} = \text{average of } \dfrac{n}{2}\text{th and } \left(\dfrac{n}{2}+1\right)\text{th values} & n \text{ even} \end{cases}$$

Example 2.2

People not familiar with classical music might tend to believe that a composer's instructions for playing a particular piece are so specific that the duration would not depend at all on the performer(s). However, there is typically plenty of room for interpretation, and orchestral conductors and musicians take full advantage of this. We went to the website ArkivMusic.com and selected a sample of 12 recordings of

Beethoven's stunningly beautiful Symphony No. 9 (the "Chorale"), and found the following durations (min) listed in increasing order:

62.3 62.8 63.6 65.2 65.7 66.4 67.4 68.4 68.8 70.8 75.7 79.0

Figure 2.2 is a dotplot of the data:

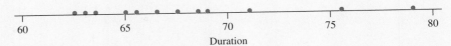

Figure 2.2 Dotplot of the data from Example 2.2

Since $n = 12$ is even, the sample median is the average of the $n/2 = $ sixth and $(n/2 + 1) = $ seventh values from the ordered list:

$$\tilde{x} = \frac{66.4 + 67.4}{2} = 66.90$$

Note that if the largest observation 79.0 had not been included in the sample, then the resulting sample median for the $n = 11$ remaining observations would have been the single middle value 67.4 [the $(n + 1)/2 = $ sixth ordered value—i.e., the sixth value in from either end of the ordered list]. The sample mean is $\bar{x} = \Sigma x_i = 816.1/12 = 68.01$, a bit more than a full minute larger than the median. The mean is pulled out a bit relative to the median because the sample "stretches out" somewhat more on the upper end than on the lower end.

The largest observation or even the largest two or three observations in Example 2.2 can be increased by an arbitrary amount without impacting $\tilde{x}$. Similarly, decreasing several of the smallest observations by any amount does not affect the median. In contrast to $\bar{x}$, the median is impervious to many outliers.

Trimmed Means

A trimmed mean is a compromise between $\bar{x}$ and $\tilde{x}$; it is less sensitive to outliers than the mean but more sensitive than the median. The observations are again first ordered from smallest to largest. Then a trimming percentage $100r\%$ is chosen, where r is a number between 0 and .5. Suppose that $r = .1$, so the trimming percentage is 10%. Then if $n = 20$, 10% of 20 is 2; the 10% trimmed mean results from deleting (trimming) the largest two and the smallest two observations, and then averaging the remaining 16 values. Notice that the trimming percentage specifies the number of observations to be deleted from *each* end of the ordered list. The sample mean is a 0% trimmed mean, whereas the median is a trimmed mean corresponding to the largest possible trimming percentage (e.g., a 45% trimmed mean when $n = 20$).

Example 2.3 Consider the following 20 observations, ordered from smallest to largest, each representing the lifetime (hr) of a certain type of incandescent lamp:

| 612 | 623 | 666 | 744 | 883 | 898 | 964 | 970 | 983 | 1003 |
| 1016 | 1022 | 1029 | 1058 | 1085 | 1088 | 1122 | 1135 | 1197 | 1201 |

The sample mean is $\bar{x} = 19{,}299/20 = 965.0$, and $\tilde{x} = (1003 + 1016)/2 = 1009.5$. The 10% trimmed mean is

$$\bar{x}_{tr(10)} = \frac{19{,}299 - 612 - 623 - 1197 - 1201}{16} = 979.1$$

The effect of trimming here is to produce a central value that is somewhat larger than the mean yet considerably below the median. Similarly, the 20% trimmed mean averages the middle 12 values to obtain $\bar{x}_{tr(20)} = 999.9$, which is even closer to the median. The various measures of center are illustrated in the dotplot of Figure 2.3.

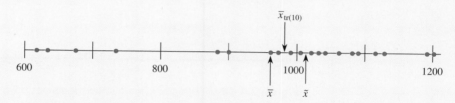

Figure 2.3 Dotplot of lifetimes and measures of center for Example 2.3

Statisticians generally recommend a trimming percentage between 5% and 25%. Notice that $(r)(n)$ may not be a whole number; if $r = .10$ and $n = 25$, then $(r)(n) = 2.5$. Eliminating two observations from each end gives a trimming percentage of 8%, whereas eliminating three observations gives 12%. The resulting two $\bar{x}_{tr}$'s can then be averaged to obtain the 10% trimmed mean. More generally, a trimmed mean for any trimming percentage can be obtained by interpolation.

Measures of Center for Distributions

The primary measure of center for a discrete distribution is the mean value, and both the mean value and the median are frequently used measures for continuous distributions.

Discrete Distributions

Plastic parts manufactured using an injection molding process may exhibit one or more defects, including sinks, scratches, black spots, and so on. Let x represent the number of defects on a single part, and suppose the distribution of x is as follows:

| x: | 0 | 1 | 2 | 3 | 4 |
| $p(x)$: | .80 | .14 | .03 | .02 | .01 |

A picture of the distribution appears in Figure 2.4. Where is this distribution centered? That is, what is the mean or long-run average value of x? A first thought might be to simply average the five possible values of x to obtain a mean value of 2.0. But this entails giving the same weight to each possible value, whereas the distribution indicates that $x = 0$ occurs much more frequently than any of the other values. So what is needed is a weighted average of x values.

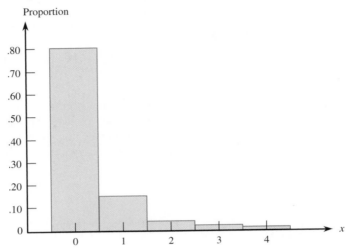

Figure 2.4 Distribution of x, the number of defects on a manufactured plastic part

DEFINITION

> The **mean value** (alternatively, expected value) **of a discrete variable x**, denoted by μ_x or just μ [alternatively, $E(x)$] is given by
>
> $$\mu_x = \sum x \cdot p(x)$$
>
> where the summation is over all possible x values.

Example 2.4   We return now to the plastic part scenario introduced at the outset of this subsection. The mean value of x, the number of defects on a part, is

$$\mu_x = \sum_{x=0}^{4} x \cdot p(x)$$
$$= 0(p(0)) + 1(p(1)) + 2(p(2)) + 3(p(3)) + 4(p(4))$$
$$= (0)(.80) + (1)(.14) + (2)(.03) + (3)(.02) + (4)(.01)$$
$$= .30$$

When we consider the population of all such parts, the population mean value of x is .30. Alternatively, .30 is the long-run average value of x when part after part is monitored. It can also be shown that the histogram of the distribution of Figure 2.4 will balance on the tip of a fulcrum placed on the horizontal axis only if the tip is at .30; μ is the balance point of the distribution.

In Example 2.4, μ is not a possible value of x. In the same way, if x is the number of children in a household, the population mean value of x might be 1.7 even though there are no households with 1.7 children.

In Chapter 1, we introduced two important types of discrete distributions, the binomial distribution and the Poisson distribution. The binomial distribution models the number of "successes" in a group of n items when conditions of individual items are independent of one another and the long-run proportion of successes is π (a number between 0 and 1). The mean value of x is

$$\mu_x = \sum_{x=0}^{n} x\, \frac{n!}{x!(n-x)!}\, \pi^x (1-\pi)^{n-x}$$

The summation looks very intimidating, but fortunately some algebraic manipulation yields an extremely simple result.

> If x is a binomial variable with parameters n = group size and π = success proportion, then $\mu = n\pi$.

Thus if $n = 10$ and $\pi = .8$, $\mu = (10)(.8) = 8$; we "expect" eight of the ten items to be successes, a very intuitive result.

When x is a Poisson variable with parameter λ,

$$\mu_x = \sum_{x=0}^{\infty} x\,\frac{e^{-\lambda}\lambda^x}{x!} = \sum_{x=1}^{\infty} x\,\frac{e^{-\lambda}\lambda^x}{x!}$$

$$= \lambda \sum_{x=1}^{\infty} \frac{e^{-\lambda}\lambda^{x-1}}{(x-1)!}$$

If we now let $y = x - 1$, the range of summation is from $y = 0$ to ∞:

$$= \lambda \sum_{y=0}^{\infty} \frac{e^{-\lambda}\lambda^y}{y!}$$

$$= \lambda \cdot (\text{sum of a Poisson mass function}) = \lambda(1) = \lambda$$

> Let x be a Poisson variable with parameter λ. The mean value of x is λ itself.

Suppose, for example, that x is the number of burnt potato chips in a 13-oz bag. If x has a Poisson distribution with parameter $\lambda = 2.5$, then $\mu_x = 2.5$; the population mean number of burnt chips per bag is 2.5.

Continuous Distributions

A distribution for a continuous variable x is specified by a density function $f(x)$ whose graph is a smooth curve. To obtain μ, we replace summation in the discrete case by integration and replace the mass function $p(x)$ by the density function.

DEFINITION

The **mean value** (or expected value) **of a continuous variable** x with density function $f(x)$ is given by

$$\mu_x = \int_{-\infty}^{\infty} x \cdot f(x)\, dx$$

Just as μ in the discrete case is the balance point for the histogram corresponding to $p(x)$, in the continuous case μ is the balance point for the density curve corresponding to $f(x)$.

Example 2.5 The distribution of the amount of gravel (tons) sold by a particular construction supply company in a given week is a continuous variable x with density function

$$f(x) = 1.5(1 - x^2) \qquad 0 \le x \le 1$$

($f(x) = 0$ outside the interval from 0 to 1). The density curve is shown in Figure 2.5. Knowledge of the mean value of x will help the company decide on a price for the gravel:

$$\mu_x = \int_{-\infty}^{\infty} xf(x)\, dx = \int_{0}^{1} x[1.5(1 - x^2)]\, dx$$

$$= 1.5\int_{0}^{1} (x - x^3)\, dx = 1.5 \left(\frac{x^2}{2} - \frac{x^4}{4} \right)\Big|_{0}^{1} = .375$$

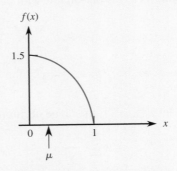

Figure 2.5 The density curve and mean value for Example 2.5

In Chapter 1, we introduced the normal distribution with parameters μ and σ. The symmetry of the associated density curve about μ certainly suggests that μ is the mean value, and this is indeed the case:

$$\int_{-\infty}^{\infty} xf(x)\, dx = \int_{-\infty}^{\infty} (\mu + x - \mu)f(x)\, dx$$

$$= \mu\int_{-\infty}^{\infty} \frac{1}{\sqrt{2\pi}\sigma} e^{-(x-\mu)^2/(2\sigma^2)}\, dx + \int_{-\infty}^{\infty} (x - \mu) \frac{1}{\sqrt{2\pi}\sigma} e^{-(x-\mu)^2/(2\sigma^2)}\, dx$$

$$= \mu + \frac{1}{\sqrt{2\pi}} \int_{-\infty}^{\infty} ye^{-y^2/2}\, dy \qquad \text{using } y = \frac{x - \mu}{\sigma}$$

The latter integral is zero because the integrand $g(y)$ is an odd function ($g(-y) = -g(y)$), which gives the desired result.

A lognormal variable x is one for which $\ln(x)$ has a normal distribution with mean value μ. That is, $\mu_{\ln(x)} = \mu$. Therefore, it might seem that $\mu_x = e^\mu$, but this is not the case. It can be shown that

$$\mu_x = e^{\mu + \sigma^2/2}$$

In Example 1.19 of Chapter 1, $\mu = .353$ and $\sigma = .754$, from which we calculate $e^\mu = 1.42$ whereas

$$\mu_x = e^{.353 + .5(.754)^2} = 1.89$$

The mean value of a Weibull variable is a somewhat complicated expression involving the parameters α and β. Consult the chapter references for details.

μ and $\bar{x}$

If $x_1, \ldots, x_n$ have been randomly selected from some population or process distribution with mean value μ, then the sample mean $\bar{x}$ gives a *point estimate* for μ. In Example 2.1, we calculated $\bar{x} = 16.36$, so a reasonable educated guess for the population mean water-absorption percentage is 16.36%. Estimation—both point (a single number) and interval—will be discussed in Chapter 7.

The Median of a Distribution

Just as the sample median $\tilde{x}$ separates the sample into two equal halves, the **median $\tilde{\mu}$ of a continuous distribution** divides the area under the density curve into two equal halves. The defining condition is

$$\int_{-\infty}^{\tilde{\mu}} f(x)\, dx = .5$$

Example 2.6   (Example 2.5 continued) The median for the distribution of weekly gravel sales satisfies

$$\int_0^{\tilde{\mu}} 1.5(1 - x^2)\, dx = 1.5\left(x - \frac{x^3}{3}\right)\Big|_0^{\tilde{\mu}} = .5$$

Using c in place of $\tilde{\mu}$, we have the cubic equation $1.5(c - c^3/3) = .5$, whose solution is $c = \tilde{\mu} = .347$. We previously calculated the mean as $\mu_x = .375$, which is somewhat larger than the median because the distribution is positively skewed (see Figure 2.5).

Figure 2.6 shows the relationship between the mean and the median for various types of unimodal distributions or (smoothed) histograms. The median of a discrete distribution can also be defined; see one of the chapter references for details.

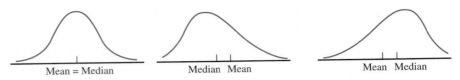

Figure 2.6 The relationship between the mean and the median for a continuous distribution or smoothed histogram

Just as the sample mean gives a point estimate of the population mean μ, the sample median $\tilde{x}$ gives a point estimate of the population median. If the population distribution is symmetric (as is any normal distribution), both $\bar{x}$ and $\tilde{x}$ are estimates of the same population characteristic, namely, the point of symmetry. The issue of which estimate to use will be addressed in Section 7.1.

Section 2.1 Exercises

1. The May 1, 2009, issue of *The Montclarian* reported the following sales figures ($ 1000s) for a sample of homes in Alameda, California, that were sold the previous month:

590	815	575	608	350	1285
408	540	555	679		

   a. Calculate and interpret the sample mean and median.
   b. Suppose the sixth observation had been 985 rather than 1285. How would the mean and median change?
   c. Calculate a 20% trimmed mean by first trimming the two smallest and two largest observations.
   d. Calculate a 15% trimmed mean.

2. Exposure to microbial products, especially endotoxin, may affect human vulnerability to allergic diseases. The article "Dust Sampling Methods for Endotoxin—An Essential but Underestimated Issue" (*Indoor Air*, 2006: 20–27) considered various issues associated with determining endotoxin concentration. The following data on concentration (EU/mg) in settled dust for one sample of urban homes and another of farm homes was kindly supplied by the authors of the article.

U:	6.0	5.0	11.0	33.0	4.0	5.0
	80.0	18.0	35.0	17.0	23.0	
F:	4.0	14.0	11.0	9.0	9.0	8.0
	4.0	20.0	5.0	8.9	21.0	9.2
	3.0	2.0	0.3			

   a. Determine the sample mean for each sample. How do they compare?
   b. Determine the sample median for each sample. How do they compare? Why is the median for the urban sample so different from the mean for that sample?
   c. Calculate the trimmed mean for each sample by deleting the smallest and largest observation. What are the corresponding trimming percentages? How do the values of these trimmed means compare to the corresponding means and medians?

3. The production of Bidri is a traditional craft of India. Bidriware (bowls, vessels, and so on) is cast from an alloy containing primarily zinc along with some copper. Consider the following observations on copper content (%) for a sample of Bidri artifacts in London's Victoria and Albert Museum ("Enigmas of Bidri," *Surface Engr.*, 2005: 333–339), which are listed in increasing order:

2.0	2.4	2.5	2.6	2.6	2.7	2.7
2.8	3.0	3.1	3.2	3.3	3.3	3.4
3.4	3.6	3.6	3.6	3.6	3.7	4.4
4.6	4.7	4.8	5.3	10.1		

   a. Construct a stem-and-leaf display of the data. How does it suggest that the sample mean and median will compare?
   b. Calculate the values of the sample mean and median. *Hint:* $\Sigma x_i = 95.0$.

c. By how much could the largest observation, 10.1, be increased without affecting the value of the sample median? By how much could this value be decreased without affecting the value of the sample median?

4. Suppose that after computing $\bar{x}_n$ based on n sample observations $x_1, \ldots, x_n$, another observation x_{n+1} becomes available. What is the relationship between the mean of the first n observations, the new observation, and the mean of all $n+1$ observations? The mean of the 10 observations in Exercise 1 is 640.5. If an 11th property had sold at a price of 780, what would be the mean sale price for all 11 properties?

5. In the article "Evaluation of Optimal Power Options for Base Transceiver Stations of Mobile Telephone Networks Cameroon" (*Solar Energy*, 2012: 2935–2949), researchers recorded site specific information for remote telecommunications stations throughout Cameroon. The following observations are daily energy demand readings (kWh) for 12 stations:

| 17.76 | 23.44 | 24.58 | 26.99 | 27.23 | 30.77 |
| 31.79 | 35.57 | 36.59 | 36.59 | 40.51 | 59.31 |

Without doing any computation, how do you think the sample mean compares to the sample median? What would you report as representative, or typical, of the daily energy demand for these stations? What prompted your choice?

6. Blood pressure values are often reported to the nearest 5 mmHg (100, 105, 110, and so on). Suppose the actual blood pressure values for nine randomly selected individuals are

| 118.6 | 127.4 | 138.4 | 130.0 | 113.7 |
| 122.0 | 108.3 | 131.5 | 133.2 | |

a. What is the median of the *reported* blood pressure values?
b. Suppose the blood pressure of the second individual is 127.6 rather than 127.4 (a small change in a single value). How does this affect the median of the reported values? What does this say about the sensitivity of the median to rounding or grouping in the data?

7. An experiment to study the lifetime (hr) for a certain type of component involved putting ten components into operation and observing them for 100 hours. Eight of the components failed during that period, and those lifetimes were recorded. Denote the lifetimes of the two components still functioning after 100 hours by 100+. The resulting sample observations were 48, 79, 100+, 35, 92, 86, 57, 17, 100+, and 29. Which of the measures of center discussed in this section can be calculated, and what are the values of those measures? *Note:* The data from this experiment is said to be "censored on the right"; patient lifetimes in medical experimentation are sometimes obtained in this way.

8. A target is located at the point 0 on a horizontal axis. Let x be the landing point of a shot aimed at the target, a continuous variable with density function $f(x) = .75(1 - x^2)$ for $-1 \leq x \leq 1$. What is the mean value of x?

9. Let x denote the amount of time for which a book on 2-hour reserve at a college library is checked out by a student, and suppose that x has density function $f(x) = .5x$ for $0 < x < 2$.
a. What is the mean value of x? Why is the mean value not 1, the midpoint of the interval of positive density?
b. What is the median of this distribution, and how does it compare to the mean value?
c. What proportion of checkout times are within one-half hour of the mean time? What proportion are within one-half hour of the median time?

10. Let x have a uniform distribution on the interval from a to b, so the density function of x is $f(x) = 1/(b - a)$ for $a \leq x \leq b$. What is the mean value of x?

11. The weekly demand for propane gas (1000s of gallons) at a certain facility is a continuous variable with density function

$$f(x) = \begin{cases} 2\left(1 - \dfrac{1}{x^2}\right) & 1 \leq x \leq 2 \\ 0 & \text{otherwise} \end{cases}$$

Determine both the mean value and the median. In the long run, in what proportion of weeks will the value of x be between the mean value and the median?

12. Refer to Exercise 27 of Section 1.3, in which x was the number of telephone lines in use at a specified time. If $\mu = 2.64$, what are the values of $p(5)$ and $p(6)$?

13. The distribution of the number of underinflated tires x on an automobile is given in Exercise 26a(ii) of Section 1.3. Determine the mean value of x.

14. Sometimes, rather than wishing to determine the mean value of x, an investigator wishes to determine the mean value of some function of x. Suppose, for example, that a repairman assesses a fixed charge of $25 plus $40 an hour that he spends on a job. Then the revenue resulting from a job that takes x hours is $h(x) = 25 + 40x$. If x is a continuous variable,

the mean value of any function $h(x)$ is computed similarly to the way in which μ itself is computed: $\mu_{h(x)} = \int h(x) f(x)\, dx$.

a. Refer to Exercise 9. Suppose the library, in a desperate search for revenue to fund its operations, charges a student $h(x) = x^2$ dollars to check a book out on 2-hour reserve for x hours. What is the mean value of the checkout charge?

b. Suppose that $h(x) = a + bx$, a linear function of x. Show that $\mu_{h(x)} = a + b\mu$ (this is true for x continuous or discrete). If the mean value of repair time is .5 hr for the repair situation mentioned at the outset of this problem, what is the mean value of repair revenue?

2.2   MEASURES OF VARIABILITY

Reporting a measure of center gives only partial information about a data set or distribution. Different samples or distributions may have identical measures of center yet differ from one another in other important ways. For example, for a normal distribution with parameters μ and σ, the normal curve becomes more spread out as the value of σ increases. Figure 2.7 shows dotplots of three samples with the same mean and median, yet the extent of spread about the center is different for all three samples. The first sample has the largest amount of variability, the third has the smallest amount, and the second is intermediate to the other two in this respect.

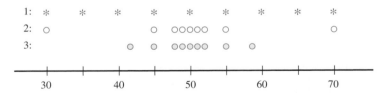

Figure 2.7 Samples with identical measures of center but different amounts of variability

Measures of Variability for Sample Data

The simplest measure of variability in a sample is the **range**, which is the difference between the largest and smallest sample values. Notice that the value of the range for sample 1 in Figure 2.7 is much larger than it is for sample 3, reflecting more variability in the first sample than in the third one. A defect of the range, though, is that it depends on only the two most extreme observations and disregards the positions of the remaining $n - 2$ values. Samples 1 and 2 in Figure 2.7 have identical ranges, yet when we take into account the observations between the two extremes, there is much less variability or dispersion in the second sample than in the first one.

Our primary measures of variability involve quantities called **deviations from the mean:** $x_1 - \bar{x}, x_2 - \bar{x}, \ldots, x_n - \bar{x}$. That is, the deviations from the mean are obtained by subtracting $\bar{x}$ from each of the n sample observations. A deviation will be positive if the observation is larger than the mean (to the right of the mean on the measurement axis) and negative if the observation is smaller than the mean. If all the deviations are small in magnitude, then all x_i's are close to the mean and there is little variability. On the other hand, if some of the deviations are large in magnitude, then some x_i's lie far from $\bar{x}$, suggesting a greater amount of variability. A simple way to combine the deviations into a single quantity is to average them (sum them and divide by n). Unfortunately, there is a major problem with this suggestion:

$$\text{sum of deviations} = \sum_{i=1}^{n} (x_i - \bar{x}) = 0$$

so that the average deviation is always zero (because $\sum_{i=1}^{n} \bar{x} = \bar{x} + \cdots + \bar{x} = n\bar{x} = \sum_{i=1}^{n} x_i$). In practice, the sum of the deviations may not be identically zero because of rounding in $\bar{x}$. The greater the decimal accuracy used in $\bar{x}$, the closer the sum will be to zero.

How can we change the deviations to nonnegative quantities so the positive and negative deviations do not counteract one another when they are combined? One possibility is to work with the absolute values of the deviations and calculate the average absolute deviation $\sum |x_i - \bar{x}|/n$. Because the absolute value operation leads to a number of theoretical difficulties, consider instead the squared deviations $(x_1 - \bar{x})^2, (x_2 - \bar{x})^2, \ldots, (x_n - \bar{x})^2$. We might now use the average squared deviation $\sum (x_i - \bar{x})^2/n$, but for several reasons we will divide the sum of squared deviations by $n - 1$ rather than n.

DEFINITIONS

The **sample variance,** denoted by s^2, is given by

$$s^2 = \frac{\sum (x_i - \bar{x})^2}{n - 1} = \frac{S_{xx}}{n - 1}$$

The **sample standard deviation,** denoted by s, is the (positive) square root of the variance:

$$s = \sqrt{s^2}$$

An alternative computational formula for s^2 is given in Exercise 18.

The unit for s is the same as the unit for each of the x_i's. If, for example, the observations are fuel efficiencies in miles per gallon (mpg), then we might have $s = 2.0$ mpg. A rough interpretation of the sample standard deviation is that it is the size of a typical or representative deviation from the sample mean within the given sample. Thus if $s = 2.0$ mpg, then some x_i's in the sample are closer than 2.0 to $\bar{x}$ whereas others are farther away; 2.0 is a representative (or standard) deviation from the mean fuel efficiency. If $s = 3.0$ for a second sample of cars of another type, a typical deviation in this sample is roughly one and one-half times what it is in the first sample, an indication of greater variability in the second sample.

Example 2.7

The website www.fueleconomy.gov contains a wealth of information about the fuel characteristics of various vehicles. In addition to EPA mileage ratings, there are many vehicles for which users have reported their own values of fuel efficiency (mpg). Consider the following sample of $n = 11$ efficiencies for the 2009 Ford Focus equipped with an automatic transmission (for this model, EPA reports an overall rating of 27 mpg—24 mpg in city driving and 33 mpg in highway driving):

Car	x_i	$x_i - \bar{x}$	$(x_i - \bar{x})^2$
1	27.3	−5.96	35.522
2	27.9	−5.36	28.730
3	32.9	−0.36	0.130
4	35.2	1.94	3.764
5	44.9	11.64	135.490
6	39.9	6.64	44.090
7	30.0	−3.26	10.628
8	29.7	−3.56	12.674
9	28.5	−4.76	22.658
10	32.0	−1.26	1.588
11	37.6	4.34	18.836
	$\Sigma x_i = 365.9$	$\Sigma(x_i - \bar{x}) = .04$	$\Sigma(x_i - \bar{x})^2 = 314.110$ $\bar{x} = 33.26$

Effects of rounding account for the sum of deviations differing slightly from zero. The numerator of s^2 is $S_{xx} = 314.110$, from which

$$s^2 = \frac{S_{xx}}{n-1} = \frac{314.110}{11-1} = 31.41, \qquad s = 5.60$$

The size of a representative deviation from the sample mean 33.26 is roughly 5.6 mpg.

Note: Of the nine people who also reported driving behavior, only three did more than 80% of their driving in highway mode; we bet you can guess which cars they drove. We haven't a clue why all 11 reported values exceed the EPA figure: Maybe only drivers with really good fuel efficiencies communicate their results.

One explanation for the use of $n - 1$ in s^2 goes back to the fact that $\Sigma(x_i - \bar{x}) = 0$. Suppose that $n = 5$ and that $x_1 - \bar{x} = -4$, $x_2 - \bar{x} = 6$, $x_3 - \bar{x} = 1$, and $x_5 - \bar{x} = -8$. Since the sum of these four deviations is -5, the remaining deviation must be $x_4 - \bar{x} = 5$ (so that the sum of all five deviations is zero). More generally, once any $n - 1$ of the deviations are available, the value of the remaining deviation is determined. The n deviations actually contain only $n - 1$ independent pieces of information about variability. Statisticians express this by saying that s^2 and s are based on $n - 1$ **degrees of freedom** (df). Many inferential procedures encountered in later chapters are based on some appropriate number of df.

The Variance and Standard Deviation of a Discrete Distribution

Let x be a discrete variable with mass function $p(x)$ and mean value μ. Just as μ itself is a weighted average of possible x values, where the weights come from the mass function, the variance is a weighted average of the squared deviations $(x - \mu)^2$ for possible x values.

DEFINITIONS

The **variance of a discrete distribution** for a variable x specified by mass function $p(x)$, denoted by σ_x^2 or just σ^2 (alternatively, $V(x)$), is given by

$$\sigma^2 = \sum (x - \mu)^2 \cdot p(x)$$

where the sum is over all possible x values. The **standard deviation** is σ, the positive square root of the variance.

If a particular x value is far from μ, resulting in a large squared deviation, it will still not contribute much to variability in the distribution if $p(x)$ is quite small. This is desirable because any x value for which $p(x)$ is quite small will be observed very infrequently in a long sequence of selections from the population or process. Just as s can be interpreted as the size of a representative deviation from the sample mean, σ can be interpreted as the size of a typical deviation from the population or process mean.

Example 2.8

Consider a computer system consisting of the computer itself, a monitor, and a printer. Let x denote the number of system components that need service while under warranty; possible x values are 0, 1, 2, and 3. Suppose that $p(0) = .532$, $p(1) = .389$, $p(2) = .076$, and $p(3) = .003$ (these come from individual component failure proportions of .2, .3, and .05 along with an assumption of component independence, so that these proportions can be multiplied as we originally did in a binomial calculation). Then $\mu = .55$ and

$$
\begin{aligned}
\sigma^2 &= \sum (x - \mu)^2 \cdot p(x) \\
&= (0 - .55)^2(.532) + (1 - .55)^2(.389) + (2 - .55)^2(.076) \\
&\quad + (3 - .55)^2(.003) \\
&= .16093 + .07877 + .15979 + .01801 = .41750
\end{aligned}
$$

from which $\sigma = .646$.

An alternative computational formula for calculating σ^2 is given in Exercise 26, which is similar to the computational formula for s^2 in Exercise 18.

Recall that the mean value of the binomial distribution based on group size n and item success proportion π is just $n\pi$. The variance is also a simple expression, though verification of this result involves some tedious manipulation of summations:

$$\sigma^2 = \sum_{x=0}^{n} (x - n\pi)^2 \frac{n!}{x!(n-x)!} \pi^x (1 - \pi)^{n-x} = n\pi(1 - \pi)$$

The standard deviation of a binomial distribution is then $\sigma = \sqrt{n\pi(1 - \pi)}$. Note that $\sigma = 0$ if $\pi = 0$ (in which case, every item is a failure, so $x = 0$ always) or $\pi = 1$ (every item a success, so $x = n$ always). The variance and standard deviation are largest when $\pi = .5$ [$\pi(1 - \pi)$ is maximized for this value], that is, when there is a 50–50 split between successes and failures. As π moves toward either 0 or 1, the variance and standard deviation decrease. If identical components are shipped in groups of size 25 and the long-run success (doesn't need warranty service) proportion is $\pi = .9$, then

$$\mu = 25(.9) = 22.5 \qquad \sigma = \sqrt{25(.9)(.1)} = \sqrt{2.25} = 1.50$$

The mean value of a Poisson distribution with parameter λ is λ itself, and this is also the variance of the distribution:

$$\sigma^2 = \sum_{x=0}^{\infty} (x - \lambda)^2 \, \frac{e^{-\lambda}\lambda^x}{x!} = \lambda$$

(Again, much summation manipulation is required.) The standard deviation is, of course, $\sqrt{\lambda}$. If the number of blemishes x on surfaces of a certain part has a Poisson distribution with parameter $\lambda = 3.5$, then the mean value is 3.5 and the standard deviation is 1.87.

The Variance and Standard Deviation of a Continuous Distribution

The variance of a continuous distribution with density function $f(x)$ is obtained by replacing summation in the discrete case by integration and substituting $f(x)$ for $p(x)$.

DEFINITIONS

The **variance of a continuous distribution** specified by density function $f(x)$ is

$$\sigma^2 = \int_{-\infty}^{\infty} (x - \mu)^2 \cdot f(x) \, dx$$

The **standard deviation** σ is again the positive square root of the variance.

Example 2.9

The distribution of x = gravel sales during a given week (tons), introduced in Example 2.5, was specified by the density function $f(x) = 1.5(1 - x^2)$ for x between 0 and 1. We found the mean value to be $\mu = .375$. The variance of the distribution is

$$\sigma^2 = \int_{0}^{1} (x - .375)^2 \cdot 1.5(1 - x^2) \, dx$$

Multiplying the factors in the integrand gives $1.5(-x^4 + .75x^3 + .859375x^2 - .75x + .140625)$. Integrating this fourth-degree polynomial term by term gives $\sigma^2 = .059375$ and $\sigma = .244$.

The Case of a Normal Distribution

The two parameters of a normal distribution were denoted by μ and σ. We have already seen that μ is in fact the mean value, and it should come as no surprise that the second parameter is the standard deviation of the distribution. That is, a bit of integration manipulation shows that

$$V(x) = \int_{-\infty}^{\infty} (x - \mu)^2 \frac{1}{\sqrt{2\pi}\sigma} e^{-(x-\mu)^2/(2\sigma^2)} dx = \sigma^2$$

Let k be some fixed positive number. Consider the area under a normal curve with parameters μ and σ that lies within k standard deviations of the mean value. That is, we wish to determine the proportion of x values that lie in the interval from $\mu - k\sigma$ to $\mu + k\sigma$. Standardizing the interval limits gives

$$\frac{\mu - k\sigma - \mu}{\sigma} = -k \qquad \frac{\mu + k\sigma - \mu}{\sigma} = k$$

Thus the desired proportion is the area under the standard normal (z) curve between $-k$ and k. This shows that the area within k standard deviations of the mean under *any* normal curve depends only on k and not on the particular normal curve under consideration. For $k = 1$, the desired proportion is the area under the z curve between -1 and 1. From Appendix Table I, this area is $.8413 - .1587 = .6826 \approx .68$. Similar calculations for $k = 2$ and $k = 3$ give $.9544$ and $.9974$, respectively. Thus for any variable x whose distribution is well approximated by a normal curve:

Approximately 68% of the values are within 1 standard deviation of the mean.
Approximately 95% of the values are within 2 standard deviations of the mean.
Approximately 99.7% of the values are within 3 standard deviations of the mean.

These three statements together are often referred to as the **empirical rule;** the name reflects the fact that histograms of a great many data sets have at least roughly the shape of a normal curve.

Other Continuous Distributions

A variable x is said to have a lognormal distribution with parameters μ and σ if $\ln(x)$ is normally distributed with mean value μ and standard deviation σ. In Section 2.1, we pointed out that the mean value of x itself is not μ. Similarly, the variance of x is not σ^2. It can be shown that

$$V(x) = e^{2\mu + \sigma^2}(e^{\sigma^2} - 1)$$

The variance of a variable having a Weibull distribution is even more complicated than the mean value; consult one of the chapter references.

σ^2 and s^2

The sample mean $\bar{x}$ is a sensible estimate (educated guess) for the value of the population or process mean μ. Similarly, the sample variance should be defined so that it gives a reasonable estimate of the population or process variance σ^2. Recall that σ^2 involves squared deviations from μ, that is, quantities of the form $(x - \mu)^2$. If the value of μ were known to an investigator, a good estimate of σ^2 based on sample observations $x_1, \ldots, x_n$ would be $\Sigma(x_i - \mu)^2/n$. It is natural to replace μ by $\bar{x}$ when the value of the former quantity is unknown. However, it can be shown that $\Sigma(x_i - \bar{x})^2 < \Sigma(x_i - \mu)^2$ unless $\bar{x} = \mu$, so $\bar{x}$ is "closer" to the sample observations than is μ. To compensate for this reduction in sum of squares, the value of the denominator n should also be reduced. According to a technical criterion called *unbiasedness*, the sample size n should be replaced by the number of df $n - 1$. The resulting sample variance s^2 will tend to provide good estimates of σ^2.

Section 2.2 Exercises

15. In the article "Mechanical Reliability of Devices Subdermally Implanted into the Young of Long-Lived and Endangered Wildlife" (*J. of Materials Engr. and Performance*, 2012: 1924–1931), researchers examined the mechanical reliability of a thin enclosure for a biotelemetry device to be subdermally implanted in young wild animals. Six enclosure specimens were subjected to puncture tests. Each specimen was placed in a test apparatus, and researchers recorded the necessary force (N) for the puncture head to cause initial cracks in the enclosure. Here is the corresponding data:

    2006.1   2065.2   2118.9
    1686.6   1966.9   1792.5

    a. Calculate $\bar{x}$ and the deviations from the mean.
    b. Use the deviations calculated in part (a) to obtain the sample variance and the sample standard deviation.
    c. Compute the sample standard deviation using a calculator or software function to confirm the accuracy of your answer in (b).

16. Return to the puncture test data given in Exercise 15.
    a. Subtract 100 from each observation to obtain a sample of transformed values. Now calculate the sample variance of these transformed values and compare it to s^2 for the original data.
    b. Consider a sample $x_1, \ldots, x_n$ and let $y_i = x_i - c$ for $i = 1, 2, \ldots, n$, where c is some specified number. Give a general argument to show that the sample variance of the y_i's is identical to that of the x_i's. *Hint:* How are $\bar{y}$ and $\bar{x}$ related?

17. Suppose the following represent quiz scores (out of 15 points) for students in two different study groups:

    Group 1: 10, 14, 8, 7, 12, 7, 11
    Group 2: 5, 8, 9.5, 8.5, 9, 9.5, 13

    a. Compute the mean and standard deviation for each group.
    b. Determine the range for each data set.
    c. Create a dotplot for each data set and ensure you use the same axis scale for each.

    d. Notice that one group exhibits the smaller standard deviation but the other exhibits the smaller range. Explain how it is possible for a data set to have the smallest standard deviation yet not have the smallest range. *Hint:* Keep in mind how standard deviation measures variability and compare the dotplots you created.

18. Traumatic knee dislocation often requires surgery to repair ruptured ligaments. One measure of recovery is range of motion (measured as the angle formed when, starting with the leg straight, the knee is bent as far as possible). The given data on postsurgical range of motion appeared in the article "Reconstruction of the Anterior and Posterior Cruciate Ligaments After Knee Dislocation" (*Amer. J. Sports Med.*, 1999: 189–197):

    154   142   137   133   122   126   135
    135   108   120   127   134   122

    a. What are the values of the sample mean and sample median?
    b. An alternative computing formula for the numerator of s^2 is:

    $$S_{xx} = \Sigma(x_i - \bar{x})^2 = \Sigma x_i^2 - \frac{1}{n}(\Sigma x_i)^2$$

    Using this formula, determine the sample variance of the data.
    Hint: $\Sigma x_i = 1695$, $\Sigma x_i^2 = 222{,}581$.

19. In the article "X-Ray Computed Tomography and Nondestructive Evaluation of Clogging in Porous Concrete Field Samples" (*J. of Materials in Civil Engr.*, 2012: 1103–1109), investigators determined the clogging percentage in porous concrete samples cored from parking lots. Porosity profiles using computed tomography scanned images were used in this study. The following represent the average porosity (%) using a gravimetric method for nine concrete cores:

    8.10   20.50   26.54   19.68   14.87
    14.36   9.19   23.55   22.27

Calculate and interpret the values of the sample mean and sample standard deviation for this data.

20. Use the alternative computing formula for S_{xx} as shown in Exercise 18 to determine the sample standard deviation for the average porosity measurements presented in Exercise 19.

21. Consider the following information on ultimate tensile strength (lb/in.) for a sample of $n = 4$ hard zirconium copper wire specimens (from "Characterization Methods for Fine Copper Wire," *Wire J. Intl.*, August 1997: 74–80):

    $\bar{x} = 76{,}831$ $s = 180$ smallest $x_i = 76{,}683$
    largest $x_i = 77{,}048$

    Determine the values of the two middle sample observations (and don't do it by successive guessing!). *Hint:* See Exercise 18 part b.

22. The federal test procedure (FTP) for determining the levels of various types of vehicle emissions is time-consuming and expensive to perform. According to the article "Motor Vehicle Emissions Variability" (*J. of the Air and Waste Mgmnt. Assoc.*, 1996: 667–675), there is a widespread belief that repeated FTP measurements on the same vehicle would yield identical (or nearly identical) results. The accompanying data is from one particular vehicle characterized as a high emitter:

    HC (gm/mi): 13.8   18.3   32.2   32.5
    CO (gm/mi): 118   149   232   236

    a. Compute the sample standard deviations for the HC and CO observations. Does the widespread belief appear to be justified?
    b. The *sample coefficient of variation* $s/\bar{x}$ (or $100s/\bar{x}$) assesses the extent of variability relative to the mean. Values of this coefficient for several different data sets can be compared to determine which data sets exhibit more or less variation. Carry out such a comparison for the given HC and CO data.

23. Suppose, as in Exercise 57 of Chapter 1, that the number of drivers traveling between a particular origin and destination during a designated time period has a Poisson distribution with $\lambda = 20$. In the long run, during what proportion of such periods will the number of drivers be

a. Within 5 of the mean value?
b. Within 1 standard deviation of the mean value?

24. Suppose that x, the number of flaws on the surface of a boiler of a certain type, has a Poisson distribution with $\lambda = 5$. For what proportion of such boilers will the number of flaws

a. Be within 1 standard deviation of the mean number of flaws?
b. Exceed the mean number of flaws by more than 2 standard deviations?

25. Let x represent the number of underinflated tires on an automobile of a certain type, and suppose that $p(0) = .4$, $p(1) = p(2) = p(3) = .1$, and $p(4) = .3$, from which $\mu = 1.8$.

a. Calculate the standard deviation of x.
b. For what proportion of such cars will the number of underinflated tires be within 1 standard deviation of the mean value? More than 3 standard deviations from the mean value?

26. Use the fact that $(x - \mu)^2 = x^2 - 2\mu x + \mu^2$ to show that $\sigma^2 = \Sigma x^2 p(x) - \mu^2$ for a discrete variable x. Then use this result to compute the variance for the variable whose distribution is given in the previous problem. *Hint:* Substitute the alternative expression for $(x - \mu)^2$ in the definition of σ^2, and break the summation into three separate terms; the argument in the continuous case involves replacing summation with integration.

27. If x has a uniform distribution on the interval from a to b $[f(x) = 1/(b - a)]$, from which $\mu = (a + b)/2$, show that $\sigma^2 = (b - a)^2/12$. If task completion time is uniformly distributed with $a = 4$ and $b = 6$, what proportion of times will be farther than 1 standard deviation from the mean value of completion time?

28. Suppose that bearing diameter x has a normal distribution. What proportion of bearings have diameters that are within 1.5 standard deviations of the mean diameter? That exceed the mean diameter by more than 2.5 standard deviations?

29. Historical data implies that 20% of all components of a certain type need service while under warranty. Suppose that whether any particular component needs warranty service is independent of whether

any other component does. If these components are shipped in batches of 25 and x denotes the number of components in a batch that need warranty service, determine the standard deviation of x and then the proportion of batches for which the number of components that need warranty service exceeds the mean number by more than 2 standard deviations.

30. If the unloading time of a forwarder in a harvesting operation is lognormally distributed with a mean value of 900 and a standard deviation of 725, what are the values of the parameters μ and σ? *Note:* An expression for the mean value of a lognormal variable is given in Section 2.1, and an expression for the variance appears in this section.

31. If component lifetime is exponentially distributed with parameter λ, obtain an expression for the proportion of components whose lifetime exceeds the mean value by more than 1 standard deviation. *Hint:* According to Exercise 26, $\sigma^2 = \int_0^\infty x^2 f(x)\, dx - \mu^2$; now use integration by parts.

32. The sample mean and sample standard deviation for the sample of $n = 100$ shear strength observations given in Exercise 17 of Section 1.2 are 5049.16 and 351.45, respectively. What percentage of the observations in the sample are within 1 standard deviation of the mean, and how does this compare to the corresponding percentage given by the empirical rule? Answer this question also for 2 standard deviations and for 3 standard deviations.

2.3 MORE DETAILED SUMMARY QUANTITIES

The median separates a data set or distribution into two equal parts, so that 50% of the values exceed the median and 50% are smaller than the median. Quartiles and percentiles give more detailed information about location of a data set or distribution by considering percentages other than 50%. In this section, we also develop another measure of spread based on the quartiles, the *interquartile range (IQR)*. The median and IQR can be used together to give a concise yet informative visual summary of sample data called a *boxplot*.

Quartiles and the Interquartile Range

The lower and upper quartiles along with the median separate a data set or distribution into four equal parts: 25% of all values are smaller than the lower quartile, 25% exceed the upper quartile, and 25% lie between each quartile and the median. This is illustrated for a continuous distribution or smoothed histogram in Figure 2.8.

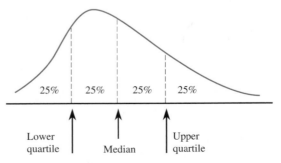

Figure 2.8 Illustrating the quartiles

Let's first consider quartiles for sample data. There are several different sensible ways to define the sample quartiles. We will use a definition that requires a minimal amount of computation; statistical computer packages actually calculate quartiles by interpolation (our quartiles are called *fourths* in some sources).

DEFINITIONS

Separate the n ordered sample observations into a lower half and an upper half; if n is an odd number, include the median $\tilde{x}$ in each half. Then

lower quartile = median of the lower half of the data
upper quartile = median of the upper half of the data

The **interquartile range (IQR),** a measure of variability that is resistant to the effect of outliers, is the difference between the two quartiles:

$$IQR = upper\ quartile - lower\ quartile$$

Example 2.10

Reconsider the flexural strength data for beams given in Example 1.2. A stem-and-leaf display of the 27 observations follows:

```
    5 | 9
    6 | 3  3  5  8  8
    7 | 0  0  2  3  4  6  7  7  8  8  9      Stem: ones digit
    8 | 1  2  7                               Leaf: tenths digit
    9 | 0  7  7
   10 | 7
   11 | 3  6  7
```

Because $n = 27$ is odd, the median $\tilde{x} = 7.7$ is included in each half of the data:

Lower half: 5.9 6.3 6.3 6.5 6.8 6.8 7.0 7.0 7.2 7.3 7.4 7.6 7.7 7.7

Upper half: 7.7 7.8 7.8 7.9 8.1 8.2 8.7 9.0 9.7 9.7 10.7 11.3 11.6 11.8

$$lower\ quartile = \frac{7.0 + 7.0}{2} = 7.0 \qquad upper\ quartile = \frac{8.7 + 9.0}{2} = 8.85$$

$$IQR = 8.85 - 7.0 = 1.85$$

Notice that if the largest observation, 11.8, were increased by any amount, the upper quartile and therefore the IQR would not be affected, whereas such an increase would change the sample variance and standard deviation. Similarly, a decrease in several of the smallest observations has no impact on the quartiles or the IQR.

The following output is from the **summary** and **IQR** commands from the R software. The former command requests that the values of various summary quantities be calculated:

```
> summary(flexural)

  Min.   1st Qu.   Median   Mean    3rd Qu.   Max.
 5.900   7.000     7.700    8.141   8.850     11.800

> IQR(flexural)
[1] 1.85
```

Minitab's reported value for the quartile Q3 is 9.000, a bit different from what R returns.

Now consider a continuous variable x whose distribution is described by a density function $f(x)$. Recall that the median $\tilde{\mu}$ results from solving the equation

$$\int_{-\infty}^{\tilde{\mu}} f(x) \, dx = .5$$

(so that half the area under the density curve lies to the left of $\tilde{\mu}$). The lower quartile q_1 and upper quartile q_u are solutions to

$$\int_{-\infty}^{q_1} f(x) \, dx = .25 \qquad \int_{q_u}^{\infty} f(x) \, dx = .25$$

Example 2.11

The exponential distribution with parameter λ has density function $\lambda e^{-\lambda x}$ for $x > 0$. For any positive number c,

$$\int_{-\infty}^{c} f(x) \, dx = \int_{0}^{c} \lambda e^{-\lambda x} \, dx = 1 - e^{-\lambda c}$$

$$\int_{c}^{\infty} \lambda e^{-\lambda x} \, dx = e^{-\lambda c}$$

Equating either of these quantities to .5 and solving for c gives $c = \tilde{\mu} = -\ln(.5)/\lambda = .693/\lambda$. Equating each of these two quantities to .25 gives

$$q_1 = -\ln(.75)/\lambda = .288/\lambda \qquad q_u = -\ln(.25)/\lambda = 1.386/\lambda$$

Suppose, for example, that times (min) between successive arrivals at a shipping terminal are exponentially distributed with $\lambda = .1$. Then $q_1 = 2.88$ min, $\tilde{\mu} = 6.93$ min, and $q_u = 13.86$ min. The upper quartile is much farther from the median than is the lower quartile because the distribution has a substantial positive skew (the mean value of x is $1/\lambda = 10$, much larger than the median).

Example 2.12

The quartiles of a normal distribution are easily expressed in terms of μ and σ. First, consider a variable z having the standard normal distribution. Symmetry of the standard normal curve about 0 implies that $\tilde{\mu} = 0$. Looking for .2500 inside Appendix Table I, we obtain the following information:

$$\text{area to the left of } -.67: \quad .2514$$
$$\text{area to the left of } -.68: \quad .2483$$

Since .25 is roughly halfway between these two tabled areas, we take $-.675$ as the lower quartile. By symmetry, .675 is the upper quartile.

It is then easily verified that if x has a normal distribution with mean value μ and standard deviation σ,

$$\text{upper quartile} = \mu + .675\sigma \qquad \text{lower quartile} = \mu - .675\sigma$$

That is, for any normal distribution, the quartiles are .675 standard deviation to either side of the mean. The interquartile range is $\mu + .675\sigma - (\mu - .675\sigma) = 1.35\sigma$. A familiar example is IQ scores in the general population, where $\mu = 100$, $\sigma = 15$, $q_l = 89.875 \approx 90$, and $q_u \approx 110$. Roughly 25% of all people have scores below 90 and roughly 25% have scores exceeding 110.

The relation $IQR = 1.35\sigma$ suggests that if the *sample* IQR is very different from $1.35s$, it is not plausible that the underlying distribution is normal. In Example 2.10, $1.35s \approx 2.2$, which is not much greater than the IQR of 1.85. A graphical technique for assessing the plausibility of a normal population or process distribution is presented in the next section.

For our purposes, it is not necessary to discuss quartiles for a discrete distribution.

Boxplots

A **boxplot** is a visual display of data based on the following five-number summary:

smallest x_i lower quartile   median   upper quartile   largest x_i

To create a boxplot, first draw a horizontal measurement scale. Then place a rectangle above this axis; the left edge of the rectangle is at the lower quartile, and the right edge is at the upper quartile (so box width = IQR). Place a vertical line segment or some other symbol inside the rectangle at the location of the median; the position of the median symbol relative to the two edges conveys information about skewness in the middle 50% of the data. Finally, draw "whiskers" out from either end of the rectangle to the smallest and largest observations. A boxplot with a vertical orientation can also be drawn by making obvious modifications in the construction process.

Example 2.13   Returning to the article on lightweight aggregates referenced in Example 2.1, the researchers also reported specific gravity measurements for all 14 LWA specimens:

1.10	1.29	1.38	1.39	1.40	1.45	1.46
1.48	1.49	1.50	1.51	1.51	1.56	1.62

The five-number summary is as follows:

Smallest $x_i = 1.10$ lower quartile = 1.39 $\tilde{x} = 1.47$ upper quartile = 1.51
Largest $x_i = 1.62$

Figure 2.9 shows the resulting boxplot.

The right edge of the box is closer to the median than is the left edge, indicating a substantial skew in the middle half of the data. The box width (IQR) is also reasonably large relative to the range of the data (distance between the tips of the whiskers).

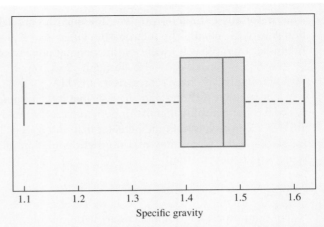

Figure 2.9 A boxplot of the LWA data generated by the R software

A boxplot is certainly more compact than a stem-and-leaf display or histogram, but it is sometimes inferior to these latter two descriptive techniques because a boxplot can mask important characteristics of the data, such as the presence of clusters. The main attraction of boxplots is that they give a quick visual comparison. A comparative or side-by-side boxplot is a very effective way of revealing similarities and differences between two or more data sets consisting of observations on the same variable.

Example 2.14   The article "Compression of Single-Wall Corrugated Shipping Containers Using Fixed and Floating Test Platens" (*J. of Testing and Evaluation*, 1992: 318–320) describes an experiment in which several different types of boxes were compared with respect to compression strength. Consider the following observations on four different types of boxes (summary quantities for this data are in good agreement with values given in the cited article):

Type of box	Compression strength (lb)					
1	655.5	788.3	734.3	721.4	679.1	699.4
2	789.2	772.5	786.9	686.1	732.1	774.8
3	737.1	639.0	696.3	671.7	717.2	727.1
4	535.1	628.7	542.4	559.0	586.9	520.0

Figure 2.10 is a comparative boxplot of this data produced by the Minitab statistical package. (Recall that Minitab uses definitions of the quartiles that differ somewhat from ours.) The most striking feature of the comparative boxplot is that strength values for the fourth type of box appear to be considerably smaller than those for the three other types; this suggests that the population mean strength for type 4 boxes is less than the mean strengths for the other three types. The differences between box types seem pretty clear-cut because within-sample variation is small relative to the

separation between sample means and medians. When this is not the case, an infer-ential method called *single-factor analysis of variance*, discussed in Chapter 9, is used to investigate differences among three or more populations or treatments.

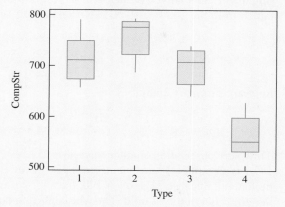

Figure 2.10 A Minitab comparative boxplot of the compressive strength data

Boxplots That Show Outliers

A boxplot can be embellished to indicate explicitly the presence of outliers.

DEFINITIONS

Any observation farther than 1.5 IQR from the closest quartile is an **outlier.** An outlier is **extreme** if it is more than 3 IQR from the nearest quartile, and it is **mild** otherwise.

Many inferential procedures are based on the assumption that the sample came from a normal distribution. Even a single extreme outlier in the sample warns the investiga-tor that such procedures should not be used, and the presence of several mild outliers conveys the same message.

Let's now modify our previous construction of a boxplot by drawing a whisker out from each end of the box to the smallest and largest observations that are *not* outliers. Each mild outlier is represented by a closed circle and each extreme outlier by an open circle. Some statistical computer packages do not distinguish between mild and extreme outliers.

Example 2.15   The National Health and Nutrition Examination Survey (NHANES), a massive annual program conducted by the National Center for Health Statistics, is a series of cross-sectional nationally representative surveys that include demographic, socioeconomic, dietary, and health-related questions. The information from the surveys is used to assess the health and nutritional status of adults and children in the United States.

One variable measured is the high-density lipoprotein (HDL) cholesterol level (mg/dl) of each survey participant. The following 30 HDL observations were obtained from the 2009–2010 NHANES data set:

$$
\begin{array}{cccccccccc}
11 & 32 & 33 & 41 & 45 & 46 & 47 & 48 & 48 & 49 \\
49 & 50 & 52 & 55 & 57 & 57 & 59 & 61 & 63 & 63 \\
66 & 67 & 71 & 71 & 71 & 72 & 73 & 76 & 111 & 144
\end{array}
$$

Relevant summary quantities are

$$\tilde{x} = 57 \qquad \text{lower quartile} = 48 \qquad \text{upper quartile} = 71$$
$$\text{IQR} = 23 \qquad 1.5\ \text{IQR} = 34.5 \qquad 3\ \text{IQR} = 69$$

Thus, any observation smaller than $48 - 34.5 = 13.5$ or larger than $71 + 34.5 = 105.5$ is an outlier. There is one outlier at the lower end of the sample and two at the upper end. Because $71 + 69 = 140$, the largest observation of 144 is an extreme outlier; the other outlier is mild. The whiskers extend out to 32 and 76, the most extreme observations that are not outliers. The resulting boxplot is in Figure 2.11.

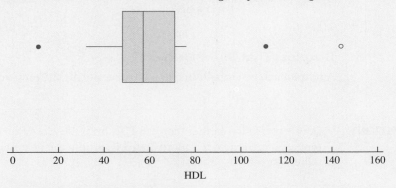

Figure 2.11 A boxplot of the HDL cholesterol data showing mild and extreme outliers

Percentiles

Let p denote a number between 0 and 1. Then the **(100p)th percentile, $\boldsymbol{\eta}_p$**—also called the **pth quantile**—separates the smallest $100p\%$ of the data or distribution from the remaining values. For example, 90% of all values lie below the 90th percentile, $\eta_{.9}$ (the .9th quantile), and only 10% of all values exceed the 90th percentile. The median is the 50th percentile, and the lower and upper quartiles are the 25th and 75th percentiles, respectively. For a continuous distribution, η_p is the solution to the equation

$$\int_{-\infty}^{\eta_p} f(x)\, dx = p$$

That is, p is the area under the density curve to the left of η_p. Figure 2.12 illustrates the definition.

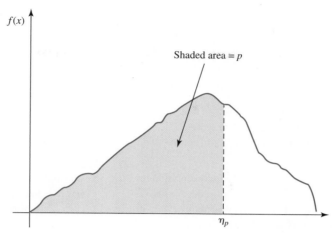

η_p

Figure 2.12 The ($100p$)th percentile of a continuous distribution

Example 2.16   Appendix Table I gives cumulative z curve areas for the standard normal distribution. To find the 90th percentile, we look for cumulative area .9000 inside the table. The entry closest to .9000 is .8997 in the 1.2 row and .08 column, so $\eta_{.9} \approx 1.28$. By symmetry, the 10th z percentile (.1th quantile) is $\eta_{.1} \approx -1.28$. It then follows that for the normal distribution with mean value μ and standard deviation σ,

$$\eta_{.9} \approx \mu + 1.28\sigma \qquad \eta_{.1} \approx \mu - 1.28\sigma$$

Once a particular z percentile is determined, the corresponding percentile for *any* normal distribution is easily calculated.

Percentiles for discrete distributions will not be needed in this book. In general, percentiles for sample data require interpolation between successive sample values. In Section 2.4, we use percentiles that correspond to the ordered sample observations. For example, if $n = 10$, we will regard the smallest sample observation as the fifth sample percentile, the second smallest observation as the 15th sample percentile, and so on.

Section 2.3 Exercises

33. Reconsider the accompanying data on postsurgical range of motion introduced in Exercise 18 of this chapter:

| 154 | 142 | 137 | 133 | 122 | 126 | 135 |
| 135 | 108 | 120 | 127 | 134 | 122 |

   a. What are the values of the quartiles? What is the value of the IQR?

   b. Construct a boxplot based on the five-number summary and comment on its features.

   c. How large or small does an observation have to be to qualify as an outlier? As an extreme outlier?

   d. By how much could the largest observation be decreased without affecting the IQR?

34. Here is a description from the R software of the strength data given in Exercise 4 from Chapter 1.

Min.	1st Qu.	Median
122.2	133.0	135.4
Mean	3rd Qu.	Max.
135.4	138.2	147.7

    a. Comment on any interesting features.
    b. Construct a boxplot of the data and comment on what you see.

35. The diameter length of contact windows used in integrated circuits is normally distributed. About 5% of all lengths exceed 3.75 μm, and about 1% of all lengths exceed 3.85 μm. What are the mean value and standard deviation of the length distribution?

36. The following data on distilled alcohol content (%) for a sample of 35 port wines was extracted from the article "A Method for the Estimation of Alcohol in Fortified Wines Using Hydrometer Baumé and Refractometer Brix" (*Amer. J. Enol. Vitic.*, 2006: 486–490). Each value is an average of two duplicate measurements.

16.35	18.85	16.20	17.75	19.58
17.73	22.75	23.78	23.25	19.08
19.62	19.20	20.05	17.85	19.17
19.48	20.00	19.97	17.48	17.15
19.07	19.90	18.68	18.82	19.03
19.45	19.37	19.20	18.00	19.60
19.33	21.22	19.50	15.30	22.25

    a. Determine the value of the IQR.
    b. Are there any outliers in the sample? Any extreme outliers?
    c. Construct a boxplot and comment on its features.
    d. By how much could the largest observation be decreased without affecting the value of the IQR?

37. Grip is applied to produce normal surface forces that compress the object being gripped. Examples include two people shaking hands and a nurse squeezing a patient's forearm to stop bleeding. The article "Investigation of Grip Force, Normal Force, Contact Area, Hand Size, and Handle Size for Cylindrical Handles" (*Human Factors*, 2008: 734–744) included the following data on grip strength (N) for a sample of 42 individuals:

16	18	18	26	33	41	54
56	66	68	87	91	95	98
106	109	111	118	127	127	135
145	147	149	151	168	172	183
189	190	200	210	220	229	230
233	238	244	259	294	329	403

Construct a boxplot that shows outliers and comment on its features.

38. A sample of 20 glass bottles of a particular type was selected, and the internal pressure strength of each bottle was determined. Consider the following partial sample information:

    median = 202.2
    lower quartile = 196.0
    upper quartile = 216.8
    three smallest observations: 125.8 188.1 193.7
    three largest observations: 221.3 230.5 250.2

    a. Are there any outliers in the sample? Any extreme outliers?
    b. Construct a boxplot that shows outliers, and comment on any interesting features.

39. A company utilizes two different machines to manufacture parts of a certain type. During a single shift, a sample of $n = 20$ parts produced by each machine is obtained, and the value of a particular critical dimension for each part is determined. The accompanying comparative boxplot is constructed from the resulting data. Compare and contrast the two samples.

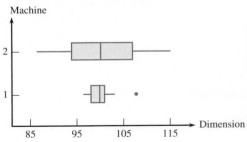

Figure for Exercise 39

40. Recall from Exercise 2 the data on the concentration (EU/mg) in settled dust for one sample of urban homes and another of farm homes:

U: 6.0   5.0   11.0   33.0   4.0   5.0
80.0   18.0   35.0   17.0   23.0

F: 4.0   14.0   11.0   9.0   9.0   8.0
4.0   20.0   5.0   8.9   21.0   9.2
3.0   2.0   0.3

a. Determine the medians, quartiles, and IQRs for the two samples.

b. Are there any outliers in either sample? Any extreme outliers?

c. Construct a comparative boxplot and use it as a basis for comparing and contrasting the two samples.

41. The authors of the article cited in Exercise 2 also provided endotoxin concentrations in dust from vacuum-cleaner dust bags:

U: 34.0 49.0 13.0 33.0 24.0 24.0 35.0 104.0
34.0 40.0 38.0 1.0

F: 2.0 64.0 6.0 17.0 35.0 11.0 17.0 13.0
5.0 27.0 23.0 28.0 10.0 13.0 0.2

Construct a comparative boxplot (which appeared in the cited paper), and compare and contrast the two samples.

42. The comparative boxplot (see below) of gasoline vapor coefficients for vehicles in Detroit appeared in the article "Receptor Modeling Approach to VOC Emission Inventory Validation" (*J. of Envir.*

Engr., 1995: 483–490). Discuss any interesting features.

43. Exercise 46 from Section 1.5 suggested a Weibull distribution with $\alpha = 5$ and $\beta = 125$ as a model for fracture strength of silicon nitride braze joints.

a. What are the quartiles of this distribution, and what is the value of the IQR?

b. Suppose that the value of β is changed to 12.5. Determine the values of the quartiles and the value of the IQR. *Note:* In essence, this amounts to dividing each observation in the population distribution by 10, because β is a "scale" parameter and changing its value stretches or compresses the x scale without changing the shape of the distribution.

44. Reconsider the lognormal distribution with $\mu = 9.164$ and $\sigma = .385$ proposed in Exercise 44 from Section 1.5 as a model for the distribution of nonpoint source load of total dissolved solids (in kg/day/km).

a. What are the values of the quartiles?

b. What is the value of the 95th percentile of the concentration distribution?

c. If μ were 10.164 rather than 9.164, would the values of the two quartiles simply increase by an identical amount?

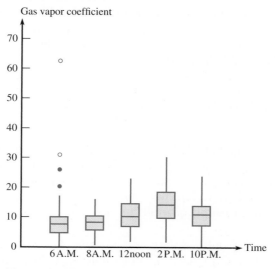

Figure for Exercise 42

2.4   QUANTILE PLOTS

An investigator frequently wishes to know whether it is plausible that a numerical sample $x_1, x_2, \ldots, x_n$ was selected from a particular type of population distribution (e.g., a normal distribution). For one thing, many inferential procedures are based on the assumption that the underlying distribution is of a specified type. The use of such procedures is inappropriate if the actual distribution differs greatly from the assumed type. Additionally, understanding the underlying distribution can sometimes give insight into the physical mechanisms involved in generating the data. An effective way to check a distributional assumption is to construct a *quantile plot* (sometimes called a *probability plot*). The essence of such a plot is that if the plot is based on the correct distribution, the points in the plot will fall close to a straight line. If the actual distribution is quite different from the one used to construct the plot, the points should depart substantially from a linear pattern.

Sample Quantiles

The details involved in constructing quantile plots differ a bit from source to source. The basis for our construction is a comparison between quantiles of the sample data and the corresponding quantiles of the distribution under consideration. Recall that for any number p between 0 and 1, the pth quantile η_p is such that area p lies to the left of η_p under the density curve. For example, Appendix Table I shows that the .9th quantile (90th percentile) for the standard normal distribution is approximately 1.28, the .1th quantile is roughly -1.28, the .8th quantile is about .84, and of course the .5th quantile (the median) is 0.

Roughly speaking, sample quantiles are defined in the same way that quantiles of a population or process distribution are defined. The .5th sample quantile should separate the smallest 50% of the sample from the largest 50%, the .9th sample quantile should be such that 90% of the sample lies below that value and only 10% above, and so on. Our interest here is only in the value of p corresponding to each of the sample observations when ordered from largest to smallest. Recall that when n is odd, the sample median or .5th quantile is the middle value in the ordered list; for example, the sixth smallest value when $n = 11$. This amounts to regarding the middle observations as being half in the lower half of the data and half in the upper half. Similarly, suppose that $n = 10$. Then if we call the third smallest value the .25th quantile, we are regarding that value as being half in the lower group (consisting of the two smallest observations) and half in the upper group (comprising the seven largest observations). This leads to the following general definition of sample quantiles:

DEFINITION

Let $x_{(1)}$ denote the smallest sample observation, $x_{(2)}$ the second smallest sample observation, . . . , and $x_{(n)}$ the largest sample observation. We take $x_{(1)}$ to be the $(.5/n)$th sample quantile, $x_{(2)}$ to be the $(1.5/n)$th sample quantile, . . . , and finally $x_{(n)}$ to be the $[(n - .5)/n]$th sample quantile. That is, for $i = 1, \ldots, n$, $x_{(i)}$ **is the** $[(i - .5)/n]$**th sample quantile.**

Thus when $n = 20$, $x_{(1)}$ is the .025th quantile, $x_{(2)}$ is the .075th quantile, $x_{(3)}$ is the .125th quantile, . . . , and $x_{(20)}$ is the .975th quantile (97.5th percentile).

A Normal Quantile Plot

Suppose now that for $i = 1, \ldots, n$, the quantities $(i - .5)/n$ are calculated and the corresponding quantiles are determined for a specified population or process distribution whose plausibility is being investigated. If the sample were actually selected from the specified distribution, the sample quantiles should be reasonably close to the corresponding distributional quantiles. That is, for $i = 1, \ldots, n$, there should be reasonable agreement between $x_{(i)}$ and the $[(i - .5)/n]$th quantile for the specified distribution. After determining the appropriate quantiles for the distribution being investigated, form the n pairs as follows:

$$\left(\left(\frac{.5}{n} \right) \text{th quantile}, x_{(1)} \right), \left(\left(\frac{1.5}{n} \right) \text{th quantile}, x_{(2)} \right), \ldots, \left(\left(\frac{n - .5}{n} \right) \text{th quantile}, x_{(n)} \right)$$

In other words, pair the smallest quantile with the smallest observation, the second smallest quantile with the second smallest observation, and so on. Each such pair can be plotted as a point on a two-dimensional coordinate system. If the first number in each pair is close to the second number, the points in the plot will fall close to a 45° line [one with slope 1 passing through the point $(0, 0)$].

For example, this program can be carried out to decide whether a normal distribution with $\mu = 100$ and $\sigma = 15$ is plausible. First the appropriate z quantiles are determined; then the desired normal quantiles are expressed in the form $\mu + $ (corresponding z quantile)σ. However, an investigator is typically not interested in knowing whether a *particular* normal distribution is plausible but instead whether *some* normal distribution is plausible. It is clearly inefficient to construct a separate normal quantile plot for each of a large number of different choices of μ and σ. Fortunately, this is not necessary because there is a linear relationship between z quantiles and those for any other normal distribution:

$$\text{quantile for normal } (\mu, \sigma) \text{ distribution} = \mu + \text{(corresponding } z \text{ quantile)} \sigma$$

DEFINITION

> A **normal quantile plot** is a plot of the (z quantile, observation) pairs. The linear relation between normal (μ, σ) quantiles and z quantiles implies that if the sample has come from a normal distribution with particular values of μ and σ, the points in the plot should fall close to a straight line with slope σ and vertical intercept μ. Thus a plot for which the points fall close to *some* straight line suggests that the assumption of a normal population or process distribution is plausible.

Note that if a straight line is fit to the points in the plot, the intercept and slope give estimates of μ and σ, respectively, though these will typically differ from the usual estimates $\bar{x}$ and s.

Example 2.17

There has been recent increased use of augered cast-in-place (ACIP) and drilled displacement (DD) piles in the foundations of buildings and transportation structures. In the article "Design Methodology for Axially Loaded Auger Cast-in-Place and Drilled

Displacement Piles" (*J. Geotech. Geoenviron. Engr.*, 2012: 1431–1441) researchers propose a design methodology to enhance the efficiency of these piles. The authors reported the following length-diameter ratio measurements based on 17 static-pile load tests on ACIP and DD piles from various construction sites. The values of p for which z percentiles are needed are $(1 - .5)/17 = .029$, $(2 - .5)/17 = .088, \ldots$, and .971.

$x_{(i)}$:	30.86	37.68	39.04	42.78	42.89	42.89	45.05	47.08	47.08
z percentile:	−1.89	−1.35	−1.05	−0.82	−0.63	−0.46	−0.30	−0.15	0.00

$x_{(i)}$:	48.79	48.79	52.56	52.56	54.8	55.17	56.31	59.94
z percentile:	0.15	0.30	0.46	0.63	0.82	1.05	1.35	1.89

Figure 2.13 shows the corresponding normal quantile plot as generated by the qqnorm function in the R software. The pattern in the plot is quite straight, indicating it is plausible that the population distribution of length-diameter ratio is normal.

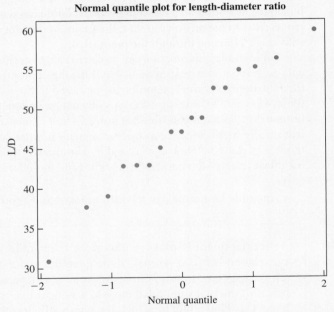

Figure 2.13 Normal quantile plot from R for the length-diameter ratio data

The judgment as to whether a plot does or does not show a substantial linear pattern is somewhat subjective. Particularly when n is small, normality should not be ruled out unless the departure from linearity is very clear-cut. Figure 2.14 displays several plots that suggest a nonnormal population or process distribution. In Section 8.4, we show how a quantitative assessment of the extent to which points in a two-dimensional plot fall close to a straight line can be used as the basis of an inferential procedure for deciding whether normality is plausible.

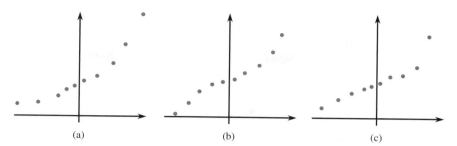

(a) (b) (c)

Figure 2.14 Quantile plots that are inconsistent with an underlying normal distribution

Minitab will automatically obtain the z percentiles in response to an "NSCORE" command, but it uses something a bit different from $(i - .5)/n$ as a basis for this calculation. Minitab also has a normal plot command in its graphics menu; the resulting plot has x on the horizontal axis and a nonlinear vertical axis constructed so that normal data should plot close to a straight line.

Plots for Other Distributions

It is easy to assess the plausibility of a lognormal population or process distribution, because to say that x is lognormally distributed is to say that $\ln(x)$ has a normal distribution. Thus one simply calculates $\ln(x_{(1)}), \ldots, \ln(x_{(n)})$ and uses these quantities in place of $x_{(1)}, \ldots, x_{(n)}$ in a normal quantile plot.

For a Weibull distribution,

$$p = \text{area to the left of } \eta_p = 1 - e^{-(\eta_p/\beta)^\alpha}$$

This implies that

$$\ln(1 - p) = -\left(\frac{\eta_p}{\beta}\right)^\alpha$$

Multiplying by -1 and taking logs again gives

$$\ln[-\ln(1 - p)] = \alpha[\ln(\eta_p) - \ln(\beta)] = \alpha \ln(\eta_p) + \gamma \quad \text{where } \gamma = -\alpha \ln(\beta)$$

Thus there is a linear relation between the logarithm of Weibull quantiles and $\ln[-\ln(1 - p)]$. This suggests that we calculate $\ln(x_{(1)}), \ldots, \ln(x_{(n)})$ and then plot the $(\ln[-\ln(1 - p)], \ln(x))$ pairs. If the plot is reasonably straight, it is plausible that the sample has come from *some* Weibull distribution.

Example 2.18

For many years it has been well established that the Weibull distribution is useful in modeling the strength of fibers used in composite materials such as carbon graphite, Kevlar, and glass. With the advent of nanotechnology where materials can be developed at miniscule levels, scientists have questioned whether the Weibull distribution is applicable to model material strength even at the nanoscale. In the article "Stochastic Strength of Nanotubes: An Appraisal of Available Data" (*Composites Sci. and Tech.*, 2005: 2380–2384) researchers reported the tensile strengths of three different types of nanotubes and assessed whether the Weibull distribution would serve as a reasonable model for each type.

The following represent the tensile strengths (in GPa) for 26 multiwall carbon nanotubes produced by chemical vapor deposition; their average diameter is roughly 97 nm. Note that the values of $p_i = (i - .5)/26$ are also given:

$x_{(i)}$:	17.4	22.3	23.7	30.0	44.2	49.3	52.7	54.8	62.1	66.2
p_i:	0.019	0.058	0.096	0.135	0.173	0.212	0.250	0.288	0.327	0.365
$x_{(i)}$:	84.9	90.1	90.3	91.1	99.5	101.6	108.5	109.5	119.1	127.0
p_i:	0.404	0.442	0.481	0.519	0.558	0.596	0.635	0.673	0.712	0.750
$x_{(i)}$:	132.9	140.8	141.0	175.0	231.8	259.7				
p_i:	0.788	0.827	0.865	0.904	0.942	0.981				

Figure 2.15 is a plot of the $(\ln[-\ln(1 - p)], \ln(x))$ pairs. Although there is some wiggling especially in the lower part of the plot, the overall pattern is reasonably straight and so the assumption of an underlying Weibull distribution for tensile strength for this type of nanotube appears to be acceptable. The article also showed that the Weibull distribution was a good fit in modeling tensile strength for the two other nanotube types discussed.

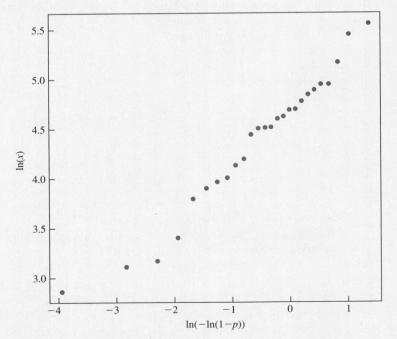

Figure 2.15 A Weibull plot of the nanotube tensile strength data

Most statistical computer packages make it easy to do the arithmetic necessary to obtain the quantities to be plotted. In addition, the Minitab graphics menu has a Weibull plot option, making it unnecessary for the user to do any arithmetic before obtaining the plot. The x values are plotted directly on the horizontal axis, and the vertical axis is constructed using a nonlinear scale so that data from a Weibull distribution should plot close to a straight line.

Plots based on other distributions can also be constructed. Consult chapter references and software packages for more information.

Section 2.4 Exercises

45. The accompanying normal quantile plot was constructed from a sample of 30 readings on tension for mesh screens behind the surface of video display tubes used in computer monitors. Does it appear plausible that the tension distribution is normal?

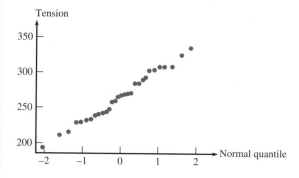

46. The following are modulus of elasticity observations for cylinders given in the article cited in Example 1.2:

37.0	37.5	38.1	40.0	40.2	40.8	41.0
42.0	43.1	43.9	44.1	44.6	45.0	46.1
47.0	62.0	64.3	68.8	70.1	74.5	

Use the quantiles for a sample of size 20 given in this section to construct a normal quantile plot, and comment on the plausibility of a normal population distribution.

47. A sample of 15 female collegiate golfers was selected, and the clubhead velocity (km/hr) of each golfer while swinging a driver was determined, resulting in the following data ("Hip Rotational Velocities During the Full Golf Swing," *J. of Sports Science and Medicine*, 2009: 296–299):

69.0	69.7	72.7	80.3	81.0
85.0	86.0	86.3	86.7	87.7
89.3	90.7	91.0	92.5	93.0

The corresponding z percentiles are

−1.83	−1.28	−0.97	−0.73	−0.52
−0.34	−0.17	0.0	0.17	0.34
0.52	0.73	0.97	1.28	1.83

Construct a normal quantile plot and a dotplot. Is it plausible that the population distribution is normal?

48. The accompanying observations are precipitation values during March over a 30-year period in Minneapolis–St. Paul.

.77	1.20	3.00	1.62	2.81	2.48
1.74	.47	3.09	1.31	1.87	.96
.81	1.43	1.51	.32	1.18	1.89
1.20	3.37	2.10	.59	1.35	.90
1.95	2.20	.52	.81	4.75	2.05

a. Construct and interpret a normal quantile plot for this data set.
b. Calculate the square root of each value and then construct a quantile plot based on this transformed data. Does it seem plausible that the square root of precipitation is normally distributed?
c. Repeat part (b) after transforming by cube roots.

49. The article "A Probabilistic Model of Fracture in Concrete and Size Effects on Fracture Toughness" (*Magazine of Concrete Res.*, 1996: 311–320) gives arguments for why fracture toughness in concrete specimens should have a Weibull distribution and presents several histograms of data that appear well fit by superimposed Weibull curves. Consider the

following sample of size $n = 18$ observations on toughness for high-strength concrete (consistent with one of the histograms); values of $p_i = (i - .5)/18$ are also given:

Obs:	.47	.58	.65	.69	.72	.74
p_i:	.0278	.0833	.1389	.1944	.2500	.3056
Obs:	.77	.79	.80	.81	.82	.84
p_i:	.3611	.4167	.4722	.5278	.5833	.6389
Obs:	.86	.89	.91	.95	1.01	1.04
p_i:	.6944	.7500	.8056	.8611	.9167	.9722

Construct a Weibull quantile plot and comment.

50. In the article "Weibull Parameter of Oil-Immersed Transformer to Evaluate Insulation Reliability on Temporary Overvoltage" (*IEEE Trans. on Dielectrics and Elec. Insul.*, 2010: 1863–1868), researchers investigated the reliability of oil-immersed transformers under various conditions. In one experiment, the researchers measured the breakdown time of the transformer oil gap under various oil flow velocities and exposure to temporary overvoltage. Consider the following breakdown time data (in s) from their experiment where an oil flow at 16 cm/s and an overvoltage of 81kV were applied.

7.2	10.0	18.0	25.0	36.0	38.0
46.0	63.0	71.0	76.0	92.0	95.0
104.0	152.0	198.0	226.0	235.0	247.0
361.0	392.0				

Construct a Weibull plot and comment on the plausibility of breakdown time having a Weibull distribution.

51. The accompanying figures show (a) a normal quantile plot of the observations on cell interdivision time (IDT) given in Exercise 16 of Section 1.2 and (b) a normal quantile plot of the logarithms of the IDTs. What do these plots suggest about the distribution of cell interdivision time?

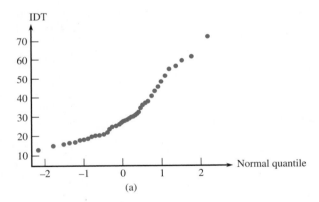

(a)

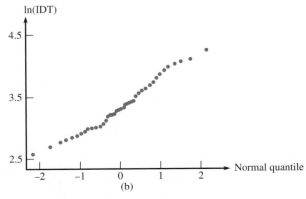

(b)

Figure for Exercise 51

52. A plot to assess the plausibility of an exponential population distribution can be based on quantiles of the exponential distribution having $\lambda = 1$ (i.e., the exponential distribution with density function $f(x) = e^{-x}$ for $x > 0$). This is because λ, like σ for a normal distribution, is a scale parameter. Consider the following failure time observations (1000s of hours) resulting from accelerated life testing of 16 integrated circuit chips of a certain type:

82.8	11.6	359.5	502.5	307.8	179.7
242.0	26.5	244.8	304.3	379.1	
212.6	229.9	558.9	366.7	204.6	

Construct a quantile plot and comment on the plausibility of failure time having an exponential distribution.

53. The article "Families of Distributions for Hourly Median Power and Instantaneous Power of Received Radio Signals" (*J. of Research for the National Bureau of Standards*, 1963: 753–762) suggests the lognormal distribution for x = hourly median power (decibels) of received radio signals transmitted between two cities. Consider the following sample of hourly median power readings:

2.7 5.4 9.7 22.8 30.5 55.7 66.2 97.3 186.5 240.0

a. Is it plausible that these observations were sampled from a normal distribution?
b. Is it plausible that these observations were sampled from a lognormal distribution?

Supplementary Exercises

54. Anxiety disorders and symptoms can often be effectively treated with benzodiazepine medications. It is known that animals exposed to stress exhibit a decrease in benzodiazepine receptor binding in the frontal cortex. The paper "Decreased Benzodiazepine Receptor Binding in Prefrontal Cortex in Combat-Related Posttraumatic Stress Disorder" (*American J. of Psychiatry*, 2000: 1120–1126) described the first study of benzodiazepine receptor binding in individuals suffering from PTSD. The accompanying data on a receptor binding measure (adjusted distribution volume) was read from a graph in the paper:

PTSD: 10 20 25 28 31 35 37 38 38
        39 39 42 46

Healthy: 23 39 40 41 43 47 51 58
         63 66 67 69 72

a. Calculate and interpret the values of the mean, median, and standard deviation for each of the two samples.
b. Calculate a trimmed mean for each sample by deleting the smallest and largest observations. What is the trimming percentage? What effect does trimming have?
c. Determine the value of the interquartile range for each sample. Does either sample contain any outliers? Any extreme outliers?

d. Construct a comparative boxplot, and comment on interesting features.
e. Would you recommend estimating the difference between the true average binding measure of PTSD individuals and the true average measure for healthy individuals using a method based on assuming that each sample was selected from a normal population distribution? Explain your reasoning.

55. A sample of 77 individuals working at a particular office was selected, and the noise level (dBA) experienced by each one was determined, yielding the following data ("Acceptable Noise Levels for Construction Site Offices," *Building Serv. Engr. Res. and Tech.*, 2009: 87–94).

55.3	55.3	55.3	55.9	55.9	55.9
55.9	56.1	56.1	56.1	56.1	56.1
56.1	56.8	56.8	57.0	57.0	57.0
57.8	57.8	57.8	57.9	57.9	57.9
58.8	58.8	58.8	59.8	59.8	59.8
62.2	62.2	63.8	63.8	63.8	63.9
63.9	63.9	64.7	64.7	64.7	65.1
65.1	65.1	65.3	65.3	65.3	65.3
67.4	67.4	67.4	67.4	68.7	68.7
68.7	68.7	69.0	70.4	70.4	71.2
71.2	71.2	73.0	73.0	73.1	73.1
74.6	74.6	74.6	74.6	79.3	79.3
79.3	79.3	83.0	83.0	83.0	

Use various techniques discussed in this chapter to organize, summarize, and describe the data.

56. Three different C_2F_6 flow rates (SCCM) were considered in an experiment to investigate the effect of flow rate on the uniformity (%) of the etch on a silicon wafer used in the manufacture of integrated circuits, resulting in the following data:

125: 2.6 2.7 3.0 3.2 3.8 4.6
160: 3.6 4.2 4.2 4.6 4.9 5.0
200: 2.9 3.4 3.5 4.1 4.6 5.1

Compare and contrast the uniformity observations resulting from these three different flow rates.

57. Consider a sample $x_1, \ldots, x_n$, and let $\bar{x}_k$ and s_k^2 denote the sample mean and variance, respectively, of the first k observations.
   a. Show that

   $$ks_{k+1}^2 = (k-1)s_k^2 + \frac{k}{k+1}(x_{k+1} - \bar{x}_k)^2$$

   b. Suppose that a sample of 15 strands of drapery yarn has resulted in a sample mean thread elongation of 12.58 mm and a sample standard deviation of .512 mm. A 16th strand results in an elongation value of 11.8. What are the values of the sample mean and sample standard deviation for all 16 elongation observations?

58. In 1997 a woman sued a computer keyboard manufacturer, charging that her repetitive stress injuries were caused by the keyboard (*Genessy v. Digital Equipment Corp.*). The jury awarded about $3.5 million for pain and suffering, but the court then set aside that award as being unreasonable compensation. In making this determination, the court identified a "normalative" group of 27 similar cases and specified a reasonable award as one within 2 standard deviations of the mean of the awards in the 27 cases. The 27 awards were (in $1000s) 37, 60, 75, 115, 135, 140, 149, 150, 238, 290, 340, 410, 600, 750, 750, 750, 1050, 1100, 1139, 1150, 1200, 1200, 1250, 1576, 1700, 1825, and 2000, from which $\Sigma x_i = 20{,}179$, $\Sigma x_i^2 = 24{,}657{,}511$. What is the maximum possible amount that could be awarded under the 2 standard deviation rule?

59. A deficiency of the trace element selenium in the diet can negatively affect growth, immunity, muscle and neuromuscular function, and fertility. The introduction of selenium supplements to dairy cows is justified when pastures have low selenium levels. Authors of the paper "Effects of Short-Term Supplementation with Selenised Yeast on Milk Production and Composition of Lactating Cows" (*Australian J. of Dairy Tech.*, 2004: 199–203) supplied the following data on milk selenium concentration (mg/L) for a sample of cows given a selenium supplement and a control sample given no supplement, both initially and after a nine-day period.

Obs	Init Se	Init Cont	Final Se	Final Cont
1	11.4	9.1	138.3	9.3
2	9.6	8.7	104.0	8.8
3	10.1	9.7	96.4	8.8
4	8.5	10.8	89.0	10.1
5	10.3	10.9	88.0	9.6
6	10.6	10.6	103.8	8.6
7	11.8	10.1	147.3	10.4
8	9.8	12.3	97.1	12.4
9	10.9	8.8	172.6	9.3
10	10.3	10.4	146.3	9.5
11	10.2	10.9	99.0	8.4
12	11.4	10.4	122.3	8.7
13	9.2	11.6	103.0	12.5
14	10.6	10.9	117.8	9.1
15	10.8	121.5		
16	8.2	93.0		

   a. Do the initial Se concentrations for the supplement and control samples appear to be similar? Use various techniques from this chapter to summarize the data and answer the question posed.
   b. Again use methods from this chapter to summarize the data and then describe how the final Se concentration values in the treatment group differ from those in the control group.

60. An inequality developed by the Russian mathematician Chebyshev gives information about the percentage of values in *any* sample or distribution that fall within a specified number of standard deviations of the mean. Let k denote any number satisfying

$k \geq 1$. Then at least $100(1 - 1/k^2)\%$ of the values are within k standard deviations of the mean.
a. What does Chebyshev's inequality say about the percentage of values that are within 2 standard deviations of the mean? Within 3 standard deviations of the mean? Within 5 standard deviations? Within 10 standard deviations?
b. What does Chebyshev's inequality say about the percentage of values that are more than 2 standard deviations from the mean? More than 3 standard deviations from the mean?
c. Suppose the distribution of slot width on a forging has a mean value of 1.000 in. and a standard deviation of .0025 in. What percentage of such forgings have a slot width that is between .995 in. and 1.005 in.? If specifications are $1.000 \pm .005$ in., what percentage of slot widths will conform to specifications?
d. Refer to part (c). What percentage of such forgings will have a slot width that is outside the interval from .995 in. to 1.005 in. (i.e., either less than .995 or greater than 1.005)? What can be said about the percentage of widths that exceed 1.005 in.?

61. Reconsider Chebyshev's inequality as stated in the previous exercise.
a. Compare what the inequality says about the percentage within 1, 2, or 3 standard deviations of the mean value to the corresponding percentages given by the empirical rule.
b. An exponential distribution with parameter λ has both mean value and standard deviation equal to $1/\lambda$. If component lifetime is exponentially distributed with a mean value of 100 hr, what percentage of these components have lifetimes within 1 standard deviation of the mean lifetime? Within 2 standard deviations? Within 3 standard deviations? Compare these to the percentages given by Chebyshev's inequality.
c. Why do you think the percentages from Chebyshev's inequality so badly understate the actual percentages in the situations of parts (a) and (b)?

62. Consider a sample $x_1, \ldots, x_n$ with mean $\bar{x}$ and standard deviation s, and let $z_i = (x_i - \bar{x})/s$. What are the mean and standard deviation of the z_i's?

63. The accompanying observations are carbon monoxide levels (ppm) in air samples obtained from a certain region:

9.3   10.7   8.5   9.6   12.2   16.6   9.2   10.5
7.9   13.2   11.0   8.8   13.7   12.1   9.8

a. Calculate a trimmed mean by trimming the smallest and largest observations, and give the corresponding trimming percentage. Do the same with the two smallest and two largest values trimmed.
b. Using the results of part (a), how would you calculate a trimmed mean with a 10% trimming percentage?
c. Suppose there had been 16 sample observations. How would you go about calculating a 10% trimmed mean?

64. Specimens of three different types of rope wire were selected, and the fatigue limit (MPa) was determined for each specimen, resulting in the accompanying data:

Type 1: 350 350 350 358 370 370 370 371
        371 372 372 384 391 391 392
Type 2: 350 354 359 363 365 368 369 371
        373 374 376 380 383 388 392
Type 3: 350 361 362 364 364 365 366 371
        377 377 377 379 380 380 392

a. Construct a comparative boxplot, and comment on similarities and differences.
b. Construct a comparative dotplot (a dotplot for each sample with a common scale). Comment on similarities and differences.
c. Does the comparative boxplot of part (a) give an informative assessment of similarities and differences? Explain your reasoning.

65. The three measures of center introduced in this chapter are the mean, median, and trimmed mean. Two additional measures of center that are occasionally used are the *midrange*, which is the average of the smallest and largest observations, and the *midhinge*, which is the average of the two quartiles. Which of these five measures of center are resistant to the effects of outliers and which are not? Explain your reasoning.

66. The capacitance (nf) of multilayer ceramic capacitors supplied by a certain vendor is normally distributed with mean value 98 and standard deviation 2. Specifications for these capacitors are 100 ± 5 nf.
    a. What proportion of these capacitors will conform to specification?
    b. Suppose that these capacitors are shipped in batches of size 20. Let x denote the number of capacitors in a batch that conform to specification. Provided that capacitances of successive capacitors are independent of one another, what kind of distribution does x have? In the long run, in what proportion of batches will at least 19 of the 20 capacitors conform to specifications? *Hint:* Think of a capacitor that conforms to specification as a "success," so x is the number of successes in the batch.

67. *Aortic stenosis* refers to a narrowing of the aortic valve in the heart. The paper "Correlation Analysis of Stenotic Aortic Valve Flow Patterns Using Phase Contrast MRI" (*Annals of Biomed. Engr.*, 2005: 878–887) gave the following data on aortic root diameter (cm) and gender for a sample of patients having various degrees of aortic stenosis:

    M: 3.7   3.4   3.7   4.0   3.9
         3.8   3.4   3.6   3.1   4.0
         3.4   3.8   3.5
    F: 3.8   2.6   3.2   3.0   4.3
         3.5   3.1   3.1   3.2   3.0

    a. Compare and contrast the diameter observations for the two genders.
    b. Calculate a 10% trimmed mean for each of the two samples and compare to other measures of center (for the male sample, the interpolation method mentioned in Section 2.1 must be used).

68. A study carried out to investigate the distribution of total braking time (reaction time plus accelerator-to-brake movement time, in ms) during real driving conditions at 60 km/hr gave the following summary information on the distribution of times ("A Field Study on Braking Responses during Driving," *Ergonomics*, 1995: 1903–1910):

    mean = 535   median = 500   mode = 500
    sd = 96   minimum = 220   maximum = 925
    5th percentile = 400   10th percentile = 430
    90th percentile = 640   95th percentile = 720

    What can you conclude about the shape of a histogram of this data? Explain your reasoning.

69. Let x denote the maximum physical stress that a unit of a certain product encounters during its lifetime. Suppose that x is normally distributed with 99th percentile = 5.33 and 10th percentile = 1.72 (suggested in the article "A Formulation of Product Reliability through Environmental Stress Testing and Screening," *J. of the Institute of Envir. Sciences*, 1994: 50–56; the unit for x was unspecified). What proportion of these units have maximum stress values exceeding 5? What proportion have maximum stress values less than 2?

70. The indoor thermal climate is an important characteristic affecting the health and productivity of workers in buildings. The paper "Adaptive Comfort Temperature Model of Air-Conditioned Buildings in Hong Kong" (*Building and Environment*, 2003: 837–852) reported data on a number of building characteristics measured during the summer and also during the winter. Consider the accompanying values of relative humidity.

    Summer: 57.18 58.11 56.53 58.61 57.40 62.64
             61.72 57.26 53.43 53.71 58.64 45.12
             47.52 54.47 55.88 51.08 53.69 54.37
             54.36 61.01 52.66 56.20 48.40 46.99
             50.63 52.40 52.20 55.95 53.77

    Winter: 52.20 41.83 55.63 54.18 54.56 56.20
             58.09 56.70 57.57 58.70 56.15 59.77
             61.58 61.81 62.48 63.31 55.57 62.25
             57.40 55.07 62.52 52.80 57.20 59.27
             54.98 58.13

    Use methods from this and the previous chapter to describe, summarize, compare, and contrast the summer and winter relative humidity data.

Bibliography

Please see the bibliography for Chapter 1.

Giancarlo Liguori/Shutterstock.com

3

Bivariate and Multivariate Data and Distributions

INTRODUCTION

Now that we have acquired some facility for working with univariate data and distributions, it's time to expand our horizons. A multivariate data set consists of observations made simultaneously on two or more variables. One important special case is that of bivariate data, in which observations on only two variables, x and y, are available. In Section 3.1, we introduce the scatterplot, a picture for gaining insight into the nature of any relationship between x and y.

Next, we discuss the correlation coefficient, which is a measure of how strongly two variables are related. In many investigations, one primary objective is to predict y from the value of x—for example, to predict yield from a chemical reaction at a particular reaction temperature. If the scatterplot shows a linear pattern, the natural strategy is to fit a straight line to the data and use it as the basis for predictions, as we do in Section 3.3. If a scatterplot shows curvature, fitting a nonlinear function, such as a quadratic or an exponential function, is appropriate; we show how this can be done in Section 3.4. Multiple regression functions, in which y is related to two or more predictor variables, are the subject of Section 3.5. Finally, Section 3.6 introduces bivariate and multivariate distributions for population or process variables. In Chapter 11, we return to this type of data and describe how formal conclusions about relationships can be drawn by using methods from statistical inference.

3.1   SCATTERPLOTS

A **multivariate data set** consists of measurements or observations on each of two or more variables. One important special case, **bivariate data,** involves only two variables, x and y. For example, x might be the distance from a particular highway and y, the lead content of the soil at that distance. When both x and y are numerical variables, each observation consists of a pair of numbers, such as (14, 5.2) or (27.63, 18.9). The first number in a pair is the value of x and the second number is the value of y.

An unorganized list of such pairs yields little information about the distribution of either the x values or the y values separately, and even less information about whether the two variables are related to one another. In Chapter 1, we saw how pictures could help make sense of univariate data. The most important picture based on bivariate numerical data is a **scatterplot.** Each observation (pair of numbers) is represented by a point on a rectangular coordinate system, as shown in Figure 3.1(a). The horizontal axis is identified with values of x and is scaled so that any x value can be easily located. Similarly, the vertical or y axis is marked for easy location of y values. The point corresponding to any particular (x, y) pair is placed where a vertical line from the value on the x axis intersects a horizontal line from the value on the y axis. Figure 3.1(b) shows the point representing the observation (4.5, 15); it is above 4.5 on the horizontal axis and to the right of 15 on the vertical axis.

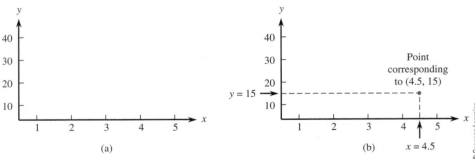

Figure 3.1 Constructing a scatterplot: (a) rectangular coordinate system for a scatterplot of bivariate data; (b) the point corresponding to the observation (4.5, 15)

Example 3.1   Visual and musculoskeletal problems associated with the use of visual display terminals (VDTs) have become rather common in recent years. Some researchers have focused on vertical gaze direction as a source of eye strain and irritation. This direction is known to be closely related to ocular surface area (OSA), so a method of measuring OSA is needed. The accompanying representative data on y = OSA (cm^2) and x = width of the palprebal fissure (i.e., the horizontal width of the eye opening, in cm) is from the article "Analysis of Ocular Surface Area for Comfortable VDT Workstation Layout" (*Ergonomics*, 1996: 877–884). The order in which observations were obtained was not given, so for convenience they are listed in increasing order of x values.

Obs:	1	2	3	4	5	6	7	8	9	10
x:	.40	.42	.48	.51	.57	.60	.70	.75	.75	.78
y:	1.02	1.21	.88	.98	1.52	1.83	1.50	1.80	1.74	1.63
Obs:	11	12	13	14	15	16	17	18	19	20
x:	.84	.95	.99	1.03	1.12	1.15	1.20	1.25	1.25	1.28
y:	2.00	2.80	2.48	2.47	3.05	3.18	3.76	3.68	3.82	3.21
Obs:	21	22	23	24	25	26	27	28	29	30
x:	1.30	1.34	1.37	1.40	1.43	1.46	1.49	1.55	1.58	1.60
y:	4.27	3.12	3.99	3.75	4.10	4.18	3.77	4.34	4.21	4.92

Thus $(x_1, y_1) = (.40, 1.02)$, $(x_5, y_5) = (.57, 1.52)$, and so on. A Minitab scatterplot is shown in Figure 3.2; we used an option that produced a dotplot of both the x values and y values individually along the right and top margins of the plot, which makes it easier to visualize the distributions of the individual variables (histograms or boxplots are alternative options).

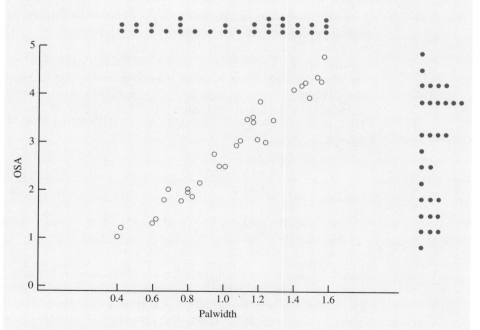

Figure 3.2 Scatterplot from Minitab for the data from Example 3.1, along with dotplots of x and y values

Here are some things to notice about the data and plot:

- Several observations have identical x values yet different y values (for example, $x_8 = x_9 = .75$, but $y_8 = 1.80$ and $y_9 = 1.74$). Thus the value of y is *not* determined solely by x but also by various other factors.
- There is a strong tendency for y to increase as x increases. That is, larger values of OSA tend to be associated with larger values of fissure width—a positive relationship between the variables.
- It appears that the value of y could be predicted from x by finding a line that is reasonably close to the points in the plot (the authors of the cited article superimposed such a line on their plot). In other words, there is evidence of a substantial (though not perfect) linear relationship between the two variables.

The horizontal and vertical axes in the scatterplot of Figure 3.2 intersect at the point $(0, 0)$. In many data sets, the values of x or y or the values of both variables differ considerably from zero relative to the range(s) of the values. For example, a study of how air conditioner efficiency is related to maximum daily outdoor temperature might involve observations for temperatures ranging from 80°F to 100°F. When this is the case, a more informative plot would show the appropriately labeled axes intersecting at some point other than $(0, 0)$.

Example 3.2   Arsenic is found in many ground waters and some surface waters. Recent research on health effects has prompted the Environmental Protection Agency to reduce allowable arsenic levels in drinking water; as a result, many water systems are no longer compliant with standards. This has spurred interest in the development of methods to remove arsenic. The accompanying data on x = pH and y = arsenic removed (%) by a particular process was read from a scatterplot in the article "Optimizing Arsenic Removal During Iron Removal: Theoretical and Practical Considerations" (*J. of Water Supply Res. and Tech.*, 2005: 545–560):

x:	7.01	7.11	7.12	7.24	7.94	7.94	8.04	8.05	8.07
y:	60	67	66	52	50	45	52	48	40

x:	8.90	8.94	8.95	8.97	8.98	9.85	9.86	9.86	9.87
y:	23	20	40	31	276	9	22	13	7

Figure 3.3 shows two Minitab scatterplots of this data. In Figure 3.3(a), the software selected the scale for both axes. We obtained Figure 3.3(b) by specifying scaling for the axes so that they would intersect at roughly the point $(0, 0)$. The second plot is much more crowded than the first one; such crowding can make it difficult to ascertain the general nature of any relationship. For example, curvature can be overlooked in a crowded plot.

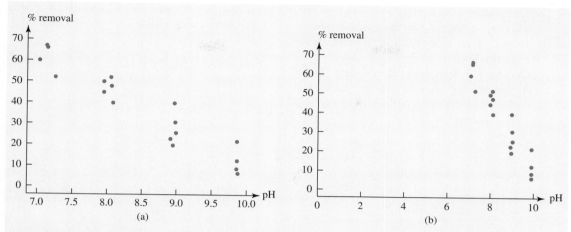

Figure 3.3 Minitab scatterplots of the data in Example 3.2

Large values of arsenic removal tend to be associated with low pH, a negative or inverse relationship. Furthermore, the two variables appear to be at least approximately linearly related, although the points in the plot would spread out somewhat about any superimposed straight line (such a line appeared in the plot in the cited article).

Section 3.1 Exercises

1. In the article "Analysis of the Thermal Properties of Air-Conditioning-Type Building Materials" (*Solar Energy*, 2012: 2967–2974), researchers investigated thermal properties of building materials that are used across a variety of climate regions. One property of interest was solar absorptance, a measure of an object's ability to absorb solar radiation. To reduce building energy consumption, it would be desirable for the building material to have higher solar absorptance in colder climates and lower solar absorptance in warmer climates. The following data (read from a graph) shows solar absorptance levels under different temperature conditions for a building material called G17S, which changes color depending on temperature, thereby allowing for variable absorptance.

Temperature (in °C): 2   9   20   28   39

Solar Absorptance: .81 .78 .69 .65 .48

Create a scatterplot for this data. How would you characterize the relationship between these two variables? Is the desired inverse relationship between temperature and absorptance evident for this material?

2. The article "Case Adaptation Method of Case-Based Reasoning for Construction Cost Estimation in Korea" (*J. Constr. Engr. Mgmt.*, 2012: 43–52) provided data on military barrack projects undertaken by the Korean Ministry of National Defense from 2004 to 2008. Two variables of interest were the floor area of a barrack and the corresponding cost (in $US). The corresponding data is given here:

Floor Area:	Cost:
382	418,930
571	609,386
618	755,489
726	660,527
802	864,438
959	1,003,495
1066	895,947
1306	1,461,549
1873	1,899,494
2460	2,331,632
3134	2,833,203
4989	4,750,468
6918	5,331,390

a. Construct stem-and-leaf displays of both floor area and cost. Comment on any interesting features.
b. Do the values of cost appear to be perfectly linearly related to the floor area values?
c. Construct a scatterplot of the data. Does it appear that cost could be accurately predicted by the value of floor area? Explain your reasoning.

3. In the article referenced in Exercise 2, the relationship between the number of beds in a barrack and the cost of the building was also investigated.

Number of Beds	Cost
22	418,930
40	609,386
40	755,489
38	660,527
24	864,438
54	1,003,495
59	895,947
98	1,461,549
106	1,899,494
142	2,331,632
190	2,833,203
68	4,750,468
392	5,331,390

Construct a scatterplot based on this data. What appears to be the nature of the relationship between these two variables? Do you notice anything peculiar in the graph?

4. Open water oil spills, such as the *Deepwater Horizon* spill of 2010, can wreak terrible consequences on the environment and be expensive to clean up. Many physical and biological methods have been developed to recover oil from water surfaces. In the article "Capacity of Straw for Repeated Binding of Crude Oil from Salt Water and Its Effect on Biodegradation" (*J. Hazard. Toxic Radioact. Waste*, 2012: 75–78), researchers examined how wheat straw could be used to extract crude oil from a water surface. An experiment was conducted in which crude oil (0 to 16.9 g) was added to 100 mL of saltwater in separate Petri dishes. Wheat straw (2 g) was then added to each dish and all dishes were shaken at 70 rpm overnight. The following data read from a graph

is based on the amount of oil added (in g) and the corresponding amount of oil recovered (in g) from wheat straw.

Oil Added	Oil Recovered
1.0	0.610
1.5	0.840
2.1	1.512
2.8	1.792
3.6	2.952
4.5	2.880
5.5	4.400
6.6	5.346
7.8	6.396
9.1	7.189
10.5	8.085
12.0	9.840
13.6	11.696
15.2	13.224
16.9	14.365

a. For each observation, determine the percentage of oil recovery by wheat straw. Is this percentage relatively constant across all observations? Was the percentage higher at certain added oil levels over others?
b. Do the values of the recovered oil appear to be perfectly linearly related to the added oil values? Why or why not?
c. Construct a scatterplot of the data. Does it appear that recovered oil could be accurately predicted by the value of added oil? Explain your reasoning.

5. The article "Objective Measurement of the Stretch-ability of Mozzarella Cheese" (*J. of Texture Studies*, 1992: 185–194) reported on an experiment to investigate how the behavior of mozzarella cheese varied with temperature. Consider the accompanying data on x = temperature and y = elongation (%) at failure of the cheese. *Note:* The researchers were Italian and used *real* mozzarella cheese, not the poor cousin widely available in the United States.

x:	59	63	68	72	74	78	83
y:	118	182	247	208	197	135	132

a. Construct a scatterplot in which the axes intersect at (0, 0). Mark 0, 20, 40, 60, 80, and

100 on the horizontal axis and 0, 50, 100, 150, 200, and 250 on the vertical axis.

b. Construct a scatterplot in which the axes intersect at (55, 100), as was done in the cited article. Does this plot seem preferable to the one in part (a)? Explain your reasoning.

c. What do the plots of parts (a) and (b) suggest about the nature of the relationship between the two variables?

6. Calcium phosphate cement is gaining increasing attention for use in bone repair applications. The article "Short-Fibre Reinforcement of Calcium Phosphate Bone Cement" (*J. of Engr. in Med.*, 2007: 203–211) reported on a study in which polypropylene fibers were used in an attempt to improve fracture behavior. The following data on x = fiber weight (%) and y = compressive strength (MPa) was provided by the article's authors.

x:	0.00	0.00	0.00	0.00	0.00	1.25
y:	9.94	11.67	11.00	13.44	9.20	9.92

x:	1.25	1.25	1.25	2.50	2.50	2.50
y:	9.79	10.99	11.32	12.29	8.69	9.91

x:	2.50	2.50	5.00	5.00	5.00	5.00
y:	10.45	10.25	7.89	7.61	8.07	9.04

x:	7.50	7.50	7.50	7.50	10.00	10.00
y:	6.63	6.43	7.03	7.63	7.35	6.94

x:	10.00	10.00
y:	7.02	7.67

Construct a scatterplot of the data. How would you describe the nature of the relationship between the two variables?

7. In surface water hydrology, a common problem is the estimation of long-term annual yield from ungauged watersheds. In the article "Generalized Mediterranean Annual Water Yield Model: Grunsky's Equation and Long-Term Average Temperature" (*J. Hydrol. Engr.*, 2011: 874–879), researchers propose a generalized water yield model for watersheds. One important watershed-specific component of the model is α, a coefficient characterizing the watershed's annual water yield response to annual precipitation. The article provided the following data from 16 California

coastal watersheds for α (in μm^{-1}) and average long-term annual temperature (T in °C):

T:	8.51	8.69	9.01	9.50	10.00	10.60	11.00	11.60
α:	.40	.42	.40	.43	.40	.38	.40	.30

T:	11.60	12.60	12.60	13.60	14.20	15.30	17.90	17.90
α:	.41	.27	.28	.19	.22	.19	.13	.09

Construct a scatterplot of the data. How would you describe the nature of the relationship between the two variables?

8. Researchers considered how the construction cost of highway resurfacing projects in Kentucky were affected by that state's asphalt price index (API) and diesel price index (DPI) among other factors. From about the mid-1990s to 2010, Kentucky's annual average API and DPI were found to be closely related to the annual average crude oil price. Based on this, the authors suggested that crude oil price could be used to predict API and DPI ("Prices of Highway Resurfacing Projects in Economic Downturn: Lessons Learned and Strategies Forward," *J. Mgmnt. Engr.*, 2012, 391–397).

Consider the following monthly API and statewide crude oil index (COI) values for California during 2010–11, obtained from the California Department of Transportation.

COI	API	COI	API
385.1	415.1	474.3	477.1
408.0	377.0	483.4	488.9
400.8	402.8	504.9	586.3
426.0	427.3	616.1	634.7
437.0	436.9	656.6	667.5
384.0	360.8	606.0	592.2
393.3	372.3	579.0	565.9
402.9	417.2	588.4	570.5
404.2	376.5	536.8	589.7
399.5	424.1	585.9	559.8
438.9	432.2	592.5	637.0
447.8	450.6	650.0	625.0

Construct a scatterplot of the data. How would you describe the nature of the relationship between the two variables? Does it seem to be the case that COI and API are closely related?

3.2   CORRELATION

A scatterplot of bivariate numerical data gives a visual impression of how strongly x values and y values are related. However, to make precise statements and draw reliable conclusions from data, we must go beyond pictures. A **correlation coefficient** (from *co-relation*) is a quantitative assessment of the strength of relationship between x and y values in a set of (x, y) pairs. In this section, we introduce the most frequently used correlation coefficient.

Figure 3.4 displays scatterplots that indicate different types of relationships between the x and y values. The plot in Figure 3.4(a) suggests a very strong *positive* relationship between x and y, that is, a strong tendency for y to increase as x increases. Figure 3.4(b) gives evidence of a substantial *negative* relationship: As x increases, there is a tendency for y to decrease (as would probably be the case for x = amount of time per week that a high school student spends watching television and y = amount of time the student spends studying). The plot of Figure 3.4(c) indicates no strong relationship between the two variables; there is no tendency for y to either increase or decrease as x increases. Finally, as illustrated in Figure 3.4(d), a scatterplot can show a strong positive (or negative) relationship through a pattern that is curved rather than linear in appearance.

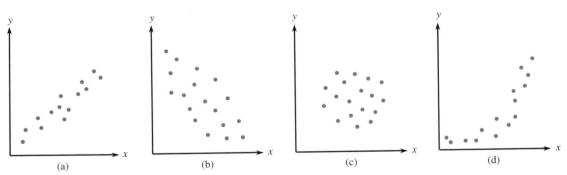

Figure 3.4 Scatterplots illustrating various types of relationships: (a) positive relationship, linear pattern; (b) negative relationship, linear pattern; (c) no relationship or pattern; (d) positive relationship, curved pattern

Pearson's Sample Correlation Coefficient

Let $(x_1, y_1), (x_2, y_2), \ldots, (x_n, y_n)$ denote a sample of (x, y) pairs. Consider subtracting $\bar{x}$ from each x value to obtain the x deviations, $x_1 - \bar{x}, \ldots, x_n - \bar{x}$, and also subtracting $\bar{y}$ from each y value to give $y_1 - \bar{y}, \ldots, y_n - \bar{y}$. Then multiply each x deviation by the corresponding y deviation to obtain products of deviations of the form $(x - \bar{x})(y - \bar{y})$.

The scatterplot in Figure 3.5(a) indicates a substantial positive relationship. A vertical line through $\bar{x}$ and a horizontal line through $\bar{y}$ divide the plot into four regions. In region I, both x and y exceed their mean values, so $x - \bar{x}$ and $y - \bar{y}$ are both positive numbers. It then follows that $(x - \bar{x})(y - \bar{y})$ is positive. The product of deviations is also positive for any point in region III, because both deviations are negative and multiplying two negative numbers gives a positive number. In each of the other two regions, one deviation is positive and the other is negative, so $(x - \bar{x})(y - \bar{y})$ is negative. Because almost all points lie in regions I and III, almost all products of deviations are positive. Thus the *sum* of products, $\Sigma(x_i - \bar{x})(y_i - \bar{y})$, will be a large positive number.

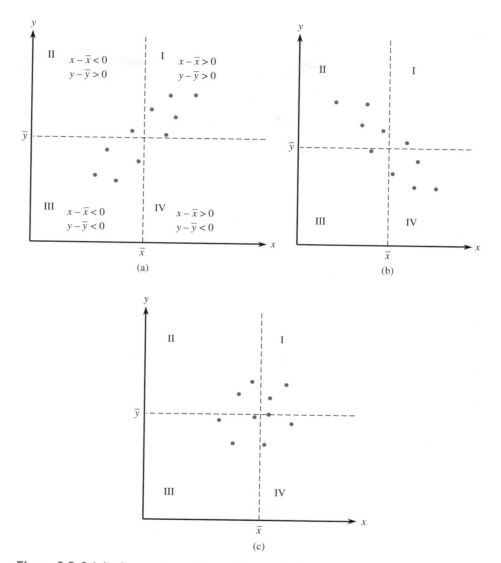

Figure 3.5 Subdividing a scatterplot according to the signs of $x - \bar{x}$ and $y - \bar{y}$: (a) a positive relation; (b) a negative relation; (c) no strong relation

Similar reasoning for the data displayed in Figure 3.5(b), which exhibits a strong negative relationship, implies that $\Sigma(x_i - \bar{x})(y_i - \bar{y})$ will be a large negative number. When there is no evidence of a strong relationship, as in Figure 3.5(c), positive and negative products of deviations tend to counteract one another, giving a value of the sum that is close to zero. In summary, $\Sigma(x_i - \bar{x})(y_i - \bar{y})$ seems to be a reasonable measure of the degree of association between the x and y values; it will be a large positive number, a large negative number, or a number close to zero according to whether there is a strong positive, a strong negative, or no strong relationship.

Unfortunately, our proposal has a serious deficiency: Its value depends on the choice of unit of measurement for both x and y. Suppose, for example, that x is height. Each x value expressed in inches will be 12 times the corresponding value expressed in feet, and the same will then be true of $\bar{x}$. It follows that the value of $\Sigma(x_i - \bar{x})(y_i - \bar{y})$ when the x unit is inches will be 12 times what it is when the unit is feet. A measure of the inherent strength of the relationship should give the same value whatever the units for the variables; otherwise our impressions may be distorted by the choice of units.

A straightforward modification of our initial proposal leads to the most popular measure of association, one that is free of the defect just alluded to and has other attractive properties.

DEFINITION

> **Pearson's sample correlation r is given by**
>
> $$r = \frac{\sum(x_i - \bar{x})(y_i - \bar{y})}{\sqrt{\sum(x_i - \bar{x})^2}\sqrt{\sum(y_i - \bar{y})^2}} = \frac{S_{xy}}{\sqrt{S_{xx}}\sqrt{S_{yy}}}$$
>
> Computing formulas for the three summation quantities are
>
> $$S_{xx} = \sum x_i^2 - \frac{\left(\sum x_i\right)^2}{n}$$
>
> $$S_{yy} = \sum y_i^2 - \frac{\left(\sum y_i\right)^2}{n}$$
>
> $$S_{xy} = \sum x_i y_i - \frac{\left(\sum x_i\right)\left(\sum y_i\right)}{n}$$

Use of the computing formulas makes all the subtraction needed to obtain the deviations unnecessary. Instead, the following five summary quantities are needed: Σx_i, Σy_i, Σx_i^2, Σy_i^2, $\Sigma x_i y_i$. The following example shows how a tabular format facilitates the calculations (we'll get to the issue of interpretation in a moment).

Example 3.3

The catch basin in a storm-sewer system is the interface between surface runoff and the sewer. A catch-basin insert is a device for retrofitting catch basins to improve their pollutant removal properties. The article "An Evaluation of the Urban Stormwater Pollutant Removal Efficiency of Catch Basin Inserts" (*Water Envir. Res.*, 2005: 500–510) reported on tests of various inserts under controlled conditions for which inflow is close to what can be expected in the field. Consider the following data, read from a graph in the article, for one particular type of insert on $x =$ amount filtered (1000s of liters) and $y = \%$ total suspended solids removed.

x:	23	45	68	91	114	136	159	182	205	228
y:	53.3	26.9	54.8	33.8	29.9	8.2	17.2	12.2	3.2	11.1

The accompanying table contains five columns for the x, y, x^2, y^2, and xy values, respectively. The sum of each column is given at the bottom of the table.

x	y	x^2	y^2	xy
23	53.3	529	2840.89	1225.9
45	26.9	2025	723.61	1210.5
68	54.8	4624	3003.04	3726.4
91	33.8	8281	1142.44	3075.8
114	29.9	12996	894.01	3408.6
136	8.2	18496	67.24	1115.2
159	17.2	25281	295.84	2734.8
182	12.2	33124	148.84	2220.4
205	3.2	42025	10.24	656
228	11.1	51984	123.21	2530.8
1251	250.6	199,365	9249.36	21,904.4
$\uparrow$	$\uparrow$	$\uparrow$	$\uparrow$	$\uparrow$
Σx_i	Σy_i	Σx_i^2	Σy_i^2	$\Sigma x_i y_i$

Then

$$S_{xx} = 199{,}365 - \frac{(1251)^2}{10} = 42{,}865,$$

$$S_{yy} = 9249.36 - \frac{(250.6)^2}{10} = 2969.3$$

$$S_{xy} = 21{,}904.4 - \frac{(1251)(250.6)}{10} = -9446$$

from which

$$r = \frac{-9446}{\sqrt{42{,}865}\,\sqrt{2969.3}} = -.837$$

Properties and Interpretation of r

1. *The value of r does not depend on the unit of measurement for either variable.* If, for example, x is height, the factor of 12 that appears in the numerator when changing from feet to inches will also appear in the denominator, so the two will cancel and leave r unchanged. The same value of r results from height expressed in inches, meters, or miles. If y is temperature, expressing values in °F, °C, or °K will give the same value of r. The correlation coefficient measures the inherent strength of relationship between two numerical variables.

2. *The value of r does not depend on which of the two variables is labeled x.* Thus if we had let $x = \%$ removed and $y =$ amount filtered in Example 3.3, the same value, $r = -.837$, would have resulted.

3. *The value of r is between -1 and $+1$.* A value near the upper limit, $+1$, is indicative of a substantial positive relationship, whereas an r close to the lower limit, -1, suggests a prominent negative relationship. Figure 3.6 shows a useful informal way to describe the

strength of relationship based on r. It may seem surprising that a value of r as extreme as $-.5$ or $.5$ should be in the "weak" category; an explanation for this is given later in the chapter.

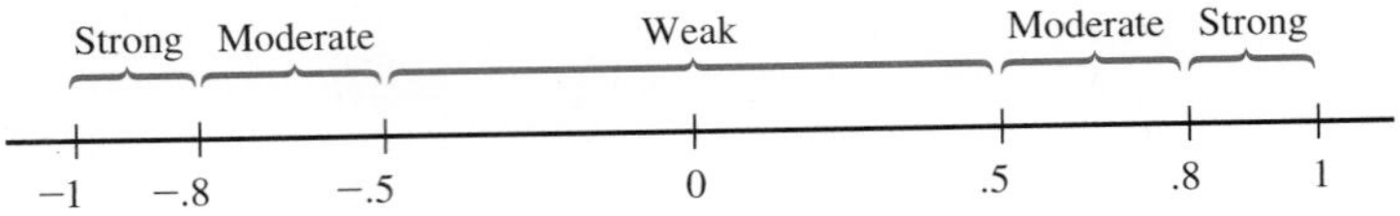

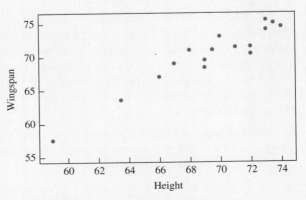

Figure 3.6 Describing the strength of relationship

4. $r = 1$ *only when all the points in a scatterplot of the data lie exactly on a straight line that slopes upward. Similarly,* $r = -1$ *only when all the points lie exactly on a downward-sloping line.* Only when there is a perfect linear relationship between x and y in the sample will r take on one of its two possible extreme values.

5. *The value of r is a measure of the extent to which x and y are **linearly** related*—that is, the extent to which the points in the scatterplot fall close to a straight line. A value of r close to zero does not rule out *any* strong relationship between x and y; there could still be a strong relationship but one that is not linear.

Example 3.4   As far back as Leonardo da Vinci, height and wingspan (measured from fingertip to fingertip between outstretched hands) were known to be closely related. For the following actual measurements (in inches) from 16 students in a statistics class notice how close the two values are.

Height:	59.0	72.0	67.0	63.5	68.0	66.0	71.0	69.0
Wingspan:	57.5	70.5	69.0	63.5	71.0	67.0	71.5	68.5

Height:	73.0	69.0	69.5	72.0	73.5	73.0	74.0	70.0
Wingspan:	74.0	69.5	71.0	71.5	75.0	75.5	74.5	73.0

The scatterplot in Figure 3.7 shows an approximately linear shape, and the point cloud is roughly elliptical. The correlation is computed to be 0.955. If the measurements were converted to centimeters, the correlation would remain unchanged.

Figure 3.7 Wingspan plotted against height

Example 3.5   The article "Quantitative Estimation of Clay Mineralogy in Fine-Grained Soils" (*J. Geotech. Geoenviron. Engr.*, 2011: 997–1008) reported on various chemical properties of natural and artificial soils. Consider the accompanying data on the cation exchange capacity (CEC, in meq/100 g) and specific surface area (SSA, in m^2/g) of 20 natural soils. A scatterplot appears in Figure 3.8.

CEC:	66	121	134	101	77	89	63	57	117	118
SSA:	175	324	460	288	205	210	295	161	314	265
CEC:	76	125	75	71	133	104	76	96	58	109
SSA:	236	355	240	133	431	306	132	269	158	303

Minitab gave the following output in response to a request for r:

```
Correlation of SSA and CEC = 0.853
```

There is evidence of a moderate to strong positive relationship.

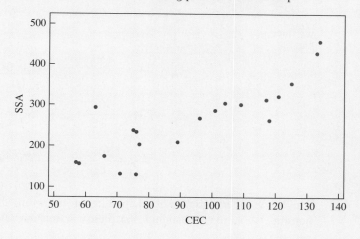

Figure 3.8 Scatterplot of the data from Example 3.5

Example 3.6   The accompanying data on y = glucose concentration (g/L) and x = fermentation time (days) for a particular brand of malt liquor was read from a scatterplot appearing in the article "Improving Fermentation Productivity with Reverse Osmosis" (*Food Tech.*, 1984: 92–96):

x:	1	2	3	4	5	6	7	8
y:	74	54	52	51	52	53	58	71

The scatterplot of Figure 3.9 (page 114) suggests a strong relationship, but *not* a linear one, between x and y. With

$$\sum x_i = 36 \quad \sum x_i^2 = 204 \quad \sum y_i = 465 \quad \sum y_i^2 = 27{,}615 \quad \sum x_i y_i = 2094$$

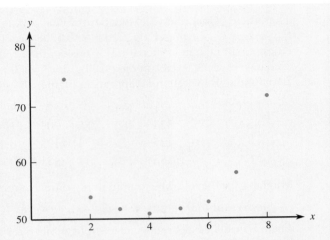

Figure 3.9 Scatterplot of the data from Example 3.6

we have

$$S_{xy} = 2094 - \frac{(36)(465)}{8} = 1.5000$$

$$S_{xx} = 204 - \frac{(36)^2}{8} = 42 \qquad S_{yy} = 586.875$$

$$r = \frac{1.500}{\sqrt{42}\sqrt{586.875}} = .0096 \approx .01$$

This shows the importance of interpreting r as measuring the extent of any *linear* relationship. We should not conclude that there is no relation whatsoever just because $r \approx 0$.

The Population Correlation Coefficient

The sample correlation coefficient r measures how strongly the x and y values in a *sample* of pairs are related to one another. There is an analogous measure of how strongly x and y are related in the entire population of pairs from which the sample $(x_1, y_1), \ldots, (x_n, y_n)$ was obtained. It is called the **population correlation coefficient** and is denoted by ρ (notice again the use of a Greek letter for a population characteristic and a Roman letter for a sample characteristic). We will never have to calculate ρ from the entire population of pairs, but it is important to know that ρ satisfies properties paralleling those of r:

1. ρ is a number between -1 and $+1$ that does not depend on the unit of measurement for either x or y, or on which variable is labeled x and which is labeled y.
2. $\rho = +1$ or -1 if and only if all (x, y) pairs in the population lie exactly on a straight line, so ρ measures the extent to which there is a linear relationship in the population.

In Chapter 11, we show how the sample characteristic r can be used to make an inference concerning the population characteristic ρ. In particular, r can be used to decide whether $\rho = 0$ (no linear relationship in the population).

Correlation and Causation

A value of r close to 1 indicates that relatively large values of one variable tend to be associated with relatively large values of the other variable. This is far from saying that a large value of one variable *causes* the value of the other variable to be large. Correlation (Pearson's or any other) measures the extent of association, but **association does not imply causation.** It frequently happens that two variables are highly correlated not because one is causally related to the other but because they are both strongly related to a third variable. Among all elementary-school children, there is a strong positive relationship between the number of cavities in a child's teeth and the size of his or her vocabulary. Yet no one advocates eating foods that result in more cavities to increase vocabulary size (or working to decrease vocabulary size to protect against cavities). Number of cavities and vocabulary size are both strongly related to age, so older children tend to have higher values of both variables than do younger ones. Among children of any fixed age, there would undoubtedly be little relationship between number of cavities and vocabulary size.

Scientific experiments can frequently make a strong case for causality by carefully controlling the values of all variables that might be related to the ones under study. Then, if y is observed to change in a "smooth" way as the experimenter changes the value of x, the most plausible explanation would be a causal relationship between x and y. In the absence of such control and ability to manipulate values of one variable, we must admit the possibility that an unidentified underlying third variable is influencing both the variables under investigation. A high correlation in many uncontrolled studies carried out in different settings can marshal support for causality—as in the case of cigarette smoking and cancer—but proving causality is often a very elusive task.

Section 3.2 Exercises

9. For each of the following pairs of variables, indicate whether you would expect a positive correlation, a negative correlation, or little or no correlation. Explain your choice.
   a. Maximum daily temperature and cooling cost
   b. Interest rate and number of loan applications
   c. Incomes of husbands and wives when both have full-time jobs
   d. Vehicle speed (mph, from 20 to 100) and fuel efficiency (mpg)
   e. Fuel efficiency and 3-year operating cost
   f. Distance from a Stanford University student's home town to campus and grade point average

10. Head movement evaluations are important because individuals, especially those who are disabled, may be able to operate communications aids in this manner. The article "Constancy of Head Turning Recorded in Healthy Young Humans" (*J. of Biomed. Engr.*, 2008: 428–436) reported data on ranges in maximum inclination angles of the head in the clockwise anterior, posterior, right, and left directions for 14 randomly selected subjects. Consider the accompanying data on average anterior maximum inclination angle (AMIA) in both the clockwise (Cl) and counterclockwise (Co) directions.

Subj:	1	2	3	4	5	6	7
Cl:	57.9	35.7	54.5	56.8	51.1	70.8	77.3
Co:	44.2	52.1	60.2	52.7	47.2	65.6	71.4

Subj:	8	9	10	11	12	13	14
Cl:	51.6	54.7	63.6	59.2	59.2	55.8	38.5
Co:	48.8	53.1	66.3	59.8	47.5	64.5	34.5

a. Construct boxplots of both the clockwise and counterclockwise direction observations, and comment on any interesting features.

b. Construct a scatterplot of the data. What does it suggest about the general nature of the relationship between Cl and Co?

c. Calculate the value of the sample correlation coefficient. Does it confirm your impression from the scatterplot?

11. Torsion during external rotation and extension of the hip may explain why acetabular labral tears occur in professional athletes. The article "Hip Rotational Velocities During the Full Golf Swing" (*J. of Sports Sci. and Med.*, 2009: 296–299) reported on an investigation in which lead hip internal peak rotational velocity (x) and trailing hip peak external rotational velocity (y) were determined for a sample of 15 golfers. Data provided by the article's authors was used to calculate the following summary quantities:

$$\sum (x_i - \bar{x})^2 = 64{,}732.83,$$

$$\sum (y_i - \bar{y})^2 = 130{,}566.96,$$

$$\sum (x_i - \bar{x})(y_i - \bar{y}) = 44{,}185.87$$

Based on this, compute the sample correlation coefficient and interpret its value. How would you characterize this correlation—as strong, moderate, or weak?

12. Historically, reinforced concrete structures used externally bonded steel plates to add strength and support. Recently, fiber reinforced polymer (FRP) plates have been used instead of steel plates because of their superior properties. In the article "Interfacial Bond Strength Characteristics of FRP and RC Substrate" (*J. of Compos.*

Constr., 2012: 35–43), investigators developed a method to mathematically model bond strength between a carbon FRP and a concrete substrate. For each of 15 carbon FRP–concrete samples, the article reported the maximum transferable load (kN) calculated by the model and compared this with the corresponding maximum transferable load (kN) as measured in the laboratory. The data is given here:

Calc:	Meas:	Calc:	Meas:
14.2	13.7	14.3	13.4
16.0	13.7	21.4	21.4
16.5	15.4	17.6	14.8
15.9	15.4	8.6	7.4
18.8	16.2	10.3	7.4
17.9	16.3	11.9	14.7
13.1	13.7	18.7	18.2
15.4	16.2		

a. Construct a scatterplot of the data. Does it seem to be the case that, in general, when the measured load is low (high), the calculated load is also low (high)? For each sample, are the two variables relatively close in value?

b. Calculate the value of the sample correlation coefficient. Does it confirm your impression from the scatterplot?

13. The article "Behavioural Effects of Mobile Telephone Use During Simulated Driving" (*Ergonomics*, 1995: 2536–2562) reported that for a sample of 20 experimental subjects, the sample correlation coefficient for x = age and y = time since the subject had acquired a driving license (yr) was .97. Why do you think the value of r is so close to 1? (The article's authors gave an explanation.)

14. An employee of an auction house has a list of 25 recently sold paintings. Eight artists were represented in these sales. The sale price of each painting is on the list. Would the correlation coefficient be an appropriate way to summarize the relationship between artist (x) and sale price (y)? Why or why not?

15. A sample of automobiles traversing a certain stretch of highway is selected. Each automobile travels at a roughly constant rate of speed, though speed does vary from auto to auto. Let x = speed and y = time needed to traverse this segment of highway. Would the sample correlation coefficient be closest to .9, .3, −3, or −.9? Explain.

16. Suppose that x and y are positive variables and that a sample of n pairs results in $r \approx 1$. If the sample correlation coefficient is computed for the (x, y^2) pairs, will the resulting value also be approximately 1? Explain.

17. Nine students currently taking introductory statistics are randomly selected, and both the first midterm exam score (x) and the second midterm score (y) are determined. Three of the students have the class at 8 A.M., another three have it at noon, and the remaining three have a night class. The resulting (x, y) pairs are as follows:

8 A.M.: (70, 60) (72, 83) (94, 85)
Noon: (80, 72) (60, 74) (55, 58)
Night: (45, 63) (50, 40) (35, 54)

a. Calculate the sample correlation coefficient for the nine (x, y) pairs.
b. Let $\bar{x}_1$ be the average score on the first midterm exam for the 8 A.M. students and $\bar{y}_1$ be the average score on the second midterm for these students. Denote the two averages for the noon students by $\bar{x}_2$ and $\bar{y}_2$, and for the night students by $\bar{x}_3$ and $\bar{y}_3$. Calculate r for these three $(\bar{x}, \bar{y})$ pairs.
c. Construct a scatterplot of the nine (x, y) pairs and another one of the three pairs of averages. Can you see why r in part (a) is smaller than r in part (b)? Does this suggest that a correlation coefficient based on averages (called an "ecological" correlation) might be misleading? Explain.

18. Suppose data is collected on two quantitative variables, x and y. Let r be the corresponding sample correlation coefficient for (x, y). The x and y values are then transformed as follows: $x' = a + bx$, $y' = c + dy$ where a, b, c, and d are constants. Let r' be the corresponding sample correlation coefficient for (x', y').
a. Show that $\bar{x}' = a + b\bar{x}$ and $\bar{y}' = c + d\bar{y}$.
b. Show that $s_{x'} = bs_x$ and $s_{y'} = ds_y$.
c. Show that $r = r'$.

3.3   FITTING A LINE TO BIVARIATE DATA

Given two numerical variables x and y, the general objective of *regression analysis* is to use information about x to draw some type of conclusion concerning y. Often an investigator wants to predict the y value that would result from making a single observation at a specified x value—for example, to predict product sales y for a sales region in which advertising expenditure x is one million dollars. The different roles played by the two variables are reflected in standard terminology: y is called the **dependent** or **response variable**, and x is referred to as the **independent, predictor,** or **explanatory variable**.

A scatterplot of y versus x frequently exhibits a linear pattern. In such cases, it is natural to summarize the relationship between the variables by finding a line that is as close as possible to the points in the plot. Before doing so, let's quickly review some elementary facts about lines and linear relationships.

Suppose a car dealership advertises that a particular type of vehicle can be rented on a one-day basis for a flat fee of $25 plus an additional $.30 per mile driven. If such a vehicle is rented and driven for 100 miles, the dealer's revenue y is

$$y = 25 + (.30)(100) = 25 + 30 = 55$$

More generally, if x denotes the distance driven in miles, then

$$y = 25 + .30x$$

That is, x and y are linearly related.

The general form of a linear relationship between x and y is $y = a + bx$. A particular relation is specified by choosing values of a and b, for example, $y = 10 + 2x$ or $y = 100 - 5x$. If we choose some x values and calculate $y = a + bx$ for each value, the points in the scatterplot of the resulting (x, y) pairs fall exactly on a straight line. The value of b, the **slope** of the line, is the amount by which y increases when x increases by 1 unit. The **vertical** or y **intercept** a is the height of the line above the value $x = 0$. The equation $y = 10 + 2x$ has slope $b = 2$, so each 1-unit increase in x results in an increase of 2 in y. When $x = 0$, $y = 10$ and the height at which the line crosses the vertical axis is 10. To draw the line corresponding to this equation, select any two x values (e.g., $x = 5$ and $x = 10$). Substitute these values into the equation to obtain the corresponding y values ($y = 20$ and $y = 30$) and thus two (x, y) points on the line. Finally, connect these two points with a straightedge.

Fitting a Straight Line

The line that gives the most effective summary of an approximate linear relation is the one that in some sense is the best-fitting line, the one closest to the sample data. Consider the scatterplot and line shown in Figure 3.10. Let's focus on the vertical deviations from the points to the line. For example,

$$\text{deviation from } (15, 47) = \text{height of point} - \text{height of line}$$

$$= 47 - [10 + 2(15)]$$

$$= 7$$

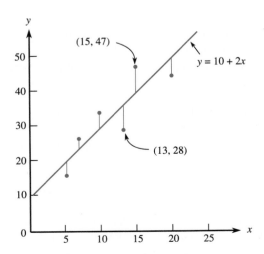

Figure 3.10 Vertical deviations from points to a line

Similarly,

$$\text{deviation from } (13, 28) = 28 - [10 + 2(13)] = -8$$

A positive deviation results from a point that lies above the chosen line, and a negative deviation from a point that lies below this line. A particular line gives a good fit if the deviations from the line are small in magnitude, that is, reasonably close to zero.

We now need a way to combine the n deviations into a single measure of fit. The standard approach is to square the deviations (to obtain nonnegative numbers) and sum these squared deviations.

DEFINITIONS

The most widely used criterion for assessing the goodness of fit of a line $y = a + bx$ to bivariate data $(x_1, y_1), \ldots, (x_n, y_n)$ is the sum of the squared deviations about the line:

$$\sum [y_i - (a + bx_i)]^2 = [y_1 - (a + bx_1)]^2 + \cdots + [y_n - (a + bx_n)]^2$$

According to the **principle of least squares,** the line that gives the best fit to the data is the one that minimizes this sum; it is called the **least squares line** or **sample regression line.**

To find the equation of the least squares line, let $g(\tilde{a}, \tilde{b}) = \sum [y_i - (\tilde{a} + \tilde{b}x_i)]^2$. Then the intercept a and slope b of the least squares line are the values of $\tilde{a}$ and $\tilde{b}$ that minimize $g(\tilde{a}, \tilde{b})$. These minimizing values are obtained by taking the partial derivative of the g function first with respect to $\tilde{a}$ and then with respect to $\tilde{b}$, and equating these two partial derivatives to zero (this is analogous to solving the single equation $f'(z) = 0$ to find the value of z that minimizes a function of a single variable). This results in the following two equations in two unknowns, called the **normal equations:**

$$n\tilde{a} + \left(\sum x_i\right)\tilde{b} = \sum y_i \qquad \left(\sum x_i\right)\tilde{a} + \left(\sum x_i^2\right)\tilde{b} = \sum x_i y_i$$

These equations are easily solved because they are linear in the unknowns (a consequence of using squared deviations in the fitting criterion).

The **slope b of the least squares line** is given by

$$b = \frac{\sum x_i y_i - \left(\sum x_i\right)\left(\sum y_i\right)/n}{\sum x_i^2 - \left(\sum x_i\right)^2/n} = \frac{S_{xy}}{S_{xx}}$$

The **vertical intercept a of the least squares line** is

$$a = \bar{y} - b\bar{x}$$

The equation of the least squares line is often written as $\hat{y} = a + bx$, where the "$\hat{\ }$" above y emphasizes that $\hat{y}$ is a prediction of y that results from the substitution of any particular x value into the equation. Notice that the numerator and denominator of b appeared previously in the formula for the sample correlation coefficient r.

Example 3.7   The cetane number is a critical property in specifying the ignition quality of a fuel used in a diesel engine. Determining this number for a biodiesel fuel is expensive and time consuming. The article "Relating the Cetane Number of Biodiesel Fuels to Their Fatty Acid Composition: A Critical Study" (*J. of Automobile Engr.*, 2009: 565–583) included the following data on $x =$ iodine value (g) and $y =$ cetane number for a sample of 14 biofuels. The iodine value is the amount of iodine necessary to saturate a sample of 100 g of oil.

x:	132.0	129.0	120.0	113.2	105.0	92.0	84.0
y:	46.0	48.0	51.0	52.1	54.0	52.0	59.0

x:	83.2	88.4	59.0	80.0	81.5	71.0	69.2
y:	58.7	61.6	64.0	61.4	54.6	58.8	58.0

The necessary summary quantities for hand calculation can be obtained by placing the x values in a column and the y values in another column and then creating columns for x^2, xy, and y^2 (the latter value is not needed at the moment but will be used shortly). Calculating the column sums gives

$$\sum x_i = 1307.5, \qquad \sum y_i = 779.2, \qquad \sum x_i^2 = 128{,}913.93,$$
$$\sum x_i y_i = 71{,}347.30, \qquad \sum y_i^2 = 43{,}745.22$$

from which

$$\bar{x} = \frac{1307.5}{14} = 93.392857, \qquad \bar{y} = \frac{779.2}{14} = 55.657143$$
$$S_{xx} = 128{,}913.93 - (1307.5)^2/14 = 6802.7693$$
$$S_{xy} = 71{,}347.30 - (1307.5)(779.2)/14 = -1424.41429$$

Thus

$$b = \frac{-1424.41429}{6802.7693} = -.20938742$$
$$a = 55.657143 - (-.20938742)(93.392857) = 75.212432$$

and the equation of the least squares line is $\hat{y} = 75.212 - .2094x$, exactly that reported in the cited article.

Figure 3.11 generated by the statistical computer package Minitab shows that the least squares line is a very good summary of the relationship between the two variables. A prediction of the cetane number when the iodine value is 100 is $\hat{y} = 75.212 - .2094(100) = 54.27$. The slope of the least squares line tells us that a decrease of roughly .209 in cetane number is associated with a 1-gram increase in iodine value.

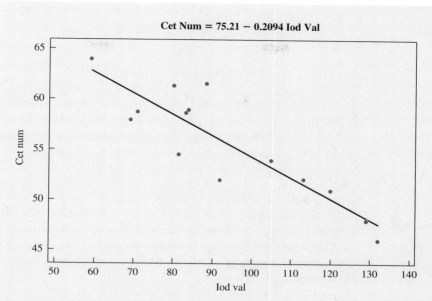

Figure 3.11 Scatterplot from Minitab for Example 3.7 with least squares line superimposed

The least squares line should not be used to make a prediction for an x value much beyond the range of the data, such as $x = 50$ or $x = 250$ in Example 3.7. The **danger of extrapolation** is that the fitted relationship (here, a line) may not be valid for such x values.

Regression

The term *regression* comes from the relationship between the least squares line and the sample correlation coefficient. Let s_x and s_y denote the sample standard deviations of the x and y values, respectively. Algebraic manipulation gives

$$b = r\left(\frac{s_y}{s_x}\right) \qquad \hat{y} = \bar{y} + r\left(\frac{s_y}{s_x}\right)(x - \bar{x})$$

If $r = 1$ and we substitute $x = \bar{x} + s_x$ (an x value 1 standard deviation above the mean x value), then $\hat{y} = \bar{y} + s_y$, which is 1 standard deviation above the mean y value. If, however, $r = .5$ and this x value is substituted, then $\hat{y} = \bar{y} + .5s_y$, which is only half a y standard deviation above the mean. More generally, when $-1 < r < 1$, for *any* x value, the corresponding predicted value $\hat{y}$ will be closer in terms of standard deviations to $\bar{y}$ than is x to $\bar{x}$; that is, $\hat{y}$ is pulled toward (regressed toward) the mean y value. This **regression effect** was first noticed by Sir Francis Galton in the late 1800s when he studied the relation between father's height and son's height; the predicted height of a son was always closer to the mean height than was his father's height.

Assessing the Fit of the Least Squares Line

How effectively does the least squares line summarize the relationship between the two variables? In other words, how much of the observed variation in y can be attributed to the approximate linear relationship and the fact that x is varying? A quantitative assessment is based on the vertical deviations from the least squares line. The height of the least squares line above x_1 is $\hat{y}_1 = a + bx_1$, and y_1 is the height of the corresponding point in the scatterplot, so the vertical deviation (residual) from this point to the line is $y_1 - (a + bx_1)$. Substituting the remaining x values into the equation gives other **predicted** (or **fitted**) **values** $\hat{y}_2 = a + bx_2, \ldots, \hat{y}_n = a + bx_n$, and the other **residuals** $y_2 - \hat{y}_2, \ldots, y_n - \hat{y}_n$ are again obtained by subtraction. A residual is positive if the corresponding point in the scatterplot lies above the least squares line and negative if the point lies below the line. It can be shown that when predicted values and residuals are based on the least squares line, $\Sigma(y_i - \hat{y}_i) = 0$, so of course the average residual is zero.

Variation in y can effectively be explained by an approximate straight-line relationship when the points in the scatterplot fall close to the least squares line—that is, when the residuals are small in magnitude. A natural measure of variation about the least squares line is the sum of the squared residuals (squaring before combining prevents negative and positive residuals from counteracting one another). A second *sum of squares* assesses the total amount of variation in observed y values.

DEFINITIONS

Residual sum of squares, denoted by **SSResid**, is given by

$$\text{SSResid} = \sum(y_i - \hat{y}_i)^2 = (y_1 - \hat{y}_1)^2 + \cdots + (y_n - \hat{y}_n)^2$$

(alternatively called *error sum of squares* and denoted by SSE).

Total sum of squares, denoted by **SSTo**, is defined as

$$\text{SSTo} = \sum(y_i - \bar{y})^2 = (y_1 - \bar{y})^2 + \cdots + (y_n - \bar{y})^2$$

Alternative notation for SSTo is S_{yy}, and a computing formula is

$$\sum y_i^2 - \frac{\left(\sum y_i\right)^2}{n}$$

A computing formula for residual sum of squares makes it unnecessary to calculate the residuals:

$$\text{SSResid} = \text{SSTo} - bS_{xy}$$

Because b and S_{xy} have the same sign, bS_{xy} is a positive quantity unless $b = 0$, so the computing formula shows that SSResid = SSTo if $b = 0$ and SSResid < SSTo otherwise.

To avoid any rounding effects, use as much decimal accuracy in b as possible when computing SSResid.

SSResid is often referred as a measure of "unexplained" variation; it is the amount of variation in y that cannot be attributed to the linear relationship between x and y. The more points in the scatterplot deviate from the least squares line, the larger the value of SSResid and the greater the amount of y variation that cannot be explained by a linear relation. Similarly, SSTo is interpreted as a measure of total variation; the larger the value of SSTo, the greater the amount of variability in the observed y_i's. The ratio SSResid/SSTo is the fraction or proportion of total variation that is *unexplained* by a straight-line relation. Subtracting this ratio from 1.0 gives the proportion of total variation that *is* explained.

DEFINITION

> The **coefficient of determination,** denoted by r^2, is given by
>
> $$r^2 = 1 - \frac{\text{SSResid}}{\text{SSTo}}$$
>
> It is the proportion of variation in the observed y values that can be attributed to (or explained by) a linear relationship between x and y in the sample. Multiplying r^2 by 100 gives the percentage of y variation attributable to the approximate linear relationship. The closer this percentage is to 100%, the more successful is the relationship in explaining variation in y.

Example 3.8
The scatterplot of the iodine value and cetane number data in Figure 3.11 portends a reasonably high r^2 value. With

$$S_{xy} = -1424.41429 \text{ (the numerator of } b) \qquad b = -.20938742$$

$$\sum y_i = 779.2 \qquad \sum y_i^2 = 43{,}745.22$$

we have

$$\text{SST} = 43{,}745.22 - (779.2)^2/14 = 377.174$$
$$\text{SSE} = 377.174 - (-.20938742)(-1424.41429) = 78.920$$

The coefficient of determination is then

$$r^2 = 1 - \text{SSE}/\text{SST} = 1 - (78.920)/(377.174) = .791$$

That is, 79.1% of the observed variation in cetane number is attributable to (can be explained by) the simple linear regression relationship between cetane number and iodine value (r^2 values are even higher than this in many scientific contexts, but social scientists would typically be ecstatic at a value anywhere near this large).

The wide availability of good statistical computer packages makes it unnecessary to hand calculate the various quantities involved in a regression analysis. Figure 3.12 shows partial Minitab output for the cetane number–iodine value data of Examples 3.7 and 3.8; the package will also provide the predicted values and residuals as well as other information on request. The formats used by other packages differ slightly from that of Minitab, but the information content is very similar. Quantities such as the standard deviations, *t*-ratios, *F*, and *P*-values are discussed in Chapter 11.

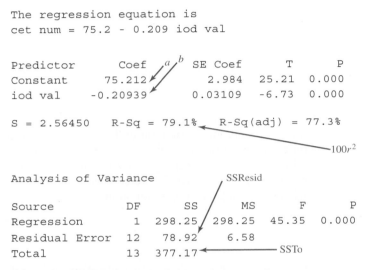

Figure 3.12 image shows:

```
The regression equation is
cet num = 75.2 - 0.209 iod val

Predictor        Coef   a  b  SE Coef        T        P
Constant       75.212              2.984    25.21    0.000
iod val       -0.20939             0.03109  -6.73    0.000

S = 2.56450   R-Sq = 79.1%   R-Sq(adj) = 77.3%
                                                      100r²

Analysis of Variance              SSResid

Source            DF      SS       MS       F        P
Regression         1   298.25   298.25   45.35    0.000
Residual Error    12    78.92     6.58
Total             13   377.17            SSTo
```

Figure 3.12 Minitab output for the regression of Examples 3.7 and 3.8

The symbol *r* was used in Section 3.2 to denote Pearson's sample correlation coefficient. It is not coincidental that r^2 is used to represent the coefficient of determination. The notation suggests how these two quantities are related:

$$(\text{correlation coefficient})^2 = \text{coefficient of determination}$$

Thus, if $r = .8$ or $r = -.8$ then $r^2 = .64$, so that 64% of the observed variation in the dependent variable can be attributed to the linear relationship. Notice that because the value of *r* does not depend on which variable is labeled *x*, the same is true of r^2. The coefficient of determination is one of the very few quantities calculated in the course of a regression analysis whose value remains the same when the role of dependent and independent variables are interchanged. When $r = .5$, we get $r^2 = .25$, so only 25% of the observed variation is explained by a linear relation. This is why values of *r* between $-.5$ and $.5$ can fairly be described as evidence of a weak relationship.

Standard Deviation About the Least Squares Line

The coefficient of determination measures the extent of variation about the best-fit line *relative* to overall variation in y. A high value of r^2 does not by itself promise that the deviations from the line are small in an absolute sense. A typical observation could deviate from the line by quite a bit, yet these deviations might still be small relative to overall y variation. Recall that in Chapter 2 the sample standard deviation $s = \sqrt{\Sigma(x - \bar{x})^2/(n - 1)}$ was used as a measure of variability in a single sample; roughly speaking, s is the typical amount by which a sample observation deviates from the mean. There is an analogous measure of variability when a line is fit by least squares.

DEFINITION

The **standard deviation about the least squares line** is given by

$$s_e = \sqrt{\frac{\text{SSResid}}{n - 2}}$$

Roughly speaking, s_e is the typical amount by which an observation deviates from the least squares line. Justification for division by $n - 2$ and the use of the subscript e are given in Chapter 11.

Example 3.9

The values of x = commuting distance and y = commuting time were determined for workers in samples from three different regions. Data is presented in Table 3.1; the three scatterplots are displayed in Figure 3.13.

For sample 1, a rather small proportion of variation in y can be attributed to an approximate linear relationship, and a typical deviation from the least squares line is roughly 4. The amount of variability about the line for sample 2 is the same as for sample 1, but the value of r^2 is much higher because y variation is much greater overall in sample 2 than in sample 1. Sample 3 yields roughly the same high value of r^2 as does sample 2, but the typical deviation from the line for sample 3 is only half that for sample 2. A complete picture of variation requires that both r^2 and s_e be computed.

Table 3.1 Data for three regions (Example 3.9)

1		2		3	
x	y	x	y	x	y
15	42	5	16	5	8
16	35	10	32	10	16
17	45	15	44	15	22
18	42	20	45	20	23
19	49	25	63	25	31
20	46	50	115	50	60

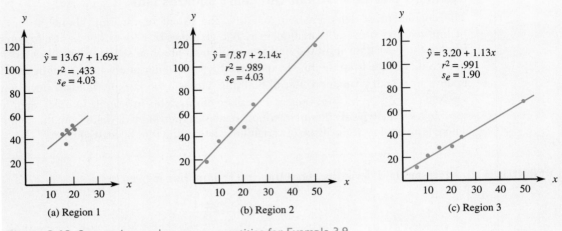

Figure 3.13 Scatterplots and summary quantities for Example 3.9

Plotting the Residuals (Optional)

It is important to have methods for identifying unusual or highly influential observations and revealing patterns in the data that may suggest how an improved fit can be achieved. A plot based on the residuals is very useful in this regard.

DEFINITION

A **residual plot** is a plot of the $(x, \text{residual})$ pairs—that is, of the pairs $(x_1, y_1 - \hat{y}_1), (x_2, y_2 - \hat{y}_2), \ldots, (x_n, y_n - \hat{y}_n)$—or of the residuals versus predicted values—the pairs $(\hat{y}_1, y_1 - \hat{y}_1), \ldots, (\hat{y}_n, y_n - \hat{y}_n)$.

A desirable plot exhibits no particular pattern, such as curvature or much greater spread in one part of the plot than in another part. Looking at a residual plot after fitting a line amounts to examining y after removing any linear dependence on x. This can sometimes more clearly show the existence of a nonlinear relationship.

Example 3.10

Consider the accompanying data (page 127) on x = height (in.) and y = average weight (lb) for American females aged 30–39 (taken from *The World Almanac and Book of Facts*). The scatterplot displayed in Figure 3.14(a) appears rather straight. However, when the residuals from the least squares line ($\hat{y} = -98.23 + 3.596x$) are plotted, substantial curvature is apparent (even though $r^2 \approx .99$). It is not accurate to say that weight increases in direct proportion to height (linearly with height). Instead, average weight increases somewhat more rapidly in the range of relatively large heights than it does for relatively small heights.

x:	58	59	60	61	62	63	64	65
y:	113	115	118	121	124	128	131	134

x:	66	67	68	69	70	71	72
y:	137	141	145	150	153	159	164

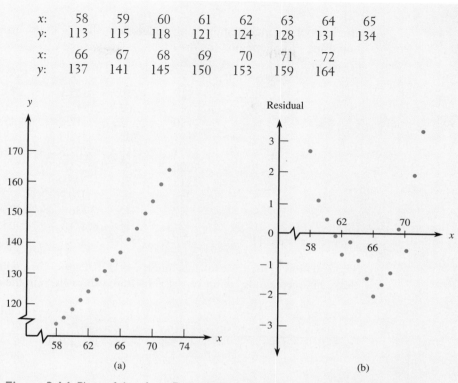

Figure 3.14 Plots of data from Example 3.10: (a) scatterplot; (b) residual plot

We also hope that there are no unusual points in the plot. A point falling far above or below the horizontal line at height zero corresponds to a large residual, which may indicate some type of unusual behavior, such as a recording error, non-standard experimental condition, or atypical experimental subject. A point whose x value differs greatly from others in the data set *may* have exerted excessive influence in determining the fitted line. One method for assessing the impact of such an isolated point on the fit is to delete it from the data set and then recompute the best-fit line and various other quantities. Substantial changes in the equation, predicted values, r^2, and s_e warn of instability in the data. More information may then be needed before reliable conclusions can be drawn.

Example 3.11   Bioaerosols are airborne particles such as bacteria or pollen that, when found in indoor environments, may cause infectious or allergic health effects. The Andersen method for determining bioaerosol concentration requires a 2–7-day incubation period. The article "Measurement of Indoor Bioaerosol Levels by a Direct Counting Method" (*J. of Envir. Engr.*, 1996: 374–378) discussed an alternative technique, the FFDC method. Consider the accompanying data, read from a plot in the cited article, on x = concentration using Andersen method (CFU/m^3) and y = concentration using FFDC method (no./m^3):

Observation	x	y	$\hat{y}$	Residual
1	119	239	225.1	13.9
2	140	262	240.3	21.7
3	150	202	247.6	−45.6
4	157	224	252.7	−28.7
5	171	255	262.8	−7.8
6	200	292	283.9	8.1
7	218	350	296.9	53.1
8	250	298	320.2	−22.2
9	272	313	336.2	−23.2
10	321	415	371.7	43.3
11	573	542	554.7	−12.7

The equation of the least squares line is $\hat{y} = 138.68 + .726x$, with $r^2 = .901$. (The slope, intercept, and r^2 differ very slightly from values given in the article.)

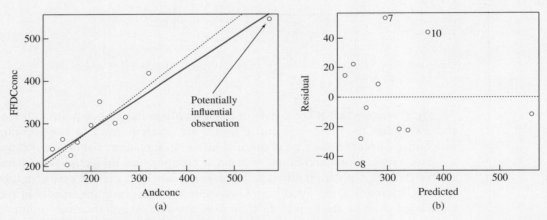

Figure 3.15 Plots from R for the bioaerosol data of Example 3.11:
(a) ——— least squares line for the full sample
    - - - least squares line when the potentially influential observation is deleted
(b) residuals versus predicted values

Figure 3.15 shows a scatterplot and a residual plot (here, residuals versus predicted values) from R (this package has excellent graphics capabilities). There is no single residual that is much larger in magnitude than the other residuals. The most striking feature here is that x_{11} is much larger than any other x value in the sample, so that (x_{11}, y_{11}) is an observation with *potentially* high influence (sometimes called a *high-leverage* observation). This point would not in fact be highly influential if it fell close to the least squares line based on just the first ten observations. However, the equation of this line is $\hat{y} = 115.09 + .850x$ with $r^2 = .757$; this r^2 value is much lower than the original value, and the slope and intercept have also changed substantially. Without the influential observation, evidence for a very strong linear relationship between concentrations assessed by the two methods is not nearly so compelling.

Resistant Lines

As Example 3.11 shows, the least squares line can be greatly affected by the presence of even a single observation that shows a large discrepancy in the x or y direction from the rest of the data. When the data set contains such unusual observations, it is desirable to have a method for obtaining a summarizing line that is resistant to the influence of these stray values. In recent years, many methods for obtaining a resistant (or robust) line have been proposed, and various statistical packages will fit such lines. Consult a statistician or a book on exploratory data analysis to obtain more information.

Section 3.3 Exercises

19. The invasive diatom species *Didymosphenia geminata* has the potential to inflict substantial ecological and economic damage in rivers. The article "Substrate Characteristics Affect Colonization by the Bloom-Forming *Didymosphenia geminata*" (*Aquatic Ecology*, 2010: 33–40) described an investigation of colonization behavior. One aspect of particular interest was whether y = colony density was related to x = rock surface area. The article contained a scatterplot and summary of a regression analysis. Here is representative data:

x:	50	71	55	50	33	58	79
y:	152	1929	48	22	2	5	35

x:	26	69	44	37	70	20	45	49
y:	7	269	38	171	13	43	185	25

a. Determine the equation of the least squares line for this data and then calculate and interpret the coefficient of determination.

b. The second observation has a very extreme y value (in the full data set consisting of 72 observations, there were 2 of these). This observation may have had a substantial impact on the form of the regression function and subsequent conclusions. Eliminate it and redo part (a). What do you conclude?

20. Electromagnetic technologies such as ground penetrating radar offer effective nondestructive sensing techniques to determine a continuous profile of a pavement structure. The propagation of electromagnetic waves through the structure depends critically on the dielectric properties of the media. However, little research has been done on the characterization of dielectric properties of asphalt mixtures. The article "Dielectric Modeling of Asphalt Mixtures and Relationship with Density" (*J. Transp. Engr.*, 2011: 104–111) reported on the dielectric response with percent air voids for various asphalt mixtures at 7-GHz frequency. The following data, kindly provided by the authors of the cited article, compares y = dielectric constant and x = air void (%) for 18 samples having 5% asphalt content:

y:	4.55	4.49	4.50	4.47	4.47	4.45
x:	4.35	4.79	5.57	5.20	5.07	5.79

y:	4.40	4.34	4.43	4.43	4.42	4.40
x:	5.36	6.40	5.66	5.90	6.49	5.70

30.79

y: 4.33   4.44   4.40   4.26   4.32   4.34
x: 6.49   6.37   6.51   7.88   6.74   7.08

a. Does a scatterplot of the data suggest it is reasonable to assume an approximate linear relationship between x and y?

b. Find the equation of the least squares line for this data and interpret its slope.

c. Determine the proportion of observed variation in the response variable that can be attributed to the approximate linear relationship between strength and fiber weight.

d. Does a residual plot indicate any deficiency in a straight line fit? Explain your reasoning.

21. For the past decade rubber powder has been used in asphalt cement to improve performance. The article "Experimental Study of Recycled Rubber-Filled High-Strength Concrete" (*Magazine of Concrete Res.*, 2009: 549–556) included on a regression of y = axial strength (MPa) on x = cube strength (MPa) based on the following sample data:

x:	112.3	97.0	92.7	86.0	102.0
y:	75.0	71.0	57.7	48.7	74.3

x:	99.2	95.8	103.5	89.0	86.7
y:	73.3	68.0	59.3	57.8	48.5

a. Does a scatterplot of the data suggest an appropriate linear relationship between x and y?

b. Obtain the equation of the least squares line and interpret its slope.

c. Calculate and interpret the coefficient of determination.

d. Roughly what is the size of a typical deviation of points in the scatterplot from the least squares line?

22. Recall the data from Exercise 4 based on amount of oil added (in g) and the corresponding amount of oil recovered (in g) from wheat straw. Suppose that we want to use the least squares line to predict the amount of oil recovered from the wheat straw based on the initial amount of oil added. Consider the accompanying output from the SAS statistical computer package.

Dependent Variable: oil_recov

Analysis of Variance

Source	DF	Sum of Squares	Mean Square	F Value	Pr > F
Model	1	289.45805	289.45805	2977.07	<.0001
Error	13	1.26398	0.09723		
C Total	14	290.72203			

Root MSE	0.31182	R-Square	0.9957	
Dep Mean	6.07513	Adj R-Sq	0.9953	
C.V.	5.13266			

Parameter Estimates

Variable	DF	Parameter Estimate	Standard Error	t Value	Pr > |t|
Intercept	1	-0.52343	0.14528	-3.60	0.0032
oil_added	1	0.87825	0.01610	54.56	<.0001

Obs	Dep Var	Predict Value	Residual
1	0.6100	0.3548	0.2552
2	0.8400	0.7939	0.0461
3	1.5120	1.3209	0.1911
4	1.7920	1.9357	-0.1437
5	2.9520	2.6383	0.3137
6	2.8800	3.4287	-0.5487
7	4.4000	4.3069	0.0931
8	5.3460	5.2730	0.0730
9	6.3960	6.3269	0.0691
10	7.1890	7.4686	-0.2796
11	8.0850	8.6982	-0.6132
12	9.8400	10.0155	-0.1755
13	11.6960	11.4207	0.2753
14	13.2240	12.8259	0.3981
15	14.3650	14.3189	0.0461

Sum of Residuals   0
Sum of Squared Residuals   1.2640

a. Write the equation of the least squares line and use it to predict the value of recovered oil when added oil is 10 g.

b. What are the values of SSResid, SSTo, r^2, and s_e? Do these values suggest that the least squares line provides an effective summary of the relationship between the two variables?

c. Construct a plot of the residuals. What does it suggest?

23. Recall the data from Exercise 6 involving x = fiber weight (%) and y = compressive strength (MPa).

a. Determine the equation of the least squares line and interpret its slope.

b. Determine the proportion of observed variation in strength that can be attributed to the approximate linear relationship between strength and fiber weight.

c. Predict the value of the compressive strength when the fiber weight percentage is 6.5.

d. Would you feel comfortable using the least squares line to predict the compressive strength when the fiber weight percentage is 25? Explain. Now predict the value of y when $x = 25$ and interpret the result.

24. By their nature, deserts are typically exposed to large amounts of solar radiation. Thus, such regions seem to be prime locations for harvesting solar energy through the installation of photovoltaic modules. These modules rely on an optical system to collect sunlight, often through some lens, so an important factor to consider would be the effect of desert sandstorms on lens performance. The authors of "Sandblasting Durability of Acrylic and Glass Fresnel Lenses for Concentrator Photovoltaic Modules" (*Solar Energy*, 2012: 3021–3025) compared the performance of sandblasted acrylic and glass Fresnel lenses used in concentrator photovoltaic modules. In the experiment, the transmittance after sandblasting of acrylic polymethylmethacrylate (PMMA) and glass Fresnel lenses were measured. The experimental data, kindly provided by the authors, compares y = reduction rate of transmittance (%) and x = sandblast momentum (g·m/s) for 14 PMMA and 8 glass substrate samples:

x_{PMMA}:	10.56	20.80	15.84	31.20	48.06
y_{PMMA}:	8.56	18.93	19.35	23.65	33.05

x_{PMMA}:	21.12	41.60	64.00	16.80	33.20
y_{PMMA}:	18.53	29.21	40.39	17.21	27.21

x_{PMMA}:	51.20	13.92	27.84	42.72
y_{PMMA}:	34.74	17.40	25.89	32.82

x_{Glass}:	35.20	52.80	105.60	52.80	70.40
y_{Glass}:	5.62	8.10	31.21	13.76	15.37

x_{Glass}:	56.00	48.00	139.20
y_{Glass}:	14.76	16.55	37.08

a. In one graph, overlay the scatterplots for the PMMA and the glass data sets and comment on any interesting features. Be sure to use different symbols for each data set.

b. Determine the equations for the least squares line for the PMMA and glass data sets. Interpret the slope for each equation.

c. For the PMMA lens, predict the reduction rate of transmittance when sandblast momentum is at 50 g·m/s. Do the same for the glass lens type.

d. Based on your results, which lens type performed better in this experiment?

25. Two important properties of a soil are its initial void ratio (e_0, a measure of soil porosity) and its compression index (C_c, an indicator of soil compressibility). The article "Consolidation and Hydraulic Conductivity of Zeolite-Amended Soil-Bentonite Backfills" (*J. Geotech. Geoenviron. Engr.*, 2012: 15–25) reported the following data (read from a graph) for the C_c and e_0 variables for sand–bentonite backfills with varying amounts and types of zeolites.

e_0:	0.988	1.018	1.058	1.070	1.085	1.145
C_c:	0.19	0.20	0.20	0.22	0.23	0.24

a. Using C_c as the response and e_0 as the explanatory variable, create the corresponding scatterplot. Do the values of C_c appear to be perfectly linearly related to the e_0 values? Explain.

b. Determine the equation of the least squares line.

c. What proportion of the observed variation in the compression index can be attributed to the approximate linear relationship between the two variables?

d. Predict the value of the compression index when the initial void ratio is 1.10. Would you feel comfortable using the least squares line to predict the compression index when the initial void ratio is .80? Explain.

26. In biofiltration of wastewater, air discharged from a treatment facility is passed through a damp porous membrane that causes contaminants to dissolve in water and be transformed into harmless products. The accompanying data on x = inlet temperature (°C) and y = removal efficiency (%) was the basis for a scatterplot that appeared in the article "Treatment of Mixed Hydrogen Sulfide

and Organic Vapors in a Rock Medium Biofilter"
(*Water Environment Research*, 2001: 426–435):

Obs	Temp	Removal %	Obs	Temp	Removal %
1	7.68	98.09	17	8.55	98.27
2	6.51	98.25	18	7.57	98.00
3	6.43	97.82	19	6.94	98.09
4	5.48	97.82	20	8.32	98.25
5	6.57	97.82	21	10.50	98.41
6	10.22	97.93	22	17.83	98.51
7	15.69	98.38	23	17.83	98.71
8	16.77	98.89	24	17.03	98.79
9	17.13	98.96	25	16.18	98.87
10	17.63	98.90	26	16.26	98.76
11	16.72	98.68	27	14.44	98.58
12	15.45	98.69	28	12.78	98.73
13	12.06	98.51	29	12.25	98.45
14	11.44	98.09	30	11.69	98.37
15	10.17	98.25	31	11.34	98.36
16	9.64	98.36	32	10.97	98.45

Calculated summary quantities are $\Sigma x_i = 384.26$, $\Sigma y_i = 3149.04$, $\Sigma x_i^2 = 5099.2412$, $\Sigma x_i y_i = 37{,}850.7762$, and $\Sigma y_i^2 = 309{,}892.6548$.

a. Does a scatterplot of the data suggest appropriateness of the simple linear regression model?

b. Determine the equation of the least square line, obtain a point prediction of removal efficiency when temperature = 10.50, and calculate the value of the corresponding residual.

c. Roughly what is the size of a typical deviation of points in the scatterplot from the least squares line?

d. What proportion of observed variation in removal efficiency can be attributed to the approximate linear relationship?

e. Personal communication with the authors of the article revealed that there was one additional observation that was not included in their scatterplot: (6.53, 96.55). What impact does this additional observation have on the equation of the least squares line and the values of s_e and r^2?

27. Consider the following four (x, y) data sets; the first three have the same x values, so these values are listed only once (from "Graphs in Statistical Analysis," *Amer. Statistician*, 1973: 17–21).

For each of these four data sets, the values of the summary quantities, Σx_i, Σy_i, and so on, are almost identical, so the equation of the least squares line($\hat{y} = 3 + .5x$), SSResid, SSTo, r^2, and s_e will be virtually the same for all four. Based on a scatterplot and a residual plot for each data set, comment on the appropriateness of fitting a straight line; include any specific suggestions for how a "straight-line analysis" might be modified or qualified.

Data set:	1–3	1	2	3	4	4
Variable:	x	y	y	y	x	y
	10.0	8.04	9.14	7.46	8.0	6.58
	8.0	6.95	8.14	6.77	8.0	5.76
	13.0	7.58	8.74	12.74	8.0	7.71
	9.0	8.81	8.77	7.11	8.0	8.84
	11.0	8.33	9.26	7.81	8.0	8.47
	14.0	9.96	8.10	8.84	8.0	7.04
	6.0	7.24	6.13	6.08	8.0	5.25
	4.0	4.26	3.10	5.39	19.0	12.50
	12.0	10.84	9.13	8.15	8.0	5.56
	7.0	4.82	7.26	6.42	8.0	7.91
	5.0	5.68	4.74	5.73	8.0	6.89

3.4   NONLINEAR RELATIONSHIPS

A scatterplot of bivariate data frequently shows curvature rather than a linear pattern. In this section, we discuss several different ways to fit a curve to such data.

Power Transformations

Suppose that the general pattern in a scatterplot is curved and monotonic — either strictly increasing or strictly decreasing. In this case, it is often possible to find a **power transformation** for x or y so that there is a linear pattern in a scatterplot of the transformed data. By a power transformation, we mean the use of exponents p and q such that the transformed values are $x' = x^p$ and/or $y' = y^q$; the relevant scatterplot is of the (x', y') pairs. Figure 3.16 displays a "ladder" of the most frequently used transformations and a guide for choosing an appropriate transformation, depending on the pattern in the original scatterplot.

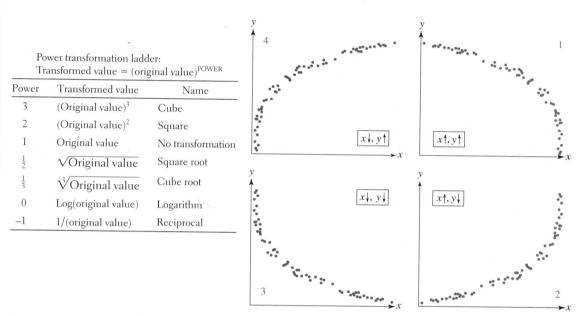

Power transformation ladder:
Transformed value = (original value)^{POWER}

Power	Transformed value	Name
3	(Original value)^3	Cube
2	(Original value)^2	Square
1	Original value	No transformation
$\frac{1}{2}$	$\sqrt{\text{Original value}}$	Square root
$\frac{1}{3}$	$\sqrt[3]{\text{Original value}}$	Cube root
0	Log(original value)	Logarithm
-1	1/(original value)	Reciprocal

Figure 3.16 Transformation ladder and guide

For example, suppose the pattern has the shape of segment 2 in Figure 3.16. Then to straighten the plot, we should use a transformation on x that is up the ladder from the no-transformation row, for example, $x' = x^2$ or x^3, or a transformation on y that is down the ladder, such as $y' = 1/y$ or $\ln(y)$ ($\log_{10}$ would produce equivalent results). A residual plot should be used to check that curvature has in fact been removed. Once a straightening transformation has been identified, a straight line can be fit to the (x', y') points using least squares. If it was not necessary to transform y, then the line provides a direct way of predicting y values: calculate x' and substitute into the equation. When y has been transformed, the line gives predictions of y' values. The transformation can then be reversed to obtain predictions of y. For example, if $x' = 1/x$ and $y' = \sqrt{y}$, the least squares line gives

$$\sqrt{y} \approx a + b/x$$

from which

$$y \approx (a + b/x)^2$$

Example 3.12 No tortilla chip aficionado likes soggy chips, so it is important to find characteristics of the production process that produce chips with an appealing texture. The following data on x = frying time (sec) and y = moisture content (%) appeared in the article "Thermal and Physical Properties of Tortilla Chips as a Function of Frying Time" (*J. of Food Processing and Preservation*, 1995: 175–189):

x:	5	10	15	20	25	30	45	60
y:	16.3	9.7	8.1	4.2	3.4	2.9	1.9	1.3

The scatterplot in Figure 3.17(a), opposite, has the pattern of segment 3 in Figure 3.16, so we must go down the ladder for x or y. A scatterplot of the $(\ln(x), \ln(y))$ pairs in Figure 3.17(b) is quite straight. A regression of $\ln(y)$ on $\ln(x)$ gives $a = 4.6384$, $b = -1.04920$, and $r^2 = .976$. The residual plot of Figure 3.17(c) shows no evidence of curvature, though there is one rather large residual.

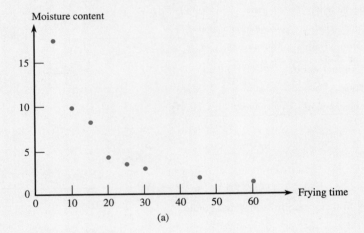

(a)

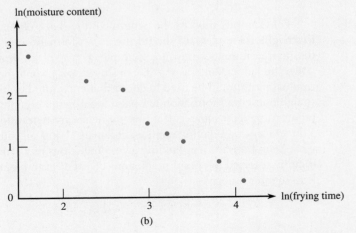

(b)

Figure 3.17 Plots of the data from Example 3.12: (a) scatterplot of the original data; (b) scatterplot of the $(\ln(x), \ln(y))$ pairs

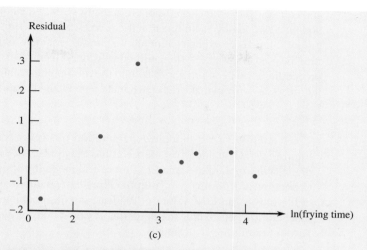

Figure 3.17 Plots of the data from Example 3.12: (c) plot of the residuals from the transformed regression

Thus $\ln(y) \approx 4.6384 - 1.04920[\ln(x)]$. Since $\ln(20) = 2.996$, a prediction of $\ln(y)$ is

$$\widehat{\ln}(y) = 4.6384 - (1.04920)(2.996) = 1.495$$

Taking the antilog of 1.495 gives a prediction of y itself: $e^{1.495} = 4.46\%$. In fact, taking the antilog of both sides of the linear equation gives an explicit nonlinear relationship between x and y:

$$y = e^{\ln(y)} \approx e^{4.6384 - 1.04920[\ln(x)]} = (e^{4.6384})(e^{-1.04920 \ln(x)}) = 103.379x^{-1.04920}$$

This is often called a *power function* relationship between x and y.

Fitting a Polynomial Function

Sometimes the general pattern of curvature in a scatterplot is not monotonic. Instead, it may be the case that as x increases, there is a tendency for y first to increase and then to decrease (like a bowl turned upside down) or for y first to decrease and then to increase. In such instances, it is reasonable to fit a quadratic function $a + b_1x + b_2x^2$, whose graph is a parabola, to the data. If the quadratic coefficient b_2 is positive, the parabola turns upward, whereas it turns downward if b_2 is negative. Just as in fitting a straight line, the principle of least squares can be employed to find the best-fit quadratic. The **least squares coefficients** a, b_1, and b_2 are the values of $\tilde{a}$, $\tilde{b}_1$, and $\tilde{b}_2$ that minimize

$$g(\tilde{a}, \tilde{b}_1, \tilde{b}_2) = \sum_i [y_i - (\tilde{a} + \tilde{b}_1x_i + \tilde{b}_2x_i^2)]^2$$

which is the sum of squared vertical deviations from the points in the scatterplot to the parabola determined by the quadratic with coefficients $\tilde{a}$, $\tilde{b}_1$, and $\tilde{b}_2$. Taking the partial derivative of the g function first with respect to $\tilde{a}$, then with respect to $\tilde{b}_1$, and finally with respect to $\tilde{b}_2$, and equating these three expressions to zero gives three equations in three unknowns. These *normal equations* are again linear in the unknowns, but because there are three rather than just two, there is no explicit elementary expression for their solution. Instead, matrix algebra must be used to solve the system numerically for each different data set. Fortunately, solution procedures have been programmed into the most popular statistical computer packages, so it is necessary only to make the appropriate request and then sit back and wait for output.

Example 3.13   The scatterplot of $y =$ glucose concentration versus $x =$ fermentation time shown in Figure 3.9 (at the end of Section 3.2) has the appearance of an upward-turning quadratic. We supplied the data to Minitab and made the appropriate regression request to obtain the accompanying output. The fitted quadratic equation appears at the top of the output, and the values of the least squares coefficients a, b_1, b_2 appear in the Coef column just below the equation. A prediction for glucose concentration when fermentation time is 4 hours is

$$\hat{y} = 84.482 - 15.875(4) + 1.7679(4)^2 = 49.27$$

```
The regression equation is
glucconc = 84.5 - 15.9 time + 1.77 timesqd
Predictor      Coef    Stdev    t-ratio        p
Constant     84.482    4.904      17.23    0.000
time        -15.875    2.500      -6.35    0.001
timesqd      1.7679   0.2712       6.52    0.001
s = 3.515    R-sq = 89.5%    R-sq (adj) = 85.3%
Analysis of Variance
SOURCE        DF       SS       MS       F        p
Regression     2   525.11   262.55   21.25    0.004
Error          5    61.77    12.35
Total          7   586.88
```

Predicted or fitted values $\hat{y}_1, \ldots, \hat{y}_n$ are obtained by substituting the successive x values $x_1, \ldots, x_n$ into the fitted quadratic equation (e.g., in Example 3.13, $\hat{y}_4 = 49.27$), and the residuals are the vertical deviations $y_1 - \hat{y}_1, \ldots, y_n - \hat{y}_n$ from the observed points to the graph of the fitted quadratic (e.g., $y_4 - \hat{y}_4 = 51 - 49.27 = 1.73$). Residual or error sum of squares and total sum of squares are defined exactly as they were previously:

$$\text{SSResid} = \sum_i (y_i - \hat{y}_i)^2 \qquad \text{SSTo} = \sum_i (y_i - \bar{y})^2$$

The Minitab output of Example 3.13 shows that SSResid $= 61.77$ and SSTo $= 586.88$. The **coefficient of multiple determination**, denoted by R^2, is now the proportion of observed y variation that can be attributed to the approximate quadratic relationship:

$$R^2 = 1 - \frac{\text{SSResid}}{\text{SSTo}}$$

The R^2 value in Example 3.13 is .895, so about 89.5% of the observed variation in glucose concentration can be attributed to the approximate quadratic relation between concentration and fermentation time.

The methodology employed to fit a quadratic is easily extended to fit a higher-order polynomial. For example, using the principle of least squares to fit a cubic equation gives a system of normal equations consisting of four equations in four unknowns. The arithmetic is best left to a statistical computer package. In practice, a cubic equation is rather rarely fit to data, and it is virtually never appropriate to fit anything of higher order than this.

Smoothing a Scatterplot

Sometimes the pattern in a scatterplot is too complex for a line or curve of a particular type (e.g., exponential or parabolic) to give a good fit. Statisticians have recently developed some more flexible methods that permit a wide variety of patterns to be modeled using the same fitting procedure. One such method is **LOWESS** (or LOESS), short for *locally weighted scatterplot smoother*. Let (x^*, y^*) denote a particular one of the n (x, y) pairs in the sample. The $\hat{y}$ value corresponding to (x^*, y^*) is obtained by fitting a straight line using only a specified percentage of the data (e.g., 25%) whose x values are closest to x^*. Furthermore, rather than use "ordinary" least squares, which gives equal weight to all points, those with x values closer to x^* are more heavily weighted than those whose x values are farther away.[1] The height of the resulting line above x^* is the fitted value $\hat{y}^*$. This process is repeated for each of the n points, so n different lines are fit (you surely wouldn't want to do all this by hand). Finally, the fitted points are connected to produce a LOWESS curve.

Example 3.14   Weighing large deceased animals found in wilderness areas is usually not feasible, so it is desirable to have a method for estimating weight from various characteristics of an animal that can be easily determined. Minitab has a stored data set consisting of various characteristics for a sample of $n = 143$ wild bears. Figure 3.18(a), opposite, displays a scatterplot of $y =$ weight versus $x =$ distance around the chest (chest girth). At first glance, it looks as though a single line obtained from ordinary least squares would effectively summarize the pattern. Figure 3.18(b) shows the LOWESS curve produced by Minitab using a span of 50% (the fit at (x^*, y^*) is determined by the closest 50% of the sample). The curve appears to consist of two straight-line segments joined together above approximately $x = 38$. The steeper line is to the right of 38, indicating that weight tends to increase more rapidly as girth does for girths exceeding 38 in.

---

[1] The weighted least squares criterion involves finding $\tilde{a}$ and $\tilde{b}$ to minimize, $\Sigma w_i[y_i - (\tilde{a} + \tilde{b}x_i)]^2$, where $w_1, \ldots, w_n$ are nonnegative weights. For example, if we take $w_5 = 0$, then (x_5, y_5) is disregarded in obtaining the fitted line. R will also fit a local quadratic in this way.

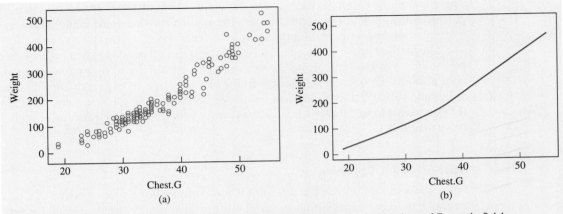

Figure 3.18 A Minitab scatterplot and LOWESS curve for the bear weight data of Example 3.14

Section 3.4 Exercises

28. Polyester fiber ropes are increasingly being used as components of mooring lines for offshore structures in deep water. The authors of the paper "Quantifying the Residual Creep Life of Polyester Mooring Ropes" (*Intl. J. of Offshore and Polar Explor.*, 2005: 223–228) used the accompanying data as a basis for studying how time to failure (hr) depended on load (% of breaking load):

x:	77.7	77.8	77.9	77.8	85.5	85.5
y:	5.067	552.056	127.809	7.611	.124	.077

x:	89.2	89.3	73.1	85.5	89.2	85.5
y:	.008	.013	49.439	.503	.362	9.930

x:	89.2	85.5	89.2	82.3	82.0	82.3
y:	.677	5.322	.289	53.079	7.625	155.299

a. Construct a scatterplot of x = load versus y = time. Would it be reasonable to characterize the relationship between the two variables to be linear?

b. Transform the response variable by computing $y' = \log(y)$. Construct a scatterplot of x and y'. Would it be reasonable to characterize the relationship between these two variables to be linear?

c. Fit a straight line to the (x, y') data. Assess the quality of the fit. Finally, based on the linear

fit, predict the value of failure time from a load of 85%.

29. The authors of "Experimental and Numerical Investigation of Bed-Load Transport Under Unsteady Flows" (*J. Hydraul. Engr.*, 2011: 1276–1282) simulated sediment yield of a gravel bed load under varying rates of water flow. The researchers wanted to mathematically model the behavior of sediment transport under such conditions and proposed a new model parameter, P_{gt}, that characterizes the unsteadiness of the water flow. Eleven simulation runs were conducted in the laboratory. For each simulation, the article reported the computed value of the unsteadiness parameter P_{gt} and the nondimensionalized total bed load, W_t^*. One aim of the study was to investigate the behavior of $y = W_t^*$ as a function of $x = P_{gt}$. Data from the experiment is given here:

x:	0.0021	0.0041	0.0045	0.0046
y:	15.4	59.0	80.9	107.5

x:	0.0049	0.0043	0.0049	0.0043
y:	313.6	163.8	857.2	40.9

x:	0.0047	0.0038	0.0046
y:	88.9	87.8	196.5

a. Would you fit a straight line to the data and use it as a basis for predicting nondimensionalized total bed load from the unsteadiness parameter? Why or why not?

b. Find a transformation that produces an approximate linear relationship between the transformed values. Then fit a line to the transformed data and use it to obtain an equation that describes approximately the relationship between the untransformed variables.

30. In the article "Sensitivity of Oklahoma Binders on Dynamic Modulus of Asphalt Mixes and Distress Functions" (*J. Mater. Civ. Engr.*, 2012: 1076–1088), researchers measured various physical characteristics of performance grade asphalt binders commonly used in Oklahoma. One important physical characteristic is dynamic shear modulus, G^* (kPa), which is the ratio of maximum shear stress to the maximum shear strain and is a measure of the stiffness or resistance of the asphalt binder to deformation under load. In one experiment, the researchers measured the dynamic shear modulus of the asphalt binder samples over a range of testing temperatures (°C). The following is the corresponding data for binder type PG64-22:

Temp:	54.4	46.1	43.3	29.4
G^*:	9.28	32.47	46.98	344.36
Temp:	21.1	12.7	4.4	
G^*:	1,030.38	4,870.00	18,300.00	

a. Construct a scatterplot of $y =$ dynamic shear modulus versus $x =$ temperature. Would it be reasonable to characterize the relationship between the two variables as approximately linear?

b. Transform only the dependent variable y so that a scatterplot of the transformed data shows a substantial linear pattern. Then fit a straight line to this data, use the line to establish an approximate relationship between x and y, and predict the dynamic shear modulus when the temperature is 35°C.

c. Plot the residuals from your linear fit in part (b) and look for any patterns that might suggest

an inappropriate choice of transformation. If necessary, return to part (b) and try a different transformation.

31. Failures in aircraft gas turbine engines due to high cycle fatigue is a pervasive problem. The article "Effect of Crystal Orientation on Fatigue Failure of Single Crystal Nickel Base Turbine Blade Superalloys" (*J. of Engr. for Gas Turbines and Power*, 2002: 161–176) gave the accompanying data and fit a nonlinear regression function in order to predict strain amplitude from cycles to failure.

Obs	Cycfail	Strampl	Obs	Cycfail	Strampl
1	1326	.01495	11	7356	.00576
2	1593	.01470	12	7904	.00580
3	4414	.01100	13	79	.01212
4	5673	.01190	14	4175	.00782
5	29,516	.00873	15	34,676	.00596
6	26	.01819	16	114,789	.00600
7	843	.00810	17	2672	.00880
8	1016	.00801	18	7532	.00883
9	3410	.00600	19	30,220	.00676
10	7101	.00575			

a. Construct scatterplots of y versus x, y versus $\ln(x)$, $\ln(y)$ versus $\ln(x)$, and $1/y$ versus $1/x$.

b. Which transformation from part (a) does the best job of producing an approximate linear relationship?

c. Use the selected transformation to predict amplitude when cycles to failure $= 5000$.

32. There has been an increasing demand for open-ended steel pipe piles to be used as deep foundations for offshore and onshore structures. When an open-ended pile is driven into the ground, a soil plug often forms within the pile. The driving resistance and the base capacity of the pile are heavily influenced by this plugging effect. As an indicator of the degree of plugging, researchers often use the plug length ratio (PLR), which is the ratio of the plug length at the end of pile installation to the length of the pile. The article "Base Capacity of Open-Ended Steel Pipe Piles in Sand" (*J. Geotech. Geoenviron. Engr.*, 2012: 1116–1128) reported the PLR and corresponding pile inner

diameter, d (mm), of nine test piles used in case studies. The data is given here:

d:	691.0	292.0	83.7	37.2	78.9
PLR:	1.00	0.82	0.76	0.44	0.76

d:	107.9	82.5	1444.0	1444.0
PLR:	0.88	0.75	1.00	1.00

a. The authors were interested in predicting PLR based on the pile inner diameter. Transform only the independent variable x so that a scatterplot of the transformed data shows a substantial linear pattern. Then fit a straight line to this data, use the line to establish an approximate relationship between x and y, and predict the plug length ratio when the pile inner diameter is 500 mm.

b. Plot the residuals from your linear fit in part (a) and look for any patterns that might suggest an inappropriate choice of transformation. If necessary, return to part (a) and try a different transformation.

33. The article "Residual Stresses and Adhesion of Thermal Spray Coatings" (*Surface Engr.*, 2005: 35–40) considered the relationship between the thickness (mm) of NiCrAl coatings deposited on stainless steel substrate and corresponding bond strength (MPa). The following data was read from a plot in the paper:

Thickness:	220	220	220	220	370
Strength:	24.0	22.0	19.1	15.5	26.3

Thickness:	370	370	370	440	440
Strength:	24.6	23.1	21.2	25.2	24.0

Thickness:	440	440	680	680	680
Strength:	21.7	19.2	17.0	14.9	13.0

Thickness:	680	860	860	860	860
Strength:	11.8	12.2	11.2	6.6	2.8

a. Is it possible to transform this data as described in this section so that there is an approximate linear relationship between the transformed variables? Why or why not?

b. Use a statistical computer package to fit a quadratic function to this data and then predict bond strength when thickness is 500. Assess the fit of the quadratic to the data.

34. The accompanying data was extracted from the article "Effects of Cold and Warm Temperatures on Springback of Aluminum-Magnesium Alloy 5083-H111" (*J. Engr. Manuf.*, 2009: 427–431). The response variable is yield strength (MPa), and the predictor is temperature (°C).

x:	−50	25	100	200	300
y:	91.0	120.5	136.0	133.1	120.8

Here is Minitab output from fitting the quadratic regression function (a graph in the cited paper suggests that the authors did this):

```
Predictor           Coef      SE Coef      T       P
Constant         111.277       2.100    52.98   0.000
temp             0.32845      0.03303     9.94   0.010
tempsqd        -0.0010050    0.0001213   -8.29   0.014

S=3.44398   R-Sq=98.1%   R-Sq(adj)=96.3%

Analysis of Variance

Source           DF       SS       MS       F       P
Regression        2   1245.39   622.69   52.50   0.019
Residual Error    2     23.72    11.86
Total             4   1269.11
```

a. What is the equation of the best-fit quadratic? Use this quadratic to predict yield strength when temperature is 110.

b. What are the values of SSResid and SSTo? Verify that these values are consistent with the value of R-sq given on the output. Do you think the fit of the quadratic is good? Explain.

3.5   USING MORE THAN ONE PREDICTOR

In Sections 3.3 and 3.4, we considered relationships between a dependent or response variable y and a single predictor, independent, or explanatory variable x. In many situations, predictions of y values can be improved and more observed y variation can be explained by utilizing information in two or more explanatory variables. Notation is a bit more complex than in the case of a single predictor. Let

$$k = \text{number of explanatory variables or predictors}$$
$$n = \text{sample size}$$

and $x_1, x_2, \ldots, x_k$ denote the k predictors, so that each observation will consist of $k + 1$ numbers: the value of x_1, the value of $x_2, \ldots,$ the value of x_k, and the value of y. Also let

$$x_{ij} = \text{value of the predictor } x_i \text{ in the } j\text{th observation}$$

so

$$\text{first observation} = (x_{11}, x_{21}, \ldots, x_{k1}, y_1)$$

$$\vdots \qquad\qquad \vdots$$

$$n\text{th observation} = (x_{1n}, x_{2n}, \ldots, x_{kn}, y_n)$$

Example 3.15   Soil and sediment adsorption, the extent to which chemicals collect in a condensed form on the surface, is an important characteristic because it influences the effectiveness of pesticides and various agricultural chemicals. The article "Adsorption of Phosphate, Arsenate, Methanearsonate, and Cacodylate by Lake and Stream Sediments: Comparison with Soils" (*J. of Environ. Qual.*, 1984: 499–504) gave the following data on y = phosphate adsorption index, x_1 = amount of extractable iron, and x_2 = amount of extractable aluminum

Observation	x_1	x_2	y
1	61	13	4
2	175	21	18
3	111	24	14
4	124	23	18
5	130	64	26
6	173	38	26
7	169	33	21
8	169	61	30
9	160	39	28
10	244	71	36
11	257	112	65
12	333	88	62
13	199	54	40

Thus the first observation is the triple $(x_{11}, x_{21}, y_1) = (61, 13, 4), \ldots,$ and the last observation is $(x_{1,13}, x_{2,13}, y_{13}) = (199, 54, 40)$.

Each observation in Example 3.15 is a triple of numbers. A scatterplot of such data would represent each observation as a point in a *three*-dimensional coordinate system, which is obviously difficult to construct or visualize. For $k > 2$, a scatterplot requires more than three dimensions! Partial information about the relationship between the variables can be obtained by forming a **scatterplot matrix.** This is just a collection of two-dimensional scatterplots, arranged in a square array, in which each variable is plotted against every other variable. The matrix gives a preliminary indication of whether any single predictor might be related to y, whether the relationship might be linear, and whether there appears to be a strong relation between any particular pair of predictors (in which case, one of them may be redundant). Figure 3.19 shows a scatterplot matrix for the adsorption data from Example 3.15. In the case $k = 2$, there are really just three plots: y versus x_1, y versus x_2, and x_2 versus x_1. Each of these plots appears twice in Figure 3.19, allowing the investigator to look across any row and see a particular variable plotted against every other variable. For example, the third row shows adsorption index versus extractable iron, followed by adsorption index versus extractable aluminum. We can see that y appears linearly related to both x_1 and x_2 and that there is not a very strong relationship between x_1 and x_2.

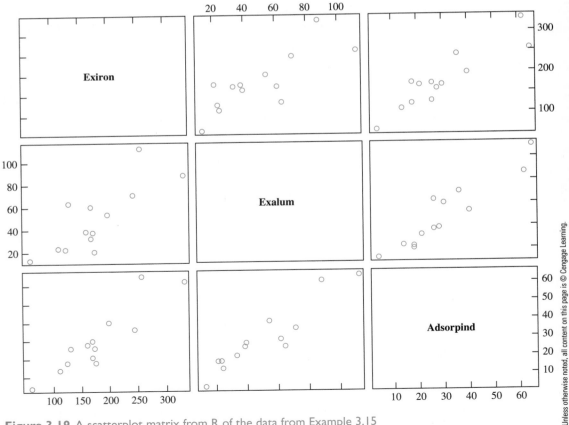

Figure 3.19 A scatterplot matrix from R of the data from Example 3.15

Fitting a Linear Function

We now consider fitting a relation of the form

$$y \approx a + b_1x_1 + b_2x_2 + \cdots + b_kx_k$$

The reasonableness of this approximation depends on patterns in the scatterplot matrix and other characteristics of the data to be considered shortly. As with bivariate data, the values of $a, b_1, \ldots, b_k$ should be selected to give the best fit. Again, the principle of least squares can be invoked: The **least squares coefficients** $a, b_1, b_2, \ldots, b_k$ are the values of $\tilde{a}, \tilde{b}_1, \ldots, \tilde{b}_k$ that minimize

$$g(\tilde{a}, \tilde{b}_1, \ldots, \tilde{b}_k) = \sum_{j=1}^{n} [y_j - (\tilde{a} + \tilde{b}_1x_{1j} + \cdots + \tilde{b}_kx_{kj})]^2$$

The $g(\ )$ function is the sum of squared deviations between observed y values and what would be predicted by $\tilde{a} + \tilde{b}_1x_1 + \cdots + \tilde{b}_kx_k$. Determination of the least squares coefficients involves multivariable calculus: Take the partial derivative of $g(\ )$ with respect to each unknown, equate these to zero to obtain a system of $k + 1$ *linear* equations in the $k + 1$ unknowns (the *normal equations*), and solve the system. The arithmetic is quite tedious, but any good statistical computer package can handle the task upon request; a regression command of some sort is usually required.

Example 3.16

(Example 3.15 continued) Figure 3.20 shows partial Minitab output from a request to fit $a + b_1x_1 + b_2x_2$ to the phosphate adsorption data using the principle of least squares. The result is

$$\hat{y} \approx -7.351 + .11273x_1 + .34900x_2 \approx -7.35 + .113x_1 + .349x_2$$

```
The regression equation is
adsorp = -7.35 + 0.113 exiron + 0.349 exalum

Predictor        Coef      Stdev    t-ratio       p
Constant       -7.351      3.485      -2.11   0.061
exiron        0.11273    0.02969       3.80   0.004
exalum        0.34900    0.07131       4.89   0.000

s = 4.379     R-sq = 94.8%    R-sq(adj) = 93.8%

Analysis of Variance

SOURCE           DF        SS        MS        F       p
Regression        2    3529.9    1765.0    92.03   0.000
Error            10     191.8      19.2
Total            12    3721.7
```

Figure 3.20 Minitab regression output for the phosphate adsorption data

A prediction of the phosphate adsorption index for an observation to be made when extractable iron is 150 and extractable aluminum is 60 is

$$\hat{y} = -7.35 + .113(150) + .349(60) = 30.54$$

We interpret $b_1 = .113$ to mean that when the amount of extractable iron increases by 1 unit *and* the amount of extractable aluminum is held fixed, we can expect the phosphate adsorption index to increase by roughly .113. A similar interpretation applies to $b_2 = .349$.

Predicted values and residuals are calculated in a manner similar to that used in the case of a single predictor. For example, $\hat{y}_1$ results from substituting $x_1 = x_{11}, x_2 = x_{21}, \ldots, x_k = x_{k1}$ (the values of the predictors for the first observation) into $\hat{y} = a + b_1 x_1 + \cdots + b_k x_k$, and the corresponding residual is $y_1 = \hat{y}_1$. From Examples 3.15 and 3.16,

$$\hat{y}_1 = -7.35 + .113(61) + .349(13) = 4.08 \qquad y_1 - \hat{y}_1 = 4 - 4.08 = -.08$$

The same two sums of squares calculated after fitting a line are relevant here:

$$\text{SSResid} = \sum (y_i - \hat{y}_i)^2 \quad \text{a measure of unexplained variation}$$

$$\text{SSTo} = \sum (y_i - \bar{y})^2 \quad \text{a measure of total variation}$$

The **coefficient of multiple determination**

$$R^2 = 1 - \frac{\text{SSResid}}{\text{SSTo}}$$

is interpreted as the proportion of observed y variation that can be explained by or attributed to the approximate linear relation between the response variable and the predictors. The value of R^2 is the first concrete indicator of whether the postulated linear relationship is indeed a good approximation. Looking at the Minitab output of Figure 3.20, we see that about 94.8% of observed variation in the phosphate adsorption index can be explained by its approximate linear relationship to extractable iron and extractable aluminum, a very impressive result. In addition, residual plots—residuals versus x_1, residuals versus $x_2, \ldots$, and residuals versus x_k—should be examined for evidence that the fitted relationship must be modified. The two residual plots of Figure 3.21 (p. 145) show no unusual pattern indicating that a modification is needed.

There is one potential difficulty with R^2: Its value can be greatly inflated by using many predictors of questionable importance when fitting the linear relationship. Suppose, for example, that y is the sale price of a home and that we have a sample of $n = 20$ homes from the region of interest. Important predictors include $x_1 =$ interior size (ft^2), $x_2 =$ lot size, $x_3 =$ age of the home, $x_4 =$ number of bedrooms, and $x_5 =$ size of the garage. Consider adding other predictors that are intuitively relatively uninformative or even frivolous: thickness of the driveway slab, diameter of a showerhead, height of the

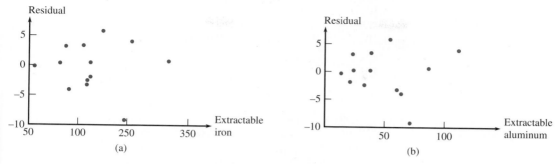

Figure 3.21 Residual plots for the adsorption data of Examples 3.15 and 3.16

doorknob on the front door, and so on. It turns out that if 19 predictors are included (one less than the number of observations), then it will virtually always be the case that $R^2 = 1$. So the goal here is not simply to obtain a set of predictors for which R^2 is large, but to obtain a large value using relatively few predictors while excluding those of marginal significance. We will further discuss this issue in Chapter 11.

Creating New Predictors from Existing Ones

The ability of predictors $x_1, \ldots, x_k$ to explain variation in y can often be considerably enhanced by having one or more predictors that are mathematical functions of the remaining predictors. As an example, let y denote the yield of a particular product resulting from a certain chemical reaction. Usually y will depend on both x_1 = reaction temperature and x_2 = reaction pressure. It may be the case, though, that using only these two predictors results in an R^2 value much less than 1, whereas including a third predictor $x_3 = x_1 x_2$ considerably increases R^2 (adding a predictor cannot possibly decrease R^2). Alternatively, using predictors x_1 and x_2 along with the two additional predictors $x_3 = x_1^2$ and $x_4 = x_2^2$ may result in most of the observed y variation being explained. Or perhaps all three of the additional predictors $x_3 = x_1 x_2$, $x_4 = x_1^2$, and $x_5 = x_2^2$ will give very impressive results. The first new variable x_3 is called an *interaction* predictor, and the other two are *quadratic* predictors (interpretations will be given in Chapter 11); the fit with all five predictors is called the *full quadratic* or *complete second-order* relationship. In fact, we used a quadratic predictor in the previous section when fitting a quadratic function to bivariate data. The two predictors there were $x_1 = x$ and $x_2 = x^2$, implying that quadratic (more generally, polynomial) regression is a special case of multiple regression.

Example 3.17   Researchers carried out a study to see how y = ultimate deflection, δ_d (mm), of reinforced ultrahigh toughness cementitious composite beams were influenced by x_1 = shear span ratio and x_2 = splitting tensile strength (MPa), resulting in the accompanying data ("Shear Behavior of Reinforced Ultrahigh Toughness Cementitious Composite Beams without Transverse Reinforcement," *J. Mater. Civ. Engr.*, 2012: 1283–1294):

x_1	x_2	x_1x_2	y
2.04	3.55	7.2420	3.11
2.04	6.07	12.3828	3.26
3.06	3.55	10.8630	3.89
3.06	6.07	18.5742	10.25
4.08	3.55	14.4840	3.11
4.08	6.16	25.1328	13.48
2.06	3.62	7.4572	3.94
2.06	6.16	12.6896	3.53
3.08	3.62	11.1496	3.36
3.08	5.89	18.1412	6.49
4.11	3.62	14.8782	2.72
4.11	5.89	24.2079	12.48
2.01	6.18	12.4218	2.82
3.02	6.18	18.6636	5.19
4.03	6.18	24.9054	8.04

Fitting $a + b_1x_1 + b_2x_2$ results in

$$\hat{y} = -9.251 + 2.322x_1 + 1.544x_2, \qquad R^2 = .576$$

Including an interaction predictor yields

$$\hat{y} = 17.279 - 6.368x_1 - 3.658x_2 + 1.707x_1x_2, \qquad R^2 = .825$$

Adding in the two quadratic predictors gives

$$\hat{y} = -34.323 - 6.568x_1 + 19.347x_2 + 1.655x_1x_2 + .058x_1^2 - 2.359x_2^2, \qquad R^2 = .845$$

General Additive Fitting

The relationships described heretofore in this section impose quite a bit of structure on how y depends on the explanatory variables. A more flexible type of relation is

$$\hat{y} = a + f_1(x_1) + f_2(x_2) + \cdots + f_k(x_k)$$

where the forms of $f_1(\ )$, . . . , and $f_k(\ )$ are left unspecified. The statistical package R, among others, will execute this **general additive fit** by calculating a and the individual $f_i(\ )$'s; one method for carrying out this latter task is based on the LOWESS technique described in Section 3.4.

Example 3.18   The ethanol data set stored in the R package contains 88 observations on variables x_1, x_2, and y obtained in an experiment in which ethanol was burned in a single-cylinder automobile test engine. The variables are

$x_1 = C =$ compression ratio of the engine
$x_2 = E =$ equivalence ratio at which the engine was run (a measure of richness of the air/ethanol mix)
$y = NO_x =$ concentration of nitric oxide and nitrogen dioxide in engine exhaust, normalized in a certain manner

Figure 3.22 shows a scatterplot matrix of the data; it appears that there is a substantial nonlinear relation between y and x_2. We asked R to obtain a general additive fit using LOWESS with a span of .75 (closest 75% of the data values) for each of the two component functions $f_1(x_1)$ and $f_2(x_2)$. Graphs of these two functions appear in Figure 3.23. Sure enough, the second graph is highly nonlinear, and there is also some nonlinearity in the first graph.

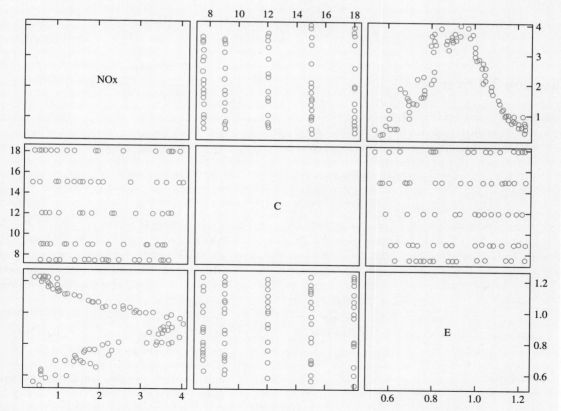

Figure 3.22 Scatterplot matrix of the ethanol data from R

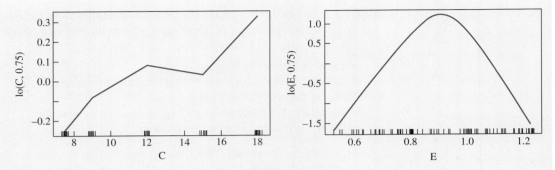

Figure 3.23 R graphs of the component functions resulting from a general additive fit to the ethanol data

The R^2 value for this fit was .873, whereas the value for the linear fit $a + b_1 x_1 + b_2 x_2$ was only .01. The reported value of the constant term a was 1.957, and the predicted value of NO_x when $C = 9.0$ and $E = 1.0$ was given by R as

$$\hat{y} = 1.957 + f_1(9.0) + f_2(1.0) = 2.743$$

Section 3.5 Exercises

35. Recently there has been increased use of stainless steel claddings in industrial settings. Claddings are used to finish the exterior walls of a building and help weatherproof the structure. To ensure the quality of claddings, it is essential to know how welding parameters impact the cladding process. The authors of "Mathematical Modeling of Weld Bead Geometry, Quality, and Productivity for Stainless Steel Claddings Deposited by FCAW" (*J. Mater. Engr. Perform.*, 2012: 1862–1872) investigated how y = deposition rate was influenced by x_1 = wire feed rate (W_f, in m/min) and x_2 = welding speed (S, in cm/min). The following 22 observations correspond to the experiment condition where applied voltage was less than 30v:

y:	2.718	3.881	2.773	3.924	2.740	3.870
x_1:	17.0	10.0	7.0	10.0	7.0	10.0
x_2:	30	30	50	50	30	30

y:	2.847	3.901	2.204	4.454	3.324	3.319
x_1:	7.0	10.0	5.5	11.5	8.5	8.5
x_2:	50	50	40	40	40	20

y:	3.423	3.242	3.385	3.420	3.380	3.402
x_1:	8.5	8.5	8.5	8.5	8.5	8.5
x_2:	60	40	40	40	40	40

y:	3.382	3.388	3.398	3.404
x_1:	8.5	8.5	8.5	8.5
x_2:	40	40	40	40

a. A least squares fit of $y = a + b_1 x_1 + b_2 x_2$ to this data gave $a = .0558$, $b_1 = .3749$, and $b_2 = .0028$. What value of deposition rate would you predict when wire feed rate = 11.5 and welding speed = 40? What is the value of the corresponding residual?

b. Residual and total sums of squares are .03836 and 5.1109, respectively. What proportion of observed variation in deposition rate can be attributed to the stated approximate relationship between deposition rate and the two predictor variables?

36. The accompanying Minitab regression output is based on data that appeared in the article "Application of Design of Experiments for Modeling

Surface Roughness in Ultrasonic Vibration Turning" (*J. of Engr. Manuf.*, 2009: 641–652). The response variable is surface roughness (mm), and the independent variables are vibration amplitude (mm), depth of cut (mm), feed rate (mm/rev), and cutting speed (m/min), respectively.

```
The regression equation is
Ra = -0.972 - 0.0312a - 0.557d - 18.3f - 0.00282v
Predictor      Coef    SE Coef        T        P
Constant    -0.9723     0.3923    -2.48    0.015
a           -0.03117    0.01864   -1.67    0.099
d            0.5568     0.3185     1.75    0.084
f           18.2602     0.7536    24.23    0.000
v            0.002822   0.003977   0.71    0.480

S=0.822059       R-Sq = 88.6%      R-Sq(adj) = 88.0%

Source        DF      SS      MS        F        P
Regression     4   401.02  100.25   148.35   0.000
Residual Error 76   51.36    0.68
Total          80  452.38
```

a. Predict the value of surface roughness when amplitude is 10, depth of cut is .5, feed rate is .25, and cutting speed is 50.

b. What proportion of observed variation in surface roughness can be explained by the approximate relationship between surface roughness and the four predictors?

37. Snowpacks contain a wide spectrum of pollutants that may represent environmental hazards. The article "Atmospheric PAH Deposition: Deposition Velocities and Washout Ratios" (*J. of Envir. Engr.*, 2002: 186–195) focused on the deposition of polyaromatic hydrocarbons. The authors proposed a multiple regression function for relating deposition over a specified time period (y, in mg/m^2) to two rather complicated predictors x_1 (mg-sec/m^3) and x_2 (mg/m^2), defined in terms of PAH air concentrations for various species, total time, and total amount of precipitation. Here is data on the species fluoranthene and corresponding output fitting $y = a + b_1x_1 + b_2x_2$ from the R software:

obs	x_1	x_2	flth
1	92017	.0026900	278.78
2	51830	.0030000	124.53
3	17236	.0000196	22.65
4	15776	.0000360	28.68
5	33462	.0004960	32.66
6	243500	.0038900	604.70
7	67793	.0011200	27.69
8	23471	.0006400	14.18
9	13948	.0004850	20.64
10	8824	.0003660	20.60
11	7699	.0002290	16.61
12	15791	.0014100	15.08
13	10239	.0004100	18.05
14	43835	.0000960	99.71
15	49793	.0000896	58.97
16	40656	.0026000	172.58
17	50774	.0009530	44.25

```
Coefficients:
                            Std.        t        Pr
              Estimate      Error     value    (>|t|)
(Intercept)    -33.46    1.490e+01   -2.246   0.0413
x_1           2.055e-03  2.945e-04    6.977   6.48e-06
x_2              29836   1.365e+04    2.185   0.0464

Residual standard error: 44.28 on 14 degrees of
freedom Multiple R-squared: 0.9234, Adjusted R-
squared: 0.9125 F-statistic: 84.39 on 2 and 14
DF, p-value: 1.546e-08

Analysis of Variance Table
Response: flth
              Df  Sum Sq  Mean Sq  F value   Pr(>F)
x_1            1  321625   321625  164.011  4.04e-09
x_2            1    9364     9364    4.775  0.04637
Residuals     14   27454     1961
```

a. Interpret the value of the coefficient of multiple determination.

b. Predict the value of deposition when $x_1 = 20,000$ and $x_2 = .001$.

c. Since $b_2 = 29,836$, is it legitimate to conclude that if x_2 increases by 1 unit while the values of the other predictors remain fixed, deposition would increase by 29,836 units? Explain your reasoning.

38. An investigation of a die-casting process resulted in the accompanying data on x_1 = furnace temperature, x_2 = die close time, and y = temperature difference on the die surface ("A Multiple-Objective Decision-Making Approach for Assessing Simultaneous Improvement in Die Life and Casting Quality in a Die Casting Process," *Quality Engr.*, 1994: 371–383).

x_1:	1250	1300	1350	1250	1300
x_2:	6	7	6	7	6
y:	80	95	101	85	92

x_1:	1250	1300	1350	1350
x_2:	8	8	7	8
y:	87	96	106	108

Use a statistical computer package to fit $y = a + b_1 x_1 + b_2 x_2$ using the least squares method. Be sure to specify all function coefficients. Also include the coefficient of multiple determination and interpret its value.

39. Use of sucrose as a carbon source for the production of chemicals is uneconomical. Beet molasses is a readily available and lower-priced substitute. The article "Optimization of the Production of β-Carotene from Molasses by Blakeslea trispora" (*J. of Chem. Tech. and Biotech.* 2002: 933–943) carried out a multiple regression analysis to relate the dependent variable y = amount of β-carotene (g/dm³) to the three predictors amount of lineolic acid, amount of kerosene, and amount of antioxidant (all g/dm³).

Obs	Linoleic	Kerosene	Antiox	Betacaro
1	30.00	30.00	10.00	0.7000
2	30.00	30.00	10.00	0.6300
3	30.00	30.00	18.41	0.0130
4	40.00	40.00	5.00	0.0490
5	30.00	30.00	10.00	0.7000
6	13.18	30.00	10.00	0.1000
7	20.00	40.00	5.00	0.0400
8	20.00	40.00	15.00	0.0065
9	40.00	20.00	5.00	0.2020
10	30.00	30.00	10.00	0.6300
11	30.00	30.00	1.59	0.0400
12	40.00	20.00	15.00	0.1320
13	40.00	40.00	15.00	0.1500
14	30.00	30.00	10.00	0.7000
15	30.00	46.82	10.00	0.3460
16	30.00	30.00	10.00	0.6300
17	30.00	13.18	10.00	0.3970
18	20.00	20.00	5.00	0.2690
19	20.00	20.00	15.00	0.0054
20	46.82	30.00	10.00	0.0640

A request to the SAS package to fit $a + b_1 x_1 + b_2 x_2 + b_3 x_3$ yielded the following output:

Dependent Variable: beta

Source	DF	Sum of Squares	Mean Square	F Value	Pr > F
Model	3	0.02352595	0.00784198	0.09	0.9648
Error	16	1.40326270	0.08770392		
C. Total	19	1.42678865			

R-Square	Coeff Var	Root MSE	beta Mean
0.016489	102.0515	0.296148	0.290195

Parameter	Estimate	Standard Error	t Value	Pr > |t|
Intercept	0.4010752535	0.38164661	1.05	0.3089
lino	0.0011095713	0.00801331	0.14	0.8916
kero	-.0032850626	0.00801331	-0.41	0.6873
anti	-.0045615514	0.01602662	-0.28	0.7796

A request to the SAS package to fit a function with predictors x_1, x_2, and x_3 as well as quadratic and interaction predictors yielded the following output:

Dependent Variable: beta

Source	DF	Sum of Squares	Mean Square	F Value	Pr > F
Model	9	1.40762342	0.15640260	81.61	<.0001
Error	10	0.01916523	0.00191652		
C. Total	19	1.42678865			

R-Square	Coeff Var	Root MSE	beta Mean
0.986568	15.08576	0.043778	0.290195

Parameter	Estimate	Standard Error	t Value	Pr > |t|
Intercept	-2.368673650	0.25095313	-9.44	<.0001
lino	0.115946557	0.00896686	12.93	<.0001
kero	0.048329827	0.00896686	5.39	0.0003
anti	0.125140001	0.01622284	7.71	<.0001
lino*kero	0.000116125	0.00015478	0.75	0.4704
lino*anti	0.000820250	0.00030956	2.65	0.0243
kero*anti	0.001002750	0.00030956	3.24	0.0089
lino*lino	-0.002108721	0.00011530	-18.29	<.0001
anti*anti	-0.009219578	0.00046120	-19.99	<.0001
kero*kero	-0.001085436	0.00011530	-9.41	<.0001

a. What is the coefficient of multiple determination for each fitted function?

b. For the fit using $a + b_1 x_1 + b_2 x_2 + b_3 x_3$, what is the predicted value of β-carotene when lineolic acid = 40, kerosene = 20, and antioxidant = 5? What is the corresponding residual?

c. For the fit with predictors x_1, x_2, and x_3 as well as quadratic and interaction predictors, what is the predicted value of β-carotene when lineolic acid = 40, kerosene = 20, and antioxidant = 5? What is the corresponding residual?

d. Note the difference in magnitude of the residuals you just computed for the two regressions. Explain how it is reasonable for one of these to have a smaller residual magnitude given the difference in coefficients of multiple determination.

40. The collapse of reinforced concrete buildings during earthquakes can result in significant loss of property and life. Often such collapses are caused by concrete column axial failure. The authors of "Rotation-Based Shear Failure Model for Lightly Confined RC Columns" (*J. Struct. Engr.*, 2012: 1267–1278) introduced a model for the deformation at onset of shear failure for a class of reinforced concrete columns. As part of the study, the authors investigated how $y =$ maximum sustained shear (V_{max}, in kN) is influenced by $x_1 =$ transverse-reinforcement yield stress (MPa) and $x_2 =$ concrete cylinder compressive strength (MPa).

y:	314.9	359.0	300.7	271.3	266.9
x_1:	469	469	469	400	400
x_2:	21.10	21.10	20.90	25.60	25.60

y:	240.2	231.3	315.8	338.1	355.9
x_1:	400	400	400	400	400
x_2:	33.10	33.10	25.70	27.60	27.60

y:	378.1	101.9	110.8	103.2	101.9
x_1:	400	46	46	365	365
x_2:	25.70	4.65	4.34	23.00	20.20

y:	120.5	111.6	219.3	213.1
x_1:	365	365	392	392
x_2:	23.00	20.20	30.70	30.70

Use a statistical computer package to fit (a) $a + b_1x_1 + b_2x_2$, (b) $a + b_1x_1 + b_2x_2 + b_3x_1x_2$, and (c) $a + b_1x_1 + b_2x_2 + b_3x_1x_2 + b_4x_1^2 + b_5x_2^2$. Be sure to specify all function coefficients. For each function, also include the coefficient of multiple determination and interpret its value.

41. A new surface finishing method has been developed for nanofinishing flat and three-dimensional workpiece surfaces. The authors of "Parametric Analysis of an Improved Ball End Magnetorheological Finishing Process" (*J. Engr. Manuf.*, 2012: 1550–1563) investigated how $y =$ percent change in surface roughness was influenced by $x_1 =$ rotational speed of tool core (N, in r/min), $x_2 =$ magnetizing current (I, in A), $x_3 =$ working gap (D, in mm).

y:	47.68	39.80	80.69	34.12	45.10
x_1:	400	500	500	600	500
x_2:	5.0	2.3	4.0	5.0	4.0
x_3:	2.00	1.50	0.66	2.00	1.50

y:	46.51	69.63	63.62	37.18	36.75
x_1:	500	500	600	668	400
x_2:	4.0	5.7	5.0	4.0	3.0
x_3:	1.50	1.50	1.00	1.50	2.00

y:	49.94	45.86	70.64	54.75	24.97
x_1:	500	500	400	600	600
x_2:	4.0	4.0	5.0	3.0	3.0
x_3:	1.50	1.50	1.00	1.00	2.00

y:	49.38	59.85	55.18	32.05	44.94
x_1:	500	400	332	500	500
x_2:	4.0	3.0	4.0	4.0	4.0
x_3:	1.50	1.00	1.50	2.34	1.50

Use a statistical computer package to fit (a) $a + b_1x_1 + b_2x_2 + b_3x_3$, (b) $a + b_1x_1 + b_2x_2 + b_3x_3 + b_4x_1x_2 + b_5x_1x_3 + b_6x_2x_3$, and (c) $a + b_1x_1 + b_2x_2 + b_3x_3 + b_4x_1x_2 + b_5x_1x_3 + b_6x_2x_3 + b_7x_1^2 + b_8x_2^2 + b_9x_3^2$. Be sure to specify all function coefficients. For each fit, also include the coefficient of multiple determination and interpret its value.

3.6   JOINT DISTRIBUTIONS

In Chapter 1, we presented several different ways to display and summarize sample data consisting of observations on a single quantitative variable x. These ideas were then extended to a population or process distribution consisting of a density function (when x is continuous) or mass function (for discrete x) and the corresponding graph. In this chapter, we have discussed bivariate and multivariate sample data. Now we consider distributions for two or more variables in a population or ongoing process.

Distributions for Two Variables

Let's initially focus on the case of two numerical variables x and y. For example, x might be the time a customer spends in a grocery checkout line (a continuous variable) and *y* the number of items purchased by the customer (a discrete variable). In practice, *x* and *y* are usually of the same type, either both discrete or both continuous. Suppose first that *x* and *y* are both discrete. Then their "joint" distribution is specified by a **joint mass function** $f(x, y)$ satisfying

$$f(x, y) \geq 0 \qquad \sum_{\text{all } (x,y)} f(x, y) = 1$$

Often there is no nice formula for $f(x, y)$. When there are only a few possible values of *x* and *y*, the mass function is most conveniently displayed in a rectangular table.

Example 3.19   A certain market has both an express checkout register and a superexpress register. Let *x* denote the number of customers queueing at the express register at a particular weekday time, and let *y* denote the number of customers in line at the superexpress register at that same time. Suppose that the joint mass function is as given in the accompanying table:

		y			
		0	1	2	3
	0	.08	.07	.04	.00
	1	.06	.15	.05	.04
x	2	.05	.04	.10	.06
	3	.00	.03	.04	.07
	4	.00	.01	.05	.06

According to the table, $f(x, y) > 0$ for only 17 (x, y) pairs. Just as in the case of a single variable, individual proportions from the mass function can be added to yield other proportions of interest. For example, (x, y) pairs for which the number of customers at the express register is equal to the number of customers at the other register are $(0, 0)$, $(1, 1)$, $(2, 2)$, and $(3, 3)$, so

$$\binom{\text{long-run proportion of}}{\text{times for which } x = y} = f(0, 0) + f(1, 1) + f(2, 2) + f(3, 3)$$

$$= .08 + .15 + .10 + .07$$
$$= .40$$

The total number of customers at these two registers will be 2 if $(x, y) = (2, 0)$, $(1, 1)$, or $(0, 2)$, so

$$\binom{\text{long-run proportion of}}{\text{times for which } x + y = 2} = f(2, 0) + f(1, 1) + f(0, 2)$$

$$= .05 + .15 + .04$$
$$= .24$$

Suppose we are presented with the joint distribution but are interested only in the distribution of x alone: the *marginal distribution of x*. In Example 3.19, we might wish to know $f_1(0)$, $f_1(1)$, $f_1(2)$, $f_1(3)$, and $f_1(4)$, the long-run proportions for various values of the first variable, x. Consider $x = 1$, which occurs when $(x, y) = (1, 0)$, $(1, 1)$, $(1, 2)$, or $(1, 3)$. Thus

$$f_1(1) = \text{long-run proportion of the time that } x = 1$$
$$= f(1, 0) + f(1, 1) + f(1, 2) + f(1, 3)$$
$$= .06 + .15 + .05 + .04 = .30$$

This is nothing more than the sum of proportions in the $x = 1$ row of the joint mass table. Adding proportions in the other rows gives the entire marginal distribution of x, whereas adding proportions in the various columns gives the marginal distribution of y, denoted by $f_2(y)$:

x:	0	1	2	3	4	y:	0	1	2	3
$f_1(x)$:	.19	.30	.25	.14	.12	$f_2(y)$:	.19	.30	.28	.23

Now let's consider the case of two continuous random variables. The distribution for a single continuous variable x is specified by a density function $f(x)$ that satisfies $f(x) \geq 0$ and $\int_{-\infty}^{\infty} f(x)\, dx = 1$. The graph of $f(x)$ is the density curve, and various proportions correspond to areas under this curve that are obtained by integrating the density function. Extending these ideas to two variables requires that we use multivariate calculus, in particular multiple integration. The joint distribution of x and y is specified by a **joint density function $f(x, y)$** that satisfies

$$f(x, y) \geq 0 \qquad \int_{-\infty}^{\infty}\int_{-\infty}^{\infty} f(x, y)\, dx\, dy = 1$$

The graph of $f(x, y)$ is a surface in three-dimensional space. The second condition indicates that the total volume under this density surface is 1.0. Suppose that x and y are reaction times by an individual to two different stimuli (e.g., two different configurations of brake lights) and that we wish to calculate the proportion of individuals for which both $.5 \leq x \leq 1$ and $.5 \leq y \leq 1$. Letting $A = \{(x, y): .5 \leq x \leq 1, .5 \leq y \leq 1\}$, a rectangular region in the x–y plane, the desired proportion is the double integral $\iint_A f(x, y)\, dx\, dy$; this is just the volume underneath the density surface that lies above the region A, as illustrated in Figure 3.24 (p. 154). Even though the region of integration is a rectangle, the integral may be quite difficult to compute if the integrand (density function) is complicated, perhaps requiring a numerical integration of some sort. When A is not a rectangle, the integration will typically be even more difficult to carry out. We are not going to do multiple integration in this text; we simply want you to be acquainted with the basic ideas of continuous distributions. To see examples of calculations, please consult one of the chapter references.

In the same way that in the discrete case the marginal distribution for either one of the variables is obtained by summing the joint mass function over values of the other variable [the row or column sums from a rectangular table of $f(x, y)$], the marginal density function $f_1(x)$ is obtained by integrating the joint density with respect to y, and $f_2(y)$ results from integrating $f(x, y)$ with respect to x.

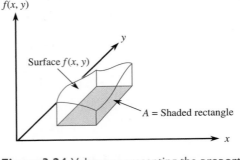

Figure 3.24 Volume representing the proportion of (x, y) in the region A

Correlation and the Bivariate Normal Distribution

Let $h(x, y)$ represent some particular function of x and y, such as $h(x, y) = x + y$ or $h(x, y) = xy$. Paralleling the definition of a mean value in the case of a single variable, the **mean** (or **expected**) **value of $h(x, y)$** is a weighted average of $h(x, y)$, with the weights given by the joint mass or density function:

$$
\mu_{h(x,y)} = \begin{cases} \sum \sum h(x, y) f(x, y) & x, y \text{ discrete} \\[2ex] \iint h(x, y) f(x, y) \, dx \, dy & x, y \text{ continuous} \end{cases}
$$

Let μ_x and μ_y denote the mean values of x and y, respectively. Then the function

$$
h(x, y) = (x - \mu_x)(y - \mu_y)
$$

is a product of x and y deviations from their mean values [like $(x - \bar{x})(y - \bar{y})$ in our discussion of sample correlation]. The mean value of this product of deviations is called the **covariance** between x and y, and the **population correlation coefficient** is

$$
\rho = \frac{\text{covariance}(x, y)}{\sigma_x \sigma_y}
$$

where σ_x and σ_y are the x and y standard deviations, respectively. This definition of ρ is very similar to the definition of the sample correlation coefficient r given in Section 3.2. You need not worry about calculating ρ, but we do want you to know that it exists and shares many properties with r. In particular,

1. ρ does not depend on the x or y units of measurement.
2. $-1 \le \rho \le 1$
3. The closer ρ is to $+1$ or -1, the stronger the *linear* relationship between the two variables.

One of the most frequently occurring bivariate distributions in statistics generalizes the univariate normal distribution introduced in Section 1.4. The **bivariate normal joint density function** is given by

$$f(x, y) = \frac{1}{2\pi\sigma_x\sigma_y\sqrt{1 - \rho^2}}\, e^{-\frac{1}{2(1 - \rho^2)}\left[\left(\frac{x - \mu_x}{\sigma_x}\right)^2 - 2\rho\left(\frac{x - \mu_x}{\sigma_x}\right)\left(\frac{y - \mu_y}{\sigma_y}\right) + \left(\frac{y - \mu_y}{\sigma_y}\right)^2\right]}$$

$$-\infty < x < \infty$$
$$-\infty < y < \infty$$

One interesting example of the use of this joint distribution appears in the article "Analysis of Size-Grouped Potato Yield Data Using a Bivariate Normal Distribution of Tuber Size and Weight" (*J. of Agric. Science*, 1993: 193–198). Figure 3.25 is a three-dimensional graph of this function for specified parameter values. The function cannot be easily integrated, so tables or numerical methods must be employed to calculate various proportions of interest. In Chapter 11, we consider an inferential procedure for drawing conclusions about ρ based on assuming that the sample was selected from a bivariate normal distribution.

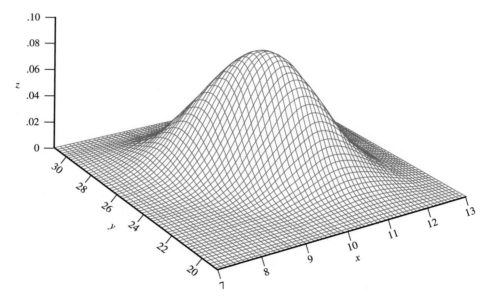

Figure 3.25 Graph of the bivariate normal density function when $\mu_x = 10$, $\sigma_x = 1$, $\mu_y = 25$, $\sigma_y = 2$, and $\rho = .5$

The Case of Independence

In general, it may be difficult to find a reasonable joint distribution for two variables x and y. The one situation in which this task is relatively straightforward is when x and y are *independent*. Intuitively, independence means that knowing the value of x does not change the distribution of y (equivalently, the distribution of y is the same for each different x value) and knowing the value of y has no bearing on the distribution of x. Look back at the joint distribution table for x and y in Example 3.19. Notice that if $x = 0$, then $y = 0$ is a possibility but not $y = 3$. However, if $x = 4$, then $y = 0$ is excluded, whereas $y = 3$ is possible. So the distribution of one variable *does* depend on the value of the other, and the variables are therefore not independent.

Let $f_1(x)$ and $f_2(y)$ denote the marginal distributions of x and y, respectively. Frequently, an investigator has enough knowledge of the situation under study to assume independence. When this is the case, the joint mass or density function must satisfy

$$(*) \qquad f(x, y) = f_1(x) \cdot f_2(y)$$

For the variables of Example 3.19 to be independent, *every* entry in the joint table would have to be the product of the row and column totals. Very importantly, once independence is assumed, one has only to select appropriate distributions for x and y separately and then use $(*)$ to create the joint distribution.

Example 3.20   A business is planning to purchase two different new vehicles, a van and a sedan. Let x denote the number of major defects on the first vehicle, and y be the number of major defects on the second one. Because the vehicles come from different manufacturers and assembly lines, an assumption of independence is reasonable. Suppose x has a Poisson distribution with $\lambda = 2$ and y has a Poisson distribution with $\lambda = 1.5$ (the marginal distributions). Then

$$f(x, y) = \left[\frac{e^{-2} \cdot (2)^x}{x!} \right] \left[\frac{e^{-1.5} \cdot (1.5)^y}{y!} \right] \qquad x = 0, 1, 2, \ldots ; y = 0, 1, 2, \ldots$$

The long-run proportion of such purchases that would result in at most one major defect for the two vehicles combined $(x + y \leq 1)$ is then $f(0, 0) + f(0, 1) + f(1, 0) = .136$.

Independence was introduced in Chapter 1 in connection with the binomial distribution. The concept will be considered further when we discuss probability in Chapter 5.

Variables x and y that have a bivariate normal distribution will be independent if $\rho = 0$, since then the joint density can be written as a product of two univariate normal densities. If the joint distribution is not bivariate normal, however, then $\rho = 0$ does not imply independence. Zero correlation means only that there is no linear relationship, whereas independence means that there is no relationship of any sort.

More Than Two Variables

Suppose that k variables $x_1, x_2, \ldots, x_k$ are under consideration. We might, for example, have a system with $k = 4$ components and let x_i be the useful lifetime of component $i (i = 1, 2, 3, 4)$. Properties satisfied by a joint mass or density function $f(x_1, \ldots, x_k)$ are analogous to those in the bivariate case. It can be quite difficult to specify a reasonable joint distribution. The *multivariate normal distribution* is frequently used when the variables are continuous. However, its density function is rather complicated. If it can be assumed that the variables are independent, then the joint distribution is again the product of the marginal distributions.

Section 3.6 Exercises

42. A large insurance agency provides services to a number of customers who have purchased both a homeowner's policy and an automobile policy. For each type of policy, a deductible amount must be specified. Let x denote the homeowner's deductible amount and y denote the automobile deductible amount for a customer who has both types of policies. The joint mass function of x and y is as follows:

$f(x, y)$		y		
		0	250	500
x	200	.20	.10	.20
	500	.05	.15	.30

a. What proportion of customers have $500 deductible amounts for both types of policies?

b. What proportion of customers have both deductible amounts less than $500?

c. What is the marginal mass function of x? What is the marginal mass function of y?

43. The joint distribution of the number of cars (x) and the number of buses (y) per signal cycle at a particular left turn lane is displayed in the accompanying table:

$f(x, y)$		y		
		0	1	2
	0	.025	.015	.010
	1	.050	.030	.020
	2	.125	.075	.050
x	3	.150	.090	.060
	4	.100	.060	.040
	5	.050	.030	.020

a. In what proportion of cycles will there be exactly one car and one bus?

b. In what proportion of cycles will there be at most one vehicle of each type?

c. In what proportion of cycles will the number of cars be the same as the number of buses?

d. What is the mean value of the number of cars per signal cycle?

e. If a bus occupies three vehicle spaces and a car occupies just one, what is the mean value of the number of vehicle spaces occupied during a signal cycle? *Hint:* Let $h(x, y) = x + 3y$.

44. Let x denote the number of major defects for a particular piece of machinery and y be the number of cosmetic flaws on this same piece. Suppose that x and y are independent variables with $f_1(x) = .80, .15$, and $.05$ for $x = 0, 1$, and 2, respectively, and $f_2(y) = .50, .25, .15, .08$, and $.02$ for $y = 0, 1, \ldots, 4$, respectively.

a. What is the joint mass function of these two variables?

b. What proportion of these machines will have no major defects or cosmetic flaws? What proportion will have at least one defect or flaw?

c. For what proportion of these machines will the number of cosmetic flaws exceed the number of major defects?

45. Refer to Exercise 42. Compute the covariance between x and y and then the value of the population correlation coefficient. Do these two variables appear to be strongly related? Explain.

Supplementary Exercises

46. Orthotropic steel bridge decks with closed ribs have been widely used in suspension bridges, cable-stayed bridges, and urban elevated expressways due to their overall light weights, ease of construction, and high load-carrying capacities. In the article "Fatigue Evaluation of Rib-to-Deck Welded Joints of Orthotropic Steel Bridge Deck" (*J. Bridge Engr.*, 2011: 492–499), researchers examine the physical properties of 22 bridge specimens. Each specimen was attached to a fatigue testing apparatus. Fatigue life was determined as the number of cycles (in millions, p. 158) at the end of the fatigue test. For each specimen, the corresponding stress range (MPa) was also recorded.

Stress:	121	71	108	99	77
Cycles:	1.257	11.250	2.240	4.030	6.650

Stress:	70	79	56	89	75
Cycles:	6.970	6.430	19.140	3.950	9.000

Stress:	95	90	110	77	64
Cycles:	2.290	4.470	2.150	10.490	19.260

Stress:	90	99	91	91	82
Cycles:	4.120	1.800	2.190	3.150	5.800

Stress:	75	79
Cycles:	5.130	5.970

a. Would you fit a straight line to the data and use it as a basis for predicting y = stress range from x = number of cycles? Why or why not?

b. Find a transformation that produces an approximate linear relationship between the transformed values. Then fit a line to the transformed data and use it to obtain an equation that describes approximately the relationship between the untransformed variables.

47. An investigation of the relationship between the temperature (°F) at which a material is treated and the strength of the material involved an experiment in which four different strength observations were obtained at each of the temperatures 100, 110, 120, 130, and 140. A scatterplot of the data showed a substantial linear pattern. The least squares line fit to the data had a slope of .500 and a vertical intercept of −25.000.

a. Interpret the value of the slope.

b. The largest strength value when temperature was 120 was 40 and the smallest was 29. What value of strength would you have predicted for this temperature, and what are the values of the residuals for the two aforementioned observations? Why do these residuals have different signs?

c. The values of SSTo and SSResid were 1060.0 and 390.0, respectively. Calculate and interpret the coefficient of determination.

48. As the air temperature drops, river water becomes supercooled and ice crystals form. Such ice can significantly affect the hydraulics of a river. The article "Laboratory Study of Anchor Ice Growth" (*J. of Cold Regions Engr.*, 2001: 60–66) described an experiment in which ice thickness (mm) was studied as a function of elapsed time (hr) under specified conditions. The following data was read from a graph in

the article: $n = 33$; $x = .17, .33, .50, .67, \ldots, 5.50$; $y = .50, 1.25, 1.50, 2.75, 3.50, 4.75, 5.75, 5.60, 7.00, 8.00, 8.25, 9.50, 10.50, 11.00, 10.75, 12.50, 12.25, 13.25, 15.50, 15.00, 15.25, 16.25, 17.25, 18.00, 18.25, 18.15, 20.25, 19.50, 20.00, 20.50, 20.60, 20.50, 19.80$.

a. The r^2 value resulting from a least squares fit is .977. Interpret this value and comment on the appropriateness of assuming an approximate linear relationship.

b. The residuals, listed in the same order as the x values, are

−1.03	−0.92	−1.35	−0.78	−0.68	−0.11	0.21
−0.59	0.13	0.45	0.06	0.62	0.94	0.80
−0.14	0.93	0.04	0.36	1.92	0.78	0.35
0.67	1.02	1.09	0.66	−0.09	1.33	−0.10
−0.24	−0.43	−1.01	−1.75	−3.14		

Plot the residuals against elapsed time. What does the plot suggest?

49. An investigation was carried out to study the relationship between speed (ft/sec) and stride rate (number of steps taken/sec) among female marathon runners. Resulting summary quantities included $n = 11$, $\Sigma(\text{speed}) = 205.4$, $\Sigma(\text{speed})^2 = 3880.08$, $\Sigma(\text{rate}) = 35.16$, $\Sigma(\text{rate})^2 = 112.681$, and $\Sigma(\text{speed})(\text{rate}) = 660.130$.

a. Calculate the equation of the least squares line that you would use to predict stride rate from speed.

b. Calculate the equation of the least squares line that you would use to predict speed from stride rate.

c. Calculate and interpret the coefficient of determination for the regression of stride rate on speed of part (a) and for the regression of speed on stride rate of part (b). How are these two related?

50. Refer to Exercise 49. Consider predicting speed from stride rate, so that the response variable y is speed. Suppose that the values of speed in the sample are expressed in meters/second. How does this change in the unit of measurement for y affect the equation of the least squares line? More generally, if each y value in the sample is multiplied by the same number c, what happens to the slope and vertical intercept of the least squares line?

51. The relationship between x = strain (in./in.) and y = stress (ksi) for an experimental alloy tension member was investigated by making an observation on stress

for each of $n = 10$ values of strain. A scatterplot of the resulting data suggested a quadratic relationship between the two variables. Employing the principle of least squares gave $\hat{y} = 88.791 + 5697.0x - 328,161x^2$ as the equation of the best-fit quadratic.

a. One observation in the sample was made when strain was .005, and the resulting value of stress was 111. What value of stress would you have predicted in this situation, and what is the value of the corresponding residual?

b. The observed values of stress were 91, 97, 108, 111, 114, 110, 112, 102, 98, and 91. Using the best-fit quadratic gave corresponding predicted values of 94.16, 98.87, 102.93, 109.07, 111.16, 113.36, 113.48, 104.22, 95.93, and 90.80, respectively. Calculate a quantitative assessment of the extent to which variation in observed stress values can be attributed to the approximate quadratic relationship between stress and strain.

c. What happens if the best-fit equation is used to predict stress when strain is .03? *Note:* The largest strain value in the sample was .017.

52. An experiment carried out to investigate the relationship between $y = $ wire bond pull strength in a semiconductor product and the two predictors $x_1 = $ wire length and $x_2 = $ die height resulted in data for which the best-fit equation according to the principle of least squares was $\hat{y} = 2.300 + 2.750x_1 + .0125x_2$.

a. Interpret the coefficients of x_1 and x_2 in the given equation.

b. The observed value of pull strength was 24.35 when wire length was 9 and die height was 100. What value of pull strength would you have predicted under these circumstances, and what is the value of the corresponding residual?

c. The values of SSTo and SSResid were 6110.2 and 123.4, respectively. Can a substantial percentage of the observed variation in strength be attributed to the postulated approximate relationship between strength and the two predictors?

53. The accompanying data resulted from an investigation of the relationship between temperature (x, in °F) and viscosity (y, in poise) for specimens of bitumen removed from tar sand deposits:

x:	750	800	700	850	590	620	650	680	710	550
y:	50	16	102	10	945	818	403	151	114	1358

a. Would a straight line fit to this data give accurate predictions of viscosity?

b. Let $x' = 1/x$ and $y' = \ln(y)$. Fit a straight line to the (x', y') data, use it as a basis for predicting viscosity when temperature is 720, and calculate a quantitative assessment of the extent to which the approximate linear relationship between x' and y' explains observed variation.

54. Ground motions resulting from an earthquake can be heavily influenced by the dynamic properties of the soils overlying bedrock. The authors of "Influence of Pore Fluid Viscosity on the Dynamic Properties of an Artificial Clay" (*J. Geotech. Geoenviron. Engr.*, 2011: 1190–1201) investigated properties of an artificial soil called *modified glyben* to study seismic soil-structure interaction. Researchers investigated the relationship between $x = $ fluid content by mass (%) and vane shear strength (kPa) for three types of modified glyben at different pore fluid viscosities (w/gw): $y' = $ vane shear strength (0% w/gw), $y'' = $ vane shear strength (25% w/gw), $y''' = $ vane shear strength (50% w/gw). The data below corresponds to a graph from the article:

x	35.0	37.5	40.0	42.5	45.0	47.5
y'	75.0	63.0	57.0	45.0	28.5	38.0
y''	52.0	41.5	38.0	35.0	20.0	16.0
y'''	33.5	24.5	22.0	19.0	13.0	10.0

a. Create the scatterplots for the pairs (x, y'), (x, y''), and (x, y'''). Does each scatterplot suggest that a linear relationship holds for the respective variables?

b. Determine the least squares regression line for each pair. For each, determine the corresponding coefficient of determination.

c. Given the slope coefficients from the regression, summarize the relationship between vane shear strength and fluid content by mass as pore fluid viscosity changes from 0%, to 25%, and to 50%.

55. Failures in aircraft gas turbine engines due to high cycle fatigue is a pervasive problem. The article "Effect of Crystal Orientation on Fatigue Failure of Single Crystal Nickel Base Turbine Blade Superalloys" (*J. of Engr. for Gas Turbines and Power*, 2002: 161–176) gave the accompanying data and fit a nonlinear regression model in order to predict strain amplitude from cycles to failure. Fit an appropriate curve, investigate the quality of the fit, and predict amplitude when cycles to failure = 5000.

Obs	Cycfail	Strampl	Obs	Cycfail	Strampl
1	1326	.01495	11	7356	.00576
2	1593	.01470	12	7904	.00580
3	4414	.01100	13	79	.01212
4	5673	.01190	14	4175	.00782
5	29,516	.00873	15	34,676	.00596
6	26	.01819	16	114,789	.00600
7	843	.00810	17	2672	.00880
8	1016	.00801	18	7532	.00883
9	3410	.00600	19	30,220	.00676
10	7101	.00575			

56. Efficient design of certain types of municipal waste incinerators requires that information about energy content of the waste be available. The authors of the article "Modeling the Energy Content of Municipal Solid Waste Using Multiple Regression Analysis" (*J. of the Air and Waste Mgmnt. Assoc.*, 1996: 650–656) kindly provided us with the accompanying data on y = energy content (kcal/kg); the three physical composition variables x_1 = % plastics by weight, x_2 = % paper by weight, and x_3 = % garbage by weight;, and the proximate analysis variable x_4 = % moisture by weight for waste specimens obtained from a certain region.

Obs	Plastics	Paper	Garbage	Water	Energy Content
1	18.69	15.65	45.01	58.21	947
2	19.43	23.51	39.69	46.31	1407
3	19.24	24.23	43.16	46.63	1452
4	22.64	22.20	35.76	45.85	1553
5	16.54	23.56	41.20	55.14	989
6	21.44	23.65	35.56	54.24	1162
7	19.53	24.45	40.18	47.20	1466
8	23.97	19.39	44.11	43.82	1656
9	21.45	23.84	35.41	51.01	1254
10	20.34	26.50	34.21	49.06	1336
11	17.03	23.46	32.45	53.23	1097
12	21.03	26.99	38.19	51.78	1266
13	20.49	19.87	41.35	46.69	1401
14	20.45	23.03	43.59	53.57	1223
15	18.81	22.62	42.20	52.98	1216
16	18.28	21.87	41.50	47.44	1334
17	21.41	20.47	41.20	54.68	1155

Obs	Plastics	Paper	Garbage	Water	Energy Content
18	25.11	22.59	37.02	48.74	1453
19	21.04	26.27	38.66	53.22	1278
20	17.99	28.22	44.18	53.37	1153
21	18.73	29.39	34.77	51.06	1225
22	18.49	26.58	37.55	50.66	1237
23	22.08	24.88	37.07	50.72	1327
24	14.28	26.27	35.80	48.24	1229
25	17.74	23.61	37.36	49.92	1205
26	20.54	26.58	35.40	53.58	1221
27	18.25	13.77	51.32	51.38	1138
28	19.09	25.62	39.54	50.13	1295
29	21.25	20.63	40.72	48.67	1391
30	21.62	22.71	36.22	48.19	1372

Using Minitab to fit a regression function with the four aforementioned variables as predictors of energy content resulted in the following output:

```
The regression equation is
enercont = 2245 + 28.9 plastics
+ 7.64 paper + 4.30 garbage - 37.4 water

Predictor    Coef     StDev     T        P
Constant     2244.9   177.9     12.62    0.000
plastics     28.925   2.824     10.24    0.000
paper        7.644    2.314     3.30     0.003
garbage      4.297    1.916     2.24     0.034
water        -37.354  1.834     20.36    0.000

s = 31.48 R-Sq = 96.4% R-Sq(adj) = 95.8%
Analysis of Variance

Source       DF    SS       MS      F        P
Regression   4     664931   166233  167.71   0.000
Error        25    24779    991
Total        29    689710
```

a. Predict the value of energy content when plastics is 17.03, paper is 23.46, garbage is 32.45, and water is 53.23. Also determine the corresponding residual.

b. What proportion of observed variation in energy content can be explained by the approximate relationship between energy content and the four predictors?

Bibliography

Kutner, M., C. Nachstein, and J. Neter, **Applied Linear Regression Models** (4th ed.), McGraw-Hill/Irwin, Burr Ridge, IL, 2004. *A comprehensive up-to-date exposition of regression and correlation analysis without overindulging in theory, though matrix algebra is rather frequently used. (This material is also* included in **Applied Linear Statistical Models**, *a longer book by the same authors.)*

Montgomery, D. C., E. A. Peck, and G. G. Vining, **Introduction to Linear Regression Analysis** (5th ed.), Wiley, New York, 2012. *A very nice treatment of regression written for engineers and physical scientists.*

4

Obtaining Data

4.1 OPERATIONAL DEFINITIONS

4.2 DATA FROM SAMPLING

4.3 DATA FROM EXPERIMENTS

4.4 MEASUREMENT SYSTEMS

INTRODUCTION

Engineering has been defined as the art of applying science and technology for the optimal conversion of the resources of nature into the uses of humankind.[1] The sciences, in turn, are grounded in mathematics, so it is natural that measurements of all kinds should play a large role in engineering and scientific practice. In this chapter, we examine some of the ways in which data is collected as well as some approaches to ensuring data quality.

Scientists and statisticians have long realized that some sets of data are definitely more useful than others, and that at the heart of data quality lies the realization that external conditions can often exert a large influence on measured values. Temperature, for example, is well known to affect the physical dimensions (length, area, etc.) of most materials, so the measured length of a thin strip of aluminum will necessarily vary depending on the ambient temperature. In an effort to control or eliminate the effects of such external or "noise" factors, engineers have developed a large number of professional **standards** whose purpose is to ensure the consistency and quality of scientific data. We will look at some specific examples of such standards in Section 4.1.

Since the early 1920s, statisticians have also addressed the problems of data quality by introducing tightly controlled data collection schemes. These schemes,

---

[1] *Encyclopedia Britannica*, 1998.

called **experimental designs** and **sampling plans,** provide methods not only for controlling or eliminating the effects of external factors but also for assessing the magnitude of their combined effect on measured data. Sampling plans also address the problem of how far we can generalize the conclusions that we draw from data. One important feature of experimental designs is the ability to study the effects of several factors simultaneously on the values of another factor, called a **response variable.** This feature is especially well suited to research and development activities. The main components of such designs are introduced in Sections 4.2 and 4.3.

The process of obtaining measurements is also vital to the eventual conclusions drawn from data. Numerous questions can be asked about measurement procedures: Can we trust a particular measuring instrument's readings? Are the readings accurate and precise? Do repeated measurements of the same object give similar results, or do the results exhibit large variation? If different people or special laboratories are involved at various stages of the measuring process, does this have an adverse effect on the quality of the data? These questions are the subject of **metrology,** the study of measurement, and are examined in Section 4.4.

4.1   OPERATIONAL DEFINITIONS

When working with data, two facts quickly emerge: (1) There are usually several ways to measure the same thing and (2) external factors can exert a large influence on our final measurements. We learn early that failing to be specific about what we want to measure can lead to endless problems and questions about how, or even whether, to use a set of data. To illustrate, suppose that you ask two people to measure the density of water. Person A might use the following method: An empty graduated cylinder is weighed and then filled with water and reweighed; the two weights are subtracted, giving the weight of the water in the cylinder; then, by reading off the water volume from the cylinder's measuring scale, the ratio of the volume to the weight is used as a measure of the water's density. Person B, however, decides to simply use a hydrometer, an instrument that directly measures the density of water. Do the two measurement methods agree? Probably not. The measurements from person A, for instance, depend on the precision and accuracy of the weighing scale used and on the person's ability to read the volume correctly from the cylinder.[2] The readings from person B depend on the precision of the hydrometer and whether it is correctly calibrated. There are additional reasons why the two measurements may not be equal. For instance, what kind of water was used? After all, pure water, freshwater, and seawater are known to have different densities. Furthermore, temperature is an important factor affecting water density (maximum water density occurs at 39.09°F). Did person A and person B measure the same sort of water at the same temperatures?

---

[2] Surface tension causes the top of the water to form a bowl-like surface, called a *meniscus*. Using the top of the meniscus leads to a different volume estimate than using the bottom of the meniscus.

As this example shows, unless you are very specific about what to measure (e.g., seawater at 50°F and 1 atmosphere of pressure) and how to measure it, data can be quite unreliable. Realizing this, the quality pioneer W. Edwards Deming recommended that, prior to collecting any set of data, one should first create an **operational definition** that spells out exactly what is to be measured and exactly how the measurements should be made. The reward for doing this is consistent, reliable data. Any two people should be able to follow the operational definition and obtain essentially the same measurements. Cognizant of the importance of operational definitions, most scientists include a Materials and Methods or Experimental Procedure section that outlines the exact procedures employed to collect the data used in a study.

Example 4.1   Automobile gasoline is a carefully balanced blend of from 8 to 15 different hydrocarbons. The resulting blends must meet up to 15 quality and environmental requirements, including standards regarding vapor pressure, boiling point, stability, color, and octane rating. The octane scale measures the degree to which a gasoline blend performs like pure isooctane (which gives the least amount of premature firing or "knock") or pure normal heptane (which produces extreme knocking). If the blend performs like a mixture of 90% isooctane and 10% heptane, it is assigned an octane rating of 90.

Because octane measurements are heavily influenced by engine speed and temperature, an operational definition must be used when assigning octane ratings. First, using a standard knock engine, the "research octane" level is measured under mild conditions (600 rpm and 120°F). Second, "motor octane" is measured under harsher conditions (900 rpm and 300°F). Finally, the "road octane" rating is calculated as the average of the research and motor octane levels. Road octane, calculated by the $(R + M)/2$ method, is the one commonly reported on gasoline station pumps.

Example 4.2   Operational definitions are often created on the job. For example, when inspecting injection-molded automobile dashboards, several types of defects can be observed, such as pinholes, creases, burn marks, and voids (hollow areas underneath the outer skin of the dashboard). To generate meaningful data about such defects, an operational definition must be created so that any two inspectors will report the same types and severity of defects. For example, we might decide to classify creases longer than 1 inch as severe, whereas creases less than one-quarter inch might be called minor. Pinholes that occur under the dashboard (not visible to passengers) could be classified differently from those that are in the passengers' field of vision. Similarly, voids with large diameters might be treated as major defects, whereas smaller voids are minor defects. Once these definitions have been established, the resulting data can be reliably used in quality control charts (Chapter 6) or other statistical methods.

Professional Standards

It often takes highly specialized knowledge to create operational definitions. Consequently, entire professional societies have arisen to create such definitions, which are then called **professional standards** or simply **standards.** One of the largest such groups is the **American Society for Testing and Materials (ASTM)**. ASTM publishes standard test methods, specifications, practices, and guides for engineers working with materials, products, systems, and services. Over 12,000 ASTM standards have now been published, and these standards are commonly adopted by government agencies for use in codes, regulations, and laws. Building codes, for example, commonly cite ASTM standards for conducting tests on structures. In the following example, notice how each step of a measurement process is carefully defined.

Example 4.3

Concrete used in construction must meet tight consistency standards. Consistency refers to the fluidity of the concrete when poured. ASTM C 143 (Standard Method for Slump of Portland Cement Concrete) is often cited in state construction codes as the required method of testing consistency.

ASTM C 143 requires that a sample of concrete be poured into a cone shaped like a megaphone (8-in. diameter at one end and 4-in. diameter at the other end). The large base of the cone is on the ground during the pour. The cone is filled one-third full and then tamped down 25 times. This procedure is repeated twice, leaving the mold full. The cement sample must come from the middle portion of the batch being poured. Next, the cone is lifted off the cement and quickly inverted and placed beside the conical pile of cement. Without the support of the cone, the height of the cement then diminishes or slumps. The distance between the top of the cone and the top of the cement is called the *slump*, and, depending on the building code used, the slump must fall within specified limits.

Other organizations, including the federal government, make extensive use of published standards. The **Code of Federal Regulations (CFR)**, for instance, is an important source of engineering standards and requirements in all federally regulated industries.

Example 4.4

The Department of Transportation (DOT) oversees the testing and rating of automobile tires. Tires are rated for treadwear, traction, and temperature resistance. These ratings are marked on the side of each tire. A treadwear rating of DOT 150, for example, means that a tire wears about one and a half times as long as a tire rated 100 on a standard government test course. Estimating the treadwear of a given brand of tire is done via regression analysis.

Because of the numerous factors that can affect treadwear (size of car, driving style, road conditions, and speed), the operational definition specified by DOT is extensive. In brief, Regulation 49CFR 575.104 (Uniform Tire Quality Grading Standards) requires that a convoy of two or four rear-wheel-drive passenger cars be driven over a 400-mile government test course in the vicinity of San Angelo, Texas. One vehicle is outfitted with special government-manufactured course-monitoring tires; the other vehicles have only test tires. Inflation pressures are specified, and each vehicle is weight-loaded to put a required test load on the tires. Wheel alignments are checked, tires are broken in for two laps (800 miles), air pressure is rechecked, and wheels are realigned. Initial tread depth, to the nearest .001 in., is measured. The convoy is then driven for 6400 miles, rotating tires every 400 miles in a specified pattern. A car's position in the convoy is also rotated. In addition, tires are also shifted from one vehicle to another every 1600 miles. Tread depth is measured every 800 miles. Finally, a regression line is fit to the nine treadwear points (one initial reading and eight readings at 800-mile intervals). The regression line is used to calculate a projected mileage for the test tires and the monitoring tires. Comparisons between the projected test tire wear and monitoring tire wear are used to assign the DOT wear rating.

Another organization that has played a major role in setting standards for various industries is the **International Organization for Standardization** (ISO). Founded in 1947, the ISO has published more than 19,500 international standards covering diverse areas such as food safety, computers, agriculture, and health care.

Example 4.5   We often assume that children's toys, once made available on the shelves of a store, are perfectly safe to use by children. Unfortunately, this is not always the case as evidenced by toy product recalls because of some hazard concern. For example, the U.S. Consumer Product Safety Commission maintains a regularly updated website that lists various hazardous toy recalls. In 2012, the ISO updated its series of toy safety standards that detail requirements and test methods for toys intended for use by children under 14 years of age; it also sets age limits for various requirements. The series contains four parts: Part 1—Safety aspects related to mechanical and physical properties; Part 2—Flammability; Part 3—Migration of certain elements; and Part 4—Swings, slides, and similar activity toys for indoor and outdoor family domestic use. Two new parts are currently under development: Part 5—Determination of total concentration of certain elements in toys; and Part 6—Toys and children's products—Determination of phthalate plasticizers in polyvinyl chloride plastics. By adopting the requirements and recommendations of the ISO safety standards, toy manufacturers can help minimize product recalls and reduce the risk of a child being injured by an unsafe toy.

Benchmarks

Operational definitions are especially appropriate for establishing industry and professional standards. However, when we want to compare several *different* products or processes, another sort of standard is needed. For these applications, **benchmarks** are the appropriate tools. Benchmarks are well-defined objects or processes whose characteristics are already explicitly known. Knowing the exact value of some characteristic in advance allows one to evaluate several products or processes by comparing how they perform against the benchmark. For example, the National Institute of Standards and Technology (NIST) keeps copies of standard physical units, such as the volt and the kilogram. These standards are the benchmarks against which the precision and accuracy of all measuring instruments are eventually compared.

Example 4.6   Benchmarks are routinely used for comparing software products. For instance, statistical software packages are evaluated for computational accuracy by using specially designed data sets whose statistical properties are precisely known. One repository of such benchmark data sets can be found at http://www.itl.nist.gov/div898/strd /index.html, a website maintained by the Information Technology Laboratory of the National Institute of Standards and Technology. This website was produced as part of the Statistical Reference Datasets Project. One of these data sets is the set of three integers 10,000,001 to 10,000,003 that is used to evaluate a software programs computation of the sample standard deviation, s. The sample standard deviation for these three values is $s = 1$, the same as for the sample 1, 2, 3.

Using this data set as a benchmark, it is possible to compare the different approaches to calculating s that are used in software packages. For instance, summing the squares of the three integers (a step used in some formulas for s) leads to inaccurate results. However, programs that use updating formulas (in which the value of s is updated as each data point is entered) are generally very accurate.

Section 4.1 Exercises

1. What is the primary difference between an operational definition and a benchmark?

2. Give an operational definition for measuring the fuel efficiency of a car. In your definition, take into account factors such as the driving speed, octane rating, distance driven, tire pressure, and driving terrain.

3. Give an operational definition for measuring the daytime temperature in a city. In your definition, take into account factors such as time of day and location.

4. To test the accuracy of a new numerical algorithm, a programmer uses the algorithm to produce the first 200 digits of the number π. The programmer checks the accuracy of 200 digits by comparing them to those in a published reference, whose accuracy has been previously verified. In this application, would

the published reference more properly be considered an operational definition or a benchmark?

5. Print speed (often measured in pages per minute, ppm) is an important property to consider when buying a printer. However, printer manufacturers measure this property in different ways, making comparison of print speeds difficult. In 2009, the ISO developed an international standard for measuring print speed. The standard, known as "ISO ppm," allows a consumer to make "apples-to-apples" comparisons of real-world print speeds under standard conditions. It is now common for the ISO ppm rating of a printer to be included in its product specifications listing. Here, would ISO ppm more properly be considered an operational definition or a benchmark?

4.2 DATA FROM SAMPLING

The data used in most applications arises from some form of **sampling.** By its very definition, a sample is simply a fraction or a part of some larger entity. Sometimes, the larger entity can be considered to be a **population,** such as the population of all electronic components made during a single workshift. At other times, the sampled entity may be a single object, such as a batch of cement, a chemical process, or a city's water supply.

The goal in all forms of sampling is to be able to draw conclusions about the larger entity based solely on our analyses of the information in a sample. For this reason, every effort is made to ensure that samples are truly representative of the thing we are sampling. Professional standards usually provide great detail on how samples are to be obtained. For example, ASTM C 172 (Standard Method of Sampling Freshly Mixed Concrete) requires that samples of fresh concrete be taken ". . . at two or more regularly spaced intervals during discharge of the middle of the batch . . ." and that the inspector should ". . . perform sampling by passing a receptacle completely through the discharge stream . . ." while taking care ". . . not to restrict the flow of the concrete . . . so as to cause segregation." Another method of assuring representative samples is based on the concept of random sampling, described later in this section.

The Advantages of Sampling

When done properly, sampling has several desirable features. Foremost among these are the savings in resources, especially time and money, that can be obtained by using samples. The economics of sampling are readily apparent because, for example, sampling and testing 20 items from a batch of 1000 items obviously involves less labor than testing the entire batch. In many cases, it is equally important to control the amount of time spent analyzing the sample data itself because production decisions often depend on tests performed on samples. In quality control applications, for instance, the decision of whether to adjust a process or to leave it alone is based on the analysis of periodic samples taken from an ongoing process. Timely test results are equally important in construction, where decisions on whether to accept a contractor's work and to proceed to the next phase of construction are based on the results of test samples.

Sometimes sampled material must be destroyed during testing. This is the case, for example, when evaluating the breaking strength of materials (e.g., metals, wood, fabrics, plastics), assessing the potency of drugs, or estimating the average lifetime of a group of electronic components. Such evaluation is called **destructive testing.** In such cases, sampling is not just an advantage, it is a necessity.

Even when testing is **nondestructive,** it still makes sense to sample. In addition to the economic benefits described previously, testing done on samples is often more reliable than testing done on entire populations. Several case studies have verified this phenomenon. The simple explanation is that testing and inspection errors begin to creep in whenever large numbers of items are tested because of inspector fatigue or differences between inspectors. With samples, more attention can be devoted to each item tested, and this almost always results in more reliable test data.

Example 4.7   The inspection and approval of metal welding in building construction can be based on nondestructive test (abbreviated NDT) methods, destructive test methods, or visual inspection. There are several NDT methods available, including magnetic particle testing, radiographic inspection, penetrant inspection, ultrasonic testing (UT), leak testing, and hardness testing. Each of these methods is based on a nondestructive examination of a sample of welded material.

Penetrant inspection, for example, involves the application of a dye (often red in color) to the welded surface. The dye penetrates any existing cracks and holes in the metal surface. After the excess dye is wiped away, only the dye in the cracks remains. To reveal these cracks, another liquid, called a *developer*, is applied to the surface. This causes the dye to come to the surface of the crack and creates a highly visible marking of each crack or hole in the weld. An experienced inspector can then make an evaluation of the quality of the weld from the number and location of these markings.

Random Sampling

Random sampling is a form of sampling used extensively in statistical methods. This technique presupposes that samples are to be obtained from some well-defined **population** of distinct items, and it provides a simple mechanism for randomly selecting items from the population to be included in a sample. The advantages of using random sampling are (1) it helps to reduce or eliminate bias in the manner in which the sampled items are chosen and (2) it enables us to make precise statements about the extent to which conclusions drawn from a sample can be applied to the entire population.

Random samples are obtained by making sure that every sample of the desired size has the same chance of being selected. This in turn implies that each item in the population has an equally likely chance of being chosen. One popular method for achieving this is to first create a list (called a sampling **frame**) of the items in a population. Next, successive positive integers are assigned to the items on the list, and then a **random number generator** is used to select a random sample of these positive integers. Random number generators can be in the form of tables, functions on handheld calculators, or commands in programming languages and statistical software. Whatever method is used, the selected integers will correspond to specific items in the sampling frame.

When sampling, we are immediately faced with a decision to sample with or without replacement. **Sampling with replacement** means that after each successive item (or integer) is selected for the random sample, the item is "replaced" back into the population and may even be selected again at a later stage. Thus, sampling with replacement allows for the possibility of having "repeats" occur in our random sample. In practice, sampling with replacement is rarely used. Instead, the more common notion of sampling is to allow only *distinct* items from the population in the sample. That is, no repeats are allowed. Sampling in this manner is called **sampling without replacement.** Although these two forms of sampling are indeed different, in most applications (i.e., when the sample size is small compared to the population size) there is little practical difference between them. Unless otherwise stated, however, we will always assume that random sampling is done without replacement.

Example 4.8	Suppose that we want to perform some electrical tests on a random sample of 5 integrated circuit chips from a package of 20 chips. Arranging the 20 chips in a horizontal line on a table is a rapid way of associating a unique integer from 1 to 20 with each chip (the leftmost chip would be labeled "1," the rightmost would be "20," and so forth). It is important to note that the particular ordering of the chips

is completely immaterial to the sampling process. All that is needed is a method for assigning integers to the chips, and horizontal positioning achieves that purpose.

Using a random number generator from a calculator or a statistical software package, we next generate a random sample of five integers from the numbers 1 through 20. When doing this, we have to decide whether to sample with replacement or without replacement. Suppose we choose to sample without replacement and that the randomly chosen integers turn out to be 4, 14, 3, 18, and 15. Then, our random sample of 5 chips would consist of the 4th, 14th, 3rd, 18th, and 15th chips, counting from left to right.

The sample size used in random sampling can sometimes change due to changes in available budgets or changes in the precision of the information required from the sample. In such cases, after already having drawn a random sample of size n from a population of N items, we may find ourselves in the position of wanting to either reduce or increase the sample size somewhat. A question then arises as to how to accomplish this. Fortunately, as the following rules illustrate, adjusting the sample size does not require that we discard the items already sampled.

Rules for Increasing or Decreasing the Size of a Random Sample[3]

1. The complement of a random sample of size n from a population of size N is itself a random sample from the population.

2. Any random subsample of a random sample is also a random sample from the population.

3. Any random subsample from the complement of a random sample is itself a random sample from the population.

4. After a random sample of size n has been selected, any random sample from its complement can be added to it to form a larger random sample from the population.

[Note: The complement of any sample is the name given to those items *not* in the sample.]

Example 4.9

Commercial and military aircraft are built using hundreds of thousands of specially designed nuts and bolts, known as "fasteners." Because these fasteners are subjected to stress, fatigue, and a host of environmental conditions, random samples of each type of fastener are routinely tested for strength requirements.

Suppose an inspector has drawn a random sample of size 10 from a box of completed fasteners and conducts torque tests on them. After testing, the inspector is informed that, in fact, a sample of size 25 is required by the customer for these fasteners. Since the fasteners remaining in the box are the complement of the original

---

[3] Wright, T., and H. Tsao, "Some Useful Notes on Simple Random Sampling," *Journal of Quality Technology*, 1985: 67–73.

sample of 10, then the inspector need only select a random sample of 15 fasteners from the box to add to the original sample. Rule 4 ensures that the group of 25 fasteners selected in this fashion qualifies as a random sample from the box.

On the other hand, suppose the inspector had originally selected a sample of size 25 but subsequently found that a sample of only 10 was needed. By simply selecting a random sample of 10 from the original 25 items, the inspector will have legitimately obtained a random sample of size 10 from the box.

Obtaining random samples often requires some ingenuity. This is especially the case when it is difficult to develop a sampling frame for the population of interest. For example, continuous processes, which are not conveniently divided into finite numbers of discrete parts, usually pose special problems when developing sampling frames. In such circumstances, it is helpful to remember that a sampling frame can also be a *procedure*, not just a list.[4]

Example 4.10   Agricultural inspectors are required to select random samples of crops for testing and evaluation. Harvested crops stored in cartons or bins, such as citrus fruit, pose special sampling problems. Although it is easy to *imagine* tagging the fruit in a bin with successive integers and applying the random number scheme to generate samples, doing so would be time-consuming and economically prohibitive. Instead, other schemes have been developed to obtain random samples in a more economical fashion. One popular technique is to select a bin of fruit at random (bins are generally easy to select by the random number method) and then follow a "random corner" method for obtaining the sample: First, one of the bin's four corners is chosen at random (a small printed table of random numbers is helpful here); then the fruit stacked in the selected corner are used to form the sample. This method relies on the reasonable assumption that the fruit were randomly mixed when packed in the bin. Choosing a corner at random has the additional benefit of not allowing human inspectors to introduce bias into the resulting data by always choosing a corner in which the fruit looks especially good (or bad).

Random Versus Nonrandom Samples

The first three chapters of this text focus primarily on the mechanics of how data is *used* to describe samples and populations. In Chapter 4, because of the importance of obtaining good (and avoiding bad) data, we look at how data is *generated* in the first place. This is one of the most important aspects of conducting a statistical study because you certainly do not want all your hard work on a problem to be negated by statements such as "your results are only as good as your data" or the well-known acronym GIGO (garbage in, garbage out).

Drawing conclusions from data always comes down to a question of trust: How reliable or trustworthy is the person, organization, or method providing the data? Statisticians address this issue by using data-gathering methods based on **random sampling** and **randomization** (see Section 4.3), techniques that then allow the use of probability calculations (Chapter 5) to *numerically* assess the reliability of the conclusions drawn

---

[4] Kish, L., *Survey Sampling*, John Wiley & Sons, New York, 1965: 53.

from the data. Such methods are objective and the only "trust" involved is in assuring that random sampling or randomization is correctly employed while gathering the data. On the other hand, with **nonrandom samples** (i.e., data *not* gathered using some sort of randomizing technique), no such probability assessments are possible and the information in such data cannot, as a rule, be generalized to larger populations.

The problems with nonrandom data go even deeper. Even with the best intentions, when trying to subjectively obtain data that we think is "representative" of a larger population, the resulting data can be badly skewed. For example, when assessing the reliability of a product, an engineer might try to ensure that the data includes examples of each kind of failure mode that the product experiences in the field. This practice automatically ignores the fact that some failure modes are usually much more prevalent than others, and inferences based on such "representative" samples may not only be unreliable, but even misleading.

So, what should you do when nonrandomly collected data arises in practice? Although it is acceptable to apply simple descriptive statistical measures to the data (e.g., means, histograms, and so forth), be aware that (1) such measures can't legitimately be generalized, and (2) the statistical techniques presented in the following chapters may not be valid when applied to such data.

Stratified Sampling

The method of random sampling can be extended to incorporate additional sources of information and to handle problems that arise when sampling from populations for which suitable sampling frames are hard to obtain. To distinguish basic random sampling (as previously described in this section) from the extended sampling schemes that rely on it, random sampling is often referred to as **simple random sampling (SRS)**.

One method for incorporating additional information is **stratified sampling.** In stratified sampling, the population of interest is first divided into several nonoverlapping subsets called **strata,** and then the SRS method is used to select a separate random sample from each of the strata. All of the strata samples are then combined into one large "stratified" sample from the population. When the strata are properly specified, stratified sampling will generally produce estimates that are *more precise* than SRS sampling.

General Rules for Choosing Strata

- Decide on a response variable y that is of interest.
- Divide the entire population into nonoverlapping groups (i.e., strata) $S_1, S_2, \ldots, S_k$, each of which is as *homogeneous* as possible.
- Decide on the sample sizes $n_1, n_2, \ldots, n_k$ to select from the k strata.
- Use SRS to obtain a sample from each stratum.

Estimating a Population Mean

Figure 4.1 illustrates the decomposition of a population into strata for estimating the mean μ of a population. Let the number of population elements that fall within these strata be denoted by $N_i\,(i = 1, 2, 3, \ldots, k)$; each stratum S_i has its own mean μ_i and standard deviation σ_i. The selection of sample sizes can be done in two steps: (1) Decide on the total sample size n that will be used, and then (2) decide how to divide n up into the strata sample sizes $n_1, n_2, n_3, \ldots, n_k$.

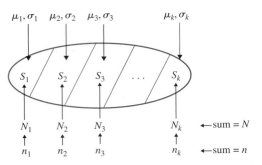

Figure 4.1 A population divided into k strata $S_1, S_2, S_3, \ldots, S_k$ of size $N_1, N_2, N_3, \ldots, N_k$

Example 4.11   Companies that produce or handle hazardous chemicals are required to apply for a National Pollutant and Discharge Elimination System (NPDES) permit from the federal government ("Measuring, Sampling, and Analyzing Storm Water," *Pollution Engr.*, Mar. 1, 1992: 50–55). The environmental concerns addressed by the NPDES permit involve the amounts of pollutants carried by storm water runoff from a company's facility to nearby public waters. Pollutant levels are estimated by taking random samples of storm water and subjecting them to chemical analysis.

Sampling runoff water is accomplished by stratifying runoff water according to the different point sources, usually water channels, that carry the runoff. Using various techniques and meters, the average velocity of water flow and the cross-sectional area of each channel are estimated. These are used to estimate total flow volumes for each point source. The flow volume can be thought of as a measure of the size N_i of the ith stratum. The total of all flow volumes represents the population size. Water samples from each point source are obtained and chemically analyzed. The total pollutant level is then calculated as a weighted average of the pollutants in each sample, weighted by the flow volume from the point source where the sample was obtained.

To choose n, we first decide on a confidence level and a **bound B on the error of estimation.** The **confidence level** (which will be discussed in greater detail in Section 7.2) is a measure of the degree of reliability, measured on a scale from 0% to 100%, that we would like to have in our final estimate of the overall population mean μ. Of course, since estimates are based on samples, 100% confidence is not possible, so confidence levels are usually restricted to large numbers (e.g., 90%, 95%, 99%, etc.) less than 100%. Estimates should also be "close enough" to the population characteristic they estimate to be useful for subsequent calculations and decision making. This requirement is achieved by specifying B, the "plus or minus" margin of error that you are willing to accept in your estimate. Finally, we let w_i denote the proportion (or weight) that the ith stratum sample represents in the total sample of n, that is, $w_i = n_i/n$ for $i = 1, 2, 3, \ldots, k$. Given the w_i's, the N_i's, the σ_i's, a confidence level of 95%, and B, it can be shown that the minimum necessary sample n for estimating the population mean μ to within a margin of error of $\pm B$ is

$$n = \frac{\displaystyle\sum_{i=1}^{k} \frac{N_i^2 \sigma_i^2}{w_i}}{N^2 \left(\dfrac{B}{1.96}\right)^2 + \displaystyle\sum_{i=1}^{k} N_i \sigma_i^2}$$

where $N = N_1 + N_2 + N_3 + \cdots + N_k$. For confidence levels other than 95%, replace 1.96 by the appropriate value from a table of standard normal curve areas.

Assuming the same per unit cost for sampling from each stratum, the optimum allocation of sample sizes (called the *Neyman allocation*) can be shown to be

$$n_i = n\left(\frac{N_i \sigma_i}{\displaystyle\sum_{i=1}^{k} N_i \sigma_i}\right) \qquad \text{where } n = \frac{\left[\displaystyle\sum_{i=1}^{k} N_i \sigma_i\right]^2}{N^2 \left(\dfrac{B}{1.96}\right)^2 + \displaystyle\sum_{i=1}^{k} N_i \sigma_i^2}$$

If, in addition, the strata standard deviations are identical, then

$$n_i = n\left(\frac{N_i}{N}\right) \qquad \text{where } n = \frac{1}{\left(\dfrac{B}{1.96}\right)^2 + \dfrac{1}{N}}$$

This is called the *proportional* allocation. Please consult one of the chapter references for the case of unequal sampling costs.

Regardless of the allocation used, the stratified estimate of the population mean μ is given by

$$\bar{x}_{str} = \bar{x}_1\left(\frac{N_1}{N}\right) + \bar{x}_2\left(\frac{N_2}{N}\right) + \bar{x}_3\left(\frac{N_3}{N}\right) + \cdots + \bar{x}_k\left(\frac{N_k}{N}\right)$$

where $\bar{x}_i$ denotes the mean of the n_i observations from stratum S_i. One of the nice features of the proportional allocation is that the resulting data is "self-weighting"; in other words, instead of calculating the stratified estimate we can simply combine the data from all the strata and calculate the ordinary sample mean of the combined data, which, only in this case, will exactly equal $\bar{x}_{str}$.

Stratified estimates of μ are usually accompanied by a measure called their **standard error** (which will be discussed more fully in Chapter 7) that can be interpreted in much the same way the sample standard deviation is interpreted. That is, if we think of all the possible stratified samples of size n that we could have selected, about 95% of the estimated means from such samples will be within about 2 standard errors of μ. For stratified sampling, the standard error is approximated by

$$s_{str} \approx \sqrt{\frac{1}{N^2}\sum_{i=1}^{k}N_i^2\left(\frac{s_i^2}{n_i}\right)\left(\frac{N_i-n_i}{N_i-1}\right)}$$

where s_i^2 is the sample variance of the n_i observations from stratum S_i.

Example 4.12   Since 1991 the USGS (U.S. Geological Survey) has conducted the National Water Quality Assessment Program (NAWQA), whose purpose is to study natural and human factors that affect water quality. One important measurement that NAWQA produces is an estimate of the percentages of a region covered by various crop types. In one study ("Validation of National Land-Cover Characteristics Data for Regional Water Quality Assessment," *Geocarto International*, Dec. vol. 10, no. 4 1995: 69–80) of the percentages of a region covered by corn crops, a region was divided into the following strata: A (irrigated crops), B (small grains and mixed crops), C (grasslands and small crops), D (wooded areas and crops), E (grasslands), and F (woods and pastures).

The region under study is first divided into smaller regions called *quadrats*, each with an area of 1 km². These subregions are then assigned to the various strata categories. Suppose that data from previous studies is used to obtain estimates of the standard deviations σ_i of the percentages of corn crops within each stratum and that this information is collected in the following table:

Stratum (S_i)	Stratum size (N_i)	Standard deviation (σ_i)
A	500	.2
B	300	.2
C	100	.3
D	50	.4
E	50	.6
F	200	.8

Since aerial photographs are used to estimate the percentage of corn coverage at a given site, the unit sampling costs will be about the same for each 1 km² subregion, so the Neyman allocation can be used. If we specify a 90% confidence level (the area under the z curve between -1.645 and $+1.645$ is .90) and a margin or error of $\pm 10\%$ (i.e., $B = .10$), then

$$n = \frac{\left[\sum_{i=1}^{k}N_i\sigma_i\right]^2}{N^2\left(\dfrac{B}{1.645}\right)^2 + \sum_{i=1}^{k}N_i\sigma_i^2}$$

$$= \frac{[500(.2)+300(.2)+100(.3)+50(.4)+50(.6)+200(.8)]^2}{1200^2\left(\dfrac{0.10}{1.645}\right)^2 + [500(.2^2)+300(.2^2)+\cdots+200(.8^2)]}$$

$$= 109.68 \approx 110 \quad \text{(rounding to the nearest integer).}$$

Using the fact that $\sum_{i=1}^{k} N_i \sigma_i = 410.0$, the Neyman allocation of $n = 110$ to the strata is

Stratum	N_i	σ_i	$N_i \sigma_i$	$n_i = n(N_i \sigma_i / \sum_{i=1}^{k} N_i \sigma_i)$
A	500	.2	100	$n_1 = 110(100/410) = 26.8 \approx 27$
B	300	.2	60	$n_2 = 110(60/410) = 16.1 \approx 16$
C	100	.3	40	$n_3 = 110(40/410) = 10.7 \approx 11$
D	50	.4	20	$n_4 = 110(20/410) = 5.4 \approx 5$
E	50	.6	30	$n_5 = 110(30/410) = 8.0 \approx 8$
F	200	.8	160	$n_6 = 110(160/410) = 42.8 \approx 43$

The next step in the study is to obtain random samples of size $n_1 = 27, n_2 = 16, \ldots, n_6 = 43$ from the respective strata and to use aerial photographs of the selected 1-km^2 regions to obtain estimates of the corn percentages in these regions. To illustrate, the following table summarizes the data from such a study:

Stratum	n_i	N_i	$\bar{x}_i$	s_i
A	27	500	.52	.18
B	16	300	.22	.23
C	11	100	.02	.35
D	5	50	.06	.45
E	8	50	.01	.64
F	43	200	.67	.78

From this data we estimate that overall percentage of the entire region that is covered by corn crops is

$$\bar{x}_{str} = .52(500/1200) + .22(300/1200) + \cdots + .67(200/1200)$$
$$= .39 \,(\text{or, } 39\%)$$

and the estimated standard deviation that accompanies this estimate is $s_{str} \approx .03$ (or, 3%).

Estimating a Population Proportion

Stratification can also be used to obtain an estimate of a population proportion π. Recall that a **population proportion** is simply the proportion of all the items in a population that have a particular attribute. In statistics, it is important to remember that the term *population proportion* refers to a proportion of the *number* of items in a population. Proportions or percentages that use *different* bases of comparison (such as in Example 4.12, where percentages of land areas were used) are treated simply as numerical data, not as proportions.

The procedure presented earlier for finding stratified estimates of a population mean can easily be converted into a procedure for estimating a population proportion. Using earlier notation, where N denotes the population size and N_i denotes the number of items in the ith stratum, S_i, the only changes in the formulas are:

1. Replace each $\bar{x}_i$ by p_i, where p_i is the *sample proportion* of items found in the sample of n_i items selected from stratum S_i.
2. Replace each σ_i by $\sqrt{\pi_i(1 - \pi_i)}$, where π_i is the proportion of items in stratum S_i that have the given attribute. Since the values of π_i

are not normally known exactly, there are various possibilities for estimating them:

a. You can approximate the π_i values based on pilot studies or on results from previous studies.

b. Or, if there is no prior information about π_i values, then be pessimistic and use $\pi_i = .5$ for each $i = 1, 2, 3, \ldots, k$ (this choice maximizes $\sqrt{\pi_i(1 - \pi_i)}$).

The stratified estimate of the population proportion π is then given by

$$p_{str} = p_1\left(\frac{N_1}{N}\right) + p_2\left(\frac{N_2}{N}\right) + p_3\left(\frac{N_3}{N}\right) + \cdots + p_k\left(\frac{N_k}{N}\right)$$

and its associated standard error is approximated by

$$s_p \approx \sqrt{\frac{1}{N^2}\sum_{i=1}^{k} N_i^2\left(\frac{N_i - n_i}{N_i}\right)\left(\frac{p_i(1 - p_i)}{n_i - 1}\right)}$$

Example 4.13   Improper handling of newly planted citrus trees can cause a defect called *benchroot*, which is the tendency for the root system to grow sideways. Benchroot eventually causes trees to be less healthy and smaller than normal, which results in smaller crops. Because citrus trees require several years of growth before reaching maximum production levels, the presence of benchroot is not apparent until years after planting. By sampling young trees shortly after planting, the extent of the benchroot problem can be estimated in time to take other measures, such as replanting selected areas.

Suppose that a citrus cooperative consists of five different farms. Using the farms as strata should increase the precision of the final sampling results since the trees within a given farm ought to be more similar to each other than to trees on other farms. The number of trees on the farms are known to be $N_1 = 2000$, $N_2 = 4000$, $N_3 = 8000$, $N_4 = 8000$, and $N_5 = 1000$. Based on records from previous plantings, the benchroot problem has affected no more than about 10% of all trees, so a value of $\pi_i = .10$ ($i = 1, 2, 3, 4, 5$) is selected for each farm. This means that $\sigma_i = \sqrt{.10(1 - .10)} = .3$ for each farm. Since the unit costs c_i ($i = 1, 2, 3, 4, 5$) of selecting and testing a tree are assumed to be equal for each farm, the Neyman allocation can be used to find the required sample size and its allocation to the strata (farms). Finally, suppose that a confidence level of 95% and an error bound of $B = .03$ (i.e., $\pm 3\%$) are chosen. Based on this information, the required sample size is

$$n = \frac{\left[\sum_{i=1}^{k} N_i\sigma_i\right]^2}{N^2\left(\frac{B}{1.96}\right)^2 + \sum_{i=1}^{k} N_i\sigma_i^2}$$

$$= \frac{[2000(.3) + 4000(.3) + 8000(.3) + 5000(.3) + 1000(.3)]^2}{20000^2\left(\frac{0.03}{1.960}\right)^2 + [2000(.3^2) + 4000(.3^2) + \cdots + 1000(.3^2)]}$$

$$= 376.92$$

which we round to $n = 377$. The following table shows the steps in allocating the total sample of 377 to the five strata (farms). After sampling, the number of trees with benchroot, x_i, is recorded for each farm. Note that we have rounded all final sample sizes to integer values.

Farm	N_i	σ_i	$N_i\sigma_i$	$n_i = n(N_i\sigma_i / \sum_{i=1}^{k} N_i\sigma_i)$	x_i
1	2000	.3	600	$n_1 = 377(600/6000) = 38$	2
2	4000	.3	1200	$n_2 = 377(1200/6000) = 75$	5
3	8000	.3	2400	$n_3 = 377(2400/6000) = 151$	8
4	5000	.3	1500	$n_4 = 377(1500/6000) = 94$	3
5	1000	.3	300	$n_5 = 377(300/6000) = 19$	2

Of the sampled trees, $x_1 = 2$, $x_2 = 5$, $x_3 = 8$, $x_4 = 3$, and $x_5 = 2$ trees were found to have the benchroot problem. Using this data, the stratified estimate of the proportion of all the 20,000 trees in the cooperative having benchroot is

$$p_{str} = p_1\left(\frac{N_1}{N}\right) + p_2\left(\frac{N_2}{N}\right) + p_3\left(\frac{N_3}{N}\right) + p_4\left(\frac{N_4}{N}\right) + p_5\left(\frac{N_5}{N}\right) = .053.$$

The reader can verify that the standard error associated with this estimate is $s_p = .011$.

Cluster Sampling

Stratified and SRS sampling are best when relatively complete lists of population elements and strata sizes are known before sampling. In some applications, however, such information is difficult or impossible to obtain. In wildlife sampling, for instance, scientists usually do not have advance knowledge of either the size of the particular population or the size of the various strata in the population. In such cases, some form of **cluster sampling** is used instead of SRS or stratified sampling. Like stratified sampling, cluster sampling requires that we first divide a population into nonoverlapping groups, called **clusters**. However, we do not need to know the number of population elements in each cluster. Instead, we simply take an SRS sample of the clusters and then measure all elements within the selected clusters. For example, the U.S. Census relies on cluster sampling when complete lists of city inhabitants are not known. A city is divided into blocks (clusters) using maps, then a random sample of these blocks is selected and *all* residences in the sampled blocks are contacted.

Example 4.14   Biologists and ecologists frequently sample geographic areas by dividing a map of a region into a collection of small square regions called **quadrats** (Ripley, B. D., *Spatial Statistics*, New York, Wiley, 2004: 102). By making sure the quadrats do not overlap, we can apply the method of cluster sampling by choosing a random sample of quadrats to investigate. In wildlife studies, for instance, the number of a given species in each of the selected quadrats is counted. Because the area of a quadrat is known, these counts are usually converted into a count per unit area, which is a measure of the abundance of the particular species per unit area.

Section 4.2 Exercises

6. Devise a procedure for selecting a random sample of words from a dictionary. Explain why your procedure guarantees that, for any n, each collection of n words has an equally likely chance of being selected.

7. Sometimes it is difficult or impossible to determine the population size before selecting a random sample. Describe how you would go about selecting a random sample of trees from a 1-square-mile area of forest.

8. Small manufactured goods are often gathered into large batches, called *lots*, for purposes of handling and shipping. Random sampling is commonly used to evaluate the quality of items in a given lot. Suppose an inspector selects a random sample of 20 items from a lot of 1000 items.

   a. Before evaluating the 20 items, the inspector decides that a sample of size 30 should be used instead. If the inspector obtains a second random sample of size 10 from the remaining 980 items, can the two samples combined be validly considered a random sample of 30 from the lot? Explain your reasoning.

   b. Suppose the inspector decides that only 15 items must be tested. Describe a method by which a valid random sample of 15 *from the lot* can be formed from the 20 items already selected.

9. Citrus trees are usually grown in orderly arrangements of rows to facilitate automated farming and harvesting practices. Suppose a group of 1000 trees is laid out in 40 rows of 25 trees each. To test the sugar content of fruit from a sample of 30 trees, researcher A suggests randomly selecting five rows and then randomly selecting six trees from each sampled row. Researcher B suggests numbering a map of the trees from 1 to 1000 and selecting a random sample (without replacement) of 30 integers from the integers 1 to 1000.

   a. Without performing any calculations, do you think that both methods are capable of generating random samples from the block of trees? Justify your answer using the rules for random samples listed in this section.

   b. Suppose that the group of trees is grown on the top and sides of a small hill. A researcher suggests that, because growing conditions (e.g., daily amounts of sunlight) are different on the four sides of the hill, the hill should be divided into four quadrants and trees should be randomly sampled from each quadrant. What is the name for this type of sampling procedure?

10. In stratified sampling, explain why it is best to choose strata such that the objects *within* any stratum are relatively homogeneous.

11. Explain how to use the =RANDBETWEEN function in Excel™ to generate a random sample from the integers 1 through 1000. Does the =RANDBETWEEN function generate samples with or without replacement?

12. A population of items is partitioned into k strata of sizes $N_1, N_2, \ldots, N_k$. Using proportional allocation, random samples of size $n_1, n_2, n_3, \ldots, n_k$ are selected from the strata and the numbers $x_1, x_2, x_3, \ldots, x_k$ of items having a specified characteristic are determined. Sample proportions $p_1, p_2, p_3, \ldots, p_k$ are then computed (i.e., $p_i = x_i/n_i$ for each i).

   a. Write an expression for the weighted average of the sample proportions, using the stratum sizes as weights.

   b. Show that the weighted average in part (a) simplifies to $(x_1 + x_2 + x_3 + \cdots + x_k)/(n_1 + n_2 + n_3 + \cdots + n_k)$.

13. Integrated circuits (ICs) consist of thousands of small circuits, electronic subcomponents (e.g., resistors), and connections. An important factor in the manufacture of ICs is the *yield*, the percentage of manufactured ICs that function correctly. Stratified sampling has recently been used to estimate the number of defects of various kinds that occur throughout an IC. The area of the IC is first divided into smaller areas (i.e., strata) and then small sample areas are selected from the strata and examined for defects. A stratified estimate of the overall proportion of defects can be used to help estimate the eventual yield of the IC manufacturing process.

    In one such study, to estimate the proportion of pinholes on an IC, its entire surface was first divided into 10 equal areas (strata), each of which was further subdivided into 1000 smaller rectangles that served as the elements to be sampled. It was also assumed that the unit costs and variances of the numbers of pinholes were equal from strata to strata.

a. Calculate the population size N.

b. Using a confidence level of 90% and a bound on the error of estimation of $B = .03$ (i.e., $\pm 3\%$), calculate the required sample size n and its allocation $n_1, n_2, n_3, \ldots, n_{10}$ to the ten strata. Round all sample sizes to the nearest integer.

c. Using the sample sizes in part (b), the results of the study showed the following numbers of pinholes per sample:

Sample #: 1 2 3 4 5 6 7 8 9 10
Pinholes: 5 4 7 6 3 9 5 6 2 8

Calculate the stratified estimate of the proportion of pinholes on the entire IC.

d. Calculate the standard error associated with the estimate in part (c).

14. Of the elements of a certain population 20% are grouped into stratum S_1 and the rest of the population elements comprise stratum S_2. Suppose that the variances of the characteristic being measured are the same for each stratum, but it costs twice as much to obtain a sampled item from stratum S_1 as it does from stratum S_2. What is the best allocation of a total sample of $n = 1000$ to these two strata?

15. When the per unit cost of sampling from stratum i is c_i, it can be shown that the optimal weights for allocating the total sample size are given by

$$w_i = \frac{\dfrac{N_i \sigma_i}{\sqrt{c_i}}}{\dfrac{N_1 \sigma_1}{\sqrt{c_1}} + \dfrac{N_2 \sigma_2}{\sqrt{c_2}} + \dfrac{N_3 \sigma_3}{\sqrt{c_3}} + \cdots + \dfrac{N_k \sigma_k}{\sqrt{c_k}}}$$

a. In the case where all unit sampling costs are equal, show that the resulting weights give the formulas for n and n_i specified by the Neyman allocation.

b. In the case where all unit sampling costs are equal and all strata variances are equal, show algebraically that the resulting weights give the formulas for n and n_i specified by the "proportional" allocation.

16. In stratified sampling, explain why the number of strata, k, should *not* exceed $n/2$, where $n = n_1 + n_2 + n_3 + \cdots + n_k$ is the total sample size and n_i denotes the number of sampled items selected from stratum S_i $(i = 1, 2, 3, \ldots, k)$.

17. In stratified sampling, what value would you use in place of 1.96 if you wanted the confidence level to be 99% rather than 95%? What is the consequence of using the higher confidence level on the necessary sample size?

4.3 DATA FROM EXPERIMENTS

The choice of a data collection method is dictated, to a large extent, by how we intend to use the data. If our work involves applying standards and codes (e.g., strength testing of concrete in commercial buildings, measuring the amount of a pollutant in a water sample, or assigning the DOT treadwear rating printed on automobile tires), then it is desirable to use operational definitions (Section 4.1) to keep tight control over every aspect of the measurement process. By doing so, we ensure that the results will be directly *comparable* to similar tests and measurements made by ourselves and others. On the other hand, if our work involves research and experimentation, then it is necessary to purposely allow some of the underlying conditions to vary so that their combined effects can be studied and understood. In this way, we can *generalize* the conclusions obtained from the data to a larger setting. This text is concerned primarily with the latter type of application: the statistical tools needed in research and experimentation.

The statistical techniques used in experimental research are collectively known as **experimental designs.** In the sciences, these tools are also referred to as the **design of experiments,** commonly abbreviated **DOE.** Experimental designs are carefully detailed plans for obtaining sample data for the purpose of understanding relationships between variables and generalizing conclusions obtained from the data. Inherent in these designs are methods for balancing the two opposing goals of comparability and generalizability mentioned in the previous paragraph.

Example 4.15   Plastic resins used in injection molding machines are designed to meet various production requirements (e.g., melting temperatures, hardness, color). Raw resins are manufactured in the form of solid plastic pellets that are subsequently melted inside an injection molding machine and then "shot" into molds.

Suppose that a company wants to test similar resins from two suppliers, A and B, to determine which one better achieves the hardness requirements for certain molded parts. One experimental approach is to test each resin two or more times using the same molding machine. By combining more than one reading for each brand, we hope to "average out" any unexpected biases that might creep into any single measurement. In such an experiment, the average hardness measurements would be directly *comparable*. That is, as long as all other experimental conditions are held constant, there would be little doubt that differences between the average hardness measurements could be attributed to differences between the two brands of resin. Figure 4.2(a) depicts this design.

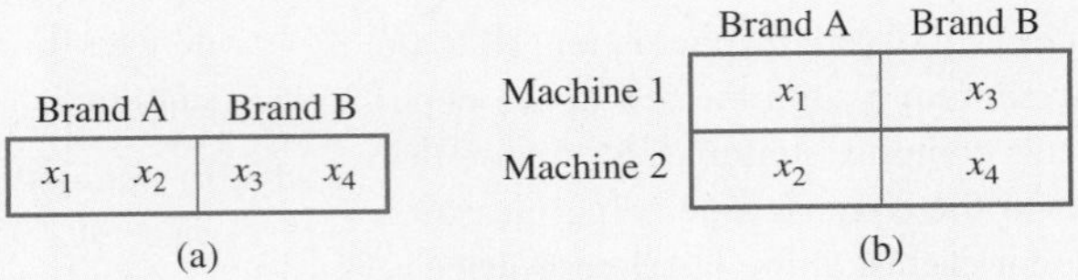

	Brand A	Brand B
Machine 1	x_1	x_3
Machine 2	x_2	x_4

Brand A		Brand B	
x_1	x_2	x_3	x_4

(a)   (b)

Figure 4.2 Experimental designs for hardness requirements: (a) one machine, two measurements per brand; (b) two machines, two measurements per brand

It is very difficult, however, to extrapolate such results to a more general setting. For instance, would the hardness measurements be significantly affected if we used several different molding machines? Figure 4.2(b) shows a simple experimental design that allows us to answer this question while simultaneously allowing us to answer the original question about differences between the two brands. The noteworthy feature of this design is that comparability between brands is maintained [by comparing the average hardness reading $(x_1 + x_2)/2$ for brand A to the average $(x_3 + x_4)/2$ for brand B], yet we can also answer questions about whether different machines influence the results [by comparing the two machine averages $(x_1 + x_3)/2$ and $(x_2 + x_4)/2$]. As this design illustrates, the key to maintaining comparability while answering questions about generalizability is to make each measurement work more than once. Note, for instance, that reading x_1 appears in the average for brand A and again in the average for machine 1. Designs such as the one in Figure 4.2(b) can easily be extended to handle more and more complex questions involving the effects of changing several test conditions.

Experimental designs are considered to be **controlled studies** because they place strict guidelines on which factors are allowed to vary and on the range of values these factors may assume. In this way, they differ from **observational studies**, in which experimenters simply observe and measure but otherwise allow all factors to vary freely. The following list shows some of the most common applications of experimental design.

Where Experimental Designs Are Used

- Studying cause-and-effect relationships
- Increasing the external validity of data
- Studying how independent variables (factors) affect a dependent variable (response)
- Studying the interrelationships among factors that affect a response
- Optimizing product and process characteristics
- Measuring experimental error

Experimental Design Terminology

Most of the concepts and terminology of experimental design were developed in the mid-1920s by the English statistician Sir Ronald Fisher while he was working at the British Agricultural Experimentation Station at Rothamsted, just outside London. Although Fisher's applications were primarily agricultural, statisticians quickly realized that the methods of experimental design were universal and soon began using them in industrial and scientific applications as well.

The object of using an experimental design is to study and quantify the effects that different test conditions have on some measurable characteristic of a product or process. For instance, experimental designs have been used for decades to analyze drilling processes. In one such study, the thrust force (lb) required to push a drill into a bar of aluminum was studied along with two explanatory variables, drill diameter (in.) and the feed rate (in./revolution) with which the drill penetrates the metal ("Design of a Metal-Cutting Drilling Experiment: A Discrete Two-Variable Problem," *Quality Engr.*, 1993: 71–98). In the language of experimental design, the thrust force is a **response variable** (also called a **dependent variable**); drill diameter and feed rate are two **factors** (also called **independent variables**) whose values are thought to explain or affect the values of the response variable. Part of the experimental process involves selecting specific factor values, called the **factor levels** (or **treatment levels**), to use in the study. In this study, five different feed rates were used (.005, .006, .009, .013, and .017 in./rev.) along with five drill sizes (.225, .250, .318, .406, and .450 in.). The final choice to be made involves the **experimental unit(s)** to which the treatments will be applied. Experimental units are the objects or material upon which the final measurements are made. In the drilling study, it was decided to use samples of a single type of aluminum alloy as the experimental units.

The particular choice of experimental units is important because it influences the range of validity of the experimental results. Generally speaking, the more variation there is between experimental units, the wider the range of validity of the experiment. By choosing a single type of aluminum alloy, for example, the results of the drilling experiment previously described are limited primarily to conclusions about drilling in aluminum. If, instead, the experimental units had consisted of different types of metals, then the experimental results would correspondingly apply to a wider range of drilling applications.

The Basic Tools of Experimental Design

Experimental designs are built from a small group of tools, each addressing specific concerns about experimental results: reducing bias, reducing experimental error, reducing the effect of external factors, and increasing the generalizability of the conclusions. What follows is an overview of these tools. Specific designs are presented in Chapter 10.

Perhaps the most familiar tool is that of **replication,** that is, making several repeated measurements at each fixed combination of factor or treatment levels. For instance, in Figure 4.2(b) of Example 4.15, suppose that we decide to make three measurements of plastic hardness at each of the four combinations of factor levels: {brand A with machine 1, brand B with machine 1, brand A with machine 2, brand B with machine 2}. The purpose of doing this is twofold: (1) Biases tend to be eliminated when several measurements are averaged and (2) the variation between repeated measurements gives a measure of **experimental error.** Experimental error is the name given to the slight differences that we expect to find between repeated experimental tests, even when we attempt to hold all test conditions constant.

The next tool, **randomization,** is somewhat less familiar than replication. Randomization requires that treatments be given to the experimental units *in random order,* or equivalently, that we assign experimental units to the various treatments in a random fashion. In Example 4.15, the experimental units are the individual containers of plastic pellets (of each brand) that are used for testing. Since we decided to use three replications for each combination of factor levels, there are a total of 12 tests to conduct (three measurements at each of the four factor combinations). Randomization requires that these 12 tests be run in *random* order. This is easy to accomplish using the methods of Section 4.2, as the next example shows.

Example 4.16   In Figure 4.2(b) of Example 4.15 (page 180), denote the four distinct treatment combinations by M1A, M1B, M2A, and M2B, where M1A stands for the combination "machine 1 and brand A," M1B stands for "machine 1 and brand B," and so forth. To run three replicate tests at each treatment combination, we first number these tests from 1 to 12 as in the following table. Next, a random sample of size 12 is chosen (without replacement) from the integers 1 through 12. Suppose, for instance, the random sample is {11, 3, 7, 1, 4, 5, 12, 2, 8, 10, 6, 9}. With this ordering, test 4 (M2A) would be the first one conducted, test 8 (M1B) would be next, and so forth. In this way, the tests will be conducted in random order.

Test #	Test conditions	Random order in which tests are conducted
1	M1A	11
2	M1A	3
3	M1A	7
4	M2A	1
5	M2A	4
6	M2A	5
7	M1B	12
8	M1B	2
9	M1B	8
10	M2B	10
11	M2B	6
12	M2B	9

Randomization is used for much the same reasons that we use random sampling (Section 4.2): to eliminate unforeseen biases from the experimental data and to lay the groundwork for the statistical inferences that we eventually draw from the experiment. The first reason is easy to understand when we again consider Example 4.15. To save time, for instance, someone might decide to run all three tests involving machine 1 and brand A sequentially, since the brand A plastic could simply be inserted in the machine three times in a row, avoiding any downtime for cleaning the machine when switching to the other brand. However, this might mean that *all* the tests with machine 1 and brand B would have to be conducted on a different day than the brand A tests. Since it is possible that environmental factors could change from one day to another or that different machine operators might be used on different days, these different conditions themselves could be responsible for substantial differences in the hardness measurements. In other words, we could no longer be confident about attributing differences between hardness measurements solely to differences between the two brands. By running the 12 tests in random order, we can avoid systematic biases such as these.

The third tool used extensively in experimental design is **blocking.** Blocking is used to screen out the effects of external factors that the experimenter suspects in advance will have a large effect on the measurements. Pharmaceutical companies, for example, use blocking when testing the effectiveness of a new drug. Because different people often differ widely in their responses to drugs, experimenters first divide the experimental subjects into homogeneous groups or **blocks.** The people in a given block are "matched" on various characteristics (e.g., blood pressure, age, gender) so that the people in any given block are very similar to one another but fairly different from the people in other blocks. The goal is to maximize the similarity of the subjects *within* each block and to maximize the differences *between* the blocks. For instance, block 1 might consist of young females with low blood pressure, block 2 could consist of middle-aged men with high blood pressure, and so forth. After the blocks are formed, the experimental treatments are applied within the blocks. For example, half of the people in block 1 would be given the new drug, whereas the other half would receive a placebo. Similarly, half the people in block 2 would receive the new drug and half would receive the placebo. In this way, when we look at a particular block, any differences in response between the two halves of the block could be attributed to the different treatments (receiving the drug or receiving the placebo), *not* to the differences between people. Without blocking, differences in the response to different treatments can often be masked by large differences between the individuals randomly selected for each treatment.

Blocking increases the sensitivity of an experiment for detecting differences between treatments. When blocking is applied in conjunction with randomization, it is possible to design experiments that are *simultaneously* sensitive to differences between the treatments studied but less sensitive to the unknown external factors that might affect the data. One popular phrase that summarizes how these tools are to be used is "block what you know, randomize what you don't."[5] In other words, try to identify known sources of variation and eliminate their effect by forming blocks. However, *within* each block, remember to assign experimental units to the treatments in a random fashion.

---

[5] Box, G. E. P., W. G. Hunter, and J. S. Hunter, *Statistics for Experimenters* (2nd ed.), John Wiley & Sons, New York, 2005: 93.

Example 4.17 The strength of concrete used in commercial construction tends to vary from one batch to another. Consequently, small test cylinders of concrete sampled from a batch are "cured" for periods up to about 28 days in temperature- and moisture-controlled environments before strength measurements are made. Concrete is then "bought and sold on the basis of strength test cylinders" (ASTM C 31 Standard Test Method for Making and Curing Concrete Test Specimens in the Field).

Suppose that we want to compare three different methods of curing concrete specimens. We know that batch-to-batch variation can be a significant factor in strength measurements. One way to compare the three methods is to use different batches of concrete as blocks in an experimental design. This is accomplished by separating each batch into three portions and then randomly assigning the portions to the three curing methods. Table 4.1 shows the data from one such test using ten batches of concrete of comparable strengths.

Table 4.1 Data from the blocked experiment of Example 4.17

| | Strength (in MPa) | | |
Batch	Method A	Method B	Method C
1	30.7	33.7	30.5
2	29.1	30.6	32.6
3	30.0	32.2	30.5
4	31.9	34.6	33.5
5	30.5	33.0	32.4
6	26.9	29.3	27.8
7	28.2	28.4	30.7
8	32.4	32.4	33.6
9	26.6	29.5	29.2
10	28.6	29.4	33.2

The purpose of blocking is to allow for fair comparisons among the three test methods. Notice, for example, that all three methods gave relatively lower values for batch 6 and higher values for batch 5. This is evidence of a difference between batches 5 and 6. By blocking, however, any differences among the batches are experienced by *all* three test methods. Consider how different things might be if we had simply assigned entire batches at random to the three test methods. By doing so, it is possible that batch 5 could be assigned to method C alone and batch 6 to method A alone, which would increase the average strength measurement for column C and decrease the average for column A. In other words, if we do not use blocking, then differences among the three test methods could be significantly influenced by the manner in which the batches of cement are assigned to the tests.

Section 4.3 Exercises _____

18. Four new word processing software programs are to be compared by measuring the speed with which various standard tasks can be completed. Before conducting the tests, researchers note that the level of a person's computer experience is likely to have a large influence on the test results. Discuss how you would design an experiment that fairly compares the word processing programs while simultaneously accounting for possible differences in users' computer proficiency.

19. What primary purpose do replicated measurements serve in an experimental design?

20. In a study of factors that affect the ability of the laser in a DVD player to read the information on a DVD, a researcher decides to examine several different photoresist thicknesses used in making the plates from which plastic DVDs are stamped. As a response variable, the researcher decides to measure the average pit depth of the holes etched on the surface of the DVD. The experiment must be conducted under a fixed budget and time constraint that allows the researcher to analyze a sample of at most 20 DVDs.
    a. Suppose that it is known that, for any fixed photoresist thickness, there tends to be little, if any, variation in the pit depths on a DVD. Which would be better: (1) an experiment with little or no replication and several photoresist thickness levels or (2) an experiment with more replication, but fewer photoresist thickness levels?
    b. Suppose it is known that, even for a fixed photoresist thickness, pit depths can vary substantially. Answer the question posed in part (a) for this situation.

21. A researcher wants to test the effectiveness of a new fuel additive for increasing the fuel efficiency (miles per gallon, mpg) of automobiles. The researcher proposes that a car be driven for a total of 500 miles and that at the end of each 100-mile segment the fuel efficiency be measured and recorded.
    a. What is the purpose of measuring efficiency every 100 miles? Why not just measure efficiency at the end of the 500-mile course?

    b. What operational definitions would you suggest that the researcher incorporate into this experiment?
    c. What changes would you make to the experiment to increase the generalizability of the experimental results?

22. In a study of the ratio of nitrogen, phosphoric acid, and potash in fertilizers, four different mixtures (M_1, M_2, M_3, M_4) of the three chemicals are to be tested for their effects on the rate of growth of grass seedlings. A square plot of land is subdivided into four equal-size square plots, each planted with the same amount, by weight, of seedlings. Before the fertilizers are applied, each square subplot is itself divided into four more squares. Two experimental methods are proposed for applying the fertilizers to the subplots. In experiment A, the four fertilizers are randomly assigned to the large subplots, whereas in experiment B, all four fertilizers are randomly assigned to the *subplots* of the four large plots. An illustration of both experimental designs follows.

M_1	M_1	M_4	M_4
M_1	M_1	M_4	M_4
M_3	M_3	M_2	M_2
M_3	M_3	M_2	M_2

Experiment A

M_1	M_4	M_3	M_2
M_3	M_2	M_4	M_1
M_2	M_1	M_2	M_3
M_4	M_3	M_1	M_4

Experiment B

    a. If care were taken to ensure that there are no significant differences in the growing conditions (soil type, irrigation, drainage, sunlight, etc.) among the four large subplots, is one of these designs preferable over the other? Why?
    b. If it is suspected that there could be significant differences in the growing conditions among the four main subplots, is one of the two designs preferable over the other? Why?

23. A complex chemical experiment is conducted and, because the amount of precipitate produced is expected to vary, the experiment is repeated several times. A lengthy lab equipment setup, followed by a tedious experimental procedure, allows the experiment to

be repeated up to six times in any given day. Consequently, one lab assistant is assigned to set up the lab equipment and then conduct six runs one day. The next day a second lab assistant conducts another six runs using the same lab setup from the previous day. What two basic experimental design principles are violated by this experimental procedure?

24. Refer to Example 4.15 and Figure 4.2(b). Suppose the hardness measurements (in Mohs) of plastics in four test runs are as follows:

	Brand A	Brand B
Machine 1	2.6	3.2
Machine 2	2.8	3.6

a. Calculate an estimate of how much plastic hardness is increased or decreased by switching from the brand A resin to the brand B resin.
b. Calculate an estimate of how much plastic hardness is increased or decreased by switching from machine 1 to machine 2.
c. Because this experiment does not provide any estimate of the experimental error expected in successive experimental runs, it is impossible to know whether the estimated change in part (a) is caused by switching brands or is simply due to experimental variation. Describe how you would improve this experiment to obtain an estimate of the experimental error.

4.4   MEASUREMENT SYSTEMS

The quality of data is affected by the type of data-gathering plan followed and the reliability of the instruments used to make required measurements. Previous sections of this chapter have dealt with concerns about data-gathering methods, especially the role of operational definitions and statistics in addressing these concerns. However, most statistical methods are not explicitly designed to address questions about the quality of the raw measurements themselves. Instead, concerns about measurement quality are usually considered separately.

The study of measurement is called **metrology**. Broadly speaking, metrology is concerned with two basic issues. The first deals with our ability to produce measurements of sufficient accuracy and precision to support any analyses based on these measurements. The second concern is **calibration**. Calibration addresses the various systematic errors that can cause an instrument's readings to be in error. A familiar example is found in common household scales, which must be "zeroed" before giving a true reading of a person's weight. If such a scale consistently gives readings that are 5 lb too high, then we say that the scale is "out of calibration" and that it has an **offset** of 5 lb. Instruments are said to be "in calibration" if they give true readings, that is, if their offset is zero. Calibrating an instrument usually requires comparing its readings to those of a similar instrument that is already known to be in calibration. In turn, these secondary instruments must themselves be calibrated by comparison with yet a higher standard until we can eventually trace all such comparisons back to the highest measurement authority—those housed within the National Institute of Standards and Technology (NIST).

Accuracy and Precision

The concepts of accuracy and precision of a measuring instrument are statistical in nature. **Accuracy** refers to the degree to which repeated measurements of a known quantity x tend to agree with x. Given several repeated measurements $x_1, x_2, x_3, \ldots, x_n$ of some known value x, we measure the accuracy of the readings by the difference between x and the *average* of the n readings:

$$\text{accuracy} = \bar{x} - x$$

Refer to the n measurement readings displayed in the histogram in Figure 4.3. We can think of accuracy as the distance between the center of the histogram (i.e., the mean) and the true value of x.

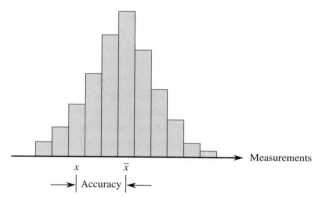

Figure 4.3 Measurement accuracy

The **precision** of an instrument describes the extent to which repeated measurements tend to agree *with one another*. They do not necessarily have to agree with the true value x that is being measured. Precision, then, is a measure of variation and is estimated by the sample standard deviation of n repeated measurements $x_1, x_2, x_3, \ldots, x_n$:

$$\text{precision} = s = \sqrt{\frac{1}{n-1}\sum (x_i - \bar{x})^2}$$

Figure 4.4 shows the various combinations of precision and accuracy that are possible in practice. The worst case occurs in Figure 4.4(a) where the measurements have a large variation (i.e., low precision) and are biased to the left of the true value of x. The best-case scenario is in Figure 4.4(d), where all the measurements are tightly packed around x (i.e., high precision and good accuracy).

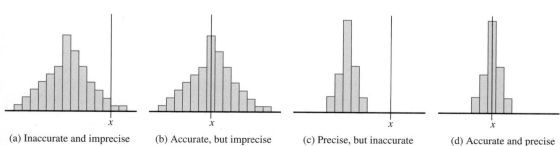

(a) Inaccurate and imprecise   (b) Accurate, but imprecise   (c) Precise, but inaccurate   (d) Accurate and precise

Figure 4.4 Precision and accuracy

Repeatability and Reproducibility

The concepts of repeatability and reproducibility refer to the amount of variation that exists between several repeated readings made by a measurement system. **Repeatability** is the amount of variation expected when almost all external sources of measurement error have been controlled and held fixed. For this reason, repeatability studies are often conducted by the same person using a single instrument to repeatedly measure a single item. Repeatability is a measure of the best that one can hope to achieve from a measuring instrument. Because the controlled environment of a repeatability study is hard to duplicate in a production environment, repeatability usually paints a very optimistic picture of measurement variation.

Repeatability is defined as the sample standard deviation, s, of several repeated measurements made under the controlled conditions described previously. When we make the additional assumption that the measurement errors follow a normal distribution, it is also common to report repeatability as $\pm 3s$, because we expect the vast majority of the readings to fall within a range of about 3 standard deviations on either side of the average reading. Because precise estimation of a population standard deviation generally requires larger sample sizes than those necessary for estimating population means, we recommend against using small sample sizes in repeatability studies. If desired, exact sample size formulas for estimating the population standard deviation can be used.

Example 4.18

In a repeatability study, a worker selects a single manufactured part and measures its length 25 times. The measurements (in.) and their sample mean and standard deviation are given in Table 4.2. The repeatability of the measuring instrument can be reported either as the standard deviation $s = .096$ in. or in terms of $\pm 3s = \pm 3(.096) = \pm .288$ in. The latter method has the intuitive interpretation that the instrument's readings generally lie within about .288 in. of the true length. For instance, if the worker measures another part and obtains a reading of 9.98 in., then the true length of that part should be somewhere between 9.692 and 10.268 in.

Table 4.2 Data for the repeatability study of Example 4.18

Repetition	Measurement (in.)	Repetition	Measurement (in.)
1	9.92	14	9.90
2	10.05	15	9.88
3	9.99	16	9.82
4	9.85	17	9.91
5	9.90	18	10.05
6	10.00	19	9.87
7	9.99	20	10.05
8	9.98	21	9.94
9	10.17	22	9.75
10	9.97	23	9.89
11	9.97	24	9.85
12	10.02	25	10.12
13	10.00		
	$\bar{x} = 9.95$	$s = .096$	

Unfortunately, the terms *repeatability* and *reproducibility* are not uniquely defined in the literature, so you may encounter alternative definitions from time to time. One popular definition of repeatability is given by the formula $k\sqrt{2}s$, which estimates the maximum difference that, with high reliability, can be expected between any *two* instrument readings. In this formula, s is the sample standard deviation and the factor k depends on the reliability level we specify. Tabled values of k, along with a detailed discussion of this form of repeatability, can be found in the article by Mandel and Lashof listed in the chapter bibliography.

As we allow more and more parts of a measurement system to vary, we move from repeatability to the concept of **reproducibility.** Reproducibility studies allow several factors to vary at the same time. In such studies, it is common to use several operators and several instruments to measure several production items. The idea is to see how the measurement system behaves in an environment more closely resembling a real production environment. Reproducibility studies are usually based on simple experimental designs that allow us to break measurement variation into distinct components that estimate the contribution of the various noise factors (different operators, different parts, etc.) to the overall measurement error. Examples of such designs are given in Chapter 10.

Interlaboratory Comparisons

Many measurements are done by laboratories specializing in complex measurement procedures. This is the case, for example, for most of the nondestructive tests mentioned in Example 4.7. For such data, our concern centers on the consistency of the results reported by different laboratories. Practically speaking, we want some assurance that if we submit the *same* sample material to laboratory A and laboratory B, then the results reported by the two laboratories will be in close agreement.

The reliability of data from different laboratories is evaluated by means of **interlaboratory comparison programs.** Professional organizations such as the American Society for Testing and Materials (see Section 4.1) run several such programs each year. For example, in the ASTM interlaboratory cross-check program for reformulated gasoline, participating laboratories are given test samples each month for measurement. The test samples are specially prepared under the direction of ASTM to ensure that each lab receives the same test material. The data from all participating laboratories is then summarized and given to the participating laboratories. In this way, each laboratory can evaluate its performance against the others and, if necessary, make changes to its measurement system.

Youden plots, introduced in 1959, are the standard technique for comparing the data from a group of laboratories (Youden, W. J., "Graphical Diagnosis of Interlaboratory Test Results," *Industrial Quality Control,* 1959: 24–28). To create these simple scatterplots, each laboratory is given two nearly identical test samples (labeled A and B) to measure. The two measurements from a given laboratory are then plotted as a single point on the Youden plot. The horizontal axis is used for the measurements of sample A and the vertical axis is used for sample B. As an aid in interpreting the plots, horizontal and vertical lines positioned at the medians of the sample A data and sample B data are included. Some typical Youden plots are shown in Figure 4.5 (page 190). The points generally fall close to a 45° line because the two samples (A and B) are similar and because each lab follows a fixed measurement procedure.

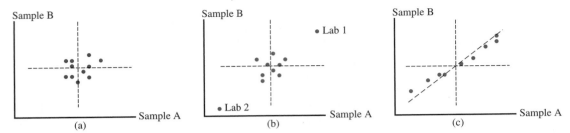

Figure 4.5 Typical Youden plots and their interpretation: (a) ideal situation with the points evenly scattered in all four quadrants; (b) laboratory 1 and laboratory 2 are using procedures that are systematically different from those used at the other labs; (c) most of the labs are following slightly different versions of the test procedure

Example 4.19   Nonsteroidal anti-inflammatory drugs (NSAIDs) are often used to reduce inflammation and relieve fever and pain. Examples of NSAIDs include ibuprofen, ketoprofen, and naproxen. In "Second Interlaboratory Exercise on Non-Steroidal Anti-Inflammatory Drug Analysis in Environmental Aqueous Samples" (*Talanta*, 2010: 1189–1196), researchers wanted to investigate interlaboratory comparisons of NSAIDs in different aqueous samples. This research was conducted to ascertain the level of interlaboratory agreement of NSAID analyses among various European laboratories and also to determine possible sources of variation. In one investigation, each of 12 laboratories measured the concentrations of ibuprofen (ng/L) in two test samples of tap water. Table 4.3 shows the data from these tests as read from a graph. The Youden plot for this data (Figure 4.6) shows many points scattered near the 45°-line, indicating that several of the laboratories are following different versions of the chemical test procedure.

Table 4.3 Ibuprofen Concentrations (ng L^{-1})

Laboratory	Sample A	Sample B
1	29.36	33.33
2	30.11	41.09
3	42.74	42.46
4	46.09	45.20
5	46.46	46.11
6	49.81	48.85
7	60.96	55.24
8	66.53	40.63
9	67.65	47.02
10	113.36	105.46
11	172.09	172.11
12	199.97	193.11

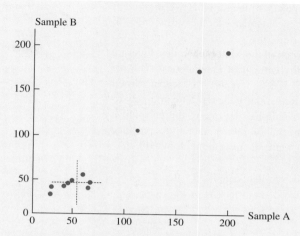

Figure 4.6 Youden plot of the data in Table 4.3

Section 4.4 Exercises

25. To estimate the accuracy and precision of an instrument that measures lengths, a .300-in. gauge block was used as a reference standard and was measured five times. The resulting measurements were: .301, .303, .299, .305, and .304. Calculate estimates of both the accuracy and the precision of the measuring instrument.

26. Calibration is the process of comparing an instrument's measurements to those of a reliable reference standard. If necessary, the instrument is adjusted to bring its measurements into agreement with the reference standard. Explain what effect calibration has on the estimated precision (not the accuracy) of a measuring instrument.

27. Many instrument makers report the accuracy of their instruments in terms of relative error as well as absolute error. The relative error in a measurement is defined as $(m - x)/x \cdot 100\%$, where m is the measured value and x is the true value. Absolute error is given by $|m - x|$.
    a. Calculate the relative errors for each of the five measurements in Exercise 25.
    b. Relative errors are often stated in terms of the maximum relative error to be expected for any

measurements within the range of an instrument. Suppose, for example, that a thermometer has a maximum relative error of ±4% over its operating range of −50°F to 150°F. What is the maximum *absolute* error you would expect in a measured reading of 70°F from this thermometer?

28. After carefully controlling all the chemical reagents and conditions during an experiment, a chemist weighs the amount of reactant produced by an experiment. The chemist weighs the reactant on an electronic balance, then reweighs the reactant five times, being careful to remove and replace the reactant on the balance between weighings.
    a. In the language of experimental design, can these six measurements be considered replications?
    b. What type of variation is measured by calculating the sample standard deviation, s, of the six measurements?

29. The melt flow index (MFI) of a polymer is defined to be the amount of the polymer (in grams) that can flow in 10 minutes through a standard die when subjected to a specified force and temperature. MFI is widely regarded as an important characteristic for

commercial polymer processing. However, there is a lack of reference standards for measuring MFI. To address this issue, the authors of "An Interlaboratory Comparison of the Melt Flow Index: Relevant Aspects for the Participant Laboratories" (*Polymer Testing*, 2007: 576–586) report MFI readings by 24 laboratories for various polypropylene and polystyrene polymer samples. For one of the polypropylene polymers, each of the 24 laboratories provided the following replicate measurements of MFI:

Laboratory	Replicate 1	Replicate 2
1	11.700	11.502
2	9.790	10.300
3	12.760	12.073
4	10.400	9.800
5	10.648	10.904
6	11.074	11.072
7	10.820	10.746
8	11.473	11.682
9	10.723	11.545

Laboratory	Replicate 1	Replicate 2
10	11.096	11.286
11	10.522	10.215
12	10.603	10.211
13	12.031	12.117
14	10.900	11.477
15	10.876	10.772
16	11.043	11.177
17	10.384	10.669
18	10.118	10.260
19	5.382	5.369
20	10.353	10.132
21	11.413	11.389
22	11.540	11.864
23	12.202	11.548
24	11.227	11.259

a. Create a Youden plot of this data.
b. What conclusions can you draw regarding the test procedures being used in these laboratories? Are there any unusual MFI measurements?

Supplementary Exercises

30. Consult a published reference, weather bureau, or Internet site to determine the operational definition used by weather forecasters when making statements like "There will be a 30% chance of rain tomorrow."

31. A common method for selecting a random sample *without replacement* from the integers 1, 2, 3, . . . , N is to generate a random sample *with replacement* (using random number tables or a software program) and then discard any duplicate numbers that appear in the sample. Use the sampling rules in Section 4.2 to justify why this procedure will produce a valid random sample *without replacement*.

32. The method of capture–recapture sampling is often used to estimate the size of wildlife populations (Thompson, S. K., *Sampling*, John Wiley & Sons, New York, 1992: 212–233). To illustrate the method, suppose an initial sample of 100 fish from a lake are caught and tagged. After releasing the fish and allowing sufficient time for them to mix with the rest of the fish in the lake, a second sample of, say, 50 fish are caught. The number of tagged fish in the second sample is counted.

a. Suppose there are five tagged fish found in the second sample. Because the samples are assumed to be random samples from the entire population of T fish, the proportion of tagged fish in the second sample should be approximately equal to the proportion of tagged fish in the population. Use this fact to estimate T, the total number of fish in the lake.

b. Generalize your result in part (a). That is, if x_{tag1} is the number of fish caught and tagged in the first sample and x_{tag2} is the number of tagged fish found in a second sample of size y, write an equation for the estimated value of T.

33. Cr(VI) is a pollutant associated with chromite ore processing. In a study of Cr(VI) concentrations, a

sampling plan was devised to estimate ambient levels of Cr(VI) in the air ["Background Air Concentrations of Cr(VI) in Hudson County, New Jersey: Implications for Setting Health-Based Standards for Cr(VI) in Soil," *J. of Air and Waste Management*, 1997: 592–597]. The authors propose using such background measurements as a basis for developing health-based standards for chromite ore processing plants.

a. In the study, background samples of air were selected to be representative of land use in the vicinity of chromite ore processing sites, but not so close that these samples would be affected by emissions from the processing plants. What role would such samples play in an experiment to subsequently evaluate emissions at chromite ore plants?

b. The authors used ASTM Standard Test Method D5281–92 when measuring the concentrations of Cr(VI). What experimental purpose does using such a standard serve?

c. Air samples were taken at two different locations, an industrial area and an undeveloped commercial site. Samples were collected at each site during six 24-hour sampling periods; wet and dry days were included. What general experimental design principles are illustrated here?

34. Youden plots are frequently used to compare two different instruments or evaluation methods. In a study of lawn mower exhaust emissions ("Exhaust Emissions from Four-Stroke Lawn Mower Engines," *J. of Air and Waste Management*, 1997: 945–950), two methods of measuring NO_x (nitrogen oxide) emission rates were compared by using both methods on

several models of gas-powered lawn mowers. The following table shows NO_x emission rates (grams/kWh) for two measuring methods: STC (similar to certification), which measures emissions for a 10-sec period, and an experimental method C6M, which is a weighted average of emission rates obtained under six different combinations of running speeds, times, and engine loads.

NO_x emission rate estimates

Lawn mower	STC	C6M
1	3.03	4.40
2	4.04	4.38
3	5.34	7.64
4	6.42	8.28
5	4.17	7.21
6	1.23	1.43
7	4.10	3.91
8	2.21	1.89
9	6.57	7.14
10	3.80	4.71
11	4.76	6.80
12	.49	.01
13	1.97	2.91
14	1.64	1.23
15	3.26	2.72
16	4.20	6.95
17	.32	.11
18	7.76	8.73
19	4.79	6.75
20	.98	1.12

a. Construct a Youden plot of this data.

b. Use the methods of Chapter 3 to fit a regression line to this data, with STC as y and C6M as x.

c. What conclusions can you draw from the results in parts (a) and (b) about the two NO_x measuring methods?

Bibliography

Box, G. E. P, W. G. Hunter, and J. S. Hunter, **Statistics for Experimenters** (2nd ed.), Wiley, New York, 2005. *Written for researchers. Emphasis is on explanation and application of experimental design techniques to real data and examples.*

Lohr, Sharon, *Sampling Design and Analysis* (2nd ed.), Duxbury, Belmont, CA, 2009. *A comprehensive survey of sampling.*

Mandel, N., and T. W. Lashof, "The Nature of Repeatability and Reproducibility," **J. of Quality Technology,** vol. 19, no. 1, 1987: 29–36. *A nice explanation and comparison of two concepts that are sometimes confused in practice.*

Thomas, G. G., **Engineering Metrology,** Wiley, New York, 1974. *Explains the various methods used to obtain measurements of different physical quantities.*

Probability and Sampling Distributions

5.1 CHANCE EXPERIMENTS

5.2 PROBABILITY CONCEPTS

5.3 CONDITIONAL PROBABILITY AND INDEPENDENCE

5.4 RANDOM VARIABLES

5.5 SAMPLING DISTRIBUTIONS

5.6 DESCRIBING SAMPLING DISTRIBUTIONS

INTRODUCTION

Chapter 5 marks a transition from purely descriptive methods to the inferential methods discussed in the remainder of this book. Beginning in this chapter, we will refer to any numerical measure calculated from sample data as a **statistic.** As you have seen in Chapters 1–3, statistics such as the sample mean, standard deviation, and correlation coefficient are useful tools for describing sets of data. Similarly, density and mass functions provide concise descriptions of populations and ongoing processes. One important question left unanswered in those chapters, however, is: How do we know what parameter values to use in a mass function or density function? For example, the Weibull density is commonly used for modeling the lifetimes of products, but how do you go about selecting *specific* numerical values for the Weibull parameters, α and β, that best describe the lifetimes of a particular product?

One way to answer such questions is to use **statistical inference,** a technique that converts the information from **random samples** (see Section 4.2) into reliable estimates of, and conclusions about, population or process parameters. Sections 5.5 and 5.6 illustrate how statistical inference works. When reading these sections, it is important to keep in mind the crucial role played by random sampling. *Without* random sampling, statistics can only provide descriptive summaries of the data itself. *With* random sampling, though, our conclusions can be reliably extended beyond the data, to the population or process from which the data arose. Figure 5.1 illustrates the difference between statistics based on ordinary data sets and statistics based on random samples.

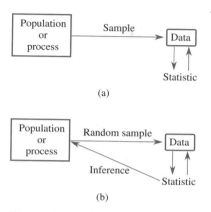

Figure 5.1 Statistical inference:
(a) descriptive statistics;
(b) inferential statistics

Drawing conclusions from samples necessarily involves some risk. Samples, after all, only give approximate pictures of populations or processes. Intuition tells us that the clarity of these pictures ought to increase as the sample size grows, but intuition fails to be more precise than that. For example, when testing a large shipment of parts for defective items, most people would agree that finding two defective items in a random sample of 10 is very different from finding 200 defectives in a random sample of 1000. Although the sample percentage (i.e., the statistic calculated from the data) is the same in both cases, the 20% defect rate in the larger sample seems much more credible than the 20% defect rate in the smaller sample. To quantify just how much more credible the information in the larger sample is, we use the tools of **probability.** Probability methods, discussed in Sections 5.1–5.4, provide the basis for measuring the amount of confidence or reliability in a statistic.

5.1   CHANCE EXPERIMENTS

The term *chance experiment* may sound self-contradictory to an engineer or a scientist. What could possibly be random or uncertain about a carefully planned scientific investigation? The answer, of course, lies in our definition of the term. A **chance experiment,** also called a **random experiment,** is simply an activity or situation whose outcomes, to some degree, depend on chance. To decide whether a given activity qualifies as a chance experiment, ask yourself the question, Will I get exactly the same result if I repeat the experiment more than once? If the answer is "no," then the experiment qualifies as a chance experiment. Under this rather wide definition, determining whether a metal part withstands a stress test, recording whether it rains tomorrow, measuring the yield of a chemical reaction, assessing the potency of a pharmaceutical product, or measuring the volume of water flowing in a drainage system all qualify as chance experiments.

Most chance experiments in the sciences arise because either (1) some natural phenomenon is at work, causing unpredictable changes in experimental outcomes, or (2) we purposely *introduce* randomness as a tool for extrapolating information from data to conclusions about populations or processes (see Section 4.3). As an example of the former, yields of chemical reactions often vary with each repetition of an experiment, no matter how hard one tries to control the conditions of the experiment. Slight differences in handling (e.g., the amount of mixing, the ambient temperature, the elapsed time of the reaction) or even in the behavior at the molecular level (e.g., Brownian motion, material flow) can induce small changes in experimental results. However chance experiments may arise, from natural forces or by statistical methodology, probability provides a structure for measuring and consistently handling uncertainty.

Events

Underlying the computations of probability is an organized system for describing and working with the outcomes of chance experiments. These outcomes can be divided into two types: (1) **simple events,** which are the individual outcomes of an experiment and, more generally, (2) **events,** which consist of collections of simple events. For instance, the chance experiment of conducting a series of stress tests on three metal parts has the eight possible outcomes *PPP, PPF, PFP, FPP, PFF, FPF, FFP,* and *FFF,* where *P* and *F* denote the test results "pass" and "fail," and the order in which the letters appear corresponds to the part number tested (e.g., *PPF* indicates that the first two parts passed the test, but the third part failed). Each of these eight outcomes is a simple event, which, taken together, form the **sample space** of the experiment.

Events are often denoted by single uppercase letters, usually from the beginning of the alphabet, much like we denote constants in formulas by lowercase letters. Single-letter names for events are very useful when applying the probability formulas in Section 5.2. Thus we might denote the event that at least two parts pass the stress test by *A*, the event that exactly 1 part passes the stress test by *B*, and so forth. Events can also be described by just listing, in brackets, the simple events that comprise them. For example, the event that at least two parts pass the stress test corresponds to the set of outcomes {*PPP, PPF, PFP, FPP*}. If we had also chosen to denote this event by the letter *A*, then we could also write *A* = {*PPP, PPF, PFP, FPP*}.

Example 5.1   Let's continue with our example of stress-testing metal parts. Suppose that we now select and test four parts. Using sequences of *P*s (for parts that pass the test) and *F*s (for parts that fail the test), the sample space of the experiment of selecting and testing four metal parts is somewhat larger than that of the experiment of selecting and testing three metal parts, discussed previously. In particular, the sample space consists of these 16 simple events: {*PPPP, PPPF, PPFP, PFPP, FPPP, PPFF, PFPF, PFFP, FPPF, FPFP, FFPP, PFFF, FPFF, FFPF, FFFP, FFFF*}. For convenience, these events are listed in order of decreasing numbers of *P*s in each four-letter sequence.

Suppose we are interested in the events A = at least two parts pass the stress test and B = at most two parts pass the stress test. In terms of simple events, we can write A and B as

$$A = \{PPPP, PPPF, PPFP, PFPP, FPPP, \underline{PPFF}, \underline{PFPF}, \underline{PFFP}, \underline{FPPF}, \underline{FPFP}, \underline{FFPP}\}$$
$$B = \{\underline{PPFF}, \underline{PFPF}, \underline{PFFP}, \underline{FPPF}, \underline{FPFP}, \underline{FFPP}, PFFF, FPFF, FFPF, FFFP, FFFF\}$$

Note that A and B have several simple events in common (shown underlined).

Example 5.2 A reasonably large percentage of C++ programs written at a particular company compile on the first run, but some do not (a compiler is a program that translates source code—in this case, C++ programs—into machine language so programs can be executed). Suppose an experiment consists of selecting and compiling C++ programs at this location one by one until encountering a program that compiles on the first run. Denote a program that compiles on the first run by S (for success) and one that does not by F (for failure). Although it may not be very likely, a possible outcome of this experiment is that the first 5 (or 10 or 20 or . . .) are F's, and the next one is an S. In other words, for any positive integer n, we may have to examine n programs before seeing the first S. The sample space is $\{S, FS, FFS, FFFS, . . .\}$, which contains an infinite number of possible outcomes. The same abbreviated form of the sample space is appropriate for an experiment in which, starting at a specified time, the gender of each newborn infant is recorded until the birth of a male is observed.

Depicting Events

Various devices have been created to help visually describe the events in a sample space. **Tree diagrams** are especially useful for depicting experiments that are conducted in a sequence of steps, such as our example of testing three metal parts. Beginning at the left, each step in the sequence is given its own set of branches, which themselves form the starting points for all branches to their right. Figure 5.2 shows a tree diagram for the experiment of selecting and testing three metal parts. Simple events are formed by following any branch of the tree diagram from the leftmost point to one of the rightmost points.

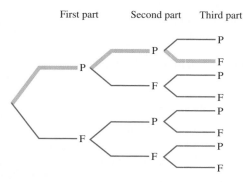

Figure 5.2 Tree diagram for the experiment of selecting and testing three metal parts (branches forming the simple event PPF are shown shaded)

Another visual device, the **Venn diagram,** is especially useful for depicting *relationships* between events. Venn diagrams are simple two-dimensional figures, often rectangles or circles, whose enclosed regions are intended to depict a collection of simple events, called *points*, in a sample space. Figure 5.3 shows a Venn diagram of several events based on Example 5.1. Events like A and B that contain points in common are depicted as overlapping regions in the diagram. Events that do not contain any common points, such as the events B = at most two parts pass the test and C = exactly three parts pass the test, are shown as nonoverlapping regions. An event that contains all the points of some other event is shown as surrounding the smaller event. For example, the event A = at least two parts pass the test contains all of the simple events in event C = exactly three parts pass the test, so C is shown inside of A in Figure 5.3.

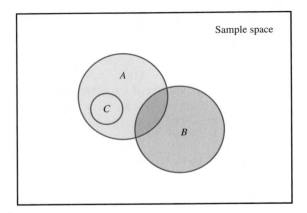

Figure 5.3 Venn diagram of the events A and B
in Example 5.1

Venn diagrams and tree diagrams are indispensable tools in many parts of probability theory, but they are not essential to conducting statistical studies. We will use these diagrams primarily as an aid for discussing certain probability concepts, but, beyond that, their use is not emphasized. The interested reader may consult texts on probability for more information on working with Venn diagrams.

Forming New Events

Simple events are fundamental to describing chance experiments, but the events that are of most interest are usually much more complex. Indeed, it is not an exaggeration to state that the majority of probability calculations involve techniques for decomposing complex events into simpler ones. One of the primary methods for creating complex events and, therefore, for unraveling them, involves the use of the words *and, or,* and *not.* The following box shows how these words are used to build new events from old ones.

DEFINITIONS

For a chance experiment and any two events A and B:

1. The event **A *or* B** consists of all simple events that are contained in either A or B. **A *or* B** can also be described as the event that *at least one* of A or B occurs.

2. The event **A *and* B** consists of all simple events common to both A and B. **A *and* B** can be described as the event that *both* A and B occur.

3. The event **A′**, called the **complement of A**, consists of all simple events that are *not* contained in A. A' is the event that A does not occur.

Example 5.3

Refer to Example 5.1, the experiment of selecting and testing four metal parts. To form the event A *or* B, we simply list all events that are in either A or B, or in both. The easiest way to do this is to list all the events in A and then add the events in B that are not duplicates of those in A. Thus

$$A \text{ or } B = \{PPPP, PPPF, PPFP, PFPP, FPPP, \underline{PPFF}, \underline{PFPF}, \underline{PFFP}, \underline{FPPF},$$
$$\underline{FPFP}, \underline{FFPP}, PFFF, FPFF, FFPF, FFFP, FFFF\}$$

For these two events, A *or* B happens to contain *all* 16 sample space points. In a similar fashion, the event A *and* B, which consists only of the underlined events in both A and B, is given by

$$A \text{ and } B = \{\underline{PPFF}, \underline{PFPF}, \underline{PFFP}, \underline{FPPF}, \underline{FPFP}, \underline{FFPP}\}$$

In this case, it is possible to give a short verbal description of the event A *and* B; namely, A *and* B = exactly two parts pass (and, hence, two fail) the stress test. Finally, the complement of event A is

$$A' = \{PFFF, FPFF, FFPF, FFFP, FFFF\}$$

A' can also be verbally described as the event that at most one part passes the test.

When two events A and B have no simple events in common, we say that they are **mutually exclusive** or **disjoint.** More intuitively, mutually exclusive events are ones that cannot occur simultaneously; the occurrence of either event precludes the occurrence of the other. In a Venn diagram, mutually exclusive events are depicted as nonoverlapping regions. As we will see in Section 5.2, probability calculations involving disjoint events are particularly simple. For this reason, we often try to decompose complex events into collections of mutually exclusive events when computing probabilities.

Several of the previous definitions can be extended to include events formed from more than two events. These definitions are given in the next box.

DEFINITIONS

Given a chance experiment and any events $A_1, A_2, A_3, \ldots, A_k$:

1. The event A_1 *or* A_2 *or* A_3 *or* ... *or* A_k consists of all the simple events that are contained in at least one of the events $A_1, A_2, A_3, \ldots,$ or A_k. It can also be described as the event that *at least one* of the events $A_1, A_2, A_3, \ldots,$ or A_k occurs.

2. The event A_1 *and* A_2 *and* A_3 *and* ... *and* A_k consists of all simple events common to *all* the events $A_1, A_2, A_3, \ldots,$ and A_k. This event can be described as the event that *all* of the events $A_1, A_2, A_3, \ldots,$ and A_k occur.

3. Several events $A_1, A_2, A_3, \ldots,$ and A_k are said to be **mutually exclusive** or **disjoint** if no two of them have any simple events in common.

Example 5.4

Sampling inspection is a common method for ascertaining the quality level of batches (called *lots*) of finished products. Sampling inspection can be used by a manufacturer to check the quality of products prior to shipment or by a customer to check the quality of incoming shipments before accepting them. In either case, sampling inspection is done by first selecting a random sample of n items from a lot and counting the number of sampled items that do not meet quality standards.

Suppose, for example, that $n = 20$ items are randomly selected from a large lot. In this situation, an event that we might be interested in is A = the sample contains at most one item that fails to meet quality standards. As you can imagine from reading the other examples in this section, the sample space of the experiment of randomly selecting and testing 20 items is prohibitively large. Even a tree diagram is of no help in depicting the simple events or the event A itself. However, relying on only verbal descriptions of the events, it is possible to decompose A into a combination of two less complex events: B = no items fail inspection and C = exactly one item fails inspection. In fact, it is not hard to see that the event B *or* C is the same as the event A. We write this as $A = B$ *or* C. Furthermore, B and C are mutually exclusive events. In Section 5.2, we show how to use this fact to more easily compute the probability that A occurs.

Section 5.1 Exercises

1. A random sample, without replacement, of three items is to be selected from a population of five items (labeled a, b, c, d, and e).
   a. List all possible different samples.
   b. List the samples that correspond to the event A = items a and c are included in the sample.
   c. List the samples that correspond to the complement of the event A in part (b).

2. An engineering firm is constructing power plants at three different sites. Define the events E_1, E_2, and E_3 as follows:

E_1 = the plant at site 1 is completed by the contract date

E_2 = the plant at site 2 is completed by the contract date

E_3 = the plant at site 3 is completed by the contract date

Draw a Venn diagram that depicts these three events as intersecting circles. Shade the region on the Venn diagram corresponding to each of the following events (redraw the Venn diagram for each question):

a. At least one plant is completed by the contract date.

b. All plants are completed by the contract date.

c. None of the plants is completed by the contract date.

d. Only the plant at site 1 is completed by the contract date.

e. Exactly one of the three plants is completed by the contract date.

f. Either the plant at site 1 or site 2 or both of the two plants are completed by the contract date.

3. Let A and B denote the events A = there are more than three defective items in a random sample of ten items and B = there are fewer than six defectives in a random sample of ten items.

a. Describe, in words, the event A *and* B.

b. Describe, in words, the event A *or* B.

c. Describe, in words, the complement of A.

4. Draw a Venn diagram depicting two events A and B that are not disjoint. Shade in the portion of this diagram that corresponds to the event A *and* B'.

5. Nuts and bolts used in aircraft manufacturing are called fasteners. To ensure that they are not loosened by vibrations during flight, some fasteners are slightly crimped so that they lock more tightly. The amount of crimping, however, must meet specific standards. To test finished fasteners, an initial inspection classifies them into two groups: those that meet standards and those that do not. Of those not meeting standards, some are completely defective and must be scrapped, whereas the rest can be run through a machine that readjusts the amount of crimping. Of the recrimped fasteners, some are corrected by the recrimping operation and pass inspection, whereas the remainder cannot be salvaged and are scrapped. Draw a tree diagram that depicts the testing and rework operations.

6. Information theory is concerned with the transmission of data, usually encoded as a stream of 0s and 1s, over communication channels. Because channels are "noisy," there is a chance that some 0s sent through the channel are mistakenly received at the other end as 1s, and vice versa. The majority of digits sent, however, are not altered by the channel. Draw a tree diagram that depicts the type of bit sent (either 0 or 1) and the type of bit received at the end of the channel.

7. Use a Venn diagram to find a simple expression for $\{A \text{ } and \text{ } B\}'$ in terms of A' and B'.

5.2   PROBABILITY CONCEPTS

Probability allows us to *quantify* the likelihood associated with uncertain events, that is, events that result from chance experiments. Generally speaking, the probability of an event can be thought of as the proportion of times that the event is expected to occur in the long run. This definition works well for experiments that can be repeated many times, such as in testing a large number of electronic components. After testing enough components, we begin to get a good idea of the chance (i.e., probability) that the next item tested will be defective or nondefective.

Probabilities are reported either as *proportions* (between 0 and 1) or as *percentages* (between 0% and 100%). To simplify computations with probabilities, the shorthand notation $P(A)$ is used to denote the probability of an event A occurring. Thus the statements $P(A) = .30$, the probability of event A occurring is .30, and the event A has a 30% chance of occurring are equivalent. As a general rule, it is best to write probabilities as proportions when performing probability calculations, converting to percentages only when it helps to interpret a probability statement.

Assigning Probabilities

Writing in his treatise *Théorie Analytique des Probabilités* (1812), mathematician and theoretical astronomer Pierre Simon de Laplace (1749–1827) stated that "at bottom, the theory of probability is only common sense reduced to calculation." With this brief statement, Laplace recognized that any rigorous definition of probability must satisfy certain commonsense requirements. For example, the probability of any event must lie between 0 and 1. This is another way of stating the obvious condition that, in any number of repetitions of an experiment, no event can occur less than 0% of the time nor more frequently than 100% of the time. In practice, this requirement provides a quick check on our probability calculations; calculated values that lie outside the interval [0, 1] are immediate signals that a mistake has occurred somewhere in the computations. Used correctly, the probability formulas given in this chapter will never yield probabilities outside the interval [0, 1].

A second self-evident requirement is that probabilities of events must not lead to logical inconsistencies. For example, it does not make sense to state that 90% of metal parts pass a stress test *and* that 20% fail the test. These two probabilities are inconsistent because we know that exactly 100%, not 110%, of the parts will either pass or fail the test. In the same vein, it would not make sense to say that 90% pass and 5% fail the test, since this implies the illogical conclusion that only 95% of all parts pass or fail the test. To avoid nonsensical statements like these, we demand that the probabilities associated with the simple events always total to exactly 1. Thus any sensible assignment of probabilities to events must satisfy the following two basic requirements:

Probability Axioms

1. The probability of any event must lie between 0 and 1. That is, $0 \leq P(A) \leq 1$ for any event A.

2. The total probability assigned to the sample space of an experiment must be 1.

Within the limits imposed by these axioms, there are several ways to determine probabilities: (1) as frequencies of occurrence, (2) from subjective estimates, (3) by assuming that events are equally likely, and (4) by using density and mass functions (see Section 5.4). Depending on the circumstances, each method has its merits. For example, when it is possible to repeat a chance experiment, the "frequentist" approach defines the probability of an event A to be the long-run ratio

$$P(A) = \frac{\text{number of times } A \text{ occurs}}{\text{numbers of times experiment is repeated}}$$

The justification for this approach is that, as the number of trials increases, we expect this ratio to stabilize and eventually approach a limiting value, which we take as our definition of $P(A)$. For example, let A be the event that a package sent within the state of California for 2nd-day delivery actually arrives within 1 day. The results from sending 10 such packages (the first 10 replications) are as follows:

Package No.	1	2	3	4	5	6	7	8	9	10
Did A occur	N	Y	Y	Y	N	N	Y	Y	N	N
Relative frequency of A	0	.5	.667	.75	.6	.5	.571	.625	.556	.5

Figure 5.4(a) shows how the relative frequency fluctuates rather substantially over the course of the first 50 replications. But as the number of replications continues to increase, Figure 5.4(b) illustrates how the relative frequency stabilizes. Using Figure 5.4(b), we would be inclined to state that $P(A)$ is close to .60.

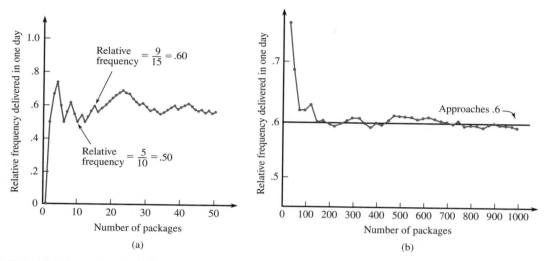

Figure 5.4 Behavior of relative frequency: (a) initial fluctuation; (b) long-run stabilization

Of course, the frequentist approach does not work when experiments cannot be faithfully replicated, as is the case with sports competitions. In these instances, subjective estimates, guided by the probability axioms, can be used to arrive at numerical probabilities that certain teams will win or lose a game. Needless to say, entire texts can and have been written comparing the various methods for assigning probabilities to events. It is not our purpose to compare each of these methods. Instead, in Section 5.4, we emphasize the technique that is most often used in statistical studies, defining probabilities by means of mass and density functions.

The Addition Rule for Disjoint Events

Probability rules, or laws, are formulas that are intended to simplify the process of calculating the probabilities of complex events. They achieve this purpose by first decomposing some event of interest into two or more less complex events whose probabilities are more easily found. The formulas then describe how to recombine the simpler probabilities to find the probability of the original event. One of the most frequently used laws is the addition rule for disjoint events, which states that the probability of the event A_1 or A_2 or A_3 or ... or A_k is simply the sum of the individual probabilities $P(A_1) + P(A_2) + P(A_3) + \cdots + P(A_k)$ *as long as all the events* $A_1, A_2, A_3, \ldots,$ *and* A_k *are mutually exclusive.* The addition rule is usually applied to an event E by first finding a collection of less complicated events $A_1, A_2, A_3, \ldots,$ and A_k that satisfy two conditions: (1) the events $A_1, A_2, A_3, \ldots,$ and A_k are disjoint and (2) $E = A_1$ or A_2 or A_3 or ... or A_k. The events $A_1, A_2, A_3, \ldots, A_k$ are sometimes said to **partition** the event E into mutually exclusive events.

The Addition Rule for Disjoint Events

Disjoint, or **mutually exclusive, events** are events that cannot occur simultaneously. For any two disjoint events A and B,

$$P(A \text{ or } B) = P(A) + P(B)$$

More generally, for any collection of disjoint events $A_1, A_2, A_3, \ldots, A_k$,

$$P(A_1 \text{ or } A_2 \text{ or } A_3 \text{ or } \ldots \text{ or } A_k) = P(A_1) + P(A_2) + P(A_3) + \cdots + P(A_k)$$

Example 5.5   Suppose that you want to find the probability that at most one item fails to meet quality standards in a random sample of $n = 20$ items from a large shipment of such items. Denote the event of interest as A = at most one item fails to meet a quality standard. In Example 5.4, we showed that A can be partitioned into the events B = no items fail inspection and C = exactly one item fails inspection. That is, we can write A = B or C, where B and C are disjoint events. According to the addition rule for disjoint events, $P(A)$ can be found by simply adding the probabilities $P(B)$ and $P(C)$, both of which are easier to find than $P(A)$. In fact, in Section 5.4 we show that the binomial mass function can be used to find both $P(B)$ and $P(C)$.

Complementary Events

The complement A' of an event A was defined in Section 5.1 to be the collection of simple events that are *not* in A. In more intuitive terms, it is helpful to think of A' as the *opposite* of A when trying to express A' in words. For example, if A is the event that at least one metal part passes a stress test, then the opposite event must be A' = no metal parts pass the stress test. Notice that we did not need to write down the sample space of the experiment to arrive at this description of A'. Consider how you might describe the complement of A'. Since A and A' are opposites, then the complement of A' is simply the event A itself, which we can write as $(A')' = A$.

Yet another way to describe the complement of an event A is to say that when A does *not* occur, then, necessarily, its complement A' *has* occurred. Viewed this way, the symbol A is somewhat like a switch that is either on (A) or off (A'). The truth-table logic you would use to describe electronic circuits can then be applied to finding complements of complex events. For instance, consider how you might go about finding the complement of the event A or B. If the event A or B does not happen, then it must be true that *both* A and B do not happen, which we can express by writing A' *and* B'. In equation form, {A or B}′ = A' *and* B'. Figure 5.5 shows how a tree diagram can be used to demonstrate the same result. The branches of the tree depict all possible combinations of the events A, B, A', and B'. The top three branches correspond to the event A or B, which implies that its complement must be the bottom branch, A' *and* B'.

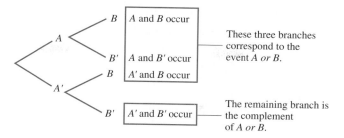

Figure 5.5 Finding the complement of the event A or B

Because an event A and its complement A′ cannot occur simultaneously, complementary events are special cases of mutually exclusive events. That is, A and A′ are disjoint. Furthermore, because we are 100% certain that exactly *one* of these events will occur, P(A or A′) = 1. Applying the formula for mutually exclusive events yields

$$1 = P(A \ or \ A') = P(A) + P(A')$$

which is called the *law of complementary events* and is usually written in the form

$$P(A) = 1 - P(A')$$

The usefulness of this simple formula lies in the fact that it is sometimes easier to find the probability of the complement A′ rather than the probability of A itself.

DEFINITION

When an event A does *not* occur, we say that its **complement**, denoted by A′, has occurred, and vice versa. The probabilities of A and A′ are related by the formula P(A) = 1 − P(A′).

Example 5.6

Refer to Example 5.5. Suppose you want to find the probability that, of the 20 items randomly selected for inspection, at least one item fails to meet quality standards. Denote this event by D = at least one item fails inspection. One approach to finding this probability is to partition D into the events $E_1, E_2, E_3, \ldots, E_{20}$, where, for each i = 1, 2, 3, . . . , 20, the E_i denotes the event that exactly i items fail inspection. Since E_1 through E_{20} are disjoint, the addition rule says that $P(D) = P(E_1) + P(E_2) + \cdots + P(E_{20})$. As mentioned in Example 5.5, the binomial mass function could then be used to find each $P(E_i)$ in this summation.

Although the addition rule will give the correct value for $P(D)$, an easier method for finding $P(D)$ is to use the law of complementary events, $P(D) = 1 - P(D')$. The complement of the event D = at least one item fails inspection is the event D' = no items fail inspection. As we will see in Section 5.4, finding $P(D')$ requires only one computation with the binomial mass function, whereas the partition method requires 20 separate computations.

The General Addition Rule

As we have seen in this section, finding the probability of an event E can be simplified considerably if it is possible to first express E in the form $E = A\ or\ B$, or more generally, in the form $E = A_1\ or\ A_2\ or\ \ldots\ or\ A_k$, where the events $A_1, A_2, \ldots, A_k$ are mutually exclusive. There are times, however, when it is not so easy to break up an event E into disjoint events. In such cases it is helpful to have another method for finding the probability of E.

The **general addition rule** is used to find the probability of an event E that can be written in the form $E = A\ or\ B$, where events A and B are not necessarily disjoint. When an event is expressed in the form $A\ or\ B$, its probability can be calculated from the following formula:

$$P(A\ or\ B) = P(A) + P(B) - P(A\ and\ B)$$

which is called the *general addition rule*. This formula can be applied to *any* two events A and B.

Although it is indeed more generally applicable than the addition rule for disjoint events, notice that the general addition rule presupposes that you are able to find the probability of the event A *and* B, which can often be just as difficult as finding $P(A\ or\ B)$. However, as you will see in Section 5.3, when A and B satisfy certain additional conditions, it is relatively easy to find $P(A\ and\ B)$.

The General Addition Rule

For any two events A and B, which need not be mutually exclusive,

$$P(A\ or\ B) = P(A) + P(B) - P(A\ and\ B)$$

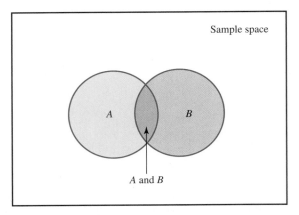

Figure 5.6 Venn diagram of A or B

Here is a simple intuitive justification for the general addition rule. Referring to the Venn diagram in Figure 5.6, imagine that the events A and B represent circular rugs on a floor and that we want to find the total floor area covered by these two rugs, analogous to determining $P(A\ or\ B)$. For the purposes of this example, think of $P(A)$ and $P(B)$ representing the floor areas covered by each rug individually. To find the total area covered by both rugs we could start by adding the areas of these two rugs, but then the floor area where the two rugs overlap has been counted twice by this simple addition. The obvious remedy is to subtract the overlapping area, represented by $P(A \text{ and } B)$, once from the sum, giving a final result of $P(A) + P(B) - P(A \text{ and } B)$. This is in essence how the general addition rule works.

Example 5.7
In a certain residential suburb, 60% of all households get Internet service from the local cable company, 80% get television service from that company, and 50% get both services from that company. If a household is randomly selected, what is the probability that it gets at least one of these two services from the company? With $A =$ {gets Internet service} and $B =$ {gets TV service}, the given information implies that $P(A) = .6$, $P(B) = .8$ and $P(A \text{ and } B) = .5$. The general addition rule now yields

$P(\text{subscribes to at least one of the two services})$
$$= P(A \text{ or } B) = P(A) + P(B) - P(A \text{ and } B) = .6 + .8 - .5 = .9$$

Section 5.2 Exercises

8. Two methods are proposed for testing a shipment of five items (call them A, B, C, D, and E). In method 1, an inspector randomly samples two of the five items and tests to see whether either item is defective. In method 2, an inspector randomly samples one item and tests it; the remaining four items are sent to a second inspector who randomly samples one item and tests it. Suppose that only item A is defective in the shipment.

   a. What is the probability that item A will be discovered by method 1?
   b. What is the probability that item A will be discovered by method 2?
   c. What general statement can you make regarding the effectiveness of the two methods? Can your statement be extended to methods involving samples of more than two items?

9. Human visual inspection of solder joints on printed circuit boards can be very subjective. Part of the problem stems from the numerous types of solder defects (e.g., pad nonwetting, knee visibility, voids) and even the *degree* to which a joint possesses one or more of these defects. Consequently, even highly trained inspectors can disagree when examining the same circuit board. The accompanying table shows the results of two inspectors who examined

the same collection of 10,000 solder joints for a particular problem:

	Number of defective solder joints found
Inspector A	724
Inspector B	751
Common to both inspectors	316

   a. How many defective solder joints were found by the two inspectors?
   b. How many defective solder joints found by inspector A were *not* found by inspector B?

10. For any collection of events $A_1, A_2, A_3, \ldots, A_k$, it can be shown that the inequality

$$P(A_1 \text{ and } A_2 \text{ and } A_3 \text{ and } \ldots \text{ and } A_k)$$
$$\geq 1 - [P(A_1') + P(A_2') + P(A_3') + \cdots + P(A_k')]$$

always holds. This inequality is particularly useful when each of the events has relatively high probability. Suppose, for example, that a system consists of ten components connected in series (cf. Example 5.8, Section 5.3) and that each component has a .999 probability of functioning without failure. What lower bound can you put on the reliability (i.e., the probability of functioning correctly) of the system built from these ten components?

11. For any collection of events A_1, A_2, A_3, . . . , A_k, it can be shown that the inequality

$$P(A_1 \text{ or } A_2 \text{ or } A_3 \text{ or } \ldots \text{ or } A_k)$$
$$\leq P(A_1) + P(A_2) + P(A_3) + \cdots + P(A_k)$$

always holds. This inequality is most useful in cases where the events involved have relatively small probabilities. For example, suppose a system consists of five subcomponents connected in series (cf. Example 5.8) and that each component has a .01 probability of failing. Find an upper bound on the probability that the entire system fails.

12. Suppose that 55% of all adults regularly consume coffee, 45% regularly consume carbonated soda, and 70% regularly consume at least one of these two types of drinks.
    a. What is the probability that a randomly selected adult regularly consumes both coffee and soda?
    b. What is the probability that a randomly selected adult doesn't regularly consume at least one of these two products?
    c. What is the probability that a randomly selected adult regularly consumes coffee but does not regularly consume soda?

5.3   CONDITIONAL PROBABILITY AND INDEPENDENCE

Conducting experimental studies is an iterative process. An initial guess or hypothesis is compared with experimental data, new hypotheses are formed, more data is gathered, and the process repeats itself until we are satisfied with the knowledge gained from experimentation. The process of adjusting our view of the world as more information is gathered can also be applied to calculating probabilities. In this context, we ask how the knowledge that a certain event B has occurred can be used to update our initial assessment of the probability that another event A will occur. Sometimes, the probability that A occurs depends heavily on whether B has occurred. In such cases, we use the methods of **conditional probability.** At other times, when the occurrence or nonoccurrence of B has no effect at all on the probability that A occurs, we say that A and B are **independent events.**

From the standpoint of probability calculations, independent events are especially easy to work with. This is one of the primary reasons that statistical methods usually incorporate some sort of random procedure, such as random sampling or randomization, as a method for ensuring that certain events will be independent.

Conditional Probability

Before shipping finished products, manufacturers routinely use automatic test equipment (ATE) to assess the functionality of products and systems. In addition to giving physical measurements of product characteristics, ATE machines can conduct a sequence of complex tests that eventually result in a final "thumbs up" or "thumbs down" determination for the item being tested. Before testing, historical process data can be used to estimate the probability that any particular item will function correctly. Suppose, for example, that such records show that 95% of the items in a certain product line perform correctly. Letting A denote the event that a randomly selected item is defect free, we can then say that $P(A) = .95$. Now consider how this estimate may change when we submit a particular item to an ATE test. Because the determinations given by ATE are good but not perfect, we will want to give a good deal of weight, but not 100%, to the ATE test result. Thus if the ATE test indicates that the item is defective, then we will definitely want to reduce our estimate of $P(A)$. Alternatively, if the item passes the ATE test, then we will revise $P(A)$ upward. In both cases, we want to update our estimate of $P(A)$ for the item being tested by factoring in the new information from the ATE test.

Let B = the item passes the ATE test. Then the **conditional probability of A given B** is denoted by **P(A|B).** Conditional probabilities are computed from the following definition:

$$P(A|B) = \frac{P(A\ and\ B)}{P(B)}$$

This formula can be justified by thinking of probability as the proportion of times that an event occurs in a large number of trials N: About $P(B) \times N$ of the trials will result in items that pass the ATE test and about $P(A\ and\ B) \times N$ of the trials will correspond to items that not only pass the test but are truly defect-free. Thus $P(A|B)$, the proportion of items that are truly defect-free out of the total number passing the ATE test, should be $P(A\ and\ B) \times N/(P(B) \times N)$, which simplifies to $P(A\ and\ B)/P(B)$.

Tree diagrams are very useful for summarizing problems that involve conditional probabilities. Figure 5.7 shows such a diagram for our ATE example. Note that conditional probabilities correspond to the branches on the tree. By writing the formula $P(A|B) = P(A\ and\ B)/P(B)$ in the form $P(A\ and\ B) = P(B)P(A|B)$, we see that the probability of taking a particular path through the diagram (from left to right) is simply the product of the probabilities of the branches that comprise that path.

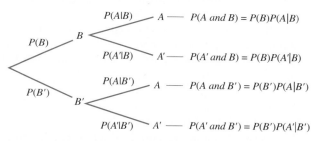

Figure 5.7 Tree diagram for depicting probabilities

DEFINITION

Let A and B be two events with $P(B) > 0$. The **conditional probability of A occurring given that event B has already occurred** is denoted by **P(A|B)** and can be calculated from the formula $P(A\ |\ B) = P(A\ and\ B)/P(B)$.

Independent Events

Conditional probability is used when the likelihood of occurrence of an event depends on whether or not another event occurs. At the other end of the spectrum are events that do not impose such restrictions on each other's chances of occurring. Two events, A and B, are said to be **independent** if the occurrence of either event has no effect whatsoever on the likelihood of occurrence of the other. This definition readily extends to any number of events.

To understand the role played by independence in probability calculations, consider the following example. To filter certain harmful particles out of a given volume of air, suppose we sequentially use two filters A and B, each of which captures a large percentage of the particles in any air passing through it. In particular, filter A allows only 5% of the particles to pass through, whereas filter B has about a 10% pass-through rate. If we begin with a fixed volume of air containing V harmful particles, then after passing through filter A, there should be $(.05)V$ particles remaining. When this air is screened through filter B, an additional 90% of the remaining particles are removed, leaving a total of $(.10)(.05)V$ particles after the two screenings. We then ask, Would it make any difference if we changed the order in which the filtering is performed? This is equivalent to asking, Do the two filters perform independently of one another? If the filters are *independent*, then we should be able to reverse the filtering procedure without changing the pass-through rates of the filters (see Figure 5.8). Thus filter B leaves $(.10)$ V particles, of which filter A then leaves $(.05)(.10)V$ particles.

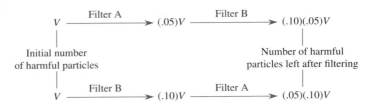

Figure 5.8 Two air filters acting independently

If we think of the pass-through rates as probabilities of the events A = filter A lets through a harmful particle and B = filter B lets through a harmful particle, then independence allows us to conclude that the proportion of particles left after applying both filters is

$$P(A \text{ and } B) = P(A)P(B) = (.05)(.10)$$

and that the order in which we apply the filters does not affect the probability of the events A and B. Independent events, then, allow us to reduce the calculation of $P(A \text{ and } B)$ to a simple multiplication. Furthermore, it is easy to see from our filter example that this multiplication formula can be extended to any number of independent filters $A_1, A_2, A_3, \ldots, A_k$, whose overall pass-through rate should be

$$P(A_1 \text{ and } A_2 \text{ and } A_3 \text{ and } \ldots \text{ and } A_k) = P(A_1)P(A_2)P(A_3) \cdots P(A_k)$$

and that the order in which the filters are applied should not affect the final probability.

DEFINITION

> Two events, A and B, are **independent events** if the probability that either one occurs is not affected by the occurrence of the other. In this case,
>
> $$P(A \text{ and } B) = P(A)P(B)$$
>
> Several events, $A_1, A_2, A_3, \ldots, A_k$, are **independent** if the probability of each event is unaltered by the occurrence of any subset of the remaining events. In this case, the product rule can be applied to any subset of the k events. That is, the probability that all the events in any *subset* occur equals the product of their individual probabilities of occurring. In particular, for all k events,
>
> $$P(A_1 \text{ and } A_2 \text{ and } A_3 \text{ and } \cdots \text{ and } A_k) = P(A_1)P(A_2)P(A_3) \cdots P(A_k)$$

Determining whether two (or more) events are independent is not quite as easy as deciding whether they are mutually exclusive. With independent events, we rely either on our intuition or on special procedures (such as random sampling and randomization). Intuition is what we generally employ when we assume that different tosses of a coin or different air filters are independent. With statistical methods, on the other hand, we rely on random sampling, not intuition, to ensure that events are independent. Practically speaking, we often assume independence when we do not know of any strong reasons why the events *should* be related. At other times, independence provides a reasonable approximation to the truth for the application at hand, but it may not be reasonable if the situation changes a little. In our filter example, for instance, independence may be a good assumption when the volume of particulate matter in the air is relatively large, but it may cease to be valid for small volumes (e.g., after being screened by one filter, the volume of particles may have dropped below the detection limit of the other filter).

Example 5.8   One branch of reliability theory, called *topological reliability*, is concerned with calculating the reliability of systems comprising several components connected in specific patterns. One common layout for components is the **series system** (Figure 5.9), in which the system operates correctly only if *each* of its subcomponents works correctly. A familiar example of such a system is a circuit with two switches, both of which must be closed for the circuit to conduct electricity. It is commonly assumed that the components are independent when performing reliability calculations.

Figure 5.9 A two-component series system, which functions correctly only if *both* components function correctly

Suppose that the switches A and B in a two-component series system are closed about 60% and 80% of the time, respectively. If we assume that the closing of switch A occurs independently of switch B, the probability that the entire circuit is closed is

$$P(\text{circuit closed}) = P(A \text{ closed } and \text{ B closed})$$
$$= P(A \text{ closed}) \, P(B \text{ closed})$$
$$= (.60)(.80) = .48$$

That is, the circuit will be closed about 48% of the time.

Combining Several Concepts

The independence of two events A and B carries over to their complements. In particular, if A and B are independent, then any pairing of A or its complement with B or its complement will also produce a pair of independent events. That is, each of the pairs of events A′ and B, A′ and B′, and A and B′ will be independent if A and B are independent (cf. Exercise 25). To see how this fact can be used, let's consider a frequently asked probability question: What is the chance that *at least one* of a set of independent events will occur? For two independent events, A and B, the event that at least one of these events occurs can be written $\{A \text{ or } B\}$. As we showed in our discussion of complementary events, the complement of $\{A \text{ or } B\}$ is the event $\{A' \text{ and } B'\}$. Therefore, using the additional knowledge that the complements of independent events must themselves be independent, we can write

P(at least one of two independent events occurs)

$$= P(A \text{ or } B) = 1 - P(A' \text{ and } B') = 1 - P(A')P(B')$$

This formula can readily be extended to any number of independent events, A_1, A_2, A_3, . . . , A_k. That is,

P(at least one of k independent events occurs)

$$= 1 - P(A'_1)P(A'_2)P(A'_3) \cdots P(A'_k)$$

The "at least one" rule has numerous applications, two of which are given in the following examples.

Example 5.9   In an example demonstrating how vendor quality affects customer quality, H. S. Gitlow and D. A. Wiesner ("Vendor Relations: An Important Piece of the Quality Puzzle," *Quality Progress*, 1988: 19–23) considered a hypothetical product consisting of 50 critical parts, any one of which, if defective, could cause the finished product to be defective. Suppose that each of these parts is purchased from a different vendor. It is therefore reasonable to assume that the condition of each part, created by a different vendor, should be *independent* of the conditions of the others. Furthermore, suppose that about 99.5% of all the parts supplied by a given vendor are good. What is the overall proportion of assembled products that can be expected to be defective?

To answer this question, let D_i denote the event that the part purchased from the ith vendor is defective, so that $P(D_i) = .005$ and $P(D'_i) = .995$. Then, the probability we seek is

$$P(\text{at least one of the 50 parts is defective}) = 1 - P(D'_1)P(D'_2)P(D'_3) \cdots P(D'_{50})$$
$$= 1 - (.995)^{50} = 1 - .7783 = .2217$$

This example demonstrates the important point that it is possible for complex systems to have high failure rates even if the quality of their individual components is relatively good.

Example 5.10   Consider the portion of an electronic circuit diagrammed in Figure 5.10. The circuit is primarily a **parallel system** (i.e., either switch A or *both* switches B and C must function if the current is to flow from left to right). The branch containing switches B and C, however, forms a **series system.** To compute the probability that a closed circuit is made between the left and right sides of the diagram, we must find the probability of the event $\{A \text{ or } \{B \text{ and } C\}\}$. Assuming that the switches function independently of one another and that they are closed with probabilities $P(A) = .80$, $P(B) = .70$, and $P(C) = .90$, we proceed as follows:

$$P(A \text{ or } (B \text{ and } C)) = P(A) + P(B \text{ and } C) - P(A \text{ and } (B \text{ and } C)) \leftarrow$$ The general addition rule applied to the events A and {B and C}

$$= P(A) + P(B)P(C) - P(A)P(B)P(C) \leftarrow$$ Since A, B, and C are independent

$$= .80 + (.70)(.90) - (.80)(.70)(.90) = .926$$

Thus the circuit is closed about 92.6% of the time. Since switch A is closed 80% of the time, the probability that the circuit is closed must certainly exceed 80%, so our answer makes sense.

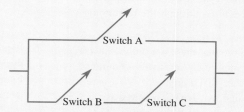

Figure 5.10 Series and parallel circuit with three switches shown in their open positions

Section 5.3 Exercises

13. Five companies (A, B, C, D, and E) that make electrical relays compete each year to be the sole supplier of relays to a major automobile manufacturer. The auto company's records show that the probabilities of choosing a company to be the sole supplier are

Supplier chosen:	A	B	C	D	E
Probability:	.20	.25	.15	.30	.10

    a. Suppose that supplier E goes out of business this year, leaving the remaining four companies to compete with one another. What are the new probabilities of companies A, B, C, and D being chosen as the sole supplier this year?

    b. Suppose the auto company narrows the choice of suppliers to companies A and C. What is the probability that company A is chosen this year?

14. Refer to the tree diagram in Figure 5.7. Suppose you want to find the probability $P(B|A)$ using the information available in the tree diagram. To do this, $P(B|A)$ must be expressed in terms of conditional probabilities, like $P(A|B)$ and $P(A'|B)$.

    a. Use the addition law to show that $P(A) = P(A \text{ and } B) + P(A \text{ and } B')$.

    b. Use the conditional probability formula to write $P(A \text{ and } B)$ in terms of $P(A|B)$ and $P(B)$. Develop a similar formula for $P(A \text{ and } B')$ in terms of $P(A|B')$ and $P(B')$.

    c. Use parts (a) and (b) to show that

    $$P(B|A) = \frac{P(A|B)P(B)}{P(A|B)P(B) + P(A|B')P(B')}$$

    This formula, known as *Bayes' theorem*, is used to "turn conditional probabilities around"; that is, it allows us to express $P(B|A)$ in terms of $P(A|B)$ and $P(A|B')$.

    d. In Figure 5.7, the probability associated with *any* path from left to right through the tree is simply the product of the probabilities of the branches. Why?

    e. Use the observation in part (d) and the conditional probability formula for $P(B|A)$ to justify Bayes' theorem.

15. In Exercise 5, suppose that 95% of the fasteners pass the initial inspection. Of those that fail inspection, 20% are defective. Of the fasteners sent to the recrimping operation, 40% cannot be corrected and are scrapped; the rest are corrected by the recrimping and then pass inspection.
    a. What proportion of fasteners that fail the initial inspection pass the second inspection (after the recrimping operation)?
    b. What proportion of fasteners pass inspection?
    c. Given that a fastener passes inspection, what is the probability that it passed the initial inspection and did not have to go through the recrimping operation?

16. In Exercise 6, suppose that there is a probability of .01 that a digit is incorrectly sent over a communication channel (i.e., that a digit sent as a 1 is received as a 0, or a digit sent as a 0 is received as a 1). Consider a message that consists of exactly 60% 1s.
    a. What is the proportion of 1s received at the end of the channel?
    b. If a 1 is received, what is the probability that a 1 was sent? *Hint:* Use the tree diagram from Exercise 6.

17. Suppose that A and B are independent events with $P(A) = .5$ and $P(B) = .6$. Can A and B be mutually exclusive events?

18. Probability calculations play an important role in modern forensic science (Aitken, C., *Statistics and the Evaluation of Evidence for Forensic Scientists*, John Wiley, New York, 1995). Suppose that a suspect is found whose blood type matches a rare blood type found at a crime scene. Let γ denote the frequency with which people in the population have this particular blood type. Assuming that people in the population are sampled at random, answer the following questions:
    a. What is the probability that a randomly chosen person from the population does not have the same blood type as that found at the crime scene?
    b. What is the probability that none of n randomly chosen people will match the blood type found at the crime scene?
    c. What is the probability that at least one person in a random sample of n people will match the blood type found at the crime scene?

    d. Suppose that $\gamma = 10^{-6}$. What is the probability that at least one person in a sample of one million will have a blood type matching that found at the crime scene?

19. In forensic science, the probability that any two people match with respect to a given characteristic (hair color, blood type, etc.) is called the *probability of a match.* Suppose that the frequencies of blood phenotypes in the population are as follows:

A	B	AB	O
.42	.10	.04	.44

    a. What is the probability that two randomly chosen people both have blood type A?
    b. Repeat the calculation in part (a) for the other three blood types.
    c. Find the probability that two randomly chosen people have matching blood types. *Note:* A person can have only one phenotype.
    d. The probability that two people do not match for a given characteristic is called *discriminating power.* What is the discriminating power for the comparison of two people's blood types in part (c)?

20. A construction firm has bid on two different contracts. Let E_1 be the event that the bid on the first contract is successful, and define E_2 analogously for the second contract. Suppose that $P(E_1) = .4$ and $P(E_2) = .3$ and that E_1 and E_2 are independent.
    a. Find the probability that both bids are successful.
    b. Find the probability that neither bid is successful.
    c. Find the probability that at least one of the bids is successful.

21. Consider a system of components connected as shown in the following figure.

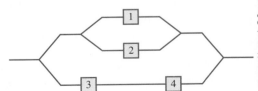

Components 1 and 2 are connected in parallel, so that their subsystem functions correctly if either component 1 or 2 functions. Components 3 and 4 are connected in series, so their subsystem works only if both components work correctly. If all components work independently of one another and P(a given component works) $= .9$, calculate the probability that the entire system works correctly.

22. The reviews editor for a certain scientific journal decides whether the review for any particular book should be short (1–2 pages), medium (3–4 pages), or long (5–6 pages). Data on recent reviews indicates that 60% of them are short, 30% are medium, and the other 10% are long. Reviews are submitted in either Word or a typesetting program called LaTeX. For short reviews, 80% are in Word, whereas 50% of medium reviews are in Word and 30% of long reviews are in Word. Suppose a recent review is randomly selected.
    a. What is the probability that the selected review was submitted in Word format?
    b. Suppose you are told the selected review was submitted in Word format. What is the probability that the review was medium in length?

23. In a certain population, 1% of all individuals are carriers of a particular disease. A diagnostic test for this disease has a 90% detection rate for carriers and a 5% detection rate for noncarriers. Suppose that the diagnostic test is applied independently to two different samples from the same randomly selected individual.
    a. What is the probability that both tests yield the same result?
    b. If both tests are positive, what is the probability that the selected individual is a carrier?

24. One of the assumptions underlying the theory of control charts (see Chapter 6) is that the successive points plotted on a chart are independent of one another.

Each point plotted on a control chart can signal either that a manufacturing process is operating correctly or that it is not operating correctly. However, even when a process is running correctly, there is a small probability, say, 1%, that a charted point will mistakenly signal that there is a problem with the process.
    a. What is the probability that at least one of ten points on a control chart signals a problem with a manufacturing process when in fact the process is running correctly?
    b. What is the probability that at least 1 of 25 points on a control chart signals a problem with a manufacturing process when in fact the process is running correctly?

25. If A and B are independent events, show that A' and B are also independent. *Hint:* Use a Venn diagram to show that $P(A'$ and $B) = P(B) - P(A$ and $B)$.

26. In October 1994, a flaw in a certain Pentium chip installed in computers was discovered that could result in a wrong answer when performing a division. The manufacturer initially claimed that the chance of any particular division being incorrect was only 1 in 9 billion, so that it would take thousands of years before a typical user encountered a mistake. However, statisticians are not typical users; some modern statistical techniques are so computationally intensive that a billion divisions over a short time period is not outside the realm of possibility. Assuming that the 1 in 9 billion figure is correct and that results of different divisions are independent of one another. What is the probability that at least 1 error occurs in 1 billion divisions with this chip?

5.4 RANDOM VARIABLES

Scientific and engineering studies rely heavily on numerical measurements derived from experiments. Indeed, it is often easy to think in terms of measurements themselves, not physical outcomes, as the end products of an experiment. Although you can imagine the various physical materials (the outcomes) that could result from repeating the chemical reaction, it is much more natural to think in terms of a numerical quantity of interest, such as the yield that might occur. Because measurements predominate in scientific studies, a mechanism is needed for extending probability concepts from the realm of simple events to the more natural scientific domain of numerical outcomes.

Random Variables

When the same numerical characteristic can conceivably be measured on any outcome of a chance experiment, we say that this quantity is a **random variable.** For instance, the measured yield of a chemical reaction is a random variable. Randomness enters the picture because we expect there to be slight unpredictable differences between each repetition of the reaction, which, in turn, will be reflected in the measured yields. There can be any number of random variables associated with a chance experiment. In a chemical reaction, any quantifiable feature associated with the reaction is a random variable (e.g., yield, density, weight, viscosity, volume, and translucence of the material produced). To make them easier to work with, random variables are usually denoted by single letters near the end of the alphabet. The yield of a chemical reaction might simply be denoted by the letter x, the density of the material by w, and so forth. The assignment of a letter to a random variable is sometimes written in the form of an equation, such as x = yield of a chemical reaction or w = density of the material produced in the reaction.

Technically speaking, the numerical values of a random variable are not the simple events of a chance experiment. Instead, a random variable is a *function* that assigns numerical values to the possible outcomes of a chance experiment, as illustrated in Figure 5.11. Notice that it is possible for more than one point in the sample space to be assigned the same real number. For instance, the random variable y = number of metal parts that pass a stress test out of three randomly selected parts assigns the number $y = 2$ to each of the sample space points PPF, PFP, and FPP.

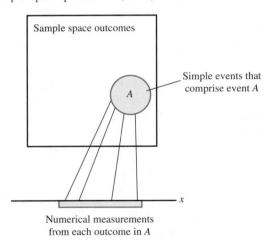

Figure 5.11 A random variable x assigns numerical values to the outcomes in the sample space

Because measurements are either **discrete** or **continuous**, random variables are also classified as **discrete random variables** or **continuous random variables.** Recall that discrete measurements generally arise when we count things, whereas continuous measurements are the result of using a measuring instrument. The yield of a chemical reaction would be a continuous random variable, whereas the number of metal parts passing a stress test would be a discrete random variable.

DEFINITIONS

A numerical characteristic whose value depends on the outcome of a chance experiment is called a **random variable**. A random variable is **discrete** if its possible values form a finite set or, perhaps, an infinite sequence of real numbers. Otherwise, a variable is **continuous** if its possible values span an entire interval of real numbers.

Events Defined by Random Variables

Although technically accurate, the description of a random variable as a function that assigns numerical values to sample space outcomes is not essential to most statistical applications. It is usually more helpful to think of random variables simply as variables whose values are likely to lie within certain ranges of the real number line. For example, the event that at least two of four randomly selected metal parts pass a stress test can simply be depicted by the numbers $x = 2$, 3, and 4 on the real number line, where $x =$ number of parts passing the stress test is a discrete random variable (Figure 5.12). In other words, we often suppress the picture of the sample space in Figures 5.11 and 5.12 and simply think of an event as a list or interval of numbers on the horizontal axis. With discrete variables, events correspond to finite or countable collections of points on the number line. For instance, the event $\{x \geq 2\}$ corresponds to the integers 2, 3, and 4 for the random variable $x =$ number of parts passing the stress test. The event $\{y \geq 2\}$ corresponds to the infinite collection of integers $y = 2$, 3, 4, ... for the variable $y =$ number of parts tested until one is found that fails the stress test. For continuous random variables, events such as $\{x > 3.21\}$, $\{x \leq 5.4\}$, or $\{18 \leq x \leq 21\}$ all refer to the real numbers contained in these intervals.

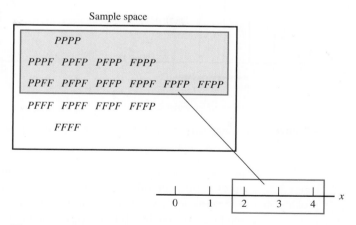

Figure 5.12 The event that at least two parts pass a stress test and the random variable $x =$ number of parts passing the stress test

The probability laws introduced in previous sections can be applied to events defined by random variables. For instance, let $x =$ length (in inches) of a randomly selected manufactured part. Then an event $\{18 \le x \le 21\}$ can, if desired, be partitioned into the disjoint events $\{18 \le x \le 21\} = \{18 \le x < 19\}$ or $\{19 \le x < 20\}$ or $\{20 \le x \le 21\}$. Notice that the particular choice of strict and inclusive inequality signs is what causes these events to be disjoint. The addition rule for disjoint events then states that $P(18 \le x \le 21) = P(18 \le x < 19) + P(19 \le x < 20) + P(20 \le x \le 21)$. Similarly, because the event $\{x > 18\}$ is the complement of the event $\{x \le 18\}$, the law of complementary events allows us to write $P(x \le 18) = 1 - P(x > 18)$.

Probability Distributions

The mechanism for assigning probabilities to events defined by random variables is to use either a mass function (for discrete random variables) or a density function (for continuous variables). In either case, we first envision an event of interest as a particular subset of the real number line. For discrete variables, the probability of the event is defined to be the sum of the mass function values that lie within the event subset. For continuous variables, the probability of an event is defined to be the area under the portion of the density curve that lies over the event on the number line. Figure 5.13 shows how a mass or density function assigns a probability to any event of interest on the real number line. When used to describe random variables, mass functions and density functions are both called **probability distributions.**

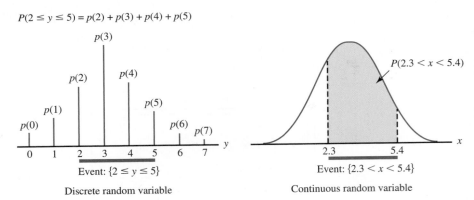

Figure 5.13 Using mass or density functions to assign probabilities to events

When a probability distribution has one of the familiar distributional forms described in Chapter 1, the methods described in that chapter can be used to find event probabilities. For example, if we believe that the length, x, of a randomly selected part can be described by a normal distribution with a mean of 20 cm and a standard deviation of 1.8 cm, then probabilities associated with x are found by standardizing, as shown in Chapter 1. Thus

$$P(18 \le x \le 21) = P\left(\frac{18 - 20}{1.8} \le z \le \frac{21 - 20}{1.8}\right)$$

$$= P(-1.11 \le z \le .56) = .5788$$

There are several ways to choose an appropriate probability distribution for describing a random variable. In the upcoming examples and chapters, we will use the following methods to justify our choices of probability distributions:

1. Examine a histogram of data, and select a familiar density or mass function whose shape approximately matches that of the histogram.
2. Use a density or mass function recommended by previous studies or professional practice.
3. Verify conditions that are known to give rise to certain mass or density functions (see binomial distributions in Section 1.6, normal distributions in Section 5.6).

Example 5.11

Examples 5.4–5.6 describe several events related to the chance experiment of randomly sampling and testing 20 items from a large shipment:

A = at most one of the sampled items fails to meet quality standards
B = none of the sampled items fails to meet quality standards
C = exactly one item fails to meet quality standards
D = at least one item fails to meet quality standards

These events can be recast in terms of the random variable x = number of items that fail to meet quality standards, as follows:

$$A \rightarrow \{x \le 1\}$$
$$B \rightarrow \{x = 0\}$$
$$C \rightarrow \{x = 1\}$$
$$D \rightarrow \{x \ge 1\}$$

Because random sampling ensures that each of the 20 selections is independent of the others, a binomial mass function is a good choice for describing probabilities associated with x (see Section 1.6). Suppose that it is known from manufacturing records that about 2% of all such items do not conform to quality standards. Using $\pi = .02$ and $n = 20$ in the formula for the binomial mass function, we calculate the probabilities of the previously described events as

$$P(x \le 1) = P(x = 0) + P(x = 1)$$

$$= \frac{20!}{0!\,20!}(.02)^0(.98)^{20} + \frac{20!}{1!\,19!}(.02)^1(.98)^{19}$$

$$= \quad 1(.98)^{20} \quad + \quad 20(.02)(.98)^{19}$$

$$= \quad \underset{\underset{P(B)}{\uparrow}}{.6676} \quad + \quad \underset{\underset{P(C)}{\uparrow}}{.2725} \quad = .9401$$

Thus if groups of 20 items are repeatedly selected, in the long run about 94% of all groups should have at most one item failing to meet standards.

The probability that at least one item fails to meet quality standards is

$$P(x \geq 1) = 1 - P(x = 0) = 1 - .6676 = .3324$$

Notice that the addition rule and the law of complementary events were used to simplify the computations of $P(x \leq 1)$ and $P(x \geq 1)$.

Mean and Variance of a Random Variable

The mean of a random variable x can be thought of as the long-run average value of x that should occur in many repeated trials of a chance experiment. Fortunately, when the probability distribution of x is known, there is no need to actually perform repeated experimental trials. Instead, we define the mean to be the mean of the population described by the mass or density function and then use the methods of Chapter 2 to compute it. The same notation μ used to describe the mean of a population is now used to denote the mean of a random variable. For a known mass function $p(x)$, the mean is defined as

$$\mu = \sum_x x p(x)$$

For a known density function $f(x)$, the mean is given by

$$\mu = \int x f(x) \, dx$$

Similarly, the variance σ^2 of a random variable is calculated from the familiar formulas in Chapter 2. The standard deviation σ of a random variable is defined to be the square root of its variance:

$$\sigma^2 = \sum (x-\mu)^2 p(x) \qquad \text{or} \qquad \sigma^2 = \int (x-\mu)^2 f(x) \, dx$$

The mean and standard deviation of a random variable frequently appear as parameters in the defining formulas for a mass or density function. For this reason, it is often necessary to obtain estimates of μ and σ before probability calculations are possible. As discussed later in the chapter, statistics such as the sample mean, $\bar{x}$, and sample standard deviation, s, are frequently used to provide such estimates.

Example 5.12   The reliability of a product at time t, denoted by $R(t)$, is defined as the probability that the product is still working correctly after t units of time (see Section 6.6). For complex products consisting of several parts and subassemblies, the time x until a product fails often follows an exponential distribution with parameter λ. In such applications, the mean of the distribution, $\mu = 1/\lambda$, is called either the **mean time between** (or **before**) **failures (MTBF)** or the **mean time to failure (MTTF)**. According to the definition of $R(t)$, the reliability can be calculated from the formula

$$R(t) = P(x > t) = \int_t^\infty \lambda e^{-\lambda x} \, dx = e^{-\lambda t}$$

Suppose that the lifetime of a certain product follows an exponential distribution with an MTBF of 10,000 hr and that we want to find the proportion of such products that fail before 20,000 hr of service. Since MTBF = $1/\lambda$, the value of λ for this distribution is $\lambda = 1/10,000 = .0001$ hr^{-1}, and the reliability function is given by

$$R(t) = P(x > t) = e^{-\lambda t} = e^{-.0001t}$$

For $t = 20,000$ hr, the reliability is then $R(20,000) = e^{-(.0001)(20,000)} = e^{-2} = .1353$. That is, about 13.53% of the products will last at least 20,000 hours. The proportion of products that do *not* last at least 20,000 hours is found by using the formula for complementary events:

$$P(\text{product fails before 20,000 hr}) = 1 - P(\text{product lasts at least 20,000 hr})$$

$$= 1 - .1353 = .8647$$

Calculations with Random Variables

Mathematical operations (e.g., addition, multiplication, exponentiation, square roots) can be applied to random variables. One reason for doing this is to reduce probability statements about one random variable to statements about a more familiar random variable whose probabilities are well known. This is what we do, for instance, when simplifying statements about a normal random variable x. By performing the arithmetic operations of subtraction (to form $x - \mu$) and division (to form $(x - \mu)/\sigma$), we eventually reduce probability statements about x to statements about the standard normal variable z, whose probabilities are easily found in tables.

Example 5.13   Resistors come in two varieties, *general purpose* (with tolerances of $\pm 5\%$ or greater) and *precision* (with tolerances of $\pm 2\%$ or less). The tolerance is the amount by which the true resistance can deviate from the stated resistance. For example, a 6.0-kilohm (kΩ) resistor with a tolerance of $\pm 10\%$ can be expected to have a measured resistance of $6.0 \pm (.10)(6.0)$, that is, from 5.4 kΩ to 6.6 kΩ. Assuming that a uniform density adequately describes the possible values of x, then the true resistance x of a randomly selected 6.0-kΩ resistor is a random variable described by the density function (see Chapter 1 for the definition of uniform densities):

$$f(x) = \begin{cases} \dfrac{1}{1.2} & \text{for } 5.4 < x < 6.6 \\ 0 & \text{otherwise} \end{cases}$$

Suppose we want to find the probability that the conductance (defined as the reciprocal of resistance) is greater than a specified amount, say, .16 siemens (S). Writing this probability statement, we can take reciprocals of both sides to find

$$P(\text{conductance} > .16) = P\left(\frac{1}{x} > .16\right) = P(x < 6.25) = .7083$$

Mathematical operations can also be applied to several random variables. In statistical applications, for example, we commonly form sums of several random variables and ask about the probability with which these sums assume various numerical values. Such calculations are greatly simplified if the random variables involved are known to be *independent* of one another. Two random variables, x and y, are said to be **independent** if the events $\{x < a\}$ and $\{y < b\}$ are independent for all possible combinations of real numbers a and b. For example, if x represents the lifetime of a randomly chosen electronic component (measured in hours of service) and y denotes the lifetime of another randomly chosen component, then, intuitively, we expect that the event $\{x < 1000\}$ should be independent of the event $\{y < 500\}$. In fact, we expect that the choice of 1000 and 500 is immaterial here, so that the events $\{x < a\}$ and $\{y < b\}$ should be independent, regardless of the values of a and b.

Independent variables commonly arise from the application of random procedures such as random sampling or randomization, or from an *assumption* of randomness. For two discrete random variables with mass functions $p_1(x)$ and $p_2(y)$, independence also means that their **joint probability** $p(a, b) = P(x = a \text{ and } y = b)$ equals the product of their mass functions; that is, $p(a, b) = p_1(a)p_2(b)$ for any combination of values of a and b. Similarly, for *continuous* random variables with densities $f_1(x)$ and $f_2(y)$, independence allows their **joint density** $f(x, y)$ to be written as a product of their individual densities: $f(x, y) = f_1(x)f_2(y)$ for any values of x and y. Now let B denote a collection of points in the $x-y$ plane. When x and y are independent, the probability that the pair (x, y) lies in B is

$$P(B) = \sum_{(x,\, y) \in B} \sum p(x, y) = \sum_{(x,\, y) \in B} \sum p_1(x)p_2(y) \qquad x, y \text{ discrete}$$

$$P(B) = \iint_B f(x, y)\, dx\, dy = \iint_B f_1(x)f_2(y)\, dx\, dy \qquad x, y \text{ continuous}$$

Example 5.14

Images displayed on computer screens consist of thousands of small regions called *picture elements*, or **pixels** for short. The intensity of the electron beam focused at a given point (x_0, y_0) on a flat screen is usually described by two independent normal random variables x and y, with means x_0 and y_0, respectively. That is, we represent the intensity of the beam by a joint density function of two independent random variables. For example, Figure 5.14 shows a graph of the joint density function describing an electron beam focused on the point $(x_0, y_0) = (30, 50)$. The standard deviations of the two normal distributions are $\sigma_x = .2$ and $\sigma_y = .2$. Because x and y are independent, we can write the joint density as the product

$$f(x, y) = f_1(x)f_2(y)$$

$$= \frac{1}{\sigma_x \sqrt{2\pi}} e^{-\frac{1}{2}\left(\frac{x - 30}{.2}\right)^2} \cdot \frac{1}{\sigma_y \sqrt{2\pi}} e^{-\frac{1}{2}\left(\frac{y - 50}{.2}\right)^2}$$

$$= \frac{1}{2\pi \sigma_x \sigma_y} e^{-\frac{1}{2}\left[\left(\frac{x - 30}{.2}\right)^2 + \left(\frac{y - 50}{.2}\right)^2\right]}$$

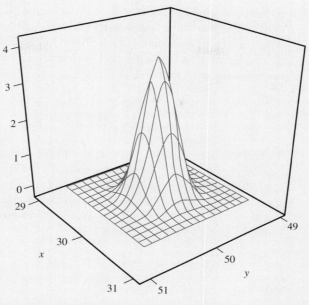

Figure 5.14 Joint density function near the point $(x_0, y_0) = (30, 50)$

The volume under this density that sits over a given region B in the x–y plane describes the proportion of time that the electron beam spends in region B. Although the joint density can be used to find the probability associated with *any* set B of points near $(30, 50)$ on the screen, the probability of some sets can be found in an easier way. For example, if we want to find the proportion of time that the beam spends in the region where $x < 29.5$ and $y < 49.6$, we can simply use the independence of x and y to obtain

$$P(x < 29.5 \text{ and } y < 49.6) = P(x < 29.5)P(y < 49.6)$$

instead of integrating the density over the region $B = \{(x, y)|x < 29.5, y < 49.6\}$. Thus

$$P(x < 29.5 \text{ and } y < 49.6) = P(x < 29.5)P(y < 49.6)$$
$$= P\left(z < \frac{29.5 - 30}{.2}\right)P\left(z < \frac{49.6 - 50}{.2}\right)$$
$$= P(z < -2.5)P(z < -2.0)$$
$$= (.0062)(.0228) = .00014$$

The proportion of time that the beam spends in this region is very small.

Example 5.15

In Sections 5.5 and 5.6, we will be concerned with sums and averages of independent random variables. Suppose, for example, that two printed circuit boards are randomly selected and tested. Let x be the number of defective computer chips found on one board; let y be the number of defectives found on the other board. Suppose the following mass functions describe x and y:

x:	1	2	3	4	y:	1	2	3	4
$p_1(x)$:	.25	.25	.25	.25	$p_2(y)$:	.25	.25	.25	.25

To find the mass function associated with the *average* number of defectives on two boards, $w = (x + y)/2$, we can use mutually exclusive events and independence to simplify each probability. For example, to find $P(w = 2.5)$, first break up the event $\{(x + y)/2 = 2.5\}$ into the disjoint events $\{x = 1\ and\ y = 4\}$, $\{x = 2\ and\ y = 3\}$, $\{x = 3\ and\ y = 2\}$, and $\{x = 4\ and\ y = 1\}$. Next, find the probabilities of these events by multiplying mass function values:

$$P(x = 1\ and\ y = 4) = P(x = 1)P(y = 4) = (.25)(.25) = .0625$$
$$P(x = 2\ and\ y = 3) = P(x = 2)P(y = 3) = (.25)(.25) = .0625$$
$$P(x = 3\ and\ y = 2) = P(x = 3)P(y = 2) = (.25)(.25) = .0625$$
$$P(x = 4\ and\ y = 1) = P(x = 4)P(y = 1) = (.25)(.25) = .0625$$

Finally, add the probabilities of these disjoint events to find $P(w = 2.5) = .2500$. Proceeding in this manner gives the mass function of the average, w:

w:	1	1.5	2	2.5	3	3.5	4
$p(w)$:	.0625	.1250	.1875	.2500	.1875	.1250	.0625

The graphs of all three mass functions are shown in Figure 5.15. Notice that the mass function of the average tends to bunch more closely around its mean than do either of its constituent mass functions, $p_1(x)$ and $p_2(y)$.

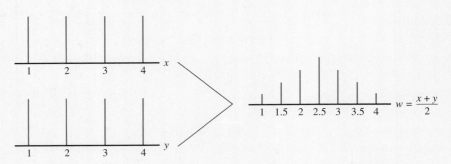

Figure 5.15 The mass function of an average of two independent random variables

Section 5.4 Exercises

27. Classify each of the following random variables as either discrete or continuous.
    a. $x =$ the number of flaws per square foot in a randomly selected sheet of fabric
    b. $y =$ the measured concentration of chemical in a solution
    c. $w =$ the proportion of oversize bolts in a randomly selected box of bolts
    d. $u =$ the number of errors per 1000 randomly selected lines of computer code
    e. $v =$ the breaking strength of a randomly selected metal bar
    f. $t =$ the lifetime of a randomly selected electronic component
    g. $x =$ the number of customer complaints in a randomly selected week

28. The probability mass function for the number x of coding errors found in 1000 randomly selected lines of computer code is given by

x:	0	1	2	3	4
$p(x)$:	.08	.15	.45	.27	.05

    a. Calculate the mean number of coding errors for all such blocks of 1000 lines of code.
    b. Calculate the variance and standard deviation of x.

29. A chemical supply company currently has in stock 100 pounds of a certain chemical, which it sells to its customers in 5-lb lots. Let x denote the number of lots ordered by a randomly selected customer, and suppose x has the following probability mass function:

x:	1	2	3	4
$p(x)$:	.2	.4	.3	.1

    a. Compute the mean number of lots ordered by a customer.
    b. Compute the variance of the number of lots ordered by a customer.
    c. Compute the expected number of pounds left after a customer's order is shipped.

30. Let x denote the number of ticketed airline passengers denied a flight because of overbooking. Suppose that x is a random variable for which

$p(x) = c(5 - x)$ for $x = 0, 1, 2, 3, 4$. Find the numerical value of c and then compute $P(x > 0)$.

31. A contractor is required by a county planning department to submit from one to five different forms, depending on the nature of the project. Let $y =$ number of forms required of the next contractor. Suppose that it is known that the probability that y forms are required is proportional to y; that is, $p(y) = ky$ for $y = 1, 2, 3, 4,$ and 5.
    a. What is the numerical value of k?
    b. What is the probability that at most three forms are required?
    c. What is the expected number of forms required?
    d. Find the standard deviation of the number of forms required.

32. Suppose that the reaction time (sec) to a certain stimulus is a continuous random variable with a density function given by

$$f(x) = \begin{cases} k/x & \text{for } 1 \le x \le 10 \\ 0 & \text{otherwise} \end{cases}$$

    a. Sketch a graph of $f(x)$.
    b. Find the numerical value of k.
    c. What is the probability that x exceeds 3?
    d. What is the probability that x lies within .25 sec of 3?

33. A printed circuit board (PCB) has 285 small holes, called "joints," into which are inserted the thin leads or "pins" emanating from electronic components soldered to the PCB (see Example 5.20). Assuming that the quality of the solder joint at any pin is independent of the quality at any other pin, a binomial mass function can be used to describe x, the number of defective solder joints. Answer the following questions, given the probability that a given solder joint is defective is .01:
    a. What are the mean and standard deviation of the number of defective solder joints on a PCB?
    b. What proportion of all PCBs are defect-free?
    c. What is the probability that a given PCB has two or more defective solder joints?

34. The Poisson mass function is often used in biology to model the number of bacteria in a solution.

Suppose a dilute suspension of bacteria is divided into several different test tubes. The number of bacteria x in a test tube has a Poisson mass function with a parameter λ that represents the mean number of bacterial cells contained in the different test tubes.

a. Express the probability that a particular test tube contains no bacteria, in terms of λ.

b. In terms of λ, what is the probability that a test tube contains at least one bacterial cell?

c. After a certain period of time, all of the test tubes are examined, and it is found that 40% of the tubes contain at least one bacterial cell. Use your answer from part (b) to estimate λ, the mean number of cells per test tube.

35. A standard procedure for testing safety glass is to drop a 1/2-lb iron ball onto a 12-in. square of glass supported on a frame ("Statistical Methods in Plastics Research and Development," *Quality Engr.*, 1989: 81–89). The height from which the ball is dropped is determined so that there is a 50% chance of breaking through the glass. A breakthrough is considered to be a failure, whereas a ball that is stopped by the glass (even if the glass cracks) is considered to be a success. Suppose that 100 sheets of safety glass are randomly selected and tested, and that no change has been made in the resin used to manufacture the glass.

a. What is the expected number of sheets that will experience a breakthrough?

b. What is the probability that 60 or more sheets will have a breakthrough?

36. The normal distribution is commonly used to model the variability expected when making measurements (Taylor, J. R., *An Introduction to Error Analysis: The Study of Uncertainties in Physical Measurements*, University Science Books, Sausalito, CA, 1997). In this context, a measured quantity x is assumed to have a normal distribution whose mean is assumed to be the "true" value of the object being measured. The precision of the measuring instrument determines the standard deviation of the distribution.

a. If the measurements of the length of an object have a normal probability distribution with a standard deviation of 1 mm, what is the probability that a single measurement will lie within 2 mm of the true length of the object?

b. Suppose the measuring instrument in part (a) is replaced with a more precise measuring instrument having a standard deviation of .5 mm. What is the probability that a measurement from the new instrument lies within 2 mm of the true length of an object?

37. Acceptance sampling is a method that uses small random samples from incoming shipments of products to assess the quality of the entire shipment. Typically, a random sample of size n is selected from a shipment, and each sampled item is tested to see whether it meets quality specifications. The number of sampled items that do not meet specifications is denoted by x. As long as x does not exceed a prespecified integer c, called the *acceptance number*, then the entire shipment is accepted for use. If x exceeds c, then the shipment is returned to the vendor. In practice, because n is usually small in comparison to the number of items in a shipment, a binomial distribution is used to describe the random variable x.

a. Suppose a company uses samples of size $n = 10$ and an acceptance number of $c = 1$ to evaluate shipments. If 10% of the items in a certain shipment are defective, what is the probability that this shipment will be returned to the vendor?

b. Suppose that a certain shipment contains no defective items. What is the probability that the shipment will be accepted by the sampling plan in part (a)?

c. Rework part (a) for shipments that are 5%, 20%, and 50% defective.

d. Let π denote the proportion of defective items in a given shipment. Use your answers to parts (a)–(c) to plot the probability of accepting a shipment (on the vertical axis) against $\pi = 0$, .05, .10, .20, and .50 (on the horizontal axis). Connect the points on the graph with a smooth curve. The resulting curve is called the *operating characteristic* (OC) curve of the sampling plan. It gives a visual summary of how the plan performs for shipments of differing quality.

38. Refer to Exercise 37. Acceptance sampling plans that use an acceptance number of $c = 0$ are given the name *zero acceptance plans*. Zero acceptance plans are not frequently used because, although they protect

against accepting shipments of inferior quality, they also tend to reject many shipments of good quality.

a. Let π denote the proportion of defective items in a shipment. Develop a general formula for the probability of accepting a shipment having $\pi \times 100\%$ defective items.

b. Plot the OC curve for the zero acceptance plan that uses sample sizes of $n = 10$.

c. For what value of π is the probability of accepting a shipment about .05?

39. Qualification exams for becoming a state-certified welding inspector are based on multiple-choice tests. As in any multiple-choice test, there is a possibility that someone who is simply guessing the answers to each question might pass the test. Let x denote the number of correct answers given by a person who is guessing each answer on a 25-question exam, with each question having five possible answers (for each question, assume only one of the five choices is correct).

a. What type of probability distribution does x have?

b. For the 25-question test, what are the mean and standard deviation of x?

c. The exam administrators want to make sure that there is a very small chance, say, 1%, that a person who is guessing will pass the test. What minimum passing score should they allow on the exam to meet this requirement?

40. When used to model lifetimes of components, a probability distribution is said to be "memoryless" if, for a component that has already lasted (without failure) for t hours, the probability that it lasts for another s hours does not depend on t. That is, $P(x \geq t + s | x \geq t) = P(x \geq s)$. Show that the exponential distribution is memoryless.

41. The concept of the median of a set of data can also be applied to the probability distribution of a random variable. If x is a random variable with density function $f(x)$, then the median of this distribution is defined to be the value $\tilde{\mu}$ for which half the area under the density curve lies to the left of $\tilde{\mu}$. That is, $\tilde{\mu}$ is the solution to the equation $\int_{-\infty}^{\tilde{\mu}} f(x) \, dx = \frac{1}{2}$.

a. Suppose the lifetime x of an electronic assembly follows an exponential distribution with an

MTBF of 500 hours (see Example 5.12 for the definition of MTBF). Find the median of this distribution. The median is the time by which half of all such assemblies will break down.

b. Is the median time to failure from part (a) larger or smaller than the mean time before failure (MTBF)?

c. From your answer to part (a), find a general formula (for any value of MTBF) for expressing the median time to failure in terms of the mean time before failure.

42. On a construction site, subcontractor A is responsible for completing the structural frame of a building. When this task is complete, subcontractor B then begins the task of installing electrical wiring and outlets. The following tables show estimated probabilities of completing each task in x days:

Framing time (days), x:	10	15	20	25	30
Probability, $p_1(x)$:	.10	.20	.30	.30	.10

Wiring time (days), y:	5	10	15	20
Probability, $p_2(y)$:	.20	.50	.20	.10

a. Calculate the expected completion time for each task.

b. Find the probability distribution of the total time for completing both tasks (assume that the framing and wiring tasks are independent).

c. What is the probability that the total time to complete both tasks is less than 35 days?

d. What is the expected time for completing both tasks?

43. Let x be the cost ($) of an appetizer and y be the cost of a main course at a certain restaurant for a customer who orders both courses. Suppose that x and y have the following joint distribution:

			y	
		10	15	20
	5	.20	.15	.05
x	6	.10	.15	.10
	7	.10	.10	.05

a. Find the probability mass function of x.

b. Find the probability mass function of y.

c. Find the probability that $x + y \leq 21$.

d. Are x and y independent?

5.5   SAMPLING DISTRIBUTIONS

The general objective of statistical inference, as we have noted in the chapter introduction, is to answer questions about the characteristics of populations and processes. In particular, we wish to be able to make statements about population and process parameters and to also accompany them by a measure of how much reliability or confidence we have in our statements. Statistical inference is based on the interplay between **random samples** (used to obtain data and calculate statistics), **sampling distributions** (which describe the behavior of such statistics), and **probability** (which gives quantitative measures of reliability about what the statistics say). In this section and the next, we show how these three tools are used in statistical inference.

The **sampling distribution** of a statistic is a mass or density function that characterizes all the possible values that the statistic can assume in repeated random samples. Depending on the particular statistic (e.g., $\bar{x}$, s, s^2, $\tilde{x}$, range, IQR), we speak of the sampling distribution of $\bar{x}$, the sampling distribution of s, and so forth. Every statistic has a sampling distribution. The sampling distribution sets the limits on which values of a statistic are likely and which are not.

DEFINITION

The **sampling distribution** of a statistic is a mass or density function that characterizes all the possible values that the statistic can assume in repeated random samples from a population or process.

How Sampling Distributions Are Used

One way to approximate the sampling distribution of a statistic is to repeatedly select a large number of random samples of size n from a given population. By calculating the value of the statistic for each sample and forming a histogram of the results, we get an approximate picture of the sampling distribution of the statistic. In turn, this picture can be used to describe the values of the statistic that are likely to occur in any random sample of size n.

Example 5.16

Suppose that we draw 1000 random samples, each of size $n = 25$, from a normal population with a mean of 50 and a standard deviation of 2. If we calculate the mean $\bar{x}$ of each sample, then the distribution of all 1000 $\bar{x}$ values gives a good approximation to the sampling distribution of $\bar{x}$. Figure 5.16 shows a histogram of the results of such an experiment. Notice that the 1000 sample means stack up around the population mean ($\mu = 50$) and that variation among the sample means is smaller than variation in the population. In particular, none of the sample means fall outside the range of 48.5 to 51.5 (i.e., none are more than 1.5 units away from μ). In fact, it also appears that very few sample means fall outside the interval 49 to 51; that is, they are generally within 1 unit of μ.

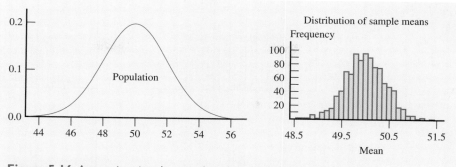

Figure 5.16 Approximating the sampling distribution of $\bar{x}$ ($n = 25$)

From the shape and location of the sampling distribution, we can begin to see which values of the sample statistic are more likely to occur than others. In this sense, the information in a sampling distribution provides a template for evaluating any sample, even future samples, from a population or process. In Figure 5.16, for instance, we can use the tails of the sampling distribution to place bounds on the values that $\bar{x}$ can assume whenever we take random samples of size 25 from a normal population with a mean of 50 and a standard deviation of 2. Going a step further, we can reasonably say that the mean affects only the location of the histogram and that the value of σ affects only the spread of the sample results. If this is so, then we now know a lot about what to expect when sampling from *any* normal population whose standard deviation is $\sigma = 2$.

Example 5.17   Refer to Example 5.16. Suppose that next week we select a *single* sample of size 25 from a normal population whose standard deviation is known to be $\sigma = 2$, but whose mean μ is *unknown* to us. If $\bar{x} = 70$ for this sample, then the results in Example 5.16 indicate that 70 is almost certainly no farther than 1.5 units away from the population mean μ and it is fairly likely that it is within 1 unit of μ. That is, we can infer that the unknown population mean is almost certainly between 68.5 and 71.5, and we can be reasonably confident that it is between 69 and 71. In this way, by using our knowledge of what the sampling distribution of $\bar{x}$ looks like, we can begin to make *inferences* about the likely values of the unknown population *parameter* μ.

General Properties of Sampling Distributions

As we noted previously, sampling distributions can be created for *any* statistic, not just $\bar{x}$. For example, Figure 5.17 shows the approximate sampling distributions of the statistics $\bar{x}$, $\tilde{x}$, s, and s^2, for the same 1000 samples of size $n = 25$ that were used to create Figure 5.16.

What about sampling from discrete populations? In particular, suppose we want to use samples of size 25 to estimate the proportion π of defectives being made by a certain process. Denoting defective items by a "1" and nondefectives by a "0," the mass function

x:	0	1
$p(x)$:	.80	.20

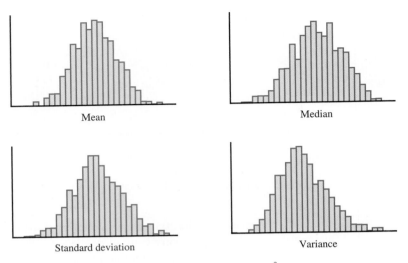

Figure 5.17 Sampling distributions of $\bar{x}$, $\tilde{x}$, s, and s^2

describes such a process in which the proportion of defective items is 20%. By calculating the sample proportion defective p for each of 1000 random samples of size $n = 25$, an approximate sampling distribution for the statistic p can be formed (Figure 5.18). Note that this distribution has many more possible values than just the values $x = 0$ and $x = 1$ in the population (each of the values 0/25, 1/25, 2/25, 3/25, . . . , 25/25 is a possible value of p). The shape of this sampling distribution is similar to the one in Figure 5.16, although it contains some gaps because only the values of p shown previously are possible to attain in a sample of size 25.

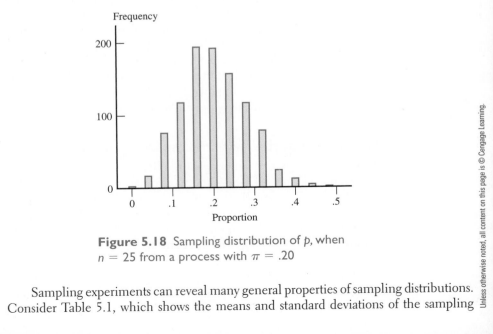

Figure 5.18 Sampling distribution of p, when $n = 25$ from a process with $\pi = .20$

Sampling experiments can reveal many general properties of sampling distributions. Consider Table 5.1, which shows the means and standard deviations of the sampling distributions in Figure 5.17 along with the actual values of the corresponding population parameters (for a normal population with $\mu = 50$, $\sigma = 2$). The similarity between the column of population parameters and the column of means of the sampling distributions leads us to conjecture that the *center* (i.e., the mean) of the sampling distribution of a statistic may, in fact, coincide with the corresponding population parameter. When this happens, we say that the statistic is **unbiased**, or that it is an **unbiased estimator** of the population parameter. As we shall see in Section 5.6, some of the most important statistics we have encountered so far are unbiased.

Table 5.1 Means and standard deviations of sampling distributions in Figure 5.17

Population parameter	Actual value	Sample mean of sampling distribution	Sample standard deviation of sampling distribution
Mean, μ	50	50.000	.418
Median	50	49.982	.515
Standard deviation, σ	2	1.9831	.2853
Variance, σ^2	4	4.0139	1.1528

Second, the standard deviations of the sampling distributions exhibit an important feature: All are smaller than the population standard deviation ($\sigma = 2$). In fact, as we shall see in Section 5.6, an even stronger statement can be made: For the majority of statistics we have studied, the variation in the sampling distribution actually decreases as the sample size increases.

Beyond these simple observations, additional questions immediately come to mind. What role does the shape of the population play in controlling the shape of the sampling distribution? What is the effect of increasing or decreasing the sample size? Exactly how is the variation exhibited by the sampling distribution related to the variation in the population? Most of these questions can be answered in *general* terms by conducting a few more sampling experiments. Such experiments provide the motivation for the following general conclusions:

General Properties of Sampling Distributions

1. The sampling distribution of a statistic often tends to be *centered* at the value of the population parameter estimated by the statistic.

2. The spread of the sampling distributions of many statistics tends to grow smaller as the sample size n increases.

3. As the sample size increases, sampling distributions of many statistics become more and more bell-shaped (more and more like normal distributions).

Finally, and perhaps most importantly, do we really have to conduct a lengthy sampling experiment every time we want to make inferences based on a statistic generated from a *single* sample? As we shall see in Section 5.6, the surprising answer is "no." In fact, the approximate shape of the sampling distribution is often known in advance, *before taking even a single sample*! Furthermore, knowing the specific shape of a sampling distribution also enables us to calculate *probabilities*, which allow us to quantify exactly what we mean by saying that, for example, the sample mean is highly likely to be within 1 unit of the population mean.

Section 5.5 Exercises

44. What primary purpose do sampling distributions serve in statistical inference?

45. Refer to Exercise 37. Suppose that a large lot of items is inspected by taking a random sample of size n and determining the number x of defective items in the sample. The result is then reported in terms of the proportion $p = x/n$ of defective items in the sample. Assume that the binomial distribution can be used to describe the behavior of the random variable x.

   a. Suppose that 5% of the items in a particular lot are defective and that a random sample of size $n = 5$ is to be taken from the lot. Calculate the probability that the sample proportion p falls within 1% of the true percent defective in the lot. That is, find $p(.05 - .01 \le p \le .05 + .01)$

   b. Answer the question in part (a) for samples of size $n = 25$.

   c. Answer the question in part (a) for samples of size $n = 100$. *Hint:* Use the normal approximation to the binomial distribution.

46. Random samples of size n are selected from a population that is uniformly distributed over the interval [10, 20]. Without sampling or performing any calculations, describe what you expect the sampling distribution of the range R (R = largest minus smallest value in a sample) to look like.

   a. For samples of size $n = 2$, what do you predict the mean of the sampling distribution of R will be?

   b. For samples of size $n = 100$, what do you predict the mean of the sampling distribution of R will be?

   c. Will the variance of the sampling distribution of R for samples of size $n = 2$ be the same or different from the variance of the sampling distribution of R for samples of size $n = 100$? Give a simple justification for your answer based on the definition of the range for $n = 2$ versus $n = 100$.

   d. What will happen to the variance of the sampling distribution of R as the sample size n increases? Give a simple justification for your answer based on the definition of the range.

47. The Food and Drug Administration (FDA) oversees the approval of both medical devices and new drugs. To gain FDA approval, a new device must be shown to perform at least as well, and hopefully better, than any similar device already on the market. Suppose a medical device company develops a new system for connecting intravenous tubes used on hospital patients. To be comparable to an already-existing product, the force required to disconnect two tubes joined by the new device must not exceed 5 lb. To estimate the maximum force required to disconnect two tubes, several tests are made. For a random sample of n connections, the forces $x_1, x_2, x_3, \ldots , x_n$ required to disconnect the tubes are recorded and the maximum, M, of the n readings is used to estimate the maximum necessary force for all such connections.

   a. Suppose that the actual distribution of forces needed to disconnect tubes can be described by a uniform distribution on the interval [2, 4]. For a sample of size $n = 2$, do you expect the mean of the sampling distribution of M to be closer to 2 or 4?

   b. Which do you expect to be larger, the mean of the sampling distribution of M for samples of size $n = 2$ or the mean of the sampling distribution of M for samples of size $n = 100$? Use the definition of the maximum of a sample to justify your answer.

   c. Will the variance of the sampling distribution of M for samples of size $n = 2$ be the same or different from the variance of the sampling distribution of M for samples of size $n = 100$? Give a simple justification for your answer based on the definition of the sample maximum for $n = 2$ versus $n = 100$.

5.6 DESCRIBING SAMPLING DISTRIBUTIONS _____

Just as histograms of sample data provide approximations to population distributions, sampling experiments (Section 5.5) furnish approximate pictures of sampling distributions. We now turn our attention to developing more precise summaries of sampling distributions. This requires a slightly deeper investigation of the role played by random sampling. For instance, Example 5.15 gives a glimpse of how random sampling and the form of the statistic $\bar{x}$ are brought together to form a more exact picture of the sampling distribution of $\bar{x}$. The essential role of random sampling is to ensure that the sampled values can be considered to be *independent*. Independence, in turn, enables us to perform the necessary probability calculations to arrive at the distribution of the statistic.

In this section, we study in some detail the exact sampling distributions of the statistics $\bar{x}$ (sample mean) and p (sample proportion). These two statistics appear in a great many statistical techniques, and their sampling distributions serve as prototypes for all other sampling distributions. In subsequent chapters, we will simply state the form of the sampling distribution that applies to a given statistical technique.

Sampling Distribution of $\bar{x}$

The sampling distribution of $\bar{x}$, also called the **sampling distribution of the mean,** is the probability distribution that describes the behavior of $\bar{x}$ in repeated random samples from a population or process. Like any distribution, the sampling distribution of $\bar{x}$ has its own unique mean and standard deviation, which we denote by $\mu_{\bar{x}}$ and $\sigma_{\bar{x}}$, respectively. The next general result relates $\mu_{\bar{x}}$ and $\sigma_{\bar{x}}$ to the population or process mean and standard deviation.

Mean and Standard Deviation of the Sampling Distribution of $\bar{x}$

Let $\bar{x}$ be the sample mean of a random sample $x_1, x_2, x_3, \ldots, x_n$ from a population or process with mean μ and standard deviation σ. Then, the mean of the sampling distribution of $\bar{x}$ coincides with μ, regardless of the sample size n. The spread of the sampling distribution, described by $\sigma_{\bar{x}}$, is equal to the population standard deviation divided by the square root of the sample size. That is,

$$\mu_{\bar{x}} = \mu \quad \text{and} \quad \sigma_{\bar{x}} = \frac{\sigma}{\sqrt{n}}$$

These equations hold regardless of the particular form of the population distribution. To emphasize the fact that it describes a sampling distribution, not a population, $\sigma_{\bar{x}}$ is also called the **standard error of $\bar{x}$,** or the **standard error of the mean.**

One of the key features of the standard error of the mean $\sigma_{\bar{x}}$ is that it decreases as the sample size increases. In fact, many statistics have this property (see Section 5.5). This makes intuitive sense, since we expect that more information ought to provide better estimates (i.e., smaller standard errors). As a result, increasing the size of a random sample has the desirable effect of increasing the probability that the estimate $\bar{x}$ will lie close to the population mean μ.

Sampling from a Normal Population

When a population follows a normal distribution, it can be shown that the sampling distribution of $\bar{x}$ is also normal, *for any sample size n*. The normality of $\bar{x}$, along with the fact that its mean $\mu_{\bar{x}}$ and standard error $\sigma_{\bar{x}}$ can be determined from μ and σ, is enough to completely characterize the sampling distribution of $\bar{x}$ in this case. As a result, with the normal distribution, probabilities of events involving $\bar{x}$ reduce to straightforward calculations. Figure 5.19 demonstrates the effect that increasing n has on the sampling distribution of $\bar{x}$.

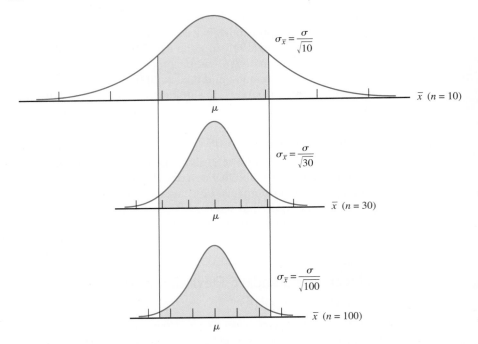

Figure 5.19 The probability that $\bar{x}$ falls within a fixed distance from μ increases as n increases

Sampling Distribution of $\bar{x}$ (Normal Population)

When a population distribution is normal, the sampling distribution of $\bar{x}$ is also normal, regardless of the size of the sample.

Example 5.18   Physical characteristics of manufactured products are often well described by normal distributions. Suppose, for example, that we want to evaluate the length (in cm) of certain parts in a production process based on the information in a random sample of five such parts. The parts are required to have a nominal length of 20 cm; past experience with this process indicates that the standard deviation is known to be $\sigma = 1.8$ cm. If we assume that the lengths can be described by a normal distribution, what is the probability that the mean of this sample will be within 2 mm of the current process mean μ? That is, what is the probability that $\bar{x}$ will lie between $\mu - 2$ and $\mu + 2$?

The solution to this type of problem lies in recognizing that the sampling distribution of $\bar{x}$ is normal with a mean of $\mu_{\bar{x}} = \mu$ and standard error of $\sigma_{\bar{x}} = \sigma/\sqrt{n} = 1.8/\sqrt{5} = .805$. To find the probability $P(\mu - 2 < \bar{x} < \mu + 2)$, we standardize, making sure to use the mean and standard error of $\bar{x}$ while doing this:

$$P(\mu - 2 < \bar{x} < \mu + 2) = P\left(\frac{\mu - 2 - \mu}{\dfrac{\sigma}{\sqrt{n}}} < z < \frac{\mu + 2 - \mu}{\dfrac{\sigma}{\sqrt{n}}} \right)$$

$$= P\left(\frac{-2}{.805} < z < \frac{2}{.805} \right) = .9868$$

That is, there is a 98.68% chance that the mean of a random sample of size $n = 5$ will be within 2 units of the population mean μ. Notice how the unknown mean μ cancels itself during the standardization. In other words, we do not need to know (or assume) a value for μ. Instead, when we select our sample of five parts, we can be relatively confident that the sample mean will be no farther than 2 cm from the true (unknown) process mean.

The Central Limit Theorem

As we have just seen, prior knowledge about the shape of a population distribution determines the sampling distribution of $\bar{x}$. Unfortunately, possessing such knowledge is more often the exception than the rule. In many applications, we are faced with sampling from populations whose distributions are, at best, only approximately understood or that sometimes deviate markedly from normality.

The remedy for this problem is to rely more heavily on the sampling process and less on our knowledge of the population. It is a fortunate and somewhat surprising fact that a complete knowledge of a population distribution is not necessary, as long as we compensate by selecting a large enough sample. *By using a moderately large sample size n, it can be shown that the sampling distribution of $\bar{x}$ is approximately normal, regardless of the particular population distribution.* This result is known as the Central Limit Theorem.

> ### The Central Limit Theorem
>
> The sampling distribution of $\bar{x}$ can be approximated by a normal distribution when the sample size n is sufficiently large, irrespective of the shape of the population distribution. The larger the value of n, the better the approximation.

Although the sampling distribution of $\bar{x}$ must reflect certain features of the population being sampled (especially its location, μ), the *shape* of the sampling distribution is primarily influenced by n. That is, as n increases, the particular shape of the population (e.g., uniform, exponential, normal, Weibull) exerts less and less influence on the shape of the sampling distribution, which becomes more and more normal in appearance. Figure 5.20 illustrates this effect for several different populations. The closer the population is to being normal, the more rapidly the sampling distribution of $\bar{x}$ approaches normality. For instance, we saw this behavior emerging in Figure 5.15, where even small samples of size $n = 2$ from a uniform population result in a sampling distribution that is already beginning to take on the characteristic normal shape.

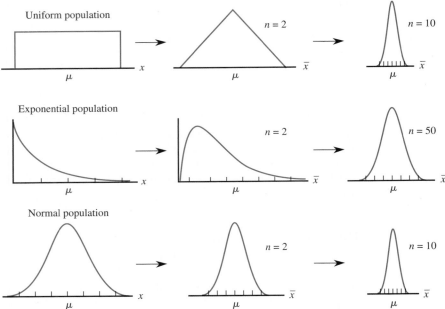

Figure 5.20 The Central Limit Theorem: The sampling distribution of $\bar{x}$ approaches a normal distribution as the sample size n increases

Many authors use $n \geq 30$ as a rough guide for what constitutes a "large enough" sample size for invoking the Central Limit Theorem. This is not a bad rule in general, but there are cases where substantially smaller values of n will suffice (e.g., with symmetric populations like the uniform and the normal), as well as cases where larger sample sizes are needed (especially for highly skewed populations). As a rule, *the less symmetric a population is, the larger the sample size will have to be to ensure normality of $\bar{x}$*. For example, in the case of an exponential population, sample sizes of 40 to 50 are often required to achieve normality.

Example 5.19 Consider the distribution shown in Figure 5.21 for the amount purchased (rounded to the nearest dollar) by a randomly selected customer at a particular gas station (a similar distribution for purchases in Britain (in pounds) appeared in the article "Data Mining for Fun and Profit," *Statistical Science*, 2000: 111–131; there were big spikes at the values 10, 15, 20, 25, and 30). The distribution is obviously quite nonnormal. We asked Minitab to select 1000 different samples, each consisting of $n = 15$ observations,

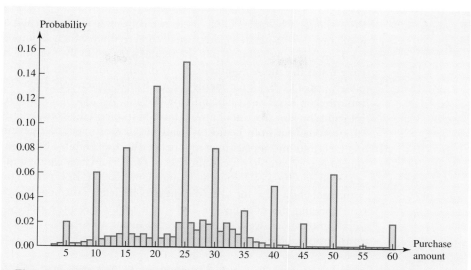

Figure 5.21 Probability distribution of x = amount of gasoline purchased ($)

and calculate the value of the sample mean $\bar{x}$ for each one. Figure 5.22 is a histogram of the resulting 1000 values; this is the approximate sampling distribution of $\bar{x}$ under the specified circumstances. This distribution is clearly approximately normal even though the sample size is not very large. A normal quantile plot based on the 1000 $\bar{x}$ values exhibits a very prominent linear pattern.

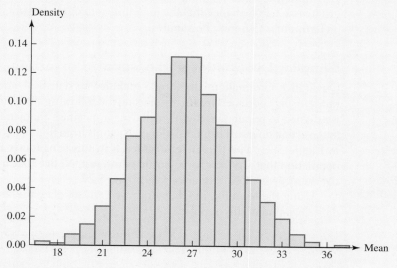

Figure 5.22 Approximate sampling distribution of the sample mean amount purchased when $n = 15$ and the population is as shown in Figure 5.21

Example 5.20

Printed circuit boards (PCBs), used in electronic equipment such as computers and appliances, are laminated cards (usually green) upon which various electronic components are mounted. One step in the manufacture of PCBs uses machines to automatically insert the metal connecting pins on the components into the appropriate hole patterns on a PCB. Components of each type (e.g., resistors, capacitors) are adhesively mounted on large paper-tape rolls and fed into the machines, which then insert them into a PCB. The amount of time it takes to insert all the components on a given PCB varies somewhat from board to board because of machine downtime for replenishing tape rolls and replacing components with broken pins. Suppose that an insertion machine can complete a certain type of PCB in an average time of 3 minutes with a standard deviation of .5 minute. If an order of 100 PCBs is run on this machine, what is the probability that the average time to complete all the boards exceeds 3.1 minutes?

Viewing the completion times as a random sample from a population with $\mu = 3$ and $\sigma = .5$, we can calculate the mean and standard error of the sampling distribution of the average completion time (of the 100 boards) as follows:

$$\mu_{\bar{x}} = \mu = 3 \quad \text{and} \quad \sigma_{\bar{x}} = \frac{\sigma}{\sqrt{n}} = \frac{.5}{\sqrt{100}} = .05$$

Because the sample size $n = 100$ is large, the Central Limit Theorem allows us to use the normal distribution to calculate the desired probability:

$$P(\bar{x} > 3.1) \approx P\left(z > \frac{3.1 - 3}{.05}\right)$$
$$= P(z > 2) = 1 - P(z \le 2) = 1 - .9772 = .0228$$

That is, there is only a 2.28% chance that the average completion time will exceed 3.1 minutes. Since $\bar{x} > 3.1$ is equivalent to $(100)\,\bar{x} > 100(3.1)$, we can also state that there is a 2.28% chance that the *total* time for completing the 100 boards will exceed 310 minutes (5 hours, 10 minutes).

Sampling Distribution of the Sample Proportion

Qualitative information can also be included in statistical studies. To do this, we first numerically code such information using the following simple device: The number "1" is assigned to population members having a specified characteristic and "0" is assigned to those that do not. The population that results from this 0–1 coding scheme is pictured in Figure 5.23. The parameter of interest in this situation is π, the proportion of the population that has the characteristic of interest. Notice that π is also the height of the bar associated with the value of 1 in Figure 5.23.

Using the formulas in Chapter 2, we can calculate the mean, variance, and standard deviation of this population:

$$\mu = \sum_x xp(x) = 0(1 - \pi) + 1 \cdot \pi = \pi$$

$$\sigma^2 = \sum_x (x - \mu)^2 p(x) = \pi(1 - \pi)$$

$$\sigma = \sqrt{\pi(1 - \pi)}$$

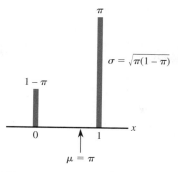

Figure 5.23 The distribution of coded values of a qualitative characteristic: "1" denotes that the specified characteristic is present; "0" indicates that it is not

Every random sample drawn from such a population will consist entirely of 0s and 1s. Suppose, for instance, that a particular sample of size 10 contains the observations $\{0, 0, 1, 1, 0, 1, 0, 0, 1, 0\}$. Then the sample mean is $(0 + 0 + 1 + 1 + 0 + 1 + 0 + 0 + 1 + 0)/10 = .40$. That is, the sample mean is simply the *proportion* of 1s in the sample. We use the notation p to denote the proportion of successes, also called the **sample proportion,** in a random sample of size n.

Since p is actually a sample mean, we can use the earlier results in this section to determine its sampling distribution. For example, the mean and standard error of the sampling distribution of p are given by

$$\mu_p = \mu = \pi \qquad \text{and} \qquad \sigma_p = \frac{\sigma}{\sqrt{n}} = \frac{\sqrt{\pi(1 - \pi)}}{\sqrt{n}} = \sqrt{\frac{\pi(1 - \pi)}{n}}$$

Furthermore, for a sufficiently large sample size n, the Central Limit Theorem indicates that the sampling distribution of p will be approximately normal. Because we record only whether each sampled item has a certain characteristic or not, large samples are often easy to come by when estimating a population proportion π. As a general rule, the accuracy of the normal approximation is best when *both* $n\pi \geq 5$ and $n(1 - \pi) \geq 5$.

Sampling Distribution of p

The mean and standard error of the sampling distribution of p are given by

$$\mu_p = \pi \qquad \text{and} \qquad \sigma_p = \sqrt{\frac{\pi(1 - \pi)}{n}}$$

In addition, for a large enough n, the sampling distribution of p is approximately normal. In general, the normal approximation is best when *both* $n\pi \geq 5$ and $n(1 - \pi) \geq 5$.

The fact that the formulas for μ_p and σ_p both contain the unknown parameter π might at first appear to negate the usefulness of the sampling distribution of p. After all, if the population proportion π is unknown, how can we possibly find $\sqrt{\pi(1-\pi)/n}$? In practice, there are two relatively simple solutions to this problem: (1) Use a predetermined value of π that describes some hypothetical value of π against which the sample data is to be compared or (2) use $\pi = 1/2$ in the formula for σ_p, which results in a conservatively large value of σ_p.

The second approach is based on the observation that $\pi(1-\pi) \leq .25$ for any value of π between 0 and 1.[1] This means that

$$\sigma_p = \sqrt{\frac{\pi(1-\pi)}{n}} \leq \sqrt{\frac{.25}{n}} = \frac{1}{2\sqrt{n}}$$

no matter what the true value of π. Thus, by choosing the sample size n large enough, $1/2\sqrt{n}$ (and hence σ_p) can be made as small as desired. This approach is commonly used in all forms of survey sampling.

Example 5.21	**Control charts** are graphs that monitor the movements in a sample statistic (such as $\bar{x}$ or p) in periodic samples taken from an ongoing process. Using the sampling distribution of the statistic as a yardstick, values of the statistic "too far" away from the center of the sampling distribution are taken to be signals of possible problems with the process. For example, a **p chart** is often used to monitor the proportion of nonconforming products in a manufacturing process. Using past data from the process, a value of π is selected as being representative of the long-run behavior of the process. Suppose, for example, that a certain process constantly generates an average of about 5% nonconforming products and that samples of size 100 are taken each day to test whether the 5% nonconformance rate has changed. On one particular day, 12 nonconforming products appear in the sample. How do we interpret this information?

Assuming that the process is behaving as it has in the past, we set $\pi = .05$. For this value of π, $n\pi = 100(.05) = 5$ and $n(1-\pi) = 100(.95) = 95$, so the condition for applying the normal approximation is met. Furthermore, the mean and standard deviation of the sampling distribution of p can be calculated:

$$\mu_p = .05 \quad \text{and} \quad \sigma_p = \sqrt{\frac{\pi(1-\pi)}{n}} = \sqrt{\frac{(.05)(1-.05)}{100}} = .0218$$

Because the sampling distribution of p is approximately normal, we can evaluate the sample proportion of $p = 12/100 = .12$ by determining how far away it is from the mean of .05. Since $(.12 - .05)/.0218 = 3.21$, we see that the value of .12 is 3.21 standard deviations above the process mean. In other words, this sample result has a very small probability of occurring *if* the process is running as usual. Our conclusion is that it is more likely that something has caused an increase in the process nonconformance rate.

---

[1] Writing $\pi(1-\pi)$ as $1/4 - (1/2 - \pi)^2$ you can see that the maximum value $1/4$ occurs when $\pi = 1/2$. Alternatively, you could use calculus, setting the derivative of $\pi(1-\pi)$ equal to 0, to find that $\pi = 1/2$ maximizes the quantity $\pi(1-\pi)$.

Section 5.6 Exercises

48. The inside diameter of a randomly selected piston ring is a random variable with a mean of 12 cm and a standard deviation of .04 cm.
    a. Where is the sampling distribution of $\bar{x}$ centered? What is the standard deviation of the sampling distribution of $\bar{x}$?
    b. Answer the questions in part (a) for sample means based on samples of size 64.
    c. Which is more likely to lie within .01 cm of 12 cm, the mean of a random sample of size 16 or the mean of a random sample of size 64?

49. A survey of the members of a large professional engineering society is conducted to determine their views on proposed changes to an ASTM measurement standard. Suppose that 80% of the entire membership favor the proposed changes.
    a. Calculate the mean and standard error of the sampling distribution of the proportion of engineers in samples of size 25 who favor the proposed changes.
    b. Calculate the mean and standard error of the sampling distribution of the proportion of engineers in samples of size 25 who do not favor the proposed changes.
    c. Calculate the mean and standard error of the sampling distribution of the proportion of engineers in samples of size 100 who favor the proposed changes.

50. A random sample of size 25 is selected from a large batch of electronic components, and the proportion of defective items in the sample is recorded. The proportion of defective items in the entire batch, however, is unknown. What is the maximum value that the standard error of the sampling distribution of the sample proportion could have?

51. Refer to Exercise 48. Assume that the distribution of piston diameters is known to be normal.
    a. Calculate the probability $P(11.99 \leq \bar{x} \leq 12.01)$ when $n = 16$.
    b. Calculate the probability $P(11.99 \leq \bar{x} \leq 12.01)$ when $n = 64$.

52. Let $x_1, x_2, x_3, \ldots, x_{100}$ denote the actual weights of 100 randomly selected bags of fertilizer.

    a. If the expected weight of each bag is 50 lb and the standard deviation of bag weights is known to be 1 lb, calculate the approximate value of $P(49.75 \leq \bar{x} \leq 50.25)$ by relying on the Central Limit Theorem.
    b. If the expected weight per bag is 49.8 lb rather than 50 lb (so that, on average, the bags are underfilled), calculate $P(49.75 \leq \bar{x} \leq 50.25)$.

53. The lifetime of a certain battery is normally distributed with a mean value of 8 hours and a standard deviation of 1 hour. There are four such batteries in a package.
    a. What is the probability that the average lifetime of the four batteries exceeds 9 hours?
    b. What is the probability that the total lifetime of the batteries will exceed 36 hours?
    c. If T denotes the total lifetime of the four batteries in a randomly selected package, find the numerical value of T_0 for which $P(T \geq T_0) = .95$.
    d. Refer to your answer to part (c). Suppose the battery manufacturer guarantees that any package of batteries that does not yield a total lifetime of T_0 hours will be replaced free of charge to the customer. If it costs the manufacturer $3.00 to replace a package of batteries (materials plus mailing to customer), calculate the expected replacement cost per package associated with a large shipment of batteries.

54. The Rockwell hardness of certain metal pins is known to have a mean of 50 and a standard deviation of 1.5.
    a. If the distribution of all such pin hardness measurements is known to be normal, what is the probability that the average hardness for a random sample of 9 pins is at least 52?
    b. What is the approximate probability that the average hardness in a random sample of 40 pins is at least 52?

55. Suppose that the sediment density (in g/cm^3) of specimens from a certain region is normally distributed with a mean of 2.65 and a standard deviation of .85 ("Modeling Sediment and Water Column Interactions for Hydrophobic Pollutants," *Water Research*, 1984: 169–174).

a. If a random sample of 25 such specimens is selected, what is the probability that the sample average sediment density is at most 3.00? Between 2.65 and 3.00?

b. How large a sample would be required to ensure that the first probability in part (a) is at least .99?

56. The number of flaws x on an electroplated automobile grill is known to have the following probability mass function:

x:	0	1	2	3
$p(x)$:	.8	.1	.05	.05

a. Calculate the mean and standard deviation of x.

b. What are the mean and standard deviation of the sampling distribution of the average number of flaws per grill in a random sample of 64 grills?

c. For a random sample of 64 grills, calculate the approximate probability that the average number of flaws per grill exceeds 1.

57. Only 2% of a large population of 100-ohm gold-band resistors have resistances that exceed 105 ohms.

a. For samples of size 100 from this population, describe the sampling distribution of the sample proportion of resistors that have resistances in excess of 105 ohms.

b. What is the probability that the proportion of resistors with resistances exceeding 105 ohms in a random sample of 100 will be less than 3%?

58. In Exercise 36, what is the probability that the average of two measurements will lie within 2 mm of the true length of the object?

59. Roughly speaking, the Central Limit Theorem says that sums of independent random variables tend to have (approximately) normal distributions. Similarly, it can be shown that products of independent *positive* random variables tend to have lognormal distributions. Recall from Section 1.5 that a random variable x is said to have a lognormal distribution with parameters μ and σ if the random variable $y = \ln(x)$ is normal with mean μ and standard deviation σ. The successive breaking of particles into finer and finer pieces, a process that can be modeled as a product of positive random variables, leads to lognormal particle size distributions. In particular, small particles suspended in the atmosphere (called aerosols) have radii that can be described by a lognormal distribution, with parameters $\mu = -2.62$ and $\sigma = .788$ (Crow, E. L., and K. Shimizu, *Lognormal Distributions: Theory and Applications*, Marcel Dekker, New York, 1988: 337).

a. Find the mean radius (in μm) of the atmospheric particles.

b. What is the probability that an atmospheric particle will have a radius exceeding .12 μm?

Supplementary Exercises

60. Figure 5.5 shows how a tree diagram can be used to verify that $\{A \text{ or } B\}' = \{A' \text{ and } B'\}$. Use a Venn diagram to prove this fact.

61. A large farming area is divided into five parcels of land of different sizes, as follows:

Parcel:	B_1	B_2	B_3	B_4	B_5
Size (acres):	15	20	25	10	20

Because crop-bearing trees are uniformly planted within each parcel, the probability that a randomly sampled tree from the farm comes from a particular parcel is assumed to be proportional to the size of the parcel.

a. What is the probability that a randomly chosen tree comes from one of the first three parcels of land?

b. What is the probability that a randomly chosen tree does not come from parcel 5?

62. A complex assembly contains 20 critical components (labeled C_1, C_2, . . .), each having a probability of .95 of functioning correctly. Each component must function correctly for the entire assembly to function. Let A denote the event that the assembly fails to function correctly and let B denote the event that component C_1 fails to function correctly.

a. Give a verbal description of the expressions $A \mid B$ and $B \mid A$.

b. Does $P(A \mid B) = P(B \mid A)$?

63. A battery-operated tool requires that each of its four batteries operate correctly to provide sufficient power to the tool. If each battery operates independently of the others and each has a .10 chance of failing over a 30-hour period of operation, what is the probability that the tool will fail sometime during the 30-hour operating period?

64. Two pumps that are connected in parallel fail independently of one another on any given day. The probability that only one pump fails is .10, and the probability that neither of the two pumps fails is .05. What is the probability that both pumps fail on a given day? *Hint:* Use a Venn diagram.

65. Find a formula for the probability that at least one of two independent events occurs. (*Hint:* If events A and B are independent, then so are the pairs of events A' and B, A and B', and A' and B'.)

66. Let x denote the number of nonzero digits in a randomly selected zip code.
a. List the possible values of the random variable x.
b. Can two or more zip codes have the same value of x?

67. A continuous random variable x has a density function of the form $f(x) = .5x$ over the interval $[0, b]$.
a. Find b.
b. Find the mean of the variable x.
c. Find the standard deviation of the variable x.

68. According to the article "Optimization of Distribution Parameters for Estimating Probability of Crack Detection" (*J. of Aircraft*, 2009: 2090–2097), the following "Palmberg" equation is commonly used to determine the probability $P_d(c)$ of detecting a crack of size c in an aircraft structure:

$$P_d(c) = \frac{(c/c^*)^\beta}{1 + (c/c^*)^\beta}$$

where c^* is the crack size that corresponds to a .5 detection probability (and thus is an assessment of the quality of the inspection process).

a. Verify that $P_d(c^*) = .5$.
b. What is $P_d(2c^*)$ when $\beta = 4$?
c. Suppose an inspector inspects two different panels, one with a crack size of c^* and the other with a crack size of $2c^*$. Again assuming $\beta = 4$ and also that the results of the two inspections are independent of one another, what is the probability that exactly one of the two cracks will be detected?

69. "Travelers" are documents that accompany a product as it sequences through various production steps. Travelers contain manufacturing instructions pertaining to the particular item or order. Suppose that each of 30 data fields on a particular traveler has a .5% chance of being filled out incorrectly. Assume that each field is independent of the others.
a. What is the probability that a given traveler will contain at least one incorrect field?
b. What is the probability that a traveler is free of errors?
c. What is the probability of finding two or more errors on a traveler?

70. A continuous signal is sent over a communication channel. The number of errors per second, x, at the receiving end of the channel has a normal distribution with a mean and standard deviation of 3 and .8 errors per second, respectively.
a. In any given 1-second period, what is the probability that no errors are transmitted?
b. Find the probability of transmitting two or more errors per second.
c. What is the probability that more than five errors per second will be transmitted?

71. Use spreadsheet (e.g., Excel™) or other software to approximate the sampling distribution of the sample mean.
a. Generate at least 100 samples of size 10 from a uniform distribution on the interval $[10, 20]$. Create a histogram of the 100 sample means, and describe the shape of the histogram.
b. Repeat part (a) by generating 100 samples of size 10 from an exponential distribution with a mean of 5.
c. Compare the shapes of the histograms in parts (a) and (b), and offer an explanation for any differences that you observe.

72. An electrical appliance uses four 1.5-volt batteries. The batteries are connected in series so that the total voltage supplied to the appliance is the sum of the voltages in the four batteries. Suppose that the actual voltage of all 1.5-volt batteries is known to have a mean of 1.5 volts and a standard deviation of .2 volt.
    a. What are the mean and standard error of the sampling distribution of the average voltage in four randomly selected 1.5-volt batteries?
    b. What is the mean of the sampling distribution of the total voltage in four randomly selected 1.5-volt batteries?

73. Five randomly selected 100-ohm resistors are connected in a series circuit. Suppose that it is known that the population of all such resistors has a mean resistance of exactly 100 ohms with a standard deviation of 1.7 ohms.
    a. What is the probability that the average resistance in the circuit exceeds 105 ohms?
    b. What is the probability that the total resistance in the circuit differs from 500 ohms by more than 11 ohms?
    c. Find the number of resistors, n, for which $p(490 \leq T \leq 510) = .95$, where T denotes the total resistance in the circuit.

74. The article "Three Sisters Give Birth on the Same Day" (*Chance*, Spring 2001, 23–25) used the fact that three Utah sisters had all given birth on March 11, 1998 as a basis for posing some interesting questions regarding birth coincidences.
    a. Disregarding leap year and assuming that the other 365 days are equally likely, what is the probability that three randomly selected births all occur on March 11? Be sure to indicate what, if any, extra assumptions you are making.
    b. With the assumptions used in part (a), what is the probability that three randomly selected births all occur on the same day?
    c. The author suggested that, based on extensive data, the length of gestation (time between conception and birth) could be modeled as having a normal distribution with mean value 280 days and standard deviation 19.88 days. The due dates for the three Utah sisters were March 15, April 1, and April 4, respectively. Assuming that all three due dates are at the mean of the distribution,

what is the probability that all births occurred on March 11? *Hint:* The deviation of birth date from due date is normally distributed with mean 0.
    d. Explain how you would use the information in part (c) to calculate the probability of a common birth date.

75. A friend who lives in Los Angeles makes frequent consulting trips to Washington, DC; 50% of the time she travels on airline #1, 30% of the time on airline #2, and the remaining 20% of the time on airline #3. For airline #1, flights are late into DC 30% of the time and late into LA 10% of the time. For airline #2, these percentages are 25% and 20%, whereas for airline #3 the percentages are 40% and 25%. If we learn that on a particular trip she arrived late at exactly one of the two destinations, what are the posterior probabilities of having flown on airlines #1, #2, and #3? *Hint:* From the tip of each first-generation branch on a tree diagram, draw three second-generation branches labeled, respectively, 0 late, 1 late, and 2 late.

76. A factory uses three production lines to manufacture cans of a certain type. The accompanying table gives percentages of nonconforming cans, categorized by type of nonconformance, for each of the three lines during a particular time period:

	Line 1	Line 2	Line 3
Blemish	15	12	20
Crack	50	44	40
Pull-tab problem	21	28	24
Surface defect	10	8	15
Other	4	8	2

During this period, line 1 produced 500 nonconforming cans, line 2 produced 400 such cans, and line 3 was responsible for 600 nonconforming cans. Suppose that one of these 1500 cans is randomly selected.
    a. What is the probability that the can was produced by line 1? That the reason for nonconformance is a crack?
    b. If the selected can came from line 1, what is the probability that it had a blemish?
    c. Given that the selected can had a surface defect, what is the probability that it came from line 1?

77. One satellite is scheduled to be launched from Cape Canaveral in Florida, and another launching is scheduled for Vandenberg Air Force Base in California. Let A denote the event that the Vandenberg launch goes off on schedule, and let B represent the event that the Cape Canaveral launch goes off on schedule. If A and B are independent events with $P(A) > P(B)$ and $P(A \ or \ B) = .626$, $P(A \ and \ B) = .144$, determine the values of $P(A)$ and $P(B)$.

78. A message is transmitted using a binary code of 0s and 1s. Each transmitted bit (0 or 1) must pass through three relays before reaching a receiver. At each relay, the probability is .20 that the bit sent is different from the bit received (a reversal). Assume that relays operate independently of one another.

Transmitter $\rightarrow$ Relay 1 $\rightarrow$ Relay 2 $\rightarrow$ Relay 3 $\rightarrow$ Receiver

   a. If a 1 is sent from the transmitter, what is the probability that a 1 is sent by all three relays?
   b. If a 1 is sent from the transmitter, what is the probability that a 1 is received by the receiver? *Hint:* Use a tree diagram.
   c. Suppose that 70% of all bits sent from the transmitter are 1s. If a 1 is received by the receiver, what is the probability that a 1 was sent? *Hint:* Use a tree diagram.

Bibliography

Devore, J. L. and K. N. Berk, **Modern Mathematical Statistics with Applications** (2nd ed.), Springer, New York, 2012. *A more mathematical treatment than given in this text, but still readable, with good examples and problems.*

Olofsson, P., **Probabilities: The Little Numbers That Rule Our Lives**, Wiley, New York, 2007. *An outstanding non-mathematical exposition, with great insights.*

Ross, S., **A First Course in Probability** (8th ed.), Wiley, New York, 2009. *A succinct mathematical treatment with good examples and problems.*

Sachs, L., **Applied Statistics: A Handbook of Techniques** (2nd ed.), Springer, New York, 1984. *A one-volume summary of statistical methods that emphasizes short summaries of essentials, easy examples, tables, notes, and detailed references for further reading.*

hilmi_m/iStockphoto.com

Quality and Reliability

6 (chapter number)

INTRODUCTION

Statistical methods for monitoring and improving the quality of manufactured goods have been around since the early 1920s when Bell Laboratories engineer W. A. Shewhart introduced the graphical **control chart** method for detecting possible problems in manufacturing processes (Sections 6.2, 6.3, and 6.5). Current applications of statistical methods of quality assurance have widened to include service industries as well as traditional manufacturing applications. Since the 1980s, there has also been a greatly increased emphasis on the use of **experimental design** techniques that seek to identify the key factors that lead to improvements in processes and products. Experimental design methods, which were briefly described in Section 4.3, are discussed in detail in Chapters 9 and 10. Although the focus in Chapter 6 is on the various control charts that have been developed to monitor existing production systems, we also include a discussion of the important topic of evaluating the reliability of finished products (Section 6.6).

The statistical tools underlying the methods of this chapter are fairly basic. Calculations of tail areas of normal distributions are used in Section 6.4 to estimate the **capability** of a production process to produce acceptable products. Control chart methods in the remaining sections are based on knowing the **sampling distribution** (Sections 5.5 and 5.6) of the various statistics used to describe the output of a production process. **Histograms** not only provide convenient summaries of process data but are also used to detect potential process problems (Section 6.1).

6.1   TERMINOLOGY

Applying statistics to a specific field, such as quality control, requires some knowledge of the jargon used in that field. The terminology introduced subsequently is used throughout the remaining sections of this chapter. In some cases, familiar statistical terms (such as *discrete* and *continuous* measurements) are given different names by quality practitioners, making it necessary to know both names when working in this field.

Specification Limits

When product designs are translated into tangible entities, it becomes necessary to precisely define the key characteristics of a product and each of its subcomponents. For manufactured products, this is done by specifying the exact physical dimensions and other quality characteristics that finished products should have. For services, specifications often take the form of rules for processing transactions or guidelines for interacting with customers. In many cases, especially in manufacturing, a single value corresponds to the most desired quality level for a given product characteristic. We refer to this value as the **nominal** or **target** value of the quality characteristic.

Practically speaking, it is almost impossible to make each unit of product identical to the next, so some flexibility is required in achieving target values. This is done by choosing **specification limits** or **tolerances** that delineate the range of measured values that we will accept as "close enough" to the target value, in the sense that products that are within the specification range should be fit for their intended use.[1] For example, car doors are made with a certain nominal width, but specification limits are necessary because doors cannot be too wide (or they may not close properly) or too narrow (or they may fail to latch correctly). Quality characteristics that have both upper and lower specification limits are said to have a **two-sided tolerance.** Those with only one specification limit have a **one-sided tolerance.** Examples of characteristics with one-sided tolerances include breaking strengths of materials, which have lower specification limits, and the level of contaminants in a water supply, which have only upper specification limits.

Nominal values and their associated specification limits are generally stated together in an abbreviated form such as 1 in. ±.005 in., which describes a characteristic with a nominal value of 1 in., a lower specification limit of .995 in., and an upper specification limit of 1.005 in. Together, the nominal value and specification limits are called the *specifications* or, more simply, the "specs." When data do not exceed the specification limits placed on them, we say that the particular process giving rise to the data is "within specifications." Otherwise, the process is said to "fail the specifications" or to be "out of spec."

---

[1] In the 1970s, *quality* was defined to be "fitness for use." Around 1983, the American Society for Quality Control (ASQC) expanded the definition to "quality is the totality of features and characteristics of a product or service that bear on its ability to satisfy given needs."

> The largest allowable value that a quality characteristic can have is called the **upper specification limit (USL)**; the smallest allowable value is called the **lower specification limit (LSL)**.

Conformance and Nonconformance

When a product or process fails to meet its specifications, there is a need to classify the seriousness of the situation. Sometimes out-of-spec conditions lead to problems that are very serious and prevent a product from ever being used. At other times, problems caused by not meeting specifications may be only cosmetic. To distinguish between these two extremes, quality practitioners have adopted the following classification scheme. Products that do not meet their specifications are called **nonconforming** and the problems or flaws in such nonconforming items are called **nonconformities.** A nonconforming product is not necessarily unfit for any use. For example, the fact that a chemical concentration is lower than its LSL does not necessarily mean that the chemical will fail to have a desired effect; it may just require a longer reaction time than if the concentration exceeds the LSL. In the garment industry, shirts that have minor nonconformities are still usable and are sold as seconds in discount stores. However, when nonconformities become so serious that a product is no longer fit for its intended use, we say that it is **defective.** A defective product can contain one or more **defects** that cause it to be classified as defective.

> A product is **nonconforming** if it has one or more **nonconformities** that cause it, or an associated product or service, not to meet a specification requirement.
> A **defective** product is one that has one or more **defects** that cause it, or an associated product or service, not to satisfy intended usage requirements.

The Process Approach

In modern quality programs, each step in a product's manufacture or each step in a service procedure is viewed as a separate **process** to be performed. Every such process has inputs (from the preceding process steps) and outputs (for use in the succeeding steps). It is common practice to use a systems diagram to depict the various process steps and their interconnections (Figure 6.1). Quality control efforts that are directed at key processes or subprocesses, with an eye to solving problems and maintaining consistent output, are called **process control activities.** When statistical methods are used for this purpose, such activities are referred to as **statistical process control (SPC).** Managing these numerous applications of SPC and other quality improvement tools usually requires well-designed implementation programs, the most popular of which are TQM (Total Quality Management) and SIX-SIGMA. Detailed descriptions of these programs can be found in many quality control textbooks.

Using statistical methods to control processes is accomplished by identifying key product characteristics, measuring them, and then converting the data into sample statistics. Both **continuous** and **discrete** measurements are used in quality control procedures. Continuous data, those obtained from measuring instruments, is also called **variables**

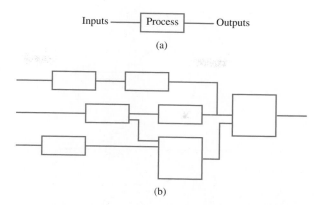

Figure 6.1 System diagrams: (a) envisioning each step in creating a product as a process; (b) products and services broken into a series of subprocesses

data in the quality professions. Discrete data, those that arise by counting things, is called **attributes data.** These names are commonly used to describe the various statistical techniques used in quality control. For example, control charts are classified as **variables control charts** or **attributes control charts,** depending on the kind of data used to form the charts (see Sections 6.2, 6.3, and 6.5).

Histograms

As shown in Figures 6.2 and 6.3, histograms are very effective tools for understanding processes that generate variables data. Because many processes tend to produce variables data that follow normal distributions, normal curves are often superimposed over such histograms. This technique is so commonly used that it is standard practice to describe a process's output by drawing a normal curve centered at the sample mean of the data, sometimes without even including the histogram of the data. When specification limits are included as well, we get a visual picture of how a process is behaving with respect to its specifications (Figure 6.2). From this figure, it is easy to see how much of the process data is nonconforming, that is, outside of the specification range.

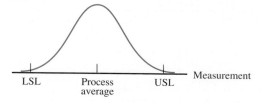

Figure 6.2 Normal curve describing process measurements

Histograms are often used to give warnings of possible process problems. A smoothly running process usually generates data whose histogram appears similar to that in Figure 6.2. Irregularities in a process are evidenced by histogram shapes that differ from a normal curve. Figure 6.3 shows some of the typical histogram shapes that can occur along with the most likely reasons for their appearance.

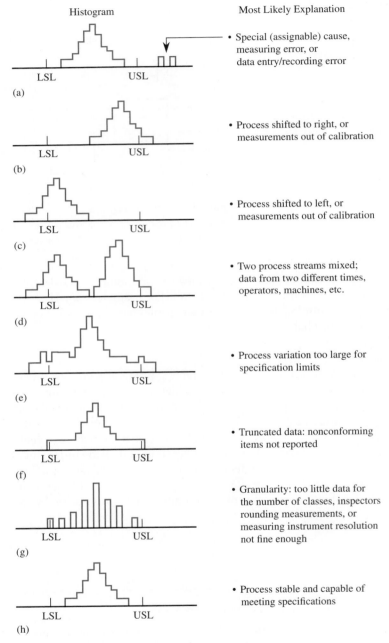

Figure 6.3 Typical histograms of process data

Section 6.1 Exercises

1. General-purpose resistors are color-coded with a sequence of four rings that identify the nominal value of the resistance (in ohms) and the plus and minus tolerance (expressed as a percentage of the nominal) to be expected in the actual resistance. For example, bands (in order) of green, blue, brown, and gold denote a resistor with a nominal value of 560 ohms and a tolerance of ±5%. What are the specification limits (in ohms) for the process that produces such resistors?

2. Determining whether structural materials conform to specifications often requires special test equipment (which can be expensive) and test procedures (which require specialized training). Consequently, independent testing and evaluation labs have arisen to perform such tests. One of the measures of the quality of the services provided by such labs is the waiting time before test results are available. Does the characteristic waiting time have a one- or a two-sided tolerance?

3. A standard legal envelope is 4 inches wide by 9.5 inches long. Normally, 8.5-inch by 11-inch pages are folded in thirds before they are inserted into such envelopes. Viewing page folding as a process whose measurable output is the width of the folded page, answer the following questions:
   a. What specification limit does the envelope size place on the page-folding process?
   b. What are the penalties for exceeding the upper specification?

4. Citrus products must have a certain sugar content, measured in degrees Brix, to be judged satisfactory to sell to grocery stores. Suppose that a certain batch of oranges fails to meet the specified Brix level. Which classification would you apply to these oranges, defective or nonconforming?

5. Measurements are to be taken on each of the following characteristics. In each case, indicate whether the resulting measurements would be classified as variables or attributes data.
   a. The number of flaws per square foot in a large sheet of metal
   b. The concentration of a chemical solution used in an electroplating process
   c. The thread diameter of a bolt
   d. The number of bolts in a batch that have oversize thread diameters
   e. The proportion of bolts in a batch that have oversize thread diameters
   f. The torque applied to an airplane wing fastener (bolts and nuts used in aerospace are called fasteners)
   g. The number of errors in 1000 lines of computer code
   h. The time between breakdowns of a certain machine
   i. The breaking strength of a molded plastic part

6. The following are measurements (in inches) of a quality characteristic with specification limits of $2.50 \pm .05$ in.:

   2.54 2.52 2.50 2.52 2.50 2.50 2.47 2.48
   2.51 2.53 2.53 2.51 2.50 2.47 2.49 2.50
   2.50 2.50 2.46 2.48 2.48 2.50 2.51 2.53
   2.51 2.53 2.53 2.52 2.47 2.51

   a. Create a histogram of the data.
   b. Estimate the mean and standard deviation of the process from which this data was taken.
   c. What percentage of these measurements falls above the USL? What percentage of the measurements falls below the LSL?
   d. Assuming that the process from which the data was taken can be described by a normal density function, what percentage of the process data is expected to fall above the USL [use your estimates from part (b)]? What percentage of the process data is expected to fall below the LSL?
   e. Explain the reason for the difference in your answers to parts (c) and (d).

7. If the measuring instrument used in Exercise 6 is out of calibration and is giving readings that are, say, .02 in. higher than the true length of an object, what is the effect on the estimated proportions of conforming and nonconforming products?

8. A cork intended for use in a wine bottle is considered acceptable if its diameter is between 2.9 cm

and 3.1 cm (so the lower specification limit is LSL = 2.9 cm, and the upper specification limit is USL = 3.1 cm).

a. If cork diameter is a normally distributed variable with mean value 3.04 cm and standard deviation .02 cm, what is the probability that a randomly selected cork will conform to specification?

b. If instead the mean value is 3.00 and the standard deviation is .05, is the probability of conforming to specification smaller or larger than it was in part (a)?

6.2   HOW CONTROL CHARTS WORK

The recognition that variation is unavoidable in every repetitive process was well understood by the early pioneers of statistical quality control. To identify, and, when possible, eliminate sources of process variation, W. A. Shewhart introduced the **control chart** method in 1924. Shewhart envisioned two types of variation that, when combined, account for all the variation in a process. The first type, **common cause variation,** is the result of the myriad imperceptible changes, or **common causes,** that occur in the everyday operation of a process. Common causes are essentially the noise in a production system and, as such, common cause variation is considered to be expected, but uncontrollable variation. Controllable variation, on the other hand, is variation for which we can find definite **assignable causes,** also called **special causes.** Assignable causes are frequently found when there are changes in brands of raw materials, turnover in the workforce, or machine wear or breakdown. Control charts are designed as a method for detecting the existence of assignable causes.

Control Charts

Control charts are constructed by taking successive samples from the output of a process, making measurements on the sampled items, and then plotting summary statistics of these results. Figure 6.4 shows a typical control chart. The samples, also called **subgroups,** of size n are taken at regular intervals of time. For each subgroup, a summary statistic is calculated and plotted (on the vertical axis) versus the subgroup number (on the horizontal axis). Any statistic of interest can be calculated, but the most commonly used are $\bar{x}$ (subgroup mean), R (subgroup range), s (subgroup standard deviation), p (proportion nonconforming), c (number of nonconformities), and u (nonconformities per unit). A control chart derives its name from the name of the particular statistic

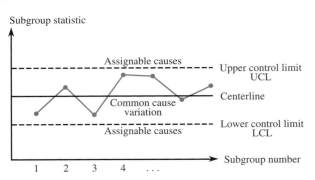

Figure 6.4 The Shewhart control chart

calculated in the subgroups. For example, an $\bar{x}$ chart (read "x bar chart") is one that monitors successive subgroup means, an R chart monitors subgroup ranges, and so forth.

The **control limits** and **centerline** of a control chart are based on the **sampling distribution** (see Sections 5.5 and 5.6) of the chart statistic. The smaller of the two control limits is called the **lower control limit (LCL)** and the larger one is called the **upper control limit (UCL)**. In the United States, control limits are set at a distance of 3 standard errors (i.e., 3 standard deviations of the subgroup statistic) from the mean of the sampling distribution. This is based on the fact that many sampling distributions closely approximate normal distributions, the majority of whose probability, about 99.73%, lies within 3 standard deviations of the mean. For example, in an $\bar{x}$ chart, the standard error $\sigma/\sqrt{n}$ of the sampling distribution of $\bar{x}$ is used to establish the control limits, which in theory would be set at $\mu \pm 3\sigma/\sqrt{n}$. In practice, of course, estimates of the process mean μ and process standard deviation σ must be used in this formula. In England and other countries, control limits are set by specifying the probability, typically around 99%, that lies under the sampling distribution curve between the control limits.

Plotted points that fall outside (i.e., above the UCL or below the LCL) are interpreted as signals of possible special causes, whereas points within the control limits are usually (but not always) associated with common cause variation, that is, the *absence* of special causes. It is also important to remember that control limits are different from specification limits, which are not plotted on a control chart.

Statistical Control

When all the points on a control chart lie between the control limits *and* when there are no other anomalous patterns in the charted points, a process is said to be in a state of **statistical control** or, more briefly, "in control." Otherwise, the process is said to be "out of control." The phrase *out of control*, which can sometimes be misinterpreted, is only a way of indicating that control chart points are behaving in a nonrandom fashion. It does not imply that the process itself is bad nor does it necessarily imply that any nonconforming products are being made. "Out of control" simply means that assignable causes are likely to be present.

When control charts were first introduced, the primary signal of an "out-of-control" condition was when one or more points were outside one of the control limits. If the sampling distribution of the subgroup statistic is approximately normal, this means that there is a probability of about .0027 (or .27%) that a control chart point will fall outside one of the control limits *when no assignable causes are present*. That is, when a process is running smoothly and no special causes are operating, there is a relatively small chance (.27%) that a control chart point will give a false positive—mistakenly signaling the presence of a special cause. On the other hand, when special causes are present, there is also a chance that the chart will fail to detect them. To increase the sensitivity of a chart for detecting special causes, while still maintaining the false positive rate at .27%, an extended set of "out-of-control" rules is often used. The "out-of-control" rules in Figure 6.5 are commonly used by quality control software to help detect the presence of special causes.

Rational Subgroups

Selecting rational subgroups is key to the proper use of control charts. The name *rational subgroup* is intended to remind us that the subgroups are chosen in a thoughtful

Each zone (A, B, C) has a width of 1 standard error.

Test 1. One point beyond Zone A

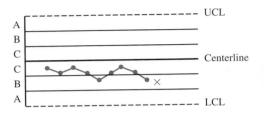

Test 2. Nine points in a row in
          Zone C or beyond

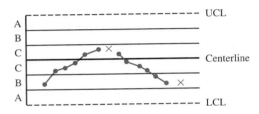

Test 3. Six points in a row steadily
          increasing or decreasing

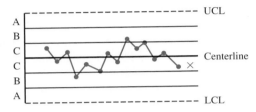

Test 4. Fourteen points in a row
          alternating up and down

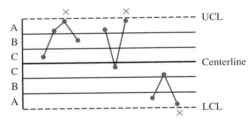

Test 5. Two out of three points in
          a row in Zone A or beyond

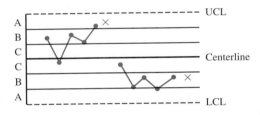

Test 6. Four out of five points in a
          row in Zone B or beyond

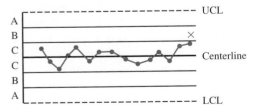

Test 7. Fifteen points in a row in
          Zone C (above and below
          centerline)

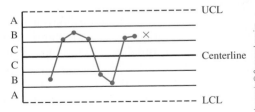

Test 8. Eight points in a row on
          both sides of centerline
          with none in Zone C

Figure 6.5 Extended list of "out-of-control" rules for Shewhart charts

manner and are *usually not random samples.* Instead, rational subgroups should be chosen in a way that maximizes the ability of the chart to detect special causes. The goal is to have the variation *within* any rational subgroup represent the common cause variation in the process. In this way, any significant variation *between* subgroups can be attributed to possible special causes. Randomness and sampling distributions enter the picture when we make the *assumption* that there are no special causes at work, in which case each rational subgroup can be considered to be a random sample from the process. That is, if a process is in control, then successive items (and subgroups of such items) should vary according to a system of random causes, which then permits us to use the properties of a sampling distribution to form control limits.

One commonly used method for forming rational subgroups is to choose subgroup elements over a fairly short span of time. The time span should be short enough so that it is unlikely for the occurrence of a special cause to overlap two subgroups. For example, if differences between raw materials are a potential source of process problems, then subgroups should be formed such that all elements in each subgroup correspond to only one type of raw material. Then, if a problem occurs when raw materials are changed, the data in all subgroups occurring after the change of materials will differ from the data in the subgroups taken before the change, and the control chart points calculated from such subgroups will have a good chance of detecting the problem.

A general strategy for deciding how to form rational subgroups is (1) to decide which causes are important to detect and which are not, then (2) to design subgroups that maximize the chance of detecting the important causes and relegate the unimportant causes to the within-subgroup variation. For instance, suppose that daily changes in temperature are known to have a small, but inconsequential, effect on the lengths of plastic parts, whereas impurities in batches of raw plastic pellets are known to have a serious effect on part lengths. If each batch of pellets lasts, say, for 4 hours of production, then subgroups of size 6 might be formed once an hour by selecting one part about every 10 minutes after a new batch of pellets is opened. In this way, each subgroup of 6 would represent a specific batch, but several different temperatures would be represented over each 1-hour collection period.

Section 6.2 Exercises

9. Two identical machines are used to make a particular metal part. The finished parts from both machines are mixed together on a conveyor system that moves the parts to a subsequent assembly operation. Consider the following two methods for generating rational subgroups for a control chart of this process:
   a. *Method 1:* Five parts per hour are sampled from the finished parts on the conveyor system each hour.
   b. *Method 2:* Before reaching the conveyor system, a sample of five parts is taken from the output of machine 1; an hour later, five parts are taken from the output of machine 2; an hour later, five parts are sampled from machine 1; and so forth.

   Which method of choosing rational subgroups would be better able to detect when one of the machines is not in statistical control?

10. When a process is in a state of statistical control, all of the points on a control chart should fall within the control limits. However, it is undesirable that all of the points should fall extremely near, or exactly on, the centerline of the control chart. Why?

11. U.S. companies commonly use 3-sigma limits to establish control limits. Some other countries (e.g., Great Britain) use control limits that are 3.09 sigmas from the chart's centerline.

   a. Using the normal distribution, what is the probability that a single control chart point falls above the UCL in a 3-sigma control chart?

   b. Using the normal distribution, what is the probability that a single control chart point falls above the UCL in a 3.09-sigma control chart?

12. Suppose that the measuring instrument used to obtain data from a certain process is out of calibration, so that each of its reported measurements is off by $+\delta$ units from the true value. What effect does this have on the signals given by the $\bar{x}$ and R charts?

13. Using the extended list of "out-of-control" rules in Figure 6.5 (page 254), determine whether the processes that give rise to the adjacent control charts appear to be in statistical control. Circle any points at which an out-of-control condition is first signaled.

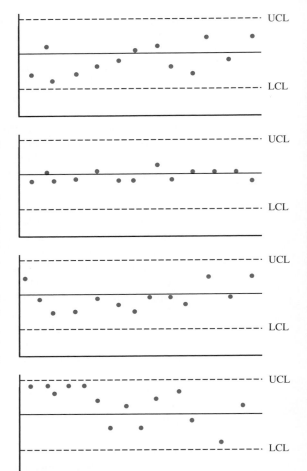

Figure for Exercise 13

6.3   CONTROL CHARTS FOR MEAN AND VARIATION

In this section, we introduce the most commonly used Shewhart charts for monitoring the mean and variation of a process. The important thing to remember about such charts is that they are generally used in pairs, one chart to track the process average and one for the process variation. Furthermore, the chart for process variation is created first because its centerline is a key ingredient in calculating the control limits for the chart that monitors the process average.

Shewhart originally used the sample mean $\bar{x}$ and the sample range R as the subgroup statistics to use in control charts for variables data. These charts, called $\bar{x}$ and R charts (read "x bar and R charts") are still among the most frequently used variables control charts. They serve as the prototype for understanding how all other Shewhart charts are intended to operate.

Theoretically, the control limits for the $\bar{x}$ chart are based on 3-sigma limits of the sampling distribution of the statistic $\bar{x}$:

$$\text{UCL} = \mu + 3\frac{\sigma}{\sqrt{n}} \qquad \text{and} \qquad \text{LCL} = \mu - 3\frac{\sigma}{\sqrt{n}}$$

where μ and σ denote, respectively, the long-run process mean and standard deviation of the process. Of course, these formulas cannot be used directly since both μ and σ must first be estimated from the available process data. The process average is estimated by the average of k successive subgroup means:

$$\bar{\bar{x}} = \frac{1}{k}\sum_{i=1}^{k}\bar{x}_i$$

The estimate is denoted by $\bar{\bar{x}}$ (read "x double bar") because it is an average of several averages; $\bar{\bar{x}}$ is also called the *grand mean* of the subgroup means. To obtain a reasonable estimate of σ, the following two-stage procedure is used. First, the chart for process variation (the R chart) is brought into statistical control. This ensures that the process variation is stable and, therefore, that the centerline of the R chart is a reliable estimate of the average range of subgroups of size n from the process. Second, this centerline is converted into an estimate of the process standard deviation σ, which is then put into the expression $\bar{\bar{x}} \pm 3\sigma/\sqrt{n}$ to obtain the approximate control limits for the $\bar{x}$ chart. Fortunately, the control limits of the R chart also turn out to be simple functions of the centerline of the R chart.

The R Chart

To construct an **R chart,** we use the data from some number, k, of successive subgroups of process measurements. It is usually recommended that about 20 to 25 subgroups be used. If possible, the same sample size n is used to form each subgroup. The centerline of the R chart is denoted by $\bar{R}$ and is calculated by averaging the sample ranges R_1, R_2, $R_3, \ldots, R_k$ of the k subgroups:

$$\bar{R} = \frac{1}{k}\sum_{i=1}^{k}R_i$$

$\bar{R}$ serves as an estimate of μ_R, the mean of the sampling distribution of the ranges (for samples of size n) from the process. Let σ_R denote the standard deviation of this sampling distribution; the 3-sigma limits, $\mu_R \pm 3\sigma_R$, are used to form the control limits for the R chart. Assuming that the process measurements can be adequately described by a normal distribution, it can be shown that the control limits for the R chart are given by

$$\text{UCL} = D_4\bar{R} \qquad \text{and} \qquad \text{LCL} = D_3\bar{R}$$

where D_3 and D_4 are constants that depend on the subgroup size, n. Values of D_3 and D_4 are found in Appendix Table XI, which lists such constants for a variety of different types of control charts.

After finding the centerline and control limits, the R chart is constructed by simply plotting the k subgroup ranges $R_i(i = 1, 2, \ldots, k)$ versus the subgroup index, i, and then drawing horizontal lines to represent the centerline $\bar{R}$ and control limits. Using the "out-of-control" rules listed in Section 6.2, we examine the R chart to see whether these k ranges seem to be in statistical control. If any out-of-control conditions are found, it is recommended that the subgroup(s) associated with these problems be eliminated and that the centerline and control limits be recalculated based on the reduced number of subgroups. When doing this, subgroups should be eliminated only if definite assignable causes can be found for the out-of-control signal associated with these subgroups. Out-of-control subgroups for which no assignable cause can be found should not be eliminated.

When the R chart is deemed to be in a state of statistical control, the centerline $\bar{R}$ can then be considered to be a reliable estimate of the average range (of samples of size n) from a normal population. This estimate can then be converted into an estimate for the process standard deviation by means of the formula

$$\hat{\sigma} = \frac{\bar{R}}{d_2}$$

where d_2 is found in the table of control chart constants (Appendix Table XI). The estimate $\hat{\sigma}$ of σ is used to calculate the control limits of the $\bar{x}$ chart and to assess the capability of the process to meet the specification limits (see Section 6.4).

The $\bar{x}$ Chart

Once the R chart is in control, the $\bar{x}$ chart is then constructed. Any subgroups that were eliminated during the construction of the R chart should automatically be eliminated from the $\bar{x}$ chart calculations. Given that we have k valid subgroups of data, whose subgroup means are denoted by $\bar{x}_1, \bar{x}_2, \bar{x}_3, \ldots, \bar{x}_k$, the centerline of the $\bar{x}$ chart is just the average of the subgroup means,

$$\bar{\bar{x}} = \frac{1}{k}\sum_{i=1}^{k}\bar{x}_i$$

as mentioned previously. The control limits are found by replacing μ and σ by the estimates $\bar{\bar{x}}$ and $\bar{R}/d_2$ in the control limit formulas:

$$\text{UCL} = \mu + 3\frac{\sigma}{\sqrt{n}} \approx \bar{\bar{x}} + 3\frac{\bar{R}/d_2}{\sqrt{n}} \quad \text{and} \quad \text{LCL} = \mu - 3\frac{\sigma}{\sqrt{n}} \approx \bar{\bar{x}} - 3\frac{\bar{R}/d_2}{\sqrt{n}}$$

Letting $A_2 = 3/d_2\sqrt{n}$, we shall now use the following estimated limits:

$$\text{UCL} \approx \bar{\bar{x}} + A_2\bar{R} \quad \text{and} \quad \text{LCL} \approx \bar{\bar{x}} - A_2\bar{R}$$

where the constant A_2 depends on the particular subgroup size, n, and is found in Appendix Table XI. These formulas show how the centerline $\bar{R}$ of the R chart directly affects the control limits of the $\bar{x}$ chart.

Example 6.1   The process of making ignition keys for automobiles consists of trimming and pressing raw key blanks, cutting grooves, cutting notches, and plating. Some of the dimensions, such as the depth of grooves and notches, are critical to the proper functioning of the keys. Table 6.1 contains measurements (in inches) of a particular groove depth on the side of each key. Due to the high volume of keys processed per hour, the sampling frequency is chosen to be five keys every 20 minutes. For convenience, the subgroup means and standard deviations are also given in Table 6.1, along with the grand mean $\bar{\bar{x}} = .007966$ and the average range $\bar{R} = .002400$. The relevant control chart constants for subgroups of size $n = 5$ are $D_4 = 2.114$, $D_3 = 0$, and $A_2 = .577$ (Appendix Table XI).

The initial estimates of the control limits for the R chart are

$$UCL = D_4\bar{R} = (2.114)(.002400) = .005074$$
$$LCL = D_3\bar{R} = (0)(.002400) = 0$$

The corresponding control chart is shown in Figure 6.6. Because there do not appear to be any out-of-control points in the chart, no subgroups need be dropped, and we can proceed immediately to the construction of the $\bar{x}$ chart.

Table 6.1 Ignition key data for Example 6.1

Subgroup number i	Groove depth (inches)					$\bar{x}_i$	R_i
1	.0061	.0084	.0076	.0076	.0044	.00682	.0040
2	.0088	.0083	.0076	.0074	.0059	.00760	.0029
3	.0080	.0080	.0094	.0075	.0070	.00798	.0024
4	.0067	.0076	.0064	.0071	.0088	.00732	.0024
5	.0087	.0084	.0088	.0094	.0086	.00878	.0010
6	.0071	.0052	.0072	.0088	.0052	.00670	.0036
7	.0078	.0089	.0087	.0065	.0068	.00774	.0024
8	.0087	.0094	.0086	.0073	.0071	.00822	.0023
9	.0074	.0081	.0086	.0083	.0087	.00822	.0013
10	.0081	.0065	.0075	.0089	.0097	.00814	.0032
11	.0078	.0098	.0081	.0062	.0084	.00806	.0036
12	.0089	.0090	.0079	.0087	.0090	.00870	.0011
13	.0087	.0075	.0089	.0076	.0081	.00816	.0014
14	.0084	.0083	.0072	.0100	.0069	.00816	.0031
15	.0074	.0091	.0083	.0078	.0077	.00806	.0017
16	.0069	.0093	.0064	.0060	.0064	.00700	.0033
17	.0077	.0089	.0091	.0068	.0094	.00838	.0026
18	.0089	.0081	.0073	.0091	.0079	.00826	.0018
19	.0081	.0090	.0086	.0087	.0080	.00848	.0010
20	.0074	.0084	.0092	.0074	.0103	.00854	.0029
						$\bar{\bar{x}} = .007966$	$\bar{R} = .002400$

The control limits for the $\bar{x}$ chart are

$$UCL = \bar{\bar{x}} + A_2\bar{R} = .007966 + (.577)(.002400) = .009351$$
$$LCL = \bar{\bar{x}} - A_2\bar{R} = .007966 - (.577)(.002400) = .006581$$

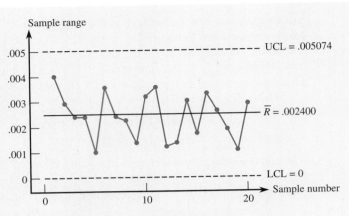

Figure 6.6 R chart for the data of Table 6.1

The $\bar{x}$ chart is shown in Figure 6.7. None of the points is outside the control units, although there is a run of eight consecutive points above the centerline (subgroups 8–15). According to the extended list of "out-of-control" rules in Section 6.2, this run of points is not quite long enough to signal an out-of-control condition.

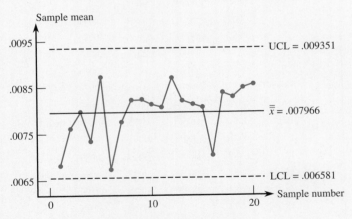

Figure 6.7 $\bar{x}$ chart for the data of Table 6.1

$\bar{x}$ and s Charts

Various alternatives to $\bar{x}$ and R charts have been proposed over the years. Because there are many different statistics available for measuring central tendency, along with several measures for variation, just about any combination of the two can be used to monitor a process average and variation. One combination that is frequently used is the $\bar{x}$ and s chart. The procedure for constructing $\bar{x}$ and s charts parallels that for $\bar{x}$ and R charts: The variation chart (i.e., the s chart) is first brought into statistical control, then the $\bar{x}$ chart is constructed using control limits formed from the centerline of the s chart.

Starting with k subgroups, each of size n, we denote the individual subgroup standard deviations by $s_1, s_2, s_3, \ldots, s_k$. Their average, $\bar{s}$, forms the centerline of the s chart:

$$\bar{s} = \frac{1}{k}\sum_{i=1}^{k} s_i$$

$\bar{s}$ is an estimate of μ_s, the mean of the sampling distribution of the sample standard deviation based on samples of size n. Following the usual 3-sigma procedure, control limits for the s chart can be shown to have the form

$$\text{UCL} = B_4\bar{s} \quad \text{and} \quad \text{LCL} = B_3\bar{s}$$

where B_3 and B_4 depend on the subgroup size, n, and are found in Appendix Table XI. In addition, to calculate the capability of a process, the standard deviation of the process measurements can be estimated by

$$\hat{\sigma} = \frac{\bar{s}}{c_4}$$

where c_4 is yet another control chart constant found in Appendix Table XI. The same extended list of "out-of-control" rules used for $\bar{x}$ and R charts can be applied to $\bar{x}$ and s charts (see Figure 6.5 on page 254).

For the $\bar{x}$ chart, the grand average of the subgroup means forms the centerline of the chart, as follows:

$$\bar{\bar{x}} = \frac{1}{k}\sum_{i=1}^{k} \bar{x}_i$$

Following the same procedure as with the $\bar{x}$ and R charts, we form the control limits for the $\bar{x}$ chart by substituting an estimate of σ into the theoretical 3-sigma limits. In this case, the estimate is $\bar{s}/c_4$, which is based on the s chart:

$$\text{UCL} = \mu + 3\frac{\sigma}{\sqrt{n}} \approx \bar{\bar{x}} + 3\frac{\bar{s}/c_4}{\sqrt{n}} \quad \text{and} \quad \text{LCL} = \mu - 3\frac{\sigma}{\sqrt{n}} \approx \bar{\bar{x}} - 3\frac{\bar{s}/c_4}{\sqrt{n}}$$

By letting $A_3 = 3/c_4\sqrt{n}$, we can write these control limits in the simpler form

$$\text{UCL} = \bar{\bar{x}} + A_3\bar{s} \quad \text{and} \quad \text{LCL} = \bar{\bar{x}} - A_3\bar{s}$$

Example 6.2

In this example, we reanalyze the key groove data of Table 6.1, this time using $\bar{x}$ and s charts. Using the average of the 20 subgroup standard deviations, $\bar{s} = .0009672$, along with the control chart constants $B_3 = 0$ and $B_4 = 2.089$ from Appendix Table XI (for subgroups of size $n = 5$), we calculate the control limits for the s chart to be

$$\text{UCL} = B_4\bar{s} = (2.089)(.0009672) = .002020$$

and

$$\text{LCL} = B_3\bar{s} = (0)(.0009672) = 0$$

The s chart, shown in Figure 6.8, does not exhibit any out-of-control conditions. With respect to the $\bar{x}$ chart, the centerline is still calculated as the average of the subgroup averages:

$$\bar{\bar{x}} = \frac{1}{k}\sum_{i=1}^{k}\bar{x}_i = .007966$$

as in Example 6.1. For subgroups of size $n = 5$, the factor $A_3 = 1.427$ is found from Appendix Table XI. This gives control limits of

$$\text{UCL} = \bar{\bar{x}} + A_3\bar{s} = .007966 + (1.427)(.0009672) = .009346$$
$$\text{LCL} = \bar{\bar{x}} - A_3\bar{s} = .007966 - (1.427)(.0009672) = .006586$$

Note that these limits are very close to the limits obtained from the R chart (UCL = .009351 and LCL = .006581). Consequently, the $\bar{x}$ chart is almost identical to that of Example 6.1, and, in particular, it gives no out-of-control signals (see Figure 6.9 below).

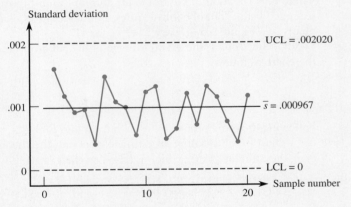

Figure 6.8 s chart for groove depth data

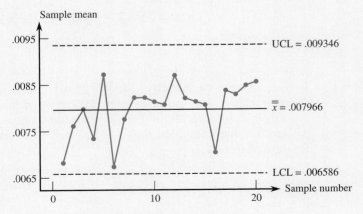

Figure 6.9 $\bar{x}$ chart for groove depth data

The choice between running a combination of $\bar{x}$ and s charts or one of $\bar{x}$ and R charts is largely a matter of personal preference. Although some authors recommend s charts in lieu of R charts because the sample standard deviation s makes more efficient use of the data than does R, the difference in efficiency is actually very small for the small subgroup sizes used in control charts, so $\bar{x}$ and R charts will almost always lead to the same conclusions as $\bar{x}$ and s charts. If control charts are done by hand, then the R chart method is definitely preferable because of the ease in calculating the ranges of small samples. If a computer is available, then computational difficulty is immaterial, and one might as well use the $\bar{x}$ and s chart approach.

Section 6.3 Exercises

14. The control limits on $\bar{x}$ charts become closer together as the subgroup size n is increased (i.e., the A_2 factor decreases as n increases). For a process that is in statistical control, does this imply that a control chart point is more likely to fall outside the control limits of an $\bar{x}$ chart based on a larger subgroup size rather than a smaller subgroup size?

15. Subgroups of four power units are selected once each hour from an assembly line, and the high-voltage output of each unit is measured. Suppose that the sum of the ranges of 30 such subgroups is 85.2. Calculate the centerline and control limits of an R chart for this data.

16. Hourly samples of size 3 are taken from a process that produces molded plastic containers, and a critical dimension is measured. Data from the most recent 20 samples is given here:

Hour	x_1	x_2	x_3	Hour	x_1	x_2	x_3
1	.36	.39	.36	11	.36	.32	.36
2	.33	.35	.30	12	.38	.47	.35
3	.51	.41	.42	13	.29	.45	.39
4	.42	.37	.34	14	.44	.38	.43
5	.39	.38	.38	15	.38	.37	.37
6	.33	.41	.45	16	.31	.43	.38
7	.43	.39	.41	17	.39	.49	.35
8	.41	.32	.32	18	.43	.36	.38
9	.37	.42	.36	19	.40	.45	.32
10	.26	.42	.32	20	.40	.40	.32

a. Construct an R chart for this data. Are any out-of-control signals indicated by this chart?

b. Construct an $\bar{x}$ chart for this data, and check for signs of special causes.

17. Refer to the data of Exercise 16.
   a. Construct an s chart for this data, and check for special causes.
   b. Construct an $\bar{x}$ chart for this data. Why are the control limits of this chart different from those in Exercise 16(b)?

18. When installing a bath faucet, it is important to properly fasten the threaded end of the faucet stem to the water-supply line. The threaded stem dimensions must meet product specifications, otherwise malfunction and leakage may occur. Authors of "Improving the Process Capability of a Boring Operation by the Application of Statistical Techniques" (*Intl. J. Sci. Engr. Research*, Vol. 3, Issue 5, May 2012) investigated the production process of a particular bath faucet manufactured in India. The article reported the threaded stem diameter (target value being 13 mm) of each faucet in 25 samples of size 4 as shown here:

Subgroup	x_1	x_2	x_3	x_4
1	13.02	12.95	12.92	12.99
2	13.02	13.10	12.96	12.96
3	13.04	13.08	13.05	13.10
4	13.04	12.96	12.96	12.97
5	12.96	12.97	12.90	13.05
6	12.90	12.88	13.00	13.05
7	12.97	12.96	12.96	12.99
8	13.04	13.02	13.05	12.97
9	13.05	13.10	12.98	12.96
10	12.96	13.00	12.96	12.99
11	12.90	13.05	12.98	12.88
12	12.96	12.98	12.97	13.02

Subgroup	x_1	x_2	x_3	x_4
13	13.00	12.96	12.99	12.90
14	12.88	12.94	13.05	13.00
15	12.96	12.96	13.04	12.98
16	12.99	12.94	13.00	13.05
17	13.05	13.02	12.88	12.96
18	13.08	13.06	13.10	13.05
19	13.02	13.05	13.04	12.97
20	12.96	12.90	12.97	13.05
21	12.98	12.99	12.96	13.00
22	12.97	13.02	12.96	12.99
23	13.04	13.00	12.98	13.10
24	13.02	12.90	13.05	12.97
25	12.93	12.88	12.91	12.90

a. Construct an R chart for this data. Are there any out-of-control signals present?

b. Construct an $\bar{x}$ chart for this data. Are there any out-of-control signals present?

c. If there are any out-of-control conditions found in parts (a) or (b), recalculate and interpret the revised $\bar{x}$ and R charts after eliminating these subgroups. (doing this assumes that assignable causes for out-of-control subgoups can be found prior to their elimination).

19. The following table gives sample means and standard deviations, each based on subgroups of six observations of the refractive index of fiber-optic cable:

Day	$\bar{x}$	s	Day	$\bar{x}$	s
1	95.47	1.30	13	97.02	1.28
2	97.38	.88	14	95.55	1.14
3	96.85	1.43	15	96.29	1.37
4	96.64	1.59	16	96.80	1.40
5	96.87	1.52	17	96.01	1.58
6	95.52	1.27	18	95.39	.98
7	96.08	1.16	19	96.58	1.21
8	96.48	.79	20	96.43	.75
9	96.63	1.48	21	97.06	1.34
10	96.50	.80	22	98.34	1.60
11	97.22	1.42	23	96.42	1.22
12	96.55	1.65	24	95.99	1.18

a. Construct an s chart for this data.

b. Construct an $\bar{x}$ chart for this data.

20. In Exercise 19, suppose that an assignable cause was found for the unusually high average refractive index in subgroup 22.

a. Recompute the control limits for both the $\bar{x}$ and s charts after removing the data from day 22.

b. Do the charts in part (a) indicate that there are any other out-of-control signals present?

21. Because processes are designed to produce products with fixed nominal dimensions, it is quite common to find that most of the variation in sample data occurs in the rightmost one or two decimal places. For example, the following data comes from a process making parts whose nominal length is required to be .254 inch:

Subgroup	x_1	x_2	x_3	Subgroup	x_1	x_2	x_3
1	.258	.254	.256	11	.253	.257	.254
2	.253	.251	.253	12	.252	.253	.258
3	.252	.258	.256	13	.258	.253	.257
4	.252	.252	.255	14	.251	.257	.256
5	.254	.252	.256	15	.256	.254	.257
6	.253	.254	.256	16	.251	.255	.253
7	.251	.257	.257	17	.252	.256	.255
8	.252	.251	.255	18	.251	.256	.253
9	.251	.255	.257	19	.253	.252	.254
10	.257	.255	.255	20	.255	.252	.253

To simplify control chart calculations for such data, practitioners often code the data by transforming the measurements into deviations from the nominal value and then multiplying by a suitable power of 10 to eliminate decimal points. In the foregoing data, for example, a reading of .258 would be converted to .258 − .254 (the deviation from the nominal value), and then multiplied by 1000. Thus .258 transforms into 4, .254 transforms into 0, .256 becomes 2, .251 becomes −3, and so forth.

a. Use the formulas for the control limits of $\bar{x}$ and R charts to explain why the signals given by charting the deviations from nominal values will always be identical to the signals given by charting the untransformed process data.

b. Transform the data in this problem as described, and then create $\bar{x}$ and R charts of the transformed data. Use the extended list of out-of-control conditions (Figure 6.5) to evaluate these charts.

c. For comparison, create the $\bar{x}$ and R charts of the untransformed data, and evaluate these charts as in part (b).

22. Three-dimensional (3D) printing is a manufacturing technology that allows the production of three-dimensional solid objects through a meticulous layering process performed by a 3D printer. 3D printing has rapidly become a time-saving and economical way to create a wide variety of products such as medical implants, furniture, tools, and even jewelry. The article "Improving the Process Capability of a Boring Operation by the Application of Statistical Techniques" (*MIT Intl. J. Mech. Engr.*, 2012: 31–38) considered the production process of metal castings by using a 3D printer manufactured by ZCorporation. Data was collected on 16 batches (each having two castings), where the outer diameter of each casting (in mm) was recorded. The target diameter of each casting was 60 mm. The corresponding data is given here:

Batch	x_1	x_2
1	59.664	59.675
2	59.661	59.648
3	59.679	59.652
4	59.665	59.654
5	59.667	59.678
6	59.673	59.657
7	59.676	59.661
8	59.648	59.651

Batch	x_1	x_2
9	59.681	59.675
10	59.655	59.672
11	59.691	59.676
12	59.682	59.651
13	59.651	59.682
14	59.668	59.685
15	59.691	59.682
16	59.661	59.673

a. Construct an R chart for this data. Are there any out-of-control signals present?

b. Construct an $\bar{x}$ chart for this data. Are there any out-of-control signals present?

23. Reconsider the data from Exercise 16.
    a. Estimate the process standard deviation.
    b. Suppose the specification limits on the process are .40 ± .08. Assuming that a normal distribution can be used to describe the process measurements, estimate the proportion of the process measurements above the USL and below the LSL.

24. Reconsider the results from Exercise 17(a).
    a. Estimate the process standard deviation.
    b. If the specification limits for the process are .40 ± .08, estimate the proportions of the process measurements above the USL and below the LSL. Compare your results to those in Exercise 23(b).

6.4 PROCESS CAPABILITY ANALYSIS

After all special causes have been identified and eliminated, a process is said to be in a state of statistical control. One of the desirable features of a controlled process is that it is *predictable*, in the sense that the process average and standard deviation are reasonably stable over time. This makes it possible to get a clear picture of how the process output compares to the requirements, or specifications, that are placed on the process. Without statistical control, it is difficult, if not impossible, to reliably evaluate the capability of a process to perform as required.

Process capability is evaluated by comparing process performance with process requirements. Since meeting specification limits is one of the most basic requirements, capability analyses usually involve specification limits somewhere in their calculations. Process data, usually from a control chart, is used to describe how a process is actually performing. Data from the chart's subgroups is used to estimate the process average and standard deviation. These, in turn, are transformed into estimates of the proportions of measurements that fall inside or outside of the specification limits. This last step requires that an *assumption* be made about the type of probability distribution that the process is thought to follow. Since many process characteristics tend to follow normal distributions, the majority of capability calculations are based on this distribution. In recent years, capability indexes have also been developed for nonnormal process data. We do not discuss the calculations required for nonnormal data, which are much more laborious than the relatively simple computations for normal processes, but we do provide references on this material for the interested reader.

Estimating the Process Mean and Variation

The best source of data for estimating process variation usually comes from the control chart used to bring the process into statistical control. In particular, variation estimates are derived from charts that monitor process variation, such as the R and s charts. As we saw in Section 6.3, depending on which variation chart is used, the process standard deviation σ is estimated by one of two formulas,

$$\hat{\sigma} = \frac{\overline{R}}{d_2} \quad \text{or} \quad \hat{\sigma} = \frac{\overline{s}}{c_4}$$

Both of these formulas are based only on the within-subgroups variation present in the data. It is also good to keep in mind that both formulas are based on the assumption that the process data follows a normal distribution. If you have reason to believe that the process is not normally distributed, then these estimates would not be appropriate.

Another method of estimating the process standard deviation is to pool the subgroup data used to make the control chart and calculate the sample standard deviation, s, of the entire set of data. For instance, rather than computing $\overline{R}/d_2$ or $\overline{s}/c_4$ from, say, 20 subgroups of size 5, you could calculate s for the combined group of 100 measurements. The reason that this is permissible is that no assignable causes should be present in the control chart data for a process that is in control and, consequently, there should be no significant difference between the subgroup-to-subgroup variation and the within-subgroup variation (as estimated by $\overline{R}/d_2$ or $\overline{s}/c_4$), which makes subgroup pooling an acceptable procedure. If a process is not in control, however, s will usually be much larger than either $\overline{R}/d_2$ or $\overline{s}/c_4$. When evaluating process capability indexes, it is often useful to know which of these methods is being used to estimate the process variation. In the ensuing discussion, we denote the estimated process average and standard deviation by $\hat{\mu}$ and $\hat{\sigma}$, regardless of the method of estimation used.

Capability studies generally use the grand mean $\overline{\overline{x}}$ of the subgroup data to estimate the process average, that is, $\hat{\mu} = \overline{\overline{x}}$. This works well for data whose distributions are fairly symmetric, such as the normal distribution. Under the assumption that a process follows a normal distribution, the 3-sigma region on either side of the process average is often called the **process spread**:

$$\text{process spread} = \hat{\mu} \pm 3\hat{\sigma}$$

This name arose from the fact that, for normal distributions, most (about 99.73%) of the process observations lie within 3 sigmas of the mean.

Nonconformance Rates

The proportion of the process measurements that fall above the upper specification limit or below the lower limit are called **nonconformance rates** or **nonconformance**

proportions. Assuming that the process measurements follow a normal distribution, we estimate these rates as follows:

$$\text{proportion above USL} = P(x > \text{USL}) \approx P\left(z > \frac{\text{USL} - \hat{\mu}}{\hat{\sigma}}\right)$$

$$\text{proportion above LSL} = P(x < \text{LSL}) \approx P\left(z < \frac{\text{LSL} - \hat{\mu}}{\hat{\sigma}}\right)$$

In these definitions, x is a normal random variable whose distribution describes the process data, and z denotes a standard normal variable. The shaded regions in Figure 6.10 show the nonconformance proportions for normally distributed process data.

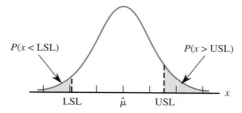

Figure 6.10 Nonconformance proportions for normally distributed process data

Nonconformance rates are usually expressed in terms of either percentages (%) or parts per million (ppm). One nonconforming (or defective) item in a collection of a million items is called one part per million, abbreviated as 1 ppm. Thus a nonconformance rate of .25% could also be expressed as 2500 ppm. For convenience, Table 6.2 shows the equivalent percentage and ppm rates for a range of values that can occur in practice. To establish small ppm rates, the standard normal table (Appendix Table I), which is traditionally limited to values of z between -3 and $+3$ or so, must be extended to accommodate the large z values required for calculations in ppm.

Table 6.2 Converting from percentage nonconforming to parts per million

Percentage (%)	Parts per million (ppm)
10.0	100,000
5.0	50,000
1.0	10,000
.1	1,000
.01	100
.001	10
.0001	1

Example 6.3

Because the control charts of the ignition key process in Example 6.1 do not indicate any out-of-control conditions, the process appears to be in statistical control. Suppose that the specification limits for the groove depth of the keys are .0072 ± .0020 inch.

Assuming that the process data is normally distributed, we can estimate the process standard deviation using the centerline of the R chart,

$$\hat{\sigma} = \frac{\bar{R}}{d_2} = \frac{.002400}{2.326} = .00103$$

Alternatively, the variation can be estimated by $\bar{s}/c_4$ from the centerline of the s chart. The nonconformance rates can then be estimated by

$$P(x > \text{USL}) \approx P\left(z > \frac{\text{USL} - \hat{\mu}}{\hat{\sigma}}\right)$$

$$= P\left(z > \frac{.0092 - .007966}{.00103}\right) = P(z > 1.20) = .1151$$

$$P(x < \text{LSL}) \approx P\left(z < \frac{\text{LSL} - \hat{\mu}}{\hat{\sigma}}\right)$$

$$= P\left(z < \frac{.0052 - .007966}{.00103}\right) = P(z < -2.69) = .0036$$

In percentage terms, we estimate that about 11.51% of the output of this process exceeds the upper specification limit, whereas only .36% is below the lower limit. This gives rise to a total percentage of 11.51% ± .36% = 11.87%, which is unacceptably high. Thus statistical control alone does not necessarily guarantee that a process will successfully meet its specification limits.

Capability Indexes

Process spread, as defined by the interval $\hat{\mu} \pm 3\hat{\sigma}$, gives a measure of how a process is currently performing. The width of this interval is $6\hat{\sigma}$. Alternatively, the distance between the specification limits, USL − LSL, provides a measure of the maximum process spread we are willing to tolerate. By comparing the two measures, it is possible to give a very succinct summary of the capability of a process to meet its specification limits. We refer to the process spread as the *actual process spread* and to USL − LSL as the *allowable process spread*. The **process capability index**, denoted by C_p, is defined by the ratio

$$C_p = \frac{\text{allowable spread}}{\text{actual spread}} = \frac{\text{USL} - \text{LSL}}{6\hat{\sigma}}$$

where $\hat{\sigma}$ is an estimate of the process standard deviation.

The C_p index is interpreted as follows. If $C_p = 1$, then the process is said to be marginally capable of meeting its specification limits. This occurs when the process is exactly centered midway between its specification limits (i.e., when $\hat{\mu} = (\text{USL} + \text{LSL})/2$ and the actual process spread uses all of the allowable spread. As you can see from Figure 6.11, this is a fairly tenuous situation since even the slightest movement of the process mean will lead to an increase in the overall nonconformance rate of the process. Normally, we would like the C_p to exceed 1, since then there is a higher likelihood that the process measurements will be able to stay within the specification limits, even if the mean wanders a little. A C_p that exceeds 1.33 (i.e., an 8-σ spread that fits within the

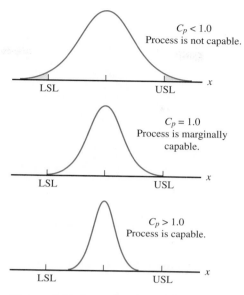

Figure 6.11 Interpretation of the process capability index C_p

specification limits) is usually considered fairly good and is commonly used as a goal by many companies. On the other hand, C_p values that are less than 1 imply that a process is not capable of meeting the specification limits.

The C_p is one of four commonly used indexes, originally invented in Japan, which are routinely used in modern quality improvement programs. The indexes derive their usefulness from the fact that they convey much information in a very simple fashion. Capability indexes also have the advantage of being *unitless* measures, making them useful for comparing related and unrelated processes alike. For example, if the copper plating thickness (in inches) from a chemical plating process has a C_p of .81, whereas the resistance (in ohms) of certain electronic components has a C_p of 2.30, then we can conclude that the electronic process is the more capable of the two, even though their measurement units, inches and ohms, are unrelated.

One drawback of the C_p is that it does not take the process location (i.e., the mean) into account. For this reason, it is often said that C_p measures only the *potential* for a process to meet its specifications. For example, Figure 6.12 shows two process distributions, both with C_p values of 2.0, one centered between the specification limits and the other located near the upper limit. Although the latter process currently has a very high nonconformance rate, it still has the potential to be capable because its C_p exceeds 1.0. However, it will realize this potential only if the process average can be moved closer to the center of the specification range.

An index that does take the process mean into account is the C_{pk} index:

$$C_{pk} = \text{minimum}\left[\frac{\text{USL} - \hat{\mu}}{3\sigma}, \frac{\hat{\mu} - \text{LSL}}{3\hat{\sigma}}\right]$$

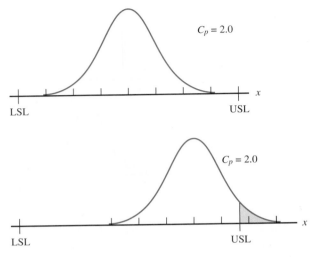

Figure 6.12 C_p is not affected by the process average

For normally distributed data, $\hat{\mu}$ is taken to be the centerline of the $\bar{x}$ chart and $\hat{\sigma}$ is chosen to be $\bar{R}/d_2$, $\bar{s}/c_4$, or perhaps the combined-subgroup estimate s mentioned previously. The k in the subscript of C_{pk} refers to the so-called k factor:

$$k = \frac{|(\text{USL} + \text{LSL})/2 - \hat{\mu}|}{(\text{USL} - \text{LSL})/2}$$

which measures the extent to which the process location $\hat{\mu}$ differs from the midpoint of the specification region. It can be shown that k lies between 0 and 1 and that C_p and C_{pk} are related by the formula

$$C_{pk} = (1 - k)C_p$$

Since $0 \leq k \leq 1$, this formula shows that C_{pk} never exceeds C_p and that $C_{pk} = C_p$ precisely when the process is centered midway between its specification limits. When used together, C_p and C_{pk} give a clear picture of process performance as well as process potential.

Example 6.4   Nonconformance rates for the groove dimension data (Table 6.1) are calculated in Example 6.3, where we concluded that the process had poor capability. The reason for the poor capability can be found by comparing the C_p and C_{pk} indexes. Using the estimates

$$\hat{\mu} = .007966 \quad \text{and} \quad \hat{\sigma} = .00103$$

from Example 6.3 along with the specification limits USL = .0092 and LSL = .0052, we calculate the k factor as follows:

$$k = \frac{|(\text{USL} + \text{LSL})/2 - \hat{\mu}|}{(\text{USL} - \text{LSL})/2} = \frac{|(.0092 + .0052)/2 - .007966|}{(.0092 - .0052)/2} = .383$$

The C_p and C_{pk} indexes are

$$C_p = \frac{.0092 - .0052}{6(.00103)} = .647$$

$$C_{pk} = (1 - k)C_p = (1 - .383)(.647) = .399$$

Neither index exceeds 1.0. The value of $C_{pk} = .399$ is a measure of how the process is currently performing with respect to meeting the specifications. The fact that C_p and C_{pk} are unequal is evidence that the process location has shifted away from the center of the specification region. However, even if the process could be adjusted so that it is centered within the specification region, a C_p of .647, which is less than 1.0, indicates that the process will still not have good capability. Clearly, attention must be focused on reducing the variation of the groove cutting process as well as bringing the process average closer to the midpoint of the specifications.

The C_p and C_{pk} indexes are used for quality characteristics that have two-sided tolerances, that is, processes with both upper and lower specification limits. Some characteristics, however, can have one-sided tolerances. The breaking strength of a material, for instance, usually has a lower specification limit, but no upper specification, since we normally want materials to have a certain minimum strength but we do not care by how much they exceed that minimum. One-sided capability indexes are used for such processes. In fact, the definitions of upper and lower capability indexes are contained within the definition of the C_{pk}. For processes having only a lower specification limit, LSL, the **lower capability index** C_{pl} is defined by

$$C_{pl} = \frac{\hat{\mu} - \text{LSL}}{3\hat{\sigma}}$$

Similarly, for processes having only an upper specification USL, the **upper capability index** C_{pu} is given by the formula

$$C_{pu} = \frac{\text{USL} - \hat{\mu}}{3\hat{\sigma}}$$

The reason $3\hat{\sigma}$ rather than $6\hat{\sigma}$ appears in the denominators is that one-sided capability indexes compare only one side of the process distribution, the upper or lower, to the corresponding upper or lower specification limit.

Even when a process has both upper and lower specification limits, calculating C_{pu} and C_{pl} is worthwhile because the smaller of the statistics indicates the direction in which the process average has shifted away from the nominal value. In fact, from the formulas it is apparent that C_{pk} is equal to the smaller of C_{pl} and C_{pu}:

$$C_{pk} = \text{minimum}\left[C_{pl}, C_{pu}\right]$$

For data that is normally distributed, it is convenient to transform C_{pu} and C_{pl} into their corresponding nonconformance rates using the following relationships:

$$P(x > \text{USL}) \approx P\left(z > \frac{\text{USL} - \hat{\mu}}{\hat{\sigma}}\right) = P(z > 3C_{pu})$$

$$P(x < \text{LSL}) \approx P\left(z < \frac{\text{LSL} - \hat{\mu}}{\hat{\sigma}}\right) = P(z < -3C_{pl})$$

Example 6.5   Using the results of Examples 6.3 and 6.4, we calculate the C_{pu} and C_{pl} indexes for the groove depth data of Table 6.1 as follows:

$$C_{pl} = \frac{\hat{\mu} - \text{LSL}}{3\hat{\sigma}} = \frac{.007966 - .0052}{3(.00103)} = .895$$

$$C_{pu} = \frac{\text{USL} - \hat{\mu}}{3\hat{\sigma}} = \frac{.0092 - .007966}{3(.00103)} = .399$$

This information could be used, if desired, to calculate C_{pk}:

$$C_{pk} = \text{minimum}[C_{pu}, C_{pl}] = \text{minimum}[.399, .895] = .399$$

Because both C_{pu} and C_{pl} are less than 1.0, we can conclude that the process is not performing well with respect to meeting *either* of its specification limits. Furthermore, the fact that C_{pl} is the smaller of the two indexes means that the process average has shifted to the right of the midpoint of the specification region.

Another way of describing capability is to estimate what proportion of the allowed process spread, USL − LSL, is "used up" by the process spread $6\hat{\sigma}$. This proportion, $6\hat{\sigma}/(\text{USL} - \text{LSL})$, is called the **capability ratio.** As you can see, the capability ratio is simply the reciprocal of the C_p index,

$$\text{capability ratio} = \frac{1}{C_p}$$

When expressed as a percentage, the capability ratio is referred to as the percentage of specification used by the process. Notice that the C_{pk} is not used in this definition. The reason is that one usually wants to know how much of the specification range is used under ideal circumstances (i.e., when the process is centered). For instance, in Example 6.4, the C_p index of the groove depth data was shown to be .647. This means that the capability ratio is $1/.647/1.55$, or in percentage terms, 155%. In other words, the process spread $6\hat{\sigma}$ uses about 155% of the allowed tolerance, which is not good. Capable processes should use up less than 100% of the allowed tolerance.

Using Capability Indexes

The interpretation of capability indexes can be complicated by the presence of measurement errors and assumptions about the process distribution. It should be stressed that the interpretations given in this section are made under the assumptions that (1) a process is in statistical control, (2) measurement errors are negligible, and (3) the process follows a normal distribution. Keep in mind that capability indexes are *statistics* that arise from taking sample data from a process. As such, capability indexes will exhibit some degree of sampling variability. That is, you should expect these indexes to vary a little with each new set of data. If desired, the amount of sampling variability can be estimated (see, for example, Kane, V. E., 1986, "Process Capability Indexes," *J. Quality Technology,* 1986: 41–52). In addition, if you suspect that the process data does not follow a normal distribution, then you will want to use indexes designed to handle nonnormal process data. The references at the end of the chapter give methods for handling such situations.

Section 6.4 Exercises

25. Why must a process be in a state of statistical control before its capability can be measured?

26. A process has a C_p index of 1.2 and is centered on its nominal value. What proportion of the specification range is used by the process measurements?

27. A computer printout shows that a certain process has a C_p of 1.6 and a C_{pk} of .9. Assuming that the process is in control, what do these indexes say about the capability of this process?

28. A process with specification limits of $5 \pm .01$ has a C_p of 1.2 and a C_{pk} of 1.0. What is the estimated process average $\bar{\bar{x}}$ from which these indexes are calculated?

29. It can be shown that the following equation always holds for processes that can be described by a normal distribution (Farnum, N. R., *Modern Statistical Quality Control and Improvement*, Duxbury, Belmont, CA, 1994: 235):

    proportion out of specification
    $$= P(z \geq 3C_{pk}) + P(z \geq 6C_p - 3C_{pk})$$

    Use this equation with the C_p and C_{pk} from Exercise 27 to estimate the proportion of the process that is not within the specification limits.

30. Use the formula given in Exercise 29 to calculate the proportion of the process that is out of specification in Exercise 28.

31. Use the data in Exercise 6 to calculate the C_p, C_{pu}, C_{pl}, and C_{pk} indexes. What do the indexes indicate about the capability of the process?

32. Using the data of Exercise 16, we estimated the process standard deviation in Exercise 23. If the specification limits for the process are $.40 \pm .08$, calculate the C_p and C_{pk} indexes. What conclusions can you draw about the capability of the process?

33. The data of Exercise 21 was analyzed by first transforming it into deviations from the nominal value and then running $\bar{x}$ and R charts on the transformed data. Suppose the specification limits on the process are $.254 \pm .01$ inch.

    a. Describe a procedure for calculating capability indexes from the transformed data.

    b. Calculate the C_p, C_{pu}, C_{pl}, and C_{pk} indexes from the transformed data.

34. Based on your analysis in Exercise 18, if the specification limits for the process are $13 \pm .2$, calculate the C_p and C_{pk} indexes. What conclusions can you draw about the capability of the process?

35. Using the data of Exercise 22, if the specification limits for the process are $60 \pm .4$, calculate the C_p and C_{pk} indexes. What conclusions can you draw about the capability of the process?

6.5 CONTROL CHARTS FOR ATTRIBUTES DATA

In quality control, counted data is called **attributes data.** Attributes data arises when we check products to see whether they possess a specified characteristic or attribute. Those that have the attribute in question are said to be **conforming** and those without it are called **nonconforming.** Attributes, or product characteristics, can be either precisely defined or fairly subjective in nature. For example, the testing of electronic devices usually results in a clear decision as to which devices are nonconforming (those that fail to function correctly) and which are conforming. Alternatively, when an injection-molded dashboard of a car is inspected for small pinholes and other blemishes, deciding which dashboards are nonconforming is a more subjective matter. Attributes measurements are frequently used in situations where variable measurements are not practical and human judgment is needed. In addition to classifying entire products as conforming or nonconforming, it is also possible to count the number of flaws or **nonconformities** on a single unit of product.

Control charts for monitoring the *proportion* of nonconforming items in subgroups are called **p charts.** Monitoring the *number* of nonconforming items is accomplished with the **np chart.** For products that are created in distinct units, such as components or appliances, the **c chart** is used to track the number of nonconformities in subgroups of such items. When products are not made in distinct units, such as reels of wire, fabric, or paper, then the **u chart** is used to monitor the number of nonconformities in specified "units" of such products.

Interpreting the Control Limits

The plotted points on control charts for nonconforming items (*p* and *np* charts) or on charts for nonconformities (*c* and *u* charts) gauge the numbers or proportions of problems found in each subgroup. Points that are above the chart's UCL indicate abnormally high levels of problems. Assignable causes for such problems should be immediately sought and eliminated. Points that fall below the LCL, however, indicate abnormally low problem levels. Although such points certainly qualify as out of control, they are also evidence that some assignable cause has brought about a temporary, but welcome, improvement in the process. In this situation, the goal changes to one of finding the assignable cause and then taking steps, not to eliminate it, but to ensure that it *continues* to exist in the future.

p and *np* Charts

The proportions of nonconforming items in successive subgroups of size *n* are plotted on *p* **charts.** If we assume that all the subgroups come from a stable process in which the true proportion of nonconforming items is π, each of the subgroup proportions p_1, $p_2, p_3, \ldots, p_k$ is a statistic whose sampling distribution (see Section 5.5) has a mean and standard deviation of

$$\mu_p = \pi \qquad \text{and} \qquad \sigma_p = \sqrt{\frac{\pi(1-\pi)}{n}}$$

In theory, the 3-sigma control limits for the *p* chart are formed by $\mu_p \pm 3\sigma_p$. In practice, of course, we must first estimate π and then substitute this estimate into the formulas.

To estimate π, the *k* subgroup proportions are averaged. This average is denoted by

$$\bar{p} = \frac{1}{k}\sum_{i=1}^{k} p_i$$

and is used as the centerline of the chart. Substituting $\bar{p}$ for π, we find the control limits for the *p* charts as

$$\text{UCL} = \bar{p} + 3\sqrt{\frac{\bar{p}(1-\bar{p})}{n}}$$

$$\text{LCL} = \bar{p} - 3\sqrt{\frac{\bar{p}(1-\bar{p})}{n}}$$

Sometimes, because of the small values of $\bar{p}$ that are encountered in practice, the LCL can be negative. When this happens, we replace the LCL by 0 since it is impossible to have negative nonconformance rates.

In the frequently occurring case where the subgroup sizes $n_1, n_2, n_3, \ldots, n_k$ are not all equal, the calculations for the centerline and control limits are modified as follows.

Letting $x_1, x_2, x_3, \ldots, x_k$ denote the *numbers* of nonconforming items in each subgroup, we estimate the centerline of the p chart by

$$\bar{p} = \frac{x_1 + x_2 + x_3 + \cdots + x_k}{n_1 + n_2 + n_3 + \cdots + n_k}$$

This formula is conveniently remembered as "the total number of nonconforming items over the total sample size." The formula is more general than the equal-samples formula and, for that reason, it is sometimes the only formula cited by some texts for estimating π. When the subgroup sizes are unequal, the control limits are calculated separately *for each subgroup*. That is, the control limits for the ith subgroup are

$$\text{UCL} = \bar{p} + 3\sqrt{\frac{\bar{p}(1 - \bar{p})}{n_i}}$$

$$\text{LCL} = \bar{p} - 3\sqrt{\frac{\bar{p}(1 - \bar{p})}{n_i}}$$

Example 6.6

Aerospace contractors and subcontractors must often demonstrate, using control charts, that their manufacturing processes are capable of meeting ever-increasing quality standards for military systems and hardware ("Department of Defense Renews Emphasis on Quality," *Quality Progress*, March 1988: 19–21). Many such systems include printed circuit board (PCB) assemblies with various electronic components soldered to them. Components are soldered in place by means of a wave solder machine, which passes the PCBs on a conveyor over a surface of liquid solder. Soldered PCBs are then connected to test stations, which electronically test the circuits and classify each board as either conforming or nonconforming. Table 6.3 contains records of the daily numbers of rejected (nonconforming) PCBs for a 30-day period. For this data,

$$\bar{p} = \frac{14 + 22 + 9 + \cdots + 12}{286 + 281 + 310 + \cdots + 289} = .054$$

Table 6.3 Daily records of numbers of tested and rejected circuit board assemblies

Day	Rejects	Tested	Proportion	Day	Rejects	Tested	Proportion
1	14	286	.049	16	15	297	.051
2	22	281	.078	17	14	283	.049
3	9	310	.029	18	13	321	.040
4	19	313	.061	19	10	317	.032
5	21	293	.072	20	21	307	.068
6	18	305	.059	21	19	317	.060
7	16	322	.050	22	23	323	.071
8	16	316	.051	23	15	304	.049
9	21	293	.072	24	12	304	.039
10	14	287	.049	25	19	324	.059
11	15	307	.049	26	17	289	.059
12	16	328	.049	27	15	299	.050
13	21	296	.071	28	13	318	.041
14	9	296	.030	29	19	313	.061
15	25	317	.079	30	12	289	.042

Since different numbers of PCBs are tested each day, the control limits for a p chart of the data are calculated separately for each subgroup:

$$\text{UCL} = \overline{p} + 3\sqrt{\frac{\overline{p}(1 - \overline{p})}{n_i}}$$

$$= .054 + 3\sqrt{\frac{(.054)(1 - .054)}{n_i}} = .054 + \frac{.6781}{\sqrt{n_i}}$$

$$\text{LCL} = \overline{p} - 3\sqrt{\frac{\overline{p}(1 - \overline{p})}{n_i}}$$

$$= .054 - 3\sqrt{\frac{(.054)(1 - .054)}{n_i}} = .054 - \frac{.6781}{\sqrt{n_i}}$$

Figure 6.13 shows the p chart for using these control limits. Note that the smaller the subgroup size n_i, the wider the control limits. Since the chart shows no signs of any out-of-control conditions, we conclude that the process is in control and currently operating at about a 5.4% nonconforming rate.

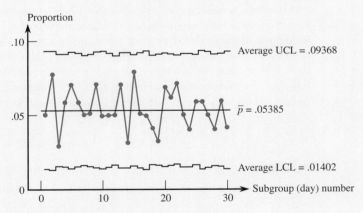

Figure 6.13 p chart of the data in Table 6.3

If you have the ability to choose constant subgroup sizes in your particular application, then the p chart calculations can be further simplified. In fact, with a constant base of comparison (i.e., constant subgroup size n), there is no need to even convert the numbers of nonconforming items into the subgroup proportions $p_1, p_2, p_3, \ldots, p_k$. Instead, we can simply plot the numbers of nonconforming items $x_1, x_2, x_3, \ldots, x_k$ on the chart. This chart is called an **np chart** because the number of nonconforming items in a subgroup is simply n times the proportion of nonconforming items.

If $x_1, x_2, x_3, \ldots, x_k$ denote the numbers of nonconforming items in k subgroups, then the centerline of the np chart is simply $n\overline{p}$, where $\overline{p}$ is calculated by either of the formulas for $\bar{p}$ used with the p chart. Similarly, the 3-sigma control limits for the np chart are found by multiplying each of the control limits of the p chart by n:

$$UCL = n\bar{p} + 3\sqrt{n\bar{p}(1 - \bar{p})}$$

$$LCL = n\bar{p} - 3\sqrt{n\bar{p}(1 - \bar{p})}$$

As in the p chart, if the LCL turns out to be negative, we replace it by 0.

Example 6.7

In complex systems, items are routed through a succession of different processes before emerging as finished products or completed services. In "build to order" systems, for example, individual orders are routed through slightly different paths from other orders, according to a customer's specific design requirements. A common method for tracking an item's progress during this journey is to attach paperwork to each order that describes the requirements for every step of the production process. These documents, often called *travelers*, are created before an order is processed. It is imperative that they be correct, since incorrect travelers are essentially recipes for nonconforming products!

To monitor the quality of such paperwork, suppose that periodic samples of 100 travelers are examined for errors, where a nonconforming document is defined to be one that contains at least one error. Table 6.4 shows data from 25 daily samples of size 100 travelers and the corresponding numbers of nonconforming ones. The total number of nonconformities in the 25 samples is 272, so $\bar{p} = 272/[(25)(100)] = .1088$ and, therefore, $n\bar{p} = 100(.1088) = 10.88$. The control limits are

$$UCL = n\bar{p} + 3\sqrt{n\bar{p}(1 - \bar{p})} = 10.88 + 3\sqrt{10.88(1 - .1088)} = 20.22$$

$$LCL = n\bar{p} - 3\sqrt{n\bar{p}(1 - \bar{p})} = 10.88 - 3\sqrt{10.88(1 - .1088)} = 1.54$$

Table 6.4 Numbers of documents containing errors in samples of 100 documents

Day	Number	Sample Size	Day	Number	Sample Size
1	10	100	14	21	100
2	12	100	15	20	100
3	10	100	16	12	100
4	11	100	17	11	100
5	6	100	18	6	100
6	7	100	19	10	100
7	12	100	20	10	100
8	10	100	21	11	100
9	6	100	22	11	100
10	11	100	23	11	100
11	9	100	24	6	100
12	14	100	25	9	100
13	16	100			

The np chart (Figure 6.14) shows one point (subgroup 14) above the UCL. Production records for day 14 should be examined for a possible assignable cause. If one is found, then subgroup 14 should be eliminated from the calculations, and an np chart with a revised centerline and control limits should be used to monitor subsequent data.

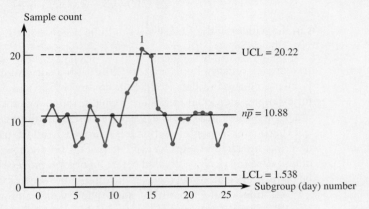

Figure 6.14 np chart from Minitab for the data of Table 6.4 (Minitab labels the first out-of-control point with a "1")

c and u Charts

Because an object can have any number of flaws, or nonconformities, it is important to establish an **inspection unit** when working with c and u charts. The inspection unit defines the fixed unit of output that will be regularly sampled and examined for nonconformities. Inspection units are often single units of product, such as a single printed circuit board or a single television. Inspection units can also be collections of items, which might be used when one examines accounting records for errors by looking at batches of 100 accounting records per day. The inspection unit is then 100 records, and the number of nonconformities for such an inspection unit is the total number of errors found in each such batch. Products are usually grouped in batches like this when the nonconformance rate is small and large samples are needed to detect nonconformities. Choosing an inspection unit is especially important with continuous processes, such as the production of long rolls of paper, wire, fabric, or metal. To count the number of surface flaws in long rolls of metal, for example, it would not be practical to look at every square foot of the metal surface. Instead, we decide on a fixed-size inspection unit, say, a 2-square-foot section of metal, and count the number of nonconformities found therein.

The number of nonconformities per unit (i.e., per inspection unit) is denoted by c. To create a **c chart**, a sample of k successive inspection units is examined, and the numbers of nonconformities $c_1, c_2, c_3, \ldots, c_k$ found in these units are counted. The centerline of the chart, denoted by $\bar{c}$, is the average

$$\bar{c} = \frac{1}{k}\sum_{i=1}^{k} c_i$$

For a stable process, the number of nonconformities, c, is modeled by a Poisson distribution. Since the mean and variance of a Poisson distribution are equal (see Chapter 2) and since the mean is estimated by the centerline $\bar{c}$, the 3-sigma control limits of the c chart are

$$\text{UCL} = \bar{c} + 3\sqrt{\bar{c}}$$
$$\text{LCL} = \bar{c} - 3\sqrt{\bar{c}}$$

As in the p and np charts, negative LCLs are replaced by 0. However, this problem can be avoided if the inspection unit is chosen so that $\bar{c}$ exceeds 9 (see Exercise 38).

Example 6.8

One measure of software quality is the number of coding errors made by programmers per 1000 lines of computer code. Using K to denote 1000, the inspection unit "a thousand lines of code" is usually abbreviated as KLOC (i.e., K Lines Of Code). The data in Table 6.5 shows the defects per KLOC obtained from weekly test logs in a software company. The average number of errors per KLOC is $\bar{c} = 134/30 = 4.467$. The upper and lower control limits of the chart are then

$$\text{UCL} = \bar{c} + 3\sqrt{\bar{c}} = 4.467 + 3\sqrt{4.467} = 10.807$$
$$\text{LCL} = \bar{c} - 3\sqrt{\bar{c}} = 4.467 - 3\sqrt{4.467} = -1.874$$

Table 6.5 Number of errors per 1000 lines of code

Week	Errors	Week	Errors
1	6	16	3
2	7	17	2
3	7	18	0
4	6	19	0
5	8	20	1
6	6	21	2
7	5	22	5
8	8	23	1
9	1	24	7
10	6	25	7
11	2	26	1
12	5	27	5
13	5	28	5
14	4	29	8
15	3	30	8

Because the LCL is negative, we reset it to 0 and then construct the c chart shown in Figure 6.15. Note that the two points at weeks 18 and 19 are touching the lower control limit and that there are several runs of points on the same side of the centerline. According to the extended "out-of-control" rules in Section 6.2, these observations do not quite qualify as out-of-control signals, but they are close. It might therefore be rewarding to conduct a small search for reasons why the error rate was so low in weeks 18 and 19.

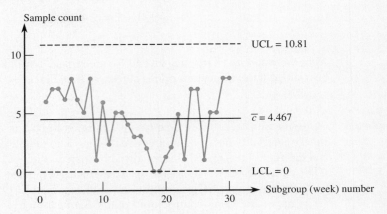

Figure 6.15 c chart for the data of Table 6.5

Sometimes it is neither possible nor convenient to use inspection units based on collections of units of a product. This is especially the case for continuous processes, such as the manufacture of sheet metal and plastic or rolls of paper, tubing, wire, and fabric. It is not convenient, nor is it necessary, to inspect entire rolls of fabric for non-conformities. Instead, smaller samples of such products are used to form control charts. This is accomplished by first deciding on an inspection unit of a specified size, such as a 2-square-foot area of fabric or, perhaps, a 2-yard section of wire. The second step is to obtain small samples of the product for testing, but these samples need not necessarily coincide with the chosen inspection unit. For example, a 4-yard section of wire might be examined for flaws on one day and a half-yard section on another. To make a fair comparison between samples of different sizes, however, we divide the number of flaws found in any sample by the number of inspection units represented in the sample. For instance, if three flaws are found in the 4-yard section of wire and we are using an inspection unit of 2 yards, the nonconformity rate would be recorded as 1.5 flaws per unit, since 4 yards represents two inspection units. Similarly, three flaws in a half-yard section of wire is more serious, because this is equivalent to 12 flaws per 2 yards, or 12 flaws per unit.

To account for variable numbers of inspection units in our subgroups, c charts are replaced by **u charts** based on the adjusted per unit rates described previously. If subgroup i contains c_i nonconformities and represents n_i inspection units, then the non-conformities per unit, u_i, is simply

$$u_i = \frac{c_i}{n_i}$$

Note that the numbers of inspection units, n_i, represented in a sample *does not have to be an integer*.

For k subgroups of such data, the statistics $u_1, u_2, u_3, \ldots, u_k$ are plotted on the u chart. The centerline on the chart is

$$\bar{u} = \frac{\text{total nonconformities in the } k \text{ subgroups}}{\text{total number of inspection units}}$$

$$= \frac{c_1 + c_2 + c_3 + \cdots + c_k}{n_1 + n_2 + n_3 + \cdots + n_k}$$

Because the subgroup size n_i usually varies from sample to sample, control limits for the u chart are computed separately for each subgroup:

$$\text{UCL} = \bar{u} + 3\sqrt{\frac{\bar{u}}{n_i}}$$

$$\text{LCL} = \bar{u} - 3\sqrt{\frac{\bar{u}}{n_i}}$$

Example 6.9

The data in Table 6.6 shows the number of flaws found in 30 samples of fabric and corresponding sizes of the samples examined (in square feet). Suppose an inspection unit of 2 square feet is used to monitor the quality of this fabric. Table 6.6 also shows the conversion of the raw nonconformity rates into the per unit rates, u_i. The u chart of this data (Figure 6.16) reveals several out-of-control points, some bad (above the UCL) and some good (below the LCL). Before this control chart can be used to monitor subsequent production, a search should be made for possible assignable causes and then appropriate actions taken. A revised chart, after eliminating out-of-control points, would then be used to monitor subsequent samples from the process.

Table 6.6 Number of flaws c_i and per unit rates u_i in 30 fabric samples

i	c_i	Sample size (ft²)	u_i	i	c_i	Sample size (ft²)	u_i
1	12	3.9	6.15	16	29	9.8	5.92
2	18	9.0	4.00	17	18	8.8	4.09
3	27	6.7	8.06	18	28	7.1	7.89
4	64	9.2	13.91	19	10	3.3	6.06
5	11	3.6	6.11	20	47	5.9	15.93
6	13	6.7	3.88	21	21	5.2	8.08
7	25	8.3	6.02	22	6	5.6	2.14
8	22	5.6	7.86	23	16	8.0	4.00
9	43	6.1	14.10	24	27	8.9	6.07
10	17	4.2	8.10	25	21	5.3	7.92
11	0	8.4	.00	26	12	3.1	7.74
12	14	6.8	4.12	27	19	6.2	6.13
13	9	4.4	4.09	28	14	4.8	5.83
14	16	5.2	6.15	29	42	8.3	10.12
15	0	7.8	.00	30	19	4.7	8.09

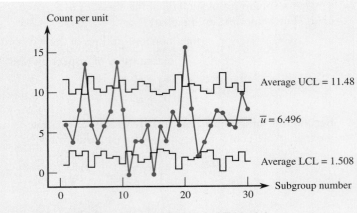

Figure 6.16 *u* chart for the data of Table 6.6

Section 6.5 Exercises

36. Explain the difference in the actions taken on a process when a point on a *p* chart exceeds the upper control limit versus the actions taken when a point falls below the lower control limit.

37. For a fixed subgroup size *n*, find the smallest value of $\bar{p}$ that will give a positive lower control limit on a *p* chart.

38. Control limits for attributes charts are never negative, and it is desirable that they be positive. For a *c* chart, what values of the centerline $\bar{c}$ will ensure that the lower control limit is positive?

39. The following data shows the number of nonconforming items found in 30 successive lots, each of size 50, of a finished product:

```
4  3  0  2  2  2  0  1  1  0
3  2  1  1  0  0  2  4  2  5
0  0  1  1  0  3  2  1  2  4
```

   a. Construct a control chart for the proportion of nonconforming items per lot.
   b. Interpret the chart in part (a).

40. On each of 25 days, 100 printed circuit boards are subjected to thermal cycling; that is, they are subjected to large changes in temperature, a procedure known to cause failures in boards with weak circuit connections. Of the boards tested, a total of 578 fail to work properly after the thermal cycling test.

   a. From this information, calculate the centerline and control limits for a *p* chart.
   b. The highest number of failures on a given day was 39 and the lowest number was 13. Would either of these points indicate an out-of-control condition?
   c. If your answer to part (b) is "yes," then eliminate the out-of-control point(s) from the data and recompute the centerline and control limits of the *p* chart.

41. After assembly and wiring of the individual keys, computer keyboards are tested by an automated test station that pushes each key several times. Daily records are kept of the number of keyboards inspected and the number that fail the inspection. Data from 25 successive manufacturing days is given here.

Day	Number tested	Number failed	Day	Number tested	Number failed
1	2186	28	11	2141	31
2	2131	21	12	2019	18
3	2158	22	13	2027	27
4	2307	14	14	2376	25
5	2262	17	15	2118	27
6	2379	27	16	2251	14
7	2069	18	17	2068	31
8	2264	20	18	2242	23
9	2383	18	19	2089	23
10	2350	19	20	2387	36

Day	Number tested	Number failed	Day	Number tested	Number failed
21	2011	38	24	2375	20
22	2059	13	25	2029	30
23	2045	11			

a. Calculate the centerline of a p chart for this data.
b. Construct the control limits for the p chart.
c. Are there any signs of out-of-control conditions in this data?

42. The following observations are the number of defects in 25 1-square-yard specimens of woven fabric (read across):

```
3  7  5  3  4  2  8  4  3  3  6  7  2
3  2  4  7  3  2  4  4  1  4  5  6
```

a. Construct a c chart for this data.
b. Check the chart for any out-of-control signals and, if necessary, eliminate such points from the data and reconstruct the c chart.

43. Off-color flaws in aspirin are caused by extremely small amounts of iron that change color when wet aspirin comes into contact with the sides of drying containers ("People: The Only Thing That Makes Quality Work," *Quality Progress*, Sept. 1988: 63–67). Such flaws are not harmful but are nonetheless unattractive to consumers. At one Dow Chemical plant, a 250-lb sample is taken out of every batch of aspirin, and the number of off-color flaws is counted. The following table shows the numbers of flaws per 250-lb sample for a period of 25 days:

Day	Number of flaws	Day	Number of flaws
1	46	10	44
2	51	11	47
3	56	12	51
4	57	13	46
5	37	14	49
6	51	15	48
7	47	16	59
8	34	17	53
9	30	18	61

Day	Number of flaws	Day	Number of flaws
19	63	23	42
20	42	24	39
21	45	25	38
22	43		

Construct an appropriate control chart for this data, and examine it for any evidence of a lack of statistical control.

44. Forty consecutive automobile dashboards are examined for signs of pinholes in the plastic molding. The numbers of pinholes found are (read across)

```
6  2  3  2  5  2  2  3  2  4  9
4  0  5  0  6  5  4  2  3  3  1
4  1  7  3  3  5  7  3  6  7  6
4  5  3  8  5  4  3
```

a. Construct a control chart for the number of pinholes per dashboard.
b. Interpret the chart in part (a).

45. Painted metal panels are examined after baking at high temperatures to harden the paint. Because the manufacturer produces panels of several different sizes, inspectors simply record the number of blemishes found along with the known area of the panel (ft^2). The following table shows the number of surface flaws found on 20 successive panels:

Panel	Number of flaws	Area of panel	Panel	Number of flaws	Area of panel
1	3	.8	11	1	.6
2	2	.6	12	3	.8
3	3	.8	13	5	.8
4	2	.8	14	4	1.0
5	5	1.0	15	6	1.0
6	5	1.0	16	12	1.0
7	10	.8	17	3	.8
8	12	1.0	18	3	.6
9	4	.6	19	5	.6
10	2	.6	20	1	.6

a. Construct a u chart for this data.
b. Examine the chart in part (a) for any out-of-control points.

6.6 RELIABILITY

Implicit in our understanding of the term *quality* is a product's ability to perform its intended function for a reasonable period of time. Unless expressly designed for short-term or one-time jobs, products that fail after only a brief period of use are not normally considered to be of high quality. In addition to applying quality improvement methods to create products, attention must also be paid to making these products last.

The field of **reliability** is concerned with the time aspect of quality. Reliability techniques are used to estimate the useful lifetime of products, to detect and fix the types of problems that occur with time, and to aid in establishing warranty, replacement, and repair policies. Directly related to reliability are issues of product safety and product liability, both of great importance to consumers and companies alike.

Failure Laws

The length of time that a product lasts until it fails, or ceases to operate correctly, is called its **lifetime.** Lifetimes are measured in terms of how a product is used. Many product lifetimes are simply measured in units of time (minutes, hours, etc.), as, for example, in a wall clock battery that begins its useful life when installed in a clock and fails sometime later when the clock stops. For items such as lightbulbs, that usually do not operate continuously, lifetimes refer to the accumulated operating time a product experiences before failure (i.e., the total number of hours during which the bulb was on). With tires, the number of miles driven is usually a better indicator of product life than simply the time that the tires have been on the car. Mechanical devices, such as springs, have lifetimes measured in **cycles** of operation, where, for example, a cycle might be defined to be one compression and release of the spring. Whatever units are used, time or cycles, we define a product's lifetime to be a measure of the total accumulated exposure to failure, often called the **time on test,** that the product experiences prior to failure.

Lifetimes are modeled as continuous random variables and, as such, their probability distributions are described by probability density functions (pdf's). Lifetimes can take on nonnegative numerical values, even zero (e.g., products that fail immediately), so density functions such as the exponential, Weibull, and lognormal are frequently used to model lifetimes. Distributions that allow negative values, such as the normal distribution, can also be used as long as their parameters are chosen in a manner that gives negligible probability to negative lifetimes. When used to model lifetimes, density functions are also called **failure laws.**

Choosing an appropriate failure law for a particular product or set of data can be done in several ways:

1. There may be a physical or mathematical reason that justifies the use of a particular density (e.g., the Central Limit Theorem justifies using the normal distribution for sums and averages).
2. Quantile plots (see Section 2.4) may show that a particular density provides a good fit to available data.
3. A failure law may have already been used by others and found to work well.

Because of the vast amount of research that has already been done on many products and materials, item (3) in the preceding list often leads to a good failure law choice. It is also useful to keep in mind the following brief list of situations that may provide the necessary justification needed in item (1):

- **Normal** failure laws often apply in situations where lifetimes are the result of a *sum* of many other variable quantities.

- **Exponential** failure laws apply to products whose current ages do not have much effect on their remaining lifetimes. This is the "memoryless" property of exponential distributions (see Exercise 61). Typical applications: fuse lifetimes, interarrival times, alpha ray arrivals, Geiger counter ticks).
- **Lognormal** failure laws work well when the degradation in lifetime is *proportional* to the previous amount of degradation (typical applications: corrosion, crack growth, diffusion, metal migration, mechanical wear).
- **Weibull** failure laws are good models for the failure time of the *weakest component* of a system (e.g., capacitor, bearing, relay, and pipe joint failures).

Example 6.10   The lognormal distribution is often used to model tread wear of tires. To fit a lognormal distribution to such data, suppose a tire manufacturer uses warranty data to estimate that the mean time to failure (measured in total miles driven) for a certain tire model is 40,000 miles with a standard deviation of 7500 miles. Denoting tire lifetimes (in miles) by a random variable x, the parameters of the lognormal distribution can be calculated using the formulas (see pages 69 and 77)

$$E(x) = e^{\mu + (\sigma^2/2)} \quad \text{and} \quad V(x) = e^{2\mu + \sigma^2}\left(e^{\sigma^2} - 1\right)$$

which can be solved for the lognormal parameters μ and σ:

$$\sigma^2 = \ln\left(1 + \frac{V(x)}{[E(x)]^2}\right) = \ln\left(1 + \frac{7500^2}{[40{,}000]^2}\right) = .034552$$

so $\sigma = .185882$ and

$$\mu = \ln(E(x)) - \frac{\sigma^2}{2} = \ln(40{,}000) - \frac{.034552}{2} = 10.57936$$

Reliability and Hazard Functions

Letting $f(x)$ be the density function (failure law) for a random variable x that describes the lifetime of a product, the **reliability at time t**, denoted $R(t)$, is the probability that the product lasts longer than time t:

$$\text{reliability at time } t = R(t) = P(T > t) = \int_t^\infty f(x)\,dx$$

Directly related to $R(t)$ is a function $Z(t)$ called the **failure rate** or **hazard function**:

$$\text{failure rate at time } t = Z(t) = \frac{f(t)}{R(t)}$$

$Z(t)$ is interpreted as the *instantaneous* rate of failure at time t, meaning that of those items that have not failed before time t, the proportion that will fail in the small interval of time from t to $t + \Delta t$ is approximately $\Delta t \cdot Z(t)$. The failure rate function is very useful for describing the manner in which failures occur.

The normal and lognormal distributions do not have closed-form expressions for either the reliability or the hazard functions; however, the exponential and Weibull distributions do have simple closed-form expressions for $R(t)$ and $Z(t)$:

	Density	$R(t)$	$Z(t)$
Exponential:	$\lambda e^{-\lambda x}(\lambda > 0)$	$e^{-\lambda t}$	λ (a constant)
Weibull:	$\dfrac{\alpha}{\beta^\alpha} x^{\alpha-1} e^{-(x/\beta)^\alpha}(\alpha, \beta > 0)$	$e^{-(t/\beta)^\alpha}$	$\dfrac{\alpha}{\beta^\alpha} t^{\alpha-1}$

(recall that the exponential is a special case of the Weibull when $\beta = 1/\lambda$ and $\alpha = 1$).

Figure 6.17 shows graphs of $Z(t)$ for various values of α (the "shape" parameter) for the Weibull distribution. Notice that for $0 < \alpha < 1$ the failure rate decreases with time, for $\alpha = 1$ (i.e., the exponential distribution) the failure rate is constant, and for $\alpha > 1$ the failure rate increases with time. In the case of the exponential distribution ($\alpha = 1$), the fact that the failure rate is constant is often interpreted as saying that products that have exponential failure law are "memoryless." That is, no matter how old such products are, their failure rates are always the same. This means, after any time t, such products are essentially "as good as new." In fact, this may be a good approximation to the behavior of items such as fuses—if a fuse has not burned out by time t, then it is probably very nearly as good as a new fuse.

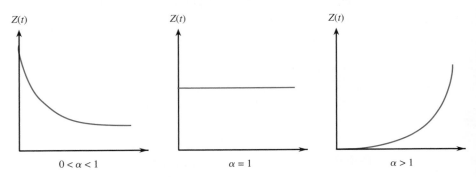

Figure 6.17 Failure rates of Weibull distributions for various values of the shape parameter α

Example 6.11   In Example 6.10, warranty data on tire failures was used to estimate the parameters $\mu = 10.57936$ and $\sigma = .185882$ of a lognormal distribution that describes tread wear (in miles). Denoting tread life by x, the reliability function for x can be calculated using the fact that $\ln(X)$ follows a normal distribution with mean μ and standard deviation σ:

$$R(t) = P(x > t) = P(\ln(x) > \ln(t)) = P\left(z > \frac{\ln(t) - \mu}{\sigma}\right) = 1 - \Phi\left(\frac{\ln(t) - \mu}{\sigma}\right)$$

where $\Phi(z)$ denotes the cumulative probability for the standard normal distribution (see Appendix Table I). Although there is no closed-form expression for $R(t)$, it is easy to use Table I or statistical software to create a graph of $R(t)$, as shown in Figure 6.18.

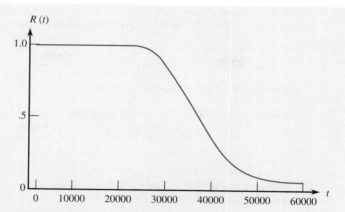

Figure 6.18 Graph of the reliability function for the lognormal distribution of Example 6.11

Similarly, the hazard function $Z(t)$ can be computed and plotted, as shown in Figure 6.19. Notice that the failure rate is an increasing function for the lognormal distribution.

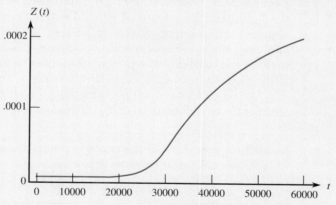

Figure 6.19 Graph of the hazard function for the lognormal distribution of Example 6.11

System Reliability

Products that consist of large assemblies of components can be at risk of failure if one or more of their individual parts fails. Studying how a product's components are connected and how this affects product lifetime is referred to as **topological** or **system reliability.**

    Systems or assemblies are usually comprised of successive levels of subsystems whose individual reliabilities are easy to estimate. By finding the subsystem reliabilities first, one can often combine these estimates into an overall estimate of product reliability. The particular combination depends on how the subsystems are connected.

    Series systems are defined to be systems whose individual components are connected end-to-end in a "series." Figure 6.20 shows a diagram of the typical series system. The main aspect of such systems is that they can only function as long as *every* component of the system functions correctly. Examples of series systems would be the tires on a vehicle, batteries in a flashlight, and the power supply and CPU in a computer.

Figure 6.20 Diagram of a series system

If we denote the reliability at time t of the ith component by $R_i(t)$, then the fundamental theorem of series systems can be summarized as follows:

> If all n components in a series system function *independently* of one another, then the reliability function $R(t)$ for the entire system is simply the product of the reliability functions of the n components. That is:
> $$R(t) = R_1(t) \cdot R_2(t) \cdot R_3(t) \cdots R_n(t)$$

Parallel systems are ones whose components function in parallel, that is, those systems that will function as long as *at least one* of their components functions correctly. Figure 6.21 shows a diagram of a typical parallel system comprised of n components. Parallel systems are often used to build **redundancy** into a product; that is, the components in parallel systems serve as "backups" for each other so that if one component fails, then the entire system will not necessarily fail. Such systems are often used to increase the reliability of a product. Examples of parallel systems include computer routing systems, pacemakers, and safety systems on airplanes.

The fundamental result for computing the reliability $R(t)$ of a parallel system in terms of the reliabilities $R_i(t)$ of its n components is

> If all n components in a parallel system function *independently* of one another, then the reliability function $R(t)$ for the entire system is given as
> $$R(t) = 1 - [1 - R_1(t)] \cdot [1 - R_2(t)] \cdot [1 - R_3(t)] \cdots [1 - R_n(t)]$$

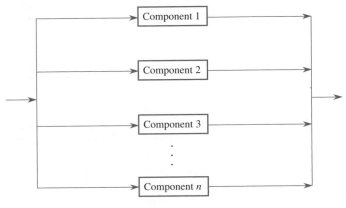

Figure 6.21 Diagram of a parallel system

The concepts of series and parallel systems can be used either separately or in combination when analyzing the reliability of complex systems. The basic method is, when possible, to break down a complex system into various combinations of series and/or parallel subsystems. The reliabilities of such subsystems can then be calculated from the theorems in this section and they, in turn, can often be combined to calculate the overall product or system reliability.

Example 6.12 **Routers** are used in the telecommunications industry to transmit data (in the form of digitized electronic signals) from one location to another. Because many important business and scientific organizations depend upon the continuous availability of data, routing systems must be highly reliable. The usual way of increasing reliability in routing systems is to include various sources of redundancy in the form of parallel subsystems. For example, Figure 6.22 shows a routing system that uses two identical routers in parallel. In addition, each router contains four different power sources (arranged in parallel) and two "supervisor cards" (also in parallel) that direct the router's actions.

Assuming all power sources are of the same kind, each with reliability function $R_p(t)$, and that they act independently of one another, the reliability of each set of four power sources is $1 - [1 - R_p(t)]^4$. Making the same assumptions for the supervisor cards [cards are independent and have a common reliability function $R_s(t)$], the reliability of each set of two cards is $1 - [1 - R_s(t)]^2$. Since the power sources in each router are connected in series to the supervisor cards, the reliability of a single router must be the product of the power source reliability and the supervisor card reliability: $\{1 - [1 - R_p(t)]^4\} \cdot \{1 - [1 - R_s(t)]^2\}$.

Since both routers are connected in parallel, the overall reliability for the routing system is $1 - [1 - \{1 - [1 - R_p(t)]^4\} \cdot \{1 - [1 - R_s(t)]^2\}^2]$. The final step would be to determine the particular form of the failure laws for the power sources and supervisor cards (e.g., exponential or Weibull) and substitute these numerical expressions into the overall reliability formula.

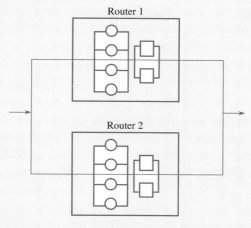

Figure 6.22 Redundancy in a routing system: two routers, four power supplies per router, two supervisor cards per router

Section 6.6 Exercises

46. Intravenous (IV) tubes that deliver liquids (drugs, saline solution, food, etc.) to medical patients are connected by inserting a plastic prong (called a *canula*) from one tub through a rubber membrane in the connector of another tube. Canula systems are needleless and therefore eliminate the possibility of needle punctures to nurses or patients when connecting IV tubes. When disconnected, the surface of the rubber membrane closes up, re-sealing the end of the tube. However, after many such connections and disconnections, the rubber membrane may eventually wear out and fail to close properly.

   Suppose the number of times (i.e., cycles) you can connect and disconnect such a system until a membrane wears out is modeled by an exponential distribution with a mean time to failure of $\mu = 500$ cycles.
   a. What is the probability that a given canula connection will last for at least 100 cycles?
   b. Find the number of cycles, t, for which the reliability equals .95 (i.e., 95%).
   c. Suppose a manufacturer of such systems wants to increase their reliability by specifying that $R(100) = .95$. What is the mean time to failure (in cycles) for such a system?

47. Reliability of mechanical springs is measured in terms of how many times (i.e., *cycles*) the spring can be compressed and released. Suppose that the life-time of a certain type of spring can be modeled by a Weibull distribution with shape parameter $\alpha = 4$ and scale parameter $\beta = 600{,}000$.
   a. Calculate the reliability at $t = 400{,}000$ cycles.
   b. Calculate the reliability at $t = 800{,}000$ cycles.
   c. Calculate the reliability at $t = \beta = 600{,}000$ cycles. Note that $R(\beta)$ is always the *same* number for *any* Weibull distribution.
   d. Write the hazard function. Is the failure rate increasing, decreasing, or flat?

48. Is the exponential distribution a reasonable one for modeling human lifetimes? To answer this question, suppose the mean lifetime is about 75 years and that lifetimes follow an exponential distribution with parameter $\lambda = 1/75 = .0133$.

   a. Use this model to calculate the percentage of people living over 150 years.
   b. Next, calculate the percentage of people living less than 10 years.
   c. Based on results in (a) and (b), do you think the exponential distribution is a good one for modeling human lifetimes?
   d. Answer part (c) using only the "memoryless" property of the exponential distribution.

49. a. Assume that a certain product can be modeled with a normal failure law having a mean lifetime of $\mu = 10$ years and standard deviation $\sigma = 2$ years. Use a spreadsheet program or other software to create a graph of the failure rate, $Z(t)$, for such a product.
   b. Based on your result in part (a), what type of failure rate (decreasing, constant, or increasing) do products with normal failure laws have?

50. Estimates of Weibull parameters can be obtained using simple linear regression (see Section 3.3). De-noting the lifetime of a product by t, the Weibull cumulative area function can be written as $F(x) = 1 - e^{-(x/\beta)\alpha}$, and it is easy to show algebraically that $\ln[\ln(1/1 - F(x))] = \alpha \ln(x) - \alpha \ln(\beta)$ (see page 93). For an ordered set of data $x_1 \le x_2 \le x_3 \le \cdots \le x_n$, we associate x_i with the $[(i - .5)/n]$th sample quantile of the Weibull distribution. That is, we use $p_i = (i - .5)/n$ in place of $F(x)$ and perform a regression of $\ln[\ln(1/1 - p_i)]$ on $\ln(x_i)$ to find estimates of β and α.
   a. Using the data of Example 2.18 (page 93), estimate the parameters of the Weibull distribution that fit this data.
   b. Compare the estimates from part (a) with those obtained in Example 7.18 (page 338).

51. RAID (Redundant Arrays of Inexpensive Disks) structures consist of various combinations of computer disks that use parallel design elements to achieve high reliability. Suppose that a RAID system consists of three disks (A, B, and C) and three "mirror" disks that contain complete copies of the data in the first three disks. Suppose each A, B, and C disk is connected in parallel to its

corresponding mirror disk and that the three such pairs of disks are connected in series.

a. Draw a diagram of this RAID system.

b. Suppose any disk has an exponential lifetime (in months) with parameter $\lambda = .025$. Calculate the reliability of this system.

Supplementary Exercises

52. When affixed to an object, each piece of paper in a pad of adhesive notepaper must stay in place but must also be easily removable. The strength of the adhesive used is a critical quality characteristic of such pads. For this type of product, does adhesive strength have a one- or a two-sided tolerance?

53. In a bottling process, a beam of light is passed through the necks of bottles passing by on a conveyor system. Underfilled bottles, which allow the beam of light to pass through, trip a sensor that routes the bottles off the conveyor system. Bottles with liquid levels above the level of the light beam do not trigger the sensor, thereby meeting the required fill specification; these bottles are then shipped to customers. Describe the shape of the distribution of fill volumes for the bottles that pass this inspection.

54. Instead of constructing $\bar{x}$ and R charts for 30 subgroups of size 4, a friend suggests the simpler alternative of calculating the standard deviation s^* of the 30 means to establish 3-sigma limits for a control chart. That is, it is suggested that the 30 means be plotted on a chart with control limits $\bar{x} \pm 3s^*$. Sample means that fall outside these control limits would indicate process problems. Explain what is wrong with this procedure.

55. A tool that drills holes in metal parts eventually wears out and periodically must be replaced. If the hole diameters drilled by this machine are monitored on a control chart, describe the type of pattern you would expect to see on the chart as the drill wears out.

56. A manufacturer of dustless chalk monitors the consistency of chalk by running an s chart on the density of chalk in subgroups of size 8. The most recent 24 such subgroups had the accompanying sample standard deviations (read across):

.204 .315 .096 .184 .230 .212 .322 .287
.145 .211 .053 .145 .272 .351 .159 .214
.388 .187 .150 .229 .276 .118 .091 .056

a. Construct an s chart based on this data.

b. Check the chart in part (a) for any out-of-control points. If there are any, eliminate them from the data and reconstruct the s chart. Repeat this process, if necessary, until there are no out-of-control signals in the s chart.

57. The deviations from nominal transformation in Exercise 21 can be used in so-called short-run processes. Even though small numbers of different-size parts are created by such processes, the deviations from the various nominal values of these parts provide information about the particular process, not the parts, that is common to all the parts. For example, consider a milling process in which metal bars of various sizes are machined to specified lengths. The size of the bars submitted to the machining process may vary from hour to hour, so there may be insufficient data to create control charts on any particular bar size. However, by subtracting the nominal value from each batch of bars, the resulting subgroups of data are sufficient to create a control chart for the milling process itself. The following table shows the raw length measurements of milled steel bars of various sizes, denoted P1, P2, P3, and P4. The nominal length for bars of type P1 is .125; for bars of type P2, .250; for P3, .375; and for P4, .500.

Subgroup	x_1	x_2	x_3	x_4	Part type
1	.251	.252	.250	.249	P2
2	.372	.378	.379	.375	P3
3	.247	.249	.254	.251	P2
4	.248	.247	.250	.252	P2
5	.249	.249	.250	.249	P2
6	.125	.127	.125	.126	P1
7	.372	.374	.375	.376	P3
8	.499	.502	.495	.503	P4
9	.124	.121	.123	.126	P1
10	.126	.126	.130	.122	P1
11	.375	.374	.378	.379	P3
12	.249	.249	.250	.247	P2
13	.250	.253	.251	.248	P2
14	.249	.250	.249	.249	P2

Subgroup	x_1	x_2	x_3	x_4	Part type
15	.252	.250	.251	.247	P2
16	.251	.249	.250	.250	P2
17	.126	.127	.122	.125	P1
18	.123	.123	.123	.128	P1
19	.252	.250	.247	.248	P2
20	.502	.496	.502	.502	P4

a. Using the nominal lengths given, convert this data into the deviations from nominal format.

b. Construct $\bar{x}$ and R charts of the transformed data in part (a). Evaluate the charts, and comment on the milling process.

58. Explain why it is possible for all the measurements in a given sample to lie within the specification limits and for the same data to yield a *nonzero* estimate of the proportion of the process data that exceeds the specification limits.

59. For a certain process, $\bar{x}$ and R charts based on subgroups of size 5 have centerlines of 14.5 and 1.163, respectively. Given that the process has specification limits of 12 and 16, calculate C_p, C_{pu}, C_{pl}, and C_{pk}.

60. For a fixed value of $\bar{p}$, how large does the subgroup size n have to be to yield a positive lower control limit on a p chart?

61. A "memoryless" system or component is one that satisfies the following property: If it has already lasted for t_1 hours, then the probability it lasts for another t_2 hours is the same as its *initial* probability of lasting t_2 hours. Prove that the exponential distribution is memoryless. That is, prove that

$P(X > t_1 + t_2 | X > t_1) = P(X > t_2)$ for a random variable X that has an exponential distribution with parameter λ.

62. Suppose that two components with reliabilities $R_1(t)$ and $R_2(t)$ are connected in series, but that the two components do *not* necessarily function independently of one another. Show in this case that $\min \{ R_1(t), R_2(t) \} \leq R_1(t)R_2(t) + \frac{1}{4}$.

63. Show that any series system consisting of two components with reliabilities $R_1(t)$ and $R_2(t)$ can never have a system reliability $R(t)$ that exceeds the reliability of its weakest link, that is, $R(t) \leq \min\{R_1(t), R_2(t)\}$.

a. Prove this under the assumption that the two components function independently of one another.

b. Then prove it in the more general case, where the two components may or may not function independently of one another.

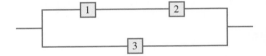

64. A small system contains three components that are connected according to the following diagram. Assuming that the components all function independently of one another, find the general expression for the system reliability $R(t)$ in terms of the component reliabilities $R_1(t)$, $R_2(t)$, and $R_3(t)$.

Bibliography

DeVor, R. E., T. Chang, and J. W. Sutherland, **Statistical Quality Design and Control** (2nd. ed.), Prentice Hall, New Jersey, 2006. *Good discussion of several of the more advanced techniques of quality control.*

Farnum, N. R., **Modern Statistical Quality Control and Improvement**, Duxbury Press, Belmont, CA, 1994. *A comprehensive overview of control charts, acceptance sampling, experimental design, metrology, and the modern approach to quality.*

Lloyd, D. K., and M. Lipow, **Reliability: Management, Methods, and Mathematics**, ASQC Press, Milwaukee, 1984. *Classic text covering all aspects of reliability. Good explanations, with many examples.*

Meeker, W. Q., and L. A. Escobar, **Statistical Methods for Reliability Data**, Wiley, New York, 1998. *Complete, modern presentation of estimation, evaluation, and graphing of reliability functions.*

Montgomery, D. C., **Introduction to Statistical Quality Control** (6th ed.), Wiley, New York, 2012. *Comprehensive and easy to read, with good examples and problems.*

Max Earey/Shutterstock.com

7

Estimation and Statistical Intervals

INTRODUCTION

The general objective of statistical inference is to use sample information as a basis for drawing various types of conclusions. In an estimation problem, we want to make an educated guess about the value of some population characteristic or parameter, such as the population mean battery lifetime μ, the proportion π of all components of a certain type that need service while under warranty, or the difference $\mu_1 - \mu_2$ between the population mean lifetimes for two different types of batteries. The simplest type of estimate is a *point estimate*, a single number that represents our best guess for the value of the parameter. Thus we might report a point estimate of 758 hours for the population mean lifetime of all brand X 100-watt lightbulbs; we are not saying that $\mu = 758$, only that sample data suggests 758 as a very plausible value for μ. Point estimation is discussed in Section 7.1.

A point estimate of a parameter almost surely differs by at least a small amount from the actual value of the parameter. That is, there is almost always at least a small error in the estimate. Our estimate, for example, may be 758 hours when μ is actually 750 hours. It would be nice if our estimates could provide some indication of precision; this is the purpose of a confidence interval (interval estimate). Such estimates, as well as several other types of intervals, are presented in Sections 7.2–7.5. Section 7.6 briefly considers several other topics relating to estimation.

7.1   POINT ESTIMATION

A **point estimate** of some parameter θ is a single number, calculated from sample data, that can be regarded as an educated guess for the value of θ. We might, for example, report 32.5 mpg as a point estimate of the population mean fuel efficiency μ for all cars of a particular type under specified conditions. Or we might decide that .350 is a point estimate for the proportion π of all individuals who would try a particular product again after using a free trial sample.

A point estimate is usually obtained by selecting a suitable statistic and calculating its value for the given sample data. For example, a natural statistic to use for estimating a population mean μ is the sample mean $\bar{x}$, and a sensible way to estimate a population variance σ^2 is to compute the value of the sample variance s^2. The statistic used to calculate an estimate is sometimes called an **estimator,** and the symbol $\hat{\theta}$ is frequently used to denote either the estimator or the resulting estimate. Thus the statement

$$\hat{\mu} = \bar{x} = 32.5$$

says that the point estimate of the population mean μ is 32.5 and that this estimate was calculated using the sample mean $\bar{x}$ as the estimator.

Example 7.1   A commonly used method of estimating the size of a wildlife population is to perform a capture/recapture experiment. Suppose a biologist wishes to estimate the number of fish in a certain lake; that is, the parameter to be estimated is the population size N. An initial sample of 100 fish is selected, each one is tagged, and the tagged fish are returned to the lake. After a time period sufficient to allow the tagged fish to mix with the other fish in the lake, a second sample of 250 fish is selected. If 25 of the fish in the recapture sample are tagged, what is a sensible estimate for N? Because 10% of the fish in the recapture sample are tagged, it is reasonable to estimate that 10% of all fish in the lake are tagged. Since we know that a total of 100 fish were initially tagged, this suggests that we use 1000 as a point estimate of N.

More generally, if M denotes the number of fish initially tagged, n the size of the recapture sample, and x the number of tagged fish in the recapture sample (so x is a random variable), the proposed estimator of N is $\hat{N} = [Mn/x]$. (The square bracket notation $[c]$ denotes the largest whole number that is at most c; this takes care of cases where Mn/x is not a whole number.)

Frequently, there is more than one estimator that can sensibly be used to calculate an estimate, as the following example shows.

Example 7.2 Consider a population of $N = 5000$ invoices. Associated with each invoice is its "book value," the recorded amount of that invoice. Let $T = \$1,761,300$ denote the known total book value. Unfortunately, some of the book values are erroneous. An audit will be carried out by randomly selecting n invoices and determining the audited (i.e., correct) value for each one. Suppose the sample gives the following results:

Invoice:	1	2	3	4	5
Book value:	300	720	526	200	127
Audited value:	300	520	526	200	157
Error:	0	200	0	0	-30

Let $\bar{y} =$ sample mean book value $= \$374.60$, $\bar{x} =$ sample mean audited value $= \$340.60$, and $\bar{e} =$ sample mean error $= \$34.00$. Each of the following estimators for the total audited (i.e., correct) value and resulting estimates is sensible:

mean per unit statistic $= N\bar{x}$; estimate $= 5000(340.60) = \$1,703,000$

difference statistic $= T - N\bar{e}$; estimate $= 1,761,300 - (5000)(34)$
$$= \$1,591,300$$

ratio statistic $= T(\bar{x}/\bar{y})$; estimate $= (1,761,300)(340.6/374.6) = \$1,601,438$

The choice among these estimates is not clear-cut. In fact, all three of the estimators have been advocated by those employing statistical methodology in auditing.

In situations where there is more than one sensible estimator available, criteria for selecting an estimator are needed. We now turn to a brief discussion of desirable properties of estimators.

Properties of Estimators

One desirable property that a good estimator should possess is that it be **unbiased.** An estimator is unbiased if, in repeated random samples, the numerical values of the estimator stack up around the population parameter that we are trying to estimate. An often-used analogy is to think of each value of an estimator as a shot fired at a target, the target being the population parameter of interest. As long as all the shots fall in a pattern with the target value in the *middle*, we say that the shots are unbiased. Notice that we do not require that any of the individual shots actually hit the target; we require only that they be centered around the target value. If the majority of the shots are centered somewhere else, then we say that they exhibit a certain amount of **bias.**

In terms of sampling distributions, an estimator is said to be unbiased if the mean of its sampling distribution coincides with the parameter that is being estimated. For instance, we know from Section 5.5 that the sampling distribution of the statistic $\bar{x}$ has a mean value of $\mu_{\bar{x}}$, which equals the mean μ of the population from which the samples are taken. Then $\bar{x}$ is said to be an estimator of the parameter μ and, because $\mu_{\bar{x}} = \mu$,

$\bar{x}$ is also an unbiased estimator of μ. In general, for any population parameter θ and any estimator $\hat{\theta}$ of that parameter, Figure 7.1 illustrates what it means for $\hat{\theta}$ to be unbiased or biased.

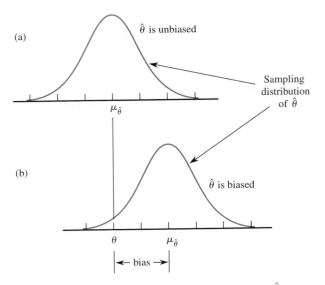

Figure 7.1 Sampling distribution of an estimator $\hat{\theta}$

DEFINITIONS

Denote a population parameter generically by the letter θ and denote any estimator of this parameter by $\hat{\theta}$. Then $\hat{\theta}$ is an **unbiased** estimator if $\mu_{\hat{\theta}} = \theta$. Otherwise, $\hat{\theta}$ is said to be biased, and the quantity $\mu_{\hat{\theta}} - \theta$. is called the **bias** of $\hat{\theta}$.

Some of the most important statistics we have studied are unbiased estimators of certain population parameters. For example, it can be shown that the sample mean $\bar{x}$ is an unbiased estimator of the population mean μ, the sample variance s^2 is an unbiased estimator of the population variance σ^2, and the sample proportion p is an unbiased estimator of the population proportion π. One important exception is the sample standard deviation s, which turns out to be a slightly biased estimator of the population standard deviation σ. Fortunately, for large samples, the amount of bias in s is negligible. For small samples from a normal population, there is a simple correction factor that can be applied to s that converts it into an unbiased statistic for estimating σ.

Unbiasedness does not imply that the estimate computed from any particular sample will coincide with the value of the parameter being estimated. Consider, for example, using the sample proportion p to estimate the population proportion π based on a sample of size $n = 25$, and suppose that $\pi = .7$. Then $\mu_p = .7$, so the sampling distribution of p is centered at .7. However, with x denoting the number of "successes" in the sample, $p = x/25 \neq .7$ for any possible value of x. That is, even though p is unbiased for estimating π, the value of the estimate calculated from any particular sample will inevitably differ from π. Nevertheless, if sample after sample is selected and the value of p calculated for each one, unbiasedness implies that the long-run average of these estimates will be the correct value, .7.

A second desirable property that estimators often possess is consistency. If $\hat{\theta}$ denotes an estimator of some population parameter θ, then $\hat{\theta}$ is said to be **consistent** if the probability that it lies close to θ increases to 1 as the sample size increases. Simply stated, consistent estimators become more and more accurate as the sample size increases. That is, as you increase n, it becomes more and more likely that such estimators will be very close to the parameter they are intended to estimate. The most common method for showing that an estimator is consistent is to show that its standard error decreases as the sample size increases. For instance, because the standard error of $\bar{x}$ is $\sigma_{\bar{x}} = \sigma/\sqrt{n}$, which must necessarily decrease as n increases, the sample mean qualifies as a consistent estimator of μ. This means that for any interval around μ, no matter how small the interval, we can eventually select n large enough so that the sampling distribution lies almost entirely within the interval. This property is illustrated in Figure 5.19. Although there are some estimators that are not consistent, such examples are fairly rare. In fact, all of the statistical applications in this text involve consistent estimators.

DEFINITION

> If the probability that an estimator $\hat{\theta}$ falls close to a population parameter θ can be made as near to 1 as desired by increasing the sample size n, then $\hat{\theta}$ is said to be a **consistent** estimator of θ.

Section 7.1 Exercises

1. A single plastic part is randomly selected from a large population of such parts. Can the length of the chosen part be considered an unbiased estimator of the average length of all the parts?

2. A random sample of ten homes in a particular area, each heated with natural gas, is selected, and the amount of gas (therms) used during January is determined for each home. The resulting observations are 103, 156, 118, 89, 125, 147, 122, 109, 138, and 99.
   a. Use an unbiased estimator to compute a point estimate of μ, the average amount of gas used by all houses in the area.
   b. Use an unbiased estimator to compute a point estimate of π, the proportion of all homes that use over 100 therms.

3. Random samples of size n are taken from a normal population whose standard deviation is known to be 5.
   a. For random samples of size $n = 10$, calculate the area under the sampling distribution curve for $\bar{x}$ between the values $\mu - 1$ and $\mu + 1$. That is, find the probability that the sample mean lies within ± 1 unit of the population mean.
   b. Repeat the probability calculation in part (a) for samples of size $n = 50$, $n = 100$, and $n = 1000$.
   c. Graph the probabilities you found in parts (a) and (b) versus their corresponding sample sizes, n. What can you conclude from this graph?

4. Random samples of n trees are taken from a large area of forest, and the proportion of diseased trees in

each sample is determined. The actual proportion of diseased trees, π, is unknown.

a. For random samples of size $n = 10$, calculate the area under the sampling distribution curve for p between the points $\pi - .10$ and $\pi + .10$. That is, find the probability that the sample proportion lies within $\pm.10$ (i.e., 10%) of the population proportion. Use the formula for the upper bound on the standard error of p (see Section 5.6) in your calculations.

b. Repeat the probability calculation in part (a) for samples of size $n = 50$, $n = 100$, and $n = 1000$. (Use the normal approximation to the binomial.)

c. Graph the probabilities you found in parts (a) and (b) versus their corresponding sample sizes, n. What can you conclude from this graph?

5. Random samples of size n are selected from a normal population whose standard deviation σ is known to be 2.

a. Suppose you want 90% of the area under the sampling distribution of $\bar{x}$ to lie within ± 1 unit of a population mean μ. Find the minimum sample size n that satisfies this requirement.

b. Repeat the calculations in part (a) for areas of 80%, 95%, and 99%.

c. Plot the sample sizes found in parts (a) and (b) versus their corresponding probabilities. What can you conclude from this graph?

6. Each of 150 newly manufactured items is examined, and the number of surface flaws per item is recorded, yielding the following data:

Number of flaws:	0	1	2	3	4	5	6	7
Observed frequency:	18	37	42	30	13	7	2	1

Let x denote the number of flaws on a randomly chosen item, and assume that x has a Poisson distribution with parameter λ.

a. Find an unbiased estimator for λ and compute the estimate using the data. *Hint:* The mean of a Poisson random variable equals λ.

b. What is the standard error of the estimator in part (a)? *Hint:* The variance of a Poisson random variable also equals λ.

7.2 LARGE-SAMPLE CONFIDENCE INTERVALS FOR A POPULATION MEAN

A point estimate, because it is a single number, by itself provides no information about the precision and reliability of estimation. Consider, for example, using the statistic $\bar{x}$ to calculate a point estimate for the true average breaking strength (g) of paper towels of a certain brand, and suppose that $\bar{x} = 9322.7$. Because of sampling variability, it is virtually never the case that $\bar{x} = \mu$. The point estimate says nothing about how close it might be to μ. An alternative to reporting a single most plausible value of the parameter being estimated is to calculate and report an entire interval of plausible values—an *interval estimate* or *confidence interval* (CI). A confidence interval is always calculated by first selecting a *confidence level*, which is a measure of the degree of reliability of the interval. A confidence interval with a 95% confidence level for the true average breaking strength might have a lower limit of 9162.5 and an upper limit of 9482.9. Then at the 95% confidence level, any value of μ between 9162.5 and 9482.9 is plausible. A confidence level of 95% implies that 95% of all samples would give an interval that includes μ, or whatever other parameter is being estimated, and only 5% of all samples would yield an erroneous interval. The most frequently used confidence levels are 95%, 99%, and 90%. The higher the confidence level, the more strongly we believe that the value of the parameter being estimated lies within the interval.

Information about the precision of an interval estimate is conveyed by the width of the interval. If the confidence level is high and the resulting interval is quite narrow, our knowledge of the value of the parameter is reasonably precise. A very wide confidence interval, however, gives the message that there is a great deal of uncertainty concerning the value of what we are estimating. Figure 7.2 shows 95% confidence intervals for true average breaking strengths of two different brands of paper towels. One of these intervals suggests precise knowledge about μ, whereas the other suggests a very wide range of plausible values.

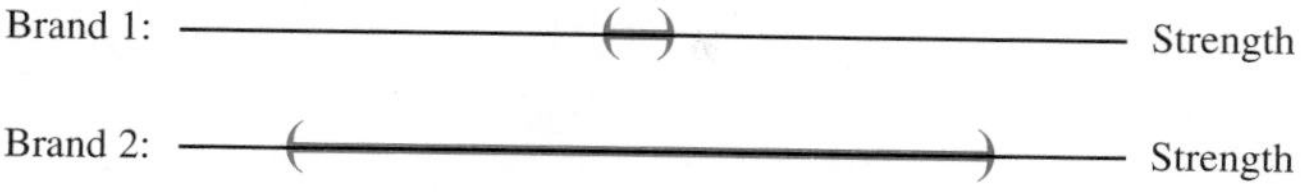

Figure 7.2 Confidence intervals indicating precise (brand 1) and imprecise (brand 2) information about μ

A Confidence Interval for μ with Confidence Level 95%

A confidence interval for a population or process mean μ is based on the following properties of the sampling distribution of $\bar{x}$:

$$\mu_{\bar{x}} = \mu \qquad \sigma_{\bar{x}} = \frac{\sigma}{\sqrt{n}}$$

When n is large, the $\bar{x}$ distribution is approximately normal (this is the Central Limit Theorem). Standardizing $\bar{x}$ by subtracting its mean value and dividing by its standard deviation gives the following standardized variable, denoted by z to emphasize that its distribution is approximately standard normal (the z curve):

$$z = \frac{\bar{x} - \mu}{\sigma/\sqrt{n}}$$

The difficulty with this standardized variable is that, in practice, the value of the population or process standard deviation σ will almost never be known to an investigator. Consider instead the standardized variable in which σ is replaced by the *sample* standard deviation s:

$$\frac{\bar{x} - \mu}{s/\sqrt{n}}$$

Because there is sampling variability in this second standardized variable both in the numerator (because of $\bar{x}$) and in the denominator (the value of s will also vary from sample to sample), it would seem as though its distribution should be more spread out than the z curve. But appearances are deceiving! It turns out that when n is large, replacement of σ by s does not add much variability; in this case, *the variable $z = (\bar{x} - \mu)/(s/\sqrt{n})$ also has approximately a standard normal distribution.*

A confidence interval with a 95% confidence level is obtained by starting with a central z curve area of .95. As Figure 7.3 illustrates, the z critical values 1.96 and -1.96 capture this area (consult Appendix Table I).

The foregoing facts justify the following probability statement:

$$P\left(-1.96 < \frac{\bar{x} - \mu}{s/\sqrt{n}} < 1.96\right) \approx .95$$

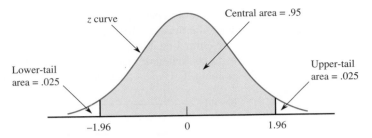

Figure 7.3 Capturing a central z curve area of .95

Now let's manipulate the inequalities inside the parentheses to isolate μ in the middle and move everything else to the two extremes. This is achieved as follows:

1. Multiply all three terms by $s/\sqrt{n}$.
2. Subtract $\bar{x}$ from all three terms (leaving only $-\mu$ in the middle).
3. Multiply by -1 (causing the direction of each inequality to reverse).

The result is $\bar{x} + 1.96(s/\sqrt{n}) > \mu > \bar{x} - 1.96(s/\sqrt{n})$, or, rewriting the terms in reverse order,

$$\bar{x} - 1.96\,\frac{s}{\sqrt{n}} < \mu < \bar{x} + 1.96\,\frac{s}{\sqrt{n}}$$

These new inequalities are algebraically equivalent to those we started with, so the probability associated with the new inequalities is also (approximately) .95. That is, think of $\bar{x} - 1.96(s/\sqrt{n})$ as the lower limit and $\bar{x} + 1.96(s/\sqrt{n})$ as the upper limit of an interval. Both of these limits involve $\bar{x}$ and s, so the values of both limits will vary from sample to sample. With a probability of approximately .95, the selected sample will be such that the value of μ is captured between these two interval limits. Substituting the values of n, $\bar{x}$, and s from any particular sample into these expressions gives a confidence interval for μ with a confidence level of approximately 95%.

A large-sample confidence interval for μ with a confidence level of (approximately) 95% has

$$\text{lower confidence limit} = \bar{x} - 1.96\,\frac{s}{\sqrt{n}}$$

$$\text{upper confidence limit} = \bar{x} + 1.96\,\frac{s}{\sqrt{n}}$$

The interval is centered at $\bar{x}$ and extends out the same distance, $1.96s/\sqrt{n}$, to each side, so it can be written in abbreviated form as

$$\bar{x} \pm 1.96\,\frac{s}{\sqrt{n}}$$

This formula is valid whatever the shape of the population distribution.

The two limits $\bar{x} \pm (1.96)s/\sqrt{n}$ can also be obtained by replacing each $<$ inside the parentheses in the probability statement by $=$ and solving the two resulting equations for μ.

Example 7.3 The alternating-current (AC) breakdown voltage of an insulating liquid indicates its dielectric strength. The article "Testing Practices for the AC Breakdown Voltage Testing of Insulation Liquids" (*IEEE Electrical Insulation Magazine*, 1995: 21–26) gave the accompanying sample observations on breakdown voltage (kV) of a particular circuit under certain conditions:

```
62  50  53  57  41  53  55  61  59  64  50  53  64  62  50  68
54  55  57  50  55  50  56  55  46  55  53  54  52  47  47  55
57  48  63  57  57  55  53  59  53  52  50  55  60  50  56  58
```

Figure 7.4 shows the output from the JMP software's Analyze/Distribution command. The boxplot of the data shows a high concentration in the middle half of the data (narrow box width). There is a single outlier at the upper end, but this value is actually a bit closer to the median (55) than is the smallest sample observation.

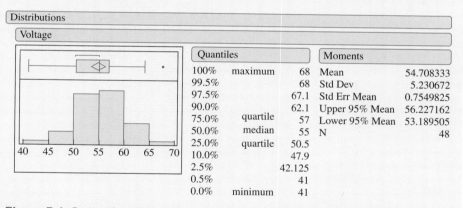

Figure 7.4 Output from JMP for the breakdown voltage data from Example 7.3

Summary quantities include $n = 48$, $\bar{x} = 54.7$, and $s = 5.23$. The 95% confidence interval is then

$$54.7 \pm 1.96 \frac{5.23}{\sqrt{48}} = 54.7 \pm 1.5 = (53.2, 56.2)$$

That is,

$$53.2 < \mu < 56.2$$

with a confidence level of approximately 95%. The interval is reasonably narrow, indicating that we have precisely estimated μ. Note that our lower and upper interval endpoints match JMP's "Lower 95% Mean" and "Upper 95% Mean," respectively.

The 95% confidence interval for μ in the foregoing example is (53.2, 56.2). It is tempting to say that there is a 95% chance that μ is between 53.2 and 56.2. Do not yield to this temptation! The 95% refers to the long-run percentage of *all* possible samples resulting in an interval that includes μ. That is, if we consider taking sample after sample from the population and use each one separately to compute a 95% confidence interval, in the long run roughly 95% of these intervals will capture μ. Figure 7.5 illustrates this for 100 samples; 93 of the resulting intervals include μ, whereas 7 do not. Without knowing the value of μ, we cannot tell whether our interval (53.2, 56.2) is one of the good 95% or the bad 5% of all intervals that might result. *The confidence level refers to the **method** used to construct the interval rather than to any particular calculated interval.*

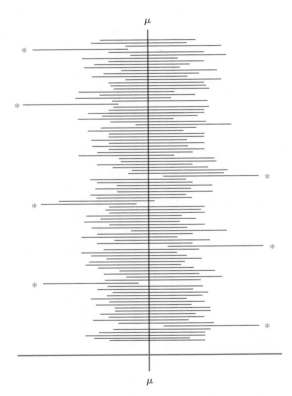

Figure 7.5 95% confidence intervals for μ from 100 different samples (* identifies an interval that does not include μ)

Other Confidence Levels and a General Formula

The confidence level of 95% was inherited from the probability .95 with which we began the derivation of the interval. This probability in turn dictated the use of the z critical value 1.96 in the confidence interval formula. It follows that if we want a confidence level of 99%, we should identify the z critical value that captures a central z curve area of .99. Figure 7.6 shows how this is done. In Appendix Table I, the closest entries for the cumulative area .9950 are .9949, in the 2.5 row and .07 column, and .9951, in the same

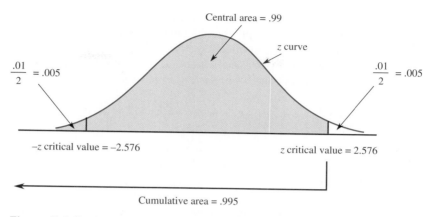

Figure 7.6 Finding the z critical value for a 99% confidence level

row and .08 column. Thus 2.576 (or 2.58, to be conservative) should be used in the CI formula in place of 1.96 to obtain the higher confidence level.

It should be clear at this point that *any* confidence level can be achieved simply by finding the z critical value that captures the corresponding z curve area. For example, it is easily verified that the interval from -1.28 to 1.28 contains above it about 80% of the area under the z curve, so using 1.28 in place of 1.96 gives a CI with confidence level 80%.

A large-sample confidence interval for a population or process mean μ is given by the formula

$$\bar{x} \pm (z \text{ critical value}) \frac{s}{\sqrt{n}}$$

As a general rule, this interval is appropriate when the sample size exceeds 30. The three most commonly used confidence levels, 90%, 95%, and 99%, use z critical values of 1.645, 1.96, and 2.576, respectively.

Why settle for 95% confidence when 99% confidence is possible? The price of a higher confidence level is that the resulting interval is wider. The width of the 95% interval is $2(1.96s/\sqrt{n})$, whereas the 99% interval has a width of $2(2.576s/\sqrt{n})$. The higher *reliability* of the 99% interval entails a loss in precision (as indicated by the wider interval). Many investigators think that a 95% confidence level gives a reasonable compromise between reliability and precision.

Choosing the Sample Size

The half-width $1.96s/\sqrt{n}$ of the 95% CI is sometimes called the **bound on the error of estimation** associated with a 95% confidence level; that is, with 95% confidence, the point estimate $\bar{x}$ will be no farther than this from μ. Before obtaining data, an investigator may wish to determine a sample size for which a particular value of the bound is achieved. For example, with μ representing the average fuel efficiency (mpg) for all cars of a certain type, the objective of an investigation may be to estimate μ to within 1 mpg with 95% confidence. More generally, suppose we wish to estimate μ to within an amount B (the specified bound on the error of estimation) with 95% confidence. This implies that $B = 1.96s/\sqrt{n}$, from which

$$n = \left[\frac{1.96s}{B}\right]^2$$

The difficulty with this formula is that calculating the value of n requires having s, which is of course not available until a sample has been selected. Instead, prior information about σ may be used as a basis for a reasonable guess for s. Alternatively, for a population distribution that is not too skewed, dividing the range (difference between the largest and smallest values) by 4 often gives a rough idea of what s might be.

Example 7.4   Refer to Example 7.3 on breakdown voltage. Suppose that the investigator believes that almost all values in the population distribution are between 40 and 70. Then $(70 - 40)/4 = 7.5$ gives a reasonable value for s. The appropriate sample size for estimating true average breakdown voltage to within 1 kV with confidence level 95% is now

$$n = \left[\frac{(1.96)(7.5)}{1}\right]^2 \approx 217$$

The sample size associated with an error bound B for any other confidence level, such as 99%, results from replacing 1.96 in the formula for n by the corresponding critical value, for example, 2.576.

One-Sided Confidence Intervals (Confidence Bounds)

The confidence intervals discussed thus far give both a lower confidence bound *and* an upper confidence bound for μ. In some circumstances, an investigator will want only one of these two types of bounds. For example, a psychologist may wish to calculate a 95% upper confidence bound for true average reaction time to a particular stimulus, or a reliability engineer may want only a lower confidence bound for true average lifetime of components of a certain type. It is easily verified that the cumulative area under the curve to the left of 1.645 is .95, implying that

$$P\left(\frac{\bar{x} - \mu}{s/\sqrt{n}} < 1.645\right) \approx .95$$

Manipulating the inequality inside the parentheses to isolate μ on one side gives the equivalent inequality $\mu > \bar{x} - 1.645s/\sqrt{n}$; the expression on the right is the desired lower confidence bound. Starting with $P(-1.645 < z) \approx .95$ and manipulating the inequality results in the upper confidence bound. A similar argument gives a one-sided bound associated with any other confidence level.

> **A large-sample upper confidence bound for μ is**
>
> $$\mu < \bar{x} + (z \text{ critical value}) \frac{s}{\sqrt{n}}$$
>
> and a **large-sample lower confidence bound for μ is**
>
> $$\mu > \bar{x} - (z \text{ critical value}) \frac{s}{\sqrt{n}}$$
>
> The three most commonly used confidence levels, 90%, 95%, and 99%, use z critical values of 1.28, 1.645, and 2.33, respectively.

Example 7.5 Recently there has been increased use of titanium and its alloys in aerospace and automotive applications. These alloys are highly durable and have a high strength-to-weight ratio. However, machining of titanium is difficult due to its low thermal conductivity. The authors of "Modelling and Multi-Objective Optimiaz-tion of Process Parameters of Wire Electrical Discharge Machining Using Non-Dominated Sorting Genetic Algorithm-II" (*J. of Engr. Manuf.*, 2012: 1186–2001), Investigated different settings that impact wire electrical discharge machining of titanium 6-2-4-2. A characteristic of interest was surface roughness (in μm) of the metal after machining. In one particular investigation a sample of 54 surface roughness observations gave a sample mean of 1.9042 μm and a sample standard deviation of .1455 μm. An upper confidence bound for true average surface roughness μ with confidence 95% is

$$1.9042 + 1.645 \frac{(.1455)}{\sqrt{54}} = 1.9042 + .0326 = 1.9368$$

That is, with a confidence level of 95%, the value of μ lies in the interval $(-\infty, 1.9368)$. Since negative values for surface roughness are not possible, we revise this interval to (0, 1.9368).

Section 7.2 Exercises

7. Assuming that n is large, determine the confidence level for each of the following two-sided confidence intervals:
   a. $\bar{x} \pm 3.09s/\sqrt{n}$
   b. $\bar{x} \pm 2.81s/\sqrt{n}$
   c. $\bar{x} \pm 1.44s/\sqrt{n}$
   d. $\bar{x} \pm s/\sqrt{n}$

8. What z critical value in the large-sample two-sided confidence interval for μ should be used to obtain each of the following confidence levels?
   a. 98%
   b. 85%
   c. 75%
   d. 99.9%

9. Discuss how each of the following factors affects the width of the large-sample two-sided confidence interval for μ:
   a. Confidence level (for fixed n and s)
   b. Sample size n (for fixed confidence level and s)
   c. Sample standard deviation s (for fixed confidence level and n)

10. Each of the following is a confidence interval for μ = true average (i.e., population mean) resonance

frequency (Hz) for all tennis rackets of a certain type:

(114.4, 115.6)   (114.1, 115.9)

a. What is the value of the sample mean resonance frequency?

b. Both intervals were calculated from the same sample data. The confidence level for one of these intervals is 90% and for the other is 99%. Which of the intervals has the 90% confidence level, and why?

11. Suppose that a random sample of 50 bottles of a particular brand of cough syrup is selected, and the alcohol content of each bottle is determined. Let μ denote the average alcohol content for the population of all bottles of the brand under study. Suppose that the resulting 95% confidence interval is (7.8, 9.4).

a. Would a 90% confidence interval calculated from this same sample have been narrower or wider than the given interval? Explain your reasoning.

b. Consider the following statement: There is a 95% chance that μ is between 7.8 and 9.4. Is this statement correct? Why or why not?

c. Consider the following statement: We can be highly confident that 95% of all bottles of this type of cough syrup have an alcohol content that is between 7.8 and 9.4. Is this statement correct? Why or why not?

d. Consider the following statement: If the process of selecting a sample of size 50 and then computing the corresponding 95% interval is repeated 100 times, 95 of the resulting intervals will include μ. Is this statement correct? Why or why not?

12. Heavy-metal pollution of various ecosystems is a serious environmental threat, in part because of the potential transference of hazardous substances to humans via food. The article "Cadmium, Zinc, and Total Mercury Levels in the Tissues of Several Fish Species from La Plata River Estuary, Argentina" (*Environmental Monitoring and Assessment*, 1993: 119–130) reported the following summary data on zinc concentration (μg/g) in the liver of fish:

Species	n	$\bar{x}$	s
Mugil liza	56	9.15	1.27
Pogonias cromis	61	3.08	1.71

a. Calculate a 95% two-sided confidence interval for population mean concentration for the *Mugil liza* species.

b. Calculate a 99% two-sided confidence interval for population mean concentration for the *Pogonias cromis* species. Why is this interval wider than the interval of part (a) even though it is based on a somewhat larger sample size?

13. Young people may feel they are carrying the weight of the world on their shoulders when in reality they are too often carrying an excessively heavy backpack. The article "Effectiveness of a School-Based Backpack Health Promotion Program" (*Work*, 2003: 113–123) reported the following data for a sample of 131 sixth graders: for backpack weight (lbs), $\bar{x} = 13.83$, $s = 5.05$; for backpack weight as a percentage of body weight, a 95% CI for the population mean was (13.62; 15.89).

a. Calculate and interpret a 99% CI for population mean backpack weight.

b. Obtain a 99% CI for population mean weight as a percentage of body weight.

c. The American Academy of Orthopedic Surgeons recommends that backpack weight be at most 10% of body weight. What does your calculation of part (b) suggest and why?

14. The article "Extravisual Damage Detection? Defining the Standard Normal Tree" (*Photogrammetric Engr. and Remote Sensing*, 1981: 515–522) discusses the use of color infrared photography in identification of normal trees in Douglas fir stands. Among data reported were summary statistics for green-filter analytic optical densitometric measurements on samples of both healthy and diseased trees. For a sample of 69 healthy trees, the sample mean dye-layer density was 1.028, and the sample standard deviation was .163.

a. Calculate a 95% two-sided CI for the true average dye-layer density for all such trees.

b. Suppose the investigators had made a rough guess of .16 for the value of s before collecting data. What sample size would be necessary to obtain an interval width of .05 for a confidence level of 95%?

15. The negative effects of ambient air pollution on children's lung function has been well established, but less research is available about the effects of

indoor air pollution. The authors of "Indoor Air Pollution and Lung Function Growth Among Children in Four Chinese Cities" (*Indoor Air*, 2012: 3–11) investigated the relationship between indoor air pollution metrics and lung function growth among children ages 6–13 years living in four Chinese cities. For each subject in the study, the authors measured an important lung-capacity index known as FEV_1, the forced volume (in ml) of air that is exhaled in 1 second. Higher FEV_1 values are associated with greater lung capacity.

Burning coal inside houses can lead to increased levels of indoor air toxins that may have negative effects on lung function. Among the children in the study, 514 came from households that use coal for cooking or heating or both. Their FEV_1 mean was 1427 with standard deviation 325. (Using a complex statistical procedure the authors went on to show that burning coal had a clear negative effect on mean FEV_1 levels.)

a. Calculate and interpret a 95% (two-sided) confidence interval for true average FEV_1 level in the population of all children from which the sample was selected.

b. Suppose the investigators had made a rough guess of 320 for the value of s before collecting data. What sample size would be necessary to obtain an interval width of 50 ml for a confidence level of 95%?

16. The article "Evaluating Tunnel Kiln Performance" (*Amer. Ceramic Soc. Bull.*, August 1997: 59–63) gave the following summary information for fracture strengths (MPa) of $n = 169$ ceramic bars fired in a particular kiln: $\bar{x} = 89.10$, $s = 3.73$.

a. Calculate a two-sided confidence interval for true average fracture strength using a confidence level of 95%. Does it appear that true average fracture strength has been precisely estimated?

b. Suppose the investigators had believed a priori that the population standard deviation was about 4 MPa. Based on this supposition, how large a sample would have been required to estimate μ to within .5 MPa with 95% confidence?

17. When the population distribution is normal and n is large, the statistic s has approximately a normal distribution with $\mu_s \approx \sigma$, $\sigma_s \approx \sigma/\sqrt{2n}$. Use this fact to develop a large-sample two-sided confidence interval formula for σ. Then calculate a 95% confidence interval for the true standard deviation of the fracture strength distribution based on the data given in Exercise 16 (the cited paper gave compelling evidence in support of assuming normality).

18. Determine the confidence level for each of the following large-sample one-sided confidence bounds:
a. Upper bound: $\bar{x} + .84s/\sqrt{n}$
b. Lower bound: $\bar{x} - 2.05s/\sqrt{n}$
c. Upper bound: $\bar{x} + .67s/\sqrt{n}$

19. The charge-to-tap time (min) for a carbon steel in one type of open hearth furnace was determined for each heat in a sample of size 36, resulting in a sample mean time of 382.1 and a sample standard deviation of 31.5. Calculate a 95% upper confidence bound for true average charge-to-tap time.

20. A Brinell hardness test involves measuring the diameter of the indentation made when a hardened steel ball is pressed into material under a standard test load. Suppose that the Brinell hardness is determined for each specimen in a sample of size 32, resulting in a sample mean hardness of 64.3 and a sample standard deviation of 6.0. Calculate a 99% lower confidence bound for true average Brinell hardness for material specimens of this type.

21. The article "Ultimate Load Capacities of Expansion Anchor Bolts" (*J. of Energy Engr.*, 1993: 139–158) gave the following summary data on shear strength (kip) for a sample of 3/8-in. anchor bolts: $n = 78$, $\bar{x} = 4.25$, $s = 1.30$. Calculate a lower confidence bound using a confidence level of 90% for true average shear strength.

7.3 MORE LARGE-SAMPLE CONFIDENCE INTERVALS _____

In Section 7.2, we used properties of the sampling distribution of $\bar{x}$ as a basis for obtaining a confidence interval formula for estimating μ when the sample size was large. In this section, we develop a large-sample interval formula for π, the proportion of individuals or objects in a population or process that possess a particular characteristic, and also for $\mu_1 - \mu_2$, the difference between two population or process means. These intervals are based on sampling distribution properties of appropriate statistics.

A Large-Sample Confidence Interval for π

Let π denote the proportion of individuals or objects in a population or process that possess a particular characteristic (the successes). For example, π might represent the proportion of all components of a certain type that do not need service while under warranty, the proportion of all computers sold at a certain store that are laptop models, or the proportion of patients suffering from a certain disease who respond favorably to a particular treatment. An inference about π will be based on a random sample of size n selected from the population or process. The natural statistic for estimating π is the sample proportion

$$p = \frac{\text{number of successes in the sample}}{n}$$

For example, if $n = 5$ and the resulting sample is SFFSS (the first, fourth, and fifth sampled individuals possess the property of interest but the second and third do not), then $p = 3/5 = .60$. The value of p is also .60 for the outcomes SSSFF and SFSFS, whereas it is .20 for the outcome FSFFF and 1 for the outcome SSSSS. When $n = 5$, the six possible values of p are 0, .2, .4, .6, .8, and 1. The larger the sample size, the more values of p are possible.

The value of π for any such population is a fixed number between 0 and 1. If, however, we select sample after sample of size n from the same population or process, the value of p will vary from sample to sample. In the case $n = 5$, a first sample might give $p = .6$, a second sample $p = .8$, a third sample $p = .6$ again, and so on. The sampling distribution of the statistic p describes this long-run variation. Consider again $n = 5$ and suppose that $\pi = .6$. Using the same reasoning that led to the binomial distribution in Chapter 1, the long-run proportion of samples with $p = 1$ (corresponding to the single outcome SSSSS) is $(.6)^5 = .078$. Similarly, there are five outcomes for which $p = .8$ (FSSSS, ..., SSSSF), and the corresponding long-run proportion is $5(.6)^4(.4) = .259$. The complete sampling distribution is

p:	0	.2	.4	.6	.8	1
Long-run proportion (probability):	.010	.077	.230	.346	.259	.078

We can then easily verify that the mean value of the statistic p is

$$\mu_p = 0(.010) + .2(.077) + \cdots + 1(.078) = .60$$

That is, the sampling distribution of p is centered exactly at the value of what the statistic is trying to estimate. This is true regardless of the values of π and n—the statistic is unbiased. Notice, however, that it is not highly likely that $p = \pi$; the sampling distribution is quite spread out about its mean value.

General properties of the sampling distribution of p:

1. $\mu_p = \pi$
2. $\sigma_p = \sqrt{\pi(1 - \pi)/n}$
3. If both $n\pi > 5$ and $n(1 - \pi) > 5$, the sampling distribution is approximately normal.

Because n is in the denominator under the square root in the expression for σ_p, the standard deviation decreases and the sampling distribution becomes more and more concentrated about π as the sample size increases. The two inequality conditions in the third property are designed to ensure that there is enough symmetry in the sampling distribution so that a normal curve with mean value π and standard deviation σ_p provides a good approximation to a histogram of the actual distribution. For example, if $n = 100$ but $\pi = .02$, there is too much (positive) skewness for the approximation to work well (much of the distribution is concentrated on the values 0, .01, .02, .03, and .04, and the rest trails out to 1, so there is almost no lower tail).

The foregoing properties allow us to form a variable having approximately a standard normal distribution when n is large:

$$z = \frac{p - \pi}{\sqrt{\pi(1 - \pi)/n}}$$

Using z^* to denote an appropriate z critical value (1.96, 1.645, etc.), we have that

$$P\left(-z^* < \frac{p - \pi}{\sqrt{\pi(1 - \pi)/n}} < z^*\right) \approx 1 - \alpha$$

As suggested earlier in the derivation of our first confidence interval for μ, consider replacing each $<$ inside the parentheses by $=$ and solving the two resulting equations for π to obtain the confidence limits. Unfortunately, these equations are not as easy to solve as were the earlier ones. This is because π appears both in the numerator and in the denominator. The equations are therefore both quadratic. Using the general formula for the solution to a quadratic equation gives the following confidence interval.

A **confidence interval for a population proportion** π is

$$\frac{p + \dfrac{z^{*2}}{2n} \pm z^* \sqrt{\dfrac{p(1 - p)}{n} + \dfrac{z^{*2}}{4n^2}}}{1 + \dfrac{z^{*2}}{n}}$$

where z^* denotes an appropriate critical value, the $-$ sign in the numerator gives the lower confidence limit, and the $+$ sign gives the upper confidence limit. The z critical values corresponding to the most frequently used confidence levels, 90%, 95%, and 99%, are 1.645, 1.96, and 2.576, respectively. A lower confidence bound for π results from using only the $-$ sign in the formula (along with the appropriate z^*), and using only the $+$ sign gives an upper confidence bound.

Although the preceding interval was derived from the large-sample distribution of p, recent research has shown that it performs well even when n is quite small. Additionally, the actual confidence level achieved by the interval is almost always quite close to the desired level corresponding to the choice of any particular z critical value. For example, using 1.96 as the z critical value implies a desired confidence level of 95%, and the actual confidence level (long-run capture percentage if the formula is used repeatedly on different samples) will almost always be roughly 95%. When n is quite large, the three terms in the CI formula involving z^* are negligible compared to the three remaining terms. In this case, the CI reduces to the traditional interval

$$p \pm (z \text{ critical value}) \sqrt{p(1 - p)/n}$$

This latter interval has the same general form as our earlier large-sample interval for μ.

Example 7.6

The article "Repeatability and Reproducibility for Pass/Fail Data" (*J. of Testing and Eval.*, 1997: 151–153) reported that in $n = 48$ trials in a particular laboratory, 16 resulted in ignition of a particular type of substrate by a lighted cigarette. Let π denote the long-run proportion of all such trials that would result in ignition. A point estimate for π is $p = 16/48 = .333$. A confidence interval for π with a confidence level of approximately 95% is

$$\frac{.333 + (1.96)^2/96 \pm 1.96\sqrt{(.333)(.667)/48 + (1.96)^2/9216}}{1 + (1.96)^2/48}$$

$$= \frac{.333 \pm .139}{1.08} = (.217, .474)$$

This interval is rather wide, indicating imprecise information about π. The traditional interval is

$$.333 \pm 1.96\sqrt{(.333)(.667)/48} = .333 \pm .133 = (.200, .466)$$

These two intervals would be in much closer agreement were the sample size substantially larger.

A Bound on the Error of Estimation

The quantity $1.96\sigma_p = 1.96\sqrt{\pi(1 - \pi)/n}$ gives a bound on the error of estimation with a 95% confidence level in the sense that in the long run, p should be within this distance of π for roughly 95% of all samples. If the desired value of the bound is B, equating this to $1.96\sigma_p$ and solving for the necessary sample size n gives

$$n = \pi(1 - \pi)\left[\frac{1.96}{B}\right]^2$$

If some other confidence level is desired, the corresponding z critical value replaces 1.96. The difficulty with using this formula is that it involves the unknown π. A conservative approach utilizes the fact that $\pi(1 - \pi)$ is largest when $\pi = .5$. The sample size resulting from this choice of π will be large enough so that the bound B is achieved with the desired confidence level no matter what the value of π.

Example 7.7

A survey is to be carried out to estimate the proportion of all registered voters in a particular state who favor certain term limits for their state legislators. How many people should be included in a random sample to estimate this proportion to within the amount .05 with 95% confidence? Substituting $\pi = .5$ in the formula for n gives

$$n = .5(1 - .5)(1.96/.05)^2 = 384.16$$

so a sample size of 385 should be used. The resulting 95% confidence interval for π will have a half-width of at most .05 regardless of the value of p. Notice that this sample size is far larger than what appeared in the previous example, which explains why that interval was so wide.

A Large-Sample Confidence Interval for $\mu_1 - \mu_2$

The symbols μ and σ have been used to denote the mean value and standard deviation, respectively, of a population, process, or treatment response distribution. When two different populations, processes, or treatments are being compared, different subscripts will be used to differentiate characteristics of the first from those of the second. Similar notation is used to distinguish between the two sample sizes, sample means, and sample standard deviations.

Notation

	Mean value	Variance	Standard deviation
Population, process, or treatment 1	μ_1	σ_1^2	σ_1
Population, process, or treatment 2	μ_2	σ_2^2	σ_2

	Sample size	Sample mean	Sample variance	Sample standard deviation
Sample from population, process, or treatment 1	n_1	$\bar{x}_1$	s_1^2	s_1
Sample from population, process, or treatment 2	n_2	$\bar{x}_2$	s_2^2	s_2

It is assumed that the observations in the first sample were obtained completely independently from those in the second sample. Notice that our notation allows for the possibility that the two sample sizes might be different. This might happen because one population, process, or treatment is more expensive to sample than the other, or perhaps because observations are "lost" in the course of obtaining data; for example, several animals receiving a first diet die (hopefully for reasons unrelated to the diet).

Example 7.8

A study was carried out to compare population mean lifetimes (hr) for two different brands of AA alkaline batteries used in a particular manner. Here, μ_1 is the mean lifetime of all brand 1 batteries and σ_1 is the population standard deviation of brand 1 lifetimes; μ_2 and σ_2 are the mean value and standard deviation for the distribution of brand 2 lifetimes. Values of the summary quantities calculated from the two resulting samples are as follows:

Brand 1: $n_1 = 50$ $\bar{x}_1 = 4.15$ $s_1 = 1.79$

Brand 2: $n_2 = 45$ $\bar{x}_2 = 4.53$ $s_2 = 1.64$

Consider estimating the difference $\mu_1 - \mu_2$. The natural statistic for estimating μ_1 is $\bar{x}_1$, and the statistic $\bar{x}_2$ gives an estimate for μ_2. The difference between the two $\bar{x}$'s then gives an estimate of the difference between the two μ's. The point estimate from the data is $4.15 - 4.53 = -.38$. That is, we estimate that, on average, brand 2 batteries last .38 hr longer than do brand 1 batteries. If the labels 1 and 2 on the two brands had been reversed, the point estimate would be .38, and the interpretation would be the same as with the original labeling.

Both $\bar{x}_1$ and $\bar{x}_2$ vary in value from sample to sample, and this will also be true of their difference. For example, repeating the study described in Example 7.8 with the same sample sizes might result in $\bar{x}_1 = 4.02$ and $\bar{x}_2 = 4.75$, giving the estimate $-.73$. Just as a confidence interval for a single μ was based on properties of the $\bar{x}$ sampling distribution, a confidence interval for $\mu_1 - \mu_2$ is derived from properties of the sampling distribution of the statistic $\bar{x}_1 - \bar{x}_2$. These properties follow from the following general results:

1. For any two random variables x and y,

$$\mu_{x-y} = \text{mean value of the difference} = \mu_x - \mu_y$$
$$= \text{difference between the two means}$$

2. If x and y are two *independent* random variables, then

$$\sigma^2_{x-y} = \text{variance of a difference} = \sigma^2_x + \sigma^2_y = \text{sum of the variances}$$

3. If x and y are independent random variables, each with a normal distribution, then the difference $x - y$ also has a normal distribution. If each variable is approximately normal, then the distribution of the difference is also approximately normal.

Properties of the Sampling Distribution of $\bar{x}_1 - \bar{x}_2$

1. $\mu_{\bar{x}_1 - \bar{x}_2} = \mu_{\bar{x}_1} - \mu_{\bar{x}_2} = \mu_1 - \mu_2$, so that $\bar{x}_1 - \bar{x}_2$ is an unbiased statistic for estimating $\mu_1 - \mu_2$.

2. $\sigma^2_{\bar{x}_1 - \bar{x}_2} = \sigma^2_{\bar{x}_1} + \sigma^2_{\bar{x}_2} = \dfrac{\sigma^2_1}{n_1} + \dfrac{\sigma^2_2}{n_2}$, from which the standard deviation of $\bar{x}_1 - \bar{x}_2$ is

$$\sigma_{\bar{x}_1 - \bar{x}_2} = \sqrt{\frac{\sigma^2_1}{n_1} + \frac{\sigma^2_2}{n_2}}$$

3. If both population distributions are normal, the sampling distribution of $\bar{x}_1 - \bar{x}_2$ is normal.

4. If both the sample sizes are large, then the sampling distribution of $\bar{x}_1 - \bar{x}_2$ will be approximately normal irrespective of the shapes of the two population distributions (a consequence of the Central Limit Theorem).

The unbiasedness of $\bar{x}_1 - \bar{x}_2$ means that the sampling distribution of this statistic is always centered at the value of what the statistic is trying to estimate. If, for example, $\mu_1 = 110$ and $\mu_2 = 100$, then the sampling distribution is centered at $110 - 100 = 10$, whereas if $\mu_1 = 100$ and $\mu_2 = 105$, the mean value of the statistic is $100 - 105 = -5$. In addition to knowing that the sampling distribution is centered at the right place, we would also like it to be highly concentrated about its center. This will be the case if the variance and standard deviation of the statistic are small. The two σ^2 values are in the numerator of the variance and the n's are in the denominator. So when there is little variability in the two population, process, or treatment distributions (small values of σ^2), the variance and standard deviation will be small even when the sample sizes are small. On the other hand, a great deal of variability in each distribution can be counteracted by increasing the sample sizes to again obtain a small variance and standard deviation (at the price of expending more resources to collect data).

Now consider using the foregoing results to standardize $\bar{x}_1 - \bar{x}_2$ when both sample sizes are large. This entails subtracting the mean value of the statistic and then dividing by its standard deviation. The standard deviation involves σ_1^2 and σ_2^2, and the values of these variances are almost never available to an investigator. Fortunately, because of the large n's, we can replace the σ^2 values by the sample variances and still end up with a z variable.

When n_1 and n_2 are both large, the standardized variable

$$z = \frac{\bar{x}_1 - \bar{x}_2 - (\mu_1 - \mu_2)}{\sqrt{\dfrac{s_1^2}{n_1} + \dfrac{s_2^2}{n_2}}}$$

has approximately a standard normal distribution (the z curve). Using this variable in the same way that z variables were used earlier to obtain confidence intervals for μ and for π gives the following large-sample confidence interval formula for estimating $\mu_1 - \mu_2$:

$$\bar{x}_1 - \bar{x}_2 \pm (z \text{ critical value})\sqrt{\frac{s_1^2}{n_1} + \frac{s_2^2}{n_2}}$$

This formula is valid irrespective of the shapes of the two underlying distributions. The three most frequently used confidence levels of 95%, 99%, and 90% are achieved by using the z critical values 1.96, 2.576, and 1.645, respectively.

Example 7.9   An experiment carried out to study various characteristics of anchor bolts resulted in 78 observations on shear strength (kip) of 3/8-in. diameter bolts and 88 observations on strength of 1/2-in. diameter bolts. Summary quantities from Minitab follow, and a comparative boxplot appears in Figure 7.7. The sample sizes, sample means, and sample standard deviations agree with values given in the article "Ultimate Load Capacities of Expansion Anchor Bolts" (*J. Energy Engr.*, 1993: 139–158). The summaries suggest that the main difference between the two samples is in where they are centered. Let's now calculate a confidence interval for the difference between true average shear strength for 3/8-in. bolts (μ_1) and true average shear strength for 1/2-in. bolts (μ_2) using a confidence level of 95%:

$$4.25 - 7.14 \pm (1.96)\sqrt{\frac{(1.30)^2}{78} + \frac{(1.68)^2}{88}} = -2.89 \pm (1.96)(.2318)$$

$$= -2.89 \pm .45 = (-3.34, -2.44)$$

Variable	N	Mean	Median	TrMean	StDev	SEMean
diam 3/8	78	4.250	4.230	4.238	1.300	0.147

Variable	Min	Max	Q1	Q3		
diam 3/8	1.634	7.327	3.389	5.075		

Variable	N	Mean	Median	TrMean	StDev	SEMean
diam 1/2	88	7.140	7.113	7.150	1.680	0.179

Variable	Min	Max	Q1	Q3		
diam 1/2	2.450	11.343	5.965	8.447		

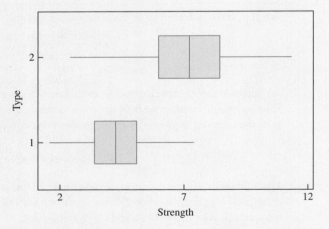

Figure 7.7 A comparative boxplot of the shear strength data

That is, with 95% confidence, $-3.34 < \mu_1 - \mu_2 < -2.44$. We can therefore be highly confident that the true average shear strength for the 1/2-in. bolts exceeds that for the 3/8-in. bolts by between 2.44 kip and 3.34 kip. Notice that if we relabel so that μ_1 refers to 1/2-in. bolts and μ_2 to 3/8-in. bolts, the confidence interval is now centered at $+2.89$ and the value .45 is still subtracted and added to obtain the confidence limits. The resulting interval is (2.44, 3.34), and the interpretation is identical to that for the interval previously calculated.

Section 7.3 Exercises

22. The American Taxpayer Relief Act of 2012 was passed by the U.S. Congress on January 1, 2013. This act helped address what became famously known as the "fiscal cliff" crisis. However, during the last months of 2012, heated debates concerning the crisis were ongoing in Congress, and there was growing concern political gridlock was preventing solution of the crisis by the end-of-year deadline. In mid-December, a *USA Today*–Gallup poll reported that only 18% of a sample of 1025 adult Americans approved of the job Congress was doing in working toward a solution to the looming fiscal cliff. Calculate a two-sided confidence interval using a 99% confidence level for the proportion of all U.S. adults who approved of the congressional handling of the crisis in December 2012.

23. TV advertising agencies face growing challenges in reaching audience members because viewing TV programs via digital streaming is increasingly popular. The Harris poll reported on November 13, 2012, that 53% of 2343 American adults surveyed said they have watched digitally streamed TV programming on some type of device.
    a. Calculate and interpret a confidence interval at the 99% confidence level for the proportion of all adult Americans who have watched streamed programming.
    b. What sample size would be required for the width of a 99% CI to be at most .05 irrespective of the value of p?

24. In a sample of 1000 randomly selected consumers who had opportunities to send in a rebate claim form after purchasing a product, 250 said they never did so ("Rebates: Get What You Deserve," *Consumer Reports*, May 2009: 7). Reasons cited for their behavior included too many steps in the process, rebate amount too small, missed deadline, fear of being placed on a mailing list, lost receipt, and doubts about receiving the money. Calculate an upper confidence bound at the 95% confidence level for the true proportion of such consumers who never apply for a rebate. Based on this bound, is there compelling evidence that the true proportion of such consumers is smaller than 1/3? Explain your reasoning.

25. The technology underlying hip replacements has changed as these operations have become more popular (more than 250,000 in the United States in 2008). Starting in 2003, highly durable ceramic hips were marketed. Unfortunately, for too many patients the increased durability has been counterbalanced by an increased incidence of squeaking. The May 11, 2008 issue of *The New York Times* reported that in one study of 143 individuals who received ceramic hips between 2003 and 2005, 10 developed squeaking problems.
    a. Calculate a lower confidence bound at the 95% confidence level for the true proportion of such hips that develop squeaking.
    b. Interpret the 95% confidence level used in part (a).

26. Researchers have developed a chemical treatment that retards the growth of trees of a certain type whose branches pose a safety threat to power lines. However, an overly severe application of the treatment can cause trees to die. In an experiment involving one particular treatment level applied to 250 trees, 38 trees died.
    a. Calculate and interpret a 95% confidence interval for the proportion of all such trees that would die if the treatment were applied at the tested level.
    b. The traditional CI for π discussed in Section 7.3 is based on the sample proportion p having approximately a normal sampling distribution, so the confidence level is only approximate rather than exact. Recent research has shown that under certain circumstances, its actual confidence level can deviate dramatically from the nominal one chosen by the investigator (e.g., the actual level may be quite different from the 95% level selected). An article by two statisticians (Agresti, A., and B. A. Coull, "Approximate Is Better Than 'Exact' for Interval Estimation of a Binomial Proportion," *The American Statistician*, May 1998: 119–126) has suggested the following remedy in the case

of a 95% confidence level: Add 2 to both the number of successes and the number of failures and then use the traditional formula. Do this for the data described in this exercise, and compare the resulting interval to the one you calculated in part (a).

27. Let π_1 and π_2 denote the proportion of successes in population 1 and population 2, respectively. An investigator sometimes wishes to calculate a confidence interval for the difference $\pi_1 - \pi_2$ between these two population proportions. Suppose random samples of size n_1 and n_2, respectively, are independently selected from the two populations, and let p_1 and p_2 denote the resulting sample proportions of successes. If the sample sizes are sufficiently large (apply the rule of thumb appropriate for a single proportion to each sample separately), the statistic $p_1 - p_2$ has approximately a normal sampling distribution with mean value $\pi_1 - \pi_2$ and standard deviation $\sqrt{\pi_1(1 - \pi_1)/n_1 + \pi_2(1 - \pi_2)/n_2}$. The estimated standard deviation of this statistic results from replacing each π under the square root by the corresponding p.
    a. Use the foregoing facts to obtain a large-sample two-sided 95% confidence interval formula for estimating $\pi_1 - \pi_2$.
    b. Is the response rate for questionnaires affected by including some sort of incentive to respond along with the questionnaire? In one experiment, 110 questionnaires with no incentive resulted in 75 being returned, whereas 98 questionnaires that included a chance to win a lottery yielded 66 responses ("Charities, No; Lotteries, No; Cash, Yes," *Public Opinion Quarterly*, 1996: 542–562). Calculate a two-sided 95% CI for the difference between the true response proportions under these circumstances. Does the interval suggest that, in fact, the values of π_1 and π_2 are different? Explain your reasoning.
    c. Recent research has shown that "coverage probability" and small-sample behavior are improved by adding one success and one failure to each sample and then using the formula you obtained in part (a). Do this for the data of part (b).

28. The article "The Effects of Cigarette Smoking and Gestational Weight Change on Birth Outcomes in Obese and Normal-Weight Women" (*Amer. J. of Public Health*, 1997: 591–596) reported on a random sample of 487 nonsmoking women of normal weight (body mass index between 19.8 and 26.0) who had given birth at a large metropolitan medical center. It was determined that 7.2% of these births resulted in children of low birth weight (less than 2500 g). The article also reported that 6.8% of a sample of 503 nonsmoking obese women (body mass index > 29) gave birth to children of low birth weight. Calculate a 95% lower confidence bound for the difference between the population proportion of normal-weight nonsmoking women and the population proportion of obese nonsmoking women who give birth to children of low birth weight. *Hint:* Refer to the previous problem.

29. Let π_1 and π_2 denote the proportions of successes in two different populations. Rather than estimate the difference $\pi_1 - \pi_2$ as described in Exercise 27, an investigator will often wish to estimate the ratio of the two π's. If, for example, $\pi_1/\pi_2 = 3$, then successes occur three times as frequently in population 1 as they do in population 2. Alternatively, if the π's refer to success proportions for two different treatments, then a ratio of 3 implies that the first treatment is three times as likely to result in a success as is the second treatment. Consider independent random samples of sizes n_1 and n_2 from the two different populations, which result in sample proportions p_1 and p_2, respectively. Also let $u =$ number of successes in the first sample and $v =$ number of successes in the second sample. When the n's are both large, the statistic $\ln(p_1/p_2)$ has approximately a normal sampling distribution with approximate mean value and standard deviation $\ln(\pi_1/\pi_2)$ and $\sqrt{(n_1 - u)/(un_1) + (n_2 - v)/vn_2)}$, respectively.
    a. Use these facts to obtain a large-sample two-sided 95% CI for $\ln(\pi_1/\pi_2)$ and a CI for π_1/π_2 itself.
    b. The article cited in Exercise 27 stated that in addition to 75 of 110 questionnaires without an incentive to respond being returned, 78 of

100 questionnaires that included a prepaid cash amount of $5 were returned. Calculate a 95% confidence interval for the ratio of the proportion of questionnaires returned when such a cash incentive is included to the proportion returned in the absence of any incentive. Does the interval suggest that such an incentive may not increase the likelihood of response?

30. A manufacturer of small appliances purchases plastic handles for coffeepots from an outside vendor. If a handle is cracked, it is considered defective and must be discarded. A very large shipment of handles is received. The proportion of defective handles, π, is of interest. How many handles from the shipment should be inspected to estimate π to within .1 with 99% confidence?

31. A manufacturer of exercise equipment is interested in estimating the proportion π of all purchasers of one of its products who still own the product two years after purchase. What sample size is required to estimate this proportion to within .05 with a confidence level of 90%?

32. Use the accompanying data to estimate with a 95% confidence interval the difference between true average compressive strength (N/mm^2) for 7-day-old concrete specimens and true average strength for 28-day-old specimens ("A Study of Twenty-Five-Year-Old Pulverized Fuel Ash Concrete Used in Foundation Structures," *Proc. Inst. Civil Engrs.*, 1985: 149–165):

7-day old: $n_1 = 68$ $\bar{x}_1 = 26.99$ $s_1 = 4.89$

28-day old: $n_2 = 74$ $\bar{x}_2 = 35.76$ $s_2 = 6.43$

33. Relative density was determined for one sample of second-growth Douglas fir 2 × 4s with a low percentage of juvenile wood and another sample with a moderate percentage of juvenile wood, resulting in the following data ("Bending Strength and Stiffness of Second-Growth Douglas Fir Dimension Lumber," *Forest Products J.*, 1991: 35–43):

Type	n	$\bar{x}$	s
Low	35	.523	.0543
Moderate	54	.489	.0450

Estimate the difference between true average densities for the two types of wood in a way that conveys information about reliability and precision.

34. Is there any systematic tendency for part-time college faculty to hold their students to different standards than full-time faculty do? The article "Are There Instructional Differences Between Full-Time and Part-Time Faculty?" (*College Teaching*, 2009: 23–26) reported that for a sample of 125 courses taught by full-time faculty, the mean course GPA was 2.7186 and the standard deviation was .63342, whereas for a sample of 88 courses taught by part-timers, the mean and standard deviation were 2.8639 and .49241, respectively.

Calculate a confidence interval at the 99% level to estimate the true mean GPA difference between full-time and part-time faculty. Does it appear that true average course GPA for part-time faculty differs from that for faculty teaching full-time? Explain your reasoning.

35. An experiment was performed to compare the fracture toughness of high-purity Ni-maraging steel with commercial-purity steel of the same type. For 32 high-purity specimens, the sample mean toughness and sample standard deviation of toughness were 65.6 and 1.4, respectively, whereas for 32 commercialpurity specimens, the sample mean and sample standard deviation were 59.2 and 1.1, respectively. Estimate the difference between true average toughness for the high-purity steel and that for the commercial steel using a lower 95% confidence bound. Does your estimate demonstrate conclusively that this difference exceeds 5? Explain your reasoning.

36. An investigator wishes to estimate the difference between population mean lifetimes of two different brands of batteries under specified conditions. If the population standard deviations are both roughly 2 hr and equal sample sizes are to be selected, what value of the common sample size n will be necessary to estimate the difference to within .5 hr with 95% confidence?

7.4 SMALL-SAMPLE INTERVALS BASED ON A NORMAL POPULATION DISTRIBUTION

Suppose we select a random sample of components of a certain type and determine the lifetime of each one, resulting in data $x_1, x_2, \ldots, x_n$. This sample can be used as a basis for calculating one of three different kinds of statistical intervals:

1. An interval of plausible values for the population mean lifetime, that is, a confidence interval for μ
2. An interval of plausible values for the lifetime of a single component of this type that you are planning to buy at some time in the near future, that is, a *prediction interval* for a single x value
3. An interval of values that includes a specified percentage, for example, 90%, of the lifetime values for components in the population, that is, a *tolerance interval* for a chosen percentage of x values in the population distribution

We have already seen how to calculate a z confidence interval for μ when n is large. In this section, we assume that the sample has been selected from a normal population distribution, and show how each of the three types of intervals can be obtained.

t Distributions and the One-Sample t Confidence Interval

When the population distribution is normal, the sampling distribution of $\bar{x}$ is also normal for any sample size n. This in turn implies that $z = (\bar{x} - \mu)/(\sigma/\sqrt{n})$ has a standard normal distribution (the z curve). The large-sample interval for μ presented in Section 7.2 was based on replacing σ by s in z; for large n, little extra variability is introduced by this substitution, so $(\bar{x} - \mu)/(s/\sqrt{n})$ also has approximately a standard normal distribution in this case. However, for small n this is no longer true. The standardized variable with s in the denominator varies much more in value from sample to sample than does the first variable. The following proposition introduces a new type of probability distribution needed for a small-sample interval.

PROPOSITION   Let $x_1, x_2, \ldots, x_n$ be a random sample from a normal distribution. Then the standardized variable

$$t = \frac{\bar{x} - \mu}{s/\sqrt{n}}$$

has a type of probability distribution called a t distribution with $n-1$ degrees of freedom (df). ∎

A passing acquaintance with properties of t distributions is important for an understanding of various inferential procedures based on these distributions.

Properties of t Distributions

1. Any particular t distribution is specified by the value of a parameter called the *number of degrees of freedom,* abbreviated df. There is one t distribution with 1 df, another with 2 df, yet another one with 3 df, and so on. The number of df for a t distribution can be any positive integer.

2. The density curve corresponding to any particular t distribution is bell-shaped and centered at 0, just like the z curve.

3. Any t curve is more spread out than the z curve.

4. As the number of df increases, the spread of the corresponding t curve decreases. Thus the most spread out of all t curves is the one with 1 df, the next most spread out is the one with 2 df, and so on.

5. As the number of df increases, the sequence of t curves approaches the z curve. (The z curve is sometimes referred to as the t curve with df $= \infty$.)

Figure 7.8 compares the z curve to several different t curves.

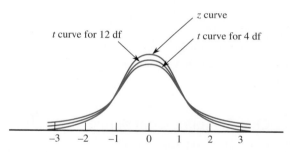

Figure 7.8 Comparison of the z curve to several t curves

Formulas for the large-sample z intervals utilized z critical values, numbers like 1.96 and 2.33, that captured certain central or cumulative areas under the z curve. Formulas for t intervals require t critical values, which play the same role for various t curves. Appendix Table IV gives a tabulation of such values. Each row of the table corresponds to a different number of df, and each column gives critical values that capture a particular central area and the corresponding cumulative area. For example, the t critical value at the intersection of the 12 df row and the .95 central area column is 2.179, so the area under the 12 df t curve between −2.179 and 2.179 is .95. The cumulative area under this t curve all the way to the left of 2.179 is the central area .95 plus the lower tail area .025, or .975. This is illustrated in Figure 7.9. The critical value 2.179 can then be used to calculate a two-sided confidence interval with a confidence level of 95%. A one-sided interval, which gives either an upper confidence bound or a lower confidence bound, with confidence level 95% necessitates going to the .95 *cumulative* area column; for 12 df, the required critical value is 1.782.

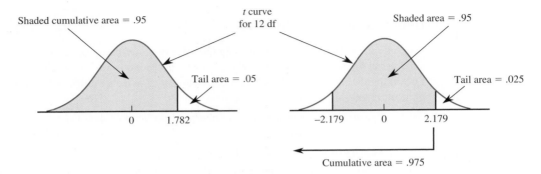

Figure 7.9 *t* critical values illustrated

As we move from left to right in any particular row of the table, the critical values increase. This is because capturing a larger central or cumulative area requires going farther out into the tail of the *t* curve. Starting with 1 df, the rows increase by 1 df until reaching 30 df, and then they jump to 40, 60, 120, and finally to ∞; this last row contains *z* critical values. Once past 30 df, there is little difference between the *t* curves and the *z* curve as far as the areas of interest to us are concerned. Rather than using the 30, 40, 60, and 120 rows or trying to interpolate, we recommend that *z* critical values be used whenever df > 30.

The large-sample *z* CI for μ was obtained by using the (approximate) standard normal variable $z = (\bar{x} - \mu)/(s/\sqrt{n})$ as the basis for a probability statement and then manipulating inequalities to isolate μ. An analogous derivation, based on the fact that $t = (\bar{x} - \mu)/(s/\sqrt{n})$ has a *t* distribution with $n - 1$ df, gives the following one-sample *t* CI.

One-Sample *t* Confidence Intervals

Let $\bar{x}$ and s be the sample mean and sample standard deviation of a random sample of size *n* from a normal population or process distribution. Then a two-sided confidence interval for the population or process mean μ has the form

$$\bar{x} \pm (t \text{ critical value}) \frac{s}{\sqrt{n}}$$

t critical values for the most frequently used confidence levels, corresponding to particular central *t* curve areas, are given in Appendix Table IV. An upper confidence bound results from replacing $\pm$ in the given formula by $+$, whereas a lower confidence bound uses $-$ in place of $\pm$. For such a one-sided interval, a *t* critical value in the cumulative area column corresponding to the desired confidence level is used.

Example 7.10 As part of a larger project to study the behavior of stressed-skin panels, a structural component being used extensively in North America, the article "Time-Dependent Bending Properties of Lumber" (*J. of Testing and Eval.*, 1996: 187–193) reported on various mechanical properties of Scotch pine lumber specimens. Consider the following observations on modulus of elasticity (MPa) obtained 1 minute after loading in a certain configuration:

10,490   16,620   17,300   15,480   12,970   17,260   13,400   13,900
13,630   13,260   14,370   11,700   15,470   17,840   14,070   14,760

Figure 7.10 shows a normal quantile plot obtained from Minitab. The straightness of the pattern in the plot provides strong support for assuming that the population distribution of modulus of elasticity is at least approximately normal.

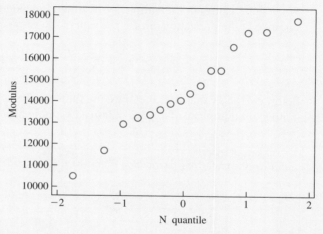

Figure 7.10 A normal quantile plot of the modulus of elasticity data

Hand calculation of the sample mean and standard deviation is simplified by subtracting 10,000 from each observation: $y_i = x_i - 10,000$. It is easily verified that $\Sigma y_i = 72,520$ and $\Sigma y_i^2 = 392,083,800$, from which $\bar{y} = 4532.5$ and $s_y = 2055.67$. Thus $\bar{x} = 14,532.5$ and $s_x = 2055.67$ (adding or subtracting the same constant from each observation does not affect variability). The sample size is 16, so a confidence interval for population mean modulus of elasticity is based on 15 df. A confidence level of 95% for a two-sided interval requires the t critical value of 2.131. The resulting interval is

$$\bar{x} \pm (t \text{ critical value}) \frac{s}{\sqrt{n}} = 14,532.5 \pm (2.131) \frac{2055.67}{\sqrt{16}}$$

$$= 14,532.5 \pm 1095.2 = (13,437.3, 15,627.7)$$

This interval is quite wide both because of the small sample size and because of the large amount of variability in the sample. A 95% lower confidence bound is obtained by using − and 1.753 in place of ± and 2.131, respectively.

A Prediction Interval for a Single x Value

Rather than wanting to estimate the population or process mean μ based on a sample $x_1, \ldots, x_n$, the individual who obtained the data may wish to use it as a basis for predicting a single x value that has not yet been observed, for example, the lifetime of the next component to be purchased, the number of calories in the next frozen dinner to be consumed, and so on. Again we assume that the underlying distribution is normal. A point prediction for x is just $\bar{x}$, which is also a point estimate for μ. An entire interval of plausible values for x is based on the prediction error $\bar{x} - x$. The value of x is of course subject to uncertainty, and so is $\bar{x}$ before obtaining data. The expected or mean value of the prediction error is

$$\mu_{(\bar{x}-x)} = \mu_{\bar{x}} - \mu_x = \mu - \mu = 0$$

Since the x_i values in the sample are assumed independent of the "future" x value, the variance of the prediction error is the *sum* of the variance of $\bar{x}$ and the variance of x:

$$\sigma^2_{(\bar{x}-x)} = \sigma^2_{\bar{x}} + \sigma^2_x = \frac{\sigma^2}{n} + \sigma^2 = \sigma^2\left(1 + \frac{1}{n}\right)$$

Statistical theory then says that if we use these results to standardize the prediction error (with s^2 used in place of σ^2), we obtain a t variable based on $n - 1$ df.

When the underlying distribution is normal, the standardized variable

$$t = \frac{\bar{x} - x}{s\sqrt{1 + \dfrac{1}{n}}}$$

has a t distribution based on $n - 1$ df. This implies that a two-sided prediction interval for x has the form

$$\bar{x} \pm (t \text{ critical value}) \cdot s\sqrt{1 + \frac{1}{n}}$$

An upper prediction bound and a lower prediction bound result from using $+$ and $-$, respectively, in place of $\pm$ and selecting the appropriate t critical value from the corresponding cumulative area column of the t table rather than the central area column.

The interpretation of a 95% prediction level is analogous to that of a 95% confidence level. If the two-sided interval is used repeatedly on different samples, in the long run about 95% of the calculated intervals will include the value of x that is being predicted. (If the samples are selected from entirely different population distributions, such as a sample of component lifetimes, then a sample of fuel efficiencies for automobiles, then a sample of service times for customers, etc., then in the long run about 95% of the intervals will include the actual values of the variables being predicted.) Notice that if the 1 under the square root in the $\pm$ factor is suppressed, the earlier confidence interval formula results. This implies that the prediction interval is wider than the confidence interval, often much wider because 1 will generally dominate $1/n$. There is a lot more uncertainty in predicting the value of a single observation x than there is in estimating a mean value μ.

Example 7.11   Reconsider the modulus of elasticity data introduced in the previous example. Suppose that one more specimen of lumber is to be selected for testing. A 95% prediction interval for the modulus of elasticity of this single specimen uses the same t critical value and values of n, $\bar{x}$, and s used in the confidence interval calculation:

$$14{,}532.5 \pm (2.131)(2055.67)\sqrt{1 + \frac{1}{16}} = 14{,}532.5 \pm 4515.5$$
$$= (10{,}017.0,\ 19{,}048.0)$$

This interval is extremely wide, indicating that there is great uncertainty as to what the modulus of elasticity for the next lumber specimen will be. Notice that the $\pm$ factor for the confidence interval is 1095.2, so the prediction interval is roughly four times as wide as the confidence interval.

Tolerance Intervals

Consider a population of automobiles of a certain type, and suppose that under specified conditions, fuel efficiency (mpg) has a normal distribution with $\mu = 30$ and $\sigma = 2$. Then since the interval from -1.645 to 1.645 captures 90% of the area under the z curve, 90% of all these automobiles will have fuel efficiency values between $\mu - 1.645\sigma = 26.71$ and $\mu + 1.645\sigma = 33.29$. But what if the values of μ and σ are not known? We can take a sample of size n, determine the fuel efficiencies, $\bar{x}$, and s, and form the interval whose lower limit is $\bar{x} - 1.645s$ and whose upper limit is $\bar{x} + 1.645s$. However, because of sampling variability in the estimates of μ and σ, there is a good chance that the resulting interval will include less than 90% of the population values. Intuitively, to have an a priori 95% chance of the resulting interval including at least 90% of the population values, when $\bar{x}$ and s are used in place of μ and σ, we should also replace 1.645 by some larger number. For example, when $n = 20$, the value 2.310 is such that we can be 95% confident that the interval $\bar{x} \pm 2.310s$ will include at least 90% of the fuel efficiency values in the population.

Let k be a number between 0 and 100. A **tolerance interval** for capturing at least $k\%$ of the x values in a normal population distribution with a confidence level 95% has the form

$$\bar{x} \pm (\text{tolerance critical value}) \cdot s$$

Tolerance critical values for $k = 90, 95$, and 99 in combination with various sample sizes are given in Appendix Table V. This table also includes critical values for a confidence level of 99% (these values are larger than the corresponding 95% values). Replacing $\pm$ by $+$ gives an upper tolerance bound and using $-$ in place of $\pm$ results in a lower tolerance bound. Critical values for obtaining these one-sided bounds also appear in Appendix Table V.

Example 7.12　Let's return to the modulus of elasticity data discussed in Examples 7.10 and 7.11, where $n = 16$, $\bar{x} = 14{,}532.5$, $s = 2055.67$, and a normal quantile plot of the data indicated that population normality was quite plausible. For a confidence level of 95%, a two-sided tolerance interval for capturing at least 95% of the modulus of elasticity values for specimens of lumber in the population sampled uses the tolerance critical value of 2.903. The resulting interval is

$$14{,}532.5 \pm (2.903)(2055.67) = 14{,}532.5 \pm 5967.6 = (8564.9, 20{,}500.1).$$

We can be highly confident that at least 95% of all lumber specimens have modulus of elasticity values between 8564.9 and 20,500.1.

　　The 95% CI for μ was (13,437.3, 15,627.7), and the 95% prediction interval for the modulus of elasticity of a single lumber specimen was (10,017.0, 19,048.0). Both the prediction interval and the tolerance interval are substantially wider than the confidence interval.

Intervals Based on Nonnormal Population Distributions

The one-sample t CI for μ is *robust* to small or even moderate departures from normality unless n is quite small. By this we mean that if a critical value for 95% confidence, for example, is used in calculating the interval, the actual confidence level will be reasonably close to the nominal 95% level. If, however, n is small and the population distribution is highly nonnormal, then the actual confidence level may be considerably different from the one you think you are using when you obtain a particular critical value from the t table. It would certainly be distressing to believe that your confidence level is about 95% when in fact it was really more like 88%! The bootstrap technique, introduced in Section 7.6, has been found to be quite successful at estimating parameters in a wide variety of nonnormal situations.

　　In contrast to the confidence interval, the validity of the prediction and tolerance intervals described in this section is closely tied to the normality assumption. These latter intervals should not be used in the absence of compelling evidence for normality. The excellent reference *Statistical Intervals*, mentioned previously, discusses alternative procedures of this sort for various other situations.

Section 7.4 Exercises

37. Determine the t critical value that will capture the desired t curve area in each of the following cases:
    a. Central area = .95, df = 10
    b. Central area = .95, df = 20
    c. Central area = .99, df = 20
    d. Central area = .99, df = 50
    e. Upper-tail area = .01, df = 25
    f. Lower-tail area = .025, df = 5

38. Determine the t critical value for a two-sided confidence interval in each of the following situations:
    a. Confidence level = 95%, df = 10
    b. Confidence level = 95%, df = 15
    c. Confidence level = 99%, df = 15
    d. Confidence level = 99%, $n = 5$
    e. Confidence level = 98%, df = 24
    f. Confidence level = 99%, $n = 38$

39. Determine the t critical value for a lower or an upper confidence bound for each of the situations described in Exercise 38.

40. According to the article "Fatigue Testing of Condoms" (*Polymer Testing*, 2009: 567–571), "tests currently used for condoms are surrogates for the challenges they face in use," including a test for holes, an inflation test, a package seal test, and tests of dimensions and lubricant quality. The investigators developed a new test that adds cyclic strain to a level well below breakage and determines the number of cycles to break. A sample of 20 condoms of one particular type resulted in a sample mean number of 1584 and a sample standard deviation of 607. Calculate and interpret a confidence interval at the 99% confidence level for the true average number of cycles to break. (*Note:* The article presented the results of hypothesis tests based on the t distribution; the validity of these depends on assuming normal population distributions.)

41. Ultra high performance concrete (UHPC) is a relatively new construction material that offers strong adhesive properties with other materials. The authors of "Adhesive Power of Ultra High Performance Concrete from a Thermodynamic Point of View" (*J. of Materials in Civil Engr.*, 2012: 1050–1058) investigated the intermolecular forces for UHPC connected to various substrates. As reported in the article, here are the work of adhesion measurements (in mJ/m^2) for five samples of UHPC adhered to steel:

107.1   109.5   107.4   106.8   108.1

a. Is it plausible that the given sample observations were selected from a normal distribution?
b. Calculate a two-sided 95% confidence interval for the true average work of adhesion for UHPC adhered to steel. Does the interval suggest that 107 is a plausible value for the true average work of adhesion for UHPC adhered to steel? What about 110?

42. The article "Measuring and Understanding the Aging of Kraft Insulating Paper in Power Transformers" (*IEEE Electrical Insul. Mag.*, 1996: 28–34)

contained the following observations on degree of polymerization for paper specimens for which viscosity times concentration fell in a certain range:

418   421   421   422   425   427
431   434   437   439   446   447
448   453   454   463   465

a. Construct a boxplot of the data and comment on any interesting features.
b. Is it plausible that the given sample observations were selected from a normal distribution?
c. Calculate a two-sided 95% confidence interval for true average degree of polymerization (as did the authors of the article). Does the interval suggest that 440 is a plausible value for true average degree of polymerization? What about 450?

43. Haven't you always wanted to own a Porsche? We investigated the Boxster (their cheapest model) and performed an online search at www.cars.com on December 30, 2012. Asking prices were well beyond our meager professorial salaries, so instead we focused on odometer readings (mileage). Here are reported readings for a sample of 16 Boxsters:

1445	25,822	26,892	29,860
35,285	47,874	49,544	64,763
72,698	75,732	84,457	91,577
93,000	109,538	113,399	137,652

A normal quantile plot supports the assumption that mileage is at least approximately normally distributed. The R software reports the following summary statistics for this data:

```
> summary(odometer, digits=6)
Min          1st Qu       Median       Mean
1445.0       33928.8      68730.5      66221.1
3rd Qu       Max
91932.8      137652.0

> sd(odometer)

37683.17
```

a. Estimate true average mileage in a way that conveys information about precision and reliability.
b. Predict the mileage for a single Porsche Boxster in a way that conveys information about precision and reliability. How does the prediction compare to the estimate calculated in part (a)?

44. A new concrete structure that experiences crack-ing within the first seven days after setting is often said to have experienced "early-age cracking." This is usually a precursor to later-age cracking and other problems that lead to an overall weak-ening of the structure. According to the article "Early-Age Cracking Tendency and Ultimate De-gree of Hydration of Internally Cured Concrete" (*J. of Materials in Civil Engr.*, 2012: 1025–1033), more than 60% of surveyed transportation agen-cies regard early-age transverse cracking to be problematic. The authors investigated the ef-fectiveness of a process known as *internal curing* to mitigate early-age cracking of bridge deck concretes.

One important mechanical property of con-crete is its modulus of elasticity (in GPa), which is the material's tendency to be deformed elasti-cally when subjected to an applied force. A higher modulus of elasticity indicates a stiffer material. As reported in the article, the following are modulus of elasticity measurements for seven specimens of internally cured concrete that have been set for one week:

27.0 25.5 28.5 34.0 31.0 34.5 32.5

a. Is it plausible that this sample was selected from a normal population distribution?
b. Estimate true average modulus of elasticity for these mixtures in a way that conveys informa-tion about precision and reliability.
c. Predict the modulus of elasticity for a single mixture in a way that conveys information about precision and reliability. How does the predic-tion compare to the estimate calculated in part (b)?

45. The article "Concrete Pressure on Formwork" (*Mag. of Concrete Res.*, 2009: 407–417) gave the following observations on maximum concrete pres-sure (kN/m^2):

33.2 41.8 37.3 40.2 36.7
39.1 36.2 41.8 36.0 35.2
36.7 38.9 35.8 35.2 40.1

a. Is it plausible that this sample was selected from a normal population distribution?

b. SAS reports the following summary information for this data:

The MEANS Procedure
Analysis Variable : pressure

Lower 95% CL for Mean	Upper 95% CL for Mean	Mean	Std Error
36.1892782	39.0373884	37.6133333	0.6639612

Calculate a two-sided 95% confidence interval for the population mean of maximum pressure and confirm the lower and upper endpoints re-ported by SAS.
c. Calculate an upper confidence bound with con-fidence level 95% for the population mean of maximum pressure.
d. Calculate an upper prediction bound with level 95% for the maximum pressure of a single obser-vation. How does the prediction compare to the estimate calculated in part (b)?

46. A study of the ability of individuals to walk in a straight line ("Can We Really Walk Straight?" *Amer. J. of Physical Anthro.*, 1992: 19–27) reported the ac-companying data on cadence (strides per second) for a sample of $n = 20$ randomly selected healthy men:

.95 .85 .92 .95 .93 .86 1.00 .92 .85 .81
.78 .93 .93 1.05 .93 1.06 1.06 .96 .81 .96

A normal quantile plot gives substantial support to the assumption that the population distribution of cadence is approximately normal. A descriptive summary of the data from Minitab follows:

Variable	N	Mean	Median	TrMean	StDev	SEMean
cadence	20	0.9255	0.9300	0.9261	0.0809	0.0181

Variable	Min	Max	Q1	Q3
cadence	0.7800	1.0600	0.8525	0.9600

a. Calculate and interpret a 95% confidence inter-val for population mean cadence.
b. Calculate and interpret a 95% prediction inter-val for the cadence of a single individual ran-domly selected from this population.
c. Calculate an interval that includes at least 99% of the cadences in the population distribution using a confidence level of 95%.

47. A sample of 25 pieces of laminate used in the manufacture of circuit boards was selected and the amount of warpage (in.) under particular conditions was determined for each piece, resulting

in a sample mean warpage of .0635 and a sample standard deviation of .0065.

a. Calculate a prediction for the amount of warpage of a single piece of laminate in a way that provides information about precision and reliability.

b. Calculate an interval for which you can have a high degree of confidence that at least 95% of all pieces of laminate result in amounts of warpage that are between the two limits of the interval.

48. A more extensive tabulation of t critical values than what appears in this book shows that for the t distribution with 20 df, the areas to the right of the values .687, .860, and 1.064 are .25, .20, and .15, respectively. What is the confidence level for each of the following three confidence intervals for the mean μ of a normal population distribution? Which of the three intervals would you recommend be used, and why?

a. $(\bar{x} - .687s/\sqrt{21}, \bar{x} + 1.725s/\sqrt{21})$
b. $(\bar{x} - .860s/\sqrt{21}, \bar{x} + 1.325s/\sqrt{21})$
c. $(\bar{x} - 1.064s/\sqrt{21}, \bar{x} + 1.064s/\sqrt{21})$

7.5 INTERVALS FOR $\mu_1 - \mu_2$ BASED ON NORMAL POPULATION DISTRIBUTIONS

In Section 7.3, we showed how to obtain a large-sample confidence interval for a difference between two population, process, or treatment means. The validity of the interval required that the two samples be selected independently of one another, and the derivation involved standardizing $\bar{x}_1 - \bar{x}_2$ to obtain a variable having approximately a standard normal distribution. In this section, we first consider two independent samples with at least one of the sample sizes being small and then an interval calculated from *paired* data.

The Two-Sample *t* Interval

The one-sample t confidence interval for μ presented in Section 7.4 can be used for any sample size n provided that the population distribution is at least approximately normal. The validity of the two-sample t interval requires that *both* population, process, or treatment response distributions be normal.

PROPOSITION Consider two normal distributions with mean values μ_1 and μ_2, respectively. Suppose a random sample of size n_1 is selected from the first distribution, resulting in a sample mean of $\bar{x}_1$ and a sample standard deviation of s_1. A random sample from the second distribution, selected independently of that from the first one, yields sample mean $\bar{x}_2$ and sample standard deviation s_2. Then the standardized variable

$$t = \frac{\bar{x}_1 - \bar{x}_2 - (\mu_1 - \mu_2)}{\sqrt{\dfrac{s_1^2}{n_1} + \dfrac{s_2^2}{n_2}}}$$

has approximately a t distribution with df estimated from the sample by the following formula:

$$df = \frac{\left[(se_1)^2 + (se_2)^2\right]^2}{\dfrac{(se_1)^4}{n_1 - 1} + \dfrac{(se_2)^4}{n_2 - 1}}$$

where $se = s/\sqrt{n}$ (Note: df should be rounded <u>down</u> to the nearest integer).

This implies that **a confidence interval for $\mu_1 - \mu_2$** in this situation is

$$\bar{x}_1 - \bar{x}_2 \pm (t\,\text{critical value})\sqrt{\frac{s_1^2}{n_1} + \frac{s_2^2}{n_2}}$$

t critical values corresponding to the most frequently used confidence levels appear in Appendix Table IV.   ■

The standardized variable in the box is identical to the one used in our previous development of the large-sample interval; it is labeled t here simply to emphasize that it now has approximately a t rather than a z distribution. The only difference between the formulas for the two intervals is that the formula here uses a t critical value instead of a z critical value. Separate normal quantile plots of the observations in the two samples can be used as a basis for checking that the normality assumption is plausible.

Example 7.13   Which way of dispensing champagne, the traditional vertical method or a tilted beer-like pour, preserves more of the tiny gas bubbles that improve flavor and aroma? The following data was reported in the article "On the Losses of Dissolved CO_2 during Champagne Serving" (*J. Agr. Food Chem.*, 2010: 8768–8775).

Temp (°C)	Type of Pour	n	Mean (g/L)	SD
18	Traditional	4	4.0	.5
18	Slanted	4	3.7	.3
12	Traditional	4	3.3	.2
12	Slanted	4	2.0	.3

Assuming the sampled distributions are normal, let's calculate confidence intervals for the difference between true average dissolved CO_2 loss for the traditional pour and that for the slanted pour at each of the two temperatures. For the 18°C temperature, the number of degrees of freedom for the interval is

$$df = \frac{\left(\frac{.5^2}{4} + \frac{.3^2}{4}\right)^2}{\frac{(.5^2/4)^2}{3} + \frac{(.3^2/4)^2}{3}} = \frac{.007225}{.00147083} = 4.91$$

Rounding down, the CI will be based on 4 df. For a confidence interval of 99%, Appendix Table IV gives t critical value = 4.604. The desired interval is

$$4.0 - 3.7 \pm (4.604)\sqrt{\frac{.5^2}{4} + \frac{.3^2}{4}} = .3 \pm (4.604)(.2915) = .3 \pm 1.3 = (-1.0, 1.6)$$

Thus, we can be highly confident that $-1.0 < \mu_1 - \mu_2 < 1.6$, where μ_1 and μ_2 are true average losses for the traditional and slant methods, respectively. Notice that this CI contains 0; so at the 99% confidence level, it is plausible that $\mu_1 - \mu_2 = 0$—that is, that $\mu_1 = \mu_2$. Note that if the 1 and 2 labels had been reversed, the resulting interval would have been $(-1.6, 1.0)$, with exactly the same interpretation.

The df formula for the 12°C comparison yields df $= .00105625/.00020208 = 5.23$. The required df is 5, and Appendix Table IV gives t critical value $= 4.032$ for a 99% CI. The resulting interval is $(.6, 2.0)$. Thus, 0 is not a plausible value for this difference. It appears from the CI that the true average loss when the slant method is used is smaller than that when the traditional method is used, so the slant method is better at this temperature. This, in fact, was the conclusion reported in the popular media.

There is a special confidence interval formula for the case of normal population distributions having $\sigma_1 = \sigma_2$. It is called the *pooled t confidence interval*; "pooled" refers to the fact that s_1 and s_2 are combined to estimate the common population standard deviation. Recent studies have shown that the behavior of this interval is rather sensitive to the assumption of equal population standard deviations. If they are not in fact the same, the actual confidence level may be quite different from the nominal level (e.g., the actual level may deviate substantially from an assumed 95% level). For this reason we recommend the use of the two-sample t interval we have described unless there is compelling evidence for at least approximate equality of the population standard deviations.

A Confidence Interval from Paired Data

Let μ_1 denote the population mean height for all married males and μ_2 represent the population height for all married females (both in inches). One way to estimate $\mu_1 - \mu_2$ would be to obtain two independent samples of heights, one for married males and the other for married females, and (assuming normality) use the two-sample t interval just discussed. Another possibility, though, is to randomly select n married couples and determine the height of the male and the female in each couple. This results in a sample of *pairs* of numerical values. The first observation might be $(69, 66)$, the second $(73, 63)$, the third $(66, 68)$, and so on. Because tall men tend to marry tall women and short men tend to marry short women, it is unreasonable to think that the two variables *height of male* and *height of female* in a married couple are independent. This invalidates the use of the two-sample t interval.

Conceptualize the entire population of pairs from which our sample was selected. For each such pair, we can subtract the second number from the first to obtain a difference value. The difference is 3 for the pair $(69, 66)$, -2 for the pair $(66, 68)$, and so on. Now let μ_d denote the population mean difference, that is, the average of all differences in the population. It can be shown that

$$\mu_d = \mu_1 - \mu_2$$

where μ_1 is the population mean value of all first numbers within pairs and μ_2 is defined similarly for all second numbers. The importance of this relationship is that if we can obtain a CI for μ_d, it will also be a CI for $\mu_1 - \mu_2$. A CI for μ_d can be calculated from the differences for pairs in the sample. In particular, if the population distribution of the differences can be assumed to be normal, then a one-sample t interval based on the sample differences is appropriate.

The Paired-t Interval

Let $\bar{d}$ and s_d denote the sample mean and sample standard deviation, respectively, for a random sample of n differences. If the distribution from which this sample was selected is normal, a confidence interval for μ_d (i.e., for $\mu_1 - \mu_2$) is given by

$$\bar{d} \pm (t\,\text{critical value})\,\frac{s_d}{\sqrt{n}}$$

The t critical value is based on $n - 1$ df. If n is large, the Central Limit Theorem ensures the validity of the interval without the normality assumption.

Example 7.14

Example 7.10 in the previous section gave data on the modulus of elasticity obtained 1 minute after loading in a certain configuration. The cited article also gave the values of modulus of elasticity obtained 4 weeks after loading for the same lumber specimens. The data is presented here.

Observation	1 minute	4 weeks	Difference
1	10,490	9110	1380
2	16,620	13,250	3370
3	17,300	14,720	2580
4	15,480	12,740	2740
5	12,970	10,120	2850
6	17,260	14,570	2690
7	13,400	11,220	2180
8	13,900	11,100	2800
9	13,630	11,420	2210
10	13,260	10,910	2350
11	14,370	12,110	2260
12	11,700	8620	3080
13	15,470	12,590	2880
14	17,840	15,090	2750
15	14,070	10,550	3520
16	14,760	12,230	2530

The normal quantile plot of the differences shown in Figure 7.11 appears to be reasonably straight, though the point on the far left deviates somewhat from a line determined by the other points. (Use of a formal inferential procedure presented in Chapter 8 indicates that it is reasonable to assume that the population distribution of the differences is approximately normal.)

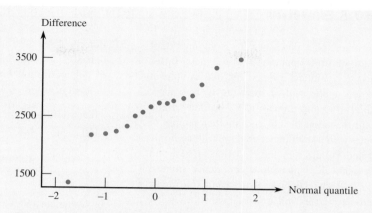

Figure 7.11 Normal quantile plot of the differences from Example 7.14

The sample consists of 16 pairs, so a 99% confidence interval based on 15 df requires the t critical value 2.947. With $\bar{d} = 2635.6$ and $s_d = 508.64$, the interval is

$$2635.6 \pm (2.947)\frac{508.64}{\sqrt{16}} = 2635.6 \pm 374.7 = (2260.9, 3010.3)$$

We can be highly confident, at the 99% confidence level, that the true average modulus of elasticity after 1 minute exceeds that after 4 weeks by between roughly 2261 MPa and 3010 MPa. This interval is rather wide, partly because of the high confidence level and partly because there is a reasonable amount of variability in the sample differences.

Although the two-sample t CI should not be used here because the 1-minute observations are not independent of the 4-week observations, the resulting interval has limits of roughly 705 and 4566. This interval is a great deal wider than the correct interval. The reason for this is that there is much less variability in the differences than there is in either the 1-minute observations or the 4-week observations ($s_d = 509$, $s_1 = 2056$, and $s_2 = 1902$).

In practice, it is frequently the case that a CI calculated from paired data is much narrower than a CI calculated from two independent samples. This is because numbers within pairs often tend to be rather similar—when one is relatively large (small), the other tends to be relatively large (small) also. The implication is that the differences will show much less variation than that in either of two independent samples. In Example 7.14, there is a natural pairing, but this is not always the case. In medical experimentation, investigators frequently create *matched pairs* by selecting patients so that within each pair, the two individuals are as similar as possible with respect to age, general physical condition, and physiological variables, such as blood pressure, heart rate, and so on. Then the differences within pairs will largely reflect the differences between the two treatments rather than extraneous variation from all other factors.

Section 7.5 Exercises

49. The firmness of a piece of fruit is an important indicator of fruit ripeness. The Magness–Taylor firmness (N) was determined for one sample of 20 golden apples with a shelf life of zero days, resulting in a sample mean of 8.74 and a sample standard deviation of .66, and another sample of 20 apples with a shelf life of 20 days, with a sample mean and sample standard deviation of 4.96 and .39, respectively. Calculate a confidence interval for the difference between true average firmness for zero-day apples and true average firmness for 20-day apples using a confidence level of 95%, and interpret the interval.

50. Anorexia nervosa (AN) is a psychiatric condition leading to substantial weight loss among women fearful of becoming overweight. The article "Adipose Tissue Distribution After Weight Restoration and Weight Maintenance in Women with Anorexia Nervosa" (*Amer. J. of Clinical Nutr.*, 2009: 1132–1137) used whole-body magnetic resonance imagery to determine various tissue characteristics for both an AN sample of individuals who had undergone acute weight restoration and maintained their weight for a year and a comparable (at the outset of the study) control sample. Here is summary data on intermuscular adipose tissue (IAT, in kg).

Condition	Sample Size	Sample Mean	Sample SD
AN	16	.52	.26
Control	8	.35	.15

    Assume that both samples were selected from normal distributions.

    a. Calculate an estimate for true average IAT under the described AN protocol; do so in a way that conveys information about the reliability and precision of the estimation.
    b. Calculate an estimate for the difference between true average AN IAT and true average control IAT; do so in a way that conveys information about the reliability and precision of the estimation. What does your estimate suggest about true average AN IAT relative to true average control IAT?

51. Refer to Exercise 42 in Section 7.4. The cited article also gave the following observations on degree of polymerization for specimens having viscosity times concentration in a higher range:

    429   430   430   431   436   437

    440   441   445   446   447

    a. Construct a comparative boxplot for the two samples, and comment on any interesting features.
    b. Calculate a 95% confidence interval for the difference between true average degree of polymerization for the middle range and that for the high range. Does the interval suggest that μ_1 and μ_2 may in fact be different? Explain your reasoning.

52. The degenerative disease osteoarthritis most frequently affects weight-bearing joints such as the knee. The article "Evidence of Mechanical Load Redistribution at the Knee Joint in the Elderly when Ascending Stairs and Ramps" (*Annals of Biomed. Engr.*, 2008: 467–476) presented the following summary data on stance duration (ms) for samples of both older and younger adults.

Age	Sample Size	Sample Mean	Sample SD
Older	28	801	117
Younger	16	780	72

    Assume that both stance duration distributions are normal.

    a. Calculate and interpret a 99% CI for true average stance duration among elderly individuals.
    b. Calculate a 99% CI for the difference between true average stance duration for the elderly and the younger individuals. Does your interval suggest that true average stance duration is larger among elderly individuals than among younger individuals?

53. Arsenic is a known carcinogen and poison. The standard laboratory procedures for measuring arsenic concentration (μg/L) in water are expensive. Consider the accompanying summary data and Minitab

output for comparing a laboratory method to a new relatively quick and inexpensive field method (from the article "Evaluation of a New Field measurement Method for Arsenic in Drinking Water Samples," *J. of Envir. Engr.*, 2008: 382–388).

```
        Two-Sample T-Test and CI
Sample  N     Mean    StDev   SE Mean
  1     3    19.70    1.10     0.64
  2     3    10.90    0.60     0.35
Estimate for difference: 8.800
95% CI for difference: (6.498, 11.102)
```

Calculate a two-sided 95% confidence interval for the difference in population means and confirm the lower and upper endpoints reported by Minitab. Based on the interval, what conclusion you can draw about the two methods? Why?

54. Suppose not only that the two population or treatment response distributions are normal but also that they have equal variances. Let σ^2 denote the common variance. This variance can be estimated by a "pooled" (i.e., combined) sample variance as follows:

$$s_p^2 = \left(\frac{n_1 - 1}{n_1 + n_2 - 2}\right)s_1^2 + \left(\frac{n_2 - 1}{n_1 + n_2 - 2}\right)s_2^2$$

($n_1 + n_2 - 2$ is the sum of the df's contributed by the two samples). It can then be shown that the standardized variable

$$t = \frac{(\bar{x}_1 - \bar{x}_2) - (\mu_1 - \mu_2)}{s_p\sqrt{\dfrac{1}{n_1} + \dfrac{1}{n_2}}}$$

has a t distribution with $n_1 + n_2 - 2$ df.
a. Use the t variable above to obtain a pooled t confidence interval formula for $\mu_1 - \mu_2$.
b. A sample of ultrasonic humidifiers of one particular brand was selected for which the observations on maximum output of moisture (oz) in a controlled chamber were 14.0, 14.3, 12.2, and 15.1. A sample of the second brand gave output values 12.1, 13.6, 11.9, and 11.2 ("Multiple Comparisons of Means Using Simultaneous Confidence Intervals," *J. of Quality Technology*, 1989: 232–241). Use the pooled t formula from part (a) to estimate the difference between true

average outputs for the two brands with a 95% confidence interval.
c. Estimate the difference between the two μ's using the two-sample t interval discussed in this section, and compare it to the interval of part (b).

55. Along any major freeway we often encounter service (or logo) signs that give information on attractions, camping, lodging, food, and gas services in advance of the off-ramp that leads to such services. These signs typically do not provide information on distances. Researchers in Virginia, with cooperation from the Virginia Department of Transportation, performed an experiment to see if the addition of distance information on the service signs would affect drivers. The results of this experiment were reported in "Evaluation of Adding Distance Information to Freeway-Specific Service (Logo) Signs" (*J. of Transp. Engr.*, 2011: 782–788).

In one investigation, the authors selected six sites along Virginia interstate highways where service signs are posted. For each site, crash data was obtained for a three-year period before distance information was added to the service signs and for a one-year period afterward. The number of crashes per year before and after the sign changes were made are given here:

Before:	15	26	66	115	62	64
After:	16	24	42	80	78	73

a. Calculate a confidence interval for the population mean difference in the number of crashes per year before and after the sign changes were made. Provide an interpretation for this interval.
b. If a seventh site were to be randomly selected among locations bearing service signs, between what values would you predict the difference in number of crashes to lie?

56. Lactation promotes a temporary loss of bone mass to provide adequate amounts of calcium for milk production. The paper "Bone Mass Is Recovered from Lactation to Postweaning in Adolescent Mothers with Low Calcium Intakes" (*Amer. J. of Clinical Nutr.*, 2004: 1322–1326) gave the following data on total body bone mineral content (TBBMC) (g) for

a sample both during lactation (L) and in the post-weaning period (P).

Subject	L	P
1	1928	2126
2	2549	2885
3	2825	2895
4	1924	1942
5	1628	1750
6	2175	2184
7	2114	2164
8	2621	2626
9	1843	2006
10	2541	2627

a. Construct a comparative boxplot of TBBMC for the lactation and postweaning periods and comment on any interesting features.

b. Estimate the difference between true average TBBMC for the two periods of concrete in a way that conveys information about precision and reliability. Does it appear plausible that the true average TBBMCs for the two periods are identical? Why or why not?

57. The paper "Quantitative Assessment of Glenohumeral Translation in Baseball Players" (*Amer. J. of Sports Med.*, 2004: 1711–1715) considered various aspects of shoulder motion for a sample of pitchers and another sample of position players [*glenohumeral* refers to the articulation between the humerus (ball) and the glenoid (socket)]. The authors kindly supplied the following data (for 19 position players and 17 pitchers) on anteroposterior translation (mm), a measure of the extent of anterior and posterior motion, for both dominant nondominant arms.

a. Estimate the true average difference in translation between dominant and nondominant arms for pitchers in a way that conveys information about reliability and precision. Interpret the resulting estimate.

b. Repeat part (a) for position players.

c. The authors asserted that "pitchers have greater difference in side-to-side anteroposterior translation of their shoulders compared with position players." Do you agree? Explain.

	Pos Dom Tr	Pos ND Tr	Pit Dom Tr	Pit ND Tr
1	30.31	32.54	27.63	24.33
2	44.86	40.95	30.57	26.36
3	22.09	23.48	32.62	30.62
4	31.26	31.11	39.79	33.74
5	28.07	28.75	28.50	29.84
6	31.93	29.32	26.70	26.71
7	34.68	34.79	30.34	26.45
8	29.10	28.87	28.96	21.49
9	25.51	27.59	31.19	20.82
10	22.49	21.01	36.00	21.75
11	28.74	30.31	31.58	28.32
12	27.89	27.92	32.55	27.22
13	28.48	27.85	29.56	28.86
14	25.60	21.95	28.64	28.58
15	20.21	21.59	28.58	27.15
16	33.77	32.48	31.99	29.46
17	32.59	32.48	27.16	21.26
18	32.60	31.61		
19	29.30	27.46		

58. Dentists make many people nervous (even more so than statisticians!). To assess any effect of such nervousness on blood pressure, the systolic blood pressure of each of 60 subjects was measured both in a dental setting and in a medical setting ("The Effect of the Dental Setting on Blood Pressure Measurement," *Amer. J. of Public Health*, 1983: 1210–1214). For each subject, the difference between dental setting pressure and medical setting pressure was computed; the resulting sample mean difference and sample standard deviation of the differences were 4.47 and 8.77, respectively. Estimate the true average difference between blood pressures for these two settings using a 99% confidence interval. Does it appear that the true average pressure is different in a dental setting than in a medical setting?

59. Antipsychotic drugs are widely prescribed for conditions such as schizophrenia and bipolar disease. The article "Cardiometabolic Risk of Second-Generation Antipsychotic Medications During First-Time Use in Children and Adolescents" (*J. of the Amer. Med. Assoc.*, 2009: 1765–1773) reported on body composition and metabolic changes for individuals who had taken various antipsychotic drugs for short periods of time.

a. The sample of 41 individuals who had taken aripiprazole had a mean change in total cholesterol (mg/dL) of 3.75, and the estimated standard error $s_d/\sqrt{n}$ was 3.878. Calculate a confidence interval with confidence level approximately 95% for the true average increase in total cholesterol under these circumstances (the cited article included this CI).

b. For the sample of 45 individuals who had taken olanzapine, the article reported (7.38, 9.69) as a 95% CI for true average weight gain (kg). What is a 99% CI?

7.6 OTHER TOPICS IN ESTIMATION (OPTIONAL)

Maximum Likelihood Estimation

Maximum likelihood estimation is a technique for automatically generating point estimators. This widely used procedure can be applied to any mass or density function, and the resulting estimators can be shown to have certain desirable statistical properties. As its name suggests, this technique is based on trying to find the value of an estimator that is most likely, given the particular set of sample data.

Example 7.15   In a random sample of ten electronic components, suppose that the first, third, and tenth components fail to function correctly when tested. Using the 0–1 coding scheme introduced in Section 5.6, we can write the data in this sample as $x_1 = 1$, $x_2 = 0$, $x_3 = 1$, $x_4 = 0, \ldots, x_{10} = 1$, where a "0" indicates that the component functioned correctly and a "1" indicates that it did not work correctly.

Since this data comes from a random sample, we can assume that the outcome involving the first item sampled is *independent* of the outcome involving the second component sampled, and so forth. Therefore, if π denotes the unknown proportion of defective components in the manufacturing process from which the sample was obtained, then the probability of getting the particular sample can be written as

$$P(x_1 = 1 \text{ and } x_2 = 0 \text{ and } x_3 = 1 \text{ and} \ldots \text{and } x_{10} = 1)$$
$$= P(x_1 = 1)\, P(x_2 = 0)\, P(x_3 = 1)\cdots P(x_{10} = 1)$$
$$= \pi(1 - \pi)\pi \cdots \pi = \pi^3(1 - \pi)^7$$

The expression $\pi^3(1 - \pi)^7$ represents the likelihood of our sample result occurring, and it is abbreviated as $L(\pi) = \pi^3(1 - \pi)^7$. We now ask, For what value of π is the observed sample most likely to have occurred? That is, we want to find the value of π that maximizes the probability $\pi^3(1 - \pi)^7$. This requires setting the derivative of $L(\pi)$ equal to 0 and solving for π. However, to simplify the calculations, we first take the natural logarithm of $L(\pi) = \pi^3(1 - \pi)^7$:

$$\ln(L(\pi)) = \ln\left[\,\pi^3(1 - \pi)^7\,\right] = 3\,\ln(\pi) + 7\,\ln(1 - \pi)$$

and then take the derivative[1]:

$$\frac{d}{d\pi}\ln(L(\pi)) = \frac{3}{\pi} - \frac{7}{1 - \pi}$$

---

[1] Since $\ln(x)$ is an increasing function of x, the value of π that maximizes $\ln(L(\pi))$ will be the same value that maximizes $L(\pi)$.

Setting this expression equal to 0 and solving for π, we find that the solution equals $3/10 = .30$. The value .30 is said to be the maximum likelihood estimate of the process proportion defective π. Notice that this estimate happens to be the ratio of the number of defective components in the sample divided by the sample size, that is, the sample proportion, p. In fact, this is true in general, regardless of the particular sample data, so we can also say that the sample proportion is a maximum likelihood estimator for a population or process proportion.

The technique in the previous example can be put into a general form that applies to any mass or density function. Let $f(x)$ denote either a mass or density function that is defined by a set of parameters $\theta_1, \theta_2, \ldots, \theta_k$. Given the data $x_1, x_2, x_3, \ldots, x_n$ in any random sample from a population whose distribution is described by $f(x)$, we form the **likelihood function**

$$L(\theta_1, \theta_2, \ldots, \theta_k) = f(x_1)f(x_2)f(x_3)\cdots f(x_n)$$

where each $f(x_i)$ is formed by simply substituting the ith data point x_i into the function $f(x)$. When $f(x)$ is a mass function, L can be interpreted as the probability that the sample result occurs. When $f(x)$ is a density, L is not a probability and, in this case, we simply call it a likelihood function.

The **maximum likelihood estimators** of the parameters $\theta_1, \theta_2, \ldots, \theta_k$ are the particular values of $\theta_1, \theta_2, \ldots, \theta_k$ that maximize the function $L(\theta_1, \theta_2, \ldots, \theta_k)$. The usual method for finding these parameter values is to treat $L(\theta_1, \theta_2, \ldots, \theta_k)$ as a function of k variables and use calculus to find the extreme points of the function. For $k = 1$, ordinary differentiation is required; for $k \geq 2$, partial derivatives are needed. Because $L(\theta_1, \theta_2, \ldots, \theta_k)$ is a product of several functions, it is usually easier to work with its natural logarithm $\ln(L(\theta_1, \theta_2, \ldots, \theta_k))$, which facilitates differentiation by converting L into a sum of functions:

$$\ln(L(\theta_1, \theta_2, \ldots, \theta_k)) = \ln(f(x_1)) + \ln(f(x_2)) + \ln(f(x_3)) + \cdots + \ln(f(x_n))$$

Because $\ln(x)$ is an increasing function of x, the values of $\theta_1, \theta_2, \ldots, \theta_k$ that maximize $\ln(L(\theta_1, \theta_2, \ldots, \theta_k))$ are the same ones that maximize $L(\theta_1, \theta_2, \ldots, \theta_k)$.

Example 7.16

The exponential distribution is commonly used to describe the lifetimes of certain products (see Example 5.12). Suppose that a sample of $n = 12$ electric appliances are tested continuously until each ceases to function. The length of time that each appliance lasted (in hours) follows:

10,502	9560	11,671	12,825	8987	7924
9508	8875	14,439	11,320	6549	10,654

To use maximum likelihood estimation to find the parameter λ of the exponential distribution that describes this data, we proceed as follows. Suppose $x_1, x_2, x_3, \ldots, x_n$

is any random sample from an exponential distribution with parameter λ. Since the exponential density function is of the form $f(x) = \lambda e^{-\lambda x}$, the likelihood function associated with the sample data is

$$L(\lambda) = f(x_1)f(x_2)f(x_3)\cdots f(x_n)$$
$$= (\lambda e^{-\lambda x_1})(\lambda e^{-\lambda x_2})(\lambda e^{-\lambda x_3})\cdots(\lambda e^{-\lambda x_n})$$
$$= \lambda^n e^{-\lambda \Sigma x_i}$$

Taking logarithms,

$$\ln(L(\lambda)) = n\ln(\lambda) - \lambda \sum x_i$$

Equating the derivative of this function to 0 and solving for λ, we find

$$\frac{d}{d\lambda}(\ln(L(\lambda))) = n/\lambda - \sum x_i = 0 \quad \text{so } \lambda = \frac{n}{\sum x_i} = \frac{1}{\bar{x}}$$

Thus the maximum likelihood estimator of λ is $\hat{\lambda} = 1/\bar{x}$. For the lifetime of the appliances, this estimate is $\hat{\lambda} = 1/10{,}234.5 = .0000977 = 9.77 \times 10^{-5}$.

Example 7.17

In Example 2.17, $n = 17$ observations on length-diameter ratio were plotted on a normal quantile plot, and it was determined that a normal distribution provides a good fit to this data. To find the maximum likelihood estimators of the parameters μ and σ^2 of the normal distribution that best fits this data, let $x_1, x_2, x_3, \ldots, x_n$ denote any random sample from a normal distribution. Then the likelihood function based on this data is as follows.

$$L(\mu, \sigma^2) = f(x_1)f(x_2)f(x_3)\cdots f(x_n)$$
$$= \left(\frac{1}{\sqrt{2\pi\sigma^2}} e^{-\frac{1}{2}\left(\frac{x_1-\mu}{\sigma}\right)^2}\right)\left(\frac{1}{\sqrt{2\pi\sigma^2}} e^{-\frac{1}{2}\left(\frac{x_2-\mu}{\sigma}\right)^2}\right)$$
$$\cdots\left(\frac{1}{\sqrt{2\pi\sigma^2}} e^{-\frac{1}{2}\left(\frac{x_n-\mu}{\sigma}\right)^2}\right)$$
$$= \left(\frac{1}{2\pi\sigma^2}\right)^{n/2} e^{-\frac{1}{2}\sum\left(\frac{x_i-\mu}{\sigma}\right)^2}$$

Taking logarithms,

$$\ln(L(\mu, \sigma^2)) = -\frac{n}{2}\ln(2\pi\sigma^2) - \frac{1}{2\sigma^2}\sum(x_i - \mu)^2$$

Since this is a function of two variables, the partial derivatives with respect to μ and σ^2 must be set to 0 and the resulting two equations solved. Omitting the details, we find the maximum likelihood estimators to be

$$\hat{\mu} = \bar{x} \qquad \hat{\sigma}^2 = \frac{1}{n}\sum (x_i - \bar{x})^2$$

Note that the first estimator, $\hat{\mu}$, is unbiased, but the second estimator, $\hat{\sigma}^2$, is slightly biased (recall that the unbiased estimator of σ^2 is the sample variance s^2, which uses a denominator of $n - 1$, not n). For the length-diameter ratio data, the maximum likelihood estimates are $\hat{\mu} = \bar{x} = 47.31$ and $\hat{\sigma}^2 = 57.153$.

Example 7.18   In Example 2.18, the following data on tensile strength for multi-wall carbon nanotubes was thought to follow a Weibull distribution:

17.4	22.3	23.7	30.0	44.2	49.3	52.7	54.8	62.1
66.2	84.9	90.1	90.3	91.1	99.5	101.6	108.5	109.5
119.1	127.0	132.9	140.8	141.0	175.0	231.8	259.7	

In general, let $x_1, x_2, x_3, \ldots, x_n$ be a random sample from a Weibull distribution with parameters α and β and density function

$$f(x) = \frac{\alpha}{\beta^\alpha} x^{\alpha-1} e^{-(x/\beta)^\alpha}$$

As in Example 7.17, the likelihood function $L(\alpha, \beta)$ is a function of two variables. So we must take partial derivatives of $\ln(L(\alpha, \beta))$, set them equal to 0, and solve the two resulting equations. Omitting the algebraic details, we find the following equations:

$$\alpha = \left[\frac{\sum x_i^\alpha \ln(x_i)}{\sum x_i^\alpha} - \frac{\sum \ln(x_i)}{n} \right]^{-1} \qquad \beta = \left(\frac{\sum x_i^\alpha}{n} \right)^{1/\alpha}$$

These two equations cannot be solved explicitly for the maximum likelihood estimates $\hat{\alpha}$ and $\hat{\beta}$. Instead, for each sample $x_1, x_2, x_3, \ldots, x_n$, the equations must be solved using an iterative numerical procedure. For the tensile strength data, the maximum likelihood estimates are $\hat{\alpha} = 1.727$ and $\hat{\beta} = 109.304$. These estimates can be obtained by using the survival package in R or by using the optimization procedure PROC NLP in SAS.

As you can see from these examples, maximum likelihood estimators are not always unbiased. In many cases, however, this bias can be removed by using a simple multiplicative correction factor. In Example 7.17, for instance, the maximum likelihood estimator of σ^2 in a normal distribution is slightly biased, but that bias can be corrected by simply multiplying the estimator by the factor $n/(n-1)$. Note that as n increases, the bias becomes negligible and the correction factor is essentially equal to 1. Beyond some slight problems with unbiasedness, maximum likelihood estimators have several properties that make them highly useful in practice. The two most important properties are listed in the following box.

Properties of Maximum Likelihood Estimators (MLEs)

1. For large n, the sampling distribution of an MLE is approximately normal and the estimator is unbiased or nearly so, with a variance smaller than that of any other estimator.

2. *Invariance property:* For any function $g(\cdot)$, if $\hat{\theta}$ is the MLE of a parameter θ, then $g(\hat{\theta})$ is the MLE of $g(\theta)$.

Example 7.19   In Example 7.16, we showed that the MLE of λ in an exponential distribution is $\hat{\lambda} = 1/\bar{x}$. Since the mean of an exponential distribution is related to λ by the equation $\mu = 1/\lambda$, the MLE of μ is simply $1/\hat{\lambda} = \bar{x}$. That is, given $g(\lambda) = 1/\lambda$, then since $\hat{\lambda}$ is the MLE for λ, $g(\hat{\lambda})$ is the MLE for $g(\lambda)$.

Density Estimation

In many applications, populations or processes can be described by normal density curves. Given a random sample of size n from a normal population, the density curve can be approximated by simply using the sample statistics $\bar{x}$ and s in place of the parameters μ and σ in the formula for the density curve:

$$f(x) \approx \frac{1}{\sqrt{2\pi s^2}} e^{-\frac{1}{2}\left(\frac{x-\bar{x}}{s}\right)^2}$$

Although this function can be graphed by itself, it is often good practice to superimpose a plot of $f(x)$ over a histogram of the sample data from which $\bar{x}$ and s were calculated. When the bars in the histogram represent densities (see p. 19), the graph of $f(x)$ will be of the same scale as the histogram, because both will have a total area of 1. When the histogram bars are simply frequencies, then $f(x)$ must be multiplied by an appropriate factor so that its area coincides with the area under the histogram. If w denotes the width of each histogram bar and there are n data points in the sample, then the total area encompassed by a frequency histogram is $w \cdot n$. Therefore, to make the approximate density function plot correctly over such a histogram, we must plot the function $w \cdot n \cdot f(x)$ instead of $f(x)$.

Kernel Density Estimation

Some populations and processes are not adequately described by common density curves. In such cases, the population density curve can be approximated by using the method of **kernel density estimation.** Creating a kernel density estimate is very similar to creating a histogram. In a histogram, the bar over any class interval can be thought of as a stack of several equal-size rectangles, each representing a single data point in that class. In the kernel density estimate, these n rectangles are replaced by n normal density curves centered at the n data points. The **kernel function** is then defined to be the average of these n normal densities. The kernel function is used as an approximation to the population density curve. Figure 7.12 illustrates this procedure on a small data set. Note that the kernel function in this figure is shown as the *sum* of the individual densities, not the average, to highlight the shape of the kernel function.

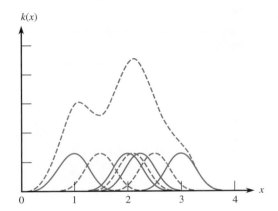

Figure 7.12 A kernel function for the data
$(1, 1.5, 1.8, 2, 2.1, 2.2, 2.5, 3)$

To put a normal density curve around each point in a set of data $x_1, x_2, x_3, \ldots, x_n$, we must determine the appropriate mean and variance to use. Let

$$s^2 = \frac{1}{n-1} \sum_{i=1}^{n} (x_i - \bar{x})^2$$

denote the sample variance of the n data points; the normal density centered at x_i has a mean and standard deviation of

$$\mu = x_i \qquad \sigma = \lambda s$$

where λ is a positive number called the **smoothing parameter** or **window width.** The smoothing parameter controls the spread of each of the normal distributions centered at the data points. These distributions have densities defined by

$$f_i(x) = \frac{1}{\lambda s \sqrt{2\pi}} e^{-\frac{1}{2}\left(\frac{x - x_i}{\lambda s}\right)^2} \qquad \text{for } -\infty < x < \infty$$

The kernel function is then given by the formula

$$k(x) = \frac{1}{n}\sum_{i=1}^{n} f_i(x) \qquad \text{for } -\infty < x < \infty$$

The effect of the smoothing constant is illustrated in the following example. Briefly, small values of λ yield kernel functions that follow the data very closely and, therefore, often have a choppy appearance similar to a histogram of the data. Larger values of λ lead to smoother-looking kernel functions.

Example 7.20   The tragedy that befell the space shuttle Challenger and its astronauts in 1986 led to a number of studies to investigate the reasons for mission failure. Attention quickly focused on the behavior of the rocket engine's O-rings. Here is data consisting of observations on x = O-ring temperature (°F) for each test firing or actual launch of the shuttle rocket engine (Presidential Commission on the Space Shuttle Challenger Accident, Vol. 1, 1986: 129–131).

31	40	45	49	52	53	57	58	58
60	61	61	63	66	67	67	67	67
68	69	70	70	70	70	72	73	75
75	76	76	78	79	80	81	83	84

The sample standard deviation of this data is $s = 12.159$. Suppose we choose a smoothing parameter of $\lambda = .5$. Starting with the leftmost point in the data, $x_1 = 31$, we then form the normal density curve with mean $\mu = 31$ and standard deviation $\sigma = \lambda s = (.5)(12.159) = 6.0795$:

$$f_1(x) = \frac{1}{6.0795\sqrt{2\pi}} e^{-\frac{1}{2}\left(\frac{x-31}{6.0795}\right)^2}$$

Proceeding to the next largest data point, $x_2 = 40$, we create a density curve with mean $\mu = 40$ and $\sigma = \lambda s = 6.0795$:

$$f_2(x) = \frac{1}{6.0795\sqrt{2\pi}} e^{-\frac{1}{2}\left(\frac{x-40}{6.0795}\right)^2}$$

After continuing in this manner through all $n = 36$ data points, we take the average of all 36 density functions to form the kernel function $k(x)$. Figure 7.13(a) shows the plot of $k(x)$ along with a histogram of the O-ring data. For comparison, Figure 7.13(b) shows a kernel function based on a value of $\lambda = .2$. Although the choice of λ is subjective, the value of $\lambda = .5$ provides a smoother fit to the data.

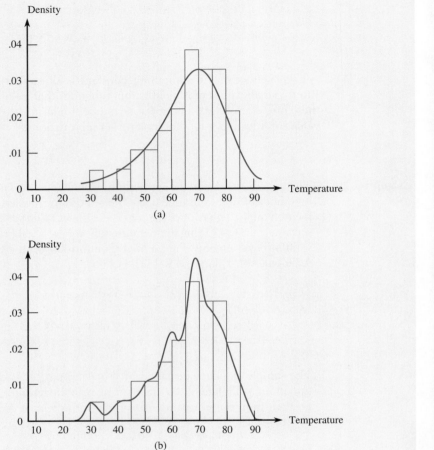

Figure 7.13 Kernel functions fit to the O-ring data of Example 7.20: (a) $\lambda = .5$; (b) $\lambda = .2$

Bootstrap Confidence Intervals

The confidence interval formulas we developed in the preceding sections require a knowledge of one or both of the following: (1) the exact distribution (e.g., the normal) of the population sampled and (2) a mathematical expression for the standard error of the statistic used to form the interval. Although requirement 1 becomes less important as the sample size increases, requirement 2 cannot be ignored. Complicating the situation further is the fact that many statistics have standard error formulas that are only approximations based on the *assumption* of normal populations. The sample correlation coefficient, r, is one such example. Even under the assumption of sampling from a bivariate normal population, the sampling distribution of r has no simple form, and formulas for the standard error of r are only approximations.

In an effort to avoid such problems, Efron ("Bootstrap Methods: Another Look at the Jackknife," *Annals of Statistics*, 1979: 1–26) introduced a computer-intensive method called the **bootstrap** method. The term *bootstrap*, a reference to the phrase "to pull oneself up by one's own bootstraps," is intended to describe the way in which bootstrap procedures approximate the sampling distribution of a statistic—by drawing large numbers of random samples *from a single sample* of data. The bootstrap is only one example of a general class of methods, called **resampling procedures,** based on the idea that there is information to be gained by sampling from a sample. The bootstrap method is described as follows:

Outline of the Bootstrap Method

1. Obtain a random sample of size n from a population or process.
2. Generate a random sample of size n, *with replacement,* from the original sample in step 1.
3. Calculate a statistic of interest for the sample in step 2.
4. Repeat steps 2 and 3 a large number of times to form the approximate sampling distribution of the statistic.

Sampling *with replacement* (see Section 4.2) is the key to the proper use of the bootstrap method. Otherwise, sampling n items *without replacement* from a set of n data values would always yield the original n items and, hence, the same calculated statistic for each sample. Much computational effort is necessary to draw repeated samples, often as many as 1000, and to calculate and compile a sampling distribution. Such effort, which would have been out of the question in the precomputer era, is a simple task for today's computers.

Understanding *how* the bootstrap works is easier than understanding *why* it works. At first glance, the procedure seems to give something for nothing. It requires no distributional assumptions about a population, needs no standard error formulas, and generates the sampling distribution of a statistic from the information in only a single sample. Roughly speaking, resampling methods work because random subsamples of a random sample are also random samples from a population (see Sampling Rules, Section 4.2). Consequently, each bootstrap sample qualifies as a genuine random sample of size n drawn, with replacement, from a population. This means that statistics calculated from such samples can properly be used to form a sampling distribution.

Bootstrap Intervals for the Mean

Bootstrap confidence intervals, also called **bootstrap percentile intervals,** for estimating a population or process are generated using the general format outlined previously. A large number, B, of bootstrap samples are randomly selected and the sample mean $\bar{x}$ is calculated for each sample. A $(1 - \alpha)100\%$ confidence interval for μ is formed by finding the upper and lower $(\alpha/2)100\%$ percentiles of the B sample means. The bootstrap procedure can be applied to large-sample and small-sample problems alike.

Choosing a value of B that makes $B(\alpha/2)$ an integer simplifies the work, because the percentiles can be found by just counting in $B(\alpha/2)$ units from both ends of the *sorted* list of sample means. Empirical studies have shown that values of B in the range of 500 to 1000 generally give good results. For the typical confidence levels used in practice (e.g., 90%, 95%, 99%), the choice $B = 1000$ will satisfy all of these requirements. In general, larger values of B should be used for larger confidence levels.

Example 7.21 In Example 7.3 (Section 7.2), the large-sample confidence interval formula $\bar{x} \pm 1.96s/\sqrt{n}$ was used to find a 95% confidence interval for the mean breakdown voltage (in kV) for a particular electronic circuit. Using the sample of $n = 48$ observations, this interval was determined to be (53.2, 56.2). For comparison, we now use this data to find a 95% bootstrap interval for the mean.

A histogram of $B = 1000$ samples, drawn with replacement from the original sample, is shown in Figure 7.14. Since $1 - \alpha = .95$, the upper and lower endpoints of the confidence interval are found by counting in $B(\alpha/2) = 1000(.05/2) = 25$ units from each end of the sorted list of 1000 sample means. For the sample means shown in Figure 7.14, the 25th largest value is 53.21 and the 975th largest value is 56.13, giving a confidence interval of (53.2, 56.1). Note how close the bootstrap interval is to the earlier interval (53.2, 56.2). This is not an accident. Bootstrap intervals usually agree closely with traditional confidence intervals when all the assumptions necessary for the traditional interval are met.

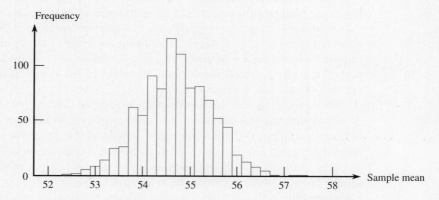

Figure 7.14 $B = 1000$ bootstrapped sample means from the data in Example 7.3

Two-Sample Bootstrap Intervals

Bootstrap methods can readily be applied to statistics based on two or more samples. The procedures are intuitive extensions of the ones illustrated previously. For the case of two independent random samples of size n_1 and n_2 (Section 7.5), each bootstrap sample consists of a *pair* of samples, one of size n_1 drawn from the first sample, the other of size n_2 drawn from the second sample. The difference of these two sample means is recorded, and the next bootstrap pair is drawn until B such differences have been obtained. The percentile method is then used to obtain the desired confidence limits. For paired data, the procedure is even easier: Select B bootstrap samples from the original n *differences* between the two paired samples of n data points. Again, we use the percentile method to form the confidence limits.

Comments

Since its inception in 1979, the bootstrap method has been successfully applied to many different situations, including regression and correlation analysis, as well as other advanced statistical procedures. During this time, computer power and availability have also dramatically increased, making the bootstrap a realistic option for data analysis. It is now relatively easy to write macros in any statistical or spreadsheet software program to carry out bootstrap computations.

As a rule, bootstrap intervals generally agree fairly well with traditional confidence interval results when the assumptions necessary for the traditional interval are met. In those cases where the assumptions are not met (e.g., when populations are not normally distributed), bootstrap intervals offer the additional advantage of giving more realistic results than traditional confidence intervals. For further reading on this subject, the book by Efron and Tibshirani entitled *An Introduction to the Bootstrap* offers a useful guide to applying the bootstrap (Efron, B., and R. J. Tibshirani, *An Introduction to the Bootstrap*, Chapman and Hall, New York, 1993).

Section 7.6 Exercises

60. Refer to Exercise 42 of Section 7.4.
    a. Use the bootstrap method to find a 95% bootstrap interval for the mean of the population from which the data of Exercise 42 was obtained.
    b. Compare your result in part (a) to the 95% confidence interval found in Exercise 42(c).

61. Refer to Exercise 46 of Section 7.4.
    a. Use the bootstrap method to find a 95% bootstrap interval for the mean of the population from which this data was obtained.
    b. Compare your result in part (a) to the 95% confidence interval found in Exercise 46(a).

62. In Exercise 14 (Section 7.2), the sample mean and standard deviation of the dye-layer density of aerial photographs of 69 forest trees were found to be 1.028 and .163, respectively. Because the raw data is not available, a researcher suggests using a computer to generate a random sample of 69 observations from a normal distribution whose mean and standard deviation are 1.028 and .163, respectively. If necessary, after obtaining the sample, the data are adjusted so that their sample mean and standard deviation coincide exactly with 1.028 and .163. A 95% bootstrap interval is then generated using this simulated data.
    a. Under what conditions will this procedure provide a reliable interval estimate?

    b. Use the procedure outlined in this exercise to generate a 95% bootstrap interval for the average dye-layer density.
    c. Compare your result in part (b) to the 95% confidence interval found in Exercise 14(a).

63. A random sample of n electronic assemblies is selected from a large shipment, and each assembly is tested on an automatic test station. The number x of assemblies that do not perform correctly is determined. Let π denote the proportion of assemblies in the entire shipment that are defective.
    a. In terms of x, what is the maximum likelihood estimator of π?
    b. Is the estimator in part (a) unbiased?
    c. What is the MLE of $(1 - \pi)^5$, the probability that none of the next five assemblies tested is defective?

64. Let x denote the proportion of an allotted time frame that a randomly selected worker spends performing a manufacturing task. Suppose the probability density function of x is

$$f(x) = \begin{cases} (\theta + 1)x^\theta & \text{for } 0 \le x \le 1 \\ 0 & \text{otherwise} \end{cases}$$

where the value of θ must be larger than 1.

a. Derive the maximum likelihood estimator of θ for a random sample of size n.

b. A random sample of ten workers yielded the following data on x: .92, .79, .90, .65, .86, .47, .73, .97, .94, .77. Use this data to obtain an estimate of θ.

65. The shear strength x of a random sample of spot welds is measured. Shear strengths (in psi) are assumed to follow a normal distribution.

   a. Find the maximum likelihood estimator of the strength that is exceeded by 5% of the population of welds. That is, find a maximum likelihood estimator for the 95th percentile of the normal distribution based on a random sample of size n. *Hint:* Determine the relationship between the 95th percentile and μ and σ, then use the invariance property of MLEs.

   b. A random sample of ten spot-weld strengths yields the following data (in psi): 392, 376, 401, 367, 389, 362, 409, 415, 358, 375. Use the result in part (a) to find an estimate of the 95th percentile of the distribution of all weld strengths.

66. Refer to Exercise 65. Suppose the strength x of another randomly selected spot weld is measured.

   a. Find a maximum likelihood estimator of the probability that x is less than 400. That is, find the MLE for $P(x < 400)$.

   b. Use the result in part (a) with the data from Exercise 65(b) to estimate $P(x < 400)$.

67. A random sample $x_1, x_2, x_3, \ldots, x_n$ is selected from a *shifted exponential distribution* whose probability density function is given by

$$f(x) = \begin{cases} \lambda e^{-\lambda(x-\theta)} & \text{for } x \geq \theta \\ 0 & \text{otherwise} \end{cases}$$

When $\theta = 0$, this probability density function reduces to the probability density function of the exponential distribution.

   a. Obtain maximum likelihood estimators of both θ and λ.

   b. In traffic flow research, *time headway* is defined to be the elapsed time between the moment that one car finishes passing a fixed point and the instant that the next car begins to pass that

point. The random variable x = time headway has been modeled by a shifted exponential distribution. For a random sample of ten headway times—3.11, .64, 2.55, 2.20, 5.44, 3.42, 10.39, 8.93, 17.82, and 1.30—use the results from part (a) to find estimates of θ and λ.

68. A specimen is weighed twice on the same scale. Let x and y denote the two measurements. Suppose x and y are independent of one another and are assumed to follow normal distributions with the same mean μ (the true weight of the specimen) and the same variance σ^2.

   a. For a random sample of n specimens, show that the maximum likelihood estimator of σ^2 is given by $(1/4n)\,\Sigma(x_i - y_i)^2$, where $(x_1, y_1), (x_2, y_2), \ldots, (x_n, y_n)$ denote the n pairs of scale measurements. *Hint:* The sample variance of two measurements z_1 and z_2 equals $(z_1 - z_2)^2/2$.

   b. Five randomly chosen specimens are weighed, yielding the following data: (3.10, 3.12), (3.52, 3.45), (4.22, 4.30), (2.98, 3.06), and (5.43, 5.38). Use the result in part (a) to find an estimate of σ^2.

69. Suppose someone suggests using a smoothing parameter of $\lambda = 2$ to create a kernel density graph. Do you expect the graph to provide a useful picture of the data? Why?

70. Refer to the data in Exercise 46 of Section 7.4.

   a. Use a smoothing parameter of $\lambda = .5$ to create a kernel density plot for this data.

   b. Repeat part (a) using a smoothing parameter of $\lambda = .3$.

   c. Which of the plots in parts (a) and (b) appears to fit the data better?

71. Suppose the smallest distance d between any two successive measurements in an ordered set of data (i.e., measurements sorted from smallest to largest) is 3 units.

   a. If s denotes the sample standard deviation of the measurements in a sample of size n, would $\lambda = d/(3s)$ lead to a kernel density graph with a choppy appearance or a smooth appearance? Why?

   b. Will values of λ that are greater than $d/(3s)$ lead to choppier- or smoother-looking kernel density estimates?

72. Refer to the data in Exercise 42 of Section 7.4.
    a. Use a smoothing parameter of $\lambda = .5$ to create a kernel density plot for this data.
    b. Repeat part (a) using a smoothing parameter of $\lambda = .3$.
    c. Which of the plots in parts (a) and (b) appears to fit the data better?

73. A kernel function is fit to the data in a sample of size n. Later, a researcher realizes that the largest observation in the sample was actually a typographical error and, because the original lab data no longer exists, this data point is removed from the sample, leaving a sample of $n - 1$ measurements.

The researcher wants to fit a new kernel function to the reduced sample of $n - 1$ data points. To produce a graph that has about the same smoothness as the original kernel function, will the value of λ have to be raised or lowered?

74. In Example 1.8 (Chapter 1), a histogram was fit to the energy consumption data (in BTUs) from a sample of 90 homes. Using this data, experiment with different values of λ until you find a value that gives a kernel density estimate that approximates the shape of the histogram of this data shown in Figure 1.7.

Supplementary Exercises

75. Exercise 4 of Chapter 1 presented a sample of $n = 153$ observations on ultimate tensile strength.
    a. Obtain a lower confidence bound for population mean strength. Does the validity of the bound require any assumptions about the population distribution? Explain.
    b. Is any assumption about the tensile strength distribution required prior to calculating a lower prediction bound for the tensile strength of the next specimen selected using the method described in this section? Explain.
    c. Use a statistical software package to investigate the plausibility of a normal population distribution.
    d. Calculate a lower prediction bound with a prediction level of 95% for the ultimate tensile strength of the next specimen selected.

76. Anxiety disorders and symptoms can often be effectively treated with benzodiazepine medications. It is known that animals exposed to stress exhibit a decrease in benzodiazepine receptor binding in the frontal cortex. The paper "Decreased Benzodiazepine Receptor Binding in Prefrontal Cortex in Combat-Related Posttraumatic Stress Disorder" (*Amer. J. of Psychiatry*, 2000: 1120–1126) described the first study of benzodiazepine receptor binding in individuals suffering from PTSD. The accompanying data on a receptor binding measure (adjusted

distribution volume) was read from a graph in the paper.

PTSD: 10, 20, 25, 28, 31, 35, 37, 38, 38, 39, 39, 42, 46

Healthy: 23, 39, 40, 41, 43, 47, 51, 58, 63, 66, 67, 69, 72

   a. Is it plausible that the population distributions from which these samples were selected are normal?
   b. Calculate an interval for which you can be 95% confident that at least 95% of all healthy individuals in the population have adjusted distribution volumes lying between the limits of the interval.
   c. Predict the adjusted distribution volume of a single healthy individual by calculating a 95% prediction interval. How does this interval's width compare to the width of the interval calculated in part (b)?
   d. Estimate the difference between the true average measures in a way that conveys information about reliability and precision.

77. The article "Quantitative MRI and Electrophysiology of Preoperative Carpal Tunnel Syndrome in a Female Population" (*Ergonomics*, 1997: 642–649) reported that $(-473.3, 1691.9)$ was a large-sample 95% confidence interval for the difference between true average

thenar muscle volume (mm³) for sufferers of carpal tunnel syndrome and true average volume for non-sufferers. Calculate a 90% confidence interval for this difference.

78. Acrylic bone cement is commonly used in total joint arthroplasty as a grout that allows for the smooth transfer of loads from a metal prosthesis to bone structure. The paper "Validation of the Small-Punch Test as a Technique for Characterizing the Mechanical Properties of Acrylic Bone Cement" (*J. of Engr. In Med.*, 2006: 11–21) gave the following data on breaking force (N):

Temp	Medium	n	$\bar{x}$	s
37°	Dry	6	325.73	34.97
37°	Wet	6	306.09	41.97

Assume that all population distributions are normal.
   a. Estimate true average breaking force in a dry medium at 37° in a way that conveys information about reliability and precision. Interpret your estimate.
   b. Estimate the difference between true average breaking force in a dry medium at 37° and true average force at the same temperature in a wet medium, and do so in a way that conveys information about precision and reliability. Then interpret your estimate.

79. An experiment was carried out to compare various properties of cotton/polyester spun yarn finished with softener only and yarn finished with softener plus 5% DP-resin ("Properties of a Fabric Made with Tandem Spun Yarns," *Textile Res. J.*, 1996: 607–611). One particularly important characteristic of fabric is its durability, that is, its ability to resist wear. For a sample of 40 softener-only specimens, the sample mean stoll-flex abrasion resistance (cycles) in the filling direction of the yarn was 3975.0, with a sample standard deviation of 245.1. Another sample of 40 softener-plus specimens gave a sample mean and sample standard deviation of 2795.0 and 293.7, respectively. Calculate a confidence interval with confidence level 99% for the difference between true average abrasion resistances for the two types of fabric. Does your interval provide convincing evidence that true average resistances differ for the two types of fabric? Why or why not?

80. As reported by the Pew Research Center's Social and Demographic Trends Project in September 2012, a survey of 6500 American households revealed that a record 19% owed student loan debt in 2010 (a sharp increase from the 15% that owed such debt in 2007).
   a. Calculate and interpret a 95% CI for the proportion of all American households in 2010 that owed student loan debt.
   b. What sample size is required if the desired width of the 95% CI is to be at most .04, irrespective of the sample results?
   c. Does the upper limit of the interval in part (a) specify a 95% upper confidence bound for the proportion being estimated? Explain.

81. Torsion during hip external rotation (ER) and extension may be responsible for certain kinds of injuries in golfers and other athletes. The article "Hip Rotational Velocities During the Full Golf Swing" (*J. of Sports Sci. and Med.*, 2009: 296–299) reported on a study in which peak ER velocity and peak IR (internal rotation) velocity (both in $\deg \cdot \sec^{-1}$) were determined for a sample of 15 female collegiate golfers during their swings. The following data was supplied by the article's authors:

Golfer	ER	IR	diff	z quan
1	−130.6	−98.9	−31.7	−1.28
2	−125.1	−115.9	−9.2	−0.97
3	−51.7	−161.6	109.9	0.34
4	−179.7	−196.9	17.2	−0.73
5	−130.5	−170.7	40.2	−0.34
6	−101.0	−274.9	173.9	0.97
7	−24.4	−275.0	250.6	1.83
8	−231.1	−275.7	44.6	−0.17
9	−186.8	−214.6	27.8	−0.52
10	−58.5	−117.8	59.3	0.00
11	−219.3	−326.7	107.4	0.17
12	−113.1	−272.9	159.8	0.73
13	−244.3	−429.1	184.8	1.28
14	−184.4	−140.6	−43.8	−1.83
15	−199.2	−345.6	146.4	0.52

   a. Is it plausible that the differences came from a normally distributed population?
   b. Estimate the true average difference in peak ER and IR velocities in a way that conveys

information about reliability and precision. Interpret the resulting estimate.

82. It is important that face masks used by firefighters be able to withstand high temperatures. In a test of one type of mask, the lenses in 11 of the 35 masks popped out at a temperature of 250°F. Calculate a lower confidence bound for the proportion of all such masks whose lenses would pop out at this temperature using both the method suggested in Section 7.3 and the method suggested in Exercise 26(b).

83. Suppose an investigator wants a confidence interval for the median $\tilde{\mu}$ of a continuous distribution based on a random sample $x_1, \ldots, x_n$ without assuming anything about the shape of the distribution.
    a. What is $P(x_1 < \tilde{\mu})$, the probability that the first observation is smaller than the median?
    b. What is the probability that *both* the first and the second observations are smaller than the median?
    c. Let $y_n = \max \{x_1, \ldots, x_n\}$. What is $P(y_n < \tilde{\mu})$? *Hint:* The condition that y_n is less than $\tilde{\mu}$ is equivalent to what about $x_1, \ldots, x_2$?
    d. With $y_1 = \min \{x_1, \ldots, x_n\}$, what is $P(\tilde{\mu} < y_1)$?
    e. Using the results of parts (c) and (d), what is $P(y_1 < \tilde{\mu} < y_n)$? Regarding (y_1, y_n) as a confidence interval for $\tilde{\mu}$, what is the associated confidence level?
    f. An experiment carried out to study the curing time (hr) for a particular experimental adhesive yielded the following observations:

    31.2 36.0 31.5 28.7 37.2
    35.4 33.3 39.3 42.0 29.9

    Referring back to part (e), determine the confidence interval and the associated confidence level.
    g. Assuming that the data in part (f) was selected from a normal distribution (is this assumption justified?), calculate a confidence interval for $\tilde{\mu}$ (which for a normal distribution is identical to μ) using the same confidence level as in part (f), and compare the two intervals.

84. Consider the situation described in Exercise 83.
    a. What is $P(x_1 < \tilde{\mu}, x_2 > \tilde{\mu}, x_3 > \tilde{\mu}, \ldots, x_n > \tilde{\mu})$, that is, the probability that only the first observation is smaller than the median and all others exceed the median?
    b. What is the probability that only x_2 is smaller than the median and all other $n - 1$ observations exceed the median?
    c. What is the probability that exactly one of the x_i's is less than $\tilde{\mu}$?
    d. What is $P(\tilde{\mu} < y_2)$, where y_2 denotes the second smallest x_i? *Hint:* $\tilde{\mu} < y_2$ occurs if either all n of the observations exceed the median or all but one of the x_i's does.
    e. With y_{n-1} denoting the second largest x_i, what is $P(\tilde{\mu} > y_{n-1})$?
    f. Using the results of parts (d) and (e), what is $P(y_2 < \tilde{\mu} < y_{n-1})$? What does this imply about the confidence level associated with the interval (y_2, y_{n-1})? Determine the interval and associated confidence level for the data given in Exercise 83.

85. Suppose we have obtained a random sample $x_1, \ldots, x_n$ from a continuous distribution and wish to use it as a basis for predicting a single new observation x_{n+1} without assuming anything about the shape of the distribution. Let y_1 and y_n denote the smallest and largest, respectively, of the n sample observations.
    a. What is $P(x_{n+1} < x_1)$?
    b. What is $P(x_{n+1} < x_1$ and $x_{n+1} < x_2)$, that is, the probability that x_{n+1} is the smallest of these three observations?
    c. What is $P(x_{n+1} < y_1)$? What is $P(x_{n+1} > y_n)$?
    d. What is $P(y_1 < x_{n+1} < y_n)$, and what does this say about the prediction level associated with the interval (y_1, y_n)? Determine the interval and associated prediction level for the curing time data given in Exercise 83.

86. The derailment of a freight train due to the catastrophic failure of a traction motor armature bearing provided the impetus for a study reported in the article "Locomotive Traction Motor Armature Bearing Life Study" (*Lubrication Engr.*, Aug. 1997: 12–19). A sample of 17 high-mileage traction motors was selected and the amount of cone penetration (mm/10) was determined both for the pinion bearing and

for the commutator armature bearing, resulting in the following data:

Motor:	1	2	3	4	5	6
Commutator:	211	273	305	258	270	209
Pinion:	226	278	259	244	273	236
Motor:	7	8	9	10	11	12
Commutator:	223	288	296	233	262	291
Pinion:	290	287	315	242	288	242
Motor:	13	14	15	16	17	
Commutator:	278	275	210	272	264	
Pinion:	278	208	281	274	268	

Calculate an estimate of the population mean difference between penetration for the commutator armature bearing and penetration for the pinion bearing, and do so in a way that conveys information about the reliability and precision of the estimate. (*Note:* A normal quantile plot validates the necessary normality assumption.) Would you say that the population mean difference has been precisely estimated? Does it look as though population mean penetration differs for the two types of bearings? Explain.

87. The article cited in Exercise 86 also included the following data on percentage of oil remaining for the commutator bearings:

71.02   86.49   81.14   84.89   87.42
84.49   82.09   80.97   69.80   89.29
86.10   86.80   83.41   60.56   88.80
86.41   86.19

Would you use the one-sample t confidence interval to estimate the population mean and median? Estimate the population median percentage of oil left using the interval suggested in Exercise 84, and determine the corresponding confidence level.

88. Wire electrical-discharge machining (WEDM) is a process used to manufacture conductive hard metal components. It uses a continuously moving wire that serves as an electrode. Coated wires have been used to substantially increase the cutting speed and precision of the process. Coating on the wire electrode allows for cooling of the wire electrode core and provides an improved cutting performance.

The article "High-Performance Wire Electrodes for Wire Electrical-Discharge Machining—A Review" (*J. of Engr. Manuf.*, 2012: 1757–1773) gave the following sample observations on total coating layer thickness (in μm) of eight wire electrodes used for WEDM:

21   16   29   35   42   24   24   25

a. Is it plausible that the given sample observations were selected from a normal distribution?
b. Calculate and interpret a 95% CI for true average total coating layer thickness in all such electrodes.
c. Predict the total coating layer thickness for a single electrode in a way that conveys information about precision and reliability.

89. Nine Australian soldiers were subjected to extreme conditions that involved a 100-min walk with a 25-lb pack when the temperature was 40°C (104°F). One of them overheated (above 39°C) and was removed from the study. Here are the rectal Celsius temperatures of the other eight at the end of the walk ("Neural Network Training on Human Body Core Temperature Data," Combatant Protection and Nutrition Branch, Aeronautical and Maritime Research Laboratory of Australia, DSTO TN-0241, 1999):

38.4   38.7   39.0   38.5   38.5   39.0   38.5   38.6

We would like to get a 95% confidence interval for the population mean.

a. Compute the t-based confidence interval of Section 7.4.

b. Use the bootstrap method to find a 95% bootstrap interval for the population mean.

c. Compare your results in parts (a) and (b).

90. Suppose that samples of size n_1, n_2, and n_3 are independently selected from three different populations. Let μ_i and σ_i ($i = 1, 2, 3$) denote the population means and standard deviations, and consider estimating $\theta = a_1\mu_1 + a_2\mu_2 + a_3\mu_3$, where the a_i's are specified numerical constants. A point estimate of θ is $\hat{\theta} = a_1\bar{x}_1 + a_2\bar{x}_2 + a_3\bar{x}_3$. When the sample sizes are all large, $\hat{\theta}$ has approximately a normal distribution with variance

$$\sigma_{\hat{\theta}}^2 = a_1^2 \cdot \frac{\sigma_1^2}{n_1} + a_2^2 \cdot \frac{\sigma_2^2}{n_2} + a_3^2 \cdot \frac{\sigma_3^2}{n_3}$$

An estimated variance $s_{\hat{\theta}}^2$ results from replacing the σ^2s; by the s^2s; $\hat{\theta}$ can then be standardized to obtain a z variable from which the confidence interval $\hat{\theta} \pm (z \text{ crit})s_{\hat{\theta}}$ is obtained. Suppose that samples of three different brands of tires with identical lifetime ratings—a store brand (1) and two national brands (2 and 3)—are selected, and the lifetime of each tire is determined, resulting in the following data:

Brand	Sample size	Sample mean	Sample standard deviation
1	40	38,376	1522
2	32	41,569	1711
3	32	42,123	1645

Calculate and interpret a confidence interval with confidence level 95% for $\theta = \mu_1 - (\mu_2 + \mu_3)/2$.

91. Recent information suggests that obesity is an increasing problem in America among all age groups. The Associated Press (October 9, 2002) reported that 1276 individuals in a sample of 4115 adults were found to be obese (a body mass index exceeding 30; this index is a measure of weight relative to height).
    a. Estimate the proportion of all American adults who are obese in a way that conveys information about the reliability and precision of the estimate.
    b. A 1998 survey based on people's own assessments revealed that 20% of all adult Americans consider themselves obese. Does the estimate of part (a) suggest that the 2002 percentage is more than 1.5 times the 1998 percentage? Explain.

92. The one-sample CI for a normal mean and PI for a single observation from a normal distribution were both based on the *central t* distribution. A CI for a particular percentile (e.g., the 1st percentile or the 95th percentile) of a normal population distribution is based on the *noncentral t* distribution. A particular distribution of this type is specified by both df and the value of the noncentrality parameter $\delta(\delta = 0$ gives the central t distribution). The key result is that the variable

$$t = \frac{\dfrac{\bar{x} - \mu}{\sigma/\sqrt{n}} - (z \text{ percentile})\sqrt{n}}{s/\sigma}$$

has a noncentral t distribution with df $= n - 1$ and $\delta = (-z \text{ percentile})\sqrt{n}$. Let $t_{.025,v,\delta}$ and $t_{.975,v,\delta}$ denote the critical values that capture lower tail area .025 and upper tail area .025, respectively, under the noncentral t curve with v df and noncentrality parameters (when $\delta = 0, t_{.025} = -t_{.975}$, since central t distributions are symmetric about 0).
    a. Use the given information to obtain a formula for a 95% confidence interval for some particular percentile of a normal population distribution.
    b. For $\delta = 6.58$ and df $= 15, t_{.025}$ and $t_{.975}$ are (from Minitab) 4.1690 and 10.9684, respectively. Use this information to obtain a 95% CI for the 5th percentile of the modulus of elasticity distribution considered in Example 7.10.

Bibliography

DeGroot, Morris, and Mark Schervish, **Probability and Statistics** (4th ed.), Addison-Wesley, Reading, MA, 2011. *A very good exposition of the general principles of statistical inference at a level somewhat above that of our book.*

Devore, Jay and Kenneth Berk, **Modern Mathematical Statistics with Applications** (2nd ed.), Springer, New York, 2012. *An excellent survey of general concepts of inference.*

Hahn, Gerald, and William Meeker, **Statistical Intervals**, Wiley, New York, 2011. *Everything you ever wanted to know about statistical intervals—confidence, prediction, tolerance, and others.*

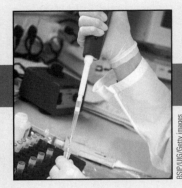

BSIP/UIG/Getty images

Testing Statistical Hypotheses

INTRODUCTION

Estimation of a parameter does not explicitly involve making a decision; instead we wish to determine the most plausible value (a point estimate) or a range of plausible values (a confidence interval). In contrast, the objective of a hypothesis-testing analysis is to decide which of two competing claims (hypotheses) is true. We have already encountered an informal situation of this sort in the context of quality control: At each time point, we used sample information to decide whether a process was out of control. The decision rule involved control limits, with the out-of-control conclusion justified only if the value of some quality statistic fell outside the limits.

In Section 8.1, we discuss the forms of hypotheses about parameters and the general nature of *test procedures* for deciding between the two relevant hypotheses. Test procedures based on t distributions are developed in Section 8.2 for testing hypotheses about a single mean μ or about the difference $\mu_1 - \mu_2$ between two means. Sections 8.3 and 8.4 introduce procedures for hypotheses about certain population proportions and population distributions. Finally, in Section 8.5, we consider a variety of issues and concepts relating to the behavior of test procedures. Hypothesis testing methods, as well as estimation methods, will be used extensively throughout the remainder of the book.

8.1   HYPOTHESES AND TEST PROCEDURES

A **statistical hypothesis,** or just **hypothesis,** is a claim or assertion either about one or more population or process characteristics (parameters) or else about the form of the population or process distribution. Here are some examples of legitimate hypotheses:

1. Parameter: π = proportion of e-mail messages emanating from a certain system that are undeliverable

   Hypothesis: $\pi < .01$

2. Parameters: μ_1 = true average lifetime for a particular name-brand tire (miles)

   μ_2 = true average lifetime for a less expensive store-brand tire

   Hypothesis: $\mu_1 - \mu_2 > 10,000$

3. Parameters: π_1 = proportion of individuals in a certain population with an AA genotype for a particular genetic characteristic

   π_2 = proportion of individuals with an Aa genotype

   π_3 = proportion of individuals with an aa genotype

   Hypothesis: $\pi_1 = .25$, $\pi_2 = .50$, $\pi_3 = .25$

4. Population distribution: $f(x)$, where x = the time between successive adjustments of a lathe process to correct for tool wear

   Hypothesis: x has an exponential distribution, that is, $f(x) = \lambda e^{-\lambda x}$ for some $\lambda > 0$

   In any hypothesis-testing problem, there are two competing hypotheses under consideration. One hypothesis might be $\mu = 1000$ and the other $\mu \neq 1000$, or we might be considering $\sigma = .10$ versus $\sigma < .10$. If it were possible to carry out a census of the entire population, we would know which of the two hypotheses is correct, but almost always our conclusion must be based on information in sample data. A **test of hypotheses** is a method for using sample data to decide between the two competing hypotheses under consideration. We initially assume that one of the hypotheses, the *null hypothesis,* is correct; this is the "prior belief" claim. We then consider the evidence (sample data), and we reject the null hypothesis in favor of the competing claim, called the *alternative hypothesis,* only if there is convincing evidence against the null hypothesis.

DEFINITIONS

> The **null hypothesis,** denoted by H_0, is the assertion that is initially assumed to be true. The **alternative hypothesis,** denoted by H_a, is the claim that is contradictory to H_0. The null hypothesis will be rejected in favor of the alternative hypothesis only if sample evidence suggests that H_0 is false. If the sample does not strongly contradict H_0, we will continue to believe in the truth of the null hypothesis. The two possible conclusions from a hypothesis-testing analysis are then *reject H_0* or *fail to reject H_0*.

Making a decision in a criminal trial is similar to what is involved in testing hypotheses. The null hypothesis, the claim initially believed to be true, is that the accused is innocent ("innocent until proven guilty"). The jury is instructed not to switch its belief to the alternative hypothesis that the accused is guilty unless there is serious and compelling evidence for reaching that conclusion. The burden of proof is on the prosecution to demonstrate conclusively from the evidence that the accused is guilty. In hypothesis testing, the burden of proof is on the alternative hypothesis; in the absence of evidence strongly contradictory to H_0 and much more consistent with H_a, we continue to believe in the null hypothesis.

The selection of the claim believed true (H_0) and the claim that will bear the burden of proof (H_a) depends on the objectives of the study. In general, if an investigator wishes to demonstrate conclusively that a particular assertion is correct, or wants to see strong evidence for an assertion before taking action, that assertion should be incorporated in H_a. Frequently in science, a researcher develops a new theory that stands in contrast to currently accepted theory. If the current theory is identified as H_0, and the new theory as H_a (the *research* hypothesis), and if H_0 can then be rejected, the investigator will have compelling evidence that the new theory is correct.

Example 8.1 Because of machining process variability, bearings produced by a certain machine do not have identical diameters. Let μ denote the true average diameter for bearings currently being produced. The machine was initially calibrated to achieve the design specification $\mu = .5$ in. However, the manufacturer is now concerned that the diameters no longer conform to this specification. That is, the hypothesis $\mu \neq .5$ must now be considered a possibility. If sample evidence suggests that this latter hypothesis is indeed correct, the production process will have to be halted while recalibration takes place. Stopping the process is quite costly, so the manufacturer wants to be sure that recalibration is necessary before this is done. Under these circumstances, a sensible choice of hypotheses is

$$H_0: \mu = .5 \quad \text{(the specification is being met, so recalibration is unnecessary)}$$
$$H_a: \mu \neq .5$$

Only compelling sample evidence would then result in H_0 being rejected in favor of H_a.

In many hypothesis-testing problems that we will consider, the null and alternative hypotheses assume particular forms. H_0 will be

$$\text{population or process characteristic} = \text{some hypothesized value}$$

H_a then results from replacing the "=" in H_0 by one of the three possible inequalities: $>, <,$ or $\neq$; the relevant inequality again depends on the research objectives. One example of this is $H_0: \sigma = .002$ versus $H_a: \sigma < .002$, where σ is the process standard deviation of bearing diameter.

Example 8.2 A pack of a certain brand of cigarettes displays the statement "1.5 mg nicotine average per cigarette by FTC method." Let μ denote the mean nicotine content per cigarette for all cigarettes of this brand. The advertised claim is that $\mu = 1.5$. People who smoke this brand would probably be disturbed if it turned out that true average nicotine content exceeded the claimed value, since excessive nicotine ingestion is a known health hazard. Suppose a sample of cigarettes of this brand is selected and the nicotine content of each cigarette is determined. Evidence from this sample against the company's claim would have to be quite strong before the accusation is made that the claim is false, since serious financial and legal consequences could ensue from any such action. This suggests that we test

$H_0: \mu = 1.5$ (the advertised claims is correct)

against the alternative hypothesis

$H_a: \mu > 1.5$ (true average nicotine level exceeds the advertised value)

and reject H_0 in favor of H_a only if sample evidence is very compelling for this conclusion.

Since the alternative hypothesis in Example 8.2 asserted that $\mu > 1.5$, it might have seemed sensible to state H_0 as the inequality $\mu \leq 1.5$. This assertion is in fact the *implicit* null hypothesis, but we will state H_0 explicitly as a claim of equality. There are several reasons for this. First of all, the development of a test procedure is most easily understood if there is a unique value of μ (or σ, or whatever other parameter is under consideration) when H_0 is true. Second, suppose sample data gives much more support to $\mu > 1.5$ than to $\mu = 1.5$. Then there would also be more support for $\mu > 1.5$ than for $\mu \leq 1.5$. If, on the other hand, $\mu = 1.5$ is much more plausible than $\mu > 1.5$ in light of the data, then $\mu \leq 1.5$ would also be deemed more plausible than $\mu > 1.5$. So the conclusion when testing $H_0: \mu = 1.5$ versus $H_a: \mu > 1.5$ should be identical to that when considering the more realistic null hypothesis $\mu \leq 1.5$ against this alternative. Similarly, whatever conclusion reached when testing $H_0: \pi = .1$ versus $H_a: \pi < .1$ would also apply to the implicit null hypothesis $H_0: \pi \geq .1$.

Errors in Hypothesis Testing

Once hypotheses have been formulated, we need a method for using sample data to determine whether H_0 should be rejected. A decision rule used for this purpose is called a **test procedure.** Just as a jury may reach the wrong verdict in a trial, there is some chance that the use of a test procedure may result in an erroneous conclusion. One incorrect conclusion in a judicial setting is for a jury to convict an innocent person, and another is for a guilty person to be set free. Similarly, there are two possible errors to consider when developing a test procedure.

DEFINITIONS

> A **type I error** is the error of rejecting H_0 when H_0 is actually true.
> A **type II error** consists of *not* rejecting H_0 when H_0 is false.

No reasonable test procedure can guarantee complete protection against either type of error; this is the price we pay for basing our inference on sample data.

Example 8.3

Suppose you have to purchase tires for your vehicle and have narrowed your choice to a certain name-brand tire and another tire sold only through a particular chain of stores. The name-brand tire is more expensive to purchase than the store-brand tire, but the extra expense would be justified if the lifetime of the former significantly exceeded that of the latter. Let μ_1 denote true average tire lifetime for the brand-name tire under specified testing conditions, and let μ_2 denote true average lifetime for the store-brand tire under these conditions. You have decided that the extra expense can be justified only if μ_1 exceeds μ_2 by more than 10,000 miles, and you want to see persuasive evidence before incurring this extra expense. The natural choice of hypotheses is then

$$H_0: \mu_1 - \mu_2 = 10{,}000$$
$$H_a: \mu_1 - \mu_2 > 10{,}000$$

A type I error here involves rejecting H_0 and purchasing the name-brand tire when its true average mileage does not exceed that of the store-brand tire by more than 10,000 miles. A type II error consists of not rejecting H_0 and purchasing the less expensive tire when the true average lifetime of the name-brand tire actually does exceed that of the store brand by more than 10,000 miles.

Recall that when sampling a population or a process, sampling variability will virtually always be present. In particular, the value of a sample mean $\bar{x}$ may be rather different from the value of μ. In the tire situation, even if $\mu_1 - \mu_2$ does equal 10,000, the name-brand tires in the sample may be unusually good and the store-brand sample unusually bad, yielding data for which H_0 should be rejected. On the other hand, perhaps $\mu_1 - \mu_2 = 12{,}000$, so H_0 is false; yet there is some chance that the store-brand sample would be unusually good and the name-brand sample not so impressive, suggesting that H_0 should not be rejected.

If a test procedure cannot offer guaranteed protection against committing either a type I error or a type II error, we would at least like the chance of making either type of error to be small.

DEFINITION

The probability of making a type I error is denoted by α and is called the **level of significance** or **significance level** of the test. Thus a test with $\alpha = .01$ is said to have a significance level of .01. This means that if H_0 is actually true and the test procedure is used repeatedly on different samples selected from the population or process, in the long run H_0 would be incorrectly rejected only 1% of the time. The probability of a type II error is denoted by β.

The ideal of $\alpha = 0$ and $\beta = 0$ cannot be achieved as long as a conclusion is to be based on sample data. The test procedures used in practice allow the user to specify the significance level α to be employed in the test. So why would someone ever select a significance level like .10 or .05 when a smaller significance level such as .01 can also be employed? Why not always select a very small value for α? The answer is that the two error probabilities are inversely related to one another. Changing the test procedure to obtain a smaller probability of making a type I error inevitably makes it more likely that a type II error will be committed if H_0 happens to be false (just as changing the rules of evidence to make it less likely that an innocent person will be convicted also makes it more likely that a guilty person will go free). If a type I error is much more serious than a type II error, a very small value of α is reasonable. When a type II error could have quite unpleasant consequences, it is better to use a larger α to keep β under control. This leads to the following general principle for specifying a test procedure:

> After thinking about the relative consequences of type I and type II errors, decide on the largest α that is tolerable for the situation under consideration. Then employ a test procedure that uses this maximum acceptable value—rather than anything smaller—as the significance level (because using a smaller level would increase β). In following this principle, we are making β as small as possible subject to keeping a clamp on α.

Thus if you decide that $\alpha = .05$ is tolerable, you should not use a test with $\alpha = .01$ or $.001$, because doing so would inflate β. The significance levels used most frequently in practice are $.05$ and $.01$ (a 1-in-20 or 1-in-100 chance of rejecting H_0 when it is true), but the level that you decide to employ should reflect the seriousness of errors in your specific situation.

Test Statistics and *P*-Values

A test of hypotheses is carried out by employing what is called a **test statistic,** the function of the data that is computed and used to decide between H_0 and H_a. Suppose, for example, that μ is the true average flexural strength of concrete beams of a certain type. These beams will not be used in a certain application unless there is strong evidence that μ exceeds 600 psi. The appropriate hypotheses then are $H_0\colon \mu = 600$ versus $H_a\colon \mu > 600$. A sample of beams will be selected, and the strength determined for each one. Obviously the value of the sample mean $\bar{x}$ will provide information about the value of μ. Recall the following properties of the sampling distribution of $\bar{x}$:

$$\mu_{\bar{x}} = \mu \quad \text{(the sampling distribution is centered at } \mu\text{)}$$

When n is large, $\bar{x}$ has approximately a normal sampling distribution (the Central Limit Theorem) with standard error

$$\sigma_{\bar{x}} = \frac{\sigma}{\sqrt{n}} \quad \text{estimated by } \frac{s}{\sqrt{n}}$$

in which case the standardized variables

$$z = \frac{\bar{x} - \mu}{\sigma/\sqrt{n}} \qquad z = \frac{\bar{x} - \mu}{s/\sqrt{n}}$$

both have an approximately standard normal distribution (the z curve).

When H_0 is true, $\mu_{\bar{x}} = 600$, whereas when H_0 is false, we expect $\bar{x}$ to exceed 600. The difference $\bar{x} - 600$ is the distance between the sample mean and what we expect it to be when H_0 is true. Consider the test statistic

$$z = \frac{\bar{x} - 600}{s/\sqrt{n}}$$

The division by $s/\sqrt{n}$ expresses the distance as some number of (estimated) standard deviations of $\bar{x}$. If, for example, $z = 3.0$, then the observed $\bar{x}$ value is 3 standard deviations larger than what would be expected were H_0 true—a result not very consistent with H_0. A z value of $.5$ results from an $\bar{x}$ value that is only half a standard deviation larger than what is expected when the null hypothesis is true; this distance is not at all contradictory to H_0.

Having decided on a test statistic and calculated its value for the given sample, we now ask the following key question: If H_0 is true, how likely is it that a test statistic value at least as contradictory to H_0 as the one obtained would result? If the likelihood of this is very small, then the test statistic value is quite extreme relative to what the null hypothesis suggests and very contradictory to H_0. On the other hand, if there is a large chance of a value at least this extreme occurring when H_0 is true, then what was observed is reasonably consistent with H_0.

DEFINITION

The **P-value**, or **observed significance level** (OSL), is the probability, calculated assuming H_0 is true, of obtaining a test statistic value at least as contradictory to H_0 as the value that actually resulted. The smaller the P-value, the more contradictory is the data to H_0. The null hypothesis should then be rejected if the P-value is sufficiently small. In particular, the following decision rule specifies a test with the desired significance level (type I error probability) α:

Reject H_0 if P-value $\le \alpha$.

Do not reject H_0 if P-value $> \alpha$.

Example 8.4

The recommended daily dietary allowance (RDA) for zinc among males older than 50 years is 15 mg/day (*World Almanac*, 1992). The article "Nutrient Intakes and Dietary Patterns of Older Americans: A National Study" (*J. of Gerontology*, 1992: M145–M150) reported the following data on zinc intake for a sample of males age 65–74 years:

$$n = 115 \qquad \bar{x} = 11.3 \qquad s = 6.43$$

Does this data suggest that μ, the average daily zinc intake for the entire population of males age 65–74, is less than the RDA? The relevant hypotheses are

$$H_0: \mu = 15$$
$$H_a: \mu < 15$$

Figure 8.1 shows a boxplot of data consistent with the given summary quantities. Roughly 75% of the sample observations are smaller than 15 (the top edge of the box is at the upper quartile). Furthermore, the observed $\bar{x}$ value, 11.3, is certainly smaller than 15, but this could be just the result of sampling variability when H_0 is true. Is it plausible that a sample mean this much smaller than what was expected if H_0 were true occurred as a result of chance variation, or is $\mu < 15$ a better explanation for what was observed?

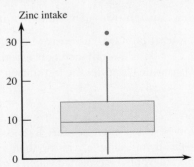

Figure 8.1 Boxplot for zinc intake data

The appropriate test statistic for testing the stated hypotheses is

$$z = \frac{\bar{x} - 15}{s/\sqrt{n}}$$

Because n is large here, when H_0 is true z has approximately a standard normal distribution (because z was formed by standardizing $\bar{x}$ using 15, the mean value of $\bar{x}$ under H_0). This implies that the P-value will be a z-curve area. The test statistic value is

$$z = \frac{\bar{x} - 15}{s/\sqrt{n}} = \frac{11.3 - 15}{6.43/\sqrt{115}} = \frac{-3.7}{.600} = -6.17$$

Values of z at least as contradictory to H_0 as this are those even smaller than -6.17 (those resulting from $\bar{x}$ values that are even farther below 15 than 11.3). Thus

P-value $= P(z < -6.17$ when H_0 is true$)$

$\qquad$ = area under the standard normal (z) curve to the left of -6.17

$\qquad \approx 0$

There is virtually no chance of seeing a z value this extreme as a result of chance variation alone when H_0 is true. If a significance level of .01 is used, then

$$P\text{-value} \approx 0 \le .01 = \alpha$$

so the null hypothesis should be rejected. Because the P-value is so small, the null hypothesis would in fact be rejected at *any* reasonable significance level, even .001 or smaller. The data is much more consistent with the conclusion that true average intake is in fact smaller than the RDA.

In Example 8.4, given that the alternative hypothesis asserted $\mu < 15$, it might seem reasonable to state H_0 as $\mu \ge 15$, previously referred to as the *implicit* null hypothesis. However, our null hypothesis is explicitly stated as a claim of equality ($H_0: \mu = 15$). On page 355 we asserted the conclusion using $H_0: \mu = 15$ versus $H_0: \mu < 15$ would be identical to that when considering $H_0: \mu \ge 15$ versus $H_a: \mu < 15$. Let us see why this is the case.

In the previous example, we tested $H_0: \mu = 15$ versus $H_a: \mu < 15$ and rejected H_0 in favor of H_a. Thus, we believe that $\mu < 15$ is a much more plausible assertion than $\mu = 15$. It follows logically that we would also believe that $\mu < 15$ is a much more plausible than the claim that $\mu = 16$, or the claim that $\mu = 17$, and so on. In other words, when we reject $H_0: \mu = 15$ in favor of $H_a: \mu < 15$, we are also *implicitly* saying that $\mu < 15$ is much more plausible than any value of μ that exceeds 15. This is why explicit consideration of the null hypothesis with a claim of equality is equivalent to considering the more realistic H_0 that includes an appropriate inequality.

Let μ_0 denote the value of μ asserted by the null hypothesis ($\mu_0 = 15$ in Example 8.4). The test statistic for testing hypotheses about μ when the sample size n is large is

$$z = \frac{\bar{x} - \mu_0}{s/\sqrt{n}}$$

When H_0 is true, this test statistic will have approximately a standard normal distribution (this will be true for *any* test statistic labeled z in this book). The P-value is then a z-curve area that depends on the inequality in H_0:

Inequality in H_0	P-value	Type of test
$>$	Area to the right of the calculated z	Upper-tailed
$<$	Area to the left of the calculated z	Lower-tailed
$\neq$	$2 \cdot$ (tail area captured by calculated z)	Two-tailed

These three cases are illustrated in Figure 8.2.

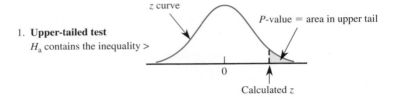

1. **Upper-tailed test**
   H_a contains the inequality $>$

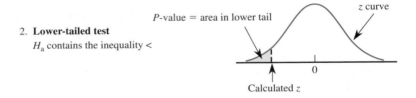

2. **Lower-tailed test**
   H_a contains the inequality $<$

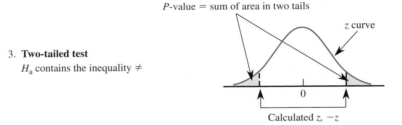

3. **Two-tailed test**
   H_a contains the inequality $\neq$

Figure 8.2 Determination of the P-value when the test statistic is z

As an example of the latter case, suppose that we are testing

$$H_0: \mu = .5 \quad \text{versus} \quad H_a: \mu \neq .5$$

where μ denotes true average bearing diameter. The large-sample test statistic is

$$z = \frac{\bar{x} - .5}{s/\sqrt{n}}$$

In this situation, values of $\bar{x}$ either much larger or much smaller than .5, corresponding to z values far from zero in *either* direction, are inconsistent with H_0 and give support to H_a. If, for example, $z = -2.76$, then

P-value = P(observing a z value at least as contradictory to H_0 as -2.76
   when H_0 is true)

$$= P(\text{either } z \leq -2.76 \text{ or } z \geq 2.76 \text{ when } z \text{ has approximately}$$
$$\text{a standard normal distribution})$$

$$= (\text{area under } z \text{ curve to the left of } -2.76)$$
$$+ (\text{area under } z \text{ curve to the right of } 2.76)$$

$$= 2(\text{area under } z \text{ curve to the left of } -2.76)$$

$$= 2(.0029) = .0058$$

The P-value would also be .0058 if $z = 2.76$. Using a significance level of .05, H_0 would be rejected because P-value $\leq \alpha$.

Section 8.1 Exercises

1. State whether each of the following assertions is a legitimate statistical hypothesis and why:
   a. $H: \sigma > 100$
   b. $H: \bar{x} = 45$
   c. $H: \tilde{\mu} \neq 2.0$
   d. $H: s \leq .50$
   e. $H: \sigma_1/\sigma_2 < 1$
   f. $H: \bar{x}_1 - \bar{x}_2 = -5.0$
   g. $H: \lambda < .01$, where λ is the parameter of an exponential distribution used to model component lifetime
   h. $H: \pi = .10$, where π is the population proportion of components that need warranty service
   i. $H: x =$ sound intensity of a certain source (decibels) has a lognormal distribution
   j. $H: x =$ rupture strength of a certain material $(10,000 \text{ N/cm}^2)$ has a Weibull distribution with $\alpha = 8$ and $\beta = 50$

2. To decide whether the pipe welds in a nuclear power plant meet specifications, a random sample of welds is to be selected and the strength of each weld (force required to break the weld) determined. Suppose a population mean strength of 100 lb/in^2 is the dividing line between welds meeting specification or not doing so. Explain why it might be better to test the hypotheses $H_0: \mu = 100$ versus $H_a: \mu > 100$ rather than $H_0: \mu = 100$ versus $H_a: \mu < 100$.

3. Many older homes have electrical systems that use fuses rather than circuit breakers. A manufacturer of 40-amp fuses wants to make sure that the true average amperage at which its fuses burn out is indeed 40. If the average amperage is lower than 40, purchasers will complain because the fuses will have to be replaced too frequently, whereas if the average exceeds 40, the manufacturer might be liable for damage to an electrical system due to fuse malfunction. After obtaining data from a sample of fuses, what null and alternative hypotheses would be of interest to the manufacturer?

4. Before agreeing to purchase a large order of polyethylene sheaths for a particular type of high-pressure, oil-filled submarine power cable, a company wants to see conclusive evidence that the population standard deviation of sheath thickness is less than .05 mm. What hypotheses should be tested, and why? In this context, what are the type I and type II errors?

5. A new design for the braking system on a certain type of car has been proposed. For the current system, the true average braking distance at 40 mph under specified conditions is known to be 120 ft. It is proposed that the new design be implemented only if sample data strongly indicates a reduction in true average braking distance for the new design. State the relevant hypotheses, and describe the type I and type II errors in the context of this situation.

6. A mixture of pulverized fuel ash and Portland cement to be used for grouting should have a true average compressive strength of more than 1300 KN/m^2. The mixture will not be used unless experimental evidence indicates conclusively that the strength specification has been met. State the relevant hypotheses, and describe the type I and type II errors in the context of this problem.

7. A regular type of laminate is currently being used by a manufacturer of circuit boards. A special laminate

has been developed in an attempt to reduce warpage. The regular laminate will be used on one sample of specimens and the special laminate on another sample; the amount of warpage will then be determined for each specimen. The manufacturer will then switch to the special laminate only if it can be demonstrated that the true average amount of warpage for that laminate is less than for the regular laminate. State the relevant hypotheses, and describe the type I and type II errors in the context of this situation.

8. a. Use the definition of a *P*-value to explain why H_0 would certainly be rejected if *P*-value = .0003.
   b. Use the definition of a *P*-value to explain why H_0 would definitely not be rejected if *P*-value = .350.

9. For which of the given *P*-values will the null hypothesis be rejected when using a test with a significance level of .05?
   a. .001   b. .021   c. .078
   d. .047   e. .156

10. For each of the given pairs of *P*-values and significance levels, state whether H_0 should be rejected.
    a. *P*-value = .084, α = .05
    b. *P*-value = .003, α = .001
    c. *P*-value = .048, α = .05
    d. *P*-value = .084, α = .10
    e. *P*-value = .039, α = .01
    f. *P*-value = .017, α = .10

11. Let μ denote the true average reaction time to a certain stimulus. A test of $H_0: \mu = 5$ versus $H_a: \mu > 5$ will be based on a large sample size so that when H_0 is true, the test statistic $z = (\bar{x} - 5)/(s/\sqrt{n})$ has approximately a standard normal distribution (the *z* curve). Determine the value of *z* and the corresponding *P*-value in each of the following cases:
    a. $n = 50, \bar{x} = 5.23, s = .89$
    b. $n = 35, \bar{x} = 5.72, s = 1.01$
    c. $n = 40, \bar{x} = 5.35, s = 1.67$

12. Newly purchased automobile tires of a certain type are supposed to be filled to a pressure of 34 psi. Let μ denote the true average pressure. A test of $H_0: \mu = 34$ versus $H_a: \mu \neq 34$ will be based on a large sample of tires so that the test statistic $z = (\bar{x} - 34)/(s\sqrt{n})$ will

have approximately a standard normal distribution when H_0 is true. Determine the value of *z* and the *P*-value in each of the following cases:
   a. $n = 50, \bar{x} = 34.43, s = 1.06$
   b. $n = 50, \bar{x} = 33.57, s = 1.06$
   c. $n = 32, \bar{x} = 33.25, s = 1.89$
   d. $n = 36, \bar{x} = 34.66, s = 2.53$

13. It is specified that a certain type of iron should contain .85 gm of silicon per 100 gm of iron (.85%). The silicon content of each of 32 randomly selected iron specimens was determined, and the accompanying Minitab output resulted from a test of the appropriate hypotheses:

Variable	N	Mean	StDev	SE Mean	Z	P-Value
sil cont	32	0.8228	0.1894	0.0335	-0.81	0.42

   a. What hypotheses were tested?
   b. What conclusion would be reached for a significance level of .05, and why? Answer the same question for a significance level of .10.

14. Lightbulbs of a certain type are advertised as having an average lifetime of 750 hours. The price of these bulbs is very favorable, so a potential customer has decided to go ahead with a purchase arrangement unless it can be conclusively demonstrated that the true average lifetime is smaller than what is advertised. A random sample of 50 bulbs was selected, the lifetime of each bulb determined, and the appropriate hypotheses were tested using Minitab, resulting in the accompanying output:

Variable	N	Mean	StDev	SEMean	Z	P-Value
lifetime	50	738.44	38.20	5.40	-2.14	0.016

   a. How can you tell from the output that the alternative hypothesis was not $H_a: \mu > 750$?
   b. What conclusion would be appropriate for a significance level of .05? A significance level of .01? What significance level and conclusion would you recommend?

15. A sample of 40 speedometers of a particular type is selected, and each speedometer is calibrated for accuracy at 55 mph, resulting in a sample mean and sample standard deviation of 53.87 and 1.36, respectively. Does this data suggest that the true average reading when speed is 55 mph is in fact something other than 55? State the relevant hypotheses, calculate the value

of the appropriate z statistic, determine the P-value, and state the conclusion for a significance level of .01.

16. To obtain information on the corrosion-resistance properties of a certain type of steel conduit, 35 specimens are buried in soil for an extended period. The maximum penetration (in mils) is then measured for each specimen, yielding a sample mean penetration of 52.7 and a sample standard deviation of 4.8. The conduits were manufactured with the specification that true average penetration be at most 50 mils. Does the sample data indicate that specifications have not been met? State the relevant hypotheses, calculate the value of the appropriate z statistic, determine the P-value, and state the conclusion for a significance level of .05.

17. Automatic identification of the boundaries of significant structures within a medical image is an area of ongoing research. The article "Automatic Segmentation of Medical Images Using Image Registration: Diagnostic and Simulation Applications" (*J. of Medical Engr. and Tech.*, 2005: 53–63) discussed a new technique for such identification.

A measure of the accuracy of the automatic region is the average linear displacement (ALD). The paper gave the following ALD observations for a sample of 49 kidneys (units of pixel dimensions).

1.38	0.44	1.09	0.75	0.66	1.28	0.51
0.39	0.70	0.46	0.54	0.83	0.58	0.64
1.30	0.57	0.43	0.62	1.00	1.05	0.82
1.10	0.65	0.99	0.56	0.56	0.64	0.45
0.82	1.06	0.41	0.58	0.66	0.54	0.83
0.59	0.51	1.04	0.85	0.45	0.52	0.58
1.11	0.34	1.25	0.38	1.44	1.28	0.51

a. Summarize and describe the data.
b. Is it plausible that ALD is at least approximately normally distributed? Must normality be assumed prior to testing hypotheses about true average ALD? Explain.
c. The authors commented that in most cases the ALD is better than or on the order of 1.0. Does the data in fact provide strong evidence for concluding that true average ALD under these circumstances is less than 1.0? Carry out an appropriate test of hypotheses.

8.2 TESTS CONCERNING HYPOTHESES ABOUT MEANS _____

In this section, we consider hypotheses either about a single population or process mean μ or about a difference $\mu_1 - \mu_2$ between two such means. Our test procedures will utilize test statistics that have either exactly or approximately a t distribution when the null hypothesis H_0 is true. This implies that the P-value for the test—the probability, calculated assuming that H_0 is true, of observing a test statistic value at least as contradictory to the null hypothesis as what was obtained—will be a t-curve tail area of some sort. The particular tail area that is relevant depends on whether the alternative hypothesis H_a contains an inequality of the form $>$, $<$, or $\neq$.

P-Values for t Tests

Inequality in H_a	Type of test	Determination of the P-value
$>$	Upper-tailed	Area under the relevant t curve to the *right* of the calculated t
$<$	Lower-tailed	Area under the relevant t curve to the *left* of the calculated t
$\neq$	Two-tailed	Twice the tail area captured by the calculated t under the relevant t curve

By the "relevant" t curve, we mean the one having the appropriate number of df. The three cases are illustrated in Figure 8.3.

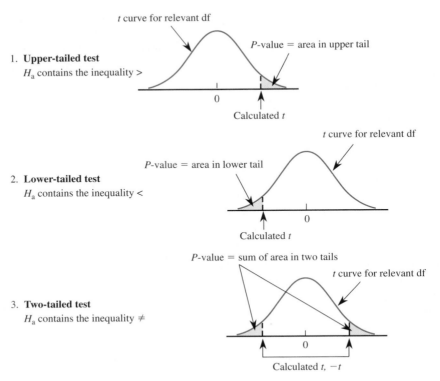

Figure 8.3 *P*-values for *t* tests: (1) upper-tailed; (2) lower-tailed; (3) two-tailed

Appendix Table VI contains a tabulation of *t*-curve upper-tail areas. Each different column of the table is for a different number of df, and the rows are for calculated values of the test statistic *t* ranging from 0.0 to 4.0 in increments of .1. For example, the number .074 appears at the intersection of the 1.6 row and the 8 df column, so the area under the 8 df curve to the right of 1.6 (an upper-tail area) is .074. Because *t* curves are symmetric, .074 is also the area under the 8 df curve to the left of −1.6 (a lower-tail area).

Suppose, for example, that a test of $H_0: \mu = 100$ versus $H_a: \mu > 100$ is based on the 8 df *t* distribution. If the calculated value of the test statistic is *t* = 1.6, then the *P*-value for this upper-tailed test is .074. Because .074 exceeds .05, we would not be able to reject H_0 at significance level .05. If the alternative hypothesis is $H_a: \mu < 100$ and a test based on 20 df yields *t* = −3.2, then Appendix Table VI shows that the *P*-value is the captured lower-tail area .002. The null hypothesis can be rejected at either level .05 or .01. Consider testing $H_0: \mu_1 - \mu_2 = 0$ versus $H_a: \mu_1 - \mu_2 \neq 0$; the null hypothesis states that the means of the two populations are identical, whereas the alternative hypothesis states that they are different without specifying a direction of departure from H_0. If the test is based on 20 df and *t* = 3.2, then the *P*-value for this two-tailed test is 2(.002) = .004. This would also be the *P*-value for *t* = −3.2. The tail area is doubled because values both larger than 3.2 and smaller than −3.2 are more contradictory to H_0 than what was calculated (values farther out in *either* tail of the *t* curve). Notice that if the calculated value of t exceeds 4.0, for all but very small df's the captured tail area is negligible. Also note that the table jumps from 40 df to 60 df to 120 df to ∞ (the z or standard normal curve). For example, for 45 df, one could either interpolate between 40 df and 60 df, or use the z-curve area as an approximation.

Tests Concerning a Single Mean

Consider testing hypotheses about the mean μ of a single population or process. The null hypothesis will be a statement of equality, such as H_0: $\mu = 100$. The alternative hypothesis H_a will contain one of three possible inequalities. A general description of the test procedures necessitates using a symbol to denote the value of μ asserted to be true by the null hypothesis. We use μ_0 to denote this **null value.** Thus the general form of the null hypothesis will be H_0: $\mu = \mu_0$, and the contradictory claim will be H_a: $\mu > \mu_0$, H_a: $\mu < \mu_0$, or H_a: $\mu \neq \mu_0$.

Suppose that the sample $x_1, \ldots, x_n$ has been randomly selected from a normal population or process distribution (recall from Chapter 2 that the plausibility of this can be checked by constructing a normal quantile plot). Then, as discussed in the development of a confidence interval for μ, the standardized variable

$$t = \frac{\bar{x} - \mu}{s/\sqrt{n}}$$

has a t distribution with $n - 1$ degrees of freedom. Our test statistic results from replacing μ by the null value μ_0. For H_0: $\mu = 100$, this gives the test statistic

$$t = \frac{\bar{x} - 100}{s/\sqrt{n}}$$

The key result is that *when the null hypothesis is true*, the test statistic has a t distribution based on $n - 1$ df; this is what justifies computing the P-value as described at the beginning of this section.

The One-Sample t Test

Null hypothesis: H_0: $\mu = \mu_0$

Test statistic: $t = \dfrac{\bar{x} - \mu_0}{s/\sqrt{n}}$

P-value: Calculated by reference to the t curve for $n - 1$ df. The test is upper-tailed when the alternative hypothesis is H_a: $\mu > \mu_0$, lower-tailed in the case H_a: $\mu < \mu_0$, and two-tailed if the alternative is H_a: $\mu \neq \mu_0$.

Assumption: $x_1, x_2, \ldots, x_n$ is a random sample from a normal population or process distribution. If n is large (usually $n > 30$ suffices), this normality assumption is no longer necessary, because the Central Limit Theorem guarantees that the $\bar{x}$ sampling distribution is approximately normal whatever the shape of the population or process distribution. The test statistic can then be denoted by z rather than t, and the P-value is obtained from the z (standard normal) curve.

Example 8.5   Glycerol is a major by-product of ethanol fermentation in wine production and contributes to the sweetness, body, and fullness of wines. The article "A Rapid and Simple Method for Simultaneous Determination of Glycerol, Fructose, and Glucose in Wine" (*American J. of Enology and Viticulture*, 2007: 279–283) includes the following observations on glycerol concentration (mg/mL) for samples of standard-quality (uncertified) white wines: 2.67, 4.62, 4.14, 3.81, 3.83. Suppose the desired concentration value is 4. Does the sample data suggest that true average concentration is something other than the desired value? The normal quantile plot in Figure 8.4 provides strong support for assuming that the population distribution of glycerol concentration is normal. Let's carry out a test of appropriate hypotheses using the one-sample *t* test with a significance level of .05.

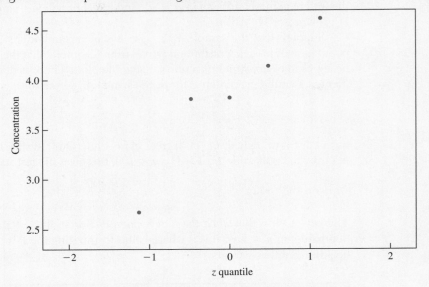

Figure 8.4 Normal quantile plot for the data of Example 8.5

Our analysis employs a sequence of steps that we advocate using for any hypothesis-testing investigation:

1. Parameter of interest: μ = true average glycerol concentration
2. Null hypothesis: $H_0: \mu = 4$
3. Alternative hypothesis: $H_a: \mu \neq 4$
4. Test statistic formula: $t = \frac{\bar{x} - 4}{s/\sqrt{n}}$ (do not substitute sample quantities yet)
5. Computation of test statistic value: $\bar{x} = 3.814$, $s = .718$, and

$$t = \frac{3.814 - 4}{.718/\sqrt{5}} = -.58 \approx -.6$$

6. Determination of the *P*-value: The test is based on $n - 1 = 4$ df. Appendix Table VI shows that the area under the 4 df curve to the right of .6 is .290.

Therefore the area under the 4 df curve to the left of $-.6$ is .290. Because the test is two-tailed, P-value $= 2(.290) = .580$.

7. Conclusion: The specified significance level is $\alpha = .05$. Since P-value $= .580 > .05 = \alpha$, we cannot reject H_0 at this (or any other reasonable) significance level. The data does not provide strong evidence for concluding that population mean glycerol concentration differs from 4. Notice that in not rejecting H_0, we may be committing a type II error (not rejecting the null hypothesis when it is false); we hope, though, we came to this conclusion for the right reason!

The R output from a request to carry out the test follows. The P-value differs slightly from ours because R uses more decimal accuracy in computing t. Thus, if H_0 were true, about 59% of all samples would yield a value of t more extreme than what we obtained. We decided not to reject H_0 because $-.58$ is not in the most extreme 5% of all t values.

```
One Sample t-test
data: concentration
t = -0.5789, df = 4, p-value = 0.5937
alternative hypothesis: true mean is not equal to 4
95 percent confidence interval: 2.921875 4.706125
sample estimates: mean of x      3.814
```

Suppose the sample size in Example 8.5 had been 45 rather than 5, with the same values of $\bar{x}$ and s. The normality assumption for glycerol concentration becomes unnecessary. The test statistic would be labeled z, and its value would be $z = -1.74$. Appendix Table I shows that the area under the z curve to the left of -1.74 is .0409, so the P-value is $2(.0409) = .0818$ and H_0 would be rejected at level .10 but not at levels .05 or .01.

Tests Concerning a Difference Between Two Means: Independent Samples

Hypothesis testing often is used as a basis for comparing two populations, processes, or treatments. For example, data might be collected to decide whether population mean fuel efficiency for a particular compact car exceeds that for a certain midsize car by more than 4 miles per gallon. Alternatively, two coatings for retarding corrosion might be available for treating a certain type of pipe. An experiment might then be carried out to decide whether the true average amount of corrosion when the first coating is used differs from the true average amount when the second coating is used; the two coatings are the treatments being studied. The same notation for the two population, process, or treatment means employed in connection with confidence intervals in the previous chapters will be used here:

μ_1 = mean of population or process 1, or the true average response when treatment 1 is applied

μ_2 = mean of population or process 2, or the true average response when treatment 2 is applied

Inferences about the value of μ_1 relative to μ_2 are based on two independently obtained random samples, one from the first population, process, or treatment and the other from the second. Let

n_1 = number of observations in the first sample

$\bar{x}_1$ = sample mean of these n_1 observations

s_1^2 = sample variance of these n_1 observations

and n_2, $\bar{x}_2$, and s_2^2 are defined analogously with respect to the second sample. Assume that both population, process, or treatment response distributions are normal. A confidence interval for the difference $\mu_1 - \mu_2$ was based on the fact that the standardized variable

$$t = \frac{\bar{x}_1 - \bar{x}_2 - (\mu_1 - \mu_2)}{\sqrt{\dfrac{s_1^2}{n_1} + \dfrac{s_2^2}{n_2}}}$$

has approximately a t distribution. Suppose the null hypothesis is $H_0: \mu_1 - \mu_2 = 4$ (i.e., the value of μ_1 is 4 larger than the value of μ_2). A test statistic results from replacing $\mu_1 - \mu_2$ in the numerator of t by the null value 4. The test statistic then has approximately a t distribution when the null hypothesis is true. The test will be upper-tailed if the alternative hypothesis is $H_a: \mu_1 - \mu_2 > 4$, lower-tailed if the alternative contains the inequality $<$, and two-tailed if $\neq$ appears in H_a.

A general description of the test procedure requires the use of a symbol for the null value; we use the Greek letter Δ for that purpose. Most frequently, in practice, $\Delta = 0$, in which case the null hypothesis says there is no difference between the two μ's.

The Two-Sample t Test

Null hypothesis: $H_0: \mu_1 - \mu_2 = \Delta$ (Δ denotes the null value, a number appropriate to the problem situation under consideration)

Test statistic: $t = \dfrac{\bar{x}_1 - \bar{x}_2 - \Delta}{\sqrt{\dfrac{s_1^2}{n_1} + \dfrac{s_2^2}{n_2}}}$

P-value: When H_0 is true, the test statistic has approximately a t distribution with

$$df = \frac{\left[(se_1)^2 + (se_2)^2\right]^2}{\dfrac{(se_1)^4}{n_1 - 1} + \dfrac{(se_2)^4}{n_2 - 1}}$$

where $se = s/\sqrt{n}$ (df should be rounded <u>down</u> to the nearest whole number). The P-value should then be calculated by reference to the corresponding t curve according to whether the test is upper-, lower-, or two-tailed.

Assumptions: The two random samples are selected independently, both from underlying normal population, process, or treatment response distributions. If the sample sizes are large (usually both $n_1 > 30$ and $n_2 > 30$ will suffice), the Central Limit Theorem implies that the normality assumption is no longer necessary. In this case, the test statistic can be denoted by z, and the *P*-value calculated by reference to the z curve.

deterioration of many municipal pipeline networks across the country is a grow-
concern. One technology proposed for pipeline rehabilitation uses a flexible lin-
hreaded through existing pipe. The article "Effect of Welding on a High-Density
yethylene Liner" (*J. of Materials in Civil Engr.*, 1996: 94–100) reported the fol-
/ing data on tensile strength (psi) of liner specimens both when a certain fusion
ocess was used and when this process was not used:

1. No fusion: 2748   2700   2655   2822   2511
          3149   3257   3213   3220   2753

$$n_1 = 10 \qquad \bar{x}_1 = 2902.8 \qquad s_1 = 277.3 \qquad se_1 = 87.69$$

2. Fused: 3027   3356   3359   3297   3125   2910   2889   2902

$$n_2 = 8 \qquad \bar{x}_2 = 3108.1 \qquad s_2 = 205.9 \qquad se_2 = 72.80$$

Figure 8.5 shows *normal probability plots* from Minitab. These plots employ a
probability scale rather than the normal quantiles discussed previously, but the critical issue is the same: Is the pattern of plotted points reasonably close to linear? There
certainly is some wiggling in these plots, but not enough to suggest that the normality assumption is implausible. Furthermore, the *P*-values that appear along with the
plots are for formal tests of the assertion that the underlying distributions are normal
(we discuss this test in Section 8.4). Because each *P*-value exceeds .1, the hypothesis
of normality cannot be rejected.

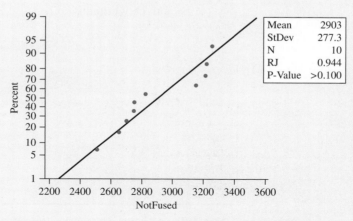

Figure 8.5 Normal probability plots from Minitab of the tensile
strength data

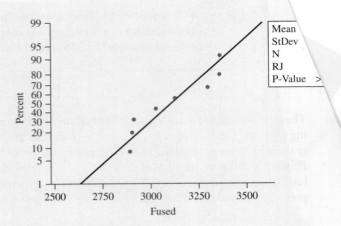

Figure 8.5 (*Continued*)

The authors of the article stated that the fusion process increased the average tensile strength. The message from the comparative boxplot of Figure 8.6 is not all that clear. Let's carry out a test of hypotheses to see whether the data supports this conclusion.

1. Let μ_1 be the true average tensile strength of specimens when the no-fusion treatment is used and μ_2 denote the true average tensile strength when the fusion treatment is used.

2. $H_0: \mu_1 - \mu_2 = 0$ (no difference in the true average tensile strengths for the two treatments)

3. $H_a: \mu_1 - \mu_2 < 0$ (true average tensile strength for the no-fusion treatment is less than that for the fusion treatment, so the investigators' conclusion is correct)

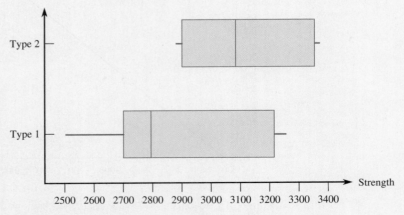

Figure 8.6 A comparative boxplot of the tensile strength data

4. The null value is $\Delta = 0$, so the test statistic is

$$t = \frac{\bar{x}_1 - \bar{x}_2}{\sqrt{\dfrac{s_1^2}{n_1} + \dfrac{s_2^2}{n_2}}}$$

5. We now compute both the test statistic value and the df for the test:

$$t = \frac{2902.8 - 3108.1}{\sqrt{\dfrac{(277.3)^2}{10} + \dfrac{(205.9)^2}{8}}} = \frac{-205.3}{113.97} = -1.8$$

$$\mathrm{df} = \frac{[(87.69)^2 + (72.80)^2]^2}{(87.69)^4/9 + (72.80)^4/7} = 15.94$$

so the test will be based on 15 df.

6. Appendix Table VI shows that the area under the 15 df t curve to the right of 1.8 is .046, so the P-value for a lower-tailed test is also .046. The following Minitab output summarizes all the computations:

```
Twosample T for nofusion vs fused

            N     Mean     StDev  SE Mean
nofusion   10     2903       277       88
fused       8     3108       206       73

95% C.I. for mu nofusion-mu fused: (-448, 38)
T-Test mu nofusion = mu fused (vs <): T= - 1.80 P = 0.046 DF=15
```

7. Using a significance level of .05, we can barely reject the null hypothesis in favor of the alternative hypothesis, confirming the conclusion stated in the article. However, someone demanding more compelling evidence might select $\alpha = .01$, a level for which H_0 cannot be rejected.

Suppose the issue in Example 8.6 had been whether fusing increased true average strength by more than 100 psi. Then the relevant hypotheses would have been $H_0: \mu_1 - \mu_2 = -100$ versus $H_a: \mu_1 - \mu_2 < -100$; that is, the null value would have been $\Delta = -100$.

Tests Concerning a Difference Between Two Means: Paired Data

A comparison of two population, process, or treatment means is often carried out by collecting data in pairs. Suppose, for example, that two different fertilizer formulations are being compared with respect to crop yield. Variation in soil characteristics, amount of precipitation, amount of sunshine, and various other factors can affect yield. To protect against this extraneous variation, an investigator could select pairs of plots (the experimental units) so that within each pair the two plots are as similar as possible with respect to any characteristics that might have a bearing on yield. Then the first fertilizer could be applied to one plot within each pair and the second formulation used on the other plot. This pairing is really a special case of blocking, as discussed in Chapter 4. The homogeneity of experimental units within each block (pair) makes it easier to detect a difference between the treatments if a difference actually exists.

Again, let μ_1 and μ_2 denote the two population, process, or treatment response means. The pairs in a sample can be viewed as having been selected from a much larger population of pairs. Now conceptualize subtracting the second number in each such pair from the first number to obtain a population of differences. If we let μ_d denote the population mean difference, it follows that

$$\mu_d = \mu_1 - \mu_2$$

This relationship implies that any hypothesis about $\mu_1 - \mu_2$ is equivalent to a hypothesis about μ_d. For example, the assertion that $\mu_1 - \mu_2 = 10$ is the same as the claim $\mu_d = 10$. But hypotheses about μ_d can be tested by using the sample differences. In particular, assuming that the underlying distribution of differences is normal, we can use a one-sample t test based on these sample differences.

The Paired t Test

Null hypothesis $H_0: \mu_d = \Delta$ (equivalent to $\mu_1 - \mu_2 = \Delta$) where μ_d denotes the population mean difference

Test statistic: $t = \dfrac{\bar{d} - \Delta}{s_d/\sqrt{n}}$, where

n = number of sample differences (pairs)

$d_1, d_2, \ldots, d_n$ = these n sample differences

$\bar{d}$ = sample mean difference

s_d = sample standard deviation of the differences

P-value: Calculated from the t curve with $n - 1$ df as described previously. The test is upper-tailed, lower-tailed, or two-tailed, depending on whether the inequality in H_a is $>$, $<$, or $\neq$, respectively.

Assumptions: The sample differences $d_1, \ldots, d_n$ have been randomly selected from a difference population having a normal distribution. If n is large, the normality assumption is not necessary; the test statistic is labeled z, and the P-value is determined from the z curve.

Example 8.7 Musculoskeletal neck-and-shoulder disorders are all too common among office staff who perform repetitive tasks using visual display units. The article "Upper-Arm Elevation During Office Work" (*Ergonomics*, 1996: 1221–1230) reported on a study to determine whether more varied work conditions would have any impact on arm movement. The accompanying data was obtained from a sample of $n = 16$ subjects. Each observation is the amount of time, expressed as a proportion of total time observed, during which arm elevation was below 30°. The two measurements from each subject were obtained 18 months apart. During this period, work conditions were changed, and subjects were allowed to engage in a wider variety of work tasks. Does the data suggest that true average time during which elevation is below 30° differs after the change from what it was before the change?

Subject:	1	2	3	4	5	6	7	8
Before:	81	87	86	82	90	86	96	73
After:	78	91	78	78	84	67	92	70
Difference:	3	−4	8	4	6	19	4	3

Subject:	9	10	11	12	13	14	15	16
Before:	74	75	72	80	66	72	56	82
After:	58	62	70	58	66	60	65	73
Difference:	16	13	2	22	0	12	−9	9

Figure 8.7 shows a normal probability plot of the 16 differences; the pattern in the plot is quite straight, supporting the normality assumption. A boxplot of these differences appears in Figure 8.8; the boxplot is located considerably to the right of zero, suggesting that perhaps $\mu_d > 0$ (note also that 13 of the 16 differences are positive and only two are negative).

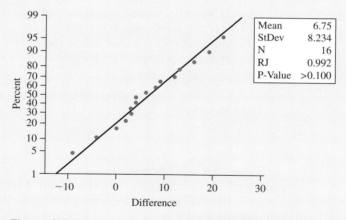

Figure 8.7 A normal probability plot from Minitab of the differences in Example 8.7

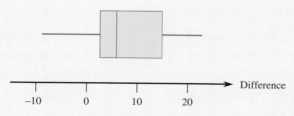

Figure 8.8 A boxplot of the differences in Example 8.7

Let's now use the recommended sequence of steps to test the appropriate hypotheses.

1. Let μ_d denote the true average difference between elevation time before the change in work conditions and time after the change.

2. $H_0: \mu_d = 0$ (there is no difference between true average time before the change and true average time after the change)

3. $H_a: \mu_d \neq 0$

4. $t = \dfrac{\overline{d} - 0}{s_d/\sqrt{n}} = \dfrac{\overline{d}}{s_d/\sqrt{n}}$

5. $n = 16$, $\Sigma d_i = 108$, $\Sigma d_i^2 = 1746$, from which $\overline{d} = 6.75$, $s_d = 8.234$, and

$$t = \frac{6.75}{8.234/\sqrt{16}} = 3.28 \approx 3.3$$

6. Appendix Table VI shows that the area to the right of 3.3 under the t curve with 15 df is .002. The inequality in H_a implies that a two-tailed test is appropriate, so the P-value is approximately $2(.002) = .004$ (Minitab gives .0051).

7. Since $.004 < .01$, the null hypothesis can be rejected at either significance level .05 or .01. It does appear that the true average difference between times is something other than zero; that is, true average time after the change is different from that before the change.

Suppose the question posed had been, Does it appear that the change in work conditions decreases true average time by more than 5? The relevant hypotheses would then be $H_0: \mu_d = 5$ versus $H_a: \mu_d > 5$, for which the test statistic is $t = (\overline{d} - 5)/(s_d/\sqrt{n})$.

In Section 8.4, we show how a test of the null hypothesis that a population distribution is normal can be based on a normal quantile or probability plot. In Section 8.5, we discuss several further aspects of hypothesis testing, including the determination of type II error probabilities for t tests.

Section 8.2 Exercises

18. Give as much information as you can about the P-value of a t test in each of the following situations:
    a. Upper-tailed test, df = 8, $t = 2.0$
    b. Lower-tailed test, df = 11, $t = -2.4$
    c. Two-tailed test, df = 15, $t = -1.6$
    d. Upper-tailed test, df = 19, $t = -.4$
    e. Upper-tailed test, df = 5, $t = 5.0$
    f. Two-tailed test, df = 40, $t = -4.8$

19. The paint used to make lines on roads must reflect enough light to be clearly visible at night. Let μ denote

the true average reflectometer reading for a new type of paint under consideration. A test of $H_0: \mu = 20$ versus $H_a: \mu > 20$ will be based on a random sample of size n from a normal population distribution. What conclusion is appropriate in each of the following situations?
    a. $n = 15$, $t = 3.2$, $\alpha = .05$
    b. $n = 9$, $t = 1.8$, $\alpha = .01$
    c. $n = 24$, $t = -.2$

20. A certain pen has been designed so that true average writing lifetime under controlled conditions

(involving the use of a writing machine) is at least 10 hours. A random sample of 18 pens is selected, the writing lifetime of each is determined, and a normal quantile plot of the resulting data supports the use of a one-sample t test.

a. What hypotheses should be tested if the investigators believe a priori that the design specification has been satisfied?

b. What conclusion is appropriate if the hypotheses of part (a) are tested, $t = -2.3$, and $\alpha = .05$?

c. What conclusion is appropriate if the hypotheses of part (a) are tested, $t = -1.8$, and $\alpha = .01$?

d. What should be concluded if the hypotheses of part (a) are tested and $t = -3.6$?

21. The true average diameter of ball bearings of a certain type is supposed to be .5 in. A one-sample t test will be carried out to see whether this is the case. What conclusion is appropriate in each of the following situations?

a. $n = 13, t = 1.6, \alpha = .05$

b. $n = 13, t = -1.6, \alpha = .05$

c. $n = 25, t = -2.6, \alpha = .01$

d. $n = 25, t = -3.9$

22. The article "The Foreman's View of Quality Control" (*Quality Engr.*, 1990: 257–280) described an investigation into the coating weights for large pipes resulting from a galvanized coating process. Production standards call for a true average weight of 200 lb per pipe. The accompanying descriptive summary and boxplot are from Minitab.

```
Variable N   Mean  Median TrMean StDev SEMean
ctg wt   30 206.73 206.00 206.81  6.35  1.16

Variable    Min    Max     Q1      Q3
ctg wt   193.00 218.00  202.75  212.00
```

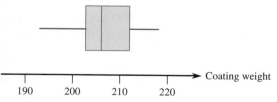

Coating weight

a. What does the boxplot suggest about the status of the specification for true average coating weight?

b. A normal quantile plot of the data was quite straight. Use the descriptive output to test the appropriate hypotheses.

23. Exercise 5 in Chapter 2 gave $n = 12$ observations on daily energy demand readings (kW h) for remote telecommunications stations throughout Cameroon, from which the sample mean and sample standard deviation are 32.59 and 10.66, respectively. Suppose the investigators had believed a priori that true average daily energy demand would be at most 30 kW h. Does the data contradict this prior belief? Assuming normality, test the appropriate hypotheses using a significance level of .05.

24. Reconsider the sample observations introduced in Exercise 15 in Chapter 2 on the required force (N) to cause initial cracks in a thin enclosure for a subdermally implanted biotelemetry device:

2006.1 2065.2 2118.9 1686.6 1966.9 1792.5

Suppose the device will not be used unless the true average required force to cause initial cracks exceeds 1800 N. Does this requirement appear to have been satisfied? State and test the appropriate hypotheses.

25. Poly(3-hydroxybutyrate) (PHB), a semicrystalline polymer that is fully biodegradable and biocompatible, is obtained from renewable resources. From a sustainability perspective, PHB offers many attractive properties though it is more expensive to produce than standard plastics. The authors of "The Melting Behaviour of Poly(3-Hydroxybutyrate) by DSC. Reproducibility Study" (*Polymer Testing*, 2013: 215–220) wanted to investigate various physical properties of PHB by using a differential scanning calorimeter (DSC).

For each of 12 PHB specimens, the authors used a DSC to measure the melting point (in °C) of the polymer, which is the temperature for 99% completion of the fusion process.

180.5 181.7 180.9 181.6 182.6 181.6
181.3 182.1 182.1 180.3 181.7 180.5

A normal probability plot of the data shows a reasonably linear pattern, so it is plausible that the population distribution of PHB melting points as measured by DSC is at least approximately normal. The sample mean and standard deviation are 181.4 and .7242, respectively. Is there compelling evidence for concluding that true average melting point exceeds 181°C? Carry out a test of hypotheses using a significance level of .05.

26. The relative conductivity of a semiconductor device is determined by the amount of impurity "doped" into the device during its manufacture. A silicon diode to be used for a specific purpose requires an average cut-on voltage of .60 V, and if this is not achieved, the amount of impurity must be adjusted. A sample of diodes was selected and the cut-on voltage was determined. The accompanying SAS out-put resulted from a request to test the appropriate hypotheses.

```
N      Mean      Std Dev         T   Prob>|T|
15  0.0453333  0.0899100  1.9527887    0.0711
```

(*Note*: SAS explicitly tests $H_0: \mu = 0$, so to test $H_0: \mu = .60$, the null value .60 must be subtracted from each x_i; the reported mean is then the average of the $(x_i - .60)$ values. Also, SAS's P-value is always for a two-tailed test.) What would be concluded for a significance level of .01? .05? .10?

27. Determine the number of degrees of freedom for the two-sample t test in each of the following situations:
    a. $n_1 = 10, n_2 = 10, s_1 = 5.0, s_2 = 6.0$
    b. $n_1 = 10, n_2 = 15, s_1 = 5.0, s_2 = 6.0$
    c. $n_1 = 10, n_2 = 15, s_1 = 2.0, s_2 = 6.0$
    d. $n_1 = 12, n_2 = 24, s_1 = 5.0, s_2 = 6.0$

28. Urban storm water can be contaminated by many sources, including discarded batteries. When ruptured, these batteries release metals of environmental significance. The article "Urban Battery Litter" (*J. of Environ. Engr.*, 2009: 46–57) presented summary data for characteristics of a variety of batteries found in urban areas around Cleveland.

Here are data on zinc mass (g) for two different brands of size D batteries:

Brand	Sample Size	Sample Mean	Sample SD
Duracell	15	138.52	7.76
Energizer	20	149.07	1.52

Assuming that both zinc mass distributions are at least approximately normal, carry out a test at significance level .05 to decide whether true average zinc mass is different for the two types of batteries.

29. Quantitative noninvasive techniques are needed for routinely assessing symptoms of peripheral neuropathies, such as carpal tunnel syndrome (CTS). The article "A Gap Detection Tactility Test for Sensory Deficits Associated with Carpal Tunnel Syndrome" (*Ergonomics*, 1995: 2588–2601) reported on a test that involved sensing a tiny gap in an otherwise smooth surface by probing with a finger; this functionally resembles many work-related tactile activities, such as detecting scratches or surface defects. When finger probing was not allowed, the sample average gap detection threshold for $n_1 = 8$ normal subjects was 1.71 mm, and the sample standard deviation was .53; for $n_2 = 10$ CTS subjects, the sample mean and sample standard deviation were 2.53 and .87, respectively. Does this data suggest that the true average gap detection threshold for CTS subjects exceeds that for normal subjects? State and test the relevant hypotheses using a significance level of .01.

30. According to the article "Fatigue Testing of Condoms" (*Polymer Testing*, 2009: 567–571), "tests currently used for condoms are surrogates for the challenges they face in use," including a test for holes, an inflation test, a package seal test, and tests of dimensions and lubricant quality. The investigators developed a new test that adds cyclic strain to a level well below breakage and determines the number of cycles to break.

The article reported that for a sample of 20 natural latex condoms of a certain type, the sample mean and sample standard deviation of the number of cycles to break were 4358 and 2218, respectively, whereas a sample of 20 polyisoprene condoms gave a sample mean and sample standard deviation of 5805 and 3990, respectively. Is there strong evidence for concluding that the true average number of cycles to break for the polyisoprene condom exceeds that for the natural latex condom by more than 1000 cycles? Carry out a test using a significance level of .01. (*Note*: The cited paper reported P-values of t tests for comparing means of the various types considered.)

31. Fusible interlinings are being used with increasing frequency to support outer fabrics and improve the shape and drape of various pieces of clothing. The article "Compatibility of Outer and Fusible Interlining Fabrics in Tailored Garments" (*Textile Res. J.*, 1997: 137–142) gave the accompanying data on extensibility (%) at 100 gm/cm for both high-quality fabric (H) and poor-quality fabric (P) specimens:

    H: 1.2 .9 .7 1.0 1.7 1.7 1.1 .9 1.7
        1.9 1.3 2.1 1.6 1.8 1.4 1.3 1.9 1.6
        .8 2.0 1.7 1.6 2.3 2.0

    P: 1.6 1.5 1.1 2.1 1.5 1.3 1.0 2.6

    a. Construct normal quantile plots to verify the plausibility of both samples having been selected from normal population distributions.
    b. Construct a comparative boxplot. Does it suggest that there is a difference between true average extensibility for high-quality fabric specimens and that for poor-quality specimens?
    c. The sample mean and standard deviation for the high-quality sample are 1.508 and .444, respectively, and those for the poor-quality sample are 1.588 and .530. Use the two-sample t test to decide whether true average extensibility differs for the two types of fabrics.

32. The article cited in Exercise 41 in Chapter 7 gave the following data on work of adhesion measurements (in mJ/m^2) for samples of ultra-high performance concrete adhered to two types of substrates:

Substrate	Observations
Steel:	107.1 109.5 107.4 106.8 108.1
Glass:	122.4 124.6 121.6 120.6 123.3

    Assuming that both samples were selected from normal distributions, carry out a test of hypotheses to decide whether the true average work of adhesion for the glass substrate is more than 12 mJ/m^2 higher than that for the steel substrate.

33. The article "The Influence of Corrosion Inhibitor and Surface Abrasion on the Failure of Aluminum-Wired Twist-on Connections" (*IEEE Trans. on Components, Hybrids, and Manuf. Tech.*, 1984: 20–25) reported data on potential drop measurements for one sample of connectors wired with alloy aluminum and another sample wired with EC aluminum. Does the accompanying SAS output suggest that the true average potential drop for alloy connections (type 1) is higher than that for EC connections (as stated in the article)? Carry out the appropriate test using a significance level of .01. In reaching your conclusion, what type of error might you have committed? *Note:* SAS reports the *P*-value for a two-tailed test.

Type	N	Mean	Std Dev	Std Error
1	20	17.4990	0.55012821	0.12301241
2	20	16.9000	0.48998389	0.10956373

Type	Variances	T	DF	Prob>|T|
1	Unequal	3.6362	37.5	0.0008
2	Equal	3.6362	38.0	0.0008

34. The article "Evaluation of a Ventilation Strategy to Prevent Barotrauma in Patients at High Risk for Acute Respiratory Distress Syndrome" (*New England J. of Medicine*, 1998: 355–358) reported on an experiment in which 120 patients with similar clinical features were randomly divided into a control group and a treatment group, each consisting of 60 of the patients. The sample mean ICU stay (days) and sample standard deviation for the treatment group were 19.9 and 39.1, respectively, whereas these values for the control group were 13.7 and 15.8.

    a. Calculate a point estimate for the difference between true average ICU stay for the treatment and control groups. Does this estimate suggest that there is a significant difference between true average stays under the two conditions?
    b. Answer the question posed in part (a) by carrying out a formal test of hypotheses. Is the result different from what you conjectured in part (a)?
    c. Does it appear that ICU stay for patients given the ventilation treatment is normally distributed? Explain your reasoning.

35. According to the article "Modelling and Predicting the Effects of Submerged Arc Weldment Process Parameters on Weldment Characteristics and Shape Profiles" (*J. of Engr. Manuf.*, 2012: 1230–1240), the submerged arc welding (SAW) process is commonly used for joining thick plates and pipes. During welding, the SAW electrode causes a slight deformation on and in the surface of the base metal. This deformation is known as the

SAW *weldment profile*; research has shown that its shape could be related to plate melting efficiency.

Authors of the article wanted to investigate how certain settings of the welding process affect macrostructure zones of the SAW weldment profile. The heat affected zone (HAZ), a band created within the base metal during welding, was of particular interest.

The article reported the impact of various SAW process settings (including current, voltage, and welding speed) on characteristics of the weldment profile. In one investigation, the SAW process was run on various current settings (A) and the depth (mm) of the HAZ was recorded. The data below is partitioned across high (525 A) and nonhigh (<525 A) current settings:

NonHigh: 1.04 1.15 1.23 1.69 1.92 1.98 2.36 2.49 2.72
          1.37 1.43 1.57 1.71 1.94 2.06 2.55 2.64 2.82

High: 1.55 2.02 2.02 2.05 2.35 2.57 2.93 2.94 2.97

Does it appear that true average HAZ depth is larger for the high current condition than for the nonhigh current condition? Carry out a test of appropriate hypotheses using a significance level of .01.

36. Which factors are relevant to the time a consumer spends looking at a product on the shelf prior to selection? The article "Effects of Base Price Upon Search Behavior of Consumers in a Supermarket" (*J. Econ. Psychol.*, 2003: 637–652) reported the following data on elapsed time (sec) for fabric softener purchasers and washing-up liquid purchasers; the former product is significantly more expensive than the latter. These products were chosen because they are similar with respect to allocated shelf space and number of alternative brands.

Product	Sample Size	Sample Mean	Sample SD
Fabric softener	15	30.47	19.15
Washing-up liquid	19	26.53	15.37

a. What if any assumptions are needed before the t inferential procedure can be used to compare true average elapsed times?

b. Carry out a test of hypotheses to decide whether the true average difference in elapsed times differs from zero.

37. Exercise 54 in Chapter 7 presented a t variable appropriate for making inferences about $\mu_1 - \mu_2$ when both population distributions are normal and, in addition, it can be assumed that $\sigma_1 = \sigma_2$.

a. Describe how this variable can be used to form a test statistic and test procedure, the *pooled t test*, for testing $H_0: \mu_1 - \mu_2 = \Delta$.

b. Use the pooled t test to test the relevant hypotheses based on the SAS output given in Exercise 33.

c. Use the pooled t test to reach a conclusion in Exercise 35.

38. The drug diethylstilbestrol was used for years by women as a nonsteroidal treatment for pregnancy maintenance, but it was banned in 1971 when research indicated a link with the incidence of cervical cancer. The article "Effects of Prenatal Exposure to Diethylstilbestrol (DES) on Hemispheric Laterality and Spatial Ability in Human Males" (*Hormones and Behavior*, 1992: 62–75) discussed a study in which ten males exposed to DES and their unexposed brothers underwent various tests. This is the summary data on the results of a spatial ability test:

exposed mean = 12.6
unexposed mean = 13.8
standard error of difference $= \dfrac{s_d}{\sqrt{n}} = .5$

Does DES exposure appear to be associated with reduced spatial ability? State and test the appropriate hypotheses using $\alpha = .05$. Does the conclusion change if $\alpha = .01$ is used?

39. Parents often urge their children to "sit up straight" when dining to practice good table manners. Although proper posture is part of maintaining good etiquette, research has shown that it can also help in reducing musculoskeletal disorders (MSDs). The authors of "Reducing Musculoskeletal Disorders Among Computer Operators: Comparison Between Ergonomics Interventions at the Workplace" (*Ergonomics*, 2012: 15711–1585) investigated the impact of a workplace intervention for reducing MSDs for computer workers. For one group of workers the intervention was in the form of a short oral presentation on how to sit; the preferred heights of chairs, tables, keyboards, and screens; and

optimal positions of the back, shoulders, elbows, and wrists.

Both an MSD score and a rapid upper limb assessment (RULA) score were obtained for each participant. The MSD score is the total number of painful body parts reported by the individual. The RULA score is a rating of the individual's posture, with lower numbers indicating better posture. Each score was determined both before and after the oral presentation intervention. (The textbook author who found this article did find that his own posture improved at least while he was typing this exercise in the manuscript.)

Measurement	Sample Size	Mean Difference (After–Before)	SD of Difference
MSD Score	21	.19	1.03
RULA Score	21	−1.52	1.56

a. Assuming that the difference in MSD scores (After–Before) is approximately normal, carry out a test at significance level .05 to decide whether true average difference in MSD scores is different from zero.

b. Assuming that the difference in RULA scores (After–Before) is approximately normal, carry out a test at significance level .05 to decide whether true average difference in RULA scores is different from zero.

c. From parts (a) and (b) you should have found that for one score the intervention had a significant impact but not for the other score. Keeping in mind what the scores measure, can you offer an explanation of why this may have occurred? (For a group of computer workers who were exposed to a more rigorous type of intervention, the article reported that intervention was beneficial for both MSD and RULA scores.)

40. The article "Selection of a Method to Determine Residual Chlorine in Sewage Effluents" (*Water and Sewage Works*, 1971: 360–364) reported the results of an experiment in which two different methods of determining chlorine content were used on samples of Cl_2-demand-free water for various doses and contact times. Observations are in mg/L.

	Sample			
	1	2	3	4
MSI method	.39	.84	1.76	3.35
SIB method	.36	1.35	2.56	3.92
	5	6	7	8
MSI method	4.69	7.70	10.52	10.92
SIB method	5.35	8.33	10.70	10.91

Does the true average content measured by one method appear to differ from that measured by the other method? State and test the appropriate hypotheses. Does the conclusion depend on whether a significance level of .05, .01, or .001 is used?

41. Shoveling is not exactly a high-tech activity but will continue to be a required task even in our information age. The article "A Shovel with a Perforated Blade Reduces Energy Expenditure Required for Digging Wet Clay" (*Human Factors*, 2010: 492–502) reported on an experiment in which each of 13 workers was provided with both a conventional shovel and a shovel whose blade was perforated with small holes. The authors of the cited article provided the following data on stable energy expenditure [kcal/kg(subject)/lb(clay)]:

Worker:	1	2	3	4	5	6	7
Conventional:	.0011	.0014	.0018	.0022	.0010	.0016	.0028
Perforated:	.0011	.0010	.0019	.0013	.0011	.0017	.0024

Worker:	8	9	10	11	12	13
Conventional:	.0020	.0015	.0014	.0023	.0017	.0020
Perforated:	.0020	.0013	.0013	.0017	.0015	.0013

Carry out a test of hypotheses at significance level .05 to see if true average energy expenditure using the conventional shovel exceeds that using the perforated shovel.

42. The article "Supervised Exercise Versus Non-Supervised Exercise for Reducing Weight in Obese Adults" (*J. Sport. Med. Phys. Fit.*, 2009: 85–90) reported on an investigation in which participants were randomly assigned either to a supervised exercise program or a control group. Those in the control group were told only that

they should take measures to lose weight. After 4 months, the sample mean decrease in body fat for the 17 individuals in the experimental group was 6.2 kg with a sample standard deviation of 4.5 kg, whereas the sample mean and standard deviation for the 17 people in the control group were 1.7 kg and 3.1 kg, respectively. Assume normality of the two body fat loss distributions (as did the investigators).

Does it appear that true average decrease in body fat is more than 2 kg larger for the experimental condition than for the control condition? Carry out a test of appropriate hypotheses using a significance level of .01.

43. The article "The Accuracy of Stated Energy Contents of Reduced-Energy, Commercially Prepared Foods" (*J. of the Amer. Dietetic Assoc.*, 2010: 116–123) presented the accompanying data on vendor-stated gross energy and measured value (both in kcal) for 10 different supermarket convenience meals):

Meal:	1	2	3	4	5	6	7	8	9	10
Stated:	180	220	190	230	200	370	250	240	80	180
Meas.:	212	319	231	306	211	431	288	265	145	228

Carry out a test of hypotheses to decide whether the true average % difference from that stated differs from zero. (*Note:* The article stated "Although formal statistical methods do not apply to convenience samples, standard statistical tests were employed to summarize the data for exploratory purposes and to suggest directions for future studies.")

8.3 TESTS CONCERNING HYPOTHESES ABOUT A CATEGORICAL POPULATION

In this section, we consider several hypothesis-testing situations involving categorical, as opposed to numerical, populations. Suppose that each individual or object in the population can be placed in one of k nonoverlapping categories. For example, systems of a particular type may consist of four components, and the failure of each system may be attributed to failure of one particular component. The four relevant categories would then be "failure of first component," . . . , "failure of fourth component." The null hypothesis will specify a particular value for each one of the category proportions (i.e., probabilities). In the system example, H_0 might specify that each of the long-run failure proportions is .25; that is, a failure is equally likely to be attributed to any one of the four components. A more complicated situation is that in which each individual or object can be categorized with respect to two different categorical factors. For example, each new automobile of a certain type might be classified with respect to color—white, black, blue, etc.—and also with respect to the type of transmission—automatic or manual. We shall consider testing the null hypothesis that categories of the first factor occur independently of those of the second, for example, that car color is independent of type of transmission, that political party registration is independent of preferred religious denomination, and so on. These tests are based on a type of probability distribution that we have not yet encountered, so we first digress from testing to introduce this distribution.

Chi-Squared Distributions

Just as with t distributions, there is not a single chi-squared distribution. Rather there is an entire family of distributions. A particular member of the family is identified by specifying some number of degrees of freedom. Thus there is one chi-squared distribution with 1 df, another with 2 df, yet another with 3 df, and so on. Curves corresponding to several different chi-squared distributions are shown in Figure 8.9. There is no density to the left of zero, so negative values of chi-squared variables are precluded. Each

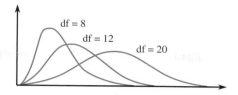

Figure 8.9 Chi-squared curves

chi-squared curve is positively skewed; as the number of df increases, the curves stretch farther and farther to the right and become more symmetric.

Our chi-squared tests are all upper-tailed, so the P-value is the area captured under a particular chi-squared curve to the right of the calculated test statistic value. The fact that t curves were all centered at zero allowed us to tabulate t-curve tail areas in a relatively compact way, with the left margin giving values ranging from 0.0 to 4.0 on the horizontal t scale and various columns displaying corresponding upper-tail areas for various df's. The rightward movement of chi-squared curves as df increases necessitates a somewhat different type of tabulation. The left margin of Appendix Table VII displays various upper-tail areas: .100, .095, .090, . . . , .005, and .001. Each column of the table is for a different value of df, and the entries are values on the horizontal chi-squared axis that capture these corresponding tail areas. For example, moving down to tail area .085 and across to the 2 df column, we see that the area to the right of 4.93 under the 2 df chi-squared curve is .085 (see Figure 8.10). To capture this same upper-tail area under the 10 df curve, we must go out to 16.54. In the 2 df column, the top row shows that if the calculated value of the chi-squared variable is smaller than 4.60, the captured tail area (the P-value) exceeds .10. Similarly, the bottom row in this column indicates that if the calculated value exceeds 13.81, the tail area is smaller than .001 (P-value < .001).

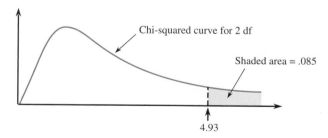

Figure 8.10 Capturing a particular upper-tail area under a chi-squared curve

Tests Based on Univariate Categorical Data

Suppose that each individual or object in a population or process can be placed in one of k nonoverlapping categories. Let

π_1 = population or long-run process proportion falling in the first category

.   .

.   .

.   .

π_k = population or long-run process proportion falling in the kth category

The π's can also be interpreted as probabilities; π_i is the probability that a randomly selected individual or object will fall in the ith category. The null hypothesis completely specifies the value of each π_i; we denote these hypothesized values by adding a subscript 0 to each π_i (as we used μ_0 to denote the null value in a test involving μ):

$$\pi_{i0} = \text{value of } \pi_i \text{ asserted to be true by the null hypothesis } (i = 1, \ldots, k)$$

As an example, suppose that the genotype for a particular genetic characteristic can be either AA, Aa, or aa ($k = 3$). The standard genetic argument in this situation implies the null hypothesis

$$H_0: \pi_1 = .25, \pi_2 = .50, \pi_3 = .25$$

The alternative hypothesis states simply that the specification in H_0 is not correct—that is, at least one of the π_{i0}'s is incorrect (because the hypothesized values add to 1.0, if a particular value is incorrect, at least one other value must also be incorrect). A test of these hypotheses will be based on a random sample taken from the population or process. Each individual or object in the sample will belong in exactly one of the k categories; thus we will have a sample consisting of univariate categorical data. For example, we might select $n = 100$ individuals and find that the first has genotype Aa, the second has genotype aa, the third and fourth both have genotype Aa, the fifth has genotype AA, and so on. Let

$$n_1 = \text{number of sampled individuals or objects falling in the first category}$$

$$\cdot \qquad\qquad\qquad\qquad \cdot$$
$$\cdot \qquad\qquad\qquad\qquad \cdot$$
$$\cdot \qquad\qquad\qquad\qquad \cdot$$

$$n_k = \text{number of sampled individuals or objects falling in the } k\text{th category}$$

The n_i values are called **observed category frequencies** or **counts.** In the genetics example with $k = 3$, we might have $n = 100$, $n_1 = 20$, and $n_2 = 53$, from which $n_3 = 100 - 20 - 53 = 27$.

  The central idea of the test procedure is to compare the observed counts with what would be expected were H_0 true. If, for example, the three hypothesized values are .25, .50, and .25, and $n = 100$, then when the null hypothesis is true,

$$\text{expected number in the first category} = n\pi_{10} = 100(.25) = 25$$

$$\text{expected number in the second category} = n\pi_{20} = 100(.50) = 50$$

$$\text{expected number in the third category} = n\pi_{30} = 100(.25) = 25$$

More generally,

$$\text{expected frequency for category } i \text{ when } H_0 \text{ is true} = n\pi_{i0} \quad (i = 1, \ldots, k)$$

That is, expected frequencies under H_0 are obtained by multiplying each hypothesized value by the sample size. Intuitively, the data supports the null hypothesis when the observed frequencies are similar to the expected frequencies. If some of the observed frequencies differ substantially from what would be expected if H_0 were true, the null hypothesis is no longer tenable.

  We now need a quantitative measure of how different the observed frequencies are from the expected frequencies, assuming H_0 is true. A first thought is to subtract each expected frequency from the corresponding observed frequency to obtain a deviation, square these deviations, and add them together. Symbolically, this would be $\Sigma(n_i - n\pi_{i0})^2$. Suppose, however, that

$$n_1 = 95, \qquad n\pi_{10} = 100$$
$$n_2 = 15, \qquad n\pi_{20} = 20$$

Then both deviations are -5, so they both contribute the same amount to our quantitative measure of discrepancy. However, the observed frequency for the first category is only 5% smaller than what was expected, whereas the observed frequency for the second category is fully 25% smaller than what we would expect if the null hypothesis were true. Our proposed measure does not reflect the fact that, on a percentage basis, the discrepancy for the second category is more sizable than that for the first category. The chi-squared test statistic takes into account percentage deviations.

The Chi-Squared Test Based on Univariate Categorical Data

Hypotheses: $H_0: \pi_1 = \pi_{10}, \ldots, \pi_k = \pi_{k0}$

H_a: the specification of π_i's in H_0 is not correct

Test statistic:
$$X^2 = \sum_{i=1}^{k} \frac{(n_i - n\pi_{i0})^2}{n\pi_{i0}} = \sum \frac{(\text{observed} - \text{expected})^2}{\text{expected}}$$

(Many sources denote this statistic by χ^2, read "chi-squared," but to avoid confusing this with a parameter we don't want to use a Greek letter.)

The smallest possible value of this test statistic is $X^2 = 0$ (when observed = expected for every category), which provides the strongest possible support for the null hypothesis. The larger the value of X^2, the stronger is the evidence against H_0.

P-value: The key result underlying the test procedure is that when H_0 is true and $n\pi_{i0} > 5$ for $i = 1, \ldots, k$ (i.e., all expected counts exceed 5), X^2 has approximately a chi-squared distribution with $k - 1$ df. The P-value is then approximately the area under the $k - 1$ df chi-squared curve to the right of the calculated X^2 value (information about tail areas for chi-squared curves appears in Appendix Table VII). If one or more expected counts is at most 5, categories should be combined in a sensible way so that the resulting expected counts are large enough.

Example 8.8   A number of psychologists have considered the relationship between various deviant behaviors and geophysical variables such as the lunar phase. The article "Psychiatric and Alcoholic Admissions Do Not Occur Disproportionately Close to Patients' Birthdays" (*Psychological Reports*, 1992: 944–946) investigated whether the chance of a patient's admission date for a particular treatment is smaller or larger than would be the case under the assumption of complete randomness. Disregarding leap year, there are 365 possible admission days, so complete randomness would imply a probability of 1/365 for each day. However, this results in far too many categories and expected counts that are too small for the chi-squared test. So the following four categories were established:

1. Within 7 days of an individual's birthday (7 days before to 7 days after)
2. Between 8 and 30 days, inclusive, from the birthday
3. Between 31 and 90 days, inclusive, from the birthday
4. More than 90 days from the birthday

Let π_i denote the true proportion of individuals in category i ($i = 1, 2, 3, 4$). Then complete randomness with respect to admission date implies that

$$\pi_1 = 15/365 = .041 \qquad \pi_2 = 46/365 = .126 \qquad \pi_3 = .329$$
$$\pi_4 = 1 - (.041 + .126 + .329) = .504$$

Thus the relevant hypotheses are

$$H_0: \pi_1 = .041, \pi_2 = .126, \pi_3 = .329, \pi_4 = .504$$

versus

$$H_a: \text{the specification of } \pi\text{'s in } H_0 \text{ is not correct}$$

The cited article gave data for $n = 200$ patients admitted for alcoholism treatment. The expected counts when H_0 is true are then

$$\text{expected count for category } 1 = n\pi_{10} = 200(.041) = 8.2$$
$$n\pi_{20} = 200(.126) = 25.2 \qquad n\pi_{30} = 200(.329) = 65.8$$
$$n\pi_{40} = 200 - (8.2 + 25.2 + 65.8) = 100.8 \; [= 200(.504)]$$

Since all expected counts exceed 5, the chi-squared test can be used. The observed counts along with their expected counterparts are as follows:

Category:	1	2	3	4
Observed:	11	24	69	96
Expected:	8.2	25.2	65.8	100.8

The value of the chi-squared statistic is thus

$$X^2 = \frac{(11 - 8.2)^2}{8.2} + \frac{(24 - 25.2)^2}{25.2} + \frac{(69 - 65.8)^2}{65.8} + \frac{(96 - 100.8)^2}{100.8}$$
$$= .96 + .06 + .16 + .23$$
$$= 1.41$$

The test is based on $k - 1 = 3$ df. The smallest entry in the 3 df column of Appendix Table VII is 6.25, corresponding to an upper-tail area of .10. Because $1.41 < 6.25$, the area captured to the right of 1.41 exceeds .10. That is, P-value $> .10$, so H_0 cannot be rejected at any reasonable significance level. Our analysis is consistent with the title of the cited article; we have no evidence to suggest that admission date is anything other than random.

Testing for Homogeneity of Several Categorical Populations

Suppose now that an investigator is interested in several different categorical populations or processes, each one consisting of the same categories. For example, there are gas stations selling four different brands of gasoline at a particular freeway interchange: Arco (A), Chevron (C), Mobil (M), and Union (U). Each station sells three different grades of gasoline: regular (R), plus (P), and super (S). The four relevant populations consist of the customers purchasing gasoline at each of the four stations, and the grades of gas are the categories. For any particular one of these four populations, there is some proportion of individuals in each of the three categories; these proportions sum to 1.0 for each population. Table 8.1 displays two different possible configurations of population proportions. In the first one, the proportion of individuals in the R category is the same for each population, and the proportion of individuals in the P category is also identical for the four populations. This, of course, implies that the proportion in the last category (S) is constant across the four populations. The populations are said to be **homogeneous with respect to the categories** when this is the case — that is, when the proportion in the first category is the same for all populations, the proportion in the second category is also identical for all populations, and so on. The second configuration in Table 8.1 corresponds to nonhomogeneous populations; the proportions in the various categories are not constant across the populations. Of course, the first configuration in Table 8.1 is not the only one for which the populations are homogeneous; any configuration for which the proportions in any particular column are identical (e.g., .7 for the first column, .2 for the second, and .1 for the third) satisfies the stated condition.

Table 8.1 Two possible configurations of proportions for four categorical populations

(a) Homogeneous populations					(b) Nonhomogeneous populations			
		Category					Category	
		R	P	S			R	P	S
	A	.50	.30	.20		A	.50	.30	.20
Population	C	.50	.30	.20	Population	C	.60	.25	.15
	M	.50	.30	.20		M	.50	.25	.25
	U	.50	.30	.20		U	.65	.25	.10

The null hypothesis that we wish to test is that the populations are homogeneous. For this purpose, we require a separate random sample from each of the populations; let's denote the corresponding sample sizes by n_1, n_2, and so on. Of the n_1 individuals or objects selected from the first population, some number will fall in the first category, some number will be in the second category, and so on. This is also the case for the samples from the other populations. The resulting category frequencies or counts can be displayed in a rectangular table called a **contingency table;** there is a row for each population and a column for each category. The row sums of these observed frequencies are the sample sizes, so they are fixed by the experimenter. Table 8.2 shows one possible set of observed frequencies when each sample size is 200.

Table 8.2 A contingency table in the case of four populations, each with three categories

| | | Category | | | |
		R	P	S	Sample size
	A	107	62	31	$n_1 = 200$
	C	95	67	38	$n_2 = 200$
Population	M	103	57	40	$n_3 = 200$
	U	98	59	43	$n_4 = 200$
Number in category		403	245	152	800

Homogeneity asserts that there is a common value of π_1, the proportion in the first category, for all populations, a common value of π_2 for all populations, and so on. If the values of these π's were known, then just as in the case of a single population, expected frequencies would result from multiplying these π's by the various sample sizes. Let's now assume that the populations are homogeneous and *estimate* the π's from the observed frequencies. Consider the frequencies in Table 8.2. Sensible estimates of π_1, π_2, and π_3 are then just the proportions of the total sample size 800 falling in the various categories:

$$\text{estimate of } \pi_1 = \text{ proportion of total sample size in first category}$$
$$= 403/800 = .50375$$

$$\text{estimate of } \pi_2 = \text{ proportion of total sample size in second category}$$
$$= 245/800 = .30625$$

$$\text{estimate of } \pi_3 = \text{ proportion of total sample size in third category}$$
$$= 152/800 = .19000$$

Multiplying these estimates by $n_1 = 200$ gives the *estimated* expected frequencies for the sample from the first population (assuming homogeneity). For example,

$$\text{estimated expected frequency for the first category in the first sample}$$

$$= 200\left(\frac{403}{800}\right) = 100.75$$

Notice that this estimated expected frequency is the product of the row total (200) and the column total (403) divided by the "grand" total (800). This is in fact the general prescription for obtaining estimated expected frequencies: (row total)(column total)/grand total. Once these have been calculated, the value of a chi-squared statistic can be obtained exactly as in the case of a single population, by summing the quantities (observed − expected)^2/expected over all cells in the contingency table.

The Chi-Squared Test for Homogeneity of Several Categorical Populations

(The word *population* may be replaced by *process* everywhere.)

Denote the number of populations by r and the number of categories for each population by k (the same k categories for all r populations).

Hypotheses: H_0: the r populations are homogeneous with respect to the categories (i.e., the proportion of each population falling in the first category is the same for all populations, the proportion falling in the second category is also the same for all populations, and so on)

H_a: the populations are not homogeneous

(for at least one of the k categories, the proportions are not identical for all populations)

Test statistic: Suppose the observed counts are displayed in a contingency table consisting of r rows, one for the sample from each population, and k columns, one for each category (an r by k table). Then the *estimated* expected frequency corresponding to any particular observed frequency (i.e., to any particular cell of the table) is computed as

$$\text{estimated expected frequency} = \frac{(\text{row total})(\text{column total})}{n}$$

where n is the sum of the individual sample sizes. The test statistic is then

$$X^2 = \sum_{\text{all } rk \text{ cells}} \frac{(\text{observed} - \text{estimated expected})^2}{\text{estimated expected}}$$

P-value: When H_0 is true and all estimated expected frequencies exceed 5, X^2 has approximately a chi-squared distribution with df $= (r-1)(k-1)$. Because any value larger than the calculated X^2 is even more contradictory to H_0, the test is upper-tailed and the P-value is approximately the area to the right of the calculated X^2 under the $(r-1)(k-1)$ chi-squared curve. If at least one estimated expected counts is at most 5, categories should be combined in a sensible way.

Example 8.9

A company packages a particular product in cans of three different sizes, each one using a different production line. Most cans conform to specifications, but a quality control engineer has identified the following reasons for nonconformance:

1. Blemish on can
2. Crack in can
3. Improper pull tab location
4. Pull tab missing
5. Other

A sample of nonconforming units is selected from each of the three lines, and each unit is categorized according to reason for nonconformity, resulting in the following contingency table data:

		Reason for nonconformity				
		Blemish	Crack	Location	Missing	Other	Sample size
	1	34	65	17	21	13	150
Production	2	23	52	25	19	6	125
line	3	32	28	16	14	10	100
	Total	89	145	58	54	29	375

Does the data suggest that the proportions falling in the various nonconformance categories are not the same for the three lines? The parameters of interest are the various proportions, and the relevant hypotheses are

H_0: the production lines are homogeneous with respect to the five nonconformance categories

H_a: the production lines are not homogeneous with respect to the categories

To calculate X^2, we must first compute the estimated expected frequencies (assuming homogeneity). Consider the first nonconformance category for the first production line. When the lines are homogeneous,

estimated expected number among the 150 selected units that are blemished

$$= \frac{(\text{first row total})(\text{first column total})}{\text{total of sample sizes}} = \frac{(150)(89)}{375} = 35.60$$

The contribution of the cell in the upper-left corner to X^2 is then

$$\frac{(\text{observed} - \text{estimated expected})^2}{\text{estimated expected}} = \frac{(34 - 35.60)^2}{35.60} = .072$$

The other contributions are calculated in a similar manner. Table 8.3 shows Minitab output for the chi-squared test. The observed count is the top number in each cell, and directly below it is the estimated expected count. The contribution of each cell to X^2 appears below the counts, and the test statistic value is $X^2 = 14.159$. All estimated expected counts exceed 5, so combining categories is unnecessary. The test is based on $(3 - 1)(5 - 1) = 8$ df. Our chi-squared table shows that the values that capture upper-tail areas of .08 and .075 under the 8 df curve are 14.06 and 14.26, respectively. Thus the P-value is between .075 and .08; Minitab gives P-value = .079. The null hypothesis of homogeneity should not be rejected at the usual significance levels of .05 or .01, but it would be rejected for the higher α of .10.

Table 8.3 Minitab output for the chi-squared test of Example 8.9

```
Expected counts are printed below observed counts
          blem      crack      loc     missing     other    Total
1          34         65        17         21         13      150
         35.60      58.00     23.20      21.60      11.60
2          23         52        25         19          6      125
         29.67      48.33     19.33      18.00       9.67
3          32         28        16         14         10      100
         23.73      38.67     15.47      14.40       7.73
Total      89        145        58         54         29      375
Chisq = 0.072 + 0.845 + 1.657 + 0.017 + 0.169 + 1.498 + 0.278 +
        1.661 + 0.056 + 1.391 + 2.879 + 2.943 + 0.018 + 0.011 +
        0.664 = 14.159
df = 8, p = 0.079
```

Testing for Independence of Two Categorical Factors in a Single Population

Rather than comparing several different categorical populations or processes, consider a *single* population or process in which each individual or object can be classified both with respect to a first categorical factor A and with respect to a second such factor B. For example, each car of a certain type manufactured in a particular year can be classified with respect to body style — two-door coupe, four-door sedan, or hatchback — and with respect to color — white, black, blue, green, or red. Suppose we take a sample of size n and classify each sampled individual or object with respect to both the A factor (style) and the B factor (color). The resulting counts can be displayed in a contingency table having a row for each category of the A factor and a column for each category of the B factor — a 3 by 5 table in the example under consideration. In this situation, neither the row nor the column totals are fixed in advance, only the sum of all counts, which equals n. The number in the upper-left corner would be the number of sampled automobiles that are both coupes and white, and so on. The null hypothesis of interest in this situation is that the two factors A and B are independent; that is, knowing the body style does not change the likelihood of a particular color and vice versa.

Although homogeneity and independence are two different scenarios, the following can be shown: (1) The estimated expected frequencies in the test of independence are calculated exactly as they were for the test of homogeneity: row total times column total divided by n; (2) X^2 is still an appropriate test statistic; (3) the test is still upper-tailed; and (4) the test is based on the same number of df as the homogeneity test.

Fisher's Exact Test

Suppose a company uses one of two methods (A and B) in the manufacture of printed circuit boards. A random sample of 15 boards is taken from the production line and each board is inspected for the existence of any major defects. The following table provides a cross-classification of the boards:

	Method A	Method B
Defects Present	7	1
Defects Absent	1	6

Consider carrying out a test of hypotheses where the null asserts that production method is independent of board condition and the alternative is that there is dependence. Here we would not be able to apply the chi-squared test due to the fact that estimated expected frequencies will not all exceed 5. In such situations the chi-squared test is known to yield unreliable results. Note in the following chi-square test output from Minitab that a warning appears concerning cells having small expected counts.

```
Expected counts are printed below observed counts
                      Method A      Method B      Total
   Defects Present        7             1            8
                         4.27          3.73
   Defects Absent         1             6            7
                         3.73          3.27

      Total               8             7           15
 Chi-Sq = 8.040, DF = 1, P-Value = 0.005
 4 cells with expected counts less than 5.
```

For a contingency table having more than two rows and two columns, if any estimated expected count is at most 5, it may be possible to consolidate some categories and generate a new contingency table whose estimated expected counts would all exceed 5. However, this option would not be available for a contingency table with two rows and two columns as the minimum number of categories for each variable has been reached. Instead of using the chi-square approach, we now introduce a different method that is popularly known as *Fisher's Exact Test*.

Recall in our example that 8 out of the 15 boards were produced using Method A and a total of 8 printed circuit boards had defects. If the null hypothesis of independence between production method and board condition is true, given that Method A accounts for 8 out of the 15 boards and that 8 out of the boards had defects, what is the probability that we would obtain results at least as extreme as what we observed? This probability is the *P*-value for Fisher's Exact Test; it can be computed explicitly by using a particular discrete distribution.

First, let us consider all possible contingency table configurations under the assumption that Method A accounts for 8 out of the 15 boards and that 8 out of the boards had defects. Figure 8.11 reveals that there are only 8 possible contingency tables. If the null hypothesis is true, it can be shown that the probability of each of the 8 possible outcomes can be determined by a discrete distribution known as the hypergeometric. Statistical software packages can readily compute probabilities from this distribution.

	A	B
Present	8	0
Absent	0	7
Prob. = .0002		

	A	B
Present	7	1
Absent	1	6
Prob. = .0087		

	A	B
Present	6	2
Absent	2	5
Prob. = .0914		

	A	B
Present	5	3
Absent	3	4
Prob. = .3046		

	A	B
Present	4	4
Absent	4	3
Prob. = .3807		

	A	B
Present	3	5
Absent	5	2
Prob. = .1828		

	A	B
Present	2	6
Absent	6	1
Prob. = .0305		

	A	B
Present	1	7
Absent	7	0
Prob. = .0012		

Figure 8.11 All possible contingency tables and corresponding hypergeometric probabilities

With all table probabilities in hand, we can now obtain *P*-value information. Our originally observed contingency table yielded 7 boards having defects manufactured by Method A. The corresponding table probability is .0087. To determine the *P*-value we need to consider other tables that would be at least as extreme than what was observed. This would include any tables having a corresponding probability that is less than or equal to .0087. From Figure 8.11 we see that only two other tables qualify (with probabilities. 0002 and .0012). Combining these probabilities, we have *P*-value = .0087 + .0002 + 0012 = .0101. Thus, at the .05 significance level we can reject the null hypothesis of independence between production method and board condition. Figure 8.12 is the corresponding output from SAS for our example:

```
                Fisher's Exact Test
--------------------------------------------------------
Cell (1,1) Frequency (F)            7
Left-sided  Pr <= F             0.9998
Right-sided Pr >= F             0.0089

Table Probability (P)          0.0087
Two-sided Pr <= P              0.0101

            Sample Size = 15
```

Figure 8.12 SAS Output for Fisher's Exact Test

From the output, the *P*-value we computed corresponds to the probability reported next to `Two-sided Pr <= P` as we were interested in testing if any type of dependence existed. As the output suggests, we can use a directional alternative for Fisher's Exact Test as well. Consult the book by Agresti cited in the chapter bibliography for more details on this test.

Section 8.3 Exercises

44. Say as much as you can about the *P*-value for a chi-squared test in each of the following situations:
   a. $X^2 = 7.5$, df $= 2$ b. $X^2 = 13.0$, df $= 6$
   c. $X^2 = 18.0$, df $= 9$ d. $X^2 = 21.3$, df $= 4$
   e. $X^2 = 5.0$, df $= 3$

45. A statistics department at a large university maintains a tutoring service for students in its introductory service courses. The service has been staffed with the expectation that 40% of its clients would be from the business statistics course, 30% from engineering statistics, 20% from the statistics course for social science students, and the other 10% from the course for agriculture students. A random sample of $n = 120$ clients revealed 52, 38, 21, and 9 from the four courses. Does this data suggest that the percentages on which staffing was based are not correct? State and test the relevant hypotheses using $\alpha = .05$.

46. Criminologists have long debated whether there is a relationship between weather and violent crime. The author of the article "Is There a Season for Homicide?" (*Criminology*, 1988: 287–296) classi-fied 1361 homicides according to season, resulting in the accompanying data. Does this data suggest that the homicide rate somehow depends on the season? State the relevant hypotheses, then test using $\alpha = .05$.

Season:	Winter	Spring	Summer	Fall
Frequency:	328	334	372	327

47. The article "Racial Stereotypes in Children's Television Commercials" (*J. of Adver. Res.*, 2008: 80–93) reported the following frequencies with which ethnic characters appeared in recorded commercials that aired on Philadelphia television stations.

Ethnicity:	African American	Asian	Caucasian	Hispanic
Frequency:	57	11	330	6

The 2000 census proportions for these four ethnic groups are .177, .032, .734, and .057, respectively. Does the data suggest that the proportions in commercials are different from the census proportions? Carry out a test of appropriate hypotheses using a significance level of .01.

48. An information retrieval system has ten storage locations. Information has been stored with the expectation that the long-run proportion of requests

for location i is given by $\pi_i = (5.5 - |i - 5.5|)/30$. A sample of 200 retrieval requests gave the following frequencies for locations 1–10, respectively: 4, 15, 23, 25, 38, 31, 32, 14, 10, and 8. Use a chi-squared test at significance level .10 to decide whether the data is consistent with the a priori proportions.

49. The article "The Gap Between Wine Expert Ratings and Consumer Preferences" (*Intl. J. of Wine Business Res.*, 2008: 335–351) studied differences between expert and consumer ratings by considering medal ratings for wines: gold (G), silver (S), or bronze (B). Three categories were then established:
    1. Rating is the same [(G,G), (B,B), (S,S)].
    2. Rating differs by one medal [(G,S), (S,G), (S,B), (B,S)].
    3. Rating differs by two medals [(G,B), (B,G)].

    The observed frequencies for these three categories were 69, 102, and 45, respectively. On the hypothesis of equally likely expert ratings and consumer ratings being assigned completely by chance, each of the 9 medal pairs has probability 1/9. Carry out an appropriate chi-squared test using a significance level of .10.

50. A random sample of smokers was obtained, and each individual was classified by both gender and age when he or she first started smoking. The data in the accompanying table is consistent with summary results reported in the article "Cigarette Tar Yields in Relation to Mortality in the Cancer Prevention Study II Prospective Cohort" (*British Med. J.*, 2004: 72–79).

		Gender	
		Male	Female
	<16	25	10
Age	16–17	24	32
	18–20	28	17
	>20	19	34

a. Calculate the proportion of males in each age category; do the same for females. Based on these proportions, does it appear there might be an association between gender and the age when an individual first smokes?
b. Carry out a test of hypotheses to decide whether there is an association between the two factors.

51. A placebo—that is, a fake medication or treatment—is well known to sometimes have a positive effect just because patients often expect the medication or treatment to be helpful. The article "Beware the Nocebo Effect" (*The New York Times*, Aug. 12, 2012) gave examples of a less familiar phenomenon: the tendency for patients informed of possible side effects to actually experience those side effects. The article cited a study reported in *The Journal of Sexual Medicine* in which a group of patients diagnosed with benign prostatic hyperplasia was randomly divided into two subgroups. One subgroup of size 55 received a compound of proven efficacy along with counseling that a potential side effect of the treatment was erectile dysfunction. The other subgroup of size 52 was given the same treatment without counseling. The percentage of the no-counseling subgroup that reported one or more sexual side effects was 15.3%, whereas 43.6% of the counseling subgroup reported at least one sexual side effect. State and test the appropriate hypotheses at significance level .05 to decide whether the nocebo effect is operating here. (*Hint*: First arrange the data into a contingency table comparing subgroup versus presence of side effects.)

52. A random sample of individuals who drive to work in a large metropolitan area was obtained, and each individual was categorized with respect to both size of vehicle and commuting distance (in miles). Does the accompanying data suggest that there is an association between type of vehicle and commuting distance?

		Commuting Distance		
		0 – <10	10 – <20	≥20
	Subcompact	6	27	19
Type of	Compact	8	36	17
vehicle	Midsize	21	45	33
	Full-size	14	18	6
		$X^2 = 14.16$		

a. Does this situation call for a test of homogeneity or a test of independence?
b. State and test the appropriate hypotheses using $\alpha = .05$.

53. We often think that occupational hazards are primarily experienced by those who work under dangerous conditions (e.g., construction workers,

law enforcement officers, dockworkers). Clearly, a dangerous job can lead to illness or death. But can the psychological stress of a work environment affect employees' overall health? This issue was investigated in the article "Are There Health Effects of Harassment in the Workplace? A Gender-Sensitive Study of the Relationships Between Work and Neck Pain" (*Ergonomics*, 2012: 147–159). The researchers wanted to identify workplace physical and psychosocial risk factors for neck pain among male and female workers. They also wanted to study the relationship between neck pain and intimidation or sexual harassment in the workplace. (Advanced statistical techniques were used to show that neck pain was significantly associated with intimidation at work among both male and female workers.)

This study was based on a representative sample (5405 men, 3987 women) of the Quebec working population. The following cross-classification table for this sample on gender versus level of neck pain is consistent with data reported in the article:

		Gender	
		Men	Women
	Never	3048	1842
Pain	Occasionally	1767	1411
	At Least Fairly Often	590	734

Does it appear that there might be an association between gender and neck pain? Carry out a test of hypotheses using the .01 significance level.

54. The article cited in Exercise 53 classified each member of the sample of workers with respect to both gender and level of work-related psychological demands. The following table is consistent with summary results reported in the article:

		Gender	
		Men	Women
	Low	1692	1324
Job Demand	Medium	1838	1352
	High	1875	1311

Does it appear that there might be an association between gender and work-related psychological demands? Carry out a test of hypotheses using the .05 significance level.

55. Children often suffer from a condition known as *tonsillitis* in which the tonsils become sore or swollen. When the condition becomes chronic, many sufferers have their tonsils surgically removed by the tonsillectomy (TE) procedure. TE is one of the most common surgeries performed in children and young adults worldwide. However, because of the invasive nature of the surgery, TE patients often experience severe postoperative complications. Tonsillotomy (TT), an alternative procedure to surgically removing the tonsils, has become increasingly popular because studies have shown it to be less invasive and to have lower risk of postoperative complications.

The article "Differences in Pain and Nausea in Children Operated on by Tonsillectomy or Tonsillotomy—a Prospective Follow-Up Study" (*J. of Advanced Nursing*, 2012) examined the differences in postoperative pain, nausea, and time of discharge in children 3–12 years of age after TE or TT. To compare differences in postoperative nausea, researchers kept track of the number of prescriptions of ondansetron (a drug to treat nausea and vomiting) that were issued to the TE and TT children. Four out of 34 TE children compared to none of the 53 TT children received such prescriptions.

a. Suppose we are interested in testing whether surgery method affects the provision of ondansetron prescriptions. Determine the estimated expected counts based on the chi-squared test method. Do all expected counts exceed 5?

b. Use Fisher's exact test to analyze this data and report the *P*-value based on a two-sided alternative (as did the authors of the cited article). If your software does not perform this test, there are many online calculators that will report the *P*-value based on this test. One such site is **http://research.microsoft.com/en-us/um /redmond/projects/mscompbio/fisherexacttest**

56. For many years, federal equal employment opportunity laws have prohibited compensation discrimination. However, according to the U.S. Equal Employment Opportunity Commission (EEOC), pay disparities continue to exist in various demographic groups. According to the EEOC website (visited on January 13, 2013), Section 10 of the EEOC

Compliance Manual describes the standards and suggested steps for investigating a charge of compensation discrimination. In the statistical analysis section, Fisher's exact test is recommended as the test of choice. The following is based on the example found in the EEOC Compliance Manual.

Suppose the employees of a particular company can be classified into one of two groups (1 and 2). There are 14 members in group 1 and 17 in group 2. Eight members of group 1 and three members of group 2 earn salaries greater than the company median salary. Use Fisher's exact test at significance level .05 to investigate whether group affiliation has an effect on salary status. (The previous exercise identifies a website that will carry out the calculations.)

8.4 TESTING THE FORM OF A DISTRIBUTION

An investigator, having obtained a sample $x_1, x_2, \ldots, x_n$ from some underlying population or process distribution, will often wish to know whether it is plausible that the underlying distribution is a member of a particular family, such as the normal family, Weibull family, or (in the case of discrete count data) the Poisson family. In this section, we first present a special test for the normal case and then show more generally how a test based on the chi-squared distribution can be carried out.

Is the Population Distribution Normal?

The validity of many inferential procedures, such as the one- and two-sample t intervals and tests presented in this chapter and in Chapter 7, requires that the underlying distribution(s) be at least approximately normal. If an assumption of normality is not justified, alternative methods of analyzing the data must be used. In Chapter 2, we suggested the use of a normal quantile plot to assess the plausibility of the underlying distribution being normal. The construction involved first determining the $(.5/n)$th quantile of the standard normal distribution, the $(1.5/n)$th quantile, the $(2.5/n)$th quantile, and so on [these are the values that separate, for $i = 1, \ldots, n$, the smallest $100((i - .5)/n)\%$ of the distribution from the remaining part]. These quantiles are then paired with the smallest sample observation, the second smallest observation, the third smallest, and so on, and the resulting pairs are plotted on a rectangular coordinate system. Normality is suggested by a plot in which the points fall reasonably close to some straight line. A plot with a substantial nonlinear pattern of some sort (e.g., curvature, or one or more points far from the line determined by the remaining points) casts doubt on population or process normality.

Some users of statistical methodology will not be comfortable with a subjective assessment of the visual evidence in a plot. After all, people may argue about what is reasonably close or what constitutes a substantial departure. Recall that in Chapter 3 we proposed the sample correlation coefficient r as a measure of the strength of any linear relationship in a bivariate sample. Consider the correlation coefficient r^* calculated from the pairs in a normal quantile plot to be our test statistic for the null hypothesis of normality. Because larger observations are paired with larger z quantiles, the points in the plot increase in height when moving from left to right. That is, the points in the plot slope upward, implying that r^* must be positive. A value of r^* quite close to 1.0 gives evidence of a very straight pattern in the plot and is thus supportive of normality. Suppose, for example, that we calculate $r^* = .962$. Then any test statistic value *smaller* than .962 is even more contradictory to the null hypothesis than what was obtained. For this reason, the test is lower-tailed; the P-value is the area under the r^* sampling distribution curve (when H_0 is true) to the left of the calculated r^*.

The test described in the next box involves a slight modification of what we have so far suggested. For technical reasons, rather than using z quantiles corresponding to $(i - .5)/n$, quantiles corresponding to $(i - .375)/(n + .25)$ are used. These alternative "plotting positions" do not greatly alter the appearance of the plot, but they have been found to improve the behavior of the test.

The Ryan–Joiner Test for Normality

(*Note:* This test is very similar to another procedure called the Wilk–Shapiro test.)

Null hypothesis: $x_1, \ldots, x_n$ comes from a normal distribution.

Alternative hypothesis: The sampled distribution is not normal.

Test statistic: r^* = the sample correlation coefficient calculated from (z quantile, observation) pairs, where the z quantiles are for proportions $(i - .375)/(n + .25), i = 1, \ldots, n$.

P-value: The sampling distribution of r^* when H_0 is true is different for each sample size n. The P-value is the area under the appropriate one of these sampling distribution curves to the left of the calculated r^*. Appendix Table XII gives, for various sample sizes, the values that capture lower-tail areas of .10, .05, and .01. Unless the calculated r^* value coincides with one of these tabulated values, one of the following four statements about the P-value can be made: (1) P-value $> .10$, (2) $.05 <$ P-value $< .10$, (3) $.01 <$ P-value $< .05$, (4) P-value $< .01$. The statistical package Minitab will give P-value information for this test upon request.

Example 8.10

The following sample of $n = 17$ observations on length-diameter ratio (LDR) measurements based on static pile load tests first appeared in Example 2.17.

Quantile:	-1.89	-1.35	-1.05	-0.82	-0.63	-0.46	-0.30	-0.15	0.00
LDR:	30.86	37.68	39.04	42.78	42.89	42.89	45.05	47.08	47.08

Quantile:	0.15	0.30	0.46	0.63	0.82	1.05	1.35	1.89
LDR:	48.79	48.79	52.56	52.56	54.8	55.17	56.31	59.94

We asked Minitab to carry out the Ryan-Joiner test, and the result appears in Figure 8.13. The test statistic value is $r^* = .990$, and Appendix Table XII gives .9549 as the critical value that captures lower-tail area .10 under the r^* sampling distribution curve when $n = 17$ and the underlying distribution is actually normal. Since .990 > .9549, we conclude that P-value = area to the left of .9881 > .10, which is what the Minitab output of Figure 8.13 reports. The P-value is larger than any reasonable significance level, so there is absolutely no reason to doubt that the length-diameter ratio is normally distributed.

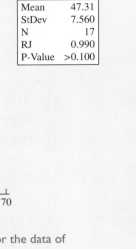

Figure 8.13 Minitab output from the Ryan-Joiner test for the data of

Chi-Squared Tests

Carrying out a chi-squared test requires that categories be established so that observed frequencies can be compared with those expected if the hypothesized family is correct. Suppose, for example, that we have observations on $x =$ number of defects for a sample of 200 automobiles. Possible values of x are 0, 1, 2, ... A reasonable null hypothesis is that x has a Poisson distribution. We might select the x value 0 as the first category, the value 1 as the second category, 2 as the third category, 3 as the fourth category, and aggregate all x values that are at least 4 as the remaining catchall category. The form of the Poisson mass function is $p(x) = e^{-\lambda}\lambda^x/x!$ for $x = 0, 1, 2, \dots$. Substituting $x = 0, 1, 2,$ and 3 and multiplying each result by $n = 200$ would give the expected frequencies for the first four categories; the last expected frequency could then be obtained by adding the first four and subtracting from 200.

However, carrying this out requires that we have a value of the parameter λ. The null hypothesis states only that the distribution is Poisson, without specifying the correct λ. So the value of λ must be estimated from the data before a test can be conducted, and the correct way to do this is to use the method of maximum likelihood introduced in Chapter 7. The estimate should be based on the grouped data (i.e., the number of observations falling in each of the five categories) rather than the individual observations, but this is virtually never done. Instead, the estimate $\hat{\lambda} = \bar{x}$ based on the full data is customarily used (this estimate is intuitively appealing because the mean value of a Poisson variable is just $\mu_x = \lambda$). Furthermore, the estimation of any parameters before calculating expected frequencies and carrying out the test reduces the number of degrees of freedom on which the test is based.

Each parameter that must be estimated from the data before calculating expected frequencies and carrying out a chi-squared test reduces the number of df for the test by one. Thus if the test is based on k categories, all (estimated) expected counts are at least 5; and if m parameters were estimated, the test is based on $k - 1 - m$ df.

For a Poisson distribution, using the five categories suggested previously would result in a test based on df $= 5 - 1 - 1 = 3$ (provided that all expected counts were at least 5). A chi-squared test for normality (not recommended because the Ryan–Joiner test, as well as other tests, have smaller type II error probabilities for the same significance level) would require estimating both μ and σ, reducing degrees of freedom by two.

Example 8.11   Consider the accompanying data on the number of *Larrea divaricata* plants found in each of $n = 48$ identically shaped sampling regions (ecologists call such regions *quadrats*), taken from the article "Some Sampling Characteristics of Plants and Arthropods in the Arizona Desert" (*Ecology*, 1962: 567–571):

Number of plants:	0	1	2	3	at least 4
Frequency:	9	9	10	14	6

The author of the article fit a Poisson distribution to this data. Suppose that the six observations in the last category were actually 4, 4, 5, 5, 6, and 6; it is easily verified that $\hat{\lambda} = \bar{x} = 2.10$ (the value reported in the article). The (estimated) expected frequency for the first category is then

$$48 \left[\frac{e^{-2.1}(2.1)^0}{0!} \right] = 5.88$$

The other four expected frequencies, calculated in the same way, are 12.34, 12.96, 9.07, and (by subtraction) 7.75. All expected frequencies exceed 5, so the test will be based on $5 - 1 - 1 = 3$ df. The test statistic value is

$$\chi^2 = \frac{(9 - 5.88)^2}{5.88} + \cdots + \frac{(6 - 7.75)^2}{7.75} = 6.31$$

The two smallest critical values in the 3 df column of our chi-squared table (Appendix Table VII) are 6.25 and 6.36, corresponding to upper-tail areas of .100 and .095, respectively. Thus the approximate P-value for the test is slightly less than .10. At a significance level of either .05 or .01, there is little reason to doubt that the distribution of the number of plants per quadrat is Poisson.

In the case of continuous data, the categories are simply class intervals. For example, we might select the following six classes: $(-\infty, 85)$, $(85, 95)$, $(95, 100)$, $(100, 105)$, $(105, 115)$, and $(115, \infty)$. After estimating any parameters, the estimated expected frequency for the fourth class would be $n \cdot \left[\int_{100}^{105} f(x)\, dx \right]$, where parameters in the density function $f(x)$ are replaced by their estimates.

Section 8.4 Exercises

57. Consider the Ryan–Joiner test for population normality.
    a. Give as much information as possible for the P-value in each of the following situations:
       i. $n = 10, r^* = .95$
       ii. $n = 10, r^* = .90$
       iii. $n = 25, r^* = .983$
       iv. $n = 25, r^* = .915$
    b. For each of the situations in part (a), state whether the null hypothesis would be rejected when using a significance level of .05.

58. The article cited in Exercise 31 of Section 8.2 gave the following observations on bending rigidity ($\mu N \cdot m$) for medium-quality fabric specimens, from which the accompanying Minitab output was obtained:

24.6	12.7	14.4	30.6	16.1	9.5	31.5	17.2
46.9	68.3	30.8	116.7	39.5	73.8	80.6	20.3
25.8	30.9	39.2	36.8	46.6	15.6	32.3	

Would you use a one-sample t confidence interval to estimate true average bending rigidity? Explain your reasoning.

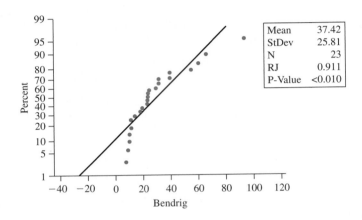

Mean	37.42
StDev	25.81
N	23
RJ	0.911
P-Value	<0.010

59. The article from which the data in Exercise 44 of Chapter 7 was obtained also gave the following data on the compressive strength (in MPa) for 7 specimens of internally cured concrete that have been set for 28 days:

38.7 40.1 40.3 47.5 48.0 56.0 61.1

Minitab gives $r^* = .953$ as the value of the correlation coefficient test statistic and reported that P-value $>.10$. Would you use the one-sample t test to test hypotheses about the value of the true average compressive strength? Why or why not?

60. The data in Exercise 40 is paired, so a paired t analysis is appropriate if it is plausible that the values of the differences were selected from a normal distribution. Based on the accompanying plot from Minitab, does this appear to be the case?

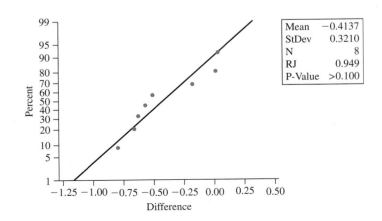

Mean	−0.4137
StDev	0.3210
N	8
RJ	0.949
P-Value	>0.100

61. The article cited in Exercise 88 of Chapter 7 gave the following observations on conductivity (% IACS) for eight wire electrodes used for wire electrical-discharge machining:

    31   28   26   24   33   65   29   29

    a. Employ software to perform a test for normality (such as the Ryan-Joiner test) using a significance level of .05.

    b. Note that there is one unusually high conductivity reading. Suppose the researchers discovered there was a recording error for this observation. Remove it and repeat part (a). How does the removal of the observation affect the test for normality?

62. In a genetics experiment, investigators examined 300 chromosomes of a particular type and counted the number of sister-chromatid exchanges on each one ("On the Nature of Sister-Chromatid Exchanges in 5-Bromodeoxyuridine-Substituted Chromosomes," *Genetics*, 1979: 1251–1264). A Poisson model was hypothesized for the distribution of the number of exchanges. Test the fit of such a model to the accompanying data by first estimating λ and then combining the frequencies for $x = 8$ and $x = 9$.

x:	0	1	2	3	4	5	6	7	8	9
Frequency:	6	24	42	59	62	44	41	14	6	2

63. In an investigation into the distribution of output tuft weight x of cotton fibers when the input weight was x_0, a truncated exponential distribution, $f(x) = (\lambda e^{-\lambda x})/(1 - e^{-\lambda x_0})$ for $0 < x < x_0$, was hypothesized ("Some Studies on Tuft Weight Distributions in the Opening Room," *Textile Res. J.*,

1976: 567–573). The mean value of this distribution is $\mu = (1/\lambda) - (x_0 e^{-\lambda x_0})/(1 - e^{-\lambda x_0})$. Replacing μ by $\bar{x}$ and λ by $\hat{\lambda}$ and solving for the latter quantity gives an estimate of λ. The expected frequencies for various categories (class intervals) can then be calculated. Use the accompanying data along with $\bar{x} = 13.086$ to decide whether the truncated exponential distribution is a plausible model ($x_0 = 70$ here).

Class:	0–<8	8–<16	16–<24
Frequency:	20	8	7

Class:	24–<32	32–<40	40–<48
Frequency:	1	2	1

Class:	48–<56	56–<64	64–<70
Frequency:	0	1	0

64. It is hypothesized that when homing pigeons are disoriented in a certain manner, they will exhibit no preference for any direction of flight after take-off (the direction x, a continuous variable, should be uniformly distributed on the interval from 0° to 360°, so $f(x) = 1/360$ on this interval). To test this, 50 pigeons were disoriented and released, resulting in the following observed directions. Use a chi-squared test based on eight classes to test the appropriate hypotheses at a significance level of .05.

171	338	238	37	92	287	203	320	88
36	131	32	61	250	99	138	155	183
201	312	89	158	206	170	204	46	323
289	141	319	242	179	249	185	277	95
46	197	251	196	326	124	350	112	37
104	290	47	310	86				

8.5 FURTHER ASPECTS OF HYPOTHESIS TESTING

Our focus in hypothesis testing thus far has been on an intuitive development of test procedures in various situations and their application to sample data. In this section, we consider several somewhat more conceptual issues: the distinction between statistical and practical significance of a test result, the interpretation and determination of type II error probabilities, a test procedure that is distribution-free in the sense that its validity does not depend on any restrictive assumptions, the relation between confidence intervals and test procedures, and a general principle for construction of test procedures.

Statistical Versus Practical Significance

Carrying out a test amounts to deciding whether the value obtained for the test statistic could plausibly have resulted when H_0 is true. If the value does not deviate too much from what is expected when the null hypothesis is true, there is no compelling reason for rejecting H_0 in favor of H_a. But suppose that the P-value is quite small, indicating a test statistic value that is quite inconsistent with H_0. One could continue to believe that H_0 is true and that such a value arose just through chance variation (a very unusual and unrepresentative sample). However, in this case a more plausible explanation for what was observed is that the null hypothesis is false and H_a is true.

When the P-value is smaller than the chosen significance level α, it is customary to say that the result is **statistically significant.** The finding of statistical significance means that, in the investigator's opinion, the observed deviation from what was expected under H_0 cannot plausibly be attributed to sampling variability alone. However, statistical significance cannot be equated with the conclusion that the true situation differs from what H_0 states in any practical sense. That is, even after the null hypothesis has been rejected, the data may suggest that there is no *practical* difference between the true value of the parameter and what the null hypothesis asserts that value to be.

Example 8.12

Samples of two different automobile braking systems were selected and the braking distance (ft) for each was determined under specified experimental conditions, resulting in the following summary information:

$$n_1 = 100 \qquad \bar{x}_1 = 120 \qquad s_1 = 5.0$$
$$n_2 = 100 \qquad \bar{x}_2 = 118 \qquad s_2 = 5.0$$

Does it appear that true average braking distance for the first system differs from that for the second system? The relevant hypotheses are $H_0: \mu_1 - \mu_2 = 0$ versus the alternative $H_a: \mu_1 - \mu_2 \neq 0$, and

$$z = \frac{\bar{x}_1 - \bar{x}_2}{\sqrt{\dfrac{s_1^2}{n_1} + \dfrac{s_2^2}{n_2}}} = \frac{120 - 118}{\sqrt{\dfrac{25}{100} + \dfrac{25}{100}}} = \frac{2}{.707} = 2.83$$

The P-value for this two-tailed z test is then $2 \cdot$ (area under z curve to the right of 2.83) = .0046. Thus the null hypothesis should be rejected at a significance level of .05 or even at .01. We say that the data is statistically significant at either of these levels. However, because of the rather large sample sizes and relatively small standard deviations, it appears that $\mu_1 - \mu_2 \approx \bar{x}_1 - \bar{x}_2 = 2.0$. From a practical point of view, a 2-foot difference in true average braking distance would appear to be relatively unimportant. This is an instance of statistical significance without any evidence of a practically significant difference.

Type II Error Probabilities

A test carried out at a specified significance level α is one for which the probability of a type I error—the probability of rejecting the null hypothesis when it is true—is the chosen α. Using a small significance level results in a test that has good protection against the commission of a type I error. However, if at the same time the likelihood of committing a type II error—not rejecting the null hypothesis when in fact it is false—is large, then the test procedure will be quite ineffective at detecting departures from the null hypothesis. For example, consider testing H_0: $\mu = 100$ versus H_a: $\mu > 100$ using a test with a significance level of $\alpha = .01$. If this test is used repeatedly on different samples selected from the population of interest and if H_0 is in fact true, in the long run only 1% of all samples will result in the incorrect rejection of the null hypothesis. Suppose, though, that the alternative $\mu = 105$ represents an important departure from the null hypothesis, but that in this situation $\beta = P(\text{type II error}) = .75$. Then if the test procedure is used over and over on different samples and in fact μ really is 105 rather than 100, in the long run only 25% of all samples will result in the rejection of H_0, whereas the other 75% of all samples will yield an incorrect conclusion. The test procedure has rather poor ability to detect a departure from the null hypothesis that has substantial practical significance. In general, it makes little sense to expend the resources necessary to acquire sample data and carry out a test if the test procedure has very poor ability to detect important departures from the null hypothesis. This is why we recommend investigating the likelihood of committing a type II error before a test with a specified α is used.

One way to determine β is to use an appropriate set of curves. Figure 8.14 shows three different β curves for a one-tailed t test (appropriate for either the alternative H_a: $\mu > \mu_0$ or the alternative H_a: $\mu > \mu_0$). Obtaining β requires that we specify an alternative value of μ (e.g., 105 in the situation considered in the previous paragraph) and also that we select a realistic value of the population or process standard deviation σ. Then we calculate the value of

$$d = \frac{|(\text{alternative value of } \mu) - \mu_0|}{\sigma}$$

the distance between the alternative value and the null value expressed as some number of population standard deviations. Thus $d = 2$ means that the alternative value of μ is 2

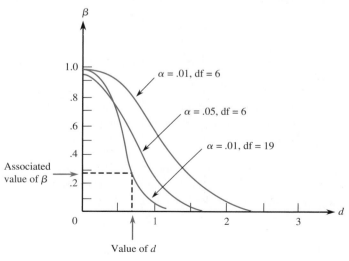

Figure 8.14 Selected β curves for the one-tailed t test

population standard deviations away from the null value. Finally, locate the value of d on the horizontal axis, move directly up to the curve for $n - 1$ df, and move over to the vertical axis to read the value of β.

The following general properties provide insight into how β behaves.

1. The larger the number of degrees of freedom, the lower is the corresponding β curve for any value of d. Because df increases as the sample size increases, we have the intuitively plausible result that β decreases as n increases.
2. The farther the alternative value of interest is from the null value, the larger the value of d. Because every curve decreases as d increases, it follows that β will be smaller for an alternative value far from what the null hypothesis asserts than for a value close to μ_0. Thus the test is more likely to detect a large departure from μ_0 than a small departure.
3. The larger the value of σ, the smaller the value of d and the larger the resulting value of β corresponding to any particular alternative value of μ. That is, the more underlying variability there is in the population or process, the more difficult it will be to detect a departure from H_0 of any given magnitude. Selecting a relatively large value of σ for the calculation gives a pessimistic value of β.

In recent years, the use of β curves has been superseded by statistical software, which is quicker and avoids the visual inaccuracies associated with the curves. In particular, Minitab will determine the **power** of the one-sample t test, where power $= 1 - \beta$, once the difference between the null value and alternative value of μ and also the value of σ have been specified (small β is equivalent to large power; a *powerful* test is one that has large power and therefore good ability to discriminate between the null hypothesis and the alternative value of μ). In addition, instead of specifying n and asking for power, the user can specify the desired power for the given difference and ask Minitab for the necessary sample size.

Example 8.13   The true average voltage drop from collector to emitter of insulated gate bipolar transistors of a certain type is supposed to be at most 2.5 volts. An investigator selects a sample of $n = 10$ such transistors and uses the resulting voltages as a basis for testing H_0: $\mu = 2.5$ versus H_a: $\mu > 2.5$ using a t test with significance level $\alpha = .05$. If the standard deviation of the voltage distribution is $\sigma = .100$, how likely is it that H_0 will not be rejected when in fact $\mu = 2.6$?

The difference value is $2.6 - 2.5 = .1$. Providing this information to Minitab along with the sample size, value of σ, and the fact that the test is upper-tailed results in power $= .8975$, from which $\beta \approx .1$. The investigator may think that this value of β is too large for such a substantial departure from H_0. When Minitab is supplied with the difference $.1$, $\sigma = .1$, and the target power of $.95$ ($\beta = .05$) for an upper-tailed test with $\alpha = .05$, the necessary sample size is returned as 13. The actual power in this case is $.9597$, whereas using $n = 12$ would result in power being somewhat below the target.

Type II error probabilities for other tests can be determined in a similar manner using appropriate statistical software.

You might ask whether there is another test procedure, based on a different test statistic (a different function of the sample data), that outperforms the one-sample t test in the sense that it has the same significance level α but smaller type II error probabilities. It turns out that there is no such test as long as the population distribution is normal. The one-sample t test is really the best possible test in this situation. Furthermore, if the population distribution is not too far from being normal, no test can improve on the one-sample t test by very much. However, if the population distribution is highly nonnormal (heavy-tailed, highly skewed, or multimodal), the t test should not be used. Then it is time to consult your friendly neighborhood statistician to see what alternative methods of analysis are available.

An Alternative Two-Sample Test for Hypotheses About $\mu_1 - \mu_2$

Unfortunately, the two-sample t test does not have the same status as does the one-sample test. The two-sample t test is an intuitively reasonable procedure that appears to protect against both type I and type II errors, but it is not known whether it is the best test in the sense described previously (smallest βs for any given α). Furthermore, if the population distributions are not normal, there are better tests available. We now give a brief description of one such test, called the *Wilcoxon rank-sum test* or alternatively the *Mann–Whitney test* (after the statisticians who discovered the procedure). The validity of the test procedure requires that both population or process distributions be continuous with the same shapes and spreads, so that the only possible difference between them is the location of the center. The two-sample t test does not require equal variances, so the new situation is more restrictive in this respect. However, the Wilcoxon test does not require normal distributions, making it more widely applicable in this sense.

The test is based on a random sample from the first distribution and another random sample, selected independently of the first one, from the second distribution. Let's take the n_1 observations in the first sample and combine them with the n_2 observations from the second sample. Suppose that there are no tied values in this combined sample (all $n_1 + n_2$ observations are distinct). We assign a **rank** to each value in the combined sample: The smallest value gets a rank of 1, the second smallest rank 2, and so on, until finally the largest value has rank $n_1 + n_2$. The following example, with $n_1 = 4$ and $n_2 = 3$, consists of observations on fuel efficiencies (mpg) for two different types of cars:

Distribution from which the observation was selected:	2	2	1	2	1	1	1
Combined sample (ordered):	27.8	29.0	29.3	29.8	31.0	32.1	33.0
Rank:	1	2	3	4	5	6	7

Consider testing $H_0: \mu_1 - \mu_2 = 0$ versus $H_a: \mu_1 - \mu_2 > 0$. The key idea behind the test is that, if the null hypothesis is true, observations from the two samples should be intermingled in magnitude, so that the ranks are intermingled. However, when μ_1 exceeds μ_2, observations in the first sample will tend to be larger than those in the second sample. In this case, the larger ranks will be assigned to sample 1 observations and the smaller ranks to the observations from sample 2. The Wilcoxon test statistic w is the sum of the ranks assigned to observations in the first sample. For the data introduced,

$$w = \text{sum of ranks for observations in sample 1} = 3 + 5 + 6 + 7 = 21$$

Because the inequality $>$ appears in H_a, values of w larger than 21 are even more contradictory to the null hypothesis than the value actually obtained. Thus

$$P\text{-value} = P(w \geq 21 \text{ when } H_0 \text{ is true})$$

Now the only set of four ranks for which $w = 21$ is the one that resulted, and the only possible w value larger than 21 is $w = 22$, which occurs when the ranks are 4, 5, 6, and 7. So

$$P\text{-value} = P(\text{ranks are 3, 5, 6, and 7 or 4, 5, 6, and 7 when } H_0 \text{ is true})$$

But when the null hypothesis is true, all seven observations have actually been selected from the same distribution, in which case any set of four ranks for the observations in the first sample has the same chance of resulting—the set 1, 2, 5, 7 or the set 1, 3, 4, 6, and so on. It is not difficult to see that there are 35 possible sets of four ranks that can be selected from the ranks 1, . . ., 7.[1] Since only two of these 35 sets have $w \geq 21$,

$$P\text{-value} = \frac{2}{35} = .0571$$

When H_0 is true and this test statistic is used repeatedly on different samples, in the long run about 5.7% of all samples will give a w value at least as contradictory to the null hypothesis as what we obtained. The P-value is small enough to justify rejection of H_0 at level .10 but not at level .05.

Unless n_1 and n_2 are quite small, it can be time-consuming to determine the sets of ranks corresponding to w values at least as extreme as what was obtained to calculate the P-value. We recommend using a statistical computer package for this purpose. The Wilcoxon test is valid whatever the nature of the two distributions as long as they are continuous with the same shapes and spreads. This test is often described as being **distribution-free** (or *nonparametric*), meaning that it is valid for a wide variety of underlying distributions rather than just one particular type of distribution. The t test is not distribution-free, because its validity is predicated on the two distributions being at least approximately normal. There are a number of other distribution-free tests in a statistician's toolbox, many of them based on ranks of the observations. The best of these tests, including the Wilcoxon test, perform almost as well as tests such as the t test that are developed with specific types of distributions in mind. That is, for the same significance level α, type II error probabilities for the distribution-free tests are not much larger than those of the best tests in various situations. Consult one of the chapter references for more information on procedures of this type.

The Relationship Between Test Procedures and Confidence Intervals

Suppose the two-sided large-sample confidence interval for a population mean μ at the 95% confidence level based on a particular sample is (103.5, 108.2). Consider using this same sample to test, at a significance level of .05, the null hypothesis $H_0\colon \mu = \mu_0$ against the two-sided alternative $H_a\colon \mu \neq \mu_0$. It is not difficult to see that if the null value μ_0 is a number in the confidence interval, such as 105 or 107.5, then the P-value will exceed .05, so H_0 cannot be rejected. If, however, μ_0 lies outside the confidence interval

---

[1] In general, there are $(n_1 + n_2)!/(n_1!)(n_2!)$ ways to select the n_1 ranks for the observations from the first sample.

(e.g., 100 or 110), then the P-value $\leq .05$ and H_0 can be rejected. In other words, the 95% interval consists precisely of all values μ_0 for which the null hypothesis H_0: $\mu = \mu_0$ cannot be rejected at a significance level of .05. This is intuitively reasonable, since the confidence interval consists of all plausible values of μ at the designated confidence level, and not rejecting H_0 means that μ_0 is plausible. The following generalization of this situation describes an important relationship between tests and confidence intervals.

Let $\hat{\theta}_L$ denote the lower confidence limit for some parameter θ and $\hat{\theta}_U$ denote the upper confidence limit, where the confidence level is $100(1 - \alpha)\%$. Consider the test procedure that rejects H_0: $\theta = \theta_0$ in favor of H_a: $\theta \neq \theta_0$ if θ_0 lies outside the interval and does not reject the null hypothesis if θ_0 falls between $\hat{\theta}_L$ and $\hat{\theta}_U$. (Notice that there is no explicit test statistic, but we still have a decision rule.) This test procedure has a significance level of α.

The result is important because a confidence interval can be used as a basis for testing hypotheses, and, by the same token, there is a confidence interval procedure corresponding to any particular test procedure. (Our discussion has focused on two-sided confidence intervals and two-tailed tests, but one-sided confidence intervals that specify a lower or an upper confidence bound give rise to one-tailed tests and vice versa.) For example, in Chapter 7 we discussed the bootstrap method for calculating confidence intervals; these intervals also form the basis for bootstrap tests of hypotheses. Similarly, the Wilcoxon rank-sum test, which was described previously, gives rise to a distribution-free confidence interval for $\mu_1 - \mu_2$. In summary, the duality between tests and confidence intervals has led to the development of many important inferential procedures.

A General Principle for Obtaining Test Procedures

The test procedures considered so far have all been developed in an ad hoc manner; an intuitively plausible test statistic was selected and its sampling distribution when H_0 is true was obtained so that the P-value could be calculated. Many frequently used test procedures can be derived using a general technique called the *likelihood ratio principle*. Suppose that the mass or density function for a single observation to be randomly selected from some population or process is $f(x; \theta)$. Recall from our discussion of maximum likelihood estimation that if the n sample observations $x_1, \ldots, x_n$ are independently selected from this distribution (a random sample), then the *likelihood* is the joint mass or density function $f(x_1; \theta) f(x_2; \theta) \cdots f(x_n; \theta)$, regarded as a function of θ. For example, for a random sample from a Poisson distribution, the likelihood would be

$$\frac{\lambda^{x_1} e^{-\lambda}}{x_1!} \cdot \ldots \cdot \frac{\lambda^{x_n} e^{-\lambda}}{x_n!} = \frac{\lambda^{\Sigma x_i} e^{-n\lambda}}{x_1! \cdots x_n!}$$

Now consider general null and alternative hypotheses of the form H_0: $\theta \in \Omega_0$ versus H_a: $\theta \in \Omega_a$ ($\in$ is read as "lies in the set. . ."). For example, Ω_0 might be the single value 10, and Ω_a might consist of all numbers except 10, whence the hypotheses are H_0: $\theta = 10$ versus H_a: $\theta \neq 10$. Now consider the following *likelihood ratio test statistic*:

$$\Lambda(\underline{x}) = \text{likelihood ratio test statistic} = \frac{\text{maximum value of likelihood for all } \theta \in \Omega_0}{\text{maximum value of likelihood for all } \theta \in \Omega_a}$$

where x is compact notation for $x_1, \ldots, x_n$. If the numerator of this statistic is much larger than the denominator (the ratio is much larger than 1.0), then there is a value of θ specified by the null hypothesis for which the observed data is a lot more likely than it would be for any value of θ specified by the alternative hypothesis. If, however, the ratio is much smaller than 1.0, there is an alternative value of θ for which the observed data is much more likely than would be the case if the null hypothesis were true. A ratio of the latter sort therefore suggests rejecting H_0 in favor of H_a. Suppose, for example, that the value of $\Lambda(x)$ is .2. Then values of this statistic *smaller* than .2 are even more contradictory to H_0 than what was obtained, implying that

$$P\text{-value} = P(\Lambda(\underline{x}) \leq .2 \text{ when } H_0 \text{ is true})$$

Suppose that the population distribution is normal and that we wish to test the null hypothesis $H_0 : \mu = \mu_0$ against one of the three alternatives considered previously. It is not at all obvious by inspection, and the argument requires a bit of tedious algebra, but it can be shown that application of the likelihood ratio principle here gives rise to the one-sample t test. So this test procedure can be derived from a general principle for test construction rather than being justified simply on intuitive grounds. This is also true of a number of test procedures to be considered in the next several chapters.

Section 8.5 Exercises

65. Let x denote the IQ of a child randomly selected from a certain large geographical region. Suppose x is known to have (approximately) a normal distribution with $\sigma = 15$. A parent group wishes to test the hypothesis $H_0 : \mu = 100$ versus $H_a : \mu > 100$, hoping to reject the null hypothesis and be able to claim that the average IQ of their children exceeds the nationwide average. The test statistic in this situation is $z = (\bar{x} - 100)/(15/\sqrt{n})$.
    a. Determine the P-value of the test for each of the following values of n when $\bar{x} = 101$ (which suggests that if there is a departure from H_0, it is of little practical significance): i. 100, ii. 400, iii. 1600, iv. 2500.
    b. At a significance level of .01, rejecting H_0 is appropriate if the P-value $\leq .01$, equivalent to $z \geq 2.33$, that is, $\bar{x} \geq 100 + (2.33)(15)/\sqrt{n}$. Determine the value of the type II error probability β when $\mu = 101$ for each of the sample sizes given in part (a). Is a large sample size likely to result in rejecting H_0 even in the absence of a practically significant departure from H_0?

66. The Charpy V-notch impact test is to be applied to a sample of 20 specimens of a certain alloy to determine transverse lateral expansion at 110°F.

To be suitable for a particular application, true average expansion should be less than 75 mils. The alloy will not be used unless there is strong evidence that the criterion has been met. Assuming a normal distribution and a test with $\alpha = .01$, what is the probability that a type II error will be committed and the alloy not used when in fact $\mu = 72$ and $\sigma = 5$? What is this probability when $\mu = 70$ and $\sigma = 5$?

67. A sample of 15 radon detectors of a particular type is to be selected, and each will be exposed to 100 pCi/L of radon. The resulting data will be used to test whether the population mean reading is in fact 100. Suppose that the reading x has a normal distribution within the population. Write a paragraph or two explaining the following Minitab output to someone who is familiar with the elements of hypothesis testing but not with type II error probabilities:

```
Testing mean = null (versus not = null)
Calculating power for mean = null + difference
Alpha = 0.01 Sigma = 1
```

Difference	Sample Size	Power
0.5	15	0.1944
0.8	15	0.5619

```
Alpha = 0.01    Sigma = 0.8
              Sample
Difference     Size      Power
    0.5         15      0.3311
    0.8         15      0.7967
Alpha = 0.01    Sigma = 0.8
              Sample    Target    Actual
Difference     Size     Power     Power
    0.5         42      0.9000    0.9047
    0.8         19      0.9000    0.9147
```

68. The article "A Study of Wood Stove Particulate Emission" (*J. of the Air Pollution Control Fed.*, 1979: 724–728) reported the following data on burn time (hr) for specimens of oak and pine. Use Wilcoxon's test at a significance level of .05 to decide whether true average burn time for oak exceeds that for pine. *Hint:* With $n_1 = 6$ and $n_2 = 8$, when H_0 is true, $P(w \geq c) = .054$ for $c = 58$ and is .010 for $c = 63$.

Pine: .98 1.40 1.33 1.52   .73 1.20

Oak: 1.72   .67 1.55 1.56 1.42 1.23 1.77 .48

69. In an experiment to compare the bond strength of two different adhesives, each adhesive was used in five bondings of two surfaces, and the force necessary to separate the two surfaces was determined for each bonding, resulting in the following data:

Adhesive 1:   229   286   245   299   250
Adhesive 2:   216   179   183   247   232

Use the Wilcoxon rank-sum test to decide whether true average bond strengths differ for the two adhesives. *Hint:* For these sample sizes, when H_0 is true, $P(w \geq c) = .048$ for $c = 36$, $.028$ for $c = 37$, and $.008$ for $c = 39$. Furthermore, when H_0 is true, the distribution of w is symmetric about $n_1(n_1 + n_2 + 1)/2$, so in this case $P(w \leq c) = .048$ for $c = 19$.

70. The confidence interval associated with Wilcoxon's rank-sum test has the following general form. First, subtract each observation in the first sample from every observation in the second sample to obtain a set of $n_1 n_2$ differences. Then the confidence interval extends from the cth smallest of these differences to the cth largest difference, where the value of c depends on the desired confidence level. In the case $n_1 = n_2 = 5$, $c = 4$ results in a confidence level of 94.4%, which is as close to 95% as can be obtained. Determine this CI for the strength data in Exercise 69.

Supplementary Exercises

71. Have you ever been frustrated because you could not get a container of some sort to release the last bit of its contents? The article "Shake, Rattle, and Squeeze: How Much Is Left in That Container?" (*Consumer Reports*, May 2009: 8) reported on an investigation of this issue for various consumer products. Suppose five 6.0-oz tubes of toothpaste of a particular brand are randomly selected and squeezed until no more toothpaste will come out. Then each tube is cut open and the amount remaining is weighed, resulting in the following data (consistent with what the cited article reported): .53, .65, .46, .50, .37. Does it appear that the true average amount left is less than 10% of the advertised net contents?
    a. Check the validity of any assumptions necessary for testing the appropriate hypotheses.
    b. Carry out a test of the appropriate hypotheses using a significance level of .05. Would your

conclusion change if a significance level of .01 had been used?
    c. Describe in context type I and II errors, and say which error might have been made in reaching a conclusion.

72. The article cited in Exercise 25 of Section 8.2 gave the following data on mass crystallinity (in %) for 12 samples of the PHB polymer:

42.97 38.81   38.83   41.03   41.25   36.99
49.57   41.77   34.50   44.77   36.92   40.48

    a. Is it plausible that the mass crystallinity for this type of polymer is normally distributed?
    b. Suppose researchers wanted to investigate whether the true average mass crystallinity exceeds 40%. Carry out a test of appropriate hypotheses using a significance level of .05.

73. The following summary data on daily caffeine consumption for a sample of adult women appeared in the article "Caffeine Knowledge, Attitudes, and Consumption in Adult Women" (*J. of Nutrition Educ.*, 1992: 179–184): $n = 47$, $\bar{x} = 215$ mg, $s = 235$ mg, range of data: 5–1176.

    a. Does it appear plausible that the population distribution of daily caffeine consumption is normal? Is it necessary to assume a normal population distribution to test hypotheses about population mean consumption? Explain your reasoning.

    b. Suppose it had previously been believed that population mean consumption was at most 200 mg. Does the given data contradict prior belief?

74. Contamination of mine soils in China is a serious environmental problem. The article "Heavy Metal Contamination in Soils and Phytoaccumulation in a Manganese Mine Wasteland, South China" (*Air, Soil, and Water Res.*, 2008: 31–41) reported that, for a sample of 3 soil specimens from a certain restored mining area, the sample mean concentration of total Cu was 45.31 mg/kg with a corresponding (estimated) standard error of the mean of 5.26. It was also stated that the China background value for this concentration was 20. The results of various statistical tests described in the article were predicated on assuming normality.

    Does the data provide strong evidence for concluding that the true average concentration in the sampled region exceeds the stated background value? Carry out a test at significance level .01.

75. In an investigation of the toxin produced by a certain poisonous snake, a researcher prepared 26 different vials, each containing 1 g of the toxin, and then determined the amount of antitoxin necessary to neutralize the toxin. The sample average amount of antitoxin necessary was found to be 1.89 mg, and the sample standard deviation was .42. Previous research had indicated that the true average neutralizing amount was 1.75 mg/g of toxin. Does the new data contradict the value suggested by prior research? State and test the relevant hypotheses using $\alpha = .05$.

76. When the population distribution is normal, it can be shown that the variable $X^2 = (n - 1)s^2/\sigma^2$ has a chi-squared distribution with $n - 1$ df. This can be used as a basis for testing $H_0: \sigma = \sigma_0$, as follows:

Replace σ^2 in X^2 by its hypothesized value σ_0^2 to obtain a test statistic. If the alternative hypothesis is $H_a: \sigma > \sigma_0$, the P-value is the area under the $n - 1$ df chi-squared curve to the right of the calculated X^2 (an upper-tailed test).

    a. To ensure reasonably uniform characteristics for a particular application, it is desired that the true standard deviation of the softening point of a certain type of petroleum pitch be at most .50°C. The softening points of ten different specimens were determined, yielding a sample standard deviation of .58°C. Assume that the distribution from which the observations were selected is normal. Does the data contradict the uniformity specification? State and test the appropriate hypotheses using $\alpha = .01$.

    b. Suppose that the investigator who performed the experiment described in part (a) had wished to test $H_0: \sigma = .70$ versus $H_a: \sigma < .70$. Can this test be carried out using the chi-squared table in this book? Why or why not?

77. Let π denote the proportion of "successes" in some population. Consider selecting a random sample of size n, and let p denote the sample proportion of successes (number of successes in the sample divided by n). Suppose we wish to test $H_0: \pi = \pi_0$. When H_0 is true and both $n\pi_0 > 5$ and $n(1 - \pi_0) > 5$, the sampling distribution of p is approximately normal with mean value π_0 and standard deviation $\sqrt{\pi_0(1 - \pi_0)/n}$. This implies that a "large-sample" test statistic is $z = (p - \pi_0)/\sqrt{\pi_0(1 - \pi_0)/n}$ (i.e., we standardize p assuming that H_0 is true); the P-value is calculated as was done in Section 8.1 for a z test concerning μ.

    Seat belts help prevent injuries in vehicle accidents, but they don't offer complete protection in extreme situations. A sample of 319 front-seat occupants involved in head-on collisions in a certain region resulted in 95 who sustained no injuries ("Influencing Factors on the Injury Severity of Restrained Front Seat Occupants in Car-to-Car Head-on Collisions," *Accident Analysis and Prevention*, 1995: 143–150). Does this data suggest that less than one-third of all such accidents result in no injuries? State and test the relevant hypotheses using a significance level of .05.

78. Some of the deadliest mass shootings in U.S. history occurred in 2012. These events led to many calls for stricter national gun control. On December 27, 2012, the Gallup organization reported that roughly 600 of 1038 American adults surveyed said they would be in favor of strengthening laws covering the sale of firearms.

    a. Does this provide strong evidence for concluding that more than 50% of the population of American adults was in favor of making laws covering the sale of firearms more strict? Conduct an appropriate test of hypotheses using a .01 significance level. (*Hint*: Read the first paragraph of the previous problem.)

    b. This poll was conducted December 19–22, just days after a mass shooting at an elementary school in Connecticut. Discuss what effects this event may have had on the poll's outcome.

79. Headability is the ability of a cylindrical piece of material to be shaped into the head of a bolt, screw, or other cold-formed part without cracking. The article "New Methods for Assessing Cold Heading Quality" (*Wire J. Intl.*, Oct. 1996: 66–72) described the result of a headability impact test applied to 30 specimens of aluminum killed steel and 30 specimens of silicon killed steel. The sample mean headability rating number for the steel specimens was 6.43 and the sample mean for aluminum specimens was 7.09. Suppose that the sample standard deviations were 1.08 and 1.19, respectively. Do you agree with the article's authors that the difference in headability ratings is significant at the 5% level?

80. The article "Two Parameters Limiting the Sensitivity of Laboratory Tests of Condoms as Viral Barriers" (*J. of Testing and Eval.*, 1996: 279–286) reported that, in brand A condoms, among 16 tears produced by a puncturing needle, the sample mean tear length was 74.0 μm, whereas for the 14 brand B tears, the sample mean length was 61.0 μm (determined using light microscopy and scanning electron micrographs). Suppose the sample standard deviations are 14.8 and 12.5, respectively (consistent with the sample ranges given in the article). The authors commented that the thicker brand B condom displayed a smaller mean tear length than the thinner brand A condom.

Is this difference in fact statistically significant? State the appropriate hypotheses and test at $\alpha = .05$.

81. Information about hand posture and forces generated by the fingers during manipulation of various daily objects is needed for designing high-tech hand prosthetic devices. The article "Grip Posture and Forces During Holding Cylindrical Objects with Circular Grips" (*Ergonomics*, 1996: 1163–1176) reported that for a sample of 11 females, the sample mean four-finger pinch strength (N) was 98.1 and the sample standard deviation was 14.2. For a sample of 15 males, the sample mean and sample standard deviation were 129.2 and 39.1, respectively.

    a. A test carried out to see whether true average strengths for the two genders were different resulted in $t = 2.51$ and P-value $= .019$. Does the appropriate test procedure described in this chapter yield this value of t and the stated P-value?

    b. Is there substantial evidence for concluding that true average strength for males exceeds that for females by more than 25 N? State and test the relevant hypotheses.

82. The article "Pine Needles as Sensors of Atmospheric Pollution" (*Environ. Monitoring*, 1982: 273–286) reported on the use of neutron-activity analysis to determine pollutant concentration in pine needles. According to the article's authors, "These observations strongly indicated that for those elements which are determined well by the analytical procedures, the distribution of concentration is lognormal. Accordingly, in tests of significance the logarithms of concentrations will be used." The given data refers to bromine concentration in needles taken from a site near an oil-fired steam plant and from a relatively clean site. The summary values are means and standard deviations of the log-transformed observations.

Site	Sample Size	Mean log concentration	Standard Deviation of log concentration
Steam plant	8	18.0	4.9
Clean	9	11.0	4.6

Let μ_1^* be the true average *log* concentration at the first site and define μ_2^* analogously for the second site.

a. Use the pooled t test (based on assuming normality *and* equal standard deviations), described in Exercise 37, to decide at significance level .05 whether the two concentration distribution means are equal.

b. If σ_1^* and σ_2^*, the standard deviations of the two log concentration distributions, are not equal, would μ_1 and μ_2, the means of the concentration distributions, be equal if $\mu_1^* = \mu_2^*$? Explain your reasoning.

83. The article cited in Exercise 78 of Chapter 7 gave additional data on breaking force (N):

Temp	Medium	n	$\bar{x}$	s
22°	Dry	6	170.60	39.08
37°	Dry	6	325.73	34.97
22°	Wet	6	366.36	34.82
37°	Wet	6	306.09	41.97

a. Is there strong evidence for concluding that true average force in a dry medium at the higher temperature exceeds that at the lower temperature by more than 100 N?

b. Is there strong evidence for concluding that true average force in a wet medium at the lower temperature exceeds that at the higher temperature by more than 50 N?

84. Long-term exposure of textile workers to cotton dust released during processing can result in substantial health problems so textile researchers have been investigating methods that will result in reduced risks while preserving important fabric properties. The accompanying data on roving cohesion strength (kN·m/kg) for specimens produced at five different twist multiples is from the article "Heat Treatment of Cotton: Effect on Endotoxin Content, Fiber and Yarn Properties, and Processability" (*Textile Research J.*, 1996: 727–738):

Twist multiple: 1.054 1.141 1.245 1.370 .481
Control strength: .45 .60 .61 .73 .69
Heated strength: .51 .59 .63 .73 .74

The authors of the cited article stated that strength for heated specimens appeared to be slightly higher on average than for the control specimens. Is the difference statistically significant? State and test the relevant hypotheses using $\alpha = .05$.

85. Tardive dyskinesia refers to a syndrome comprising a variety of abnormal involuntary movements assumed to follow long-term use of antipsychotic drugs. An experiment carried out to investigate the effect of the drug deanol also used a placebo treatment, something that resembled deanol in every way but was known to be inert and have absolutely no medical effect. The two treatments were administered for 4 weeks each in random order to 14 patients, resulting in the following total severity index scores ("Double Blind Evaluation of Deanol in Tardive Dyskinesia," *J. of the Amer. Med. Assoc.*, 1978: 1997–1998):

Patient:	1	2	3	4	5	6	7
Deanol:	12.4	6.8	12.6	13.2	12.4	7.6	12.1
Placebo:	9.2	10.2	12.2	12.7	12.1	9.0	12.4

Patient:	8	9	10	11	12	13	14
Deanol:	5.9	12.0	1.1	11.5	13.0	5.1	9.6
Placebo:	5.9	8.5	4.8	7.8	9.1	3.5	6.4

Does the data indicate that, on average, deanol yields a higher total severity index score than does the placebo treatment?

86. The authors of the article "Predicting Professional Sports Game Outcomes from Intermediate Game Scores" (*Chance*, 1992: 18–22) used statistical analysis to determine whether there was any merit to the idea that basketball games are not settled until the last quarter, whereas baseball games are "over" by the seventh inning. They also considered football and hockey. Data was collected for a sample of games of each type, selected from all games played during the 1990 season for baseball and football and during the 1990–1991 season for the other two sports. For each game, the late-game leader was determined, and it was noted whether the leader actually ended up winning the game. The leader was defined as the team ahead after three quarters in basketball and football, two periods in hockey, and seven innings in baseball. The results follow:

Sport	Leader wins	Leader loses
Basketball	150	39
Baseball	86	6
Hockey	65	15
Football	72	21

Do the four sports appear to be identical with respect to the proportion of games won by the late-game leader? State and test the appropriate hypotheses using $\alpha = .05$. Do you think your conclusion can be attributed to a single sport being an anomaly?

87. As the population ages, there is increasing concern about accident-related injuries to the elderly. The article "Age and Gender Differences in Single-Step Recovery from a Forward Fall" (*J. of Gerontology*, 1999: M444–M50) reported on an experiment in which the maximum lean angle—the furthest a subject is able to lean and still recover in one step—was determined for both a sample of younger females (21–29 years) and a sample of older females (67–81 years). The following observations are consistent with summary data given in the article:

    YF: 29 34 33 27 28 32 31 34 32 27
    OF: 18 15 23 13 12

    Does the data suggest that true average maximum lean angle for older females is more than 10 degrees smaller than it is for younger females? State and test the relevant hypotheses at significance level .10.

88. Adding computerized medical images to a database promises to provide great resources for physicians. However, there are other methods of obtaining such information, so the issue of efficiency of access needs to be investigated. The article "The Comparative Effectiveness of Conventional and Digital Image Libraries" (*J. of Audiovisual Media in Medicine*, 2001: 8–15) reported on an experiment in which 13 computer-proficient medical professionals were timed both while retrieving an image from a library of slides and while retrieving the same image from a computer database with a Web front end.

Subject:	1	2	3	4	5	6	7
Slide:	30	35	40	25	20	30	35
Digital:	25	16	15	15	10	20	7
Difference:	5	19	25	10	10	10	28

Subject:	8	9	10	11	12	13
Slide:	62	40	51	25	42	33
Digital:	16	15	13	11	19	19
Difference:	46	25	38	14	23	14

Does the true mean difference between slide retrieval time and digital retrieval time appear to exceed 10 sec? Be sure to check the validity of any assumptions on which your chosen inferential method is based.

89. The NCAA basketball tournament begins with 64 teams that are apportioned into four regional tournaments, each involving 16 teams. The 16 teams in each region are then ranked (seeded) from 1 to 16. During the 12-year period from 1991 to 2002, the top-ranked team won its regional tournament 22 times, the second-ranked team won 10 times, the third-ranked team was 5 times, and the remaining 11 regional tournaments were won by teams ranked lower than 3. Let P_{ij} denote the probability that the team ranked i in its region is victorious in its game against the team ranked j. Once the P_{ij}'s are available, it is possible to compute the probability that any particular seed wins its regional tournament (a complicated calculation because the number of outcomes in the sample space is quite large). The paper "Probability Models for the NCAA Regional Basketball Tournaments" (*The American Statistician*, 1991: 35–38) proposed several different models for the P_{ij}'s.
    a. One model postulated $P_{ij} = .5 + \lambda(i - j)$ with $\lambda = 1/32$ (from which $P_{16.1} = \lambda$, $P_{16.2} = 2\lambda$, etc.). Based on this, $P(\text{seed #1 wins}) = .27477$, $P(\text{seed #2 wins}) = .20834$, and $P(\text{seed #3 wins}) = .15429$. Does this model appear to provide a good fit to the data?
    b. A more sophisticated model has $P_{ij} = .5 + .2813625(z_i - z_j)$ where the z's are measures of relative strengths related to standard normal percentiles (percentiles for successive highly seeded teams are closer together than is the case for teams seeded lower, and .2813625 ensures that the range of probabilities is the same as for the model in part (a)). The resulting probabilities of seeds 1, 2, or 3 winning their regional tournaments are .45883, .18813, and .11032, respectively. Assess the fit of this model.

90. One way to reduce the equipment problems that occur during die casting is to apply a thin coating to the core pins. The paper "Tool Treatment Extends

Core and Pin Life" (*Die Casting Engineer*, March/April 1999: 88) reported on an experiment in which one group of core pins was coated using the traditional nitride process and a second group was coated using a new thermal diffusion process. Use the accompanying data to decide at significance level .01 whether there is strong evidence for concluding that true average lifetime for the thermal treatment is more than four times that of the nitride treatment. *Hint:* Consider the parameter $\theta = 4\mu_1 - \mu_2$ with corresponding estimator $\hat{\theta} = 4\bar{x}_1 - \bar{x}_2$. This

estimator is unbiased and normally distributed provided that the two population distributions are normal, and its variance can be determined from the fact that for any two *independent* random variables y_1 and y_2 and numerical constants a_1 and a_2, $V(a_1 y_1 + a_2 y_2) = a_1^2 V(y_1) + a_2^2 V(y_2)$.

Nitrite:	9000	20,000	10,000	20,000
	21,000	3000	4000	
Thermal:	49,000	23,000	20,000	100,000
	114,000	35,000	30,000	

Bibliography

Agresti, Alan, **An Introduction to Categorical Data Analysis** (2nd ed.), Wiley, New York, 2007. *An excellent treatment of various aspects of categorical data analysis by one of the most prominent researchers in this area.*

Devore, Jay L., **Probability and Statistics for Engineering and the Sciences** (8th ed.), Brooks/Cole -Cengage,

Belmont, CA, 2012. *A somewhat more comprehensive and slightly advanced treatment of hypothesis testing and other topics than what is presented in this book.*

See also the books by Devore et al. and by DeGroot et al. listed in the Chapter 7 bibliography.

9

The Analysis of Variance

INTRODUCTION

As we saw in Chapter 8, there is more than one way to make comparisons between two populations or processes. Choosing the best approach involves using one's technical knowledge of a problem to select an appropriate statistical technique. In some cases, the independent samples test (Section 8.2) may be the best approach. At other times, the paired-samples test (also Section 8.2) may be superior. In Chapter 9, both of these methods are extended to comparisons between more than two population means. The independent samples test generalizes to the **single-factor analysis of variance** (Section 9.2), whereas the paired-samples test generalizes to the **randomized block design** (Section 9.4). The procedures in Chapter 9 are the tip of a large statistical iceberg called **experimental design,** which is discussed further in Chapter 10.

One of the important features of the designs in Chapter 9 is that they combine the sample data from several populations into a *single* test capable of detecting when one or more of the population means differ from the rest. That is, analysis of variance tests are not conducted by simply performing the two-sample tests of Chapter 8 on all the different pairings of several populations. Only after an analysis of variance test signals a possible difference between the population means do we begin to conduct **multiple comparisons** of the populations, discussed in Section 9.3, to pinpoint the specific populations whose means differ from one another.

9.1 TERMINOLOGY AND CONCEPTS

Although the focus in this chapter is on detecting differences between several population or process *means*, the primary tool used in these tests is based on a comparison of *variances*. Consequently, the procedures in this chapter have collectively become known as the **analysis of variance.** This phrase is often shortened to the acronym **ANOVA.**

ANOVA methods are concerned with testing null hypotheses of the form

H_0: the means of several populations or processes are the same

The alternative hypothesis H_a is that at least two of the means differ from each other. Letting $\mu_1, \mu_2, \mu_3, \ldots, \mu_k$ denote k population or process means, H_0 and H_a can be written as

$$H_0: \mu_1 = \mu_2 = \mu_3 = \cdots = \mu_k$$
$$H_a: \text{at least two of the } \mu_i\text{'s are different}$$

Some typical ANOVA applications follow:

- Do four different brands of gasoline have different effects on automobile fuel efficiency? (H_0: the mean fuel efficiency (mpg) obtained is the same for all four brands.)
- Is there any difference in crop yields when five different fertilizers are used? (H_0: the mean crop yield per acre is the same for all five fertilizers.) What about using four different watering schedules? (H_0: the mean crop yield per acre is the same for each watering schedule.)
- Will three different levels of a chemical concentration have differing effects on an electroplating process? (H_0: the mean plating thickness is the same for all three concentration levels.)

In each of these examples, the populations share some common characteristic, called a **factor** or **treatment,** whose various **levels** or **treatment levels** distinguish one population from another. For example, when testing for possible differences in fuel efficiency among four brands of gasoline, the factor of interest is gasoline brand, which has four different levels, for example, brand 1, brand 2, brand 3, and brand 4. Alternatively, we could refer to gasoline brand as a treatment with the four brands representing the treatment levels. The k levels of a factor correspond to the k different populations being compared in the test of the hypothesis $H_0: \mu_1 = \mu_2 = \mu_3 = \cdots = \mu_k$.

Comparisons between populations are made by choosing a numerical quantity, called a **response variable,** that is measured for each sampled item. In the fuel efficiency study, for example, fuel efficiency (mpg) would be the response variable, and we would measure the mpg for cars selected to use gasoline of brand 1, brand 2, brand 3, and brand 4, respectively. Using this terminology, these three examples can be summarized as follows:

Factor	Levels	Response
Gasoline	Brand 1, brand 2, brand 3, brand 4	Fuel efficiency, in mpg
Fertilizer	Fertilizer 1, fertilizer 2, fertilizer 3	Crop yield, in bushels/acre
Chemical concentration	Concentration 1, concentration 2, concentration 3	Plating thickness, in mm

When an ANOVA problem is expressed in terms of a factor and a response, the goal of the study is to determine whether the different factor levels have different effects on the response variable. Think of the factor as the *independent variable* and the response as the *dependent variable*. It often helps to draw a picture, as shown in Figure 9.1, to visualize the data from an ANOVA study.

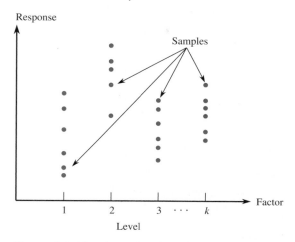

Figure 9.1 Visualizing data from an ANOVA study

Of course, populations may be characterized by several factors, not just one, and each factor can have any number of levels. Populations that represent different levels of a *single* factor are said to form a **one-way classification,** and we compare these populations using a **single-factor (one-way) ANOVA.** Characterizing populations by two factors is called a **two-way classification,** and so forth. Techniques for studying two or more factors are presented in Chapter 10.

How ANOVA Tests Work

Like the two-sample tests in Chapter 8, ANOVA procedures use just *one* test for comparing k population means. Suppose, for example, that we select random samples from each of $k = 4$ populations and present the data in a graph such as that in Figure 9.1. This is the natural extension of the independent samples situation of Section 8.2. To determine whether the population means differ, the ANOVA approach compares the variation *between* the four sample means to the inherent variability *within* each sample (see Figure 9.2). The more the sample means differ, the larger will be the between-samples variation shown at the right in Figure 9.2. The test statistic that compares these two types of variation is the *ratio* of the between-samples variation to the within-samples variation,

$$\text{test statistic} = \frac{\text{between-samples variation}}{\text{within-samples variation}}$$

Figure 9.3 shows how this test statistic behaves when there is no difference between the four means (i.e., when H_0: $\mu_1 = \mu_2 = \mu_3 = \mu_4$ is true) and when the means *do* differ (when H_0 is false). In essence, large values of the test statistic tend to support the alternative hypothesis (that some of the means differ from the others), whereas small values of the statistic support the null hypothesis.

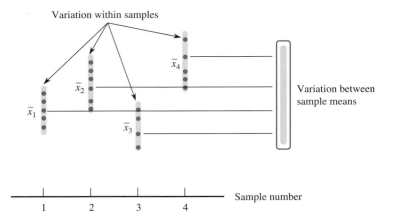

Figure 9.2 ANOVA methods compare *between-samples* variation to *within-samples* variation

F Distributions

When the hypothesis H_0: $\mu_1 = \mu_2 = \mu_3 = \cdots = \mu_k$ is true, it can be shown that the test statistic described previously follows a continuous probability distribution called an *F* **distribution.** *F* distributions arise in statistical tests that involve ratios of two variation measures, such as the ratio of between-samples variation to within-samples variation, as shown in Figure 9.3. The variation measures used in an *F* ratio are based on

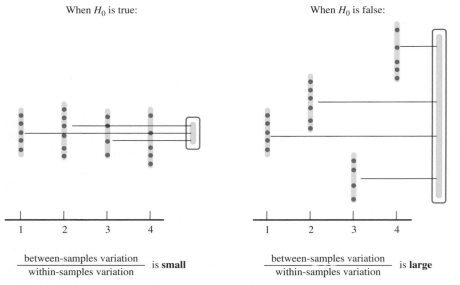

Figure 9.3 How an ANOVA test works

certain sums of squares calculated from the sample data, and each sum of squares has an associated number of degrees of freedom. The **numerator degrees of freedom,** denoted df_1, is the number of degrees of freedom associated with the sum of squares in the numerator of an F ratio. Similarly, df_2 denotes the **denominator degrees of freedom.**

There is a different F distribution for every different combination of positive integers df_1 and df_2. For example, there is an F distribution with 4 numerator degrees of freedom and 12 denominator degrees of freedom, another with 3 numerator degrees of freedom and 20 denominator degrees of freedom, and so forth. Because they are ratios of nonnegative quantities, variables that follow F distributions have only nonnegative values, and their density curves have a shape similar to that shown in Figure 9.4.

ANOVA tests, as Figure 9.3 illustrates, are upper-tailed tests. In other words, only large values of the F ratio lead to rejecting H_0; small values do not reject H_0. In terms of P-values, this means that the P-value associated with a calculated F ratio is the area under the F distribution to the *right* of the calculated F ratio. Figure 9.4 shows the P-value associated with a calculated F ratio of 9.15 based on $df_1 = 4$ and $df_2 = 6$. Tables of critical values for F distributions can be found in Appendix Table VIII.

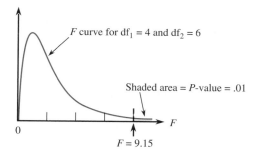

Figure 9.4 P-value for an upper-tailed F test

The F table (Table VIII) contains critical F values associated with tail areas of .10, .05, .01, and .001. To use the table with an F ratio based on $df_1 = 4$ and $df_2 = 6$, read across the top of the table to find the column with $df_1 = 4$ and read down the left side of the table to find the row with $df_2 = 6$. At the intersection of this column and row, there will be four critical values, corresponding to right-tail probabilities of .10, .05, .01, and .001. For example, a P-value of .01 is associated with an F ratio of 9.15, a P-value of .05 is associated with an F ratio of 4.53, and so forth.

Section 9.1 Exercises

1. Three types of wood (denoted A, B, and C) are being considered for use in a building project. Each type of wood differs in cost, so the builder is interested in keeping costs down as well as in selecting wood that will be strong enough. To determine whether there is a difference between the average strengths of the three types of wood, a random sample of ten beams of each type is selected and their strengths are measured.

a. What hypotheses would you test in such a study? Describe, in words, the parameters that appear in the hypotheses.

b. Suppose an ANOVA test indicates that beams of types A and B are not significantly different in strength from one another, but that both types are significantly stronger than beams of type C. If the builder's objective is to use as strong a beam as possible, what type of beam should be used?

c. Suppose the ANOVA test does not reveal any significant differences in strength between the three types of beams. If the builder must use one of the three types, which type should be chosen?

2. Suppose you have a fixed budget to allocate to the samples used in a study of the effect of the factor "chemical concentration" on the plating thickness of electroplated plastic parts. Describe in general terms how you would allocate the samples. Specifically, what information would make you want to use fewer levels of chemical concentration and, correspondingly, more plastic parts at each concentration level? Conversely, what scenario would lead you to use a larger number of concentration levels and, therefore, fewer plastic parts per concentration level? Include the two sources of variation in an ANOVA experiment in your answers.

3. In a one-way ANOVA test for comparing the mean strengths (in kilograms) of three different alloys, suppose that the measuring instrument used is out of calibration, causing it to give readings that are consistently 2.5 kilograms higher than the true measured strength. Using the general description of the techniques given in this section, explain what effect you think such data would have on the results of an ANOVA test comparing samples of the three alloys. Do you think an ANOVA test based on accurate measurements of the *same* samples of alloys will lead to a different conclusion?

4. Repeated measurements in an ANOVA study are supposed to indicate what would happen if another researcher tried to repeat your study. In particular, simply measuring the same sampled item several times, which gives repeated measurements of that item, is *not* considered to be a valid form of replication. Instead, several *different* items should each be measured once. What is the danger in using repeated measurements of the same item instead of truly replicating an experimental result? What do you expect the effect on the F statistic to be if repeated measurements of a single item are used at each level of a factor?

5. As a simple method of determining which of k factor levels maximizes the average value of a certain response variable, inexperienced researchers sometimes calculate the k sample means and then simply select the factor level having the largest sample mean. This strategy has been called the "pick the winner" approach in the literature on experimental design. Explain what is wrong with this approach and why it does not take the place of an ANOVA test.

6. Use the table of F distribution critical values (Appendix Table VIII) to find
   a. The F critical value based on $df_1 = 5$ and $df_2 = 8$ that captures an upper-tail area of .05
   b. The F critical value based on $df_1 = 8$ and $df_2 = 5$ that captures an upper-tail area of .05
   c. The F critical value based on $df_1 = 5$ and $df_2 = 8$ that captures an upper-tail area of .01
   d. The F critical value based on $df_1 = 8$ and $df_2 = 5$ that captures an upper-tail area of .01
   e. The 95th percentile of an F distribution with $df_1 = 3$ and $df_2 = 20$
   f. The probability $P(F \leq 6.16)$ for $df_1 = 6$ and $df_2 = 4$
   g. The probability $P(4.74 \leq F \leq 7.87)$ for $df_1 = 10$ and $df_2 = 5$

7. Based on your answers to Exercise 6(a)–(d), what effect does interchanging df_1 and df_2 have on the critical F value (for a fixed upper-tail area)?

8. An experiment was carried out to compare flow rates for four different types of nozzles.
   a. Samples of five type-A nozzles, six type-B nozzles, seven type-C nozzles, and six type-D nozzles were tested. ANOVA calculations yielded an F value of 3.68 with $df_1 = 3$ and $df_2 = 20$. State and test the relevant hypotheses using $\alpha = .01$.
   b. Analysis of the data using statistical software yielded a P-value of $P = .029$. Using $\alpha = .01$, what conclusion would you draw regarding the test in part (a)?

9. In a test of the hypothesis $H_0: \mu_1 = \mu_2 = \mu_3 = \mu_4$, samples of size 6 were selected from each of four populations, and an F statistic value of 4.12 was calculated (using the methods in the next section). The appropriate degrees of freedom for the F distribution in this exercise are $df_1 = 3$, $df_2 = 20$. Using $\alpha = .05$, conduct the test to determine whether you can conclude that there are differences between μ_1, μ_2, μ_3, and μ_4.

9.2 SINGLE-FACTOR ANOVA

Notation and Formulas

One way to test the null hypothesis H_0: $\mu_1 = \mu_2 = \mu_3 = \cdots = \mu_k$ is to compare the means of random samples selected from each of the k populations specified by H_0. This method of sampling is the basis of the **completely randomized design.** The sample sizes $n_1, n_2, n_3, \ldots,$ and n_k do not have to be equal. Let x_{ij} denote the measured response for the jth item in a sample from the ith population. The following notation will be used in our computations:

ANOVA Notation

Sample sizes: $n_1, n_2, n_3, \ldots, n_k$

Sample means: $\bar{x}_1, \bar{x}_2, \bar{x}_3, \ldots, \bar{x}_k$

Sample variances: $s_1^2, s_2^2, s_3^2, \ldots, s_k^2$

Total sample size: $n = n_1 + n_2 + n_3 + \cdots + n_k$

Grand average: $\bar{\bar{x}} =$ average of all n responses

The double bar in the notation for the grand average is meant to imply that $\bar{\bar{x}}$ is an average of averages. More accurately, $\bar{\bar{x}}$ is a *weighted* average of the k sample means:

$$\bar{\bar{x}} = \left(\frac{n_1}{n}\right)\bar{x}_1 + \left(\frac{n_2}{n}\right)\bar{x}_2 + \left(\frac{n_3}{n}\right)\bar{x}_3 + \cdots + \left(\frac{n_k}{n}\right)\bar{x}_k$$

where the weights $n_1/n, n_2/n, n_3/n, \ldots, n_k/n,$ sum to 1 (because $n = n_1 + n_2 + n_3 + \cdots + n_k$). Alternatively, $\bar{\bar{x}}$ can also be thought of as the sample mean of the combined group of n response values.

With this notation, the **treatment sum of squares** (denoted **SSTr**) and **error sum of squares** (denoted **SSE**) are defined as

$$\text{SSTr} = n_1(\bar{x}_1 - \bar{\bar{x}})^2 + n_2(\bar{x}_2 - \bar{\bar{x}})^2 + n_3(\bar{x}_3 - \bar{\bar{x}})^2 + \cdots + n_k(\bar{x}_k - \bar{\bar{x}})^2$$

$$\text{SSE} = \sum_{j=1}^{n_1}(x_{1j} - \bar{x}_1)^2 + \sum_{j=1}^{n_2}(x_{2j} - \bar{x}_2)^2 + \sum_{j=1}^{n_3}(x_{3j} - \bar{x}_3)^2$$

$$+ \cdots + \sum_{j=1}^{n_k}(x_{kj} - \bar{x}_k)^2$$

SSTr and SSE form the basis of the **between-samples variation** and **within-samples variation** described in Section 9.1. Before these quantities are used to conduct the hypothesis test, however, they must be adjusted to take into account the effects of sample sizes. This is done later in the section.

SSE can also be written in the form

$$\text{SSE} = (n_1 - 1)s_1^2 + (n_2 - 1)s_2^2 + (n_3 - 1)s_3^2 + \cdots + (n_k - 1)s_k^2$$

which more clearly shows how SSE combines or pools the information in the k sample variances $s_1^2, s_2^2, s_3^2, \ldots, s_k^2$. Together, these two sources of variation comprise the **total sum of squares** (denoted **SST**). That is,

$$SST = SSTr + SSE$$

where SST is the sum of squared deviations from the grand mean:

$$SST = \sum_{i=1}^{k} \sum_{j=1}^{n_i} (x_{ij} - \bar{\bar{x}})^2$$

Hypothesis Tests

Until now, our ANOVA formulas have been merely arithmetic constructs. To put these ingredients together to form a statistical procedure, we must be willing to make a few assumptions about the populations being studied. ANOVA tests, in particular, are based on the following assumptions:

ANOVA Assumptions

1. All of the k population variances are equal (i.e., $\sigma_1^2 = \sigma_2^2 = \sigma_3^2 = \cdots = \sigma_k^2$)
2. Each of the k populations follows a normal distribution.

These assumptions are identical to those for the two-sample equal-variance procedures in Exercise 54 of Chapter 7 and Exercise 37 of Chapter 8, which, in fact, are just special cases (when $k = 2$) of the more general single-factor ANOVA test we are currently discussing. If the ratio of the largest sample variance to the smallest one does not exceed 4 by very much, then Assumption 1 is plausible. And for very small sample sizes, this rule is conservative, so 4 can be replaced by 6. Formal test procedures can be found in the chapter references. Assumption 2 can be checked by examining normal quantile plots of each sample or, if sample sizes are quite small, a single quantile plot of the deviations $x_{ij} - \bar{x}_i$ calculated separately within each sample.

When sampling from normal populations, each sum of squares (such as SST, SSTr, and SSE) has its own unique number of **degrees of freedom.** Furthermore, just as SST can be decomposed into the sum of SSTr and SSE, the degrees of freedom associated with these sums of squares also decompose in a similar fashion. In a one-way classification, the **total degrees of freedom** (associated with SST) is $n - 1$, which equals the sum of $k - 1$ (the **degrees of freedom for treatments**) plus $n - k$ (the **degrees of freedom for error**)[1]:

Sums of Squares and Their Degrees of Freedom (Single-factor ANOVA)

Decomposition of sums of squares: **SST   = SSTr + SSE**

Decomposition of degrees of freedom: $n - 1 = k - 1 + n - k$

---

[1] The total degrees of freedom is always $n - 1$, regardless of the ANOVA test we use. However, the *other* sums of squares have df values that depend on the particular test. For example, the df for SSE in a one-way ANOVA is different from the df for SSE in a two-way ANOVA.

Our purpose in finding the degrees of freedom is to convert the sums of squares into **mean squares** by dividing each sum of squares by its associated df. Thus we define

$$\text{Mean square for treatments (between-samples)} = \text{MSTr} = \frac{\text{SSTr}}{k-1}$$

$$\text{Mean square error (within-samples)} = \text{MSE} = \frac{\text{SSE}}{n-k}$$

MSTr and MSE serve as our measures of the between-samples and within-samples variation described in Section 9.1. All of this information is usually organized into an **ANOVA table** (Figure 9.5). The ANOVA table is arranged in column form to emphasize the fact that the sums of squares and degrees of freedom *sum* to SST and $n-1$, respectively.

Source of variation	df	SS	MS	F	P-value
Between samples (treatments)	$k-1$	SSTr	MSTr	MSTr/MSE	
Within samples (error)	$n-k$	SSE	MSE		
Total variation	$n-1$	SST			

Figure 9.5 ANOVA table for the one-way classification

The entry in the F column of the ANOVA table is the test statistic value

$$F = \frac{\text{MSTr}}{\text{MSE}}$$

which is used to test the hypothesis $H_0: \mu_1 = \mu_2 = \mu_3 = \cdots \mu_k$. This F distribution has $(k-1, n-k)$ degrees of freedom since the numerator in MSTr/MSE has df $= k-1$ and the denominator has df $= n-k$. As we mentioned in Section 9.1, the test procedure is always right-tailed; that is, the P-value associated with an F statistic is equal to the area to the *right* of the statistic under the appropriate F density curve. We reject H_0 whenever the P-value of the test statistic F is less than or equal to the desired significance level α. Software packages usually include the P-value in the table.

One-Way ANOVA F Test (Significance level α)

Null hypothesis:	$H_0: \mu_1 = \mu_2 = \mu_3 = \cdots = \mu_k$
Alternative hypothesis:	At least two μ's are different.
Test statistic:	$F = \dfrac{\text{MSTr}}{\text{MSE}}$
P-value:	P is the area under the F density with $(k-1, n-k)$ degrees of freedom to the right of the calculated F.
Decision:	Reject H_0 if P-value $\leq \alpha$.

Example 9.1   Numerous factors contribute to the smooth running of an electric motor ("Increasing Market Share Through Improved Product and Process Design: An Experimental Approach," *Quality Engineering*, 1991: 361–369). In particular, it is desirable to keep motor noise and vibration to a minimum. To study the effect that the brand of bearing has on motor vibration, five different motor bearing brands were examined by installing each type of bearing on different random samples of six motors. The amount of motor vibration (measured in microns) was recorded when each of the 30 motors was running. The data for this study is given in Table 9.1. Because each sample of six motors was selected independently of the other samples, this is a completely randomized design with the factor brand at five levels (brand 1, brand 2, . . . , brand 5). Determining whether the bearing brands have different effects on the response variable (motor vibration) can be accomplished with a one-way ANOVA test. The null hypothesis is $H_0: \mu_1 = \mu_2 = \mu_3 = \mu_4 = \mu_5$, where μ_i average vibration (in microns) for motors using bearings of brand i. We use a significance level of .05 to conduct this test.

Table 9.1 Vibration (in microns) in five groups of electric motors with each group using a different brand of bearing

	Brand 1	Brand 2	Brand 3	Brand 4	Brand 5
	13.1	16.3	13.7	15.7	13.5
	15.0	15.7	13.9	13.7	13.4
	14.0	17.2	12.4	14.4	13.2
	14.4	14.9	13.8	16.0	12.7
	14.0	14.4	14.9	13.9	13.4
	11.6	17.2	13.3	14.7	12.3
Mean:	13.68	15.95	13.67	14.73	13.08
St. dev.:	1.194	1.167	.816	.940	.479

ANOVA Table

Source	df	SS	MS	F
Factor	4	30.88	7.72	8.45
Error	25	22.83	.913	
Total	29	53.71		

The ANOVA calculations proceed as follows. The sum of all $n = 30$ values in the data is 426.7, so the grand mean is $\bar{\bar{x}} = 426.7/30 = 14.22$. Alternatively, we could use the sample means to find $\bar{\bar{x}}$:

$$\bar{\bar{x}} = \left(\frac{n_1}{n}\right)\bar{x}_1 + \left(\frac{n_2}{n}\right)\bar{x}_2 + \left(\frac{n_3}{n}\right)\bar{x}_3 + \cdots + \left(\frac{n_5}{n}\right)\bar{x}_5$$

$$= \left(\frac{6}{30}\right)(13.68) + \left(\frac{6}{30}\right)(15.95) + \left(\frac{6}{30}\right)(13.67) + \left(\frac{6}{30}\right)(14.73)$$

$$+ \left(\frac{6}{30}\right)(13.08)$$

$$= 14.22$$

Furthermore,

$$\text{SSTr} = n_1(\bar{x}_1 - \bar{\bar{x}})^2 + n_2(\bar{x}_2 - \bar{\bar{x}})^2 + n_3(\bar{x}_3 - \bar{\bar{x}})^2 + \cdots + n_5(\bar{x}_5 - \bar{\bar{x}})^2$$

$$= 6(13.68 - 14.22)^2 + 6(15.95 - 14.22)^2 + \cdots + 6(13.08 - 14.22)^2$$

$$= 30.88$$

and

$$\begin{aligned}
\text{SSE} &= (n_1 - 1)s_1^2 + (n_2 - 1)s_2^2 + (n_3 - 1)s_3^2 + \cdots + (n_5 - 1)s_5^2 \\
&= (6 - 1)(1.194)^2 + (6 - 1)(1.167)^2 + \cdots + (6 - 1)(.479)^2 \\
&= 22.83
\end{aligned}$$

Putting these results into the formulas for MSTr and MSE, we find

$$\text{MSTr} = \frac{\text{SSTr}}{k - 1} = \frac{30.88}{5 - 1} = 7.72$$

$$\text{MSE} = \frac{\text{SSE}}{n - k} = \frac{22.83}{30 - 5} = .913$$

which yields the test statistic value

$$F = \frac{\text{MSTr}}{\text{MSE}} = \frac{7.72}{.913} = 8.45$$

Using Appendix Table VIII for the F distribution with $(k - 1, n - k) = (5 - 1, 30 - 5) = (4, 25)$ degrees of freedom, we find that the P-value associated with the test statistic $F = 8.45$ is less than .001. Since this P-value is smaller than the prescribed α of .05, we can reject the hypothesis that all five means are equal and conclude that the type of motor bearing used *does* have a significant effect on motor vibration. In particular, a visual inspection of the sample means in Table 9.1 suggests that brand 5 is the best choice for reducing vibration. In Section 9.3, we present a statistical procedure to sort out which brands are indeed the better ones to use.

Section 9.2 Exercises

10. Five brands of raw materials are tested for their effect on a process yield. Random samples of size 10 are used for each of the materials. Complete the following ANOVA table for this experiment:

Source of variation	df	SS	MS	F
Brand	___	___	___	15.32
Error	___	___	.64	
Total variation	___	___		

11. An experiment was carried out to compare electrical resistivity for six different low-permeability concrete bridge deck mixtures. There were 26 measurements on concrete cylinders for each mixture; these were obtained 28 days after casting. The entries in the accompanying ANOVA table are based on information in the article "In-Place Resistivity of Bridge Deck Concrete Mixtures" (*ACI Materials J.*,

2009: 114–122). A partial ANOVA table for this data follows:

Source	df	Sum of Squares	Mean Square	F
Mixture	___	___	___	___
Error	___	___	13.929	
Total	___	5664.415		

a. Fill in the missing entries in the ANOVA table.
b. State the null and alternative hypotheses of interest in this experiment.
c. Use $\alpha = .05$ to carry out the hypothesis test in part (b).

12. Super duplex stainless steels (SDSS) are iron-based alloys that offer an excellent combination of toughness and mechanical strength. Such alloys are useful for many applications in the chemical and petrochemical industries. Recent research

has shown that the pulsed current gas tungsten arc welding (PCGTAW) process offers superior SDSS welds compared to other methods. The authors of "Optimization of Experimental Conditions of the Pulsed Current GTAW Parameters for Mechanical Properties of SDSS UNS S32760 Welds Based on the Taguchi Design Method" (*J. of the Air and Waste Mgmt. Assoc.*, 2012: 1978–1988) researched the impact of different PCGTAW process parameters on mechanical properties of the welds of a particular SDSS. One investigation focused on seeing how pulse current (A) of the PCGTAW affects the toughness (J) of the SDSS welds. Here are experimental results for toughness measurements under three pulse current settings:

Pulse Current: 100 100 100 120 120 120 140 140 140
Toughness: 39 47 44 52 56 53 40 46 42

Use $\alpha = .05$ to conduct the test for whether there are any differences in the true average weld toughness that may be attributable to the different pulse currents.

13. The article "Influence of Contamination and Cleaning on Bond Strength to Modified Zirconia" (*Dental Materials*, 2009: 1541–1550) reported on an experiment in which 50 zirconium-oxide disks were divided into 5 groups of 10 each. Then a different contamination/cleaning protocol was used for each group. The following summary data on shear bond strength (MPa) appeared in the article:

Treatment:	1	2	3	4	5
Sample mean:	10.5	14.8	15.7	10.0	21.6
Sample sd:	4.5	6.8	6.5	6.7	6.0

a. State the hypotheses of interest in this experiment.
b. Using a significance level of .01, can you conclude that there is a difference between the mean shear bond strength of the five groups?

14. In "Investigation on Machining Performance of Inconel 718 Under High Pressure Cooling Conditions" (*J. of Mech. Engr.*, 2012: 683–690), researchers varied selected high-pressure jet-assisted (HPJA) machining parameters for the nickel-based alloy Inconel 718 and investigated their effect on tool wear.

In one experiment, the researchers machined six specimens of Inconel 718 at each of three different HPJA coolant pressure levels (.6, 10, and 30 MPa) and recorded the corresponding average tool flank wear (ATFW), a combination of abrasive and depth of cut notch wear:

Pressure
.6: 145.00 158.14 157.32 409.42 143.00 135.50
10:   75.00 113.82 76.02 378.65 61.58 183.39
30:   94.03 65.90 102.31 131.62 53.12 108.41

Consider conducting an ANOVA test to see if there are any differences in the true mean ATFW caused by the different coolant pressures. The validity of an ANOVA test depends on the extent to which the two fundamental ANOVA assumptions (normal populations; equal population variances) are satisfied.
a. Create a single normal probability (quantile) plot based on the deviations of the sample data from the sample mean for each of the three samples. Does the assumption of normality appear to hold?
b. The assumption of equal population variances is plausible if the ratio of the largest sample variance to the smallest sample variance is not much more than 4. Is it plausible that the population variances are approximately equal?

15. It is common practice in many countries to destroy (shred) refrigerators at the end of their useful lives. In this process, material from insulating foam may be released into the atmosphere. The article "Release of Fluorocarbons from Insulation Foam in Home Appliances During Shredding" (*J. of the Air and Waste Mgmt. Assoc.*, 2007: 1452–1460) gave the following data on foam density (g/L) for each of two refrigerators produced by four different manufacturers:

Manufacturer:	1	1	2	2	3	3	4	4
Foam Density:	30.4	29.2	27.7	27.1	27.1	24.8	25.5	28.8

Does it appear that true average foam density is not the same for all these manufacturers? State and test the relevant hypotheses using a significance level of $\alpha = .05$. Summarize your analysis in an ANOVA table.

16. According to "Evaluating Fracture Behavior of Brittle Polymeric Materials Using an IASCB Specimen" (*J. of Engr. Manuf.*, 2013: 133–140), researchers have recently proposed an improved test for the investigation of fracture toughness of brittle polymeric

materials. The authors applied this new fracture test to the brittle polymer polymethylmethacrylate (PMMA), more popularly known as Plexiglas, which is widely used in commercial products.

The test was performed by applying asymmetric three-point bending loads on PMMA specimens and varied the location of one of the three loading points to determine its effect on fracture load. In one experiment, three loading point locations based on different distances (mm) from the center of the specimen's base were selected, resulting in the following fracture load data (kN):

Distance	Fracture Load			
42	2.62	2.99	3.39	2.86
36	3.47	3.85	3.77	3.63
31.2	4.78	4.41	4.91	5.06

Here is the corresponding Minitab ANOVA table:

```
One-way ANOVA: Fracture versus Distance

Source  DF   SS      MS      F      P
Dist.    2  6.7653  3.3826  48.58  0.000
Error    9  0.6267  0.0696
Total   11  7.3920
```

a. Use your calculator to confirm Minitab's computations.

b. At a significance level of .01, can you conclude there is a difference among true average fracture loads for the three loading point locations?

c. Returning to the Minitab output, note that the number reported under P corresponds to the P-value. Is the P-value exactly zero? What does it mean when Minitab reports 0.000?

17. In an experiment to study the possible effects of four different concentrations of a chemical on heights of newly grown plants, suppose that an ANOVA test is conducted and that plant height is measured in inches. At a later date, the experimenter decides that plant heights should have been measured in centimeters instead of inches. After multiplying the data in the original samples by 2.54 (1 in. = 2.54 cm), the experimenter wants to know what effect this data conversion will have on the conclusions drawn from the ANOVA test.

a. Use the formulas for SSTr, SSE, SST, MSE, and MSTr to discuss the effect that changing

from inches to centimeters has on the ANOVA calculations.

b. Based on your conclusions in part (a), what general statement can you make about the effect of changing units of measure in an ANOVA test?

18. The accompanying summary data on skeletal-muscle citrate synthase activity (nmol/min/mg) appeared in the article "Impact of Lifelong Sedentary Behavior on Mitochondrial Function of Mice Skeletal Muscle" (*J. of Gerontology*, 2009: 927–939):

	Young	Old Sedentary	Old Active
Sample size	10	8	10
Sample mean	46.68	47.71	58.24

Suppose that the total sum of squares for the experimental data is SST = 2116.81.

a. Construct an ANOVA table for this experiment.

b. Using $\alpha = .05$, can you conclude that true average activity differs for the three groups?

19. A study was conducted to determine whether certain physical properties of asphalt are related to portions of a gel permeation chromatogram of the asphalt ("Methodology for Defining LMS Portion in Asphalt Chromatogram," *J. of Materials in Civil Engr.*, 1997: 31–39). To determine whether certain bands or slices of the chromatogram can be used to distinguish different aging conditions in asphalt, samples of grade AC-10 asphalt were sampled from several sources and artificially aged, some samples for 5 hours and others for 24 hours. Another group of samples was not aged. The following table shows the percentage area of the same slice of the chromatograms of these samples (i.e., area of the slice as a percentage of the entire chromatogram):

	Age of asphalt		
	0 hours	5 hours	24 hours
Mean	3.43	3.18	3.22
Standard deviation	.22	.13	.11
Sample size	6	6	6

Can you conclude (using $\alpha = .05$) that there is a difference between the means for the three age categories?

20. To assess the reliability of timber structures and related building design codes, many researchers have studied strength factors of structural lumber. In one such study ("Size Effects in Visually Graded Softwood Structural Lumber," *J. of Materials in Civil Engr.*, 1995: 19–29), three species of Canadian softwood were analyzed for bending strength. Because the amount of bending depends on the width and length of a board and the particular stress applied, the board dimensions were kept the same in each of the three wood species. Wood samples were selected from randomly selected sawmills, and, according to ASTM Standard D 4761, each sample was conditioned by kiln and air drying to achieve approximately a 15% moisture content. The results of the experiment are given here.

	Bending strength (in MPa)
Douglas-Fir:	65 46 52 39 41 44
Species Hem-Fir:	45 48 32 30 47 50
Spruce Pine- Fir:	42 38 30 28 39 40

Using a significance level of 5%, conduct an ANOVA test to determine whether there is a difference in the mean bending strengths among the three types of wood.

21. Pegged mortise and tenon joints have been used to build wooden structures for centuries. Since the mid-1960s, there has been renewed interest in this method of timber connection because of its inherent strength compared with other methods of connection. In a recent study of the bearing strength of white oak dowels, a random sample of white oak boards was used to create several pegs, from which a random sample of pegs was drawn ("Characterization of Bearing Strength Factors in Pegged Timber Connections," *J. of Structural Engr.*, 1997: 326–332). To determine whether the bearing strength of a peg is affected by the direction with which forces are applied to the peg, three different peg orientations were used in the study: 0°, 45°, and 90°. The pegs were randomly assigned to one of the three orientations, and a stress measurement (in MPa) was recorded for each peg:

Sample number	Peg orientation		
	0°	45°	90°
1	17.7	22.0	19.3
2	17.4	18.7	20.8
3	17.1	20.5	27.5
4	17.3	19.5	19.6
5	16.8	17.4	19.3
6	22.4	22.0	22.3
7	22.3	19.4	22.9
8	20.4	18.3	19.6

a. Conduct an ANOVA test to determine whether the mean bearing strength of the pegs is affected by the orientation of the pegs in the joint connection (use $\alpha = .05$).

b. Would you say the test results in part (a) are favorable or unfavorable for the practice of using wooden pegs in timber connections?

22. Friction in machining processes generates high cutting temperatures that ultimately lead to wear and thermal damage of cutting tools. Fluid is traditionally used to reduce cutting temperature, but this can lead to environmental pollution, health hazards, and higher production costs. An alternative and novel process known as *dry cutting* uses no cooling liquids and has shown great promise for the machining industry to produce components in an economical and ecologically desirable manner.

Within the dry cutting device an interchangeable cooling structure is placed near the cutting tip. The authors of "Design and Analysis of an Internally Cooled Smart Cutting Tool for Dry Cutting" (*J. of Engr. Manuf.*, 2012: 585–591) investigated how various physical characteristics of the cooling compartment affect cutting temperature. Data from one experiment that compared thickness of the cooling structure (mm) to the corresponding cutting temperature (K) is given here:

Thickness	Temperature		
0.5	425.60	426.95	424.30
1.0	415.38	415.04	418.71
1.5	416.91	418.84	418.63

Using a significance level of 5%, can it be concluded that there is a difference among the true mean temperature measurements for the three structure thicknesses?

23. In Exercise 3, a measuring instrument that was out of calibration was used to measure strengths (in kg) of three different alloys. Use the formulas in Section 9.2 to give a more specific answer to the question posed in Exercise 3. That is,
    a. Using the formulas for SSTr, SSE, SST, MSE, MSTr, and F, describe the exact effect the

calibration problem in Exercise 3 will have on the entries in the ANOVA table.
    b. Based on your conclusions in part (a), what general statement can you make about the effect of calibration problems in measuring the response variable of a single-factor ANOVA test?

24. Check the validity of the two fundamental ANOVA assumptions for the data in Exercise 21 by following the steps stated in Exercise 14.

9.3   INTERPRETING ANOVA RESULTS

Effects Plots

A useful way to summarize the results of an ANOVA test is to create a graph showing, on average, how a response variable changes as the levels of the independent variable change. Such graphs are called **effects plots** because they depict the effect of changing the levels of the independent variable. For a factor with k levels, this amounts to simply plotting the sample averages $\bar{x}_1, \bar{x}_2, \bar{x}_3, \ldots, \bar{x}_k$ versus the integers $1, 2, 3, \ldots, k$. To make the graph easier to read, the k averages are also joined by straight-line segments. By following these line segments from point to point, we get a clearer picture of the relationship between the response and independent variables in an experiment.

Statistical software programs often include effects plots to accompany ANOVA calculations. When you look at such printouts, remember that effects plots depict only the between-samples variation in the experiment. They do not show the within-samples variation and, consequently, cannot be used as substitutes for ANOVA tests. Technically, effects plots are only used *after* an ANOVA test shows that the independent variable is statistically significant. Even then, effects plots give a general picture and do not conclusively indicate which factor levels are truly distinct from others. Making that determination requires the use of the multiple comparisons procedures presented later in this section.

Example 9.2

In Example 9.1, we compared five different brands of motor bearings to find out which brands, if any, are better for reducing motor vibration. Because the ANOVA test in that example shows that the factor "brand" is statistically significant, it is permissible to create the effects plot for the five sample means given in Table 9.1. This plot (Figure 9.6) clearly shows that the sample from brand 5 gives the lowest average vibration. However, this fact still does not allow us to conclusively say that brand 5 is the best. It might prove to be the case, for instance, that brands 1 and 3 are about the same as brand 5 in their effectiveness for reducing vibration, even though the effects plot shows that their sample means are slightly higher than the mean for brand 5. Example 9.3 further clarifies the results of this study.

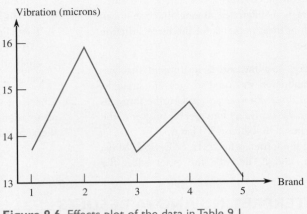

Figure 9.6 Effects plot of the data in Table 9.1

Multiple Comparisons: Tukey's Method

If the F statistic in a single-factor ANOVA test is not significant, then we have no statistical evidence for concluding that the mean response differs at any of the k treatment levels. Depending on the particular problem at hand, a nonsignificant ANOVA test result can be as important as a significant result. Suppose, for instance, that the test in Example 9.1 had turned out to be nonsignificant, so that no statistically significant differences were detected between the five brands of bearings. As long as we are confident that a sufficient amount of data was used to ensure the reliability of the experimental results, a nonsignificant test result would allow us to freely choose any of the five brands to use in producing electric motors. This would be very useful information because the decision of which brand to use could then be based on other considerations, such as a brand's reliability or unit cost.

If the F statistic in an ANOVA test is significant, however, we must do further testing before drawing conclusions. The most common method for doing this involves the use of a **multiple comparisons procedure.** There are several such procedures in the statistics literature. The one we present, called **Tukey's procedure,** was developed by Princeton statistician John Tukey, who is better known to scientists and engineers for inventing the fast Fourier transform (FFT) method and for introducing the term *bit* as a shortened version of *binary digit.*

Tukey's procedure allows us to conduct separate tests to decide whether $\mu_i = \mu_j$ for each pair of means in an ANOVA study of k population means. The method is based on the selection of a "family" significance level, α, that applies to the *entire* collection of pairwise hypothesis tests. For example, when using the Tukey procedure with a significance level of, say, 5%, we are assured that there is at most a 5% chance of obtaining a false positive among the entire set of pairwise tests. That is, there is at most a 5% chance of mistakenly concluding that two population means differ when, in fact, they are equal. This is very different from simply conducting all the pairwise tests as individual tests, each at a $\alpha = .05$, which can result in a high probability of finding false positives among the pairwise tests.

Consider first the case of equal sample sizes. Tukey's procedure is based on comparing the distance between any two sample means, $|\bar{x}_i - \bar{x}_j|$, to a threshold value T that depends on α as well as on the MSE from the ANOVA test. The formula for T is

$$T = q_\alpha \sqrt{\frac{\text{MSE}}{n_i}}$$

where n_i is the size of the sample drawn from each population. The value of q_α is found from a table of right-tail values of a statistic, q, that follows the **Studentized range distribution.** A table of the values of q_α is given in Appendix Table IX. The Studentized range distribution is a probability distribution that depends on a pair of degrees of freedom (k, m), where

$k =$ number of population means to be compared

$m =$ error degrees of freedom

$\quad = n - k$, for single-factor ANOVA

$n = n_1 + n_2 + n_3 + \cdots + n_k$

$\quad =$ total number of observations used in the ANOVA study

To determine whether two means μ_i and μ_j differ, we simply compare $|\bar{x}_i - \bar{x}_j|$ to T. If $|\bar{x}_i - \bar{x}_j|$ exceeds T, then we conclude that $\mu_i \neq \mu_j$. Otherwise, we cannot conclude that there is a difference between the two means.

Tukey's Procedure for Equal Sample Sizes

1. Select a family significance level α at which to conduct the hypothesis tests.

2. Compute $T = q_\alpha \sqrt{\dfrac{MSE}{n_i}}$.

3. Conclude that $\mu_i \neq \mu_j$ if $|\bar{x}_i - \bar{x}_j| > T$.

4. Use bars to connect each pair of means $\bar{x}_i$ and $\bar{x}_j$ for which $|\bar{x}_i - \bar{x}_j|$ does *not* exceed T in Step 3. The corresponding means μ_i and μ_j of such pairs are not considered to differ statistically from one another.

One easy method for keeping track of the results of all these pairwise tests is to arrange the sample means $\bar{x}_1, \bar{x}_2, \bar{x}_3, \ldots, \bar{x}_k$ in increasing order, plot the ordered means along a horizontal line, and then draw horizontal bars connecting pairs of means that are *no* farther than T units apart. These connecting bars are usually drawn in several rows beneath the corresponding means to keep the diagram uncluttered. The bars show which population means do not significantly differ from one another. Likewise, means that are not connected by a bar *do* differ significantly from one another. Figure 9.7 illustrates how this graphical procedure would be used to summarize the multiple comparisons of an ANOVA test using $k = 4$ populations.

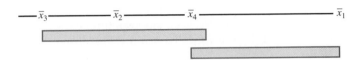

Figure 9.7 Using bars to connect means that do not significantly differ for $\bar{x}_1 = 5$, $\bar{x}_2 = 2$, $\bar{x}_3 = 1$, $\bar{x}_4 = 3$, and critical value $T = 2.5$

Example 9.3   Because the ANOVA test in Example 9.1 is significant, it is necessary to conduct a multiple comparisons procedure to delineate exactly which of the five bearing brands are better than the others. Using Tukey's procedure with a family significance level of $\alpha = .05$, we calculate the critical distance between sample means to be

$$T = q_\alpha \sqrt{\frac{\text{MSE}}{n_i}} = (4.15)\sqrt{\frac{.913}{6}} = 1.62$$

where q_α is based on $(k, n - k) = (5, 25)$ degrees of freedom and is approximated by interpolating between the values of $q_{.05}(5, 24)$ and $q_{.05}(5, 30)$ found in Appendix Table IX. The pairwise distances between the five sample means in Table 9.1 are then compared to T:

Samples	Distance	T	Conclusion
1, 2	$\|13.68 - 15.95\| = 2.27 > 1.62$		μ_1 differs from μ_2
1, 3	$\|13.68 - 13.67\| = .01 < 1.62$		
1, 4	$\|13.68 - 14.73\| = 1.05 < 1.62$		
1, 5	$\|13.68 - 13.08\| = .60 < 1.62$		
2, 3	$\|15.95 - 13.67\| = 2.28 > 1.62$		μ_2 differs from μ_3
2, 4	$\|15.95 - 14.73\| = 1.22 < 1.62$		
2, 5	$\|15.95 - 13.08\| = 2.87 > 1.62$		μ_2 differs from μ_5
3, 4	$\|13.67 - 14.73\| = 1.06 < 1.62$		
3, 5	$\|13.67 - 13.08\| = .59 < 1.62$		
4, 5	$\|14.73 - 13.08\| = 1.65 > 1.62$		μ_4 differs from μ_5

The information from these ten tests is summarized in Figure 9.8 by arranging the five sample means in ascending order and then drawing rows of bars connecting the pairs whose distances do *not* exceed $T = 1.62$. Starting at the left, the top row connects the means that do not significantly differ from μ_5; the next row shows the means that do not differ from μ_4; etc. Using this diagram along with the effects plot (Figure 9.6), we can now summarize what is happening in the ANOVA test. Although brand 5 has the lowest mean, it does not significantly differ from brands 1 and 3 in its effect on vibration. We can conclude, however, that brand 5 is definitely better than brands 2 and 4. Thus the choice of which bearing brand is best has been narrowed to brands 1, 3, and 5. If we are satisfied that the average vibration levels produced by these three brands are acceptable for use in the motors, then the choice could be further narrowed by considering additional factors, such as unit cost and reliability. Figure 9.9 shows the SAS output from the application of Turkey's procedure.

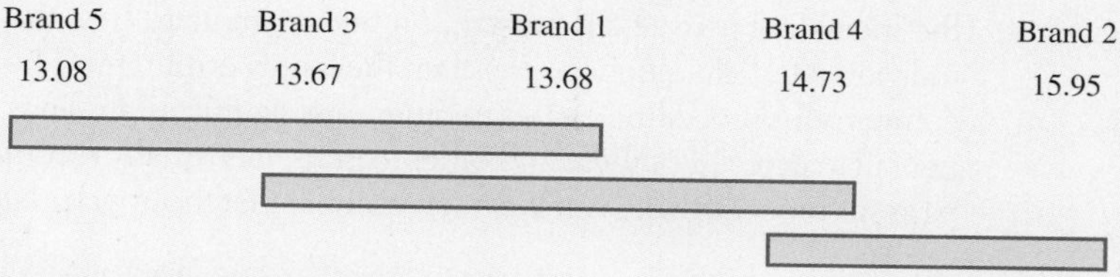

Brand 5	Brand 3	Brand 1	Brand 4	Brand 2
13.08	13.67	13.68	14.73	15.95

Figure 9.8 Summarizing the ten comparisons from Tukey's procedure for the data of Example 9.1

Alpha = 0.05 df = 25 MSE = 0.913533
Critical Value of Studentized Range = 4.15336
Minimum Significant Difference = 1.6206

Means with the same letter are not significantly different.

Tukey Grouping			Mean	N	Brand
		A	15.9500	6	2
		A			
B		A	14.7333	6	4
B					
B		C	13.6833	6	1
B		C			
B		C	13.6667	6	3
		C			
		C	13.0833	6	5

Figure 9.9 Turkey's Method in SAS

Experimental designs that use the same sample size for each treatment level are called **balanced designs**, whereas those with different sample sizes for some treatment levels are called **unbalanced designs**. For an unbalanced design, the Tukey procedure is often run by choosing the minimum of the numbers $n_1, n_2, n_3, \ldots, n_k$ to use in the calculation of the critical value T. This leads to a slightly larger value of T than necessary for multiple comparisons; consequently, this practice is considered a conservative procedure. That is, differences between sample means that exceed T would surely remain significant if larger values of n_i were to be used in the calculation of T. Other modifications of Tukey's procedure include using $T_{ij} = q_\alpha \sqrt{(MSE/2)(1/n_i + 1/n_j)}$ in place of $q_\alpha \sqrt{MSE/n_i}$ when comparing two sample means based on unequal sample sizes.

One question that sometimes arises when first encountering multiple comparisons procedures is: Why not simply conduct such procedures at the outset and bypass the step of conducting the ANOVA test? One answer is that most multiple comparison procedures tend to be not quite so powerful as the ANOVA test for detecting differences between means. The main reason for this is that, faced with a large number of pairwise hypothesis tests, multiple comparisons procedures attempt to avoid the problem of making too many type I errors (i.e., falsely detecting differences between means that are, in fact, equal) by using family significance levels. These family significance levels essentially put more demands on the individual pairwise tests than we might normally do if we were comparing only one pair of means, not several. By controlling the overall, or family, error rate of all the tests, each of the individual pairs tested must pass a higher standard (i.e., the significance levels for each individual test are much smaller than the family error rate). The end result is that a multiple comparisons procedure can sometimes miss significant findings that the ANOVA test would not fail to detect. For these and other reasons, it is usually recommended that multiple comparisons procedures be run *after* determining that the appropriate ANOVA test is significant.

Multiple Comparisons to a Control: Dunnett's Method

Many scientific studies involve comparisons of several treatment populations to a fixed **control population.** For example, in tests for levels of contaminants in water, water samples taken downstream from an industrial discharge source are usually compared to a control sample of water taken upstream from the source. Many biological studies compare the potential effects of drugs or other treatments on treated samples to a control sample that is not treated. In such studies, we are mainly interested in the comparisons between the $k - 1$ treatment means and the mean of a single control sample, but we are not necessarily interested in making *all* possible pairwise comparisons between samples. Multiple comparisons procedures, such as Tukey's method, which take into account all possible pairwise comparisons, are usually too conservative for applications involving control groups. Consequently, alternative procedures, such as **Dunnett's method**, are used when only comparisons to a control are desired.

The steps in Dunnett's method are similar to those in Tukey's method except that the critical value T is computed as

$$T = t_\alpha(k - 1, n - k)\sqrt{\text{MSE}(1/n_i + 1/n_c)}$$

where n_c denotes the sample size used in the control group and n_i is the sample size of the treatment group being compared to the control. The critical value $t_\alpha(k - 1, n - k)$, called **Dunnett's t,** is based on $(k - 1, n - k)$ degrees of freedom, where n is the total of the sample sizes used in the experiment. Values of $t_\alpha(k - 1, n - k)$ can be found in Appendix Table X. Instead of making all $k(k - 1)/2$ possible pairwise comparisons, Dunnett's method involves only $k - 1$ comparisons of the $k - 1$ treatment means to the single control group mean.

Example 9.4	To illustrate Dunnett's method, we reconsider the data of Example 9.1. Suppose that bearings of brand 2 are currently used to manufacture the electric motors and that we want to compare each of four competing brands to brand 2. To conduct such a test of $k = 5$ means, we would use Dunnett's method and compare the $k - 1 = 4$ treatment samples (brands 1, 3, 4, and 5) to the control sample (brand 2). Because the sample sizes are equal, the same T value would be used for all four comparisons. Using a family significance level of a $\alpha = .05$, we find

$$T = t_\alpha(k - 1, n - k)\sqrt{\text{MSE}\left(\frac{1}{n_i} + \frac{1}{n_c}\right)} = (2.61)\sqrt{(.913)\left(\frac{1}{6} + \frac{1}{6}\right)} = 1.440$$

where the value of $t_{.05}(4, 25)$ is found in Appendix Table X to be approximately 2.61. The four comparisons to the control sample yield the following results:

Samples	Distance	T	Conclusion
2, 1	$\|15.95 - 13.68\| = 2.27 > 1.440$		μ_1 differs from μ_2
2, 3	$\|15.95 - 13.67\| = 2.28 < 1.440$		μ_3 differs from μ_2
2, 4	$\|15.95 - 14.73\| = 1.22 < 1.440$		
2, 5	$\|15.95 - 13.08\| = 2.87 < 1.440$		μ_5 differs from μ_2

There is no need to create a bar diagram as in Tukey's method because all four comparisons are being made to a single population, brand 2. The results of the test show that brands 1, 3, and 5 each differ significantly from brand 2, so we are free to choose among these three brands when considering a replacement for brand 2.

Fixed and Random Effects

Factor or treatment levels used in an experiment arise in essentially two ways, each of which forces a different interpretation on the results of an ANOVA test. Sometimes, the factor levels chosen may be the only ones of interest to us. This would be the case, for instance, if the five brands of motor bearings studied in Examples 9.1–9.4 are the only brands of such motor bearings currently available in the market. In this situation, our conclusions pertain only to these five brands and to comparisons between them. A factor whose levels are the only ones of interest in an experiment is called a **fixed factor,** and ANOVA models based on such factors are said to be **fixed effects models.** Alternatively, the levels of a factor may be only a *sample* from a larger population of possible levels. When this is the case, we call the factor a **random factor,** and ANOVA models based on such factors are called **random effects models.** For example, if the five brands studied in Example 9.1 are only a sample from a large population of possible brands, then "brand" would be considered a random factor.

It is important to understand the difference between fixed and random factors for two reasons: (1) The computations of F ratios for testing whether a factor is significant usually depend on whether the factor is fixed or random, and (2) the interpretation of the ANOVA results differs for the two types of factor. The first fact is especially important when working with multifactor models (see Chapter 10). With more than one factor in a model, some factors may be random whereas others are fixed. In such cases, the F ratios for random factors are often calculated differently from F ratios for fixed factors. Fortunately, though, for single-factor ANOVA models, it turns out that the statistical test procedure is identical for either the random or the fixed effects model.

For example, the single-factor ANOVA test of Example 9.1 would be conducted in exactly the same manner, regardless of whether the factor "brand" was considered to be fixed or random. The interpretations, though, would differ in the following ways. If "brand" is a fixed factor, then we would report the ANOVA results by pointing out the significant differences between the population means, $\mu_1, \mu_2, \mu_3, \ldots,$ and μ_k. Furthermore, the conclusions of the study would not be extended beyond these populations. If "brand" is a random factor, however, then the purpose of the study is to extrapolate the ANOVA findings to the larger population from which the factor levels are chosen. In particular, we are interested in estimating how much of the variability in the sample results is due to the variability between the various brands in the population (from which the five brands in the study were selected) and how much is due to the experimental error. These two **components of variance** sum to the total variation, σ^2:

$$\sigma^2 = \sigma_\tau^2 + \sigma_\varepsilon^2$$

where σ_τ^2 denotes the variability in the population from which the treatment levels are chosen and σ_ε^2 is the experimental, or within-samples, error. In the random effects model, the hypotheses we test are $H_0: \sigma_\tau^2 = 0$ versus $H_a: \sigma_\tau^2 > 0$. For the case of equal sample sizes, estimates of σ_τ^2 and σ_ε^2 are given by the formulas

$$\hat{\sigma}_\varepsilon^2 = \text{MSE}$$
$$\hat{\sigma}_\tau^2 = \frac{\text{MSTr} - \text{MSE}}{n_i}$$

Example 9.5   The study of nondestructive forces and stresses in materials furnishes important information for efficient engineering design. The paper "Zero-Force Travel-Time Parameters for Ultrasonic Head-Waves in Railroad Rail" (*Materials Evaluation*, 1985: 854–858) reports on a study of travel time for a certain type of wave that results from longitudinal stress of rails used for railroad track. Three measurements were made on each of six rails randomly selected from a population of rails. The investigators used random effects ANOVA to decide whether some of the variation in travel time could be attributed to "between-rail variability." The data for this experiment and the corresponding ANOVA table appear in Table 9.2.

The error variance is estimated by $\hat{\sigma}_\varepsilon^2 = \text{MSE} = 16.17$, and the estimated variation in the population of rails is $\hat{\sigma}_\tau^2 = (\text{MSTr} - \text{MSE})/n_i = (1862.1 - 16.17)/3 = 615.31$. Furthermore, since the F ratio of 115.2 is highly significant (i.e., it has a low P-value), we can conclude that the differences between rails are an important source of travel-time variability.

Table 9.2 Wave travel times (in nanoseconds)

	Observations			Sample mean
Rail 1	55	53	54	54.00
Rail 2	26	37	32	31.67
Rail 3	78	91	85	84.67
Rail 4	92	100	96	96.00
Rail 5	49	51	50	50.00
Rail 6	80	85	83	82.67

ANOVA Table

Source	df	SS	MS	F
Treatments	5	9310.5	1862.1	115.2
Error	12	194.0	16.17	
Total	17	9504.5		

Section 9.3 Exercises

25. Explain why creating an effects plot does not take the place of performing an ANOVA test.

26. Refer to the data from Exercise 18.
    a. Create an effects plot of the data.
    b. Use Tukey's multiple comparisons procedure to determine which groups differ from one another with respect to CS activity.

27. An experiment to compare the wall coverage area of five different brands of yellow interior latex paint used 4 gallons of each brand. The sample means of the coverage areas (in ft^2/gal) for the five brands were: 462.0, 512.8, 437.5, 469.3, 532.1. The MSE was 272.8 and the computed F statistic for the ANOVA test was found to be significant at $\alpha = .05$. Use Tukey's test (at $\alpha = .05$) to investigate the pairwise differences between the coverage areas of the five brands of paint.

28. In Exercise 27, suppose that the third sample mean is 427.5 (instead of 437.5). Use Tukey's procedure to see which population averages can be considered different from one another ($\alpha = .05$). Use the method of placing bars under those means that are not statistically different from one another. Write a short sentence summarizing your conclusions. Assume the MSE remains the same as in Exercise 27.

29. Repeat Exercise 28 for the case where the sample means are 462.0, 502.8, 427.5, 469.3, 532.1 (i.e., the second and third sample means have been changed from their original values in Exercise 27).

30. Refer to the data from Exercise 19.
    a. Construct an effects plot for this data.
    b. Use Tukey's method with $\alpha = .05$ to determine which age categories differ from each other.
    c. Suppose that asphalt that is not aged is taken to be a control group. Use Dunnett's method with $\alpha = .05$ to decide whether one or both of the aged asphalt groups differ from the control group.

31. Exercise 11 described an experiment in which 26 resistivity observations were made on each of six different concrete mixtures. The article cited there gave the following sample means: 14.18, 17.94, 18.00, 18.00, 25.74, 27.67. Apply Tukey's method using $\alpha = .05$ to identify significant differences, and describe your findings (use MSE = 13.929).

32. In Exercise 16, samples of three different loading points were tested to determine whether there were differences among their average fracture loads.
    a. Draw an effects plot for the data.
    b. Using $\alpha = .05$, apply Tukey's method to determine which if any of the loading points differ from the others.

33. Using a significance level of $\alpha = .05$, apply Tukey's method to the data of Exercise 12. Is there a pulse current that seems to be the best choice to yield maximum average toughness?

9.4 RANDOMIZED BLOCK EXPERIMENTS

Using one's knowledge about a problem, whether it comes from technical experience or common sense, often helps guide the choice of an experimental design. For instance, consider how additional knowledge might affect a comparison of the fuel efficiency (measured in miles per gallon) of several different brands of gasoline. Our first inclination might be to conduct this study as a completely randomized design (Section 9.2) involving the hypothesis $H_0: \mu_1 = \mu_2 = \mu_3 = \cdots = \mu_k$, where μ_i = average mpg obtained using brand i. However, there is a potential problem: Experience tells us that compact cars get better fuel efficiency (higher mpg) than mid-size cars, which, in turn, are more efficient than luxury cars. So what would happen if our random sampling happened to produce a disproportionate number of compact cars in sample 1 (cars that use brand 1)? Clearly, the average mpg of the cars in such a sample would probably be higher than the average mpg for the other samples, even if brand 1 were the *worst* of the four in terms of average fuel efficiency. In fact, it is easy to imagine many scenarios in which the sample means might reflect more about the particular sizes of the automobiles chosen than about the efficiency of the gasoline brands. To avoid such problems, we should use an experimental design that ensures that each brand of fuel is tested on the same range of car sizes.

External influences, such as car size, can be thought of as additional factors to be included in an experimental design. There is usually no need to test such external factors for statistical significance. Either common sense or technical knowledge tells us that they are influential, and our reason for considering them is to make sure that they do not invalidate conclusions about the factor in which we are truly interested (e.g., brand of fuel).

The effect of such external influences can be eliminated, or at least substantially reduced, by using them as **blocks** in an experiment. Blocks are groups of items in a population that have similar characteristics, such as the block of compact cars and the block of mid-size cars. By making sure that a member of each block is included in each of the samples, we can eliminate the effect of external factors on the differences between average responses for the factor we are studying.

For example, to eliminate the influence of car size in the fuel efficiency study, we could select a range of car sizes, call them $B_1, B_2, B_3, \ldots, B_b$, and then make sure that each gasoline brand is used on a car from each of these blocks. Denoting the levels of the factor "gasoline brand" by $A_1, A_2, A_3, \ldots, A_a$, we can summarize the data from such an experiment in matrix form (Figure 9.10).

The design in Figure 9.10, called a **randomized block design**, is the natural extension of the paired-samples test of Section 8.2. In Figure 9.10, blocks of homogeneous experimental units take the place of the data pairs of Section 8.2. Notice, for instance, that the observations in any two rows of this matrix are paired because each level of the blocking factor is represented in each row. Just as in the paired-samples test, the effect of the different blocks is subtracted out when calculating the difference between any two row means (i.e., the differences in the average responses for the levels of factor A).

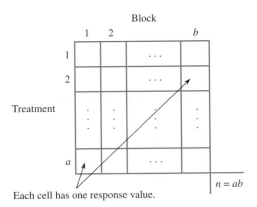

Each cell has one response value.

Figure 9.10 Data layout for a randomized block design

Sums of Squares

In a randomized block design, the total variation SST in the response variable decomposes into three terms, one representing the variation due to the differences in treatment levels (SSTr), one representing the variation between the block means (SSB), and the error term (SSE, which accounts for all other variation):

$$SST = SSTr + SSB + SSE$$

SST, SSTr, and SSB are computed from the formulas shown in the following box:

Sums of Squares Formulas for a Randomized Block Experiment

x_{ij} = observation on factor level i in block j

$$SST = \sum_{i=1}^{a} \sum_{j=1}^{b} (x_{ij} - \bar{\bar{x}})^2$$

$$SSTr = b \sum_{i=1}^{a} (\bar{A}_i - \bar{\bar{x}})^2, \text{ where } \bar{A}_i \text{ denotes the mean of the data the } i\text{th row}$$

$$SSB = a \sum_{j=1}^{b} (\bar{B}_j - \bar{\bar{x}})^2, \text{ where } \bar{B}_j \text{ denotes the mean of the data in the } j\text{th column}$$

The remaining term, SSE, is computed by rewriting the ANOVA decomposition as
$$SSE = SST - SSTr - SSB$$

Hypothesis Tests

Under the usual ANOVA assumptions of normal populations and equal variances, the **total degrees of freedom** associated with SST is $n - 1$, where $n = a \cdot b$. The degrees of freedom for treatments and the blocking factor B are $a - 1$ and $b - 1$, respectively. The remaining degrees of freedom, $(a - 1)(b - 1)$, are associated with the error term:

ANOVA decomposition: **SST** = **SSTr** + **SSB** + **SSE**

Degrees of freedom: $ab - 1 = (a - 1) + (b - 1) + (a - 1)(b - 1)$

The **mean squares** are given by

$$\textbf{MSTr (treatments)} = \frac{SSTr}{a - 1}$$

$$\textbf{MSB (blocks)} = \frac{SSB}{b - 1}$$

$$\textbf{MSE (error)} = \frac{SSE}{(a - 1)(b - 1)}$$

The hypothesis tests for a randomized block design are summarized in the following box:

Randomized Block Test (Significance level α)

Hypothesis	Test statistic	Decision
H_0: there is no treatment effect	$F = \dfrac{MSTr}{MSE}$ with $[a - 1, (a - 1)(b - 1)]$ df	Reject H_0 if $P \le \alpha$

H_0: there is no block effect

$$F = \frac{MSB}{MSE}$$

Reject H_0 if $P \le \alpha$

with $[(b-1),(a-1)(b-1)]$ df

(In each case, the alternative hypothesis is that the particular effect *does* exist.)

Example 9.6   The applications of statistics to crop studies, which began in the 1920s, frequently makes use of a particular blocking variable, the *plot*. As farmers have long known, different plots of land have unique combinations of water, sunlight, and soil chemicals, each having a significant effect on crop growth and yield. Oranges, for example, are so sensitive to different amounts of sunlight that it is a well-known fact that the sweetest oranges come from the south side of the tree.[2]

In a study of different rootstocks for orange trees, four different varieties of rootstock are tested by planting each variety on the same ten plots of land.[3] The numbers of oranges produced by these trees are recorded in Figure 9.11. In this study, the factor A = "variety" has $a = 4$ levels. The blocking factor B = "plot" has $b = 10$ levels.

					Block (plot)							
		1	2	3	4	5	6	7	8	9	10	Average:
Treatment (variety)	1	11	12	10	10	10	9	10	10	10	12	10.4
	2	12	12	10	10	10	9	10	10	10	12	10.5
	3	14	15	12	13	12	12	13	13	14	16	13.4
	4	12	13	10	10	11	9	11	12	11	14	11.3
Average:		12.25	13.0	10.5	10.75	10.75	9.75	11.0	11.25	11.25	13.5	

Figure 9.11 Number of oranges per tree (in 100s) for Example 9.6

The grand average of the 40 values is $\bar{\bar{x}} = 11.4$, and the ANOVA calculations for this data are as follows:

$$SSTr = b\sum_{i=1}^{a}(\bar{A}_i - \bar{\bar{x}})^2$$
$$= 10[(10.4 - 11.4)^2 + (10.5 - 11.4)^2 + (13.4 - 11.4)^2 + (11.3 - 11.4)^2] = 58.2$$

$$SSB = a\sum_{j=1}^{b}(\bar{B}_j - \bar{\bar{x}})^2$$
$$= 4[(12.25 - 11.4)^2 + (13.0 - 11.4)^2 + \cdots + (13.5 - 11.4)^2] = 49.1$$

---

[2] McPhee, J., *Oranges*, Farrar, Straus, Giroux, New York, 1967, p. 8.
[3] Oranges, like roses, are grown by grafting plants with desirable characteristics onto the root structure of *another* plant whose root system is known to be resistant to disease and other problems.

$$\text{SST} = \sum_{i=1}^{a}\sum_{j=1}^{b}(x_{ij} - \bar{\bar{x}})^2$$

$$= (11 - 11.4)^2 + (12 - 11.4)^2 + (10 - 11.4)^2 + \cdots + (11 - 11.4)^2$$
$$+ (14 - 11.4)^2 = 113.6$$

By subtraction, SSE = SST − SSTr − SSB = 113.6 − 58.2 − 49.1 = 6.3. All of this information is summarized in the ANOVA table:

Source of variation	df	SS	MS	F
Treatments (variety)	$a - 1 = 3$	58.2	19.4	83.15
Blocks (plots)	$b - 1 = 9$	49.1	5.456	23.38
Error	$(a - 1)(b - 1) = 27$	6.3	.2333	
Total variation	$ab - 1 = 39$	113.6		

Using a significance level of a $\alpha = .05$, we can conclude that the different varieties do have different mean yields since F = MSTr/MSE = 83.15 has a P-value smaller than .05. We can also conclude that the different plots have differing effects on yield since F = MSB/MSE = 23.38 also has a P-value smaller than .05, although this conclusion only confirms our original belief about the effect of different plots.

When H_0 is rejected, Tukey's method can be applied to identify significant differences among treatments.

Section 9.4 Exercises

34. A pharmaceutical company wants to begin testing a drug designed to reduce blood pressure. The company wants to test the drug by measuring the blood pressures of two samples of people, those who take the drug for a prescribed period of time and those who do not take this drug (or any other medications during the test period). Because researchers know that several human characteristics (e.g., age, weight, diet, exercise) may have considerable effects on the experimental results, they want to run their experiment as a randomized block design. Using the characteristics mentioned, describe how the researchers should go about creating the blocks for such an experiment.

35. A consumer protection organization wants to compare the annual power consumption of five different brands of dehumidifiers. Because power consumption depends on the prevailing humidity level, each brand was tested at four different humidity levels, ranging from moderate to heavy humidity. For each brand, a sample of four humidifiers was randomly assigned, one each, to the four humidity levels. The resulting annual power consumption (in kilowatt-hours) is given in the following table:

Brand	1	2	3	4
		Humidity level		
1	685	792	838	875
2	722	806	893	953
3	733	802	880	941
4	811	888	952	1005
5	828	920	978	1023

a. Using $\alpha = .01$, can you conclude that there is a difference between the power consumptions of the five brands?

b. Using $\alpha = .01$, can you conclude that there are differences in power consumption between the levels of the blocking factor "humidity"? Does this result support the experimenters' use of humidity as a blocking factor?

36. A certain county uses three assessors to determine the values of residential properties. To see whether the three assessors differ in their assessments, five houses are selected and each assessor is asked to determine the market value of each house. Let A denote the factor "assessors" and B denote the blocking factor "houses." An ANOVA calculation revealed that SSA = 11.7, SSB = 113.5, and SSE = 25.6.

    a. Using $\alpha = .05$, test the hypothesis that there are no differences between the average values reported by the three assessors.

    b. Based on the ANOVA results, was the use of houses as a blocking factor warranted in this study?

37. The article "A Software-Based Resource Selection Process in Competitive Network Environment Using ANOVA (A Case Study)" (*Intl. J. of Comp. Appl.*, 2012: 17–21) reported on a study in which three types of lathes were compared. Each of three operators used each of the lathes for the equivalent of a full workday shift. For each shift, the researchers recorded the percentage of acceptable products manufactured by the operator. The data from the experiment is given here:

		Lathe Brand		
		1	2	3
	1	86	86	88
Operator	2	85	86	91
	3	82	83	85

    a. Using the three operators as blocks, can you conclude that there is a difference among the percent of acceptable products due to lathes? (Use $\alpha = .05$)

    b. Can you conclude that the different operators have differing effects on product acceptability rate? (Use $\alpha = .05$).

38. In the article "The Effects of a Pneumatic Stool and a One-Legged Stool on Lower Limb Joint Load and Muscular Activity During Sitting and Rising" (*Ergonomics*, 1993: 519–535), the following data is given on the effort (measured on the Borg scale) required by a subject to arise from sitting on four different stools. Because it was suspected that different people could exhibit large differences in effort, even from the same type of stool, a sample of nine people was selected and each was tested on all four stools:

		Subject								
		1	2	3	4	5	6	7	8	9
Type of stool	A	12	10	7	7	8	9	8	7	9
	B	15	14	14	11	11	11	12	11	13
	C	12	13	13	10	8	11	12	8	10
	D	10	12	9	9	7	10	11	7	8

    a. Using a significance level of $\alpha = .05$, can you conclude that there is a difference in the average effort required to rise from each type of stool?

    b. Do the differences in rising effort that the researchers expected seem to be confirmed by the data?

39. To assess the potential risks associated with failure of a particular process, investigators often perform a *failure modes and effects analysis* (FMEA). An FMEA identifies opportunities for failure, known as *failure modes*, in a given process. Each mode is assessed with a numeric score based on (1) severity of the consequences of failure, (2) likelihood of failure occurrence, and (3) likelihood that failure would not be detected. The product of these scores is the *risk priority number* (RPN) for the mode. Modes having the highest RPN values are usually given the highest priority in carrying out further analyses.

    The article "Continuous Quality Improvement in Investment Castings: An Experimental Study using a Modified FMEA Approach Called FEAROM" (*Eur. J. of Sci. Res.*, 2012: 308–325) reported on a study that compared four design methods (M_1, M_2, M_3, M_4) in preproduction trials of the upper range for a particular casting valve. The design methods are applied by human operators, which introduces potential operator-to-operator variation in RPN values. To account for this, each of the four design methods was used (in random order) by all 21 individuals in the study.

    The data was analyzed by the R software, giving the following output. Note that the format of the ANOVA table in R is very similar to the one we use, except R eliminates the row of "totals" and uses the word *residuals* instead of *error*. The column labeled 'Pr(>F)' represents *P*-value.

```
           Df Sum Sq Mean Sq F-value Pr(F)
DESIGN     ?  519515    ?       ?      ?
PERSON     ?    ?      5023     ?     0.445
Residuals  ?  293009    ?
```

a. Fill in the missing values in the table above.
b. Using $\alpha = .05$, can it be concluded that there is a difference in the true average RPN among the four design methods?
c. Do the person-to-person differences in RPN seem to be confirmed by the data? Explain.

40. In the study described in Exercise 20, the wood grade is known to affect wood strength. To incorporate this information, three wood grades were studied: SS (select structural), grade 2, and grade 3. Wood grades are determined by visual inspection. The following table shows bending strengths from testing wood samples of each type and grade:

		Wood grade		
		SS	Grade 2	Grade3
	Douglas Fir	65	43	41
Species	Hem-Fir	45	38	32
	Spruce-Pine-Fir	42	35	30

a. Using the three wood grades as blocks, can you conclude that there is a difference between the mean bending strengths of the three species of wood? (use $\alpha = .05$.)
b. Can you conclude that there are differences between the mean bending strengths for the three grades of wood? (use $\alpha = .05$.)
c. Suppose that wood with a large bending strength is needed for a particular structure and that any wood grade is acceptable. Which type and grade of wood is best for such a structure?

d. Explain why your conclusions about wood types in this experiment differ from the conclusions reached in Exercise 20.

41. Example 4.15 (Chapter 4) describes a randomized block experiment for comparing three different methods (A, B, and C) of curing concrete. Different batches of concrete are used as the blocks in the experiment. For convenience, the data from Table 4.1 is repeated here:

Batch	Method A	Method B	Method C
	Strength (in MPa)		
1	30.7	33.7	30.5
2	29.1	30.6	32.6
3	30.0	32.2	30.5
4	31.9	34.6	33.5
5	30.5	33.0	32.4
6	26.9	29.3	27.8
7	28.2	28.4	30.7
8	32.4	32.4	33.6
9	26.6	29.5	29.2
10	28.6	29.4	33.2

a. Using a significance level of 5%, can you conclude that there is a difference in mean concrete strength between the three curing methods?
b. Can you conclude that there are differences between the batch means? (Use $\alpha = .05$.)
c. Suppose that you ignore the fact that the batches are blocks in this experiment and that you simply run a one-factor ANOVA test, treating the three columns of data as three random samples. Using a significance level of .05, what conclusion do you reach regarding the differences between the three curing methods?

Supplementary Exercises

42. The authors of "Statistical Analysis and Optimization Study on the Machinability of BerylliumCopper Alloy in Electro Discharge Machining" (*J. of*

Engr. Manuf., 2012: 1847–1861) investigated the machinability of berylliumcopper alloy in an electro discharge machining (EDM) process. The accompanying data resulted from an EDM process using an oil dielectric medium where researchers applied four different EDM pulse times (μs) and recorded the corresponding material removal rate (MRR, in mm^3/s).

		MRR			
	20	0.1797	0.3353	0.4073	0.7548
Pulse	40	0.2433	0.3830	0.5625	0.7258
Time	60	0.2338	0.3372	0.5552	0.7453
	80	0.1341	0.2806	0.5502	0.8212

Use $\alpha = .05$ to conduct the test for whether there are any differences in the true average MRR that may be attributable to the different pulse times.

43. The lumen output was determined for three different brands of 60-watt soft-white light bulbs, with eight bulbs of each brand tested. From the resulting lumen measurements, the following sums of squares were computed: SSE = 4773.3 and SSTr = 591.2.
    a. State the hypotheses of interest. Describe, in words, the parameters that appear in the hypotheses.
    b. Compute each of the entries in the ANOVA table for this experiment.
    c. Using $\alpha = .05$, can you conclude that there are any differences between the average lumen outputs for the three brands?

44. In the study described in Exercise 12, the authors also investigated how pulse current affects the hardness of the SDSS welds. Hardness is measured in HV (known as the *Vickers number*; higher values indicate harder metals).

Pulse Current: 100 100 100 120 120 120 140 140 140

Hardness: 326 296 312 245 273 276 299 296 282

Use $\alpha = .05$ to conduct the test for whether there are any differences in the true average weld hardness attributable to the different pulse currents.

45. In the special case where $df_1 = 1$, the right-tail areas associated with an F distribution are related to similar areas under a t distribution's density curve.

In particular, it can be shown that $F_\alpha = (t_{\alpha/2})^2$, for an F distribution with $df_1 = 1$ and any value of df_2 and for a t distribution with $df = df_2$. The subscripts α and $\alpha/2$ on F_α and $t_{\alpha/2}$ denote right-tail areas of α and $\alpha/2$ under the density curves for the F and t distributions, respectively.
    a. Verify this relationship by looking up $F_{.05}(df_1 = 1, df_2 = 10)$ and $t_{.025}(df = 10)$ in the F and t tables, Appendix Tables VIII and IV, respectively.
    b. For $\alpha = .05$, the values of $t_{\alpha/2}$ approach $z_{\alpha/2} = 1.96$ as the degrees of freedom increase. What limit does $F_{.05}(df_1 = 1, df_2)$ approach as df_2 increases?

46. Consider the following data on plant growth after the application of five different types of growth hormone:

		Data		
A	13	17	7	14
B	21	13	20	17
C	18	15	20	17
D	7	11	18	10
E	6	11	15	8

    a. Perform the F test for this single-factor ANOVA at $\alpha = .05$.
    b. Apply Tukey's procedure to this data with $\alpha = .05$. Compare your results to the conclusion obtained in part (a).

47. Consider a single-factor ANOVA in which samples of size 5 each are measured at each of three levels of a certain factor. The means of the three samples are 10, 12, and 20. Find a value of SSE that satisfies the following two requirements:
    (1) The calculated F statistic is larger than the tabled value of F for $\alpha = .05$, $df_1 = 2$, and $df_2 = 12$, so the hypothesis $H_0: \mu_1 = \mu_2 = \mu_3$ is rejected at $\alpha = .05$.
    (2) When Tukey's procedure is applied, none of the three μ_i's can be said to differ from one another (again using $\alpha = .05$).

48. For the data referenced in Exercise 39, the article reported that there was a difference in RPN means for the four design methods (M_1, M_2, M_3, M_4). Perform a post hoc analysis by applying Tukey's procedure

(as the authors did) using the following output from the SAS software:

```
Alpha = 0.05 df = 60 MSE = 4883.488
Critical Value of Studentized Range =
            3.73709
Minimum Significant Difference = 56.989
```

Means with the same letter are not significantly different.

Tukey Grouping	Mean	N	trt
A	336.00	21	M2
A			
A	301.00	21	M4
B	171.43	21	M3
B			
B	155.71	21	M1

49. In Exercise 47, suppose that the three sample means are 10, 15, and 20. Can you now find a value of SSE that satisfies the two conditions in Exercise 47?

50. Helmet-mounted displays (HMDs) are computer displays that are presented on see-through screens attached to the helmets of helicopter pilots. HMDs are normally employed to aid night flights. In a study of HMDs, researchers tested Apache helicopter pilots to determine whether the presence of in-flight vision problems has an effect on a pilot's ability to focus the HMD panel. Thirteen pilots were divided into two groups: those who experience certain in-flight vision problems and those who do not. Subjects were asked to set the focus of the HMD for a fixed test pattern, and their focus settings were then measured with a dioptometer ("Oculomotor Responses with Aviator Helmet-Mounted Displays and Their Relation to In-Flight Symptoms," *Human Factors*, 1995: 699–710). The data from one such experiment is given here:

In-Flight symptom tested: Distance misperception (measurements are in diopters)

	Symptom present	Symptom absent
Sample size	9	4
Sample mean	−.83	−.70
Sample standard deviation	.172	.184

a. Using $\alpha = .01$, conduct an ANOVA test to determine whether there is a difference in the average focus settings between the two groups of pilots.

b. Which test procedure in Chapter 8 could have been used on this data in place of the ANOVA test in part (a)?

c. Conduct the appropriate test you identified in part (b), using $\alpha = .01$, and compare your answer to the answer in part (a).

51. The results on the effectiveness of line drying on the smoothness of fabric were studied in the paper "Line-Dried vs. Machine-Dried Fabrics: Comparison on Appearance, Hand, and Consumer Acceptance" (*Home Econ. Research J.*, 1984: 27–35). Smoothness scores were given for nine types of fabric and five different drying methods. Because the different types of fabric were expected to have large differences in smoothness, regardless of drying method, each of the five drying methods was used on five samples of each fabric type. The smoothness scores for this experiment were as follows:

	Drying method				
Fabric type	**1**	**2**	**3**	**4**	**5**
Crepe	3.3	2.5	2.8	2.5	1.9
Double knit	3.6	2.0	3.6	2.4	2.3
Twill	4.2	3.4	3.8	3.1	3.1
Twill mix	3.4	2.4	2.9	1.6	1.7
Terry	3.8	1.3	2.8	2.0	1.6
Broadcloth	2.2	1.5	2.7	1.5	1.9
Sheeting	3.5	2.1	2.8	2.1	2.2
Corduroy	3.6	1.3	2.8	1.7	1.8
Denim	2.6	1.4	2.4	1.3	1.6

a. Construct an ANOVA table for this experiment.

b. Using a significance level of .05, can you conclude that there is a difference between the mean smoothness scores for the five drying methods?

52. A consumer protection organization carried out a study to compare the electricity usage for four

different types of residential air-conditioning systems. Each system was installed in five homes and the monthly electricity usage (in kilowatt-hours) was measured for a particular summer month. Because of the many differences that can exist between residences (e.g., floor space, type of insulation, type of roof, etc.), five different groups of homes were identified for study. From each group of homes of a similar type, four homes were randomly selected to receive one of the four air-conditioning systems. The resulting data is given in the table.

		Type of home				
		1	2	3	4	5
Air-conditioning system	1	116	118	97	101	115
	2	171	131	105	107	129
	3	138	131	115	93	110
	4	141	141	115	93	99

a. Construct an ANOVA table for this experiment.
b. Using a significance level of .05, can you conclude that there is a difference between the monthly mean kilowatt-hours of electricity used by the four types of air conditioners?

Bibliography

Montgomery, D. C., **Design and Analysis of Experiments** (8th ed.), Wiley, New York, 2012. *The first half of the book gives a good introduction to statistical inference and the analysis of variance method. The remaining chapters give an equally readable account of the experimental design techniques described in Chapter 10.*

Ott, R.L. and M. Longnecker, **Introduction to Statistical Methods and Data Analysis** (6th ed.), Cengage Learning, Belmont, CA, 2008. *A practitioner's guide to analysis of variance and experimental design. Emphasizes applications, calculations, and interpretation of results.*

10

Experimental Design

INTRODUCTION

Methods of **experimental design** are used to evaluate the effects of several different treatments on a response variable. In the field of agronomy, where experimental design techniques were first applied in the 1920s, different fertilizer blends (the treatments) were applied to a crop in an effort to find the particular blend that maximized crop yield (the response). The essential statistical ideas underlying experimental design lie in the commonsense notion that the usefulness of the conclusions drawn from an experiment will critically depend on how the experiment is conducted.

Scientific applications of experimental design methods are often called **design of experiments** (abbreviated **DOE**). Furthermore, the designs discussed in this chapter are from a special class called **factorial designs.** The multifactor designs presented in this chapter are an extension of the single-factor designs discussed in Chapter 9. Consequently, the terminology in Section 10.1 builds on that already introduced in Chapter 9. Sections 10.2 and 10.3 show how to conduct factorial experiments and how to interpret the results from such experiments.

Throughout the chapter, the statistical tool of **analysis of variance (ANOVA)** is used to analyze the data from experiments and to make decisions about whether a given factor has a significant impact on a response variable.

In addition, the graphical tools of **effects plots** and **probability plots** provide very simple, yet powerful methods for visually summarizing the results of an experiment and for sorting out factors that are influential from those that are not. Effects plots, which were introduced in Section 9.3, are discussed in Section 10.2, and probability (quantile) plots, first introduced in Section 2.4, are used throughout Sections 10.4 and 10.5.

Sections 10.4 and 10.5 deal with a class of factorial designs called 2^k **designs.** These designs have been widely used in industrial and scientific applications. Because each factor in a 2^k design is restricted to only two levels, the resulting statistical analyses are simplified, making these designs very intuitive and easy to use.

10.1   TERMINOLOGY AND CONCEPTS

Much of the terminology of experimental design has already been introduced in Sections 4.3 and 9.1. Recall from those discussions that a **response variable,** or more simply, a **response,** is a measurable characteristic of a product or process that we would like to study. The object of the study is to determine the extent to which various **factors** (also called **independent variables**) affect the values of the response variable. Experiments are carried out by simply changing the **levels** of each factor and then measuring whether, and by how much, the response changes. **Experimental designs** are specific procedures that stipulate exactly how each factor is to be varied to obtain the most information from the experimental data.

One of the most surprising things to come out of Fisher's original work on experimental design in the 1920s was the realization that the intuitive one-factor-at-a-time approach to experimentation has several disadvantages. One-factor-at-a-time experiments are conducted by allowing one factor to vary at a time, keeping the levels of all other factors fixed while doing so. By successively testing each factor in this manner, an experimenter hopes to determine both the individual and combined effects that the factors have on a response variable. Fisher pointed out how inefficient the one-factor-at-a-time approach is and suggested that better experiments could be designed by using **factorial designs** along with the statistical tools of **randomization, replication,** and **blocking** (Section 4.3).

There are several major difficulties with one-factor-at-a-time experiments: (1) They require more experimental runs than do the factorial designs discussed in this chapter; (2) they are incapable of detecting how the interplay between two or more factors influences a response variable; and (3) they usually cannot detect the specific levels of each factor that will optimize a response variable. In short, one-factor-at-a-time experiments fail to achieve most of the important goals of an experimenter.

We illustrate each of these shortcomings by reconsidering the discussion from Example 4.15. In that example, two factors, the particular injection molding machine used (machine 1 or machine 2) and the brand of plastic pellets used (brand A or brand B) in the machines were thought to affect the hardness of molded plastic parts. In the terminology of experimental design, machine and brand are the factors and plastic hardness is the response variable. Using the one-factor-at-a-time approach, an experimenter might conduct a series of six tests, as shown in Figure 10.1.

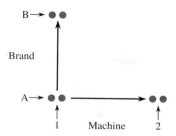

Figure 10.1 Experimental runs in a one-factor-at-a-time experiment

The arrows in the figure indicate the direction in which the factor is varied (e.g., test runs for brand are first done for brand A and machine 1). Two experimental runs are made at each fixed combination of factor settings to help increase the precision of the estimates derived from the data. Employing repeated measurements, or **replication,** is an intuitive method often used in experiments to reduce errors introduced by outside factors that can bias experimental results. Along the horizontal axis in Figure 10.1, the experimenter holds the brand factor fixed (i.e., only brand A is used) and allows the machine factor to vary. Then, holding the machine factor fixed (at machine 1), the brand factor is varied as shown on the vertical axis. A total of six experimental measurements are made using this one-factor-at-a-time method. To estimate the effect of changing from machine 1 to machine 2, the experimenter can compare the average of the two response values for machine 1 with the average of the two responses for machine 2. The difference between these two averages is a measure of how much the response changes when the machine factor is varied. Similarly, the difference in the two averages associated with brands A and B can be used to measure the effect of varying the brand factor.

Figure 10.1 highlights one of three problems with one-factor-at-a-time experiments mentioned above: the inability of this design to capture all the information about the interplay between factors. Suppose, for illustration, that the plastic of brand B works about the same as brand A does in machine 1, but that brand B works significantly better in machine 2 than in machine 1. If so, such information would not be seen in the results of the experiment shown in Figure 10.1. Instead, the data from the one-factor-at-a-time experiment would show that there was very little effect on hardness when changing the brand factor, since brand B is evaluated only on machine 1. From those results, an experimenter would incorrectly conclude that changing plastic brands has little effect on the hardness of the molded parts. As you can see from Figure 10.1, this potential problem is caused by the fact that the one-factor-at-a-time approach does not include any experimental runs using plastic of brand B on machine 2.

In contrast, the designs introduced in this chapter are constructed to expressly take into account the possibility of significant interplay between factors. In statistics, such interplay between factors is called **interaction.** Two or more factors are said to **interact** if, as described in the previous paragraph, the magnitude of a factor's effect on the response variable depends on the particular level(s) of the other factor(s) in the experiment.

In our example, the effect of changing plastic brands on plastic hardness was negligible when machine 1 was used, but the brand effect becomes substantial when machine 2 was used. Thus there is an interaction between brand and machine. Interactions between factors are discussed in more detail in Section 10.2.

Figure 10.2 shows an experimental design that does allow for detecting such an interaction, if it exists. This design is an example of the factorial designs discussed throughout the chapter. One of the important features of such designs is that experimental tests are conducted at many, if not all, combinations of the levels of the factors. In particular, note that the design in Figure 10.2 includes a test measurement for the combination of machine 2 with plastic brand B. If there is an interaction between the two factors, this design will be able to detect it.

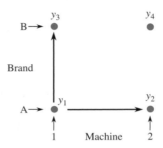

Figure 10.2 A factorial design using two factors

Another significant feature of the design in Figure 10.2 is that only one measurement is made at each of the combinations of factor levels, which means that a total of four experimental runs are needed. This brings up the question of whether this four-run experiment is capable of estimating the factor effects with the same precision as the one-factor-at-a-time experiment, in which each factor effect is estimated as the difference between two averages, each based on two measurements. To answer this question, we denote the four test measurements in Figure 10.2 by y_1, y_2, y_3, and y_4. First consider the factor *machine*. The difference $y_2 - y_1$ estimates the change in plastic hardness caused by changing machines when brand A is used on both machines. Similarly, the difference $y_4 - y_3$ estimates the effect of changing machines when brand B is used on both. Therefore, by averaging these two estimates, we obtain a more precise estimate of the effect of changing machines:

$$\text{machine effect} = \frac{1}{2}\left[(y_2 - y_1) + (y_4 - y_3)\right]$$

By rearranging this expression, we can write the machine effect in the form

$$\text{machine effect} = \frac{1}{2}(y_2 + y_4) - \frac{1}{2}(y_1 + y_3)$$

which shows that the machine effect is estimated by the difference between two averages, each based on two measurements, just as is done in the one-factor-at-a-time experiment that uses six experimental runs. Thus the four-run factorial not only is able to achieve the same degree of precision as the one-factor-at-a-time experiment but also does so with *fewer* experimental runs. Using the same reasoning, we can show that the factor *brand* is also measured with the same precision and can be written

$$\text{brand effect} = \frac{1}{2}(y_3 + y_4) - \frac{1}{2}(y_1 + y_2)$$

As the preceding paragraph illustrates, factorial experiments are more efficient than one-at-a-time experiments. In fact, as you will see in later sections, the efficiency of factorial designs compared to one-factor-at-a-time experiments increases as more and more factors are included in an experiment. As Figure 10.2 shows, factorial experiments achieve their efficiency by using the data more than once. Note, for example, that the same four data values in Figure 10.2 are used in *both* of the effects estimates described in the previous paragraph. Cuthbert Daniel, one of the pioneers in applying factorial designs to industrial processes, describes this feature of factorial designs as "making each piece of data work twice," an expression originally credited to the statistician W. J. Youden.[1]

To demonstrate that one-factor-at-a-time experiments do not generally yield the optimum settings for each factor, it is helpful to imagine what would happen if we were fortunate enough to know the exact relationship between the factors and the response variable. Suppose, for instance, that such information is available for a particular response value y and two factors whose measured values are denoted by x_1 and x_2. Thus we can find the exact value of y associated with any two values of x_1 and x_2 and, therefore, create a graph of y versus x_1 and x_2. Such a graph is called a **response surface.** Figure 10.3 is an idealized example of a response surface, which illustrates how the percentage yield y of a process might be related to the levels of two factors known to affect process yield. From this graph, it is easy to find the particular values of x_1 and x_2 that will maximize the percentage yield y. In a real experiment, of course, the shape of the response surface is unknown, and the experimenter's goal is to come as close as possible to the settings of x_1 and x_2 that optimize the response variable.

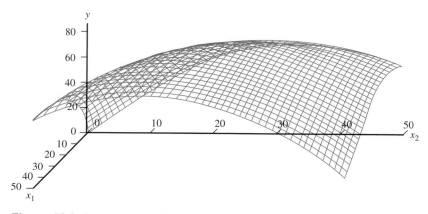

Figure 10.3 A response surface of process yield y (in percent) versus the values x_1 and x_2 of two factors

[1] Daniel, C., *Application of Statistics to Industrial Experimentation*, John Wiley & Sons, New York, 1976: 3.

Another way to summarize the information in a response surface is to create a **contour plot.** Contour plots are similar to two-dimensional topographical maps in that they consist of a series of lines in the plane connecting all points (x_1, x_2) with a common y value. For example, by connecting all points (x_1, x_2) whose associated y value is 90%, a **contour** line is formed in the plane. By comparing the contours associated with other y values, the reader can then form a mental image of how the height of the response surface changes. Figure 10.4 shows a contour plot created from the response surface of Figure 10.3. Notice how much easier the contour plot makes the task of finding the x_1 and x_2 coordinates of the point where the surface achieves its maximum. For this reason, we will now use the contour plot to illustrate why one-factor-at-a-time experiments generally fail to find the optimum factor settings.

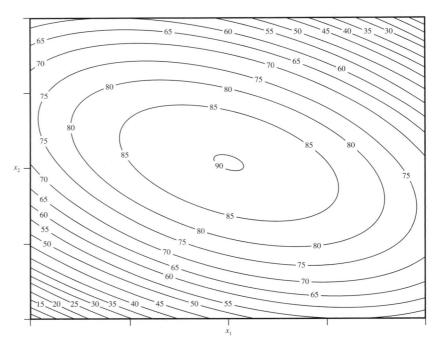

Figure 10.4 Contour plot of the response surface in Figure 10.3

Figures 10.5 and 10.6 show two different experimental strategies that could be followed in a one-factor-at-a-time experiment. Suppose, for illustration, that a process is currently running with the two factors set at the values associated with point A in the figures. Starting with Figure 10.5, suppose that an experimenter begins by varying the values of x_1 (keeping x_2 fixed) and tries to maximize the process yield. As Figure 10.5 shows, the best value of x_1 occurs near point B in the figure. Next, keeping x_1 fixed at its value from point B, the experimenter then varies x_2 until its optimum value is found near point C. The experimenter would conclude that both factors had been optimized and that the best process yield possible is about 86%.

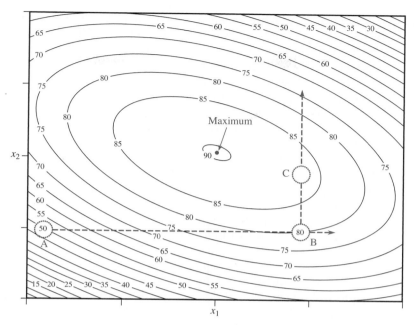

Figure 10.5 Contour plot of a one-factor-at-a-time experiment: changing x_1 to a new value, then changing x_2

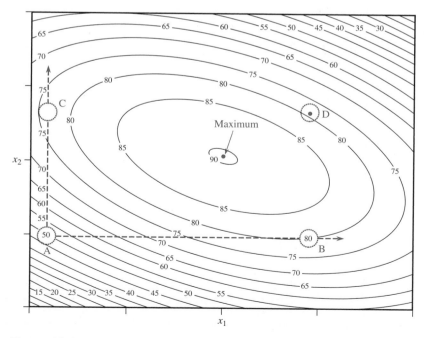

Figure 10.6 Contour plot of a one-factor-at-a-time experiment: separate searches for x_1 and x_2 values are combined

Alternatively, an experimenter could employ the strategy shown in Figure 10.6, in which x_1 is first varied until point B is found, x_1 is returned to its original value from point A, and then x_2 is varied until its best value is found at point C. Putting these results together, the experimenter might surmise that the best combination of x_1 and x_2 is at point D, which uses the x_1 coordinate from point B along with the x_2 coordinate from point C. This time, the experimenter concludes that the optimum process yield is about 79%. In both cases, the experimenter has indeed improved the process yield, but in neither case has the optimum yield been located.

In practice, one-factor-at-a-time procedures usually require that several experiments be conducted to ascertain the approximate location of the points B and C illustrated in Figures 10.5 and 10.6. Thus not only do such experiments generally fail to pinpoint optimal factor settings but several repeated tests are also needed to do so. By comparison, factorial experiments require much less experimentation and usually come closer to achieving the goal of finding optimum factor settings. To see why this happens, consider Figure 10.7, which shows the results of running a factorial experiment near the starting point A. Based on the size of the response values at the four corners of this factorial design, it is readily apparent that the experimenter should move in the direction indicated by the arrow in Figure 10.7. By repeating this process at points B and C, the experimenter quickly determines the optimum factor settings.

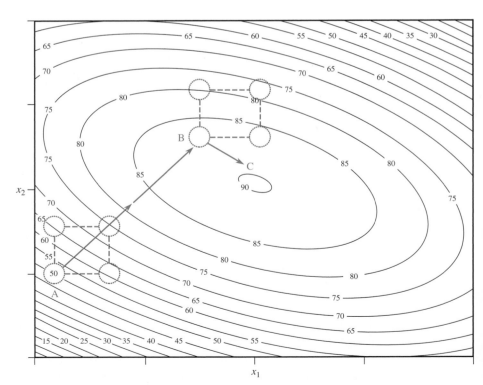

Figure 10.7 Using factorial designs to search for optimum factor settings

Section 10.1 Exercises

1. What statistical purpose does replication serve in an experimental design?

2. Factors A and B are thought to have an effect on a certain response value, y. The following table contains data on the response variable measured at each combination of the two levels of factors used in a study:

		Factor A level	
		1	2
Factor B level	1	5.2	7.4
	2	4.0	6.3

a. Calculate an estimate of the effect of changing factor A from level 1 to level 2.

b. Calculate an estimate of the effect of changing factor B from level 1 to level 2.

3. Suppose that the response surface for a two-factor experiment can be described by the function $f(x, y) = e^{(-1/2)[(x-2)^2+(y-5)^2]}$.

a. Use a computer package to create a graph of the response surface.

b. From the graph in part (a), determine the approximate coordinates of the point (x, y) at which the response surface is at its maximum.

c. Find an equation that describes the typical contour of the response surface.

d. Sketch some of the contours using your answer to part (b). From this sketch, determine the approximate coordinates of the point at which the response surface is at its maximum from these contours.

4. Suppose that the response surface for a two-factor experiment can be described by the function $f(x, y) = e^{-(x-y)^2}$.

a. Use a computer package to create a graph of the response surface.

b. Find an equation that describes the contours of the response surface.

c. Sketch some of the contours using the equation(s) in part (b). Using these results, determine from these contours the approximate coordinates of the point(s) at which the response surface is at its maximum.

10.2 TWO-FACTOR DESIGNS

In a **two-factor design,** two factors (labeled A and B in the ensuing discussion) are specified along with the number of levels of each factor, which are denoted by a and b, respectively. For example, suppose that we want to expand the motor vibration study described in Example 9.1 to include two factors, A = brand of bearing used in the motor and B = material used for the motor casing. If we decide to use five bearing brands and three types of casing material, then a = 5 and b = 3 for such an experiment.

A two-factor design is often denoted as an **$a \times b$ design** (read "a by b design"). The design in the previous paragraph would therefore be called a 5 × 3 design. In addition to allowing us to quickly read the number of levels for each factor, this notation reminds us of multiplication (e.g., 5 × 3 = 15), because the product of a and b happens to be the number of distinct treatments (i.e., different combinations of factor levels) created by the two factors. Thus in the motor vibration study, there are 15 distinct combinations of bearing brand and casing material that must be included in the experiment. Although it is certainly possible to conduct any number of tests at each factor–level combination, it simplifies the calculations if we choose the *same* number of items for each such treatment. Designs that use the same number of samples for each factor–level combination are called **balanced designs.** We will use the letter r (which stands for *repeated measures* or *replicates*) to denote the common sample size selected from each factor–level combination in a balanced design.

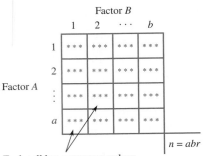

Each cell has r response values.

Figure 10.8 Data layout for a balanced two-factor design with r replicates per cell

The most convenient way to keep track of the information in a two-way design is to display it in matrix form, as shown in Figure 10.8. Each of the $a \times b$ combinations has its own **cell** in which r response values are recorded. With r values in each of the $a \times b$ cells, the total number of experimental runs is $n = a \times b \times r$.

Main Effects and Interactions

Graphs of the average response versus the factor levels can reveal much about the influence the factors have on a response variable. Such graphs are called **effects plots** since they illustrate the effect that changing the levels of a factor has on the response variable. Recall that effects plots for one-factor experiments were first introduced in Section 9.3.

One simple rule governs all effects plots: A *plotted point corresponding to any factor level (or factor–level combination) is simply the average of all response values in which that factor level (or factor–level combination) is present.* The following example illustrates the process of creating effects plots from the data in a two-way design matrix. Suppose the data for a 3×2 design is as follows:

		Factor B		
		1	2	
	1	10, 14	18, 14	$14 = (10 + 14 + 18 + 14)/4$
Factor A	2	23, 21	16, 20	$20 = (23 + 21 + 16 + 20)/4$
	3	31, 27	21, 25	$26 = (31 + 27 + 21 + 25)/4$

$$21 = \frac{10 + 14 + \cdots + 27}{6} \qquad 19 = \frac{18 + 14 + \cdots + 25}{6}$$

In the margins of the matrix, we have included the averages of all the responses in the rows and columns. For instance, the average response for the first row is 14, which is the average of all four numbers in that row. Notice that these four numbers each correspond to the first level of factor A, which we will denote by A_1 in the graphs that follow. Also, because we have used a *balanced* design, each level of B is included an *equal* number of times in these four numbers, which is what makes the average response of 14 a good representation of what to expect at level A_1. As you can see, each level of B is also represented in the four numbers used to find the average responses for levels A_2 and A_3.

Following our general rule for computing effects averages, we compute the average responses for B_1 and B_2 from six numbers, because each column (i.e., each level of B) in the matrix contains a total of six measurements.

By plotting the average response versus the levels of a factor, we obtain a graph of the **main effect** of that factor. In Figure 10.9, for example, the plots of the main effects of factors A and B in our example show that the average response tends to increase as factor A changes from level A_1 to A_2 to A_3, whereas changing factor B from B_1 to B_2 has the effect of decreasing the average response from 21 to 19. Plotting both the A and B main effects on the same graph allows you to easily compare the magnitudes—of the A and B effects. In those cases where the average response stays relatively constant from level to level (e.g., if the average response had been 20 at *all three* levels of A), we say that a factor has **no main effect.**

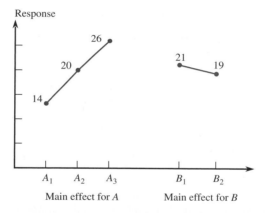

Figure 10.9 Plotting the main effects of factors A and B

From Figure 10.9, we can see that going from level A_1 to level A_3 has the net effect of raising the average response by 12 units (i.e., from 14 at A_1 to 26 at A_3) and that going from B_1 to B_2 lowers the average response by 2 units (from 21 at B_1 to 19 at B_2). Looking at this figure, it is tempting to want to treat A and B separately, by simply choosing desirable settings first for A, then for B. If this were always the case, it would make the results of a two-way experiment exceedingly easy to interpret. Unfortunately, things are not always that simple.

It is possible, as noted in Section 10.1, that two (or more) factors do not act independently of one another. Two factors are said to **interact** when the effect of changing the levels of one factor *depends* on the particular level of the other factor. This is the case in our example. The following calculations show that the effect of changing factor A depends on the particular setting of factor B:

Effect of changing A (B fixed at B_1)

$$\left.\begin{array}{ll} A_1 \text{ and } B_1 & (10+14)/2 = 12 \\ A_2 \text{ and } B_1 & (23+21)/2 = 22 \\ A_3 \text{ and } B_1 & (31+27)/2 = 29 \end{array}\right\}17$$

Effect of changing A (B fixed at B_2)

$$\left.\begin{array}{ll} A_1 \text{ and } B_2 & (18+14)/2 = 16 \\ A_2 \text{ and } B_2 & (16+20)/2 = 18 \\ A_3 \text{ and } B_2 & (21+25)/2 = 23 \end{array}\right\}7$$

Notice that the effect of going from A_1 to A_3 is an increase of 17 units when B is at level B_1, whereas the corresponding increase is only 7 units when B is at level B_2. Thus the effect of changing the levels of A seems to depend on the particular level of B.

Like main effects, such **two-factor interaction effects** can also be plotted. This can be done as shown in Figure 10.10 by overlaying separate graphs, one for each level of factor B. Alternatively, the two values of B could be used on the horizontal axis with three overlaid graphs (one for each level of A). The presence of interaction between two factors is indicated by graphs that either cross one another or, more generally, are not parallel. Parallel graphs, as depicted in Figure 10.11, are a sign of *no* interaction between the factors. Why?

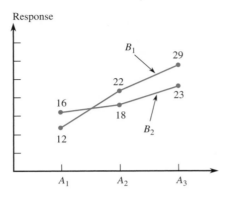

Figure 10.10 A two-factor interaction plot

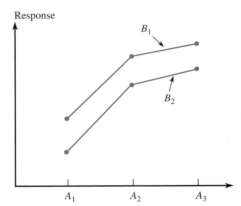

Figure 10.11 A plot showing no interaction between factors A and B

Keep in mind that effects plots do *not* take the place of statistical tests. You should always run an ANOVA test first to determine which of the effects are significant and which are not. It may turn out, for example, that the interaction effect is not statistically significant, in which case you can interpret the main effects without worrying about factor interactions. At other times, you may discover that a factor that you initially thought was important turns out to have no significant effect on the response variable.

When statistical testing shows that an interaction effect is significant, then the results of the experiment must be interpreted by examining the interaction plots, *not* the main effects plots. When interactions exist, the conclusions drawn from the main effects plots may or may not agree with those drawn from the interaction plots. On the other hand, if the interaction between factors is not significant, then you can simply examine and interpret the main effects plots. For instance, in our example, neither the main effect for factor B nor the interaction effect is significant at $\alpha = .05$ (see Exercise 6). This means that we need only examine the main effects plot for factor A. If the goal of the study is, say, to maximize the response value, then the main effects plot suggests that we set factor A at level 3. Because the main effect for factor B is not significant and because the interaction between A and B is not significant, choosing either level of B should give substantially the same response value.

ANOVA Formulas

All ANOVA procedures share a common goal: to analyze the total variation (SST) in a response variable by breaking it into identifiable sources of variation. This is accomplished by defining a separate sum of squares for each source of variation and then decomposing SST into a sum of these components. Such formulas are called **ANOVA decompositions.**

The general ANOVA decomposition for a two-factor analysis of variance is

$$SST = SSA + SSB + SS(AB) + SSE$$

where SSA, SSB, and SS(AB) denote the sums of squares associated with factor A, factor B, and the AB interaction, respectively. SSE, the error or residual sum of squares, represents the variation from all sources of variation *other* than A, B, and their interaction. The formulas for these sums of squares are given in the following box.[2] Note that once SST, SSA, SSB, and SSE are computed, SS(AB) can easily be found by rewriting the ANOVA decomposition as

$$SS(AB) = SST - SSA - SSB - SSE$$

Sums of Squares Formulas (Balanced Two-Way ANOVA)

$$SSA = br \sum_{i=1}^{a} (\bar{A}_i - \bar{\bar{y}})^2$$

$$SSB = ar \sum_{j=1}^{b} (\bar{B}_j - \bar{\bar{y}})^2$$

SST = sum of squared deviations of all individual response values y_{ijk} from the grand average, $\bar{\bar{y}}$

$$= \sum_{i=1}^{a} \sum_{j=1}^{b} \sum_{k=1}^{r} (y_{ijk} - \bar{\bar{y}})^2$$

SSE = sum of squared deviations of response values y_{ijk} from the corresponding cell means, $\bar{y}_{ij}$

$$= \sum_{i=1}^{a} \sum_{j=1}^{b} \sum_{k=1}^{r} (y_{ijk} - \bar{y}_{ij})^2$$

$$SS(AB) = SST - SSA - SSB - SSE$$

where

y_{ijk} = kth observation when A is at level i and B is at level j

a = number of levels of factor A

b = number of levels of factor B

r = number of replications per cell

$\bar{A}_i$ = average of all response values associated with the ith level of factor A

$\bar{B}_j$ = average of all response values associated with the jth level of factor B

Hypothesis Tests

We now proceed to find the degrees of freedom and mean squares associated with each source of variation. The **total degrees of freedom** is $n - 1$, where $n = abr$. The degrees of freedom associated with a *factor* is simply its number of levels minus 1, and the

---

[2] In the precomputer era, shortcut formulas were often used instead of the formulas we have given. The interested reader may consult other texts for these formulas.

degrees of freedom for an **interaction** term is the *product* of the degrees of freedom of the corresponding factors. The **error degrees of freedom** equals $ab(r-1)$. Decomposition of degrees of freedom mimics that for sums of squares:

ANOVA decomposition:
$$SST = SSA + SSB + SS(AB) + SSE$$

Degree of freedom:
$$abr - 1 = (a-1) + (b-1) + (a-1)(b-1) + ab(r-1)$$

By dividing each sum of squares by its degrees of freedom, we form the **mean squares:**

$$MSA = \frac{SSA}{a-1} \qquad MS(AB) = \frac{SS(AB)}{(a-1)(b-1)}$$

$$MSB = \frac{SSB}{b-1} \qquad MSE = \frac{SSE}{ab(r-1)}$$

These are used to form the F ratios used in our hypothesis tests. In a two-way ANOVA, we can conduct separate tests for the presence of each main effect and the interaction effect. In each such test, the null hypothesis is that the effect does *not* exist, and the alternative hypothesis is that the effect *is* present. To conclude, for example, that the factor A (or B) effect is present means that the average response *differs* at different levels of A (or B). The following box summarizes the test procedures for a two-factor ANOVA. An ANOVA table (Figure 10.12) provides the most convenient way to summarize these results.

Two-Way ANOVA Tests (Significance Level α)

To test these hypotheses:	Test statistic	Degrees of freedom for P-value determination
H_{0A}: There is no main effect for A	$F = \dfrac{MSA}{MSE}$	$a - 1,\ ab(r-1)$
H_{0B}: There is no main effect for B	$F = \dfrac{MSB}{MSE}$	$b - 1,\ ab(r-1)$
H_{0AB}: There is no AB interaction effect	$F = \dfrac{MS(AB)}{MSE}$	$(a-1)(b-1),\ ab(r-1)$

Reject H_0 if P-value $\leq \alpha$ (In each case, H_a is that the particular effect *does* exist.)
If H_{0AB} is rejected, then the interaction plot takes precedence over the main effects plots when interpreting the effects of A and B.

Source of variation	df	SS	MS	F
Factor A	$a - 1$	SSA	MSA	MSA/MSE
Factor B	$b - 1$	SSB	MSB	MSB/MSE
AB interaction	$(a-1)(b-1)$	SS(AB)	MS(AB)	MS(AB)/MSE
Error	$ab(r-1)$	SSE	MSE	
Total variation	$abr - 1$	SST		

Figure 10.12 ANOVA table for the two-way classification

Technically speaking, the statistical tests just described are based on a **fixed effects model,** in which the particular levels of A and B are assumed to be the *only* ones of interest in the study. If, on the other hand, we think of the levels only as *samples* from all of the possible levels of A and B, then a **random effects model** should be used. Recall that the distinction between fixed and random effects was first introduced in Section 9.3. Although the distinction between fixed and random factors does not alter the ANOVA calculations for one-factor experiments (Chapter 9), this situation changes for multifactor designs. In particular, the calculation of F ratios for random effects models and **mixed models** (one factor fixed, the other random) are slightly different from those of the fixed effects models. These topics are beyond the scope of our introductory discussion. Throughout this chapter, we consider all factors in a design to be fixed factors.

Example 10.1   Refer to Example 9.1, where we examined the possible causes of electric motor vibration. Suppose that we have identified two product characteristics (factors) that are thought to influence the amount of vibration (the response, measured in microns) of running motors: factor A = the brand of bearing used in the motor and B = the material used for the motor casing. Figure 10.13 shows the data from an experiment in which a = 5 brands of bearings were tested along with b = 3 types of casing material (steel, aluminum, and plastic). Two motors (r = 2) were constructed and tested for *each* of the ab = 5 · 3 = 15 combinations of bearing brand and casing type, giving a total sample size of abr = 5 · 3 · 2 = 30.

Factor B
(casing material)

		1	2	3	Averages:
	1	13.1, 13.2	15.0, 14.8	14.0, 14.3	14.07
	2	16.3, 15.8	15.7, 16.4	17.2, 16.7	16.35
Factor A (brand)	3	13.7, 14.3	13.9, 14.3	12.4, 12.3	13.48
	4	15.7, 15.8	13.7, 14.2	14.4, 13.9	14.62
	5	13.5, 12.5	13.4, 13.8	13.2, 13.1	13.25
Averages:		14.39	14.52	14.15	

Figure 10.13 Data on electric motor vibration for Example 10.1

Before proceeding with the ANOVA calculations, it is instructive to look at the margins of the data array in Figure 10.13. In particular, note that there appears to be very little difference between the average responses for the three levels of factor B, which is a preliminary indication that factor B may have little or no effect on reducing vibration.

The sum of all 30 response values is 430.6, so the grand average is $\bar{\bar{y}} = 430.6/30 = 14.353$, which, with the row averages $\overline{A}_1 = 14.07, \overline{A}_2 = 16.35, \overline{A}_3 = 13.48, \overline{A}_4 = 14.62$, and $\overline{A}_5 = 13.25$ gives

$$SSA = b \cdot r \sum_{i=1}^{a} (\overline{A}_i - \bar{\bar{y}})^2$$
$$= 3 \cdot 2 \big[(14.07 - 14.353)^2 + (16.35 - 14.353)^2 + (13.48 - 14.353)^2$$
$$+ (14.62 - 14.353)^2 + (13.25 - 14.353)^2 \big]$$
$$= 6 [6.118125] = 36.709$$

Similarly, the column averages $\overline{B}_1 = 14.39, \overline{B}_2 = 14.52, \overline{B}_3 = 14.15$ yield

$$SSB = a \cdot r \sum_{j=1}^{b} (\overline{B}_j - \bar{\bar{y}})^2$$
$$= 5 \cdot 2 \big[(14.39 - 14.353)^2 + (14.52 - 14.353)^2 + (14.15 - 14.353)^2 \big]$$
$$= 10 [.070467] = .705$$

The total sum of squares is the sum of the squared differences of all 30 values from $\bar{\bar{y}}$:

$$SST = (13.1 - 14.353)^2 + \cdots + (13.1 - 14.353)^2 = 50.655$$

whereas SSE is the sum of the squared differences of each response value from its own cell mean,

$$SSE = \big[(13.1 - 13.15)^2 + (13.2 - 13.15)^2 \big]$$
$$+ \big[(15.0 - 14.9)^2 + (14.8 - 14.9)^2 \big] + \cdots$$
$$= 1.670$$

By subtraction, $SS(AB) = SST - SSA - SSB - SSE = 50.655 - 36.709 - .705 - 1.670 = 11.571$. These results, along with their associated degrees of freedom and mean squares, are summarized in the following ANOVA table:

Source of variation	df	SS	MS	F
Factor A (bearing brand)	$5 - 1 = $ 4	36.709	36.709/4 = 9.177	9.177/.1113 = 82.45
Factor B (casing material)	$3 - 1 = $ 2	.705	.705/2 = .353	.353/.1113 = 3.17
AB interaction	$(5-1)(3-1) = $ 8	11.571	11.571/8 = 1.446	1.446/.1113 = 12.99
Error	$5 \cdot 3(2-1) = $ 15	1.670	1.670/15 = .1113	
Total variation	$5 \cdot 3 \cdot 2 - 1 = $ 29	50.655		

At a significance level of $\alpha = .05$, let's first test for the presence of any interactions. Because the P-value for $F = 12.99$ (based on $df_1 = 8$, $df_2 = 15$) is less than .001, H_{0AB} must be rejected. It appears that there is interaction between the two factors. Therefore we should consider the corresponding effects plot (see Figure 10.14) to draw conclusions. Although the casing material does not have a significant effect by itself, it *does* influence the A main effect (because the AB interaction is significant). The lowest vibration occurs for bearing brand A_3, but *only* if casing B_3 (plastic casing) is used with A_3.

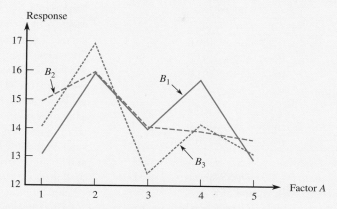

Figure 10.14 Effects plot for Example 10.1

Section 10.2 Exercises

5. Why do parallel line segments in effects plots indicate that there is no interaction between two factors?

6. In the example discussed on page 454, perform the necessary hypothesis tests to show that neither factor B nor the two-factor AB interaction is significant (using $\alpha = .05$).

7. A fixed effects model is used to analyze two factors, each of which has five levels. Three replicated measurements are available for each combination of factor levels. Complete the following ANOVA table for this experiment:

Source of variation	df	SS	MS	F
Factor A	___	20	___	___
Factor B	___	___	___	8.1
AB interaction	___	___	___	___
Error	___	___	2	
Total variation	___	200		

8. A chemical engineer conducts an experiment to test the effects of gas flow rate (factor A) and liquid flow rate (factor B) on the gas film heat transfer coefficient (in Btu/hr ft^2). Four levels of each factor are used in the study, and two replications are conducted at each combination of factor levels:

		Factor B			
		1	2	3	4
	1	200, 211	226, 219	240, 249	261, 250
	2	278, 267	312, 324	330, 337	381, 375
Factor A	3	369, 355	416, 402	462, 457	517, 524
	4	500, 487	575, 593	645, 632	733, 718

a. Is there evidence of a significant interaction between the two factors? Use $\alpha = .01$.

b. Use $\alpha = .01$ to test the hypothesis that gas flow rate has no effect on the heat transfer coefficient.

c. Use $\alpha = .01$ to test the hypothesis that liquid flow rate has no effect on the heat transfer coefficient.

9. The following data was obtained in an experiment to investigate whether the yield from a certain chemical process depends on either the chemical formulation of the input materials or the mixer speed, or on both factors:

		Speed		
		60	70	80
		189.7	185.1	189.0
	1	188.6	179.4	193.0
		190.1	177.3	191.0
Formulation				
		165.1	161.7	163.3
	2	165.9	159.8	166.6
		167.6	161.6	170.3

A statistical software package gave these results: SS(formulation) = 2253.44, SS(speed) = 230.81, SS(interaction) = 18.58, and SSE = 71.87.

a. Does there appear to be interaction between the two factors? (Use $\alpha = .05$.)

b. Does the yield appear to depend on either the formulation or the speed? (Use $\alpha = .05$.)

10. Draw an interaction plot for the data of Exercise 9.

11. Lightweight aggregate asphalt mix has been found to have lower thermal conductivity, which is desirable, than a conventional mix would have. The article "Influence of Selected Mix Design Factors on the Thermal Behavior of Lightweight Aggregate Asphalt Mixes" (*J. of Testing and Eval.*, 2008: 1–8) reported on an experiment in which various thermal properties of mixes were determined. Three different binder grades were used in combination with three different coarse aggregate contents (%), with two observations made for each such combination, resulting in the conductivity data (W/m·K) given here:

		Coarse Aggregate Content (%)		
		38	41	44
Asphalt	PG58	.835, .845	.822, .826	.785, .795
Binder	PG64	.855, .865	.832, .836	.790, .800
Grade	PG70	.815, .825	.800, .820	.770, .790

a. Test for the presence of interaction between the two factors. Use $\alpha = .01$.

b. Use $\alpha = .01$ to test the hypothesis that coarse aggregate content has no effect on thermal conductivity.

c. Use $\alpha = .01$ to test the hypothesis that asphalt binder grade has no effect on thermal conductivity.

12. Factorial designs have been used to study productivity of software engineers ("Experimental Design and Analysis in Software Engineering," *Software Engineering Notes*, 1995: 14–16). Suppose that an experiment is conducted to study the time it takes to code a software module. Factors that may affect the coding time are the size of the module and whether the programmer has access to a library of previously coded submodules. Module size is studied at two levels, large and small, whereas access to a library of submodules is either available or not. After running a two-factor design on sample modules, suppose that the interaction between module size and library access is found to be significant.

a. If the goal is to reduce coding time, describe the conclusions you can draw from the experiment if the interaction plot looks like this:

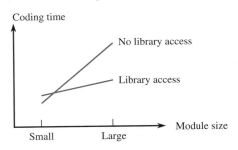

b. What possible reasons can you give for an interaction plot that looks like the following one?

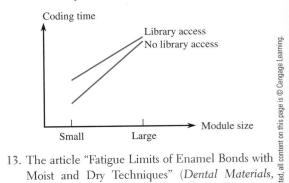

13. The article "Fatigue Limits of Enamel Bonds with Moist and Dry Techniques" (*Dental Materials*, 2009: 1527–1531) described an experiment to investigate the ability of adhesive systems to bond to mineralized tooth structures. The response variable is shear bond strength (MPa), and two different

adhesives—Adper Single Bond Plus (SBP) and OptiBond Solo Plus (OBP)—were used in combination with two different surface conditions. The accompanying data was supplied by the authors of the article. The first 12 observations came from the SBP-dry treatment, the next 12 from the SBP-moist treatment, the next 12 from the OBP-dry treatment, and the last 12 from the OBP-moist treatment.

```
SBP-Dry    56.7 57.4 53.4 54.0 49.9 49.9
           56.2 51.9 49.6 45.7 56.8 54.1
SBP-Moist  49.2 47.4 53.7 50.6 62.7 48.8
           41.0 57.4 51.4 53.4 55.2 38.9
OBP-Dry    38.8 46.0 38.0 47.0 46.2 39.8
           25.9 37.8 43.4 40.2 35.4 40.3
OBP-Moist  40.6 35.5 58.7 50.4 43.1 61.7
           33.3 38.7 45.4 47.2 53.3 44.9
```

a. Construct a comparative boxplot of the data on the four different treatments and comment.
b. Carry out an appropriate analysis of variance and state your conclusions (use a significance level of .01 for any tests). Include any graphs that provide insight.
c. If a significance level of .05 is used for the two-way ANOVA, the interaction effect is significant (just as in general different glues work better with some materials than with others). So now it makes sense to carry out a one-way ANOVA on the four treatments SBP-D, SBP-M, OBP-D, and OBP-M. Do this and identify significant differences among the treatments.

14. Experiments often have more than one response value of interest. In the article "Towards Improving the Properties of Plaster Moulds and Castings" (*J. Engr. Manuf.*, 1991: 265–269), a study was undertaken to determine the effects of carbon fiber (in %) and sand addition (in %) on two response variables, casting hardness and wet-mold strength.

Sand addition (%)	Carbon fiber addition (%)	Casting hardness	Wet-mold strength
0	0	61.0	34.0
0	0	63.0	16.0
15	0	67.0	36.0
15	0	69.0	19.0
30	0	65.0	28.0
30	0	74.0	17.0
0	.25	69.0	49.0
0	.25	69.0	48.0
15	.25	69.0	43.0
15	.25	74.0	29.0
30	.25	74.0	31.0
30	.25	72.0	24.0
0	.50	67.0	55.0
0	.50	69.0	60.0
15	.50	69.0	45.0
15	.50	74.0	43.0
30	.50	74.0	22.0
30	.50	74.0	48.0

a. Construct an ANOVA table for the effects of these factors on wet-mold strength. Test for the presence of significant effects using $\alpha = .05$.
b. Construct an ANOVA table for the effects of these factors on casting hardness. Test for the presence of significant effects using $\alpha = .05$.
c. From your results in parts (a) and (b), what levels of each factor would you select to maximize wet-mold strength? What factor levels would you choose to maximize casting hardness?

10.3 MULTIFACTOR DESIGNS

The two-factor designs of Section 10.2 can be extended to include any number of factors A, B, C, D, . . . , each with its own number of levels a, b, c, d, . . . , and so on. These **factorial designs,** as they are called, require that experimental runs be made at all possible combinations of the factor levels. As in the two-factor case, the total sample size for a factorial design is the product of the number of factor levels times the number of **replicates,** r. A four-factor experiment, for example, would require $n = a \cdot b \cdot c \cdot d \cdot r$ sample measurements. Needless to say, sample sizes can grow rapidly as more and more factors are included in an experiment, a problem that is addressed in Sections 10.4 and 10.5 of this chapter.

The "×" notation used to describe two-factor designs also provides compact descriptions of multifactor designs. For instance, a $3 \times 2 \times 2$ factorial design is one that has three factors A, B, and C, with $a = 3$ levels of A, $b = 2$ levels of B, and $c = 2$ levels of C. Figure 10.15 shows a data layout for such a design. Note that the number of replicates, r, is not included in this notation. To indicate that repeated measurements have been made at each factor–level combination, we simply state this fact when referring to the design, for example, a replicated $3 \times 2 \times 2$ design, or a $3 \times 2 \times 2$ design with r replicates.

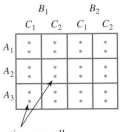

r observations per cell

Figure 10.15 Data layout for a $3 \times 2 \times 2$ factorial design (three levels of factor A, two levels of B, two levels of C)

Main Effects and Interactions

In multifactor designs, main effects and two-factor interactions are interpreted in the same manner as in two-factor designs (Section 10.2). With more than two factors, however, the opportunity arises to incorporate even higher-order interactions, such as the interaction between three or more factors. Notationally, a three-factor interaction between factors A, B, and C is denoted by either ABC or $A \times B \times C$, a four-factor interaction is denoted by either $ABCD$ or $A \times B \times C \times D$, and so forth. The basic rule for interpreting interaction terms is the same as in Section 10.2: If an interaction is statistically significant, then each of the component factors' effects will depend on the particular combination of the other factors in the interaction term. For instance, the presence of a significant ABC interaction means that the effect of factor A is different at different settings of factors B and C, or, equivalently, that the effect of B depends on the levels of A and C, or that the effect of C depends on the levels of A and B.

Operationally, the presence of an interaction term indicates that we must first look at interaction plots, not main effects plots, when interpreting the experimental results. However, factors that are *not* involved in significant interaction terms may be interpreted by simply examining their main effects plots. For example, suppose that an ANOVA test reveals that both the main effect for factor A and the BC interaction are significant. This means that we can look at the main effects plot for A when deciding on the best setting for A, but the BC interaction plot must be consulted when determining the best settings for factors B and C.

ANOVA Formulas

ANOVA decompositions for factorial designs contain a sum of squares term for every possible main effect and interaction. For example, a factorial design based on factors A, B, and C gives rise to three main effects terms (A, B, and C), three two-factor interactions

(AB, AC, and BC), and one three-factor interaction (ABC). The sum of squares for an interaction term is denoted by putting the interaction term in parentheses after the SS notation. Thus SS(AB) denotes the sum of squares associated with the AB interaction, and so on. The ANOVA decomposition for a three-factor model is given by

$$\text{SST} = \text{SSA} + \text{SSB} + \text{SSC} + \text{SS(AB)} + \text{SS(AC)} + \text{SS(BC)} + \text{SS(ABC)} + \text{SSE}$$

where SST (the total sum of squares) measures the total variation in the response data and SSE (the error sum of squares) is the variation from all sources *other* than the factors included in the experiment.

For a three-factor design including every possible main effect and interaction, computational formulas for sums of squares are given in the following box. The key is to start by computing the sums of squares of main effects and then use these results to find sums of squares for the two-factor interactions. Similarly, the sums of squares of the two-factor interactions are used to find SS(ABC). Although the patterns evident in these formulas can be extended to the case of four or more factors, in practice one usually relies on statistical software to perform the calculations.

Sum of Squares Formulas (Balanced Three-Factor ANOVA)

Let G denote the grand total of all $n = abcr$ response values.

$$\text{SSA} = \sum_{i=1}^{a} \frac{A_i^2}{bcr} - \frac{G^2}{abcr}, \text{ where } A_i = \text{sum of all data for } i\text{th level of factor } A$$

$$\text{SSB} = \sum_{j=1}^{b} \frac{B_j^2}{acr} - \frac{G^2}{abcr}, \text{ where } B_j = \text{sum of all data for } j\text{th level of factor } B$$

$$\text{SSC} = \sum_{k=1}^{c} \frac{C_k^2}{abr} - \frac{G^2}{abcr}, \text{ where } C_k = \text{sum of all data for } k\text{th level of factor } C$$

$$\text{SS(AB)} = \sum_{i=1}^{a} \sum_{j=1}^{b} \frac{AB_{ij}^2}{cr} - \frac{G^2}{abcr} - \text{SSA} - \text{SSB}, \text{ where } AB_{ij} = \text{sum of all data for } i\text{th level of } A \text{ and } j\text{th level of } B$$

$$\text{SS(AC)} = \sum_{i=1}^{a} \sum_{k=1}^{c} \frac{AC_{ik}^2}{br} - \frac{G^2}{abcr} - \text{SSA} - \text{SSC}, \text{ where } AC_{ik} = \text{sum of all data for } i\text{th level of } A \text{ and } k\text{th level of } C$$

$$\text{SS(BC)} = \sum_{j=1}^{b} \sum_{k=1}^{c} \frac{BC_{jk}^2}{ar} - \frac{G^2}{abcr} - \text{SSB} - \text{SSC}, \text{ where } BC_{jk} = \text{sum of all data for } j\text{th level of } B \text{ and } k\text{th level of } C$$

$$\text{SS(ABC)} = \text{SST} - \text{SSA} - \text{SSB} - \text{SSC} - \text{SS(AB)} - \text{SS(AC)} - \text{SS(BC)} - \text{SSE}$$

SST = sum of squared deviations of all $n = abcr$ response values from the grand average of the data

SSE = sum over all cells of squared deviations of cell entries from corresponding cell means

---

For a three-factor design restricted to main effects and two-factor interactions (i.e., a design that excludes the ABC interaction), we can determine the sums of squares for total variation, main effects, and two-factor interactions using the computational formulas in the foregoing box. However, now that we are excluding the ABC term, the SSE must necessarily change. It is no longer just the sum of squared cell deviations (shown in the foregoing table), but it will now be increased by an amount equal to SS(ABC). The correct ANOVA decomposition will now be given by

$$SST = SSA + SSB + SSC + SS(AB) + SS(AC) + SS(BC) + SSE^*$$

Rearrangement of this decomposition yields an expression for the new SSE:

$$SSE^* = SST - SSA - SSB - SSC - SS(AB) - SS(AC) - SS(BC).$$

Comparing SSE^* to the SSE for the full three-factor model, we see that the error term now includes the ABC contribution in the sense that $SSE^* = SSE + SS(ABC)$.

Similarly, if we want to restrict the model to only main effects, the ANOVA decomposition becomes

$$SST = SSA + SSB + SSC + SSE^{**}$$

from which

$$SSE^{**} = SST - SSA - SSB - SSC.$$

Again, the SSE^{**} term has simply absorbed the SSE for the full three-factor model and all sums of squares of the terms omitted from this model. This also happens when terms from a two-factor design are dropped (cf. page 457) or, in general, from a model having any number of factors, including multiple regression models (discussed in Chapter 11).

Hypothesis Tests

Hypothesis tests concerning main effects and interactions are based on the familiar ANOVA assumption that the response values at *each* fixed factor–level combination follow a normal distribution and that the variances of these distributions are the *same*, regardless of the particular factor–level combination. From these assumptions, a separate degrees of freedom and mean square can be computed for each source of variation. **The total degrees of freedom** is $n - 1$, where n is the total number of experimental runs. The degrees of freedom for each main effect equals its number of levels minus 1 and the degrees of freedom for any interaction term is simply the product of the degrees of freedom for its component factors. The **mean square** associated with any main effect or interaction equals its sum of squares divided by its degrees of freedom. All of this information is summarized in the form of an ANOVA table. For example, Figure 10.16 shows the general form of the ANOVA table for a three-factor design.

Source of variation	df	SS	MS	F
A	$a - 1$	SSA	MSA	MSA/MSE
B	$b - 1$	SSB	MSB	MSB/MSE
C	$c - 1$	SSC	MSC	MSC/MSE
AB	$(a - 1)(b - 1)$	SS(AB)	MS(AB)	MS(AB)/MSE
AC	$(a - 1)(c - 1)$	SS(AC)	MS(AC)	MS(AC)/MSE
BC	$(b - 1)(c - 1)$	SS(BC)	MS(BC)	MS(BC)/MSE
ABC	$(a - 1)(b - 1)(c - 1)$	SS(ABC)	MS(ABC)	MS(ABC)/MSE
Error	$abc(r - 1)$	SSE	MSE	
Total variation	$abcr - 1$	SST		

Figure 10.16 ANOVA table for a factorial design with three factors, A, B, and C

Example 10.2

Over the past decade researchers and consumers have shown increased interest in renewable fuels such as biodiesel, a form of diesel fuel derived from vegetable oils and animal fats. According to www.fueleconomy.gov, compared to petroleum diesel, the advantages of using biodiesel include its nontoxicity, biodegradability and lower greenhouse gas emissions. One popular biodiesel fuel is fatty acid ethyl ester (FAEE). The authors of "Application of the Full Factorial Design to Optimization of Base-Catalyzed Sunflower Oil Ethanolysis" (*Fuel*, 2013: 433–442) performed an experiment to determine optimal process conditions for producing FAEE from the ethanolysis of sunflower oils. In one study, the effects of three process factors on FAEE purity (%) were investigated.

Factor	Factor name	Factor levels
A	Reaction Temperature	25°C, 50°C, 75°C
B	Ethanol-to-oil molar ratio	6:1, 9:1, 12:1
C	Catalyst loading	.75 wt.%, 1.00 wt.%, 1.25 wt.%

Table 10.1 shows the data from this $3 \times 3 \times 3$ experiment. Note that there are $r = 2$ repeated tests run at each combination of factor levels. Figure 10.17 shows the resulting ANOVA table. All effects except the BC and ABC interaction effects are significant at $\alpha = .05$. Because some interaction terms are significant, the interaction plots must be examined when drawing conclusions about the factor effects.

Plots of all two-factor interactions are shown in Figure 10.18, along with the main effects plots for the three factors. Suppose we are interested in maximizing the value of the response variable, FAEE purity. Looking at the interaction plots, the combination of factor levels that best accomplishes this objective is $A = 75°C$, $B = 12:1$, and $C = 1.25\%$. In this example, the conclusions from the interaction plots agree with the conclusions that we would have drawn from inspecting the main effects plots.

Table 10.1 Purity (%) of fatty acid ethyl ester

		Ratio							
		6:1			9:1			12:1	
Loading →	.75	1.00	1.25	.75	1.00	1.25	.75	1.00	1.25
25	81.07	88.71	95.42	81.54	89.12	96.32	86.07	92.05	97.02
	82.22	87.61	94.06	82.82	86.49	95.45	87.73	91.72	96.16
50	87.31	89.52	94.68	87.99	90.05	96.44	89.61	90.32	98.30
	87.94	88.75	95.45	88.98	90.42	96.47	89.02	90.61	96.62
75	90.66	91.60	93.65	92.14	92.55	97.41	92.88	96.12	97.66
	91.87	92.34	95.73	92.22	97.06	97.08	93.30	97.41	97.59
↑									
Temp									

Source	df	SS	MS	F	P
A	2	215.38	107.69	112.07	.0000
B	2	74.51	37.26	38.77	.0000
C	2	602.72	301.36	313.60	.0000
AB	4	13.45	3.36	3.50	.0200
AC	4	107.41	26.85	27.94	.0000
BC	4	4.37	1.09	1.14	.3598
ABC	8	12.47	1.56	1.62	.1649
Error	27	25.95	.961		
Total	53	1056.26			

Figure 10.17 ANOVA for the data of Table 10.1

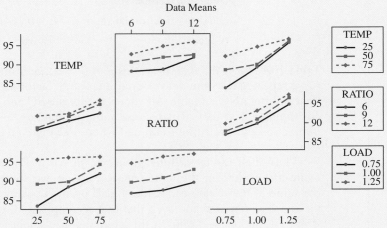

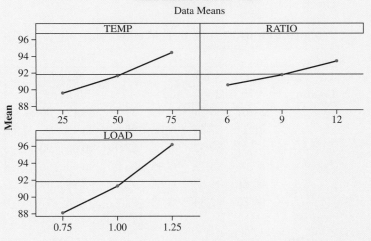

Figure 10.18 Two-factor interaction plots and main effects plots for Example 10.2

Based on years of empirical evidence, the results of a factorial experiment usually show that, over the range of factor levels studied, only a few factors are significant and even fewer interaction terms are significant. When *all* main effects and interactions are significant, the experimenter should carefully examine how the test runs were conducted to make sure that correct procedures were followed. Recall from Section 4.3 that the proper method of conducting repeated tests is to *completely* replicate the experimental conditions for each test.

For example, in Example 10.2, the two repeated tests made at $A = 25°C$, $B = 6:1$ molar ratio, and $C = .75$ wt.% catalyst loading should be conducted by resetting the apparatus used in the first test, substituting a new sunflower oil sample using the specified molar ratio and catalyst loading, allowing the temperature to change and be reset to $25°C$, and *then* running the second test. If, instead, the experimenter simply leaves the apparatus from the first test in place and immediately conducts a second test, then the variation between the two FAEE purity responses is more likely to be a measure of the repeatability of the purity measurement system. It will not truly capture the experimental error we would expect for any sunflower oil sample under the conditions $A = 25°C$, $B = 6:1$ molar ratio, and $C = .75$ wt.% catalyst loading. Test runs that are incorrectly conducted by simply taking two successive measurements usually result in underestimating the experimental error MSE, thereby artificially increasing the F ratios on which hypothesis tests are based.

Section 10.3 Exercises

15. Highly precise finishing methods are important for the manufacturing of ultraprecision optical parts but conventional polishing methods have proven to be unsatisfactory. Magnetic abrasive finishing (MAF), a relatively new technology that uses abrasive particles surrounded by magnets that generate a magnetic field around the polishing area, has drawn attention as an alternative finishing method. The authors of "Run-to-Run Process Control of Magnetic Abrasive Finishing Using Bonded Abrasive Particles" (*J. of Engr. Manuf.*, 2012: 1963–1975) examined the impact of MAF process control parameters on finishing outcomes. To see whether average surface roughness (R_a) is affected by the abrasive size (A), abrasive quantity (B), and quill gap (C), an experiment using three sizes, three quantities, and three gaps was performed, with two replicates at each of the factor combinations. The resulting sums of squares were SSA = 210.67, SSB = 132.17, SSC = 2586.35, SS(AB) = 57.48, SS(AC) = 636.84, SS(BC) = 875.00, SS(ABC) = 888.52, SSE = 5416.67 and SST = 10,803.70.

a. Construct an ANOVA table for this data.

b. Test to see whether any interaction effects are significant at $\alpha = .05$.

c. Test to see whether any main effects are significant at $\alpha = .05$.

16. Factorial designs have been used in forestry to assess the effects of various factors on the growth behavior of trees. In one such experiment, researchers thought that healthy spruce seedlings should bud sooner than diseased spruce seedlings ("Practical Analysis of Factorial Experiments in Forestry," *Canadian J. of Forestry*, 1995: 446-461). In addition, before planting, seedlings were also exposed to three levels of pH to see whether this factor has an effect on virus uptake into the root system. The following table shows data from a 2×3 experiment to study both factors:

		pH		
		3	**5.5**	**7**
	Diseased	1.2, 1.4, 1.0, 1.2, 1.4	.8, .6, .8, 1.0, .8	1.0, 1.0, 1.2, 1.4, 1.2
Health status				
	Healthy	1.4, 1.6, 1.6, 1.6, 1.4	1.0, 1.2, 1.2, 1.4, 1.4	1.2, 1.4, 1.2, 1.2, 1.4

The response variable is an average rating of five buds from a seedling. The ratings are 0 (bud not broken), 1 (bud partially expanded), and 2 (bud fully expanded).

a. Using a significance level of 5%, conduct an ANOVA test for this data. Indicate which factors are significant and whether the interaction term is significant.

b. Create an effects plot for the factors that were found to be significant in part (a).

c. What conclusions can you draw regarding the effects of the two factors on bud rating?

17. The output of a continuous extruding machine that coats steel pipe with plastic was studied as a function of thermostat temperature profile (A, at three levels), type of plastic (B, at three levels), and the speed (C, at three levels) of the rotating screw that forces the plastic through a tube-forming die. Two replications were obtained at each factor–level combination, yielding a total of 54 observations. The sums of squares were $SSA = 14{,}144.44$, $SSB = 5{,}511.27$, $SSC = 244{,}696.39$, $SS(AB) = 1{,}069.62$, $SS(AC) = 62.67$, $SS(BC) = 331.67$, $SSE = 3127.50$, and $SST = 270{,}024.33$.

a. Construct an ANOVA table for this experiment.

b. Use the appropriate F ratios to show that none of the two- or three-factor interactions is significant at $\alpha = .05$.

c. Which main effects are significant at $\alpha = .05$?

18. To see whether the force in drilling is affected by the drilling speed (A), feed rate (B), or material used (C), an experiment using four speeds, three rates, and two materials was performed, with two replicate samples drilled at each combination of levels of the three factors. A software package was used to obtain the sums of squares for the experimental data: $SSA = 19{,}149.73$, $SSB = 2{,}589{,}047.62$, $SSC = 157{,}437.52$, $SS(AB) = 53{,}238.21$, $SS(AC) = 9{,}033.73$, $SS(BC) = 91{,}880.04$, $SSE = 56{,}819.50$, and $SST = 2{,}983{,}164.81$.

a. Construct an ANOVA table for this experiment, and identify significant effects using $\alpha = .01$.

b. Is there any single factor that appears to have no effect on thrust force? If so, how would you go about choosing the level of this factor that would minimize thrust force?

19. An experiment was conducted to investigate how the length of steel bars is affected by the time of day (A),

heat treatment applied (B), and machine used (C). The three times were 8:00 A.M., 11:00 A.M., and 3:00 P.M. Two types of heat and four machines were used. The data from this $3 \times 2 \times 4$ factorial design is given in the following table. *Note:* Data is coded as $1000(\text{length} - 4.380)$; this does not affect the analysis.

	B_1			
	C_1	C_2	C_3	C_4
A_1	6, 9, 1, 3	7, 9, 5, 5	1, 2, 0, 4	6, 6, 7, 3
A_2	6, 3, 1, −1	8, 7, 4, 8	3, 2, 1, 0	7, 9, 11, 6
A_3	5, 4, 9, 6	10, 11, 6, 4	−1, 2, 6, 1	10, 5, 4, 8

	B_2			
	C_1	C_2	C_3	C_4
A_1	4, 6, 0, 1	6, 5, 3, 4	−1, 0, 0, 1	4, 5, 5, 4
A_2	3, 1, 1, −2	6, 4, 1, 3	2, 0, −1, 1	9, 4, 6, 3
A_3	6, 0, 3, 7	8, 7, 10, 0	0, −2, 4, −4	4, 3, 7, 0

a. Construct an ANOVA table for this data.

b. Test to see whether any interaction effects are significant at $\alpha = .05$.

c. Test to see whether any main effects are significant at $\alpha = .05$.

20. The deposition of thick protective coatings on substrates can be facilitated by laser cladding, in which an alloy powder is melted on the substrate surface. Experiments were conducted to determine how three processing parameters, laser power (A), scanning velocity (B), and powder flow rate (C) affect the coating hardness. ("Laser Cladding: An Experimental Study of Geometric Form and Hardness of Coating Using Statistical Analysis," *J. of Engr. Manuf.*, 2006: 1549–1554). Each factor had three levels, and there was one observation at each factor-level combination. The following corresponds to the ANOVA table from the article; only main effects and two-factor interactions were considered there:

SOURCE	DF	SS	MS	F
A	?	?	?	63.24
B	?	2034.74	?	?
C	?	?	480.26	?
AB	?	?	?	6.48
AC	?	729.04	?	?
BC	?	?	115.26	?
Error	?	?	104.26	
Total	?	?		

a. Fill in the missing entries in the table.

b. Identify significant effects using $\alpha = .01$.

21. Recently, nickel titanium (NiTi) shape memory alloy (SMA) has become widely used in medical devices. This is attributable largely to the alloy's shape memory effect (material returns to its original shape after heat deformation), superelasticity, and biocompatibility. An alloy element is usually coated on the surface of NiTi SMAs to prevent toxic Ni release. The alloy element is coated by laser cladding, a technique first described in Exercise 20.

The authors of "Parametrical Optimization of Laser Surface Alloyed NiTi Shape Memory Alloy with Co and Nb by the Taguchi Method" (*J. of Engr. Manuf.*, 2012: 969–979) conducted a study to see whether the percent by weight of nickel in the alloyed layer is affected by carbon monoxide powder paste thickness (A, at three levels), scanning speed (B, at three levels), and laser power (C, at three levels). One observation was made at each factor-level combination (*Note:* Thickness column headings were incorrect in the cited article):

Power	Speed	Paste Thickness .2	.3	.4
600	600	38.64	35.13	19.20
	900	38.16	34.24	26.23
	1200	37.54	33.46	30.44
700	600	36.56	35.91	34.62
	900	39.16	33.10	28.71
	1200	37.06	31.78	21.50
800	600	39.44	40.42	37.21
	900	39.34	37.64	35.65
	1200	39.30	34.97	32.50

a. Construct an ANOVA table for this experiment, including all main effects and two-factor interactions (as did the authors of the cited article).

b. Use the appropriate F ratios to show that none of the two-factor interactions is significant at $\alpha = .05$.

c. Which main effects are significant at $\alpha = .05$?

22. A four-factor factorial design was used to investigate the effect of fabric (A), type of exposure (B), level of exposure (C), and fabric direction (D) on the extent of color change as measured by a spectrocolorimeter (from "Accelerated Weathering of Marine Fabrics," *J. Testing and Eval.*, 1992: 139–143). Two observations were made at each combination of the factor levels. The resulting mean squares were MSA = 2,207.329, MSB = 47.255, MSC = 491.783, MSD = .44, MS(AB) = 15.303, MS(AC) = 275.446, MS(AD) = .470, MS(BC) = 2.141, MS(BD) = .273, MS(CD) = .247, MS(ABC) = 3.714, MS(ABD) = 4.072, MS(ACD) = .767, MS(BCD) = .280, and MSE = .977. Perform an analysis of variance using $\alpha = .01$ for all tests, and summarize your conclusions.

23. One property of automobile air bags that contributes to their ability to absorb energy is the permeability of the woven material used to construct the air bags. Understanding how permeability is influenced by various factors is important for increasing effectiveness. In one study, the effects of three factors were studied: temperature (A), fabric denier (B), and air pressure (C). Two specimens were measured at each factor-level combination ("Analysis of Fabrics Used in Passive Restraint Systems—Airbags," *J. of the Textile Institute*, 1996: 554–571).

	Temperature 8° 17.2	34.4	103.4	50° 17.2	34.4	103.4	75° 17.2	34.4	103.4
Pressure →									
420-D	73	157	332	52	125	281	37	95	276
	80	155	332	51	118	264	31	106	281
630-D	35	91	288	16	72	169	30	91	213
	43	98	271	12	78	173	41	100	211
840-D	125	234	477	90	149	338	102	170	307
	111	233	464	100	155	350	98	160	311

↑
Denier

a. Construct an ANOVA table for this data.

b. Test to see whether any interaction effects are significant at $\alpha = .01$.

c. Test to see whether any main effects are significant at $\alpha = .01$.

10.4 2^k DESIGNS

The minimum number of experimental runs needed for a factorial experiment can increase rapidly as more factors are added to an experiment. Recall, for instance, that to study factor A at three levels, factor B at two levels, and factor C at four levels, a minimum of $3 \times 2 \times 4 = 24$ runs are needed, one run for each different combination of factor levels. If each test run is replicated r times, then the total number of runs will further increase by a factor of r. As a consequence, the cost of resources needed to conduct a factorial experiment can quickly become prohibitive.

One method of combating the problem of extremely large numbers of runs is to use only *two* levels of each of the factors of interest. Using this approach to study k different factors, each having only two levels, the minimum number of experimental runs needed is $2 \cdot 2 \cdot 2 \cdots 2 = 2^k$, which is the reason such experiments are called 2^k **factorial designs.** These designs are very popular in the research and development of products and processes, not only because they require smaller sample sizes but also because the associated statistical analyses are exceedingly simple and, if necessary, can even be done by hand.

Coding Schemes and the Design Matrix

It is convenient to use coding schemes to describe the factor levels in a 2^k experiment. Two such schemes are in common use, one based on $+$ and $-$ signs, the other based on using lowercase English letters. The $+$ and $-$ sign scheme is particularly useful for simplifying the computations needed for the analysis of a 2^k design. The other coding scheme is better for compactly describing the particular combinations of factor levels used in an experiment. It is useful to understand both methods.

In the $+$ and $-$ coding method, $+1$ is used to denote one level of a factor, often called the **high level,** whereas -1 is used to denote the **low level.** For example, if a factor such as temperature is studied at the two levels 60°F and 100°F, then we would use -1 to code 60°F and $+1$ to code 100°F. Although it does not matter which factor level is assigned $+1$ or -1, for *numerical* factors (such as temperature) it is usually best to assign the $+1$ coding to the numerically larger of the two levels. For *qualitative* factors, such as the brand of raw material used, it does not matter which brand is assigned the $+1$ or -1 code. When creating models that relate the factors to the response variable, the actual factor settings are called the **uncoded factor levels.**

		Uncoded factor levels	Coded factor levels
Factor A *temperature*	$\longrightarrow$	60°F	-1 (low level of A)
		100°F	$+1$ (high level of A)

The ± 1 coding scheme provides a quick method for listing all 2^k experimental runs. Using capital letters $A, B, C, \ldots$, to denote the names of the k factors in an experiment, we form k columns of $+1$ and -1 values according to the following rule:

Creating the Design Matrix for a 2^k Experiment

Column 1:   Starting with -1, create a column of length 2^k by alternating -1 and $+1$ values.

Column 2:   Create a column of alternating blocks of *two* -1 values and *two* $+1$ values.

Column 3:   Create a column of alternating blocks of *four* -1 values and *four* $+1$ values.

Column 4:   Create a column of alternating blocks of *eight* -1 values and *eight* $+1$ values.

$$\vdots \qquad\qquad\qquad \vdots$$

Continue in this manner, using block sizes that are successive powers of 2, until all k columns have been formed.

When these columns are placed side by side, they form the **design matrix** of the experiment, in which each row specifies a particular combination of factor settings. That is, each row constitutes one of the 2^k experimental test runs. The order in which these runs are listed in the design matrix is called **Yates standard order** after Frank Yates, a colleague of Fisher's who helped develop the methodology of factorial designs.

Example 10.3   For a 2^3 experiment based on the factors A, B, and C, the eight experimental runs in Yates standard order are as follows:

Run	A	B	C
1	-1	-1	-1
2	$+1$	-1	-1
3	-1	$+1$	-1
4	$+1$	$+1$	-1
5	-1	-1	$+1$
6	$+1$	-1	$+1$
7	-1	$+1$	$+1$
8	$+1$	$+1$	$+1$

The alternative coding scheme used with 2^k designs is based on lowercase letters $a, b, c, d, \ldots$, which are intended to denote the *high* levels of the corresponding factors $A, B, C, D, \ldots$. To denote a particular experimental run, we form a string of lowercase letters, showing which factors in the run are set to their high levels. Letters are omitted for factors that are set to their low levels. For instance, in a 2^3 experiment with factors A, B, and C, the combination of letters ab refers to the test run in which both A and B are set at their high levels and C is set at its low level. Similarly, the letter b denotes the run with B high and both A and C low. The notation (1) is used for the one test run in which *all* factors are set to their low levels.

Example 10.4　Using the letter coding method, the eight test runs of the 2^3 experiment in Example 10.3 are coded as follows. The table shows the letter codes that correspond to runs that have been written in Yates standard order:

Run	A	B	C	Letter code
1	-1	-1	-1	(1)
2	$+1$	-1	-1	a
3	-1	$+1$	-1	b
4	$+1$	$+1$	-1	ab
5	-1	-1	$+1$	c
6	$+1$	-1	$+1$	ac
7	-1	$+1$	$+1$	bc
8	$+1$	$+1$	$+1$	abc

Conducting an Experiment

Yates's method for generating the columns of the design matrix provides a quick and organized method for laying out the factor–level combinations of a 2^k experiment. However, when it comes to actually performing the experimental tests, *test runs should be conducted in random order*. **Randomization** of experimental runs, first discussed in Section 4.3, helps reduce the possible effects of unknown factors on the test results.

　　To see why randomization is used, suppose that we begin to conduct the runs in a 2^3 experiment in standard order (as in Example 10.3) but that unforeseen problems occur and only half the runs can be performed in one day, the remaining runs being postponed until later in the week. Because the runs are not randomized, factor C is always at its low level during the first day of testing. Later in the week, the remaining runs will be conducted when C is at its high level *and* when other external conditions may possibly have changed. Consequently, any effect that C has on the response will be commingled with the effects of changing conditions during the week. If statistical tests eventually show that factor C has a significant effect on the response, the experimenter will not be able to tell whether this effect is really caused by factor C or, instead, if it is caused by changes in *other* conditions that might have arisen between the two days of testing. If the test runs had been randomized, there would have been a much smaller chance that such external factors could systematically influence the test results. For instance, it is highly unlikely that a randomized run sequence would have resulted in having C always at its low level during the first half of the runs.

Example 10.5　To randomize the test runs in a 2^3 experiment, first find the total number of runs required, including replicated runs. For example, if we decide to conduct two replicate runs for each factor–level combination, then a total of $N = r2^k = 2 \times 2^3 = 16$

test runs must be conducted. Using a random number generator in a statistical computer program or spreadsheet, choose a random sample of size N, *without replacement*, from the integers 1, 2, 3, . . . , N. Assign the first random number chosen to the first row in the design matrix, the second random number to the second row, and so forth. These numbers indicate the order in which the tests are to be conducted.

Suppose, for instance, that the random sample of 1, 2, 3, . . . , 16 turns out to be

$$8 \quad 6 \quad 10 \quad 16 \quad 4 \quad 15 \quad 7 \quad 3 \quad 14 \quad 1 \quad 11 \quad 12 \quad 5 \quad 13 \quad 2 \quad 9$$

Proceeding down the rows of the design matrix, we write the first set of eight random numbers. Returning to the top row, we record the second set of eight random numbers. According to this randomization, the experimenter should begin by conducting run 2, followed by run 7, then run 8, run 5, run 5, run 2, and so forth. The response value measured at each run is recorded in the row corresponding to its test number.

Run	A	B	C	Run order		Responses
1	−1	−1	−1	8	14	y_{11}, y_{12}
2	+1	−1	−1	6	1	y_{21}, y_{22}
3	−1	+1	−1	10	11	y_{31}, y_{32}
4	+1	+1	−1	16	12	y_{41}, y_{42}
5	−1	−1	+1	4	5	y_{51}, y_{52}
6	+1	−1	+1	15	13	y_{61}, y_{62}
7	−1	+1	+1	7	2	y_{71}, y_{72}
8	+1	+1	+1	3	9	y_{81}, y_{82}

Calculating Effects Estimates

Main effects and two-factor interaction effects can be plotted for a 2^3 experiment in exactly the same manner as described for general factorial designs in Sections 10.2 and 10.3. Furthermore, in 2^k designs, it is also possible to calculate a *numerical* estimate for each main effect and interaction effect. Main effects and two-factor interaction effects are defined as follows:

DEFINITIONS

The **main effect of a factor** is the average response value for all test runs at the high level of the factor minus the average response value for runs at the low level of the factor.

The **two-factor interaction effect** is one-half of the difference between the main effects of one factor calculated at the two levels of the other factor.

These definitions are best understood by considering a numerical example. The 2^4 experiment in Table 10.2 shows four process variables used in the first stages of an industrial chemical reaction. The response variable is the percentage of a critical chemical that is converted during the first stage of the reaction. (We thank Eric Ziegel of AMOCO Corp. for providing this data.)

Table 10.2 2^4 experiment for studying the effects of four factors on the percent yield of a chemical reaction

Factor	Low level	High level
A, Pressure(psi)	14.0	20.0
B, Steam ratio	7.5	11.5
C, Throughput rate	.52	.66
D, Temperature (°F)	1150	1200

Run	A	B	C	D	y
1	−1	−1	−1	−1	27.22
2	1	−1	−1	−1	25.19
3	−1	1	−1	−1	23.23
4	1	1	−1	−1	18.93
5	−1	−1	1	−1	25.32
6	1	−1	1	−1	22.61
7	−1	1	1	−1	26.80
8	1	1	1	−1	20.20
9	−1	−1	−1	1	44.53
10	1	−1	−1	1	42.44
11	−1	1	−1	1	43.78
12	1	1	−1	1	37.66
13	−1	−1	1	1	42.16
14	1	−1	1	1	38.97
15	−1	1	1	1	48.85
16	1	1	1	1	42.05

The main effect for factor D in this experiment is calculated as follows:

$$\text{main effect for } D = \frac{1}{8}(44.53 + 42.44 + 43.78 + \cdots + 42.05)$$

$$- \frac{1}{8}(27.22 + 25.19 + 23.23 + \cdots + 20.20)$$

$$= 42.56 - 23.69 = 18.87$$

That is, we estimate that changing factor D from its low to its high level results in an increase of about 18.87 in the response variable. Figure 10.19 shows the main effect graph for factor D. Note that the vertical distance (dashed line) in this graph is the numerical value of the main effect. The sloped line connecting the two average response values shows whether the effect is increasing or decreasing the response value as we change from the low to the high level of the factor.

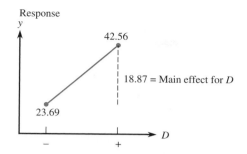

Figure 10.19 Main effect for factor D
for the data of Table 10.2

The interaction effect between two factors A and B is denoted by writing either AB or $A \times B$. Both notations are found in the literature. The interaction between three factors A, B, and C is written either as ABC or $A \times B \times C$; four-factor interactions are written $ABCD$ or $A \times B \times C \times D$, and so forth. To illustrate the calculation of a two-factor interaction effect, consider the BC interaction for the experiment in Table 10.2. Figure 10.20 shows the BC interaction graph created by plotting the average response values for all four combinations of levels of B and C. Each plotted point is now the average of four data points, not eight. For instance, the point where B is low and C is low is the average of the data points 27.22, 25.19, 44.53, and 42.44. With B on the horizontal axis, the pairs of points with the same level of C are joined by line segments. These two lines show the main effect of changing B from low to high *while holding each level of C fixed*. As you can see from the graph, the effect of changing B from low to high is very different for the two levels of C. The BC interaction is defined to be one-half of the difference between the main effect for B with C at its high level and the main effect for B with C at its low level:

$$BC \text{ interaction effect} = \frac{1}{2}\left[(34.48 - 32.27) - (30.90 - 34.85)\right] = 3.08$$

B effect when C is held at $+1$ B effect when C is held at -1

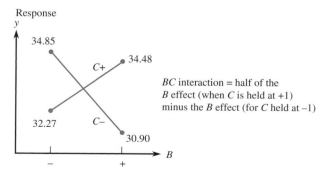

Figure 10.20 BC interaction effect for the data of Table 10.2

We leave it as an exercise for the reader to show that the calculation of an interaction effect does not depend on the order in which the factors appear. That is, the BC

and *CB* interaction effects are exactly the same and are both treated as measures of the same two-factor interaction.

The definitions of higher-order interaction effects become more complex as the number of factors increases. For example, the three-factor *ABC* interaction is defined to be one-half of the difference between the *AB* interaction values calculated at the two levels of *C*. As is the case with all interaction calculations, this definition is symmetric in the sense that the *ABC* interaction can also be calculated by using the difference between the *BC* interactions at the two levels of *A* or by using the difference between the *AC* interaction values at the two levels of factor *B*.

Fortunately, there is a much simpler method for calculating interactions of any order. Starting with the design matrix, we first create additional columns by forming all possible products (two at a time, three at a time, etc.) of columns in the design matrix. It is convenient to append these columns to the right of the design matrix. For example, an *AB* column is formed by multiplying the corresponding entries in columns *A* and *B*, an *ABC* column is formed by multiplying across the rows of *A*, *B*, and *C*, and so forth. Next, a **contrast** is calculated for each column in the extended matrix by multiplying the signs in a particular column by the column of response values and then summing. In the case where there are repeated runs at each factor–level setting (as illustrated in Example 10.5), the column signs are multiplied by the *total* of the responses at each factor–level combination. Each contrast is given the name of the column from which it is constructed. For instance, in a 2^3 design, there will be contrasts for *A*, *B*, *C*, *AB*, *AC*, *BC*, and *ABC*. The final step is to divide each contrast by *half* the number of runs:

$$\text{effect estimate} = \frac{\text{contrast}}{r2^{k-1}} = \frac{\text{contrast}}{\text{half the number of runs}}$$

The resulting values will be the estimates for each main effect and each interaction effect.

Example 10.6   To calculate all main effects and interaction effects for the 2^4 design in Table 10.2, the design matrix (in Yates standard order) is first extended to include all possible products of columns. For illustration, the *BC* and *ABC* columns are shown here. As part of Exercise 25, the reader should fill in the remaining columns.

Run	A	B	C	D	AB	AC	AD	BC	BD	CD	ABC	ABD	ACD	BCD	ABCD	y
1	−1	−1	−1	−1				+1			−1					27.22
2	1	−1	−1	−1				+1			+1					25.19
3	−1	1	−1	−1				−1			+1					23.23
4	1	1	−1	−1				−1			−1					18.93
5	−1	−1	1	−1				−1			+1					25.32
6	1	−1	1	−1				−1			−1					22.61
7	−1	1	1	−1				+1			−1					26.80
8	1	1	1	−1				+1			+1					20.20

Run	A	B	C	D	AB	AC	AD	BC	BD	CD	ABC	ABD	ACD	BCD	ABCD	y
9	−1	−1	−1	1	1					+1		−1				44.53
10	1	−1	−1	1	1					+1		+1				42.44
11	−1	1	−1	1	1					−1		+1				43.78
12	1	1	−1	1	1					−1		−1				37.66
13	−1	−1	1	1	1					−1		+1				42.16
14	1	−1	1	1	1					−1		−1				38.97
15	−1	1	1	1	1					+1		−1				48.85
16	1	1	1	1	1					+1		+1				42.05

As a check on your calculations, each column in the extended design matrix should consist of exactly half +1s and half −1s. By multiplying each entry in an effect column by the corresponding entry in the response column and summing, we obtain the contrast for each effect. For instance, the A, BC, and ABC contrasts are

$$A \text{ contrast} = -27.22 + 25.19 - 23.33 + 18.93 - 25.32 + 22.61$$
$$-26.80 + 20.20 - 44.53 + 42.44 - 43.78 + 37.66$$
$$-42.16 + 38.97 - 48.85 + 42.05$$
$$= -33.84$$

$$BC \text{ contrast} = -27.22 + 25.19 - 23.23 - 18.93 - 25.32 - 22.61$$
$$+26.80 + 20.20 + 44.53 + 42.44 - 43.78 - 37.66$$
$$-42.16 - 38.97 + 48.85 + 42.05$$
$$= 24.62$$

$$ABC \text{ contrast} = -27.22 + 25.19 + 23.23 - 18.93 + 25.32 - 22.61$$
$$+26.80 + 20.20 - 44.53 + 42.44 + 43.78 - 37.66$$
$$+42.16 - 38.97 - 48.85 + 42.05$$
$$= -1.20$$

Because the total number of test runs is 16, each contrast is divided by 8 (half the number of runs) to obtain the effect estimates:

$$\text{Main effect for A} = \frac{-33.84}{8} = -4.23$$
$$BC \text{ interaction effect} = \frac{24.62}{8} = 3.08$$
$$ABC \text{ interaction effect} = \frac{-1.20}{8} = -.15$$

Analyzing a 2^k Experiment

Having obtained estimates of all main effects and interaction effects, we must now use a statistical procedure to sort the important effects from the unimportant ones. The particular procedure used depends on whether the experiment is replicated. If only one response value is measured for each test run, then there is no replication of test runs; consequently, no estimate of experimental error is available. In this commonly occurring situation, the recommended procedure is to create a normal quantile or probability plot of the effects and fit, by eye, a straight line through the "small" effects, that is, the effects with magnitudes close to zero. Only the effects that do *not* fall on or near the straight line are considered to be the important ones. The effects falling near the line are thought to be due to experimental error or "noise." To date, there is no universally agreed-upon method for deciding which group of "small" effects to fit with a straight line. Fortunately, though, decades of empirical studies have shown that the nonsignificant effects usually comprise the majority of the plotted points, so fitting an appropriate straight line is usually fairly easy.

Example 10.7   The 15 effects $A, B, C, D, AB, \ldots, ABCD$ for the 2^4 experiment in Table 10.2 are shown in a normal quantile plot in Figure 10.21. As expected, many of the small effects tend to fall very close to a straight line (fit by eye). The effects that fall off the line appear to be A, D, AB, BD, and BC, although it is possible the last three may be close enough to the line to ignore. Based on these results, we tentatively propose that these five effects are the only ones that matter in the experiment. In particular, factor D (temperature) has a large positive effect on the response variable (% of chemical converted). Specifically, changing D from its low to high level causes an increase in the percentage of chemical converted. The situation for the other factors is not as clear since the three smaller interaction terms, AB, BD, and BC, are potentially significant, which means that their interaction plots must be examined before deciding on the best settings for these factors.

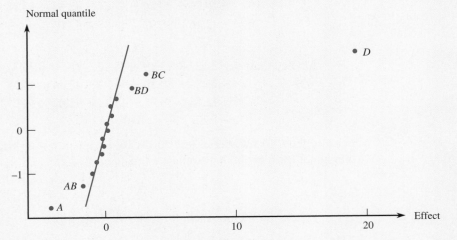

Figure 10.21   Normal quantile plot of the effects (response is % conversion)

Another method for separating the important effects from the others is to *assume* that certain higher-order effects are nonsignificant and to use these effects to obtain an estimate of the experimental error. This procedure is based on decades of empirical evidence suggesting that main effects and two-factor interaction effects are usually the most important ones in an experiment. Given a choice, the method based on normal probability plotting is usually more reliable than simply making assumptions about the outcomes of an experiment. However, when a normal probability plot suggests that the higher-order interactions may indeed be insignificant, there is more justification for combining these effects to calculate an SSE.

Suppose that m effects, call them $E_1, E_2, E_3, \ldots, E_m$, are thought to be insignificant. To create an estimate of the variance of any effect, we form the average of the squared effect values:

$$\text{variance of any effect} \approx \frac{1}{m}\sum_{i=1}^{m} E_i^2$$

This estimate has m degrees of freedom associated with it. Consequently, confidence intervals for the remaining effects in the experiment can be constructed by using the following formula:

$$\text{confidence interval for effect } E: \ E \pm (t \text{ critical value})\sqrt{\frac{1}{m}\sum_{i=1}^{m} E_i^2} \quad (\text{df} = m)$$

Example 10.8

The normal quantile plot in Figure 10.21 shows that only a few of the main effects and two-factor interactions are likely to be significant for the experiment in Table 10.2. Consequently, it is reasonable to assume that at least the three-factor and four-factor interactions are negligible and can be safely used to derive confidence intervals for the remaining 10 effects. Combining these $m = 5$ effects allows us to approximate the variance of any effect as follows:

	Effect name	Effect estimate	Squared effect
1	ABC	−.150	$(-.150)^2$
2	ABD	−.185	$(-.185)^2$
3	ACD	.150	$(.150)^2$
4	BCD	.748	$(.748)^2$
5	ABCD	.255	$(.255)^2$
			Sum = .70375

variance of any effect $\approx .70375/5 = .14075$

We can then determine the 95% confidence interval for an effect E as follows:

$$E \pm t_{\alpha/2}\sqrt{\frac{1}{m}\sum_{i=1}^{m} E_i^2} = E \pm (t \text{ critical value for 5 df})\sqrt{.14075}$$

$$= E \pm (2.571)(.3752)$$

$$= E \pm .965$$

For instance, a 95% confidence interval for the D effect is $18.87 \pm .965$. Since this interval does *not* contain 0, we conclude that the D effect is significantly different from 0. Similarly, a 95% interval estimate for the BC interaction is $3.08 \pm .965$, which indicates that the BC interaction is also significant. The same effects identified by the normal quantile plot (A, D, AB, BC, and BD) turn out to be the only significant effects identified when we assume that the three- and four-factor interactions are negligible.

When test runs are replicated, that is, when $r \geq 2$, then a 2^k experiment can be analyzed using ANOVA techniques. The sum of squares for any main effect or interaction effect can easily be computed from the effect's contrast:

$$\text{sum of squares for an effect} = \frac{\text{contrast}^2}{r2^k} = \frac{\text{contrast}^2}{\text{total number of runs}}$$

The error sum of squares, SSE, can be computed in two ways: (1) by calculating the total sum of squares, SST, for the data and then subtracting the sums of squares of the effect estimates or (2) directly, by finding the error variation for each of the 2^k test runs. Both methods are illustrated in Example 10.9.

Example 10.9

Compact discs (CDs) and digital video discs (DVDs) are manufactured by the same process. First, a master disc is created by baking a photosensitive material on a round glass plate. Next, timed pulses from a laser beam etch digital signals in a tight spiral on the plate. The plate is then "developed" to reveal a sequence of surface "pits" that encode the digital information. Master plates are electroplated to produce metal "stampers," which, when placed in plastic injection molding machines, press thousands of copies of the final disc.

Some of the factors that affect the mastering stage are listed here, along with the factor levels that were used in a 2^3 experiment on compact discs. The goal of the experiment was to minimize an electronic response called "jitter," which is a measure of how well the CD can be read by a CD-ROM device. The factor "linear velocity" is a measure of the speed with which the laser travels in a slowly increasing spiral path as it burns the pits in the photosensitive material.

Factor	Low level	High level
Laser power	90%	110%
Developing time	20 sec	30 sec
Linear velocity	1.20	1.30

Data from two replicated runs of the 2^3 experiment is given in Table 10.3. The 16 test runs were conducted in random order, but the data is presented in the table in Yates standard order. By extending the design matrix (see Example 10.6) and applying the columns of "+" and "−" signs to the column of response *totals*, we obtain the contrasts, effects, and sums of squares listed below Table 10.3.

Table 10.3 Data for the 2^3 experiment in Example 10.9

Run	A	B	C	Response values	
1	−1	−1	−1	34	40
2	1	−1	−1	26	29
3	−1	1	−1	33	35
4	1	1	−1	21	22
5	−1	−1	1	24	23
6	1	−1	1	23	22
7	−1	1	1	19	18
8	1	1	1	18	18

Effect	Symbol	Contrast	Effect estimate	Effect sum of squares
Laser power	A	−47.0	$-47.0/8 = -5.875$	$(-47.0)^2/16 = 138.063$
Linear velocity	B	−37.0	$-37.0/8 = -4.625$	$(-37.0)^2/16 = 85.563$
Developing time	C	−75.0	$-75.0/8 = -9.375$	$(-75.0)^2/16 = 351.563$
Laser power × Linear velocity	AB	−5.0	$-5.0/8 = -.625$	$(-5.0)^2/16 = 1.563$
Laser power × Developing time	AC	41.0	$41.0/8 = 5.125$	$(41.0)^2/16 = 105.063$
Linear velocity × Developing time	BC	−1.0	$-1.0/8 = -.125$	$(-1.0)^2/16 = .063$
Laser power × Linear velocity × Developing time	ABC	7.0	$7.0/8 = .875$	$(7.0)^2/16 = 3.063$

The total sum of squares SST can be found by calculating the sample variance of all 16 measurements and then multiplying by 15. Thus SST $= 15(6.8869)^2 = 711.441$. Subtracting all of the effects sums of squares from SST gives the value of SSE $= 711.441 - 138.063 - 85.563 - \cdots - .063 - 3.063 = 26.5$. Alternatively, the error variation can be calculated separately for each run by finding $\sum_{i=1}^{r}(y_i - \bar{y})^2$ and then summing all $2^3 = 8$ results:

Run	$\sum_{i=1}^{r}(y_i - \bar{y})^2$
1	18.0
2	4.5
3	2.0
4	.5
5	.5
6	.5
7	.5
8	.0
	SSE = 26.5

Finally, converting SSE to MSE by dividing by the error degrees of freedom $(r - 1)2^k$, we have the following ANOVA table from Minitab:

```
Source   DF        SS          MS           F        P
A         1    138.063     138.063      41.68    0.000    *
B         1     85.563      85.563      25.83    0.000    *
C         1    351.563     351.563     106.13    0.000    *
A*B       1      1.563       1.563       0.47    0.512
A*C       1    105.063     105.063      31.72    0.000    *
B*C       1      0.063       0.063       0.02    0.894
A*B*C     1      3.063       3.063       0.92    0.364
Error     8     26.500       3.313
Total    15    711.441
```

At a significance level of $\alpha = .01$, this table shows that the significant effects are A, B, C, and the AC interaction. Because B does not appear to interact with A or C (i.e., neither the AB nor the BC interaction is significant), we can immediately conclude that increasing the linear velocity will, on average, cause the response variable to decrease by about 4.625 units. Because the AC interaction is significant, it is necessary to examine the AC interaction plot before deciding on the proper settings for A and C (Figure 10.22). From the plot, we see that the settings that minimize the response variable are $A = -1$ and $C = +1$. In this example, the conclusions from the interaction plot do not agree with those from the main effects plots, which would have (incorrectly) indicated that both A and C should be set at their -1 levels.

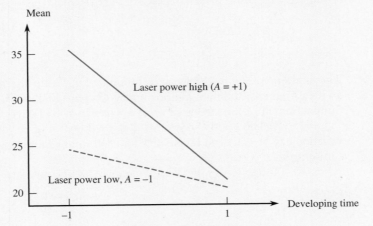

Figure 10.22 Laser power × developing time interaction plot for Example 10.9

Fitting a Model

After the important effects have been identified, it is often useful to write an equation for predicting the response value. Although in other applications this task would require the methods of regression analysis (Chapters 3 and 11), the special arrangement of factor levels in a factorial design makes it especially easy to find prediction equations. All that is needed is to define k predictor variables, one for each factor in a 2^k experiment. These predictor variables, also called **indicator variables** or **dummy variables,** use the same $+1$ and -1 coding that we used to form the design matrix. For example, the indicator variable for factor A is denoted by x_A and is defined as follows:

$$x_A = \begin{cases} +1 & \text{When A is at its high level} \\ -1 & \text{When A is at its low level} \end{cases}$$

In the same fashion, indicator variables x_B, x_C, ... are defined for the remaining factors in the experiment. Interaction terms are represented by products of the indicator variables for the factors comprising the interaction term. For instance, the AB interaction term in a prediction equation is represented by the product $x_A x_B$, the ABC interaction by the product $x_A x_B x_C$, and so forth.

To write a prediction equation based on certain main effects and interactions, we simply include the associated indicator variables (or products of indicators) accompanied by coefficients that are exactly *half* of the corresponding effect estimates. For instance, if factor A is to be included in the equation, then we include the term cx_A, where the coefficient c equals one-half of the estimated main effect for A. As another example, to include the AB interaction term, we would write cx_Ax_B, where the constant c is now one-half of the estimated AB interaction effect. In this manner, a separate term is included for each effect that we choose to put in the model. These terms, along with the grand average of all the data, are then added together to form the desired prediction equation.

Example 10.10   In Example 10.9, our analysis showed that the important effects are A, B, C, and the AC interaction. The corresponding effect estimates are -5.875 (A), -4.625 (B), -9.375 (C), and 5.125 (AC). Furthermore, the grand average of all 16 data points in the experiment is 25.313. Using indicator variables x_A, x_B, and x_C, the prediction equation based on A, B, C, and AC is

$$\substack{\text{predicted} \\ \text{value of } y} = \hat{y} = 25.313 - 2.938x_A - 2.313x_B - 4.688x_C + 2.563x_Ax_C$$

This equation can then be used to find predicted values of the response variable for selected values of factors A, B, and C. For example, suppose that we set A high and both B and C low. This corresponds to the choice $x_A = +1$, $x_B = -1$, $x_C = -1$. Substituting these values into the prediction equation, we get

$$\hat{y} = 25.313 - 2.938(+1) - 2.313(-1) - 4.688(-1) + 2.563(+)(-1)$$
$$= 26.813$$

Notice that the predicted value of 26.813 agrees reasonably well with the average of the two response values (26 and 29) that were measured at this combination of factor settings.

Prediction equations are used for several purposes: (1) to generate diagnostic checks on the adequacy of the chosen model, (2) to create response surface and contour plots, and (3) to establish factor settings that lie between the $+1$ and -1 levels. Discussing all of these applications is beyond the scope of our presentation. However, Example 10.11 illustrates how the prediction equation can help in choosing factor settings.

Example 10.11   Based on our analysis of the compact disc experiment in Examples 10.9 and 10.10, the prediction equation

$$\hat{y} = 25.313 - 2.938x_A - 2.313x_B - 4.688x_C + 2.563x_Ax_C$$

should provide an adequate description of how the response variable is affected by the factors A (laser power), B (linear velocity), and C (developing time). To increase the speed with which discs are manufactured, the compact disc company would like to set the linear velocity (factor B) as fast as possible, while shortening the developing time as much as possible. Within the range of factor values studied in this experiment, this means that they would like to operate the mastering process at the high setting for B and the low setting for C. Given this situation, what setting should they choose for the laser power if the goal is to minimize the response variable "jitter"?

Substituting $x_B = +1$ and $x_C = -1$ into the prediction equation and collecting terms, we find that

$$\hat{y} = 25.313 - 2.938x_A - 2.313(+1) - 4.688(-1) + 2.563x_A(-1)$$

or

$$\hat{y} = 27.688 - 5.501x_A$$

From this equation, we see that minimizing the response value y can be accomplished by making x_A as large as possible. Within the range of values studied in the experiment, the best setting for x_A should be $x_A = +1$; that is, laser power should be set at its high level of 110%. When this is done, the value of the response variable should be about $27.688 - 5.501(+1) = 22.187$. If the value of 22.187 is small enough to satisfy customer requirements for jitter, then the company can proceed to use these factor settings. If not, then it can further reduce the jitter by choosing the settings $x_A = -1, x_B = +1$, and $x_C = +1$ as in Example 10.9, even though these settings will necessarily increase the production time for each master disc (since developing time, C, will now be at its high level).

Section 10.4 Exercises

24. Write the design matrix (in Yates standard order) for a complete 2^3 experiment. Denote the response measurements associated with the runs as $y_1, y_2, \ldots, y_8$.
    a. Using the definition that the BC interaction is one-half the difference between the main effect for B with C at its high level and the main effect for B at its low level, write the formula for the BC interaction in terms of the data $y_1, y_2, \ldots, y_8$.
    b. Reversing the order of the factors, repeat the calculation in part (a) for the CB interaction.
    c. Show that the formulas in parts (a) and (b) are equivalent.

25. Fill in the remaining columns of contrasts for the 2^4 design in Example 10.6.

26. Polyolefin blends and composites can often improve the strength of existing polymers. In a study to determine which blends lead to increased material strength, composites of isotactic polypropylene (PP) and linear low-density polyethylene (LLDPE) were mixed with red mud (RM) particles ("Application of Factorial Design of Experiments to the Quantitative Study of Tensile

Strength of Red Mud Filled PP/LLDPE Blends," *J. of Materials Science Letters*, 1996: 1343–1345). The factors studied were the ratio of PP to LLDPE and the amount of red mud particles (in parts per hundred parts of resin). The levels at which these factors were studied are given in the following table:

	Lower level	Upper level
PP/LLDPE ratio	.25	4
RM particles	4	10

Composites made with each combination of factor levels were strength tested, with the following results:

Run	PP/ LLPDE ratio	RM particles	Strength (in MPa) Replication 1	Strength (in MPa) Replication 2
1	4	10	19.3	20.2
2	.25	10	8.1	9.7
3	4	4	20.3	24.5
4	.25	4	10.4	11.8

a. Calculate the main effects and the two-factor interaction effect for this experiment.

b. Create the ANOVA table for the experiment. Which factors appear to have an effect on strength? (Use $\alpha = .05$.)

c. Draw the main effects and interaction effects plots for the factors identified in part (b).

d. Which settings (high or low) of the factors in part (b) lead to maximizing the strength of a composite?

e. Using the important effects identified in part (b), write a model for predicting strength of a composite.

27. The following data resulted from a study of the dependence of welding current on three factors: welding voltage (A), wire feed speed (B), and tip-to-workpiece distance (C). Two levels of each factor were used, with two replicate observations made at each combination of factor levels.

Test run	Response values
(1)	200.0, 204.2
a	215.5, 219.5
b	272.7, 276.9
ab	299.5, 302.7
c	166.6, 172.6
ac	186.4, 192.0
bc	232.6, 240.8
abc	253.4, 261.6

a. Create the ANOVA table for this experiment.

b. At $\alpha = .01$, which effects appear to be important?

28. The article "Effect of Cutting Conditions on Tool Performance in CBN Hard Turning" (*J. of Manuf. Processes*, 2005: 10–16) reported the accompanying data, from a 2^3 design, on cutting speed (m/s), feed (mm/rev), depth of cut (mm), and tool life (min). Perform an ANOVA to investigate two-factor interactions and main effects.

Obs	Cut spd	Feed Cut	Depth	Life
1	1.21	0.061	0.102	27.5
2	1.21	0.168	0.102	26.5
3	1.21	0.061	0.203	27.0
4	1.21	0.168	0.203	25.0
5	3.05	0.061	0.102	8.0
6	3.05	0.168	0.102	5.0
7	3.05	0.061	0.203	7.0
8	3.05	0.168	0.203	3.5

29. As with many dried products, sun-dried tomatoes can exhibit an undesirable discoloration during the drying and storage process. A replicated 2^3 experiment was conducted in an effort to optimize color by considering storage time, temperature, and packaging type ("Use of Factorial Experimental Design for Analyzing the Effect of Storage Conditions on Color Quality of Sun-Dried Tomatoes," *Sci. Res. and Essays*, 2012: 477–489). In the following table, higher values of the response variable (based on chromaticity measurements) are associated with higher color quality:

Run	Storage time	Storage temp	Packaging	Color Quality Replication 1	Color Quality Replication 2
1	−	−	−	2.38	2.40
2	+	−	−	2.38	2.40
3	−	+	−	2.42	2.40
4	+	+	−	2.31	2.29
5	−	−	+	2.38	2.40
6	+	−	+	2.38	2.40
7	−	+	+	1.94	1.94
8	+	+	+	1.93	1.92

a. Calculate all main effects and two-factor interaction effects.

b. Construct an ANOVA table and use it as a basis for deciding which factors appear to affect color quality (Use $\alpha = .01$).

c. Create main effects and interaction effects plots for the factors identified in part (b).

d. Which settings (high or low) of the factors in part (b) lead to maximizing color quality?

30. Self-consolidating concrete (SCC) is a highly flowable product that can easily fill heavily congested reinforcement areas. Despite its low viscosity, SCC also maintains high stability to prevent segregation. The authors of "Effect of SCC Mixture Composition on Thixotropy and Formwork Pressure" (*J. Mater. Civ. Engr.*, 2012: 876–888) conducted a study to determine the effect of three mixture parameters—base material slump flow (A), sand-to-total aggregate ratio by volume (B), and relative content of coarse aggregate (C)—on characteristics of the resulting SCC mixtures. The following table gives the coded factor levels along with values of the time

(s) required for the SCC mixture to reach 500-mm slump flow.

Run	Slump	S/A	Coarse	Time
1	−1	−1	−1	1.71
2	−1	−1	1	3.19
3	−1	1	−1	1.75
4	−1	1	1	3.06
5	1	−1	−1	.88
6	1	−1	1	2.44
7	1	1	−1	1.34
8	1	1	1	3.37

a. Calculate all main effects and interaction effects.

b. Create a probability plot of the effects from part (a). Which effects appear to be important?

c. Which settings (high or low) of the factors in part (b) lead to maximizing the response variable? Which settings lead to minimizing the value of the response variable?

d. Determine a model equation relating time needed to reach 500-mm slump flow to the effects identified in part (b).

31. Combustion experiments of medium crude oil were conducted to determine which of three factors (oxygen partial pressure, oxygen flow rate, and oxygen molar concentration) affect various aspects of the combustion process. ("Factorial Analysis of *In Situ* Combustion Experiments," *Trans. of the Institution of Chemical Engineers*, 1991: 237–244). Two response variables, combustion time (in hours) and coke burnoff (in grams/hour), were studied using a full 2^3 design with no replications:

Run	Partial pressure	Flow rate	Molar concentration	Combustion time	Coke burnoff
1	−1	−1	−1	10.6	5.73
2	1	−1	−1	11.2	5.70
3	−1	1	−1	24.4	3.05
4	1	1	−1	20.3	2.87
5	−1	−1	1	9.2	5.57
6	1	−1	1	7.0	5.87
7	−1	1	1	14.3	3.13
8	1	1	1	17.5	3.05

a. For the response variable combustion time, calculate all main effects and interaction effects for this experiment.

b. Create a probability plot of the effects in part (a). Which effects appear to be important?

c. Which settings (high or low) of the factors in part (b) lead to maximizing combustion time? Which settings lead to minimizing combustion time?

d. Determine a model equation relating combustion time to the effects identified in part (b).

e. Repeat parts (a)–(d) for the response variable coke burnoff.

32. Impurities in the form of iron oxides lower the economic value and usefulness of industrial minerals, such as kaolins, to ceramic and paper-processing industries. A 2^4 experiment was conducted to assess the effects of four factors on the percentage of iron removed from kaolin samples ("Factorial Experiments in the Development of a Kaolin Bleaching Process Using Thiourea in Sulphuric Acid Solutions," *Hydrometallurgy*, 1997: 181–197). The factors and their levels are displayed in the following table:

Factor	Description	Units	Low level (−1)	High level (+1)
A	H_2SO_4	M	.10	.25
B	Thiourea	g/l	0.0	5.0
C	Temperature	°C	70	90
D	Time	min	30	150

The data from an unreplicated 2^4 experiment is given in the table below:

Test run	Iron extraction (%)	Test run	Iron extraction (%)
(1)	7	d	28
a	11	ad	51
b	7	bd	33
ab	12	abd	57
c	21	cd	70
ac	41	acd	95
bc	27	bcd	77
abc	48	abcd	99

a. Calculate all main effects and two-factor interaction effects for this experiment.

b. Create a probability plot of the effects. Which effects appear to be important?

c. Which settings (high or low) of the factors in part (b) lead to maximizing the percentage of iron extracted?

d. Write a model for predicting iron extraction percentage from the factors identified in part (b).

33. An unreplicated 2^5 experiment was performed to determine which factors affect the percent of arsenic removed from contaminated water by electrocoagulation (EC) ("Prediction of Arsenic Removal by Electrocoagulation: Model Development by Factorial Design," *J. Hazard. Toxic Radioact. Waste*, 2011: 48–54). The factors and corresponding levels are shown here along with the resulting data.

Factor	Description	Units	Low level (−1)	High level (+1)
A	Time	s	30	120
B	Current	amp	.6	3.0
C	EC area	cm^2	57	91.2
D	Volume	L	1	3
E	Arsenic	mg/L	.23	1.18

Test run	Removal (%)	Test run	Removal (%)	Test run	Removal (%)	Test run	Removal (%)
(1)	48.70	d	35.70	e	57.20	de	36.40
a	86.50	ad	59.60	ae	81.00	ade	52.50
b	89.10	bd	69.10	be	85.10	bde	61.00
ab	97.00	abd	89.10	abe	96.90	abde	89.30
c	58.30	cd	37.00	ce	57.60	cde	47.50
ac	84.80	acd	64.80	ace	78.80	acde	55.90
bc	90.90	bcd	71.70	bce	87.30	bcde	58.50
abc	95.20	abcd	93.90	abce	97.10	abcde	89.00

a. Calculate all main effects and two-factor interaction effects.

b. Create a probability plot of the effects. Three effects in particular should appear to be important; what are they?

c. Which settings (high or low) of the factors in part (b) lead to maximizing the percentage of arsenic extracted?

d. Develop a model equation for predicting arsenic removal percentage from the factors identified in part (b).

10.5 FRACTIONAL FACTORIAL DESIGNS

The two experiments analyzed in Section 10.4 exhibit a phenomenon commonly found in 2^k designs: Only a few main effects and interactions are important. Most of the effects, especially the higher-order interactions, tend not to be significant. Early researchers quickly devised methods for taking advantage of this situation. One such method was discussed in Section 10.3, where higher-order interaction effects are sometimes assumed to be negligible and are then pooled to form an estimate of the experimental error for testing the remaining effects in an experiment. Another procedure that relies on the scarcity of significant interaction effects is the method of **fractional factorial designs** discussed in this section.

An important reason for using fractional factorial designs is that *full* factorial designs, in which all 2^k tests are conducted at least once, expend a large amount of resources in estimating interaction terms. That is, as the number of factors k increases, the ratio of the number of main effects to the total number of effects shrinks rapidly in a 2^k design. Table 10.4 (page 490) illustrates how quickly this ratio declines. For instance, in a full 2^6 experiment with 64 test runs, only 9.5% of the effects calculated are main effects. The remaining 90.5% of the estimates are devoted to interaction effects, many of which are not likely to be of statistical or practical importance. Because simple models based on main effects and, perhaps, some two-factor interactions tend to predominate in actual applications, full 2^k designs can be somewhat inefficient for studying large numbers of experimental factors.

Table 10.4   Percentage (rounded) of main effect estimates in a full 2^k design

Number of factors	Number of main effects	Number of interaction effects	Total number of effects ($2^k - 1$)	Percentage of main effects
1	1	0	1	100
2	2	1	3	67
3	3	4	7	43
4	4	11	15	27
5	5	26	31	16
6	6	57	63	9.5
7	7	120	127	5.5
8	8	247	255	3.1
9	9	502	511	1.8
10	10	1013	1023	1.0

Creating a Fraction of a 2^k Design

To reduce the problem of estimating large numbers of possibly unimportant interaction effects, fractional factorial designs are created by replacing some of the higher-order interaction terms by additional experimental factors. For example, suppose that we want to study four factors, A, B, C, and D, but that we want to use 8 test runs rather than the 16 runs required by a full 2^4 design. To do this, first write down the extended design matrix for the *full* 2^3 design (i.e., the 2^k design with 8 runs):

A	B	C	AB	AC	BC	ABC
-1	-1	-1	$+1$	$+1$	$+1$	-1
$+1$	-1	-1	-1	-1	$+1$	$+1$
-1	$+1$	-1	-1	$+1$	-1	$+1$
$+1$	$+1$	-1	$+1$	-1	-1	-1
-1	-1	$+1$	$+1$	-1	-1	$+1$
$+1$	-1	$+1$	-1	$+1$	-1	-1
-1	$+1$	$+1$	-1	-1	$+1$	-1
$+1$	$+1$	$+1$	$+1$	$+1$	$+1$	$+1$

Next, since the highest-order interaction is least likely to be important, replace the ABC column by the letter D. This is abbreviated by writing $D = ABC$. Then erase all remaining interaction columns to obtain the design matrix:

A	B	C	D
-1	-1	-1	-1
$+1$	-1	-1	$+1$
-1	$+1$	-1	$+1$
$+1$	$+1$	-1	-1
-1	-1	$+1$	$+1$
$+1$	-1	$+1$	-1
-1	$+1$	$+1$	-1
$+1$	$+1$	$+1$	$+1$

This four-column matrix is the design matrix of a fractional factorial design based on four factors. In fact, these 8 test runs correspond to certain rows in the full 2^4 design, as shown (shaded) here.

Run	A	B	C	D
1	-1	-1	-1	-1
2	$+1$	-1	-1	-1
3	-1	$+1$	-1	-1
4	$+1$	$+1$	-1	-1
5	-1	-1	$+1$	-1
6	$+1$	-1	$+1$	-1
7	-1	$+1$	$+1$	-1
8	$+1$	$+1$	$+1$	-1
9	-1	-1	-1	$+1$
10	$+1$	-1	-1	$+1$
11	-1	$+1$	-1	$+1$
12	$+1$	$+1$	-1	$+1$
13	-1	-1	$+1$	$+1$
14	$+1$	-1	$+1$	$+1$
15	-1	$+1$	$+1$	$+1$
16	$+1$	$+1$	$+1$	$+1$

Because the 8 test runs comprise only a fraction of the 16 runs required in a full 2^4 design, we say that the 8-run experiment is a **fractional factorial experiment.** Furthermore, since this design uses only half of the 16 runs, we say that it is a *half fraction* of the full factorial design based on four factors.

All of the information about the 8-run design can be compactly summarized using the following notation system. The particular fractional factorial design we have created is denoted as a 2^{4-1} **design.** This notation carries the following information:

1. The design has 8 test runs (because $2^{4-1} = 2^3 = 8$).
2. Four factors are studied in the experiment.
3. Each factor has two levels.
4. One factor (factor D) has been added to a *full* design based on 8 runs.
5. The design uses a fraction, $1/2^1$, of the runs of a full 2^k design.

In general, any fractional factorial design can be described by the notation 2^{k-p}, which is intended to convey that

1. The design has a total of 2^{k-p} test runs.
2. k factors are studied in the experiment.
3. Each factor has two levels.
4. p factors have been added to a *full* design based on 2^{k-p} runs.
5. The design uses a fraction, $1/2^p$, of the runs of a full 2^k design.

The general procedure for creating a fractional factorial design is similar to that in the previous example: First, create the extended design matrix for a *full* design based on $k - p$ test runs, and then rename p of the interaction columns with the p additional factors. It is convenient to use sequential capital English letters to denote the factors. As we will see subsequently, the choice of which columns to replace with the additional factors is important and cannot simply be made arbitrarily.

Example 10.12   Suppose that you want to study five factors using only 8 test runs. How do you create a fractional factorial design to accomplish this? First, start with the *full* 2^3 design (i.e., the full 2^k design that has 8 runs). Write the column headings of the extended design matrix: A, B, C, AB, AC, BC, and ABC. Finally, choose two of the interaction columns, say, ABC and AC, and assign the additional two factors, D and E, to these columns. Denote this column assignment by writing $D = ABC$ and $E = AC$. Because we are adding two factors (D and E) to a full design based on three factors (A, B, and C), this design is called a 2^{5-2} fractional factorial. To create the design matrix for this particular 2^{5-2} experiment, first write the design matrix in Yates standard order for the 2^3 experiment with factors A, B, and C. Then append columns D and E. The entries in D are found by multiplying the entries of columns A, B, and C. Similarly, the entries in column E are found by multiplying the entries of columns A and C. Exercise 35 asks you to write this design matrix.

Finding the Alias Structure

The reward for using fractional factorial designs is a substantial reduction in the required number of test runs. It stands to reason, however, that there is also a price to pay. After all, how can a 2^{4-1} design with 8 test runs be expected to give exactly the same quality of information about four factors that a full 2^4 design with 16 runs can? What is lost in a fractional design is the ability to clearly distinguish some of the effects from one another. To illustrate, consider the 2^{4-1} design created previously by the assignment $D = ABC$. We immediately see that the D effect and the ABC effect cannot be distinguished from one another because the *same* column of $+1$s and -1s in the design matrix is used to compute both the ABC and D effects. Consequently, D and ABC are said to be **aliases** of one another. We also say that the D effect is **confounded** with the ABC effect. Of course, the reason that we chose to alias D with the ABC column in the first place was that we hoped the ABC effect would be negligible. If this turns out to be the case, then we will have obtained a main effect estimate for D using only 8 runs.

Unfortunately, the assignment $D = ABC$ induces even more confounding than you might first imagine. Consider, for example, the AB and the CD interactions. In Exercise 36, you are asked to show, by multiplying the appropriate columns, that the AB and CD columns are identical. Thus not only are D and ABC aliased but AB and CD are also aliased. In fact, there are many sets of aliased effects generated by our original choice of $D = ABC$. The entire set of aliases in a fractional factorial design is called the **alias structure** of the design.

As is the case with all the other aspects of 2^k designs, there is a fairly easy method for writing down the alias structure of a fractional design. This method depends on some simple observations about multiplying columns of $+1$s and -1s:

1. First, the letter I denotes the column consisting entirely of $+1$s.
2. Note that any column multiplied by itself yields column I. For example, $A \cdot A = A^2 = I, B \cdot B = B^2 = I$, and so forth.
3. Multiplying column I by any other column does not change the column. For example, $A \cdot I = I \cdot A = A$.

Using these facts, we can obtain the alias structure of any fractional factorial as follows:

Finding the Alias Structure of a Fractional Factorial

1. First, write the p assignments of additional factors in equation form. These p equations are called the **design generators.**
2. Multiply each generator from Step 1 by its left side to put each generator into the form $I = w$, where w is a "word" composed of several letters representing particular experimental factors (e.g., $D = ABC$ becomes $I = ABCD$). It is also possible to create words with "$-$" signs, such as $D = -ABC$. If this is done, the resulting design will use a different fraction of the runs from the full 2^k design.
3. Letting $I = w_1, I = w_2, \ldots, I = w_p$ denote the p design generators from Step 2, form all possible products of the words w_i (one at a time, two at a time, three at a time, etc.). Use the fact that squares of factors can be eliminated (e.g., $A^2 = I$ and multiplying by I does not change anything). There will be a total of 2^p words formed. This collection is called the **defining relation** of the design.
4. Multiply each word in the defining relation by all $2^k - 1$ effects based on k factors. Use the fact that squares of factors cancel out to simplify the products. The result is called the **alias structure** of the design.

As the following examples show, finding the alias structure is not as complicated a task as the procedure may indicate.

Example 10.13   Let's determine the alias structure of the 2^{4-1} design where D is aliased with ABC. The generator of this design is $D = ABC$. Multiplying both sides by D gives $D \cdot D = D(ABC)$, or $I = ABCD$. Since there is only one "word" in this equation, the defining relation is also of the form $I = ABCD$. Multiplying each of the $2^4 - 1$ effects by the relation $I = ABCD$ yields the following:

Effect	Aliases	Effect	Aliases
A	$= BCD$	BD	$= AC$
B	$= ACD$	CD	$= AB$
C	$= ABD$	ABC	$= D$
D	$= ABC$	ABD	$= C$
AB	$= CD$	ACD	$= B$
AC	$= BD$	$ABCD$	$= A$
AD	$= BC$	$ABCD$	$= I$
BC	$= AD$		

There is a lot of repetition in this list. Eliminating duplicate equations, we can summarize the alias structure of the design as follows:

A	$= BCD$	AB	$= CD$
B	$= ACD$	AC	$= BD$
C	$= ABD$	AD	$= BC$
D	$= ABC$	$ABCD$	$= I$

The alias structure can be summarized as follows: (1) Each main effect is aliased with a three-factor interaction; (2) all two-factor interactions are aliased with one another; and (3) the single four-factor interaction is aliased with the grand average of the data.

Example 10.14

The 2^{5-2} design of Example 10.12 provides a better illustration of how the defining relation is formed. Recall that the design generators in that example are $D = ABC$ and $E = AC$. Writing these in the form $I = ABCD$ and $I = ACE$, we can see that the defining relation is formed from the "words" $ABCD$ and ACE and all possible products of these words. Since there is only one such product, namely, $(ABCD)(ACE) = A^2BC^2DE = BDE$, the defining relation is $I = ACE = BDE = ABCD$. Multiplying each of the $2^5 - 1$ effects through by the defining relation gives the following alias structure (Exercise 37):

I	$= ACE$	$=$	BDE	$=$	$ABCD$	
A	$= CE$	$=$	BCD	$=$	$ABDE$	
B	$= DE$	$=$	ACD	$=$	$ABCE$	
C	$= AE$	$=$	ABD	$=$	$BCDE$	
D	$= BE$	$=$	ABC	$=$	$ACDE$	
E	$= AC$	$=$	BD	$=$	$ABCDE$	
AB	$= CD$	$=$	ADE	$=$	BCE	
AD	$= BC$	$=$	ABE	$=$	CDE	

Notice that each main effect is now aliased with at least one two-factor interaction as well as higher-order interactions in this design.

Analyzing a Fractional Factorial Experiment

Fractional factorial designs are also called **screening designs** because they are used to separate the few important effects from the many unimportant effects in the early stages of experimentation. Because of their emphasis on studying a large number of factors with as small a number of runs as possible, replicated fractional factorials are fairly rare. It is much more likely to find fractional designs run using only one test run for each combination of factor levels. Therefore, normal quantile or probability plots are generally used to analyze fractional designs. In those fortunate cases where replicated test runs are available, ordinary ANOVA tests can be used to distinguish the important effects from the others.

To begin the analysis of an unreplicated fractional design, first compute all 2^{k-p} effects (this includes the grand average) associated with the design. Then construct a normal plot of all effects *except* the grand average. Analyze the plot in the usual fashion by fitting a straight line, by eye, through the effects with small magnitudes. Finally, use the alias structure to formulate the model that is most likely to explain the pattern in the plot. One common practice is to opt for main effects and two-factor interactions rather than higher-order effects when formulating a tentative model.

Example 10.15 Pyrometallurgical processes are normally used to extract manganese from raw mineral ores, but alternative methods based on chemical reactions are currently being studied. One such method, based on reductive chemical leaching, uses sucrose in a solution of sulfuric acid to extract manganese dioxide ("Fractional Factorial Experiments in the Development of Manganese Dioxide Leaching by Sucrose in Sulfuric Acid Solutions," *Hydrometallurgy*, 1994: 215–230). In this investigation, five factors were studied to determine their effect on the percentage of manganese dioxide, MnO_2, obtained from the leaching process (Table 10.5, page 496).

A 2^{5-1} design with generator $E = ABCD$ was used. From the data in Table 10.5, a normal quantile plot of the effects was created (Figure 10.23, page 496). From this plot, it appears that only factors A (sucrose concentration), B (particle size of ore), and E (sulfuric acid concentration) have a significant effect on the percentage of MnO_2 extracted by the leaching process. None of the interaction terms appears to be significant. The effect estimates are

Factor	Main effect
A (sucrose)	10.69
B (size)	11.19
E (H_2SO_4)	−32.69

From these results, we can conclude that raising the sucrose concentration and using ores of larger particle size tend to increase the MnO_2 yield. In addition, because raising the sulfuric acid concentration tends to *reduce* the yield, it would be better to use the lower concentration. We divide the effects by 2 to obtain the model coefficients. In addition, we can write a model for predicting the percentage yield, y, given the (coded) values of the variables x_A, x_B, and x_E.

$$\hat{y} = 42.97 + 5.35x_A + 5.60x_B - 16.35x_E$$

The fact that lowering the sulfuric acid concentration has such a large effect on yield suggests that further experiments be conducted with even lower H_2SO_4 levels.

Table 10.5 2^{5-1} design for studying the effects of five factors on percentage yield of a chemical process

Factor	Factor name	Low level	High level
A	Sucrose (g/L)	5	10
B	Ore particle size (μm)	90–125	200–300
C	Mixing rate (min^{-1})	150	200
D	Temperature (°C)	30	50
E	Sulfuric acid (M)	1	2

Run	Sucrose	Particle size	Agitation	Temperature	H$_2$SO$_4$	Yield %
1	−1	−1	−1	−1	1	14.0
2	1	−1	−1	−1	−1	56.0
3	−1	1	−1	−1	−1	63.5
4	1	1	−1	−1	1	38.0
5	−1	−1	1	−1	−1	48.0
6	1	−1	1	−1	1	25.5
7	−1	1	1	−1	1	26.5
8	1	1	1	−1	−1	81.0
9	−1	−1	−1	1	−1	45.0
10	1	−1	−1	1	1	25.0
11	−1	1	−1	1	1	24.0
12	1	1	−1	1	−1	51.5
13	−1	−1	1	1	1	18.0
14	1	−1	1	1	−1	67.5
15	−1	1	1	1	−1	62.0
16	1	1	1	1	1	42.0

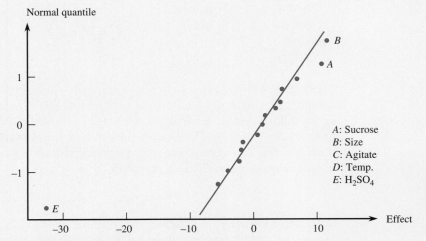

Figure 10.23 Normal quantile plot of the effects (response is yield %)

Section 10.5 Exercise

34. In a 2^{7-3} fractional factorial design,
    a. How many factors are being studied?
    b. How many experimental runs are required (assuming no replications)?
    c. What fraction of the runs of a full 2^7 design are used by this experiment?

35. Fill in all the columns in the design matrix for the 2^{5-2} design of Example 10.12.

36. A 2^{4-1} design is specified by setting $D = ABC$.
    a. Fill in the columns of the design matrix for this fractional factorial design.
    b. By multiplying the appropriate columns in the design matrix from part (a), show that AB and CD contrasts are identical.

37. Using the design generators $I = ABCD$ and $I = ACE$, verify all the entries in the alias structure of the 2^{5-2} design of Example 10.14.

38. A quarter-fraction of a 2^7 experiment (factors A, B, . . . , G) is constructed using the design generators $ABCDE = F$ and $CDE = G$.
    a. How many experimental runs (assuming no replications) must be conducted?
    b. Write down the alias structure for this design.

39. A fractional factorial experiment with 16 test runs was conducted to determine the effects of several factors on the antioxidant capacity in carotenoid extracts of the bacterium *Thermus filiformis* ("Evaluation of Biomass Production, Carotenoid Level and Antioxidant Capacity Produced by *Thermus Filiformis* Using Fractional Factorial Design," *Braz. J. Microbiol.*, 2012: 126–134). The variables studied were temperature (at 65°C and 75°C), pH (at 7 and 8), tryptone (at 5 and 10 g/L), yeast extract (at 5 and 10 g/L), and Nitsch's trace elements (2 and 5 mL/L). The Nitsch's trace elements factor was aliased with the highest-order interaction term.
    a. What are k and p for this 2^{k-p} design?
    b. Determine the alias structure of the design.
    c. Suppose that it is reasonable to assume that all interactions consisting of three or more factors are negligible. In this case, will any of the estimates of the remaining effects be aliased with one another?

40. Metal "leads" that protrude from electronic components often have their bases sealed with glass to protect against moisture ingress. Fractures in the glass can be caused by bending or twisting the leads and by large thermal changes. In an experiment designed to evaluate how different factors affect the peak stress applied to a glass seal, the following factors and factor levels were studied ("A Fractional Factorial Numerical Technique for Stress Analysis of Glass-to-Metal Lead Seals," *J. of Electronic Packaging*, 1994: 98–104):

Factor	Description	Low level (L, in.)	High level (H, in.)
s	Half the distance between neighboring leads	.025	.35
w_{lead}	Horizontal width of lead	.010	.020
h_{lead}	Distance from package base to center of lead	.127	.381
r_{port}	Radius of port in package for lead seal	.4572	.5588
t_{wall}	Wall thickness of package	.030	.050

The design matrix for the study was

Run	s	w_{lead}	h_{lead}	r_{port}	t_{wall}
1	L	L	L	L	L
2	L	L	L	H	H
3	L	L	H	L	H
4	L	L	H	H	L
5	L	H	L	L	H
6	L	H	L	H	L
7	L	H	H	L	L
8	L	H	H	H	H
9	H	L	L	L	H
10	H	L	L	H	L
11	H	L	H	L	L
12	H	L	H	H	H
13	H	H	L	L	L
14	H	H	L	H	H
15	H	H	H	L	H
16	H	H	H	H	L

a. Find k and p for this 2^{k-p} design.
b. Determine the alias structure of this design.

41. In an effort to reduce the variation in copper plating thickness on printed circuit boards, a fractional factorial design was used to study the effect of three factors—anode height (up or down), circuit board orientation (in or out), and anode placement (spread or tight)—on plating thickness ("Characterization of Copper Plating Process for Ceramic Substrates," *Quality Engr.*, 1990: 269–284). The following factor combinations were run:

Anode height	Board orientation	Anode placement	Thickness variation
−	−	−	11.63
−	+	+	3.57
+	−	+	5.57
+	+	−	7.36

a. Find k and p for this 2^{k-p} design.
b. Determine the alias structure of this design.
c. Calculate estimates of the effects for this experiment.
d. Assuming that the AB interaction is negligible, use this information to obtain an estimate of SSE and perform hypothesis tests for both main effects. (Use $\alpha = .05$.)
e. From the results in part (d), which factors have a significant effect on plating thickness variation?
f. If the objective of the study is to minimize the variation in plating thickness, what setting of each factor do you recommend?

42. Lateritic nickel ore deposits are an important source of nickel. Atmospheric acid leaching (AL) has grown in popularity as a method to extract nickel from such deposits. In the AL process, a high concentration of ferric iron may remain in the leach solution which would diminish the purity of the desired nickel. A study was conducted to investigate how five AL process factors impact iron removal efficiency (%) from leach solutions. These factors were pH (2 versus 4), temperature (25°C and 85°C), neutralizing agents [15% (W/W) MgO and 25% (W/W) $CaCO_3$], Fe/Ni ratio (6 versus 18), and stirring speed (200 and 500 rpm) ("The Effect of Iron Precipitation Upon Nickel Losses from Synthetic Atmospheric Nickel Laterite Leach Solutions: Statistical Analysis and Modelling," *Hydrometallurgy*, 2011: 140–152).

Here is data from the resulting fractional factorial experiment:

pH	Temp	Agents	Ratio	Speed	Iron Removal (%)
−	−	−	+	+	29.19
+	−	−	−	+	84.72
−	+	−	−	−	95.25
+	+	−	+	−	96.08
−	−	+	+	−	49.89
+	−	+	−	−	87.92
−	+	+	−	+	89.22
+	+	+	+	+	96.17

a. What are k and p for this 2^{k-p} design?
b. Determine the alias structure of this design. *Hint:* Each of the last two design columns is a product of two of the initial three columns.
c. Calculate estimates of the effects for this study.
d. Create a normal probability plot for the effects determined in part (b) and identify any effects that appear to be important.

43. Exercise 39 described a half-fraction of a factorial experiment in which the Nitsch's trace elements factor was aliased with the highest-order interaction term. The response variable, antioxidant capacity, was measured in percent protection against singlet oxygen [$O_2(^1\Delta_g)$]. The cited article reported the following data:

Temp	pH	Yeast	Tryptone	Nitsch	%Prot
−	−	−	−	+	51.5
+	−	−	−	−	85.1
−	+	−	−	−	46.1
+	+	−	−	+	49.0
−	−	+	−	−	33.6
+	−	+	−	+	82.9
−	+	+	−	+	57.1
+	+	+	−	−	71.9
−	−	−	+	−	34.4
+	−	−	+	+	42.7
−	+	−	+	+	31.4
+	+	−	+	−	64.8
−	−	+	+	+	4.3
+	−	+	+	−	40.4
−	+	+	+	−	48.9
+	+	+	+	+	60.5

a. Calculate estimates of the various effects.

b. Suppose that additional experimentation shows that only those effects whose magnitudes exceed 15 are important. Which factors or interactions have a significant effect on percent protection?

c. Create an effects plot for the important effects identified in part (b).

d. If the objective of the study is to maximize percent protection, what setting of each factor do you recommend?

Supplementary Exercises

44. The following data was used to investigate whether the compressive strength of concrete depends on the type of capping material used or on type of curing method used. The numbers in the matrix are *totals*, each based on three replications. In addition, SSE = 4716.67 and SST = 35,954.31 for this data.

		Curing method				
		1	2	3	4	5
Capping material	1	1847	1942	1935	1891	1795
	2	1779	1850	1795	1785	1626
	3	1806	1892	1889	1891	1756

a. Construct an ANOVA table for this experiment.

b. Using $\alpha = .01$, test to see whether either factor or their interaction is significant. Describe your conclusions from these tests.

45. In an experiment to assess the effects of curing time (factor A) and type of mix (factor B) on the compressive strength of concrete cylinders, three different curing times were used in combination with four different mixes, with three replicate observations obtained for each of the 12 factor–level combinations. The resulting sums of squares were SSA = 30,763.0, SSB = 34,185.6, SSE = 97,436.8, and SST = 205,966.6.

a. Construct an ANOVA table for this experiment.

b. Using $\alpha = .05$, can you conclude that there is a significant interaction between the two factors?

c. Test, at $\alpha = .05$, the hypothesis that factor A has no effect on compressive strength.

d. Test, at $\alpha = .05$, the hypothesis that factor B has no effect on compressive strength.

46. The authors of the article cited in Exercise 15 also performed an experiment to see whether the maximum peak to valley profile height (R_{max}) is affected by the abrasive size (A), abrasive quantity (B), and quill

gap (C); the experiment involved three sizes, three quantities, and three gaps, with two replicates at each of the factor combinations. The resulting sums of squares were SSA = 12,209.77 SSB = 19,641.09 SSC = 367,688.98 SS(AB) = 8721.72 SS(AC) = 40,008.11 SS(BC) = 44,347.01 SS(ABC) = 94,554.41 SSE = 334,393.64 and SST = 921,564.7275.

a. Construct an ANOVA table for this data.

b. Test to see whether any interaction effects are significant at $\alpha = .05$.

c. Test to see whether any main effects are significant at $\alpha = .05$.

47. Exercise 20 described an experiment involving three processing parameters: laser power (A), scanning velocity (B), and powder flow rate (C). Another experiment considered how depth penetration of the cladding layer is affected by these same factors. Each factor had three levels and there was one observation at each factor combination. Here is the ANOVA table from the article, which only considered main effects and two-factor interactions:

SOURCE	DF	SS	MS	F
A	?	?	?	162.38
B	?	0.080570	?	?
C	?	?	0.130195	?
AB	?	?	?	0.56
AC	?	0.145137	?	?
BC	?	?	?	0.76
Error	?	?	0.006387	
Total	?	?		

a. Fill in the missing entries in the table.

b. Identify significant effects using $\alpha = .01$

48. The article "An Assessment of the Effects of Treatment, Time, and Heat on the Removal of Erasable Pen Marks" (*J. Testing and Eval.*, 1991: 394–397) reports the following sums of squares for the response

variable "degree of removal of marks" (larger values of this variable are associated with more complete removal of marks): SSA = 39.171, SSB = .665, SSC = 21.508, SS(AB) = 1.432, SS(AC) = 15.953, SS(BC) = 1.382, SS(ABC) = 9.016, and SSE = 115.820. Four different laundry treatments (factor A), three different types of pen (factor B), and six different fabrics (factor C) were used in the experiment. Three observations were obtained for each combination of the factor levels. Perform an analysis of variance using $\alpha = .01$ for all tests, and state your conclusions.

49. The article cited in Exercise 21 also reported on another experiment in which the authors investigated whether the percent by weight of nickel in the alloy layer is affected by niobium powder paste thickness (A, at three levels), scanning speed (B, at three levels), and laser power (C, at three levels). One observation was made at each factor-level combination, yielding the accompanying data (*Note:* Thickness column headings were incorrect in the cited article):

		Paste Thickness		
Power	Speed	.2	.3	.4
700	600	17.14	20.16	18.73
	900	24.75	17.19	26.54
	1200	18.78	18.80	21.42
800	600	26.55	13.03	18.92
	900	19.96	29.37	21.41
	1200	26.66	19.80	22.01
900	600	33.33	27.65	28.71
	900	37.33	28.81	23.22
	1200	34.98	26.40	15.44

a. Construct an ANOVA table for this experiment including only main effects and two-factor interactions (as did the authors of the cited article).
b. Use the appropriate F ratios to show that none of the two-factor interactions are significant at $\alpha = .05$.
c. Which main effects are significant at $\alpha = .05$?

50. Even under the increased levels of security sought by current airport security practices, airports try to assure rapid processing of individuals through security checkouts. In an experiment designed to find combinations of factors that will minimize travelers' processing times at security checkpoints, three factors were studied: the number of ticket checkers (2 or 3), the number of X-ray machines (1 or 2), and the number of metal detectors (1 or 2) ("Operation of Airport Security Checkpoints Under Increased Threat Conditions," *J. of Transp. Engr.*,1996: 264–269). Each of the possible combinations of these factors was studied by using eight separate random samples of 67 travelers. The processing times (in seconds) are summarized in the table below.

a. Calculate all main effects and interaction effects for this experiment.
b. Pool the standard deviations of the replicated runs to find a value for SSE.
c. Using the SSE from part (b), determine which effects are significant (at $\alpha = .05$).
d. Which settings (high or low) of the factors in part (c) lead to minimizing processing time?
e. What is the best way to staff a security checkpoint if management wants to limit the number

					Processing time	
Test	Ticket checkers	X-ray machines	Metal detectors	Number of replicates	Mean	Standard deviation
1	2	2	2	67	39.10	1.29
2	3	2	1	67	46.50	4.30
3	2	2	1	67	50.56	5.41
4	3	2	2	67	35.07	1.05
5	2	1	2	67	93.37	37.75
6	3	1	1	67	90.55	33.52
7	2	1	1	67	97.70	34.79
8	3	1	2	67	88.86	37.58

of employees to five per checkpoint? *Note:* X-ray machines and metal detectors each require one operator.

f. Is the disparity in magnitudes of the standard deviations a possible cause for concern in this experiment?

51. Shea tree oxidation experiments were conducted to determine which of three factors (reaction time, air pressure, reaction temp.) affect various aspects in converting the woody biomass into a renewable biofuel. Optimal enzymatic conversion of the Shea tree into ethanol occurs when the cellulose content is maximized and lignin content is minimized ("Optimization of Pretreatment Conditions Using Full Factorial Design and Enzymatic Convertibility of Shea Tree Sawdust," *Biomass and Bioenergy*, 2013: 130–138). The response variable *lignin removal* (g/kg) was studied using a full 2^3 design with no replication:

Run	Time	Pressure	Temp	Lignin
1	-1	-1	-1	30
2	1	-1	-1	110
3	-1	1	-1	241
4	1	1	-1	192
5	-1	-1	1	116
6	1	-1	1	201
7	-1	1	1	230
8	1	1	1	191

a. Calculate all main effects and interaction effects for this experiment.

b. Create a probability plot of the effects in part (a).

c. Suppose that additional experimentation shows that only those effects whose magnitudes exceed 40 are important. Which factors or interactions have a significant effect on lignin removal?

d. Draw an effects plot for the important effects identified in part (c).

e. Suppose that additional experiments show that the AB and BC interactions are not significant. If the objective of the study is to maximize lignin removal, what setting of each factor do you recommend?

52. In an automated chemical coating process, the speed with which objects on a conveyor belt are passed through a chemical spray (belt speed), the amount of chemical sprayed (spray volume), and the brand of chemical used (brand) are factors that may affect the uniformity of the coating applied. A replicated 2^3 experiment was conducted in an effort to increase the coating uniformity. In the following table, higher values of the response variable are associated with higher surface uniformity:

Run	Spray volume	Belt speed	Brand	Surface uniformity Replication 1	Replication 2
1	$-$	$-$	$-$	40	36
2	$+$	$-$	$-$	25	28
3	$-$	$+$	$-$	30	32
4	$+$	$+$	$-$	50	48
5	$-$	$-$	$+$	45	43
6	$+$	$-$	$+$	25	30
7	$-$	$+$	$+$	30	29
8	$+$	$+$	$+$	52	49

a. Calculate all main effects and two-factor interaction effects for this experiment.

b. Create the ANOVA table for this experiment. Which factors appear to have an effect on surface uniformity? (Use $\alpha = .01$).

53. A half-fraction of a 2^5 experiment is used to study the effects of heating time (A), quenching time (B), drawing time (C), position of heating coils (D), and measurement position (E) on the hardness of steel castings. The following data was obtained:

Test run	Obs	Test run	Obs
a	70.4	acd	66.6
b	72.1	ace	67.5
c	70.4	ade	64.0
d	67.4	bcd	66.8
e	68.0	bce	70.3
abc	73.8	bde	67.9
abd	67.0	cde	65.9
abe	67.8	$abcde$	68.0

Assuming that second- and higher-order interactions are negligible, conduct tests (at $\alpha = .01$) for the presence of main effects.

Bibliography

Box, G. E. P., W. G. Hunter, and J. S. Hunter, **Statistics for Experimenters** (2nd ed.), Wiley, New York, 2005. *This is one of the definitive texts on industrial experimental design, with emphasis on 2^k designs and fractional factorial designs.*

Daniel, C., **Applications of Statistics to Industrial Experimentation,** Wiley, New York, 1976. *A classic text that briefly, yet eloquently, explains 2^k and fractional factorial designs from the point of view of the practitioner. The author's considerable experience in applying these designs makes it a very valuable reference.*

Montgomery, D. C., **Design and Analysis of Experiments** (8th ed.), Wiley, New York, 2012. *This book gives complete coverage of experimental designs, including general factorials, blocking, 2^k designs, fractional factorial designs, and more. Rigorous treatment, good examples, and easy to read.*

Myers, R.H., D.C. Montgomery, and C.M. Anderson-Cook, **Response Surface Methodology: Process and Product Optimization Using Designed Experiments** (3rd ed.), Wiley, New York, 2009. *Easy-to-read presentations of response surface analysis and factorial designs. Includes some of the most recent developments and tools in experimental design.*

11

Inferential Methods in Regression and Correlation

INTRODUCTION

Regression and correlation were introduced in Chapter 3 as techniques for describing and summarizing data consisting of observations on a dependent or response variable y and one or more independent variables. We first focused on the case of a single independent variable x and suggested constructing a scatterplot of sample data $(x_1, y_1), \ldots, (x_n, y_n)$ to gain preliminary insight into the nature of any relationship between the two variables. When the scatterplot exhibits a linear pattern, a line fit to the data by the principle of least squares provides a convenient summary of the approximate relationship; the coefficient of determination r^2 describes what proportion of the total variation in the observed y values can be attributed to this relation. Substituting a particular x value into the linear equation results in a *point prediction* for the value of y that would be observed if one more observation were made at this particular x value.

When data is available on k independent variables $(k \geq 2)$ $x_1, \ldots, x_k$, the same line of reasoning leads to a best-fit prediction equation having the general form $\hat{y} = a + b_1 x_1 + \cdots + b_k x_k$ and a value of the coefficient of multiple determination R^2. Again, a point prediction of y results from substituting specified values of the x_i's into the prediction equation.

In this chapter, we introduce probabilistic models as a way of describing situations where there is uncertainty in the value of y even after the values of selected predictor variables have been specified. Such models are then used to test various hypotheses of interest and to calculate both confidence intervals for mean y values and prediction intervals for individual y values to be observed at some future time. We also show how the sample correlation coefficient r can be used to test hypotheses about the *population correlation coefficient ρ*.

11.1 REGRESSION MODELS INVOLVING A SINGLE INDEPENDENT VARIABLE

A *deterministic relationship* between two variables x and y is one in which the value of y is completely and uniquely determined, with no uncertainty, by the value of x. Such a relationship can be described using traditional mathematical notation: $y = f(x)$, where $f(x)$ is a specified function, such as $10 + 2x$, $5e^{-.2x}$, or $100 - 4/\sqrt{x}$. In many engineering and science applications, it is unreasonable to assume that the variables of interest are deterministically related. For example, there is presumably a strong relationship between x = engine horsepower and y = time to go from 0 mph to 60 mph. Yet it is possible for two different engines with the same x values to result in two different values of y, so that the value of the latter variable is not determined solely by the value of the former.

A description of the relations between variables x and y that are not deterministically related can be given by specifying a *probabilistic model*. The general form of an **additive probabilistic model** allows y to deviate from $f(x)$ by a random amount. The model equation is

$$y = \text{deterministic function of } x + \text{random deviation}$$
$$= f(x) + e$$

(The random deviation e is sometimes referred to as a *random error*.) Consider graphing $f(x)$ on a two-dimensional rectangular coordinate system. If we fix x at the value x^* and make an observation on y, in the absence of the random deviation, the resulting point (x^*, y) would fall exactly on the graph. However, if $e > 0$, the point falls above the graph, whereas $e < 0$ implies that the point falls below the graph. So the role of the random deviation is to allow observed points to deviate from the graph of the deterministic function by random amounts.

Occasionally, some sort of theoretical argument will suggest an appropriate choice of $f(x)$. Most frequently, though, a scatterplot of the data is used for this purpose. When the plot shows a linear pattern, it is natural to take $f(x)$ to be a linear function, resulting in what is called the *simple linear regression model.*

DEFINITIONS

> The **simple linear regression model** assumes that there is a line with slope β and vertical or y intercept α, called the **true** or **population regression line.** When a value of the independent variable x is fixed and an observation on the dependent variable y is made, the variables are related by the model equation
>
> $$y = \alpha + \beta x + e$$
>
> Without the random deviation e, all points would fall exactly on the population regression line. We shall assume that for any fixed x value, e has a normal distribution with mean value 0 ($\mu_e = 0$) and standard deviation σ ($\sigma_e = \sigma$). We also assume that the random deviations $e_1, e_2, \ldots, e_n$ associated with different observations are independent of one another.

Figure 11.1 shows several observations in relation to the population regression line.

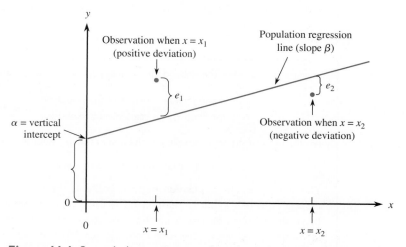

Figure 11.1 Several observations resulting from the simple linear regression model

Randomness in e implies that y itself is subject to uncertainty. The foregoing assumptions about the distribution of e imply that the distribution of y values in repeated sampling satisfies certain properties. Consider y when x equals some fixed value x^*, and let

$$\mu_{y \cdot x^*} = \text{mean or expected value of } y \text{ when } x = x^*$$

$$\sigma^2_{y \cdot x^*} = \text{variance of } y \text{ when } x = x^*$$

$$\sigma_{y \cdot x^*} = \text{standard deviation of } y \text{ when } x = x^*$$

For example, if x = engine horsepower and y = time to go from 0 to 60 mph, then

$\mu_{y \cdot 200}$ = mean time to go from 0 to 60 mph when horsepower is 200

$\sigma_{y \cdot 200}$ = standard deviation of time when horsepower is 200

Because α and β in the equation $y = \alpha + \beta x^* + e$ are fixed numbers, so is $\alpha + \beta x^*$. Taking the mean value on both sides of this equation then gives (since $\mu_e = 0$)

$$\mu_{y \cdot x^*} = \alpha + \beta x^*$$

which is just the height of the population regression line above the value $x = x^*$. Similarly, taking the variance on both sides of the equation and using the fact that the variance of a constant is zero gives

$$\sigma^2_{y \cdot x^*} = \sigma^2_e = \sigma^2 \qquad \sigma_{y \cdot x^*} = \sigma$$

That is, for a given value of x, the amount of variability in y is the same as the amount of variability in e, which in turn is the amount of variability about the population line. Finally, e is assumed to have a normal distribution, and the sum of a constant $\alpha + \beta x^*$ and a normally distributed variable itself has a normal distribution. Thus the distribution of y for any fixed x value is normal.

For any fixed x value, the dependent variable y has a normal distribution with

$$\frac{\text{mean } y \text{ value}}{\text{for fixed } x} = \frac{\text{height of the population}}{\text{regression line above } x} = \alpha + \beta x$$

(so the population regression line is the line of mean y values) and

$$\text{standard deviation of } y \text{ for fixed } x \text{ value} = \sigma$$

The slope β of the population regression line is the mean or expected change in y associated with a 1-unit increase in x. The value of σ determines the extent to which (x, y) observations deviate from the population regression line—roughly speaking, it is the size of a "typical" deviation from the line. Most or even all of the (x, y) observations will fall quite close to the population line when σ is close to 0, but when σ is large there are likely to be some large deviations from the line. Finally, independence of the e's corresponding to different observations implies that the different y's are independent.

The key features of the model are illustrated in Figures 11.2 and 11.3. The three normal curves in Figure 11.2 have identical spreads because the amount of variability in y is the same at each x value.

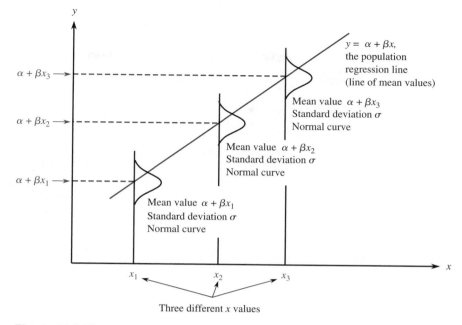

Figure 11.2 The simple linear regression model

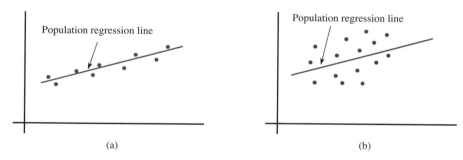

Figure 11.3 Data from the simple linear regression model: (a) σ small; (b) σ large

Example 11.1

Recently the use of granite in construction and as an ornamental material has grown in popularity. However, due to its textural properties, granite is a difficult material to process by traditional machining methods. Abrasive waterjet (AWJ) is an advanced cutting process that has shown promise in improving granite machining. The authors of "Performance of Abrasive Waterjet in Granite Cutting: Influence of the Textural Properties" (*J. of Materials in Civil Engr.*, 2012: 944–949) examined the effect of textural properties on the cutting performance of AWJ. The article suggested the simple linear regression model as a way to relate $y =$ AWJ cut depth (mm) to $x =$ granite grain size (mm).

Suppose that the parameter values for the actual model (as suggested by data in the cited article) are

$$\beta = -.4 \qquad \alpha = 25.5 \qquad \sigma = .9 \text{ mm}$$

Then for any particular fixed x value, y is normally distributed with

$$\textbf{mean value} = \mu_{y \cdot x} = 25.5 - .4x$$

$$\textbf{standard deviation} = \sigma_{y \cdot x} = .9$$

For example, when $x = 5$, AWJ cut depth has mean value $= 25.5 - .4(5) = 23.5$ mm. Because $23.5 \pm 2\sigma = 21.7$ and 25.3, roughly 95% of all AWJ cut depths made when granite grain size is 5 mm will be between these limits. The slope $\beta = -.4$ is the mean decrease in AWJ cut depth associated with a 1-mm increase in granite grain size. Thus, if we make one observation on AWJ cut depth when $x = 5$ and another when $x = 6$, we expect the former cut depth to exceed the latter by .4 mm (but the actual difference in y values will almost always be either larger or smaller than this because observations will deviate from the population line).

In practice, the judgment as to whether the simple linear regression model is appropriate is virtually always based on sample data and a scatterplot. The plot should show a linear rather than a curved pattern, and the vertical spread of points should be relatively homogeneous throughout the range of x values. Figure 11.4 shows plots with three different patterns, only one of which is consistent with the model.

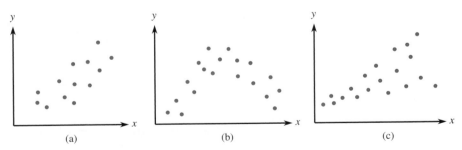

Figure 11.4 Some commonly encountered patterns in scatterplots: (a) consistent with the simple linear regression model; (b) suggests a nonlinear probabilistic model; (c) suggests that variability in y changes with x

Estimating Model Parameters

Estimates of the three parameters α, β, and σ^2 (or σ) are based on n sample observations $(x_1, y_1), (x_2, y_2), \ldots, (x_n, y_n)$ assumed to have been obtained independently according to the simple linear regression model; that is, $y_1 = \alpha + \beta x_1 + e_1$, $y_2 = \alpha + \beta x_2 + e_2$, and so on. Denote estimates of the intercept and slope by a and b, respectively. These estimates come from applying the principle of least squares introduced in Chapter 3; the least squares line has smaller sum of squared vertical deviations than does any other line.

> The least squares estimates of the slope β and intercept α of the population regression line are the slope and intercept, respectively, of the least square line, given by
>
> $$b = \text{point estimate of } \beta = \frac{S_{xy}}{S_{xx}}$$
>
> $$a = \text{point of estimate of } \alpha = \bar{y} - b\bar{x}$$
>
> where
>
> $$S_{xy} = \sum x_i y_i - \left(\frac{\sum x_i \sum y_i}{n}\right)$$
>
> $$S_{xx} = \sum x_i^2 - \frac{(\sum x_i)^2}{n}$$
>
> The estimate of the population regression line is then just the least squares line
>
> $$\hat{y} = a + bx$$

Let x^* denote some particular value of the predictor variable x. Then $a + bx^*$ has two different interpretations:

1. It is a point estimate of the mean y value when $x = x^*$ (i.e., of $\alpha + \beta x^*$).
2. It is a point prediction of an individual y value to be observed when $x = x^*$.

Example 11.2   Variations in clay brick masonry weight have implications not only for structural and acoustical design but also for design of heating, ventilating, and air conditioning systems. The article "Clay Brick Masonry Weight Variation" (*J. of Architectural Engr.*, 1996: 135–137) gave a scatterplot of $y =$ mortar dry density (lb/ft^3) versus mortar air content (%) for a sample of mortar specimens, from which the following representative data was read:

x:	5.7	6.8	9.6	10.0	10.7	12.6	14.4	15.0	15.3
y:	119.0	121.3	118.2	124.0	112.3	114.1	112.2	115.1	111.3
x:	16.2	17.8	18.7	19.7	20.6	25.0			
y:	107.2	108.9	107.8	111.0	106.2	105.0			

The scatterplot of this data in Figure 11.5 certainly suggests the appropriateness of the simple linear regression model; there appears to be a substantial negative linear relationship between air content and density, one in which density tends to decrease as air content increases.

The values of the summary statistics required for calculation of the least squares estimates are

$$\sum x_i = 218.1 \qquad \sum y_i = 1693.6 \qquad \sum x_i^2 = 3577.01$$

$$\sum x_i y_i = 24{,}252.54 \qquad \sum y_i^2 = 191{,}672.90$$

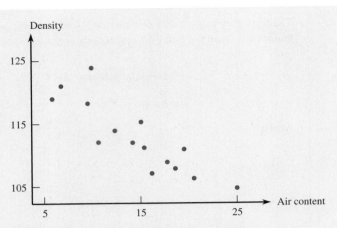

Figure 11.5 Scatterplot of the data from Example 11.2

from which

$$S_{xy} = 24{,}252.54 - \frac{(218.1)(1693.6)}{15} = -372.404000$$

$$S_{xx} = 3577.01 - \frac{(218.1)^2}{15} = 405.836000$$

$$b = \frac{-372.404000}{405.836000} = -.917622 \approx -.9176$$

$$a = \frac{1693.6}{15} - (-.917622)\left(\frac{218.1}{15}\right) = 126.248889 \approx 126.25$$

The equation of the estimated regression line (the least squares line) is then

$$\hat{y} = 126.25 - .9176x$$

Substitution of the air content value 12.0 into this equation gives $\hat{y} = 115.24$, which can be interpreted either as a point estimate of the mean dry density for all specimens whose air content is 12% or as a prediction for the dry density of a single mortar specimen whose air content is 12%.

Inferences based on the fitted model require that the error standard deviation be estimated. The estimate is based on calculating the vertical deviations from the estimated regression line. First, the **predicted** or **fitted values** are obtained by substituting the x values from the sample into the equation of the estimated regression line: $\hat{y}_1 = a + bx_1, \hat{y}_2 = a + bx_2$, and so on. The **residuals** are then the differences between the observed y values and the predicted y values: $y_1 - \hat{y}_1, \ldots, y_n - \hat{y}_n$. These are the vertical deviations from the points in the scatterplot to the estimated regression line (least squares line). Squaring and summing these residuals gives **residual** or **error sum of squares**, denoted by either **SSResid** or by **SSE**:

$$\text{SSResid} = \sum (y_i - \hat{y}_i)^2 = S_{yy} - bS_{xy}$$

Each sum of squares in statistics has associated with it a specified number of degrees of freedom. In simple linear regression, SSResid is based on $n - 2$ df, because before SSResid can be calculated, the two parameters α and β must be estimated, resulting in a loss of 2 df (just as in the case of a single sample, estimating μ by $\bar{x}$ gives the sum of squares $\Sigma(x_i - \bar{x})^2$ based on $n - 1$ df). The statistic for estimating the third model parameter σ^2 is the **mean square error,** obtained by dividing error SS by its df:

$$\text{estimate of } \sigma^2 = s_e^2 = \frac{\text{SSResid}}{n - 2}$$

$$\text{estimate of } \sigma = s_e = \sqrt{s_e^2}$$

Roughly speaking, s_e is the size of a typical deviation in the sample from the estimated regression line.

In Chapter 3, SSResid was interpreted as a measure of the variation in observed y values not explained by the approximate linear relationship between x and y. We also introduced total sum of squares

$$\text{SSTo} = S_{yy} = \sum(y_i - \bar{y})^2 = \sum y_i^2 - \frac{\left(\sum y_i\right)^2}{n}$$

interpreted as a measure of total variation in the observed y values. In the present context, the **coefficient of determination**

$$r^2 = 1 - \frac{\text{SSResid}}{\text{SSTo}}$$

is interpreted as the proportion of observed y variation that can be attributed to (or, equivalently, explained by) the simple linear regression model relationship between y and x. The closer r^2 is to 1.0, the better the model explains the y variation. The difference between SSTo and SSResid is itself a sum of squares, called **regression sum of squares,** which is interpreted as explained variation:

$$\text{SSRegr} = \text{SSTo} - \text{SSResid} \qquad r^2 = \frac{\text{SSRegr}}{\text{SSTo}}$$

Example 11.3

Let's reconsider the data on $x =$ air content and $y =$ mortar dry density from Example 11.2. The first predicted value and residual are

$$\hat{y}_1 = 126.248889 - .917622(5.7) = 121.0184$$

$$y_1 - \hat{y}_1 = 119.0 - 121.0184 = -2.0184$$

(The negative residual implies that the point (5.7, 119.0) lies below the estimated regression line.) The relevant sums of squares are

$$\text{SSTo} = S_{yy} = 191{,}672.90 - (1693.6)^2/15 = 454.1693$$

$$\text{SSResid} = S_{yy} - bS_{xy} = 454.1693 - (-.917622)(-372.4040) = 112.4432$$

from which the coefficient of determination is

$$r^2 = 1 - \frac{112.4432}{454.1693} = .752$$

Thus roughly 75% of the observed variation in density can be attributed to the simple linear regression model relationship between density and air content. SSResid is based on $15 - 2 = 13$ df, and the estimates of the "error" variance and standard deviation are

$$s_e^2 = \frac{112.4432}{13} = 8.6495 \qquad s_e = 2.941$$

Figure 11.6 shows output from the SAS software package. Values on the output agree quite closely with our hand calculations.

Dependent Variable: DENSITY

Analysis of Variance

Source	DF	Sum of Squares	Mean Square	F Value	Prob>F
Model	1	341.72606	341.72606	39.508	0.0001
Error	13	112.44327	8.64948		
C Total	14	454.16933			

Root MSE	2.94100	R-square	0.7524
Dep Mean	112.90667	Adj R-sq	0.7334
C.V.	2.60481		

Parameter Estimates

Variable	DF	Parameter Estimate	Standard Error	T for H0: Parameter=0	Prob > |T|
INTERCEP	1	126.248889	2.25441683	56.001	0.0001
AIRCONT	1	-0.917622	0.14598888	-6.286	0.0001

Obs	Dep Var DENSITY	Predict Value	Residual
1	119.0	121.0	-2.0184
2	121.3	120.0	1.2909
3	118.2	117.4	0.7603
4	124.0	117.1	6.9273
5	112.3	116.4	-4.1303
6	114.1	114.7	-0.5869
7	112.2	113.0	-0.8351
8	115.1	112.5	2.6154
9	111.3	112.2	-0.9093
10	107.2	111.4	-4.1834
11	108.9	109.9	-1.0152
12	107.8	109.1	-1.2894
13	111.0	108.2	2.8283
14	106.2	107.3	-1.1459
15	105.0	103.3	1.6917

Sum of Residuals	0
Sum of Squared Residuals	112.4433
Predicted Resid SS (Press)	146.4144

Figure 11.6 SAS output for the data of Example 11.3

Exponential Regression

A scatterplot of data obtained in a scientific or engineering investigation will often show curvature rather than a linear pattern. The scatterplot of Figure 11.7 shows a *monotonic* pattern, a tendency for y to decrease as x increases (alternatively, y might tend to increase as x increases). In this case, an exponential regression model may be a reasonable way to relate y to x. The model equation is multiplicative rather than additive:

$$y = \alpha e^{\beta x} \cdot \varepsilon, \ \varepsilon > 0$$

(The multiplicative random deviation is denoted by ε to avoid confusion with the base e of the natural logarithm system, whose value is approximately 2.7182818.) The *population regression function* is $\alpha e^{\beta x}$. When $\varepsilon > 1$, the point (x, y) lies above the graph of the regression function, and $\varepsilon < 1$ implies that the point lies below the graph. Now consider the percentage change in the population regression function when x increases by 1:

$$100 \frac{\left[\alpha e^{\beta(x+1)} - \alpha e^{\beta x}\right]}{\alpha e^{\beta x}} = 100(e^{\beta} - 1)$$

a constant not dependent on x. In simple linear regression, when x increases by 1 unit, on average y will increase by a constant amount β; in this case, when x increases by 1 unit, on average y will increase (or decrease, if $\beta < 0$) by a constant percentage.

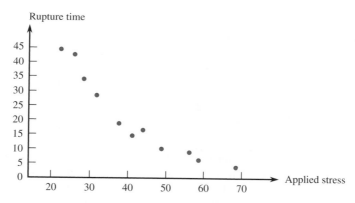

Figure 11.7 A scatterplot consistent with an exponential regression model (y = time to rupture a brass specimen, x = applied stress)

Let's now take the logarithm of both sides of the model equation:

$$y' = \ln(y) = \ln(\alpha) + \beta x + \ln(\varepsilon) = \alpha' + \beta' x + \varepsilon'$$

where $\alpha' = \ln(\alpha)$, $\beta' = \beta$, and $\varepsilon' = \ln(\varepsilon)$. This is exactly the equation for simple linear regression. Thus to say that y and x are related via the exponential regression model is the same as saying that $\ln(y)$ and x are related by the simple linear regression model (provided that $\ln(\varepsilon)$ is normally distributed, which is equivalent to ε itself having a lognormal distribution). In particular, using the previous formulas for the slope and intercept of the least squares line on the $(x_i, \ln(y_i))$ pairs gives point estimates of β and ln(α), respectively. A point estimate of α results from taking the antilog of the estimate for ln(α). Figure 11.8 shows the result of transforming the y values in Figure 11.7 by logs and then fitting the simple linear regression model. The r^2 value from this regression is obviously very high, so the simple linear regression model explains virtually all of the observed variation in ln(time to rupture).

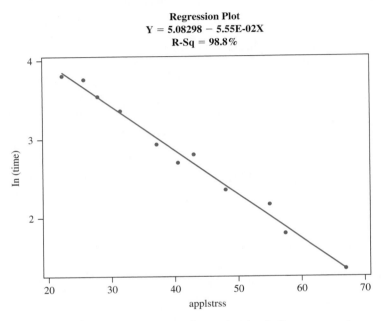

Regression Plot
Y = 5.08298 − 5.55E-02X
R-Sq = 98.8%

Figure 11.8 Minitab output from fitting the simple linear regression model to the (x, ln(y)) pairs resulting from the data of Figure 11.7

The key point here is that making a transformation [transformed y = ln(y)] results in the simple linear regression model. There are many other models nonlinear in y or x for which a transformation on one or both of the variables recaptures the simple linear regression model. Parameters of the original model can then be estimated in a relatively straightforward way.

Section 11.1 Exercises

1. The flow rate y (m³/min) in a device used for air-quality measurement depends on the pressure drop x (in. of water) across the device's filter. Suppose that for x values between 5 and 20, the two variables are related according to the simple linear regression model with true regression line $y = -.12 + .095x$.
   a. What is the expected (i.e., true average) change in flow rate associated with a 1-in. increase in pressure drop? Explain.

   b. What change in flow rate can be expected when pressure drop increases from 10 in. to 15 in.?
   c. What is the expected (i.e., true average) flow rate when the pressure drop is 10 in.? When the pressure drop is 15 in.?
   d. Suppose that $\sigma = .025$ and consider making repeated observations on flow rate when the pressure drop is 10 in. What is the long-run

proportion of observed flow rates that will exceed .835 [that is, what is $P(y > .835$ when $x = 10)$]?

2. In a certain chemical process the reaction time y (hr) is known to be related according to the simple linear regression model to the temperature x (°F) in the chamber in which the reaction takes place. The model equation is $y = 5.00 - .01x + e$, with $\sigma = .075$.

a. What is the true average change in reaction time associated with a 1°F increase in temperature? A 10°F increase in temperature?

b. What is the true average reaction time when temperature is 200°F? When temperature is 250°F?

c. What is $P(2.4 < y < 2.6$ when $x = 250)$? If an investigator makes five independent experimental runs, each for a temperature of 250°F, what is the probability that all five observed reaction times are between 2.4 and 2.6?

3. Let V be the vapor pressure of water (mm Hg) at a specific temperature T (°K). The Clausius–Clapeyron equation from physical chemistry suggests that $y = \ln(V)$ is related to $x = 1/T$ according to the simple linear regression model.

a. What is the implied probabilistic relationship between V and T?

b. If the coefficients in the simple linear regression model are $\alpha = 20.607$ and $\beta = -5200.762$, what would you predict for the value of vapor pressure when temperature is 300?

4. The article "Characterization of Highway Runoff in Austin, Texas, Area" (*J. of Envir. Engr.*, 1998: 131–137) gave a scatterplot, along with the least squares line, of $x =$ rainfall volume (m^3) and $y =$ runoff volume (m^3) for a particular location. The accompanying values were read from the plot:

x:	5	12	14	17	23	30	40	47
y:	4	10	13	15	15	25	27	46

x:	55	67	72	81	96	112	127
y:	38	46	53	70	82	99	100

a. Does a scatterplot of the data support the use of the simple linear regression model?

b. Calculate point estimates of the slope and intercept of the population regression line.

c. Calculate a point estimate of the true average runoff volume when rainfall volume is 50.

d. Calculate a point estimate of the error standard deviation σ.

e. What proportion of the observed variation in runoff volume can be attributed to the simple linear regression relationship between runoff and rainfall?

5. The bond behavior of reinforcing bars is an important determinant of strength and stability. The article "Experimental Study on the Bond Behavior of Reinforcing Bars Embedded in Concrete Subjected to Lateral Pressure" (*J. of Materials in Civil Engr.*, 2012: 125–133) reported the results of one experiment in which the researchers applied varying levels of lateral pressure on 21 concrete cube specimens, each with an embedded 16-mm plain steel round bar, and measured the corresponding bond capacity. Due to differing concrete cube strengths (f_{cu}, in MPa), the applied lateral pressure was equivalent to a fixed proportion of the specimen's f_{cu} ($0, .1f_{cu}, \ldots, .6f_{cu}$). Also, since bond strength can be heavily influenced by the specimen's f_{cu}, bond capacity was expressed as the ratio of bond strength (MPa) to $\sqrt{f_{cu}}$.

Pressure:	0	0	0	.1	.1	.1	.2
Ratio:	0.123	0.100	0.101	0.172	0.133	0.107	0.217

Pressure:	.2	.2	.3	.3	.3	.4	.4
Ratio:	0.172	0.151	0.263	0.227	0.252	0.310	0.365

Pressure:	.4	.5	.5	.5	.6	.6	.6
Ratio:	0.239	0.365	0.319	0.312	0.394	0.386	0.320

a. Does a scatterplot of the data support the use of the simple linear regression model?

b. Calculate point estimates of the slope and intercept of the population regression line.

c. Calculate a point estimate of the true average bond capacity when lateral pressure is $.45f_{cu}$.

d. Calculate a point estimate of the error standard deviation σ.

6. A study reported in the article "The Effects of Water Vapor Concentration on the Rate of Combustion of an Artificial Graphite in Humid Air Flow" (*Combustion and Flame*, 1983: 107–118) gave data on $x =$ temperature of a nitrogen–oxygen mixture (1000s of °F) under specified conditions and

y = oxygen diffusivity. Summary quantities are

$$n = 9 \quad \sum x_i = 12.6 \quad \sum y_i = 27.68$$

$$\sum x_i^2 = 18.24 \quad \sum x_i y_i = 40.968$$

$$\sum y_i^2 = 93.3448$$

a. Assuming that the variables are related by the simple linear regression model, determine the equation of the estimated regression line.

b. Calculate a point estimate of mean diffusivity when temperature is 1.5. How does this point estimate compare to a point prediction of the diffusivity value that would result from making one more observation when temperature is 1.5?

c. Estimate the error standard deviation σ.

d. Calculate and interpret the coefficient of determination.

7. Timber piles are often used to buttress multiple-span simply supported (MSSS) bridges that are commonly found in rural areas. The authors of "Bridge Timber Piles Load Rating under Eccentric Loading Conditions" (*J. Bridge Engr.*, 2012: 700–710) examined the effect of various geometric and structural characteristics on the critical rating (an overall structural assessment score) of MSSS bridges. The article reported the following data (read from a graph) for x = timber pile length (m) and y = critical rating for a particular timber profile at various damage levels.

x = Timber pile length (m):	7.32	7.93	8.54	9.14	9.75
y' = Critical rating (damage = 0%):	59.09	54.79	49.74	44.11	37.99
y'' = Critical rating (damage = 20%):	57.52	52.63	44.28	33.85	25.74
y''' = Critical rating (damage = 40%):	43.94	30.70	19.12	9.77	2.48

a. Create the scatterplots for the pairs (x, y'), (x, y''), and (x, y'''). Does each scatterplot suggest that a simple linear regression model holds for the respective variables?

b. For each pair, calculate point estimates of the slope and intercept of the respective population regression line and determine the corresponding coefficients of determination.

c. Given the slope coefficients from the regression, summarize the relationship between critical rating and pile length as timber damage changes from 0%, to 20%, and to 40%.

d. Calculate a point estimate of the error standard deviation σ for each of the pairs. How do these point estimates change as timber damage increases from 0% to 20% and then to 40%?

8. Exercise 30 in Section 3.4 gave data on x = testing temperature and y = dynamic shear modulus for a particular asphalt binder type. A scatterplot of x and $y' = \log(y)$ shows a substantial linear pattern, suggesting that these variables are related by the simple linear regression model.

a. What probabilistic model for relating y = dynamic shear modulus to x = testing temperature is implied by the simple linear regression relationship between x and y'?

b. Summary quantities calculated from the data are

$$n = 7 \quad \sum x_i = 211.4 \quad \sum y_i' = 40.64$$

$$\sum x_i^2 = 8449.68 \quad \sum (y_i')^2 = 282.58$$

$$\sum x_i y_i' = 917.48$$

Calculate estimates of the parameters for the model in part (a), and then obtain a point prediction of dynamic shear modulus when temperature is 35°F.

9. The authors of the article "Long-Term Effects of Cathodic Protection on Prestressed Concrete Structures" (*Corrosion*, 1997: 891–908) presented a scatterplot of y = steady-state permeation flux ($\mu A/cm^2$) versus x = inverse foil thickness (cm^{-1}); the substantial linear pattern was used as a basis for an important conclusion about material behavior. This is the Minitab output from fitting the simple linear regression model to the data.

```
The regression equation is
flux = -0.398 + 0.260 invthick

Predictor    Coef     Stdev    t-ratio      p
Constant    -0.3982   0.5051    -0.79    0.460
invthick     0.26042  0.01502   17.34    0.000

s = 0.4506   R-sq = 98.0%   R-sq(adj) = 97.7%

Analysis of Variance

Source      DF     SS       MS       F       p
Regression   1   61.050   61.050   300.64   0.000
Error        6    1.218    0.203
Total        7   62.269
```

Obs.	inv-thick	flux	Fit	Stdev. Fit	Residual	St. Resid
1	19.8	4.3	4.758	0.242	-0.458	-1.20
2	20.6	5.6	4.966	0.233	0.634	1.64
3	23.5	6.1	5.722	0.203	0.378	0.94
4	26.1	6.2	6.399	0.182	-0.199	-0.48
5	30.3	6.9	7.493	0.161	-0.593	-1.41
6	43.5	11.2	10.930	0.236	0.270	0.70
7	45.0	11.3	11.321	0.253	-0.021	-0.06
8	46.5	11.7	11.711	0.271	-0.011	-0.03

a. Interpret the estimated slope and the coefficient of determination.
b. Calculate a point estimate of true average flux when inverse foil thickness is 23.5.
c. Predict the value of flux that would result from a single observation made when inverse foil thickness is 45.
d. Verify that the sum of the residuals is zero and that squaring and summing the residuals results in the value of SSResid given in the output.

11.2 INFERENCES ABOUT THE SLOPE COEFFICIENT β _____

The slope β of the population regression line is the true average change in the dependent variable y associated with a 1-unit increase in the independent variable x. The slope of the least squares line, b, gives a point estimate of β. A confidence interval is a more effective way to estimate a parameter than is a point estimate, because it gives information about reliability (via the confidence level) and precision (from the width of the interval). Recall that the development of the one-sample t confidence interval for a population mean μ was based on properties of the sampling distribution of the statistic $\bar{x}$: $\mu_{\bar{x}} = \mu$, $\sigma_{\bar{x}} = \sigma/\sqrt{n}$, and that $\bar{x}$ is normally distributed when the population itself has a normal distribution. These results in turn implied that the standardized variable

$$t = \frac{\bar{x} - \mu}{s/\sqrt{n}}$$

has a t distribution with $n - 1$ degrees of freedom, from which the interval estimate $\bar{x} \pm$ (t critical value) $(s/\sqrt{n})$ emerges.

In the same way that the statistic $\bar{x}$ varies in value from sample to sample, the statistic b does also. For example, if the slope of the population regression line is actually $\beta = 25.0$, a first sample might result in $b = 24.2$, a second in an estimate of 26.5, a third in 25.4, and so on.

Properties of the Sampling Distribution of b

1. $\mu_b = \beta$ (The sampling distribution is always centered at the value of what the statistic is trying to estimate; that is, b is an unbiased statistic.)

2. $\sigma_b = \sigma/\sqrt{S_{xx}}$, where $S_{xx} = \Sigma(x_i - \bar{x})^2 = \Sigma x_i^2 - (\Sigma x_i)^2/n$.

The estimated standard deviation of b results from replacing σ by its estimate s_e:

$$s_b = s_e/\sqrt{S_{xx}}$$

3. b is normally distributed (because e in the model equation is assumed to have a normal distribution).

The smaller the value of σ_b, the more precisely β will tend to be estimated. Because σ is in the numerator, the less variability there is about the population line, the smaller is the standard deviation of b and the more concentrated is its sampling distribution. The value of σ is, of course, not under our control. However, we may be able to have an impact on the value of S_{xx}. Because this quantity is in the denominator of σ_b, the larger its value, the smaller is the value of the standard deviation. Since S_{xx} is a measure of how much the x_i values in the sample spread out, the implication is that spreading out the values of the independent variable tends to give a more precise estimate than if these values were quite close together. Intuitively, if the sample x_i values were highly concentrated, very small changes in the resulting y_i's might substantially affect the slope of the least squares line, whereas such changes would have little effect on the slope if the x_i's were quite spread out. So if the investigator can select the x values at which observations will be made (frequently not possible in social science and business scenarios), they should be spread out as much as possible while still preserving the approximate linearity of the relationship between x and y.

A Confidence Interval for the Slope Parameter

Just as in the case of $\bar{x}$ and μ, the foregoing properties allow us to form a t variable, which then gives rise to the desired confidence interval.

The standardized variable

$$t = \frac{b - \beta}{s_b}$$

has a t distribution based on $n - 2$ df. This in turn implies that a confidence interval for β is

$$b \pm (t \text{ critical value})s_b$$

Appendix Table IV contains t critical values corresponding to the most frequently used confidence levels.

Example 11.4
Let's reconsider the data on air content and mortar dry density introduced in Examples 11.2 and 11.3. In this context, β is the average or expected change in dry density associated with an increase of 1% in air content. We previously calculated $S_{xx} = 405.836000$, $b = -.918$, and $s_e = 2.941$, from which the estimated standard deviation (standard error) of b is

$$s_b = \frac{2.941}{\sqrt{405.836}} = .1460$$

The confidence interval is based on $n - 2 = 15 - 2 = 13$ df, and the corresponding t critical value for a confidence level of 95% is 2.160. The confidence interval is

$$-.918 \pm (2.160)(.1460) = -.918 \pm .315 = (-1.233, -.603)$$

With a high degree of confidence, we estimate that an average decrease in density of between .603 lb/ft^3 and 1.233 lb/ft^3 is associated with a 1% increase in air content (at least for air content values between roughly 5 and 25%, corresponding to the x values in our sample). The interval is reasonably narrow, indicating that the slope of the population line has been precisely estimated. Notice that the interval includes only negative values, so we can be quite confident of the tendency for density to decrease as air content increases.

Looking back to the SAS output of Figure 11.6, we find the value of s_b in the Parameter Estimates table as the second number in the Standard Error column. All of the widely used statistical packages include this estimated standard error in output. There is also an estimated standard error for the statistic a, from which a confidence interval for the intercept α of the population regression line can be calculated.

Testing Hypotheses About β

The form of the null hypothesis when testing hypotheses about a population or process mean μ was $H_0: \mu = \mu_0$, where the symbol μ_0 ("mu naught") represented the value of μ asserted to be true by the null hypothesis, or simply the null value. The test statistic resulted from using the null value to standardize $\bar{x}$: $t = (\bar{x} - \mu_0)/(s/\sqrt{n})$. Let β_0 denote the null value when testing hypotheses about β. Then when $H_0: \beta = \beta_0$ is true, the statistic $t = (b - \beta_0)/s_b$ has a t distribution based on $n - 2$ df. The P-value for the test (probability of obtaining a value of b more contradictory to H_0 than the value actually obtained from the given sample) is then a t curve tail area whose computation depends on the nature of the inequality in the alternative hypothesis.

Null hypothesis: $H_0: \beta = \beta_0$

Test statistic: $t = \dfrac{b - \beta_0}{s_b}$, which is based on $n - 2$ df

Alternative hypothesis	Type of test	P-value
$H_a: \beta > \beta_0$	Upper-tailed	Area under the $n - 2$ df t curve to the right of the calculated t
$H_a: \beta < \beta_0$	Lower-tailed	Area under the $n - 2$ df t curve to the left of the calculated t
$H_a: \beta \neq \beta_0$	Two-tailed	Twice the area under the $n - 2$ df t curve to the right of the calculated $|t|$

Upper-tail areas captured under various t curves are given in Appendix Table VI. Because t curves are symmetric about zero, these are also lower-tail areas.

In practice, the most frequently tested null hypothesis is $H_0: \beta = 0$. When the slope of the population regression line is zero, there is no useful linear relationship between x and y. The usual alternative hypothesis is $H_a: \beta \neq 0$, according to which there *is* a useful linear relationship between the two variables. A test of these two hypotheses is often referred to as the **model utility test** in simple linear regression. Unless H_0 can be rejected at a reasonably small significance level, the simple linear regression model should not be used as a basis for making various inferences (e.g., for predicting y from knowledge of x). In practice, the model will generally be judged useful by this test when r^2 is reasonably large. On occasion, the alternatives $H_a: \beta > 0$ or $H_a: \beta < 0$ may be of interest; the former says that there is in fact a *positive* linear relationship between the two variables (a tendency for y to increase linearly as x increases). The test statistic in all three cases is the *t*-ratio, b/s_b.

Example 11.5

The presence of hard alloy carbides in high chromium white iron alloys results in excellent abrasion resistance, making them suitable for materials handling in the mining and materials processing industries. The accompanying data on x = retained austentite content (%) and y = abrasive wear loss (mm^3) in pin wear tests with garnet as the abrasive was read from a plot in the article "Microstructure-Property Relationships in High Chromium White Iron Alloys" (*Intl. Materials Reviews*, 1996: 59–82).

x:	4.6	17.0	17.4	18.0	18.5	22.4	26.5	30.0	34.0
y:	.66	.92	1.45	1.03	.70	.73	1.20	.80	.91
x:	38.8	48.2	63.5	65.8	73.9	77.2	79.8	84.0	
y:	1.19	1.15	1.12	1.37	1.45	1.50	1.36	1.29	

A scatterplot of the data (not shown) suggests that the simple linear regression may specify a useful relationship between these two variables. Is this indeed the case? Let's base our analysis on the SAS output in Figure 11.9.

```
Dependent Variable: ABRLOSS

                        Analysis of Variance

                            Sum of         Mean
     Source        DF      Squares        Square      F Value     Prob>F

     Model          1      0.63690       0.63690       15.444     0.0013
     Error         15      0.61860       0.04124
     C Total       16      1.25551

          Root MSE        0.20308     R-square      0.5073
          Dep Mean        1.10765     Adj R-sq      0.4744
          C.V.           18.33410

                        Parameter Estimates

                     Parameter     Standard    T for H0:
     Variable   DF    Estimate        Error    Parameter=0    Prob > |T|

     INTERCEP    1    0.787218     0.09525879      8.264        0.0001
     AUSTCONT    1    0.007570     0.00192626      3.930        0.0013
```

Figure 11.9 SAS output from a simple linear regression of the data in Example 11.5

The parameter of interest is β, the average change in wear loss associated with a 1% (i.e., 1-unit) increase in austentite content. The relevant hypotheses are

$H_0: \beta = 0$ (the model is not useful)

$H_a: \beta \neq 0$ (there is a useful linear relationship between the variables)

The test statistic is the model utility t-ratio $t = b/s_b$. From the Parameter Estimates table in Figure 11.9,

$$b = .007570 \qquad s_b = .00192626 \qquad t = \frac{.007570}{.00192626} = 3.93 \approx 3.9$$

The two-tailed test is based on $n - 2 = 15$ df. In Appendix Table VI, the area under the 15 df t curve to the right of 3.9 is .001, so the P-value for the test is roughly .002. Figure 11.9 gives this P-value as .0013 (so the area to the right of 3.93 must be about .00065). Clearly the P-value is smaller than either .05 or .01. H_0 can obviously be rejected in favor of the conclusion that there is a useful linear relationship. Notice that the r^2 value is .507, which is not terribly impressive. But as long as n is not too small, the model will be judged useful even when r^2 is moderate to small.

The article's authors asserted that "increasing the austentite content leads to greater wear rates with garnet as the abrasive." The implied alternative hypothesis is $H_a: \beta > 0$ (a positive linear relationship). The P-value for this upper-tailed test is about .001 (more exactly, .00065), which clearly supports the authors' contention.

Regression and ANOVA

An alternative to the t test for model utility is based on the decomposition of total sum of squares into regression or model sum of squares and error sum of squares:

$$\text{SSTo} = \text{SSRegr} + \text{SSResid}$$

where df $= 1$ for SSRegr and df $= n - 2$ for SSResid. The two mean squares are then MSRegr $=$ SSRegr$/1$ and MSResid $=$ SSResid$/(n - 2)$, and the F ratio is given by $F = $ MSRegr$/$MSResid. The calculations are usually summarized in an ANOVA table, as shown in Table 11.1.

Table 11.1 ANOVA table for simple linear regression

Source of variation	df	Sum of squares	Mean square	F	P-value
Model					Area to right of
(Regression)	1	SSRegr	SSRegr	MSRegr/MSResid	calculated F
Error	$n - 2$	SSResid	SSResid/$(n - 2)$		
Total	$n - 1$	SSTo			

Looking at the ANOVA table on the SAS output of Figure 11.9, we see that the calculated F ratio for the data of Example 11.5 is $F = 15.444$, and the corresponding P-value (the area under the F curve with 1 numerator and 15 denominator df to the right of 15.444) is .0013. That this P-value is identical to the P-value for the model utility t test is no accident: It can be shown that $t^2 = F$ [$(3.930)^2 = 15.444$ in Example 11.5], and the distribution of the square of a t variable with ν df is the F distribution with 1 numerator and ν denominator df. However, in multiple regression, the test for model utility is an F test, and t tests are used for another purpose.

Correlation Revisited

The sample correlation coefficient r was introduced in Chapter 3 as a measure of the extent of linear association between values of x and y in a sample. An analogous measure for the entire population from which the sample of pairs was selected is called the **population correlation coefficient** and is denoted by ρ. The most important properties of r are also satisfied by ρ; in particular, $-1 \leq \rho \leq 1$, so the closer ρ is to 1 or -1, the stronger the linear relationship within the population. The value $\rho = 0$ indicates the complete absence of any linear relationship in the population. Even if $\rho = 0$, the value of r will usually differ somewhat from zero because of sampling variability—r is a statistic and its value will vary from sample to sample in the same way that $\bar{x}$ and b do. It is therefore important to have a formal test of the null hypothesis that $\rho = 0$. The usual test procedure assumes that $(x_1, y_1), \ldots, (x_n, y_n)$ have been randomly selected from a bivariate normal population distribution (introduced in Section 3.6). This assumption is difficult to check. A partial assessment of plausibility is based on constructing one normal quantile plot of the x's and another of the y's. A nonlinear pattern in either plot is a warning of implausibility.

A Test for Linear Association in a Bivariate Normal Population

Null hypothesis: $H_0: \rho = 0$

Test statistic: $\quad t = \dfrac{r\sqrt{n-2}}{\sqrt{1-r^2}} \quad$ where $r = \dfrac{S_{xy}}{\sqrt{S_{xx}S_{yy}}}$

When H_0 is true, the test statistic has a t distribution based on $n-2$ df, so a P-value is computed as was done for previous t tests. In particular, the usual alternative hypothesis is $H_a: \rho \neq 0$ (no linear association, positive or negative, in the population), for which the test is two-tailed and the P-value is twice the tail area captured by the calculated $|t|$.

Example 11.6   Neurotoxic effects of manganese are well known and are usually caused by high occupational exposure over long periods of time. In the fields of occupational hygiene and environmental hygiene, the relationship between lipid peroxidation, which is responsible for deterioration of foods and damage to live tissue, and occupational exposure has not been previously reported. The article "Lipid Peroxidation in Workers Exposed to Manganese" (*Scand. J. Work and Environ. Health*, 1996: 381–386) gave data on $x =$ manganese concentration in blood (ppb) and $y =$ concentration (μmol/L) of malondialdehyde, which is a stable product of lipid peroxidation, both for a sample of 22 workers exposed to manganese and for a control sample of 45 individuals. The value of r for the control sample was .29, from which

$$t = \frac{(.29)\sqrt{45 - 2}}{\sqrt{1 - (.29)^2}} \approx 2.0$$

The corresponding P-value for a two-tailed t test based on 43 df is roughly .052 (the cited article reported only that P-value $>.05$). We would not want to reject the assertion that $\rho = 0$ at either significance level .01 or .05. For the sample of exposed workers, $r = .83$ and $t \approx 6.7$, clear evidence that there is a linear relationship in the entire population of exposed workers from which the sample was selected.

The hypothesis $H_0: \beta = 0$ for the model utility test in regression also asserts that there is no linear relationship between x and y. Although it is certainly not obvious by inspection, it can be shown that the t-ratio b/s_b is algebraically identical to the t statistic in the previous box for testing $\rho = 0$. The value of the latter statistic is easier to compute, since it requires only r and not any of the calculations appropriate for regression.

Test procedures for $H_0: \rho = \rho_0$ when $\rho_0 \neq 0$ are rather complicated, as is the procedure for obtaining a confidence interval for ρ. Please consult one of the chapter references for further information.

Section 11.2 Exercises

10. Exercise 4 of Section 11.1 gave data on x = rainfall volume and y = runoff volume (both in m^3). Use the accompanying Minitab output to decide whether there is a useful linear relationship between rainfall and runoff, and then calculate a confidence interval for the true average change in runoff volume associated with a 1-m^3 increase in rainfall volume.

```
The regression equation is
runoff = -1.13 + 0.827 rainfall
Predictor     Coef    Stdev  t-ratio      p
Constant    -1.128    2.368    -0.48  0.642
rainfall   0.82697  0.03652    22.64  0.000
s =5.240 R-sq = 97.5% R-sq(adj) = 97.3%
```

11. In the same way that b/s_b is the t-ratio for testing $H_0: \beta = 0$, the t-ratio a/s_a is appropriate for testing $H_0: \alpha = 0$, where s_a is the estimated standard deviation of the statistic a and the test is again based on $n - 2$ df. The null hypothesis says that the vertical intercept of the population line is zero, so that the line passes through the origin $(0, 0)$. Carry out this test using the information given in Exercise 10.

12. Use the computer output given in Exercise 9 of the previous section to decide whether the simple linear regression model specifies a useful relationship between flux and inverse foil thickness.

13. Exercise 22 (Section 3.3) of Chapter 3 gave SAS output from a regression of amount of oil recovered from wheat straw on amount of oil added.
    a. Does the simple linear regression model appear to specify a useful relationship between these two variables? State the relevant hypotheses, and carry out a test in two different ways.
    b. If the roles of the two variables were reversed, so that the amount of oil recovered from wheat straw was the independent variable, what would be the value of the t-ratio for testing model utility? (Answer without actually carrying out another regression analysis, and explain your reasoning.)

14. Exercise 20 (Section 3.3) of Chapter 3 presented data on y = dielectric constant and x = air void (%)

for 18 asphalt mixture samples having 5% asphalt content. The following R output is from a simple linear regression of y on x:

```
                          Std.       t       Pr
              Estimate    Error    value   (>|t|)
(Intercept)   4.858691  0.059768  81.293   <2e-16
AirVoid      -0.074676  0.009923  -7.526  1.21e-06
Residual standard error: 0.03551 on 16
    degrees of freedom
Multiple R-squared: 0.7797,
    Adjusted R-squared: 0.766
F-statistic: 56.63 on 1 and 16 DF, p-value: 1.214e-06
Analysis of Variance Table
Response: Dielectric
            DF    Sum Sq   Mean Sq  F value    Pr(>F)
AirVoid      1  0.071422  0.071422   56.635  1.214e-06
Residuals   16  0.020178  0.001261
```

a. What are the values of SSRegr, SSResid, and SSTo?

b. Determine and interpret the value of r^2 for this regression. What is the corresponding value of r? Note that the sign of r can be determined based on the output.

c. Use the output to calculate a confidence interval with a confidence level of 95% for the slope β of the population regression line and interpret the resulting interval.

d. Suppose it had previously been believed that when air void increased by 1 percent, the associated true average change in dielectric constant would be at least $-.05$. Does the sample data contradict this belief? State and test the relevant hypotheses.

15. Suppose that the unit of measurement for y = wear loss in Example 11.5 is changed from mm^3 to in^3, which amounts to multiplying each y value by the same conversion factor c. How does this change affect the value of the t-ratio for testing model utility? Explain your reasoning.

16. The value of the sample correlation coefficient is .722 for the $n = 14$ observations on average anterior maximum inclination angle (AMIA) in both the clockwise (Cl) and counterclockwise (Co) directions given in Exercise 10 (Section 3.2) of Chapter 3. Carry out a test at significance level .05 to decide whether these two variables are linearly related in the population from which the data was selected (assuming that the population distribution is bivariate normal).

17. A sample of $n = 13$ steel specimens was selected, and the values of x = nickel content and y = percentage austentite were determined, resulting in

$$\sum(x_i - \bar{x})^2 = 1.183 \quad \sum(y_i - \bar{y})^2 = .05080$$
$$\sum(x_i - \bar{x})(y_i - \bar{y}) = .2073$$

Does there appear to be a *positive* linear relationship between these two variables in the sampled population? State and test the relevant hypotheses.

18. In what was surely an unpleasant data collection experience, the article "Annual Variations of Odor Concentrations and Emissions from Swine Gestation, Farrowing, and Nursery Buildings" (*J. of the Air and Waste Mgmnt.*, 2011: 1361–1368) reported on monthly odor concentrations and emission rates from a Canadian swine farm for a period of one year. One study objective was to identify possible relationships, if any, between odor and presence of other gases such as ammonia (NH_3), hydrogen sulfide (H_2S), carbon dioxide (CO_2), and methane (CH_4). Identifying such relationships would be helpful in that the gas concentration could be used as an odor indicator.

a. A scatterplot of the $n = 32$ observations on y = odor concentration (OU/m^3) and x = H_2S concentration (ppb) suggested the plausibility of a positive linear relationship. The coefficient of determination for the simple linear regression of y on x was .58. State and test the relevant hypotheses to see if the message from the scatterplot can be confirmed.

b. A scatterplot of the $n = 32$ observations on y = odor concentration (OU/m^3) and x = CH_4 concentration (ppm) also suggested the plausibility of a positive linear relationship. The coefficient of determination for the simple linear regression of y on x was 0.33. State and test the relevant hypotheses to see if the message from the scatterplot can be confirmed.

19. How does lateral acceleration—side forces experienced in turns that are largely under driver control—affect nausea as perceived by bus passengers? The article "Motion Sickness in Public Road Transport: The Effect of Driver, Route, and Vehicle" (*Ergonomics*, 1999: 1646–1664) reported data on x = motion sickness dose (calculated in accordance with a British standard for evaluating

similar motion at sea) and y = reported nausea (%). Relevant summary quantities are

$$n = 17 \quad \sum x_i = 221.1 \quad \sum y_i = 193$$

$$\sum x_i^2 = 3056.69 \quad \sum x_i y_i = 2759.6$$

$$\sum y_i^2 = 2975$$

Values of dose in the sample ranged from 6.0 to 17.6.

a. Assuming that the simple linear regression model is valid for relating these two variables (this is supported by the raw data), calculate and interpret an estimate of the slope parameter that conveys information about the precision and reliability of estimation.

b. Does it appear that there is a useful linear relationship between these two variables?

c. Would it be sensible to use the simple linear regression model as a basis for predicting % nausea when dose = 5.0? Explain your reasoning.

d. When Minitab was used to fit the simple linear regression model to the raw data, the observation (6.0, 2.50) was flagged as possibly having a substantial impact on the fit. Eliminate this observation from the sample and recalculate the estimate of part (a). Based on this, does the observation appear to be exerting an undue influence?

20. Mineral mining is one of the most important economic activities in Chile. Mineral products are frequently found in saline systems composed largely of natural nitrates. Freshwater is often used as a leaching agent for the extraction of nitrate, but the Chilean mining regions have scarce freshwater resources. An alternative leaching agent is seawater. The authors of "Recovery of Nitrates from Leaching Solutions Using Seawater" (*Hydrometallurgy*, 2013: 100–105) evaluated the recovery of nitrate ions from discarded salts using freshwater and seawater leaching agents. Tests were performed in salt columns irrigated at the same rate for a period of more than 150 hours.

Here is data on x = leaching time (h), y_{fw} = nitrate extraction percentage (freshwater), and y_{sw} = nitrate extraction percentage (seawater):

x:	25.5	31.5	37.5	43.5	49.5	55.5
y_{fw}:	25.7	43.2	55.3	62.9	68.6	73.2
y_{sw}:	26.4	40.1	50.2	57.4	62.7	67.3

x:	61.5	67.5	73.5	79.5	85.5	91.5
y_{fw}:	76.7	79.4	81.8	83.7	85.1	86.5
y_{sw}:	71.4	74.7	77.8	80.3	82.3	84.1

x:	97.5	103.5	109.5	115.5	121.5	127.5
y_{fw}:	87.7	88.6	89.6	90.5	90.7	91.2
y_{sw}:	85.5	86.6	87.9	89.0	89.9	90.6

x:	133.5	139.5	145.5	151.5	157.5
y_{fw}:	91.9	92.5	93.1	93.9	94.7
y_{sw}:	91.2	91.8	92.3	92.8	93.3

a. Construct scatterplots of y_{fw} versus x and y_{sw} versus x. Note the nonlinearity of the plots. Would it be reasonable to describe the patterns in both plots as curved and monotonic?

b. In Section 3.4, we described how a power transformation can be applied to create a linear pattern in the transformed data. Using the transformation $x' = 1/x$, construct scatterplots of y_{fw} versus x', and y_{sw} versus x'. For each set of pairs, calculate point estimates of the slope and intercept of the respective population regression line.

c. Does the simple linear regression model appear to specify a useful relationship between either dependent variable and x' in part (b)? State and test the relevant hypotheses.

d. The researchers concluded that the freshwater and seawater leaching agents yield similar nitrate extraction efficiencies. Using the regression models from part (b), calculate a point estimate of true nitrate extraction percentage when leaching time is 150 hours. Are the two estimates similar?

11.3 INFERENCES BASED ON THE ESTIMATED REGRESSION LINE ___

Once the simple linear regression model has been judged useful by the model utility test discussed in Section 11.2, the estimated model can be used as the basis for further inferences. Let x^* denote a particular value of the independent or predictor variable x. In this section, we show how to obtain a confidence interval for the mean y value when

$x = x^*$ and also how to calculate a prediction interval for the value of a single y to be observed at some time in the future when $x = x^*$. For example, x might be the tensile force applied to a steel specimen (1000s of lb) and y the resulting amount of elongation (thousandths of an inch). Then we might wish to calculate a confidence interval (interval of plausible values) for the average amount of elongation for all specimens to which a tensile force of 5000 lb is applied (so $x^* = 5$). Alternatively, we might subject a single specimen to a force of 5000 lb and wish to calculate a prediction interval (interval of plausible values) for the resulting amount of elongation.

Recall that substituting a particular value x^* into the equation of the estimated regression line gives a number $\hat{y} = a + bx^*$ that has two different interpretations: It can be regarded either as a point estimate of the mean y value when $x = x^*$ or as a point prediction of the y value that would result from making a single observation when x has this value. Because the point estimate and point prediction are single numbers, they convey no information about the reliability or precision of estimation or prediction. An interval gives information about reliability through its confidence or prediction level (e.g., 95%) and about precision from the width of the interval.

Before we obtain sample data, both a and b are subject to sampling variability—that is, they are both statistics whose values will vary from sample to sample. Suppose, for example, that $\alpha = 50$ and $\beta = 2$. Then a first sample of (x, y) pairs might give $a = 52.35$, $b = 1.895$, a second sample might result in $a = 46.52$, $b = 2.056$, and so on. It follows that $\hat{y} = a + bx^*$ itself varies in value from sample to sample, so it is a statistic. If the intercept and slope of the population line are the aforementioned values 50 and 2, respectively, and $x^* = 10$, then this statistic is trying to estimate the value $50 + 2(10) = 70$. The estimate from a first sample might be $52.35 + 1.895(10) = 71.30$, from second sample might be $46.52 + 2.056(10) = 67.08$, and so on. In the same way that a confidence interval for β was based on properties of the sampling distribution of b, a confidence interval for a mean y value in regression is based on properties of the sampling distribution of the statistic $\hat{y}$.

Properties of the Sampling Distribution of $a + bx^*$

Let x^* denote a particular value of the independent variable x. Then the sampling distribution of the statistic $\hat{y} = a + bx^*$ has the following properties:

1. The mean value of this statistic is $\alpha + \beta x^*$, so the sampling distribution is centered at the value that the statistic is attempting to estimate (i.e., the statistic is unbiased for estimating $\alpha + \beta x^*$).

2. The standard deviation of the statistic is

$$\sigma_{\hat{y}} = \sigma\sqrt{\frac{1}{n} + \frac{(x^* - \bar{x})^2}{S_{xx}}}$$

The *estimated* standard deviation of the statistic $\hat{y}$, which we denote by $s_{\hat{y}}$, results from replacing σ in this expression by its estimate, s_e.

3. The assumptions that any particular random deviation e in the model equation is normally distributed and that different deviations are independent of one another imply that $\hat{y}$ itself is normally distributed.

The values of both $\sigma_{\hat{y}}$ and $s_{\hat{y}}$ increase as the value of $(x^* - \bar{x})^2$ gets larger. That is, these standard deviations increase in value as the specified value x^* deviates farther from $\bar{x}$, the center of the x values for the sample observations. Thus the farther x^* is from $\bar{x}$, the less precisely $\hat{y}$ tends to estimate $\alpha + \beta x^*$.

A Confidence Interval for a Mean y Value

In the same way that a confidence interval for the slope was based on the t variable $t = (b - \beta)/s_b$, a confidence interval here is based on a standardized variable having a t distribution.

The standardized variable

$$t = \frac{\hat{y} - (\alpha + \beta x^*)}{s_{\hat{y}}}$$

has a t distribution based on $n - 2$ df, where

$$s_{\hat{y}} = s_e \sqrt{\frac{1}{n} + \frac{(x^* - \bar{x})^2}{S_{xx}}}$$

This implies that a **confidence interval for $\alpha + \beta x^*$**, the mean y value when $x = x^*$, is

$$\hat{y} \pm (t \text{ critical value}) s_{\hat{y}}$$

The t critical values corresponding to the usual confidence levels are given in Appendix Table IV; a value from the $n - 2$ df row of this table should be used.

Example 11.7 Corrosion of steel reinforcing bars is the most important durability problem for reinforced concrete structures. Carbonation of concrete results from a chemical reaction that lowers the pH value by enough to initiate corrosion of the rebar. Representative data on x = carbonation depth (mm) and y = strength (MPa) for a sample of core specimens taken from a particular building follow (read from a plot in the article "The Carbonation of Concrete Structures in the Tropical Environment of Singapore," *Magazine of Concrete Res.*, 1996: 293–300):

x:	8.0	15.0	16.5	20.0	20.0	27.5	30.0	30.0	35.0
y:	22.8	27.2	23.7	17.1	21.5	18.6	16.1	23.4	13.4

x:	38.0	40.0	45.0	50.0	50.0	55.0	55.0	59.0	65.0
y:	19.5	12.4	13.2	11.4	10.3	14.1	9.7	12.0	6.8

A scatterplot of the data (see Figure 11.11 on p. 529) gives strong support to use of the simple linear regression model. Relevant quantities are as follows:

$$\sum x_i = 659.0 \qquad \sum x_i^2 = 28{,}967.50 \qquad \bar{x} = 36.61111 \qquad S_{xx} = 4840.7778$$

$$\sum y_i = 293.2 \qquad \sum x_i y_i = 9293.95 \qquad \sum y_i^2 = 5335.76$$

$$b = -.297561 \qquad a = 27.182936 \qquad \text{SSResid} = 131.2402$$

$$r^2 = .766 \qquad s_e = 2.8640$$

Let's now calculate a confidence interval, using a 95% confidence level, for the mean strength for all core specimens having a carbonation depth of 45 mm—that is, a confidence interval for $\alpha + \beta(45)$. The interval is centered at

$$\hat{y} = a + b(45) = 27.18 - .2976(45) = 13.79$$

The estimated standard deviation of the statistic $\hat{y}$ is

$$s_{\hat{y}} = 2.8640\sqrt{\frac{1}{18} + \frac{(45 - 36.6111)^2}{4840.7778}} = .7582$$

The 16 df t critical value for a 95% confidence level is 2.120, from which we determine the desired interval to be

$$13.79 \pm (2.120)(.7582) = 13.79 \pm 1.61 = (12.18, 15.40)$$

The narrowness of this interval suggests that we have reasonably precise information about the mean value being estimated. Remember that if we recalculated this interval for sample after sample, in the long run about 95% of the calculated intervals would include $\alpha + \beta(45)$. We can only hope that this mean value lies in the single interval that we have calculated.

Figure 11.10 shows Minitab output resulting from a request to fit the simple linear regression model and calculate confidence intervals for the mean value of strength at depths of 45 mm and 35 mm. The intervals are at the bottom of the output; note that the second interval is narrower than the first, because 35 is much closer to $\bar{x}$ than is 45. Figure 11.11 (on page 529) shows a Minitab scatterplot with (1) curves corresponding to the confidence limits for each different x value and (2) prediction limits, to be discussed shortly. Notice how the curves get farther and farther apart as x moves away from $\bar{x}$.

```
The regression equation is
strength = 27.2 - 0.298 depth
Predictor       Coef     Stdev    t-ratio        p
Constant       27.183    1.651      16.46    0.000
depth         -0.29756  0.04116     -7.23    0.000
s = 2.864    R-sq = 76.6%   R-sq(adj) = 75.1%

Analysis of Variance
SOURCE        DF       SS       MS        F       P
Regression     1    428.62   428.62    52.25   0.000
Error         16    131.24     8.20
Total         17    559.86
   Fit   Stdev.Fit        95.0% C.I.          95.0% P.I.
13.793       0.758   (12.185, 15.401)    (7.510, 20.075)
   Fit   Stdev.Fit        95.0% C.I.          95.0% P.I.
16.768       0.678   (15.330, 18.207)   (10.527, 23.009)
```

Figure 11.10 Minitab regression output for the data of Example 11.7

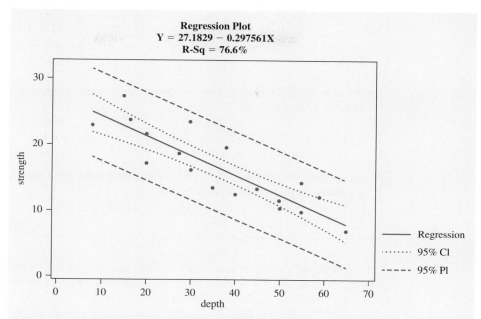

Figure 11.11 Minitab scatterplot with confidence intervals and prediction intervals for the data of Example 11.7

A Prediction Interval for a Single y Value

Suppose an investigator is contemplating making a single observation on the dependent variable y at some future time when x has the value x^*. Let y^* denote the resulting future observation. The point prediction for y^* is $\hat{y} = a + bx^*$, and this is also the point estimate for $\alpha + \beta x^*$, the mean y value when $x = x^*$. Consider the errors of estimation and prediction:

$$\text{estimation error} = \text{estimate} - \text{true value} = \hat{y} - (\alpha + \beta x^*)$$

$$\text{prediction error} = \text{prediction} - \text{true value} = \hat{y} - y^*$$

The estimation error is the difference between a random quantity ($\hat{y}$) and a fixed quantity, whereas the prediction error is the difference between *two* random quantities. This implies that there is more uncertainty associated with making a prediction than with estimating a mean y value. The mean value of the prediction error is

$$\mu_{\hat{y}-y^*} = \mu_{\hat{y}} - \mu_{y^*} = \alpha + \beta x^* - (\alpha + \beta x^*) = 0$$

Furthermore, $\hat{y}$ and y^* are independent of one another, because the former is based on the sample data and the latter is to be observed at some future time. This implies that

$$\sigma^2_{\hat{y}-y^*} = \sigma^2_{\hat{y}} + \sigma^2_{y^*} = \sigma^2\left[\frac{1}{n} + \frac{(x^* - \bar{x})^2}{S_{xx}}\right] + \sigma^2$$

The standard deviation of the prediction error is the square root of this expression, and the *estimated* standard deviation results from replacing σ^2 by s_e^2. Using these results to standardize the prediction error gives a t variable from which the prediction interval is obtained.

The standardized variable

$$t = \frac{\hat{y} - y^*}{\sqrt{s_e^2 + s_{\hat{y}}^2}}$$

has a t distribution with $n - 2$ df. This implies that a **prediction interval for a future y value y^* to be observed when $x = x^*$** is

$$\hat{y} \pm (t \text{ critical value}) \sqrt{s_e^2 + s_{\hat{y}}^2}$$

Without s_e^2 under the square root in the prediction interval formula, we would have the confidence interval formula. This implies that the prediction interval (PI) is wider than the confidence interval (CI)—often much wider because s_e^2 is frequently much larger than $s_{\hat{y}}^2$. The prediction level for the interval is interpreted in the same way that a confidence level was previously interpreted. If a prediction level of 95% is used in calculating interval after interval from different samples, in the long run about 95% of the calculated intervals will include the value y^* that is being predicted. Of course, we will not know whether the single interval that we have calculated is one of the good 95% until we have observed y^*.

Example 11.8 Let's return to the carbonation depth–strength data of Example 11.7 and calculate a 95% prediction interval for a strength value that would result from selecting a single core specimen whose carbonation depth is 45 mm. Relevant quantities from that example are

$$\hat{y} = 13.79 \qquad s_{\hat{y}} = .7582 \qquad s_e = 2.8640$$

For a prediction level of 95% based on $n - 2 = 16$ df, the t critical value is 2.120, exactly what we previously used for a 95% confidence level. The prediction interval is then

$$13.79 \pm (2.120) \sqrt{(2.8640)^2 + (.7582)^2} = 13.79 \pm (2.120)(2.963)$$

$$= 13.79 \pm 6.28 = (7.51, 20.07)$$

Plausible values for a single observation on strength when depth is 45 mm are (at the 95% prediction level) between 7.51 MPa and 20.07 MPa. The 95% confidence interval for mean strength when depth is 45 was (12.18, 15.40). The prediction interval is much wider than this because of the extra $(2.8640)^2$ under the square root. Figure 11.10, the Minitab output in Example 11.7, shows this interval as well as the confidence interval.

Simultaneous Intervals

Suppose we wish to calculate a confidence interval for the mean y value or a prediction interval for a future y value both when $x = x_1^*$ and also when $x = x_2^*$, two different values of the predictor variable. If the confidence or prediction level for each individual interval is 95%, then the joint or simultaneous level of confidence for both intervals will be smaller than 95%. For example, from Examples 11.7 and 11.8, we can be 95% confident that a y value to be observed when $x = 45$ will be in the interval (7.51, 20.07) and also 95% confident that a y value to be observed when $x = 35$ will lie in the interval (10.53, 23.01). The degree of confidence in the simultaneous statements

$$7.51 < \text{1st } y < 20.07, \qquad 10.53 < \text{2nd } y < 23.01$$

must be less than 95%. It is very difficult to say exactly what the degree of simultaneous confidence is, because the two intervals are not based on independent data sets [if they were, the simultaneous level would be $100(.95)^2 \approx 90\%$]. What can be said is that the simultaneous confidence level will be *at least* $100(1 - 2(.05))\%$, that is, at least 90%. More generally, if k different intervals are calculated, each using a confidence or prediction level of $100(1 - \alpha)\%$, then the simultaneous confidence or prediction level for all k intervals will be at least $100(1 - k\alpha)\%$. Thus if three different 99% confidence intervals were computed, the simultaneous confidence level would be at least 97%. There is a special table of t critical values for which the simultaneous level for k intervals is at least 95% ($k = 2, 3, 4, \ldots$) and another such table for at least 99%; the tabulated numbers are called *Bonferroni t critical values* after the mathematician whose inequality justifies the "at least" statement. If more than two or three of these intervals are calculated, they will have to be quite wide to guarantee at least the desired level.

Section 11.3 Exercises

21. Mist (airborne droplets or aerosols) is generated when metal-removing fluids are used in machining operations to cool and lubricate the tool and workpiece. Mist generation is a concern to OSHA, which has recently lowered substantially the workplace standard. The article "Variables Affecting Mist Generation from Metal Removal Fluids" (*Lubrication Engr.*, 2002: 10–17) gave the accompanying data on $x =$ fluid flow velocity for a 5% soluble oil (cm/sec) and $y =$ the extent of mist droplets having diameters smaller than 10 μm (mg/m^3):

x:	89	177	189	354	362	442	965
y:	.40	.60	.48	.66	.61	.69	.99

    a. The investigators performed a simple linear regression analysis to relate the two variables. Does a scatterplot of the data support this strategy?

    b. What proportion of observed variation in mist can be attributed to the simple linear regression relationship between velocity and mist?

    c. The investigators were particularly interested in the impact on mist of increasing velocity from 100 to 1000 (a factor of 10 corresponding to the difference between the smallest and largest x values in the sample). When x increases in this way, is there substantial evidence that the true average increase in y is less than .6?

    d. Estimate the true average change in mist associated with a 1 cm/sec increase in velocity, and do so in a way that conveys information about precision and reliability.

22. Phenolic compounds are found in the effluents of coal conversion processes, petroleum refineries,

herbicide manufacturing, and fiberglass manufacturing. These compounds are toxic, carcinogenic, and have contributed over the past decades to environmental pollution of aquatic environments. In one study reported in "Photolysis, Biodegradation, and Sorption Behavior of Three Selected Phenolic Compounds on the Surface and Sediment of Rivers" (*J. of Envir. Engr.*, 2011: 1114–1121), the authors examined the sorption characteristics of three selected phenolic compounds. The following data on y = sorbed concentration ($\mu g/g$) and x = equilibrium concentration ($\mu g/mL$) of 2, 4-Dinitrophenol (DNP) in a particular natural river sediment was read from a graph in the article.

x: 0.11 0.13 0.14 0.18 0.29 0.44 0.67 0.78 0.93
y: 1.72 2.17 2.33 3.00 5.17 7.61 11.17 12.72 14.78

a. Calculate point estimates of the slope and intercept of the population regression line.
b. Using the simple linear regression model fit to this data, confirm that $\hat{y}$ = 3.404, $s_{\hat{y}}$ = .107 when x = .2, and $\hat{y}$ = 6.616, $s_{\hat{y}}$ = .088 when x = .4. Explain why $s_{\hat{y}}$ is larger when x = .2 than when x = .4.
c. Calculate a confidence interval with a confidence level of 95% for the true average DNP sorbed concentration of all river sediment specimens using an equilibrium concentration of .4.
d. Calculate a prediction interval with a prediction level of 95% for the DNP sorbed concentration of a single river sediment specimen using an equilibrium concentration of .4.
e. If a 95% CI is calculated for true average DNP sorbed concentration when equilibrium concentration is .2, what will be the simultaneous confidence level for both this interval and the interval calculated in part (c)?

23. Refer to Exercise 6 of Section 11.1.
a. Predict oxygen diffusivity for a single observation to be made when temperature is 1500°F, and do so in a way that conveys information about reliability and precision.
b. Would a prediction interval for diffusivity when temperature is 1200°F using the same prediction level as in part (a) be wider or narrower than the interval of part (a)? Answer without computing this second interval.

24. The simple linear regression model provides a very good fit to the data on rainfall and runoff volume given in Exercise 4 of Section 11.1. The equation of the least squares line is $\hat{y}$ = −1.128 + .82697x, r^2 = .975, and s_e = 5.24. Use the fact that $s_{\hat{y}}$ = 1.44 when rainfall volume is 40 m^3 to predict runoff in a way that conveys information about reliability and precision. Does the resulting interval suggest that precise information about the value of runoff for this future observation is available? Explain your reasoning.

25. The article "Root Dentine Transparency: Age Determination of Human Teeth Using Computerized Densitometric Analysis" (*Amer. J. of Physical Anthro.*, 1991: 25–30) reported on an investigation of methods for age determination based on tooth characteristics. A single observation on y = age (yr) was made for each of the following values of x = % of root with transparent dentine: 15, 19, 31, 39, 41, 44, 47, 48, 55, 64. Consider the following six intervals based on the resulting data: (i) a 95% CI for mean age when x = 35; (ii) a 95% PI for age when x = 35; (iii) a 95% CI for mean age when x = 42; (iv) a 95% PI for age when x = 42; (v) a 99% CI for mean age when x = 42; (vi) a 99% PI for age when x = 42. Without computing any of these intervals, what can be said about their relative widths?

26. During oil drilling operations, components of the drilling assembly may suffer from sulfide stress cracking. The article "Composition Optimization of High-Strength Steels for Sulfide Cracking Resistance Improvement" (*Corrosion Sci.*, 2009: 2878–2884) reported on a study in which the composition of a standard grade of steel was analyzed. The following data on y = threshold stress (% SMYS) and x = yield strength (MPa) was read from a graph in the article (which also included the equation of the least squares line).

x:	635	644	711	708	836	820	810
y:	100	93	88	84	77	75	74

x:	870	856	923	878	937	948
y:	63	57	55	47	43	38

a. Does a scatterplot support the use of the simple linear regression model for relating y to x?

b. What proportion of observed variation in stress can be attributed to the approximate linear relationship between the two variables?

c. Determine a 90% confidence interval for the true average threshold stress of all similar steel specimens whose yield strength is 800 MPa.

d. Determine a 90% prediction interval for the threshold stress of a single steel specimen whose yield strength is 800 MPa.

27. Milk is an important source of protein. How does the amount of protein in milk from a cow vary with milk production? The article "Metabolites of Nucleic Acids in Bovine Milk" (*J. of Dairy Science*, 1984: 723–728) reported the accompanying data on x = milk production (kg/day) and y = milk protein (kg/day) for Holstein-Friesan cows:

x:	42.7	40.2	38.2	37.6	32.2	32.2	28.0
y:	1.20	1.16	1.07	1.13	.96	1.07	.85

x:	27.2	26.6	23.0	22.7	21.8	21.3	20.2
y:	.87	.77	.74	.76	.69	.72	.64

Relevant calculated values include $S_{xx} = 762.012$, $b = .024576$, $a = .175576$, SSTo $= .48144$, and SSResid $= .02120$.

a. Does the simple linear regression model specify a useful relationship between production and protein?

b. Estimate true average protein for all cows whose production is 30 kg/day; use a confidence interval with a confidence level of 99%. Does the resulting interval suggest that this mean value has been precisely estimated? Explain your reasoning.

c. Calculate a 99% prediction interval for the protein from a single cow whose production is 30 kg/day.

28. Obtain an expression for s_a, the estimated standard deviation of the intercept of the least squares line. Then use the fact that $t = (a - \alpha)/s_a$ has a t distribution with $n - 2$ df to test $H_0: \alpha = 0$ for the data in Exercise 27 (this null hypothesis says that the population regression line passes through the origin). *Hint:* When $x = 0$, $\hat{y} = a + b(0) = a$, and we have a general expression for $s_{\hat{y}}$.

11.4 MULTIPLE REGRESSION MODELS

The regression models considered thus far have involved relating the dependent or response variable y to a single independent or predictor variable x. But it is virtually always the case that a model relating y to two or more predictors will explain more variation and provide better predictions than will a model with just a single predictor. For example, we should be able to predict fuel efficiency of a car more precisely from knowing *both* engine size and weight of the car than from knowing only one of these variables. Let k denote the number of predictor variables to be used in a model, and denote the predictors themselves by $x_1, x_2, \ldots, x_k$ (previously $x_1, x_2, \ldots$ represented various values of the single variable x, whereas now they represent different variables). For example, let y be the concentration of a certain chemical contaminant in an industrial worker's bloodstream. Then we might use the four predictors

x_1 = number of years of exposure to the contaminant
x_2 = number of years since the last exposure
x_3 = age of the worker
x_4 = a quantitative index of body mass

It is almost never true that the value of y is completely and uniquely determined by values of $x_1, \ldots, x_k$. A probabilistic relationship is obtained by starting with some deterministic function $f(x_1, \ldots, x_k)$ and adding (or perhaps multiplying by) a random deviation e to incorporate uncertainty due to various other factors.

DEFINITIONS

A **general additive multiple regression model,** which relates a dependent variable y to k predictor variables $x_1, x_2, \ldots, x_k$, is given by the model equation

$$y = \alpha + \beta_1 x_1 + \beta_2 x_2 + \cdots + \beta_k x_k + e$$

The random deviation e is assumed to be normally distributed with mean value 0 and variance σ^2 for any particular values of the predictors, and the e's resulting from different observations are assumed to be independent of one another. The β_i's are called **population regression coefficients,** and the deterministic portion $\alpha + \beta_1 x_1 + \cdots + \beta_k x_k$ is the **population regression function.**

Let $x_1^*, x_2^*, \ldots, x_k^*$ denote particular values of the predictors. Then the model equation and assumptions about e imply that

$$(\text{mean } y \text{ value when } x_1 = x_1^*, \ldots, x_k = x_k^*) = \alpha + \beta_1 x_1^* + \cdots + \beta_k x_k^*$$

$$(\text{variance of } y \text{ when } x_1 = x_1^*, \ldots, x_k = x_k^*) = \sigma^2$$

As in simple linear regression, if σ^2 is quite close to 0, any particular observed y value will tend to be quite near its mean value. When σ^2 is large, many of the y observations may deviate substantially from their mean y values.

The slope coefficient β in simple linear regression was interpreted as the mean change in y associated with a 1-unit increase in the value of x. Each population regression coefficient in multiple regression has a similar interpretation. For example, β_2 is the mean change in y associated with a 1-unit increase in x_2 *provided that the values of the remaining predictors $x_1, x_3, \ldots, x_k$ are held fixed.*

Example 11.9

Cardiorespiratory fitness is widely recognized as a major component of overall physical well-being. Direct measurement of maximal oxygen uptake (VO_2max) is the single best measure of such fitness, but direct measurement is time-consuming and expensive. It is therefore desirable to have a prediction equation for VO_2max in terms of easily obtained quantities. Consider the variables

$$y = VO_2\text{max (L/min)} \qquad x_1 = \text{weight (kg)} \qquad x_2 = \text{age (yr)}$$

$$x_3 = \text{time necessary to walk 1 mile (min)}$$

$$x_4 = \text{heart rate at the end of the walk (beats/min)}$$

Here is one possible model, for male students, consistent with the information given in the article "Validation of the Rockport Fitness Walking Test in College Males and Females" (*Research Quarterly for Exercise and Sport*, 1994: 152–158):

$$y = 5.0 + .01x_1 - .05x_2 - .13x_3 - .01x_4 + e \qquad \sigma = .4$$

The population regression function is

$$\text{mean } y \text{ valued for fixed } x_1, \ldots, x_k = 5.0 + .01x_1 - .05x_2 - .13x_3 - .01x_4$$

For individuals whose weight is 76 kg, age is 20 yr, walk time is 12 min, and heart rate is 140 beats/min,

$$\text{mean value of VO}_2\text{max} = 5.0 + .01(76) - .05(20) - .13(12) - .01(140)$$
$$= 1.80 \text{ L/min}$$

With $2\sigma = .80$, it is quite likely (a probability of roughly .95) that an actual y value observed when the x_i's are as stated will be within .80 of the mean value, that is, in the interval from 1.00 to 2.60.

The value $\beta_2 = -.05$ is interpreted as the average change in VO_2max (here a decrease) associated with a 1-year increase in age while weight, walk time, and heart rate are all held fixed. The three other β_i's associated with predictors have similar interpretations.

A Special Case: Polynomial Regression

Consider again the case of a single independent variable x, and suppose that a scatterplot of the n sample (x, y) pairs has the appearance of Figure 11.12. The simple linear regression model is clearly not appropriate. It does, however, look as though a parabola, the graph of a quadratic function $y = \alpha + \beta_1 x + \beta_2 x^2$, would provide a very good fit to the data for appropriately chosen values of α and the β_i's. Because no quadratic would give a perfect fit, we need a probabilistic model that allows observed points to deviate from the parabola. Adding a random deviation e to the quadratic function gives such a model:

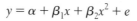

$$y = \alpha + \beta_1 x + \beta_2 x^2 + e$$

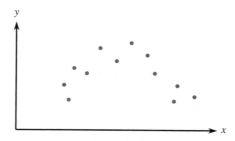

Figure 11.12 A scatterplot consistent with a quadratic regression model

If we rewrite this equation with $x_1 = x$ and $x_2 = x^2$, a special case of the general multiple regression model with $k = 2$ results. Notice that one of the two predictors is a mathematical function of the other one: $x_2 = (x_1)^2$. In general, in a multiple regression model, it is perfectly legitimate to have one or more of the k predictors that are mathematical functions of other predictors. For example, we will shortly discuss models that include an *interaction* predictor of the form $x_3 = x_1 x_2$, a product of two other predictors. In particular, the general polynomial regression model begins with a single independent variable x and creates predictors $x_1 = x, x_2 = x^2, \ldots, x_k = x^k$ for some specified value of k.

DEFINITIONS

The **kth-degree polynomial regression model**

$$y = \alpha + \beta_1 x + \beta_2 x^2 + \cdots + \beta_k x^k + e$$

is a special case of the general additive multiple regression model with $x_1 = x$, $x_2 = x^2, \ldots, x_k = x^k$. The population regression function is

$$\text{mean } y \text{ value for fixed } x = \alpha + \beta_1 x + \cdots + \beta_k x^k$$

The most important special case other than simple linear regression ($k = 1$) is the **quadratic regression model**

$$y = \alpha + \beta_1 x + \beta_2 x^2 + e$$

This model replaces the line of mean y values in simple linear regression with a parabolic curve of mean values $\alpha + \beta_1 x + \beta_2 x^2$. If $\beta_2 < 0$, the curve opens downward, as in Figure 11.13(a), whereas it opens upward when $\beta_2 > 0$. A less frequently encountered case is that of cubic regression, in which $k = 3$.

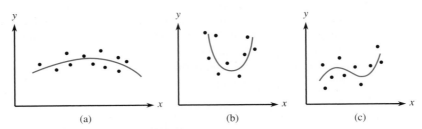

Figure 11.13 Polynomial regression models: (a) quadratic regression model with $\beta_2 < 0$; (b) quadratic regression model with $\beta_2 > 0$; (c) cubic regression model with $\beta_3 > 0$

Example 11.10

Researchers have examined a variety of climatic variables in an attempt to gain an understanding of the mechanisms that govern rainfall runoff. The article "The Applicability of Morton's and Penman's Evapotranspiration Estimates in Rainfall-Runoff Modeling" (*Water Resources Bull.*, 1991: 611–620) reported on a study in which data on x = cloud cover and y = daily sunshine (hr) was gathered from a number of different locations. The authors used a cubic regression model to relate these variables. Suppose that the actual model equation for a particular location is

$$y = 11 - .400x - .250x^2 + .005x^3 + e$$

Then the regression function is

(mean daily sunshine for given cloud cover x) $= 11 - .400x - .250x^2 + .005x^3$

For example,

(mean daily sunshine when cloud cover is 4) $= 11 - .400(4) - .250(4)^2 + .005(4)^3$
$$= 5.72$$

If $\sigma = 1$, it is quite likely that an observation on daily sunshine made when $x = 4$ would be between 3.72 and 7.72 hr.

The interpretation of β_i given previously for the general multiple regression model is not legitimate in polynomial regression. This is because all predictors are functions of x, so $x_i = (x)^i$ cannot be increased by 1 unit while the values of all other predictors are held fixed. In general, the interpretation of regression coefficients requires extra care when some predictor variables are mathematical functions of other variables.

Interaction Between Variables

Suppose that an industrial chemist is interested in the relationship between product yield (y) from a certain reaction and two independent variables, x_1 = reaction temperature and x_2 = pressure at which the reaction is carried out. The chemist initially proposes the relationship

$$y = 1200 + 15x_1 - 35x_2 + e$$

for temperature values between 80 and 100 in combination with pressure values ranging from 50 to 70. The population regression function $1200 + 15x_1 - 35x_2$ gives the mean y value for any particular values of the predictors. Consider this mean y value for three different particular temperature values:

$x_1 = 90$:   mean y value $= 1200 + 15(90) - 35x_2 = 2550 - 35x_2$
$x_1 = 95$:   mean y value $= 2625 - 35x_2$
$x_1 = 100$:   mean y value $= 2700 - 35x_2$

Graphs of these three mean y value functions are shown in Figure 11.14(a). Each graph is a straight line, and the three lines are parallel, each with a slope of -35. Thus irrespective of the fixed value of temperature, the average change in yield associated with a 1-unit increase in pressure is -35.

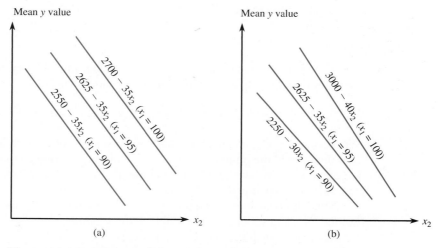

Figure 11.14 Graphs of the mean y value for two different models:
(a) $1200 + 15x_1 - 35x_2$; (b) $-4500 + 75x_1 + 60x_2 - x_1x_2$

Since chemical theory suggests that the decline in average yield when pressure x_2 increases should be more rapid for a high temperature than for a low temperature, the chemist now has reason to doubt the appropriateness of the proposed model. Rather than the lines being parallel, the line for a temperature of 100 should be steeper than the line for a temperature of 95, and that line in turn should be steeper than the line for $x_1 = 90$. A model that has this property includes, in addition to predictors x_1 and x_2, a third predictor variable, $x_3 = x_1 x_2$. One such model is

$$y = -4500 + 75x_1 + 60x_2 - x_1 x_2 + e$$

for which the population regression function is $-4500 + 75x_1 + 60x_2 - x_1 x_2$. This gives

$$\text{mean } y \text{ value when temperature is } 100 = -4500 + (75)(100) + 60x_2 - 100x_2$$
$$= 3000 - 40x_2$$
$$\text{mean value when temperature is } 95 = 2625 - 35x_2$$
$$\text{mean value when temperature is } 90 = 2250 - 30x_2$$

These are graphed in Figure 11.14(b), where it is clear that the three slopes are different. Now each different value of x_1 yields a line with a different slope, so the average change in yield associated with a 1-unit increase in x_2 depends on the value of x_1. When this is the case, the two variables are said to *interact*.

DEFINITION

If the change in the mean y value associated with a 1-unit increase in one independent variable depends on the value of a second independent variable, there is **interaction** between these two variables. Denoting the two independent variables by x_1 and x_2, we can model this interaction by including as an additional predictor $x_3 = x_1 x_2$, the product of the two independent variables.

The general equation for a multiple regression model based on two independent variables x_1 and x_2 that also includes an interaction predictor is

$$y = \beta_0 + \beta_1 x_1 + \beta_2 x_2 + \beta_3 x_3 + e \qquad \text{with } x_3 = x_1 x_2$$

When x_1 and x_2 do interact, this model will usually give a much better fit to resulting data than would the no-interaction model. Failure to consider a model with interaction too often leads an investigator to conclude incorrectly that the relationship between y and a set of independent variables is not very substantial.

More than one interaction predictor can be included in the model when more than two independent variables are available. If, for example, three independent variables x_1, x_2, and x_3 are available, one possible model is

$$y = \alpha + \beta_1 x_1 + \beta_2 x_2 + \beta_3 x_3 + \beta_4 x_1 x_2 + \beta_5 x_1 x_3 + \beta_6 x_2 x_3 + e$$

One could even include a three-way interaction $x_7 = x_1 x_2 x_3$, although in practice this is rarely done. In applied work, quadratic predictors such as x_1^2 and x_2^2 are often included to model a curved relationship between y and several independent variables. A frequently used model with $k = 5$ based on two independent variables x_1 and x_2 is the **full quadratic** or **complete second-order model**

$$y = \alpha + \beta_1 x_1 + \beta_2 x_2 + \beta_3 x_1 x_2 + \beta_4 x_1^2 + \beta_5 x_2^2 + e$$

This model replaces the straight lines of Figure 11.14 with parabolas (each one is the graph of the population regression function as x_2 varies when x_1 has a particular value). Starting with four independent variables $x_1, \dots, x_4$, one could create a model with four quadratic predictors and six two-way interaction predictor variables. Clearly, a great many different models can be created from just a small number of independent variables. In Section 11.6 we briefly discuss methods for selecting one model from a number of competing models.

Qualitative Predictor Variables

Thus far we have explicitly considered the inclusion of only quantitative (numerical) predictor variables in a multiple regression model. Using simple numerical coding, qualitative (categorical) variables, such as bearing material (aluminum or copper/lead) or type of wood (pine, oak, or walnut), can also be incorporated into a model. Let's first focus on the case of a dichotomous variable, one with just two possible categories — male or female, U.S. or foreign manufacture, and so on. With any such variable, we associate a **dummy** or **indicator variable** x whose possible values 0 and 1 indicate which category is relevant for any particular observation.

Example 11.11   The article "Estimating Urban Travel Times: A Comparative Study" (*Trans. Res.*, 1980: 173–175) described a study relating the dependent variable $y = $ travel time between locations in a certain city and the independent variable $x_2 = $ distance between locations. Two types of vehicles, passenger cars and trucks, were used in the study. Let

$$x_1 = \begin{cases} 1 & \text{if the vehicle is a truck} \\ 0 & \text{if the vehicle is a passenger car} \end{cases}$$

One possible multiple regression model is

$$y = \alpha + \beta_1 x_1 + \beta_2 x_2 + e$$

The mean value of travel time depends on whether a vehicle is a car or a truck:

$$\text{mean time} = \alpha + \beta_2 x_2 \qquad \text{when } x_1 = 0 \quad \text{(cars)}$$

$$\text{mean time} = \alpha + \beta_1 + \beta_2 x_2 \qquad \text{when } x_1 = 1 \quad \text{(trucks)}$$

The coefficient β_1 is the difference in mean times between trucks and cars with distance held fixed; if $\beta_1 > 0$, on average it will take trucks longer to traverse any particular distance than it will for cars.

A second possibility is a model with an interaction predictor:

$$y = \alpha + \beta_1 x_1 + \beta_2 x_2 + \beta_3 x_1 x_2 + e$$

Now the mean times for the two types of vehicles are

$$\text{mean time} = \alpha + \beta_2 x_2 \qquad \text{when } x_1 = 0$$
$$\text{mean time} = \alpha + \beta_1 + (\beta_2 + \beta_3) x_2 \qquad \text{when } x_1 = 1$$

For each model, the graph of the mean time versus distance is a straight line for either type of vehicle, as illustrated in Figure 11.15. The two lines are parallel for the first (no-interaction) model, but in general they will have different slopes when the second model is correct. For this latter model, the change in mean travel time associated with a 1-mile increase in distance depends on which type of vehicle is involved—the two variables "vehicle type" and "travel time" interact. Indeed, data collected by the authors of the cited article suggested the presence of interaction.

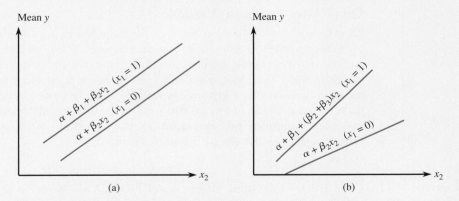

Figure 11.15 Regression functions for models with one dummy variable (x_1) and one quantitative variable x_2: (a) no interaction; (b) interaction

You might think that the way to handle a three-category situation is to define a single numerical variable with coded values such as 0, 1, and 2 corresponding to the three categories. This is incorrect, because it imposes an ordering on the categories that is not necessarily implied by the problem context. The correct approach to incorporating three categories is to define *two* different dummy variables. Suppose, for example, that y is the lifetime of a certain cutting tool, x_1 is cutting speed, and there are three brands of tool being investigated. Then let

$$x_2 = \begin{cases} 1 & \text{if a brand A tool is used} \\ 0 & \text{otherwise} \end{cases} \qquad x_3 = \begin{cases} 1 & \text{if a brand B tool is used} \\ 0 & \text{otherwise} \end{cases}$$

When an observation on a brand A tool is made, $x_2 = 1$ and $x_3 = 0$, whereas for a brand B tool, $x_2 = 0$ and $x_3 = 1$. An observation made on a brand C tool has $x_2 = x_3 = 0$, and it is not possible that $x_2 = x_3 = 1$ because a tool cannot simultaneously be both brand

A and brand B. The no-interaction model would have only the predictors x_1, x_2, and x_3. The following interaction model allows the mean change in lifetime associated with a 1-unit increase in speed to depend on the brand of tool:

$$y = \alpha + \beta_1 x_1 + \beta_2 x_2 + \beta_3 x_3 + \beta_4 x_1 x_2 + \beta_5 x_1 x_3 + e$$

Construction of a picture like Figure 11.14 with a graph for each of the three possible (x_2, x_3) pairs gives three nonparallel lines (unless $\beta_4 = \beta_5 = 0$).

More generally, incorporating a categorical variable with c possible categories into a multiple regression model requires the use of $c - 1$ indicator variables (e.g., five brands of tools would necessitate using four indicator variables). Thus even one categorical variable can add many predictors to a model.

Nonlinear Multiple Regression Models

Many nonlinear relationships can be put in the form of our basic additive model equation by transforming one or more of the variables. For example, taking the logarithm on both sides of the multiplicative exponential model equation

$$y = \beta_0 e^{\beta_1 x_1 + \beta_2 x_2 + \cdots + \beta_k x_k} \cdot \varepsilon, \ \varepsilon > 0$$

gives an equation of the desired form [with $\alpha = \ln(\beta_0)$]. An appropriate transformation could be suggested by theory or by various plots of the data, such as those to be discussed in Section 11.6. There are also relationships that cannot be linearized by means of a transformation, necessitating more complex methods of analysis. Consult one of the chapter references for more information.

Section 11.4 Exercises

29. A trucking company considered a multiple regression model for relating the dependent variable $y =$ total daily travel time for one of its drivers (hours) to the predictors $x_1 =$ distance traveled (miles) and $x_2 =$ the number of deliveries made. Suppose that the model equation is
$$y = -.800 + .060x_1 + .900x_2 + e$$
    a. What is the mean value of travel time when distance traveled is 50 miles and three deliveries are made?
    b. How would you interpret $\beta_1 = .060$, the coefficient of the predictor x_1? What is the interpretation of $\beta_2 = .900$?
    c. If $\sigma = .5$ hour, what is the probability that travel time will be at most 6 hours when three deliveries are made and the distance traveled is 50 miles?

30. Consider the regression model $y = -6.50 + .250x_1 + .600x_2 - .150x_3 + .160x_4 + e$, where $y =$ gasoline yield (% of crude oil), $x_1 =$ crude oil gravity (°API), $x_2 =$ crude oil vapor pressure (PSIA), $x_3 =$ crude oil ASTM 10% point (°F), and $x_4 =$ gasoline end point (°F).
    a. Interpret the population regression coefficients β_1 and β_3.
    b. What is the mean yield when $x_1 = 40$, $x_2 = 5$, $x_3 = 230$, and $x_4 = 360$?

31. High-alumina refractory castables have been extensively investigated in recent years because of their significant advantages over other refractory brick of the same class: lower production and application costs, versatility, and performance at high temperatures. The authors of "Processing of Zero-Cement Self-Flow Alumina Castables" (*The Amer. Ceramic Soc. Bull.*, 1998: 60–66) proposed a quadratic regression model to describe the relationship between $x =$ viscosity (MPa · sec) and $y =$ free flow (%). Suppose the actual model is $y = -296 + 2.20x - .003x^2 + e$.

a. Graph the true regression function $y = -296 + 2.20x - .003x^2$ for x values between 350 and 485.

b. Would mean free flow percentage be higher for a viscosity value of 450 or 470?

c. What is the change in mean free flow percentage when the viscosity increases from 450 to 460? From 460 to 470?

32. Let y = wear life of a bearing, x_1 = oil viscosity, and x_2 = load. Suppose that the multiple regression model relating life to viscosity and load is

$$y = 125.0 + 7.750x_1 + .0950x_2 - .0090x_1x_2 + e$$

a. What is the mean value of life when viscosity is 40 and load is 1100?

b. When viscosity is 30, what is the change in mean life associated with an increase of 1 in load? When viscosity is 40, what is the change in mean life associated with an increase of 1 in load?

33. Let y = sales at a fast-food outlet (1000s of \$), x_1 = number of competing outlets within a 1-mile radius, x_2 = population within a 1-mile radius (1000s of people), and x_3 be an indicator variable that equals 1 if the outlet has a drive-up window and 0 otherwise. Suppose that the true regression model is

$$y = 10.00 - 1.2x_1 + 6.8x_2 + 15.3x_3 + e$$

a. What is the mean value of sales when the number of competing outlets is 2, there are 8000 people within a 1-mile radius, and the outlet has a drive-up window?

b. What is the mean value of sales for an out-let without a drive-up window that has 3 competing outlets and 5000 people within a 1-mile radius?

11.5   INFERENCES IN MULTIPLE REGRESSION

We now assume that a dependent or response variable y is related to k independent, predictor, or explanatory variables $x_1, \ldots, x_k$ via the general additive multiple regression model

$$y = \alpha + \beta_1 x_1 + \cdots + \beta_k x_k + e$$

discussed in Section 11.4. Estimation of model parameters and other inferences are based on a sample of n observations, each one consisting of $k + 1$ numbers: a value of x_1, a value of $x_2, \ldots,$ a value of x_k, and a value of y. As in simple linear regression, the principle of least squares is used to estimate the population regression coefficients $\alpha, \beta_1, \ldots, \beta_k$. The least squares estimates $a, b_1, b_2, \ldots, b_k$ are chosen to minimize the sum of squared deviations:

$$\sum_{\text{all obs}} [y - (a + b_1 x_1 + \cdots + b_k x_k)]^2$$

As described in Section 3.5, the minimization requires taking $k + 1$ partial derivatives, equating these to zero to obtain a system of linear equations (the *normal* equations), and solving this system for the estimates. There are formulas for the least squares estimates, but the only sensible way to express them is to use the branch of mathematics called matrix algebra. Fortunately, this is not necessary for our purposes; these formulas have been programmed into all of the most popular statistical computer packages. When using any particular package, it is necessary only to enter the data, make an appropriate request, and know how to find the estimates on the output. The estimated regression equation $\hat{y} = a + b_1 x_1 + \cdots + b_k x_k$ can then be used to estimate a mean y value or predict a single y value.

Example 11.12   The article "How to Optimize and Control the Wire Bonding Process: Part II (*Solid State Technology*, Jan. 1991: 67–72) described an experiment carried out to assess the impact of the variables x_1 = force (g), x_2 = power (mW), x_3 = temperature (°C), and x_4 = time (ms) on y = ball bond shear strength (g). The following data[1] was generated to be consistent with the information given in the article:

Observation	Force	Power	Temperature	Time	Strength
1	30	60	175	15	26.2
2	40	60	175	15	26.3
3	30	90	175	15	39.8
4	40	90	175	15	39.7
5	30	60	225	15	38.6
6	40	60	225	15	35.5
7	30	90	225	15	48.8
8	40	90	225	15	37.8
9	30	60	175	25	26.6
10	40	60	175	25	23.4
11	30	90	175	25	38.6
12	40	90	175	25	52.1
13	30	60	225	25	39.5
14	40	60	225	25	32.3
15	30	90	225	25	43.0
16	40	90	225	25	56.0
17	25	75	200	20	35.2
18	45	75	200	20	46.9
19	35	45	200	20	22.7
20	35	105	200	20	58.7
21	35	75	150	20	34.5
22	35	75	250	20	44.0
23	35	75	200	10	35.7
24	35	75	200	30	41.8
25	35	75	200	20	36.5
26	35	75	200	20	37.6
27	35	75	200	20	40.3
28	35	75	200	20	46.0
29	35	75	200	20	27.8
30	35	75	200	20	40.3

A statistical computer package gave the following least squares estimates:

$$a = -37.48 \quad b_1 = .2117 \quad b_2 = .4983 \quad b_3 = .1297 \quad b_4 = .2583$$

---

[1] From the book *Statistics Engineering Problem Solving* by Stephen Vardeman, an excellent exposition of the territory covered by our book, albeit at a somewhat higher level.

Thus we estimate that .1297 gm is the average change in strength associated with a 1-degree increase in temperature when the other three predictors are held fixed; the other estimated coefficients are interpreted in a similar manner.

The estimated regression equation is

$$\hat{y} = -37.48 + .2117x_1 + .4983x_2 + .1297x_3 + .2583x_4$$

A point prediction of strength resulting from a force of 35 g, power of 75 mW, temperature of 200 degrees, and time of 20 ms is

$$\hat{y} = -37.48 + (.2117)(35) + (.4983)(75) + (.1297)(200) + (.2583)(20)$$

$$= 38.41 \text{ g}$$

This is also a point estimate of the mean value of strength for the specified values of force, power, temperature, and time.

Substituting the values of the predictors from the successive observations into the equation for the estimated regression gives the **predicted** or **fitted values** $\hat{y}_1, \hat{y}_2, \ldots, \hat{y}_n$. For example, since the values of the four predictors for the last observation in Example 11.12 are 35, 75, 200, and 20, respectively, the corresponding predicted value is $\hat{y}_{30} = 38.41$. The **residuals** are the differences $y_1 - \hat{y}_1, \ldots, y_n - \hat{y}_n$. The last residual in Example 11.12 is $40.3 - 38.41 = 1.89$. The closer the residuals are to zero, the better the job our estimated equation is doing in predicting the y values corresponding to values of the predictors in our sample. Squaring these residuals and summing gives **residual** or **error sum of squares** $\sum(y_i - \hat{y}_i)^2$, denoted by SSResid. The number of df associated with SSResid is $n - (k + 1)$. The explanation is that the $k + 1$ parameters $\alpha, \beta_1, \ldots, \beta_k$ have to be estimated from the data before SSResid can be calculated, resulting in a loss of this many df (in simple linear regression, $k = 1$ so df $= n - 2$). The variance σ^2 of a random deviation e in the model equation is estimated by $s_e^2 = \text{SSResid}/[n - (k + 1)]$, and s_e is the estimate of σ. For the data of Example 11.12, SSResid $= 665.12$, so $665.12/[30 - (4 + 1)] = 26.60$ and the estimated standard deviation is 5.16. We estimate that, roughly speaking, the size of a typical deviation of y from its mean value will be about 5.2 g.

Model Utility

A very important quantity introduced in Section 3.5 is the **coefficient of multiple determination**, R^2, given by

$$R^2 = 1 - \frac{\text{SSResid}}{\text{SSTo}} \qquad \text{where SSTo} = \sum(y_i - \bar{y})^2$$

R^2 is interpreted as the proportion of variation in the observed y values that can be attributed to (or explained by) the model relationship between y and the predictors. The closer R^2 is to 1, the more effectively the model has explained variation in y by relating it to the predictors. The coefficient of multiple determination for the data of Example 11.12 is .714, so somewhat more than 70% of the observed variation in strength can be attributed to the model relationship between strength and the four predictors force, power, temperature, and time.

The value of R^2 cannot decrease when an extra predictor is added to the model, and it will generally increase. Furthermore, the value of R^2 can almost always be made very close to 1 simply by using a model whose number of predictors is quite close to the sample size, even if many of these predictors are "frivolous" in the sense that they would contribute only marginally to explaining variation in y. Because R^2 can be misleading in this way, a quantity called **adjusted R^2** is included on multiple regression output from most statistical computer packages. It is defined by

$$\text{adjusted } R^2 = 1 - \frac{\text{SSResid}/[n - (k + 1)]}{\text{SSTo}/(n - 1)} = 1 - \left[\frac{n - 1}{n - (k + 1)}\right]\frac{\text{SSResid}}{\text{SSTo}}$$

Replacing the expression in brackets on the far right by 1 gives R^2 itself. Since this expression is less than 1, the adjusted R^2 is smaller than R^2. This downward adjustment will be small when R^2 is reasonably high and this has been achieved by using a model with relatively few predictors compared to the sample size. For example, adjusted R^2 for the model fit in Example 11.12 is .668, which is not all that much smaller than R^2 itself. The adjustment will be more dramatic when R^2 is not so high or when k is large relative to n.

High values of R^2 and adjusted R^2 certainly suggest that the model fit is a useful one. But how large should these values be before we draw this conclusion? It is desirable to have a formal test procedure so that we will not be led astray by intuition. Recall that the null hypothesis for the model utility test in simple linear regression was that $\beta = 0$; its interpretation was that there is no useful linear relationship between y and the single predictor x. Here, the null hypothesis states that there is no useful linear relationship between y and *any* of the k predictors included in the model. The test is based on F distributions, which were first encountered in Chapter 9 in connection with the analysis of variance.

The Model Utility F Test in Multiple Regression

Null hypothesis: $H_0: \beta_1 = \beta_2 = \cdots = \beta_k = 0$

Alternative hypothesis: $H_a:$ at least one among $\beta_1, \ldots, \beta_k$ is not zero

Test statistic: $F = \dfrac{R^2/k}{(1 - R^2)/[n - (k + 1)]} = \dfrac{\text{MSRegr}}{\text{MSResid}}$

where

$$\text{MSResid} = \text{SSresid}/[(n - (k + 1)]$$
$$\text{MSRegr} = \text{SSRegr}/k$$
$$\text{SSRegr} = \text{SSTo} - \text{SSResid}$$

The larger the value of R^2, the larger the value of F will be, implying that the test is upper-tailed (as were F tests in ANOVA). When H_0 is true, the test statistic has an F distribution based on k numerator and $n - (k + 1)$ denominator df. The P-value for the test is the area under the corresponding F curve to the right of the calculated value of F. Partial information about this P-value can be obtained from the table of F critical values given in Appendix Table VIII. As usual, the null hypothesis is rejected if the P-value is less than or equal to the chosen significance level.

A large value of R^2 is no guarantee that the model will be judged useful by the F test. If k is large relative to n, F will not exceed 0 by a great deal and the P-value will not be very small.

Example 11.13   Returning to the bond shear strength data of Example 11.12, a model with $k = 4$ predictors was fit, so the relevant hypotheses are

$$H_0: \beta_1 = \beta_2 = \beta_3 = \beta_4 = 0$$

$$H_a: \text{at least one of these four } \beta\text{'s is not zero}$$

Figure 11.16 shows output from the JMP statistical package. The values of the estimated coefficients, s_e (Root Mean Square Error), R^2, and adjusted R^2 agree with those given previously.

Response: strength

Summary of Fit

RSquare	0.713959
RSquare Adj	0.668193
Root Mean Square Error	5.157979
Mean of Response	38.40667
Observations (or Sum Wgts)	30

Parameter Estimates

Term	Estimate	Std Error	t Ratio	Prob>|t|
Intercept	-37.47667	13.09964	-2.86	0.0084
force	0.2116667	0.210574	1.01	0.3244
power	0.4983333	0.070191	7.10	<.0001
temp	0.1296667	0.042115	3.08	0.0050
time	0.2583333	0.210574	1.23	0.2313

Whole-Model Test

Analysis of Variance

Source	DF	Sum of Squares	Mean Square	F Ratio
Model	4	1660.1400	415.035	15.6000
Error	25	665.1187	26.605	Prob>F
C Total	29	2325.2587		<.0001

Figure 11.16 Multiple regression output from JMP for the data of Example 11.12

The value of the model utility F ratio is

$$F = \frac{R^2/k}{(1 - R^2)/[n - (k + 1)]} = \frac{.713959/4}{.286041/(30 - 5)} = 15.60$$

This value also appears in the F Ratio column of the ANOVA table in Figure 11.16. The largest F critical value for 4 numerator and 25 denominator df in our F table is 6.49, which captures an upper-tail area of .001. Thus P-value < .001. The ANOVA table in the JMP output (Figure 11.16) shows that P-value < .0001. This is a highly significant result. The null hypothesis should be rejected at any reasonable significance level. We conclude that there *is* a useful linear relationship between y and *at least one* of the four predictors in the model. This does not mean that all four predictors are useful; we will say more about this subsequently.

Inferences About an Individual β_i

Just as the value of the estimated slope coefficient b in simple linear regression varies from sample to sample, so too does the value of any estimated coefficient b_i in multiple regression. That is, b_i is a statistic, therefore it has a sampling distribution. It can be shown that the sampling distribution is normal (a consequence of the assumption that the random deviation e is normally distributed and that the various deviations are independent of one another). The mean value of the statistic b_i is β_i. That is, the sampling distribution is always centered at the value of what the statistic is trying to estimate, so the statistic is unbiased. We denote the estimated standard deviation of b_i by s_{b_i}; the formulas for these estimated standard deviations are complicated, but their values will be available on output from all of the most popular statistical computer packages. In the JMP output of Figure 11.16, the estimated standard deviations are shown in the Std Error column right next to the estimated coefficients. These quantities are the basis for calculating confidence intervals and testing hypotheses.

The standardized variable

$$t = \frac{b_i - \beta_i}{s_{b_i}}$$

has a t distribution based on $n - (k + 1)$ df. This implies that a confidence interval for β_i is

$$b_i \pm (t \text{ critical value}) s_{b_i}$$

The test statistic for $H_0: \beta_i = \text{hypothesized value}$ is

$$t = \frac{b_i - \text{hypothesized value}}{s_{b_i}}$$

The test is upper-, lower-, or two-tailed, depending on whether the inequality in H_a is $>$, $<$, or $\neq$. In practice, the most frequently tested null hypothesis is $H_0: \beta_i = 0$. The interpretation of H_0 is that as long as all the other predictors are retained in the model, the predictor x_i provides no useful information about y.

Example 11.14

The JMP output of Figure 11.16 gives $b_2 = .498333$, $s_{b_2} = .070191$, and error df $= n - (k + 1) = 25$. The t critical value for a confidence interval for β_2 with a confidence level of 95% is 2.060. The confidence interval is

$$.498333 \pm (2.060)(.070191) \approx .498 \pm .145 = (.353, .643)$$

We therefore estimate with a high degree of confidence that, when the value of power is increased by 1 mw while force, temperature, and time are all held fixed, the associated change in average strength will be between .353 gm and .643 gm.

Example 11.15

In Example 3.15 from Section 3.5, we gave a data set consisting of 13 observations on the variables $y =$ adsorption, $x_1 =$ extractable iron, and $x_2 =$ extractable aluminum. Figure 11.17 is the Minitab output from fitting the model $y = \alpha + \beta_1 x_1 + \beta_2 x_2 + \beta_3 x_3 + e$, where $x_3 = x_1 x_2$ is an interaction predictor.

Judging from the P-value of .000 for the model utility test, the fitted model specifies a very useful relationship between y and the predictors. Provided that iron and aluminum are retained in the model, does the interaction predictor appear to provide useful information about adsorption? The relevant hypotheses are

$$H_0: \beta_3 = 0$$

$$H_a: \beta_3 \neq 0$$

The test statistic is the t-ratio b_3/s_{b_3}, with value $.0005278/.0006610 = .80$. Our table of t curve tail areas shows that the area under the $13 - (3 + 1) = 9$ df curve to the right of .8 is .222 (see Appendix Table VI), so the P-value for the two-tailed test is .444 (.445 according to Minitab). The null hypothesis should not be rejected at any reasonable significance level. It is very plausible that $\beta_3 = 0$, from which we conclude that the interaction predictor does not appear to provide useful information beyond what is provided by the predictors iron and aluminum.

```
The regression equation is
adsorp = -2.37 + 0.0828 iron + 0.246 alum + 0.000528 ir*al

Predictor         Coef       Stdev    t-ratio        p
Constant        -2.368       7.179      -0.33    0.749
iron           0.08279     0.04818       1.72    0.120
alum            0.2460      0.1481       1.66    0.131
ir*al         0.0005278   0.0006610      0.80    0.445

s = 4.461    R-sq = 95.2%    R-sq(adj) = 93.6%

Analysis of Variance

SOURCE        DF        SS        MS        F        p
Regression     3    3542.6    1180.9    59.34    0.000
Error          9     179.1      19.9
Total         12    3721.7
```

Figure 11.17 Minitab output for Example 11.15

More Intervals

Because the individual estimated coefficients vary from sample to sample, so will the value of $\hat{y} = a + b_1 x_1 + \cdots + b_k x_k$ for fixed values of $x_1, \ldots, x_k$. Properties of the sampling distribution of the statistic $\hat{y}$ can be used to obtain both a confidence interval for a mean y value and a prediction interval for a single y value when the predictors have specified values. Both intervals are based on $n - (k + 1)$ df and have the same general form as in the case of simple linear regression. The CI for a mean y value is

$$\hat{y} \pm (t \text{ critical value}) s_{\hat{y}}$$

and the PI for a single as-yet-unobserved y value is

$$\hat{y} \pm (t \text{ critical value}) \sqrt{s_e^2 + s_{\hat{y}}^2}$$

where $s_{\hat{y}}$ is the estimated standard deviation of the statistic $\hat{y}$. The PI is always wider than the corresponding CI.

Example 11.16

Figure 11.18 shows Minitab output from fitting the model, using only the predictors x_1 and x_2, to the adsorption data referred to in Example 11.15. About 95% of the observed variation in adsorption can be attributed to the model relationship. The P-value for model utility is .000, confirming the utility of the chosen model. The P-values corresponding to t-ratios for the two β coefficients are .004 and .000, respectively, indicating that neither of these predictors should be deleted from the model when the other one is retained. That is, both predictors appear to provide useful information about y. The last line of the output gives estimation and prediction information when $x_1 = 200$ and $x_2 = 40$. The values of $\hat{y}$ and $s_{\hat{y}}$ are 29.16 and 1.76, respectively. The limits of both a 95% CI for mean adsorption and a 95% PI for a single adsorption value are also displayed. Notice how much wider the PI is than the CI. Even with a very high R^2 value, there is still a reasonable amount of uncertainty involved in predicting a single value of adsorption.

```
The regression equation is
strength = -7.35 + 0.113 iron + 0.349 alum

Predictor       Coef      Stdev    t-ratio        p
Constant      -7.351      3.485      -2.11    0.061
iron         0.11273    0.02969       3.80    0.004
alum         0.34900    0.07131       4.89    0.000

s = 4.379    R-sq = 94.8%    R-sq(adj) = 93.8%

Analysis of Variance

SOURCE         DF         SS         MS         F        p
Regression      2     3529.9     1765.0     92.03    0.000
Error          10      191.8       19.2
Total          12     3721.7


  Fit   Stdev.Fit       95.0% C.I.         95.0% P.I.
29.16        1.76   (25.24, 33.07)     (18.64, 39.67)
```

Figure 11.18 Minitab output for Example 11.16

Eliminating a Group of Predictors

The null hypothesis in the model utility test asserts that none of the predictors is useful. The usefulness of a single predictor can be assessed using a t-ratio. Sometimes an investigator will want to know whether any of the predictors in some specified group provides useful information. Let

$$g = \text{number of predictors in the group under investigation}$$

The relevant hypotheses are

H_0: all β's corresponding to the g predictors in the group have value 0

H_a: at least one of the β's referred to in H_0 is not 0

The alternative hypothesis is interpreted as saying that at least one predictor in the group does provide useful information about y.

The test is carried out by fitting two different models: the "full" model, consisting of all predictors (those in the group of interest and those not being considered for deletion), and the "reduced" model, which contains only those predictors not in the specified group. This results in an SSResid(full) value and an SSResid(red) value. The former SSResid cannot be larger than the latter, because it results from adding extra predictors (those in the group) without deleting anything. The usefulness of at least one predictor in the group is suggested by an SSResid(full) value that is a good deal smaller than the SSResid(red) value, because much less variation is left unexplained by the full model than by the reduced model. The test statistic is

$$F = \frac{[\text{SSResid(red)} - \text{SSResid(full)}]/g}{\text{SSResid(full)}/[n - (k + 1)]}$$

The test is upper-tailed and is based on the F distribution having g numerator df and $n - (k + 1)$ denominator df.

Example 11.17	For the bond shear strength data given in Example 11.12, the model with the four predictors force, power, temperature, and time gave SSResid = 665.12, $R^2 = .714$, and adjusted $R^2 = .668$. Now consider as the full model the complete second-order model containing not only x_1–x_4 but also 4 quadratic predictors and 6 interaction predictors, for a total of 14 predictors. The estimated regression equation is

$$\text{strength} = -1 - 2.30\text{force} - .08\text{power} + .836\text{temp} - 3.99\text{time}$$
$$+ .0240\text{for}^*\text{pow} - .0093\text{for}^*\text{temp} + .0755\text{for}^*\text{time}$$
$$- .00467\text{pow}^*\text{temp} + .0237\text{pow}^*\text{time}$$
$$+ .0007\text{temp}^*\text{tim} + .0152\text{forsqd} + .00130\text{powsqd}$$
$$- .00011\text{tempsqd} - .0078\text{timesqd}$$

with SSResid(full) = 426.93, $R^2 = .816$, adjusted $R^2 = .645$, and P-value = .002 for the model utility F test. Should any of the second-order predictors be retained in the model? The relevant null hypothesis is

$$H_0: \beta_5 = \beta_6 = \cdots = \beta_{14} = 0$$

whereas the alternative hypothesis states that at least one of these β's is not zero (that is, there is at least one useful second-order predictor). The number of predictors in the subset being considered for deletion is $g = 10$, which is numerator df; denominator df is $30 - (14 + 1) = 15$. The test statistic value is

$$F = \frac{(665.12 - 426.93)/10}{426.93/15} = .84$$

for which P-value $> .10$. The null hypothesis should not be rejected at any reasonable significance level. None of the second-order predictors appears to provide useful information beyond what is contained in the four first-order predictors.

In Chapter 3, we discussed briefly fitting more general kinds of functions to bivariate or multivariate data using the LOWESS technique or a general additive relationship. Inferential techniques for fits of these types are still in the developmental stages. Confidence intervals, for example, can be calculated using the bootstrap method presented in Chapter 7. Statistical packages such as R and SAS will do this sort of thing without much difficulty. Please consult a more advanced reference for details. In the next section, we consider further aspects of regression modeling, including checking model adequacy and variable selection.

Section 11.5 Exercises

34. The article "Validation of the Rockport Fitness Walking Test in College Males and Females" (*Research Quarterly for Exercise and Sport*, 1994: 152–158) recommended the following estimated regression equation for relating $y = \mathrm{VO_2max}$ (L/min, a measure of cardiorespiratory fitness) to the predictors $x_1 = $ gender (female $= 0$, male $= 1$), $x_2 = $ weight (lb), $x_3 = $ 1-mile walk time (min), and $x_4 = $ heart rate at the end of the walk (beats/min):

$\hat{y} = 3.5959 + .6566x_1 + .0096x_2 - .0996x_3 - .0080x_4$

a. How would you interpret the estimated coefficient $b_3 = -.0996$?

b. How would you interpret the estimated coefficient $b_1 = .6566$?

c. Suppose that an observation made on a male whose weight was 170 lb, walk time was 11 min, and heart rate was 140 beats/min resulted in $\mathrm{VO_2max} = 3.15$. What would you have predicted for $\mathrm{VO_2max}$ in this situation, and what is the value of the corresponding residual?

d. Using SSResid $= 30.1033$ and SSTo $= 102.3922$, what proportion of observed variation in $\mathrm{VO_2max}$ can be attributed to the model relationship?

35. Exercise 35 of Section 3.5 gave data on $x_1 = $ wire feed rate, $x_2 = $ welding speed, and $y = $ deposition rate of a welding process. Minitab output from fitting the multiple regression model with x_1 and x_2 as predictors is given here.

```
The regression equation is
DepRate = 0.0558 + 0.375 FeedRate + 0.00278 WeldSpd

Predictor      Coef      Stdev    t-ratio      p
Constant    0.05580    0.07836      0.71    0.485
FeedRate    0.374917   0.007476    50.15    0.000
WeldSpd     0.002775   0.001121     2.47    0.023

s = 0.0448530   R-sq = 99.3%   R-sq(adj) = 99.2%

Analysis of Variance

SOURCE        DF      SS       MS         F        p
Regression     2   5.0726   2.5363   1260.71   0.000
Error         19   0.0382   0.0020
Total         21   5.1108
```

a. Carry out the model utility test.

b. Calculate and interpret a 95% confidence interval for β_2, the population regression coefficient of x_2.

c. When $x_1 = 11.5$ and $x_2 = 40$, the estimated standard deviation of $\hat{y}$ is $s_{\hat{y}} = .02438$. Calculate a 95% confidence interval for true average deposition rate for the given values of x_1 and x_2.

d. Calculate a 95% prediction interval for the deposition rate resulting from a single experimental run with $x_1 = 11.5$ and $x_2 = 40$.

36. Exercise 37 of Section 3.5 gave R output for a regression of $y =$ deposition over a specified time period on two complex predictors x_1 and x_2 defined in terms of PAH air concentrations for various species, total time, and total amount of precipitation. Use the output in that exercise to answer the following:

a. Does there appear to be a useful linear relationship between y and at least one of the predictors?

b. The estimated standard deviation of $\hat{y}$ when x_1 is 20,000 and x_2 is .002 is $s_{\hat{y}} = 21.7$. Calculate a 95% confidence interval for the mean value of deposition under these circumstances.

c. Fitting the model with predictors x_1 and x_2 gave SSResid $= 27,454$, whereas fitting with x_1, x_2, and $x_3 = x_1x_2$ resulted in SSResid $= 20519$. Using $\alpha = .01$, can we conclude that the x_1x_2 term adds useful information to a 'reduced' model containing only x_1 and x_2? Note: when $g = 1$, the resulting F test gives the same conclusion as the t-test for whether a single variable (here, x_1x_2) contributes useful information to a model.

37. The article "Analysis of the Modeling Methodologies for Predicting the Strength of Air-Jet Spun Yarns" (*Textile Res. J.*, 1997: 39–44) reported on a study carried out to relate yarn tenacity (y, in g/tex) to yarn count (x_1, in tex), percentage polyester (x_2), first nozzle pressure (x_3, in kg/cm^2), and second nozzle pressure (x_4, in kg/cm^2). The estimate of the constant term in the corresponding multiple regression equation was 6.121. The estimated coefficients for the four predictors were $-.082$, $.113$, $.256$, and $-.219$, respectively, and the coefficient of multiple determination was .946.

a. Assuming that the sample size was $n = 25$, state and test the appropriate hypotheses to decide whether the fitted model specifies a useful linear relationship between the dependent variable and at least one of the four model predictors.

b. Again using $n = 25$, calculate the value of adjusted R^2.

c. Calculate a 99% confidence interval for true mean yarn tenacity when yarn count is 16.5, yarn contains 50% polyester, first nozzle pressure is 3, and second nozzle pressure is 5 if the estimated standard deviation of predicted tenacity under these circumstances is .350.

38. A regression analysis carried out to relate $y =$ repair time for a water filtration system (hr) to $x_1 =$ elapsed time since the previous service (months) and $x_2 =$ type of repair (1 if electrical and 0 if mechanical) yielded the following model based on $n = 12$ observations: $\hat{y} = .950 + .400x_1 + 1.250x_2$. In addition, SSTo $= 12.72$, SSResid $= 2.09$, and $s_{b_2} = .312$.

a. Does there appear to be a useful linear relationship between repair time and the two model predictors? Carry out a test of the appropriate hypotheses using a significance level of .05.

b. Given that elapsed time since the last service remains in the model, does type of repair provide useful information about repair time? State and test the appropriate hypotheses using a significance level of .01.

c. Calculate and interpret a 95% confidence interval for β_2.

d. The estimated standard deviation of a prediction for repair time when elapsed time is 6 months and the repair is electrical is .192. Predict repair time under these circumstances by calculating a prediction interval with a 99% prediction level. Does the resulting interval suggest that the estimated model will give an accurate prediction? Why or why not?

39. The accompanying data on $x =$ frequency (MHz) and $y =$ power (W) for a certain laser configuration was read from a graph in the article "Frequency Dependence in RF Discharge Excited Waveguide CO$_2$ Lasers" (*IEEE J. of Quantum Electronics*, 1984: 509–514):

x:	60	63	77	100	125	157	186	222
y:	16	17	19	21	22	20	15	5

Fitting a quadratic regression model to this data yielded the following summary quantities: $a = -1.5127$, $b_1 = .391902$, $b_2 = -.00163141$, SSResid $= .29$, SSTo $= 202.87$, and $s_{b_2} = .00003391$.

a. Why is b_2 negative rather than positive?

b. What proportion of observed variation in output power can be attributed to the model relationship between power and frequency?

c. Carry out a test of hypotheses to decide whether the quadratic regression model is useful.

d. Carry out a test of hypotheses to decide whether the quadratic predictor should be retained in the model.

e. When $x = 150$, the estimated standard deviation of $\hat{y}$ is $s_{\hat{y}} = .1410$. Calculate a 99% confidence interval for true average power when frequency is 150, and also a 99% prediction interval for a single output power observation to be made when frequency is 150.

40. The article "Sensitivity Analysis of a 2.5 kW Proton Exchange Membrane Fuel Cell Stack by Statistical Method" (*J. of Fuel Cell Sci. and Tech.*, 2009: 1–6) used regression methodology to investigate the relationship between fuel cell power (W) and the independent variables $x_1 = H_2$ pressure (psi), $x_2 = H_2$ flow (stoc), $x_3 =$ air pressure (psi), and $x_4 =$ airflow (stoc).

    Here is the Minitab output from fitting the model with the aforementioned independent variables as predictors (also fit by the authors of the cited article):

```
Predictor    Coef     SE Coef      T       p
Constant    1507.3     206.8     7.29    0.000
x1          -4.282     4.969    -0.86    0.407
x2           7.46      62.11     0.12    0.907
x3          -0.9162    0.6227   -1.47    0.169
x4          90.60      24.84     3.65    0.004

s = 4.6885  R-sq = 59.6%  R-sq(adj) = 44.9%

SOURCE       DF       SS       MS      F       p
Regression    4      40048    10012   4.06    0.029
Res.Error    11      27158    2469
Total        15      67206
```

a. Does there appear to be a useful relationship between power and at least one of the predictors? Carry out a formal test of hypotheses.

b. Fitting the model with predictors x_3, x_4, and the interaction x_3x_4 gave $R^2 = .834$. Does this model appear to be useful? Can an F test be used to compare this model to the model of part (a)? Explain.

c. Fitting the model with all 4 predictors as well as all second-order interactions gave $R^2 = .960$

(this model was also fit by the investigators). Does it appear that at least one of the interaction predictors provides useful information about power over and above what is provided by the first-order predictors? State and test the appropriate hypotheses using a significance level of .05.

41. The article "The Undrained Strength of Some Thawed Permafrost Soils" (*Canadian Geotechnical J.*, 1979: 420–427) reported the following data on undrained shear strength of sandy soil (y, in kPa), depth (x_1, in m), and water content (x_2, in %):

Obs	Depth	Watcont	Shstren
1	8.9	31.5	14.7
2	36.6	27.0	48.0
3	36.8	25.9	25.6
4	6.1	39.1	10.0
5	6.9	39.2	16.0
6	6.9	38.3	16.8
7	7.3	33.9	20.7
8	8.4	33.8	38.8
9	6.5	27.9	16.9
10	8.0	33.1	27.0
11	4.5	26.3	16.0
12	9.9	37.8	24.9
13	2.9	34.6	7.3
14	2.0	36.4	12.8

Fitting the model with predictors x_1 and x_2 only gave SSResid $= 894.95$, whereas fitting the complete second-order model with predictors x_1, x_2, x_1^2, x_2^2, and x_1x_2 resulted in SSResid $= 390.64$. Carry out a test at significance level .01 to decide whether at least one of the second-order predictors provides useful information about shear strength.

42. Soluble dietary fiber (SDF) can provide health benefits by lowering blood cholesterol and glucose levels. The article "Effects of Twin-Screw Extrusion on Soluble Dietary Fiber and Physicochemical Properties of Soybean Residue" (*Food Chemistry*, 2013: 884–889) reported the following data on $y =$ SDF content (%) in soybean residue and the three predictors $x_1 =$ extrusion temperature (in °C), $x_2 =$ feed moisture (in %), and $x_3 =$ screw speed (in rpm) of a twin-screw extrusion process.

Obs	x_1	x_2	x_3	y
1	35	110	160	11.13
2	25	130	180	10.98
3	30	110	180	12.56
4	30	130	200	11.46
5	30	110	180	12.38
6	30	110	180	12.43
7	30	110	180	12.55
8	25	110	160	10.59
9	30	130	160	11.15
10	30	90	200	10.55
11	30	90	160	9.25
12	25	90	180	9.58
13	35	110	200	11.59
14	35	90	180	10.68
15	35	130	180	11.73
16	25	110	200	10.81
17	30	110	180	12.68

a. The authors fit the complete second-order model with predictors x_1, x_2, x_3, x_1^2, x_2^2, x_3^2, $x_1 x_2$, $x_1 x_3$, and $x_2 x_3$, which resulted in SSResid = .215 and SSTo = 16.798. Determine the corresponding values of R^2 and adjusted R^2.

b. If we include in the model only the predictors x_1, x_2, and x_3, the corresponding SSResid = 11.428. Carry out a test at significance level .01 to decide whether at least one of the second-order predictors provides useful information about SDF content.

43. The use of high-strength steels (HSS) rather than aluminum and magnesium alloys in automotive body structures reduces vehicle weight. However, HSS use is still problematic because of difficulties with limited formability, increased springback, difficulties in joining, and reduced die life. The article "Experimental Investigation of Springback Variation in Forming of High Strength Steels" (*J. of Manuf. Sci. and Engr.*, 2008: 1–9) included data on y = springback from the wall opening angle and x = blank holder pressure (BHP). Three different material suppliers and three different lubrication regimens (no lubrication, lubricant 1, and lubricant 2) were also utilized.

a. What predictors would you use in a model to incorporate supplier and lubrication information in addition to BHP?

b. The accompanying Minitab output resulted from fitting the model of part (a) (the articles authors also used Minitab; amusingly, they employed a significance level of .06 in various tests of hypotheses). Does there appear to be a useful relationship between the response variable and at least one of the predictors? Carry out a formal test of hypotheses.

c. When BHP is 1000, material is from supplier 1, and no lubrication is used, $s_{\hat{y}}$ = .524. Calculate a 95% PI for the springback that would result from making an additional observation under these conditions.

d. From the output, it appears that lubrication regimen may not be providing useful information. A regression with the corresponding predictors removed resulted in SSResid = 48.426. What is the coefficient of multiple determination for this model, and what would you conclude about the importance of the lubrication regimen?

e. A model with predictors for BHP, supplier, and lubrication regimen, as well as predictors for interactions between BHP and both supplier and lubrication regiment, resulted in SSResid = 28.216 and R^2 = .849. Does this model appear to improve on the model with just BHP and predictors for supplier? Use α = .05.

```
Predictor        Coef  SE Coef        T      p
Constant      21.5322   0.6782    31.75  0.000
BHP        -0.0033680 0.0003919    -8.59  0.000
Supp1_1      -1.7181   0.5977     -2.87  0.007
Supp1_2      -1.4840   0.6010     -2.47  0.019
Lub_1        -0.3036   0.5754     -0.53  0.602
Lub_2         0.8931   0.5779      1.55  0.133

s = 1.18413   R-sq = 77.5%   R-sq(adj) = 73.8%

SOURCE        DF       SS      MS      F      p
Regression     5  144.915  28.983  20.67  0.000
Res.Error     30   42.065   1.402
Total         35  186.980
```

44. Coir fiber, derived from coconut, is an eco-friendly material with great potential for use in construction. The article "Seepage Velocity and Piping Resistance of Coir Fiber Mixed Soils" (*J. of Irrig. and Drainage Engr.*, 2008: 485–492) included several multiple regression analyses. The article's authors kindly provided the accompanying data on x_1 = fiber content (%), x_2 = fiber length (mm), x_3 = hydraulic gradient (no unit provided), and y = seepage velocity (cm/sec).

Obs	cont	lngth	grad	vel	Obs	cont	lngth	grad	vel
1	0.0	0	0.400	0.027	26	1.5	50	1.141	0.058
2	0.0	0	0.716	0.050	27	1.5	50	1.474	0.082
3	0.0	0	0.925	0.080	28	1.5	50	1.581	0.112
4	0.0	0	1.098	0.099	29	1.5	50	1.983	0.144
5	0.0	0	1.226	0.107	30	1.0	25	0.462	0.028
6	0.0	0	1.427	0.140	31	1.0	25	0.705	0.059
7	0.0	0	1.709	0.178	32	1.0	25	0.987	0.084
8	0.0	0	1.872	0.200	33	1.0	25	1.154	0.101
9	0.5	50	0.380	0.022	34	1.0	25	1.479	0.150
10	0.5	50	0.774	0.040	35	1.0	25	1.786	0.194
11	0.5	50	1.056	0.060	36	1.0	25	1.957	0.218
12	0.5	50	1.329	0.111	37	1.0	40	0.419	0.030
13	0.5	50	1.598	0.158	38	1.0	40	0.705	0.050
14	0.5	50	1.799	0.188	39	1.0	40	0.979	0.068
15	1.0	50	0.410	0.026	40	1.0	40	1.226	0.091
16	1.0	50	0.577	0.038	41	1.0	40	1.470	0.126
17	1.0	50	0.748	0.049	42	1.0	40	1.744	0.168
18	1.0	50	0.927	0.060	43	1.0	60	0.436	0.034
19	1.0	50	1.090	0.070	44	1.0	60	0.650	0.051
20	1.0	50	1.239	0.088	45	1.0	60	0.889	0.068
21	1.0	50	1.496	0.111	46	1.0	60	1.222	0.093
22	1.0	50	1.744	0.134	47	1.0	60	1.477	0.112
23	1.0	50	1.915	0.145	48	1.0	60	1.726	0.139
24	1.5	50	0.444	0.014	49	1.0	60	1.983	0.173
25	1.5	50	0.821	0.037					

a. Here is output from fitting the model with the three x_i's as predictors:

```
Predictor          Coef     SE Coef        T       p
Constant       -0.002997    0.007639    -0.39   0.697
fib cont       -0.012125    0.007454    -1.63   0.111
fib lngth     -0.0003020    0.0001676   -1.80   0.078
hyd grad        0.102489    0.004711    21.76   0.000

s = 0.0162355   R-sq = 91.6%   R-sq(adj) = 91.1%

Source          DF    SS        MS        F       p
Regression       3  0.129898  0.043299  164.27  0.000
Residual Error  45  0.011862  0.000264
Total           48  0.141760
```

How would you interpret the number $-.0003020$ in the Coef column on the output?

b. Does fiber content appear to provide useful information about velocity provided that fiber length and hydraulic gradient remain in the model? Carry out a test of hypotheses at $\alpha = .05$.

c. Fitting the model with just fiber length and hydraulic gradient as predictors gave the estimated regression coefficients $a = -.005315$, $b_1 = -.0004968$, and $b_2 = .102204$ (the t-ratios for these two predictors are both highly significant). In addition, $s_{\hat{y}} = .00286$ when fiber length $= 25$ and hydraulic gradient $= 1.2$. Is there convincing evidence that true average velocity is something other than .1 in this situation? Carry out a test using a significance level of .05.

d. Fitting the complete second-order model (as did the article's authors) resulted in SSResid $= .003579$. Does it appear that at least one of the second-order predictors provides useful information over and above what is provided by the three first-order predictors? Test the relevant hypotheses at $\alpha = .05$.

11.6   FURTHER ASPECTS OF REGRESSION ANALYSIS

This last section surveys a variety of issues in regression analysis, including diagnostic checks for model adequacy, identification of unusual observations, selection of a good group of predictors from a candidate pool, problems associated with a strong linear relationship among the predictors, and a model appropriate when y is a 0–1 variable corresponding to a success–failure dichotomy.

Checking Model Adequacy

In Section 11.5, we presented inferential methods based on the general additive multiple regression model. These methods are appropriate only if the model assumptions, for example, normality of the random deviation e in the model equation, are satisfied. Checks of model adequacy are usually based on the residuals, and in particular various plots involving these or related quantities. Recall that the residuals are the differences $y_1 - \hat{y}_1, y_2 - \hat{y}_2, \ldots, y_n - \hat{y}_n$ between observed and predicted y values. Before the data is obtained, each one of these residuals is subject to randomness; we do not know a priori whether any particular residual will be -3.2, 5.7, 0, or any other possible value. If the correct model has been fit, the mean value of any particular residual is zero. Unfortunately, the amount of variability in any particular residual will depend on the values of the predictors at which the corresponding observation is made. This can make it difficult to compare the various residuals to one another. One remedy for this difficulty is to standardize the residuals:

$$\text{standardized residual} = \frac{\text{residual} - 0}{\text{estimated standard deviation of residual}}$$

$$= \frac{\text{residual}}{\text{estiamted standard deviation of residual}}$$

The most popular statistical packages will produce these standardized residuals on request.

Example 11.18   The adsorption data introduced in Example 3.15, repeated here, is used in several examples in the previous section. The residuals are based on the model with the two predictors $x_1 =$ iron content and $x_2 =$ aluminum content.

Obs	Iron	Aluminum	Adsorption	Residual	Estimated standard deviation	Standardized residual
1	61	13	4	−.06305	3.64425	−.01730
2	175	21	18	−1.70661	3.72079	−.45867
3	111	24	14	.46130	4.02690	.11455
4	124	23	18	3.34477	4.04931	.82601
5	130	64	26	−3.64064	3.50644	−1.03827
6	173	38	26	.58585	4.14741	.14126
7	169	33	21	−2.21821	4.09222	−.54206
8	169	61	30	−2.99022	4.07688	−.73346
9	160	39	28	3.70238	4.18323	.88505
10	244	71	36	−8.93520	4.03193	−2.21611
11	257	112	65	4.29026	2.98776	1.43595
12	333	88	62	1.09857	2.99775	.36647
13	199	54	40	6.07079	4.18560	1.45040

Notice that the estimated standard deviations for the 11th and 12th observations are much smaller than those of most other observations. This is because the x_1 and x_2 values for these two observations are quite far from the center of the data. This is analogous to the least squares line in simple linear regression being pulled toward an observation whose x value is far to the left or right of the other x values; there is less variability in the corresponding residual than for the other observations. The only unusually large residual here is for the 10th observation; because the standardized residual is -2.22, the residual -8.94 is more than 2 standard deviations smaller than what would be expected if the correct model had been fit.

We previously advocated the use of a normal quantile plot to check a normality assumption. In regression, we suggest that the assumption of normally distributed random deviations be investigated by constructing a normal quantile plot of the standardized residuals. A reasonably linear pattern in this plot suggests that normality is plausible.

Example 11.19 Figure 11.19 shows a normal quantile plot of the standardized residuals for the adsorption data given in Example 11.18. The straightness of the plot casts little doubt on the assumption that the random deviation e is normally distributed.

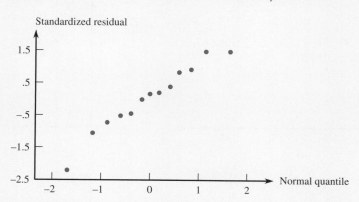

Figure 11.19 A normal quantile plot of the standardized residuals from Example 11.18

Another model assumption is that the variance σ^2 of a random deviation is a constant; that is, it does not depend on the values of the predictors. This can be checked by plotting the standardized residuals against each predictor in turn—one plot of standardized residuals versus x_1, another of standardized residuals versus x_2, and so on. Ideally, the points in each of these plots should appear randomly placed with no discernible pattern. If there is any marked tendency for the points in one of these plots to spread out substantially more at one end than the other, the constant variance assumption is suspect. Remedial action must be taken, and the advice of a statistician should be sought. Additionally, if there is substantial curvature in a plot, the population regression function in the chosen model has been incorrectly specified. It would then be necessary to try transforming one or more of the variables or introducing new predictors, for example, quadratic predictors. Some statisticians suggest replacing plots of the standardized residuals (or residuals) versus each predictor by a single omnibus plot of the standardized residuals (or residuals) versus the predicted values ($\hat{y}$'s). Again, any marked deviation from randomness is a call for remedial action. A plot of $\hat{y}$ versus y gives a visual impression of how well the model is predicting for the observations in the sample. The closer the points in this plot are to a 45° line, the better the predictions; the vertical deviations from this line are just the residuals. Finally, if the observations were obtained in time sequence, the standardized residuals should be plotted in time order to see whether there is an effect over time. Such an effect might indicate that the e's for successive observations are not independent, necessitating a more complex model.

Example 11.20 Figure 11.20 shows the suggested plots for the adsorption data. Given that there are only 13 observations in the data set, there is not much evidence of a pattern in any of the first three plots other than randomness. The point at the bottom of each of these three plots corresponds to the observation with the large residual. We will say more about such observations subsequently. For the moment, there is no compelling reason for remedial action.

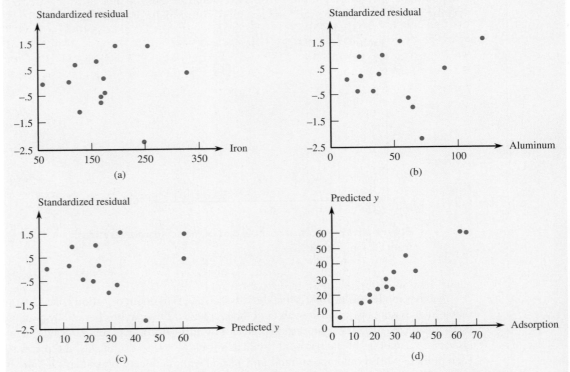

Figure 11.20 Diagnostic plots for the adsorption data: (a) standardized residual versus x_1; (b) standardized residual versus x_2; (c) standardized residual versus $\hat{y}$; and (d) $\hat{y}$ versus y

Identifying Unusual Observations

Two different types of unusual observations can occur in a regression data set: those with extreme y values and those with extreme values of the predictors (the x_i's). Extreme y values are indicated by standardized residuals quite different from zero. Minitab, for example, will flag any observation for which a standardized residual exceeds 2 in absolute value. There is one such observation in the adsorption data set.

One way to recognize an observation whose predictor values are unusual relies on the fact that each predicted value is a linear function of all the observed y values:

$$\hat{y}_1 = h_{11}y_1 + h_{12}y_2 + \cdots + h_{1n}y_n$$
$$\hat{y}_2 = h_{21}y_1 + h_{22}y_2 + \cdots + h_{2n}y_n$$
$$\vdots$$

The h_{ij} coefficients depend only on the values of the predictors for the various observations and not on the resulting y values. The coefficient h_{11} is the weight given to y_1 in computing the corresponding predicted value, and an analogous interpretation applies to $h_{22}, \ldots, h_{nn}$. Intuitively, a large value of h_{ii} for any particular i identifies an observation that is heavily weighted in calculating the corresponding predicted value. The first observation is said to have **high leverage**—high *potential* influence—if h_{11} is large relative to the other h_{ii}'s. The influence is only potential because whether an observation is actually influential depends on its y value as well as the values of the predictors. Minitab will flag any observation whose h_{ii} exceeds $3(k + 1)/n$ ($\sum h_{ii} = k + 1$, so an observation is flagged if its h_{ii} is three times the average of all the h_{ii}'s). The h_{ii}'s for the adsorption data are as follows:

.308 .278 .154 .145 .359 .103 .127 .133 .088 .152 .535 .531 .087

Since $3(k + 1)/n = 3(3)/13 = .692$, no observation can be characterized as having high leverage.

A commonly used strategy for assessing the impact of an "unusual" observation—either a large standardized residual or high leverage—is to remove the observation from the data set and refit the same model using the remaining observations. If any of the calculated quantities, such as the b_i's, R^2, and s_e, change substantially from their values before deletion of the unusual observation, the regression analysis is unstable. When the observation with the large standardized residual was removed from the adsorption data, estimated coefficients and other quantities changed very little. When large changes do occur, one possibility is to use a "robust" fitting technique for which estimated coefficients are not so heavily affected by unusual observations as they may be for a least squares fit. Consult one of the chapter references for more information on these matters.

Model Selection

An investigator has obtained data on a response variable y and a "candidate pool" of p predictors (some of which may be mathematical functions of others, such as interaction or quadratic predictors) and wishes to fit a multiple regression model. Frequently, some of these p predictors are only weakly related to y or contain information that duplicates information provided by some of the other predictors. So the issue is how to select a subset of predictors from the candidate pool to obtain an effective model.

One type of model selection strategy involves fitting all possible models, computing one or more summary quantities from each fit, and comparing these quantities to identify the most satisfactory model. With p predictors in the pool, there are 2^p possible models when the model that contains none of the predictors is counted (because there are two possibilities for each predictor, it could be included in the model or not included). When p exceeds 5, it is obviously time-consuming to sit in front of a computer and explicitly request that each possible model in turn be fit. Several of the most powerful statistical computer packages have an all-subsets option, which will give limited output from several of the best (according to criteria discussed shortly) models of each different size. Once the field has been narrowed, the fit of each finalist can then be examined in more detail. Minitab can be used for this purpose as long as $p \leq 31$ (for $p = 31$, over 2 billion models are under consideration).

Suppose that p is small enough for the all-subsets option to be feasible. What criteria can be used to select a winner? An obvious and appealing choice is the coefficient of multiple determination, R^2. Certainly for two models containing the same number of predictors, if the corresponding R^2 values are quite different, the model with the larger value should be preferred to the one with the smaller value. However, using R^2 as a basis for choosing between models that contain different numbers of predictors is not so straightforward. The reason is that adding a predictor to a model can never result in a decrease in R^2; there is almost always an increase, though it may be quite small. In particular, let

$$R_i^2 = \text{largest } R^2 \text{ value for any model containing } i \text{ predictors } (i = 1, 2, \ldots, p)$$

Then $R_1^2 \leq R_1^2 \leq \cdots \leq R_p^2$. The objective then is not simply to find the model with the largest R^2 value; the model with all p predictors from the candidate pool does that. Instead, we should look for a model that contains relatively few predictors but has a large R^2 value. The model should be such that no other model containing more predictors yields much of an improvement in R^2 value. Suppose, for example, that $p = 5$ and that

$$R_1^2 = .427 \qquad R_2^2 = .733 \qquad R_3^2 = .885 \qquad R_4^2 = .898 \qquad R_5^2 = .901$$

The best three-predictor model seems to be a good choice, since it substantially improves on the best one- and two-predictor models, whereas very little is gained by using more than three predictors.

A small increase in R^2 resulting from the addition of a predictor to a model may be offset by the increased complexity of the new model and the reduction in df associated with SSResid (resulting in less precise estimates and predictions). This is the rationale for adjusted R^2, which can either decrease or increase when a predictor is added to the model. We can then think of identifying the model whose adjusted R^2 is largest and then consider only this model and any others whose adjusted R^2 values are nearly as large.

When considering models containing some fixed number of predictors, for example, $k = 8$, there may be several different models whose R^2 and adjusted R^2 values are rather close to one another. By focusing only on the model with the highest values of these two criterion measures, we may miss out on other good models that are easier to interpret and use for estimation and prediction. For this reason, most all-subsets procedures allow the analyst to specify some number of models c of each given size (e.g., $c = 3$) for which output should be provided.

One other criterion for model selection that has been used with increasing frequency in recent years is *Mallows' CP*. Let μ_i denote the mean or expected value of y_i, which is the value of the response variable for the ith observation in our sample. Then after fitting any particular model, $\hat{y}_i$ calculated from the fit provides an estimate of μ_i, and the total expected estimation error for all observations in the data set is $\sum E[(\hat{y}_i - \mu_i)^2]$. Mallows' CP is an estimate of this total expected estimation error normalized in a certain way. It is desirable to choose a model for which CP is small. One additional consideration is that to protect against possible biases in estimates of population regression coefficients, it is desirable to have $CP \approx k + 1$ when the model under consideration has k predictors.

Example 11.21

The bond shear strength data introduced in Section 11.5 contains values of four different independent variables x_1–x_4. We found that the model with only these four variables as predictors was useful and that there was no compelling reason to consider the inclusion of second-order predictors. Figure 11.21 is the Minitab output that results from a request to identify the two best models of each given size.

```
Response is strength                      f  p
                                          o  o  t  t
                                          r  w  e  i
                          Adj.            c  e  m  m
      Vars   R-sq   R-sq     C-p      s   e  r  p  e
        1    57.7   56.2    11.0   5.9289  X
        1    10.8    7.7    51.9   8.6045     X
        2    68.5   66.2     3.5   5.2070  X  X
        2    59.4   56.4    11.5   5.9136  X        X
        3    70.2   66.8     4.0   5.1590  X  X     X
        3    69.7   66.2     4.5   5.2078  X  X  X
        4    71.4   66.8     5.0   5.1580  X  X  X  X
```

Figure 11.21 Output from Minitab's Best Subsets option

The best two-predictor model, with predictors power and temperature, seems to be a very good choice on all counts: R^2 is significantly higher than for models with fewer predictors yet almost as large as for any larger models, adjusted R^2 is almost at its maximum for this data, and CP is small and close to $2 + 1 = 3$.

The choice of a "best" model in Example 11.21 seemed reasonably clear-cut. This is often not the case. More typically, there will be several different models that are more or less equally appealing in terms of the criteria discussed here. These finalists would then have to be examined in more detail to choose the best model.

If the number of predictors in the candidate pool is too large or if suitable software is not available, an alternative to an *all-subsets* or *best regression* approach is to use an *automatic selection* procedure. The most easily understood such procedure is *backward elimination*. First, fit the model containing all predictors in the candidate pool, then eliminate predictors one by one until at some point all remaining predictors seem important. This involves looking at the *t*-ratios b_i/s_{b_i} on all coefficients for predictors in the model at each stage of the process. The obvious candidate for elimination is the predictor corresponding to the t-ratio closest to zero. The most frequently used rule of thumb in practice is to stop eliminating predictors when all t-ratios either exceed 2 or are less than -2. Some packages use F ratios, which are the squares of t-ratios, with a cutoff of 4.

Example 11.22

Figure 11.22 shows Minitab output from the backward elimination procedure applied to the bond shear strength data (this was done within Minitab's Stepwise option). At the first stage, the t-ratio closest to zero was 1.01 for the β coefficient corresponding to the predictor force. Since this t-ratio is between -2 and 2, force is eliminated (this would also have been the case if the t-ratio had been -1.01). At the next stage, the model with the three remaining predictors was fit. The predictor time now qualifies for elimination, since the corresponding t-ratio 1.23 is closest to 0 and between -2 and 2. When the model with the two remaining predictors is fit, both the corresponding t-ratios exceed 2 in absolute value, and the procedure is terminated. The resulting model is the same one that we suggested previously based on all-subsets considerations.

```
Response is strength on 4 predictors,
with N = 30
        Step         1        2        3
    Constant    -37.48   -30.07   -24.90

       force      0.21
     T-Ratio      1.01

       power     0.498    0.498    0.498
     T-Ratio      7.10     7.10     7.03

        temp     0.130    0.130    0.130
     T-Ratio      3.08     3.08     3.05

        time      0.26     0.26
     T-Ratio      1.23     1.23

           S      5.16     5.16     5.21
        R-Sq     71.40    70.24    68.52
```

Figure 11.22 Backward elimination output from Minitab

Another automatic selection procedure is *forward selection*, in which predictors from the candidate pool are added to the model one-by-one until, at a certain point, none of the predictors not already added appears useful. Suppose, for example, that $p = 10$ and that x_1, x_6, and x_9 have already been added. Then at the next stage, each four-predictor model that includes these three predictors along with one of the predictors not yet added would be fit (e.g., the model with predictors x_1, x_6, x_9, and x_2). The t-ratios for the coefficients corresponding to not-yet-entered predictors are the basis for deciding whether to enter at least one more predictor or to terminate. A ± 2 cutoff is frequently employed. Clearly, forward selection involves fitting many more models than is the case with backward elimination; for example, at the first stage, all p one-predictor models must be fit to decide whether at least one predictor should enter.

A variation on this is *stepwise regression,* in which predictors are added one-by-one with the option of deleting a predictor at some later stage that was added previously. The justification for this variation is that a predictor that earlier seemed important may become redundant once several other predictors have been entered into the model. The models identified by backward elimination, forward selection, and stepwise regression may not be the same. Furthermore, none of these automatic selection procedures may identify the model selected using the all-subsets criteria. Our recommendation is that all-subsets be used in preference to any of the automatic selection procedures whenever possible.

Multicollinearity

When the values of the single predictor x in a simple linear regression analysis are all quite close to one another, s_b will usually be quite large, indicating that the slope coefficient β has been imprecisely estimated. The analogous situation in multiple regression is referred to as *multicollinearity.* When the model to be fit includes the k predictors $x_1, \ldots, x_k,$ there is said to be multicollinearity if there is a strong linear relationship between *these predictors* (so multicollinearity has nothing to do with the response variable y). Severe multicollinearity leads to poorly estimated population regression coefficients and various other problems. The most straightforward way to recognize the presence of multicollinearity is to fit k different regression models, each of which has one of the x variables as the dependent variable and the other $k - 1$ predictors as the independent variables (e.g., if $k = 5$, there would be five regressions, the first with x_1 as the dependent variable, the second with x_2 playing this role, and so on). If one or more of the resulting R^2 values is close to 1, multicollinearity exists. If you use Minitab to regress y against the k predictors, a warning message will appear if any of these R^2's exceeds .99, and the package will not allow you to include all predictors if any R^2 exceeds .9999. Many analysts would be more conservative and say that multicollinearity is a problem if any R^2 exceeds .9.

When values of the predictor variable are under the control of the experimenter, as was the case in the bond shear strength example, a careful choice of values will preclude multicollinearity from arising. It is, however, often a problem in social science or business applications, where data results simply from observation rather than from intervention by an investigator. Statisticians have proposed various remedies for the problems associated with multicollinearity, but a discussion would take us beyond the scope of this book (after all, we want to leave something for your next statistics course!).

Logistic Regression

The simple linear regression model is appropriate for relating a quantitative response variable y to a quantitative predictor x. Suppose that y is a dichotomous variable with possible values 1 and 0 corresponding to success and failure. Let $\pi = P(S) = P(y = 1)$. Frequently, the value of π will depend on the value of some quantitative variable x. For example, the probability that a car needs warranty service of a certain kind might well depend on the car's mileage, or the probability of avoiding an infection of a certain type might depend on the dosage in an inoculation. Instead of using just the symbol π for the success probability, we now use $\pi(x)$ to emphasize the dependence of this probability on the value of x. The simple linear regression equation $y = \alpha + \beta x + e$ is no longer appropriate, for taking the mean value on each side of the equation gives

$$\mu_y = 1 \cdot \pi(x) + 0 \cdot (1 - \pi(x)) = \pi(x) = \alpha + \beta x$$

Whereas $\pi(x)$ is a probability and therefore must be between 0 and 1, $\alpha + \beta x$ need not be in this range.

Instead of letting the mean value of y be a linear function of x, we now consider a model in which some function of the mean value of y is a linear function of x. In other words, we allow $\pi(x)$ to be a function of $\alpha + \beta x$ rather than $\alpha + \beta x$ itself. A function that has been found quite useful in many applications is the **logit function,**

$$\pi(x) = \frac{e^{\alpha + \beta x}}{1 + e^{\alpha + \beta x}}$$

Figure 11.23 shows a graph of $\pi(x)$ for particular values of α and β with $\beta > 0$. As x increases, the probability of success increases. For β negative, the success probability would be a decreasing function of x.

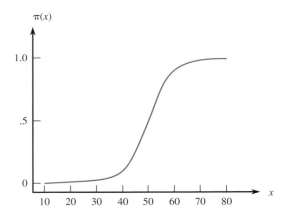

Figure 11.23 A graph of a logit function

Logistic regression means assuming that $\pi(x)$ is related to x by the logit function. Straightforward algebra shows that

$$\frac{\pi(x)}{1 - \pi(x)} = e^{\alpha + \beta x}$$

The expression on the left-hand side is called the *odds*. Suppose, for example, that $\pi(60)/[1 - \pi(60)] = 3$. Then when $x = 60$ a success is three times as likely as a failure. We now see that the logarithm of the odds is a linear function of the predictor. In particular, the slope parameter β is the change in the log odds associated with a 1-unit increase in x. This implies that the odds itself changes by the multiplicative factor e^{β} when x increases by 1 unit.

Fitting the logistic regression to sample data requires that the parameters α and β be estimated. This is usually done using the maximum likelihood technique described in Chapter 7. The details are quite involved, but fortunately the most popular statistical computer packages will do this on request and provide quantitative and pictorial indications of how well the model fits.

Example 11.23

Here is data on launch temperature and the incidence of failure for O-rings in 24 space shuttle launches prior to the *Challenger* disaster of 1986.

Temperature	Failure	Temperature	Failure
53	Y	70	Y
57	Y	72	N
63	N	73	N
66	N	75	N
67	N	75	Y
67	N	76	N
67	N	76	N
68	N	78	N
69	N	79	N
70	N	80	N
70	Y	81	N
70	N		

Figure 11.24 shows Minitab output for a logistic regression analysis and a graph of the estimated logit function from the R software. We have chosen to let π denote the probability of failure. The graph of decreases as temperature increases because failures tended to occur at lower temperatures than did successes. The estimate of β is $b = -.232$, and the estimated standard deviation of b is $s_b = .1082$. Provided that n is large enough, and we assume it is in this case, b has approximately a normal distribution. If $\beta = 0$ (temperature does not affect the likelihood of O-ring failure), $z = b/s_b$ has approximately a standard normal distribution. The value of this z-ratio is -2.14, and the P-value for a two-tailed test is .032 (some packages report a chi-square value, which is just z^2, with the same P-value). At significance level .05, we reject the null hypothesis of no temperature effect.

The estimated odds of failure for any particular temperature value x is

$$\frac{\pi(x)}{1 - \pi(x)} = e^{15.0429 - .232163x}$$

This implies that the odds ratio, the odds of failure at a temperature of $x + 1$ divided by the odds of failure at a temperature of x, is

$$\frac{\pi(x+1)/[1 - \pi(x+1)]}{\pi(x)/[1 - \pi(x)]} = e^{-.232163} = .7928$$

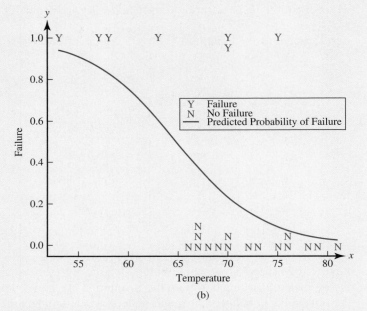

```
Binary Logistic Regression: Failure versus Temp

Logistic Regression Table
```

					Odds Ratio	95% Lower	CI Upper
Predictor	Coef	SE Coef	z	p			
Constant	15.0429	7.37862	2.04	0.041			
Temp	-0.232163	0.108236	-2.14	0.032	0.79	0.64	0.98

(a)

(b)

Figure 11.24 (a) Logistic regression output from Minitab for Example 11.23; (b) graph of estimated logistic function from R

The interpretation is that for each additional degree of temperature, we estimate that the odds of failure will decrease by a factor of .79 (21%). A 95% CI for the true odds ratio also appears on the output.

The launch temperature for the *Challenger* mission was only 31°F. This temperature is much smaller than any value in the sample, so it is dangerous to extrapolate the estimated relationship. Nevertheless, it appears that O-ring failure is virtually a sure thing for a temperature this small.

Our treatment of logistic regression modeling can be extended in an obvious way to incorporate more than one predictor. The probability of success π is now a function of the predictors $x_1, x_2, \ldots, x_k$:

$$\pi(x_1, \ldots, x_k) = \frac{e^{\alpha + \beta_1 x_1 + \cdots + \beta_k x_k}}{1 + e^{\alpha + \beta_1 x_1 + \cdots + \beta_k x_k}}$$

Simple algebra yields an expression for the odds:

$$\frac{\pi(x_1, \ldots, x_k)}{1 - \pi(x_1, \ldots, x_k)} = e^{\alpha + \beta_1 x_1 + \cdots + \beta_k x_k}$$

The interpretation of β_i $(i = 1, \ldots, k)$ is analogous to the interpretation for β given in the logit function containing only a single predictor x. That is, for $i = 1, \ldots, k$, the following argument shows that the odds changes by the multiplicative factor $e^{\beta i}$ when x_i increases by 1 unit and all other predictors remain fixed.

$$\frac{\pi(x_1, \ldots, x_i + 1, \ldots, x_k)}{1 - \pi(x_1, \ldots, x_i + 1, \ldots, x_k)} = e^{\alpha + \beta_1 x_1 + \cdots \beta_i(x_i + 1) + \cdots + \beta_k x_k}$$

$$= e^{\alpha + \beta_1 x_1 + \cdots \beta_i x_i + \cdots + \beta_k x_k + \beta_i}$$

$$= \frac{\pi(x_1, \ldots, x_k)}{1 - \pi(x_1, \ldots, x_k)} e^{\beta_i}$$

Again, statistical software must be used to estimate parameters, calculate relevant standard deviations, and provide other inferential information.

Example 11.24

Data was obtained from 189 women who gave birth during a particular period at the Baystate Medical Center in Springfield, Massachusetts, in order to identify factors associated with low birth weight. The accompanying Minitab output resulted from a logistic regression in which the dependent variable indicated whether (1) or not (0) a child had low birth weight (<2500 g), and predictors were weight of the mother at her last menstrual period, age of the mother, and an indicator variable for whether (1) or not (0) the mother had smoked during pregnancy.

Logistic Regression Table

Predictor	Coef	SE Coef	z	p	Odds Ratio	95% CI Lower	Upper
Constant	2.06239	1.09516	1.88	0.060			
Wt	-0.01701	0.00686	-2.48	0.013	0.98	0.97	1.00
Age	-0.04478	0.03391	-1.32	0.187	0.96	0.89	1.02
Smoke	0.65480	0.33297	1.97	0.049	1.92	1.00	3.70

It appears that age is not an important predictor of low birth weight, provided that the two other predictors are retained. The other two predictors do appear to be informative. The point estimate of the odds ratio associated with smoking status is 1.92 (ratio of the odds of low birth weight for a smoker to the odds for a nonsmoker); at the 95% confidence level, the odds of a low-birth-weight child could be as much as 3.7 times higher for a smoker than what it is for a nonsmoker.

Please see one of the chapter references for more information on logistic regression, including methods for assessing model effectiveness and adequacy.

We have reached the end of our exposition, but hopefully this is not the end of your statistical education. Our hope is that you have enjoyed the journey through statistics thus far and that you will find many opportunities to apply the concepts and methods in the near future. Enjoy!!

Section 11.6 Exercises

45. Reconsider the data on $x =$ inverse thickness and $y =$ flux from Exercise 9 of Section 11.1. The values of the standardized residuals from a simple linear regression analysis and the corresponding normal quantiles follow:

x:	19.8	20.6	23.5	26.1
Standardized residual:	-1.20	1.64	.94	$-.48$
z quantile:	$-.85$	1.43	.85	$-.47$

x:	30.3	43.5	45.0	46.5
Standardized residual:	-1.41	.70	$-.06$	$-.03$
z quantile:	-1.43	.47	$-.15$	.15

a. Does it appear plausible that the random deviations in the simple linear regression model equation are normally distributed?

b. Construct a plot of the standardized residuals versus x and comment.

46. Exercise 41 of Section 11.5 gave data on $y =$ shear strength of a soil specimen, $x_1 =$ depth, and $x_2 =$ water content. The data is presented again, along with the standardized residuals and corresponding normal scores obtained from fitting the complete second-order model.

Obs	Shstren	Depth	Watcont	Stresid	NQuant
1	14.7	8.9	31.5	-1.50075	-1.20448
2	48.0	36.6	27.0	.53889	.89743
3	25.6	36.8	25.9	$-.52893$	$-.65862$
4	10.0	6.1	39.1	$-.17350$	$-.26585$
5	16.0	6.9	39.2	.33350	.45321
6	16.8	6.9	38.3	.04076	$-.08767$
7	20.7	7.3	33.9	$-.41791$	$-.45321$
8	38.8	8.4	33.8	2.16543	1.70991
9	16.9	6.5	27.9	.22720	.26585
10	27.0	8.0	33.1	.43788	.65862
11	16.0	4.5	26.3	.19601	.08767
12	24.9	9.9	37.8	$-.90858$	$-.89743$
13	7.3	2.9	34.6	-1.53399	-1.70991
14	12.8	2.0	36.4	1.02146	1.20448

a. Construct a normal quantile plot of the standardized residuals to see whether it is plausible that the random deviations in the fitted model come from a normal distribution.

b. Plot the standardized residuals against depth and against water content, and comment on the plots.

47. The accompanying table shows the smallest value of SSResid for each number of predictors k ($k = 1, 2, 3, 4$) for a regression problem in which $y =$ cumulative heat of hardening in cement, $x_1 =$ % tricalcium aluminate, $x_2 =$ % tricalcium silicate, $x_3 =$ % aluminum ferrate, and $x_4 =$ % dicalcium silicate:

Number of predictors k	Predictors	SSResid
1	x_4	880.85
2	x_1, x_2	58.01
3	x_1, x_2, x_3	49.20
4	x_1, x_2, x_3, x_4	47.86

In addition, $n = 13$ and SSTo $= 2715.76$.

a. Use the criteria discussed in the text to recommend the use of a particular model.

b. Would the forward selection method of model selection have considered the best two-predictor model? Explain your reasoning.

48. A study carried out to investigate the relationship between a response variable relating to pressure drops in a screen-plate bubble column and the predictors $x_1 =$ superficial fluid velocity, $x_2 =$ liquid viscosity, and $x_3 =$ opening mesh size resulted in the accompanying data (top of page 569; "A Correlation of Two-Phase Pressure Drops in Screen-Plate Bubble Column," *Canad. J. of Chem. Engr.*, 1993: 460–463).

The standardized residuals and h_{ii} values resulted from the model that included the three independent variables as predictors. Are there any unusual observations?

Data for Exercise 48

Obs	Velocity	Viscosity	Mesh size	Response	Standardized residual	h_{ii}
1	2.14	10.00	.34	28.9	2.01721	.202242
2	4.14	10.00	.34	26.1	1.34706	.066929
3	8.15	10.00	.34	22.8	.96537	.274393
4	2.14	2.63	.34	24.2	1.29177	.224518
5	4.14	2.63	.34	15.7	−.68311	.079651
6	8.15	2.63	.34	18.3	.23785	.267959
7	5.60	1.25	.34	18.1	.06456	.076001
8	4.30	2.63	.34	19.1	.13131	.074927
9	4.30	2.63	.34	15.4	−.74091	.074927
10	5.60	10.10	.25	12.0	−1.38857	.152317
11	5.60	10.10	.34	19.8	−.03585	.068468
12	4.30	10.10	.34	18.6	−.40699	.062849
13	2.40	10.10	.34	13.2	−1.92274	.175421
14	5.60	10.00	.55	22.8	−1.07990	.712933
15	2.14	112.00	.34	41.8	−1.19311	.516298
16	4.14	112.00	.34	48.6	1.21302	.513214
17	5.60	10.10	.25	19.2	.38451	.152317
18	5.60	10.10	.25	18.4	.18750	.152317
19	5.60	10.10	.25	15.0	−.64979	.152317

49. The article "Anatomical Factors Influencing Wood Specific Gravity of Slash Pines and the Implications for the Development of a High-Quality Pulpwood" (*TAPPI*, 1964: 401–404) reported the results of an experiment in which 20 specimens of slash pine wood were analyzed. A primary objective was to relate wood specific gravity (*y*) to various other wood characteristics. Consider the accompanying data (top of page 570) on *y* and the predictors x_1 = number of fibers/mm² in springwood, x_2 = number of fibers/mm² in summerwood, x_3 = springwood %, x_4 = % springwood light absorption, and x_5 = % summerwood light absorption.

Based on the accompanying Minitab output, which model(s) would you recommend investigating in more detail?

```
                                       s s s s s
                                       p u p u u
                                       f f p l l
                          R-Sq         i i e a a
Vars  R-Sq  (adj)   C-p        s  b b r b b
 1    56.4  53.9  10.6  0.021832        X
 1    10.6   5.7  38.5  0.031245              X
 1     5.3   0.1  41.7  0.032155     X
 2    65.5  61.4   7.0  0.019975          X   X
 2    62.1  57.6   9.1  0.020950      X X
 2    60.3  55.6  10.2  0.021439  X       X
 3    72.3  67.1   4.9  0.018461  X       X   X
 3    71.2  65.8   5.6  0.018807          X X X
 3    71.1  65.7   5.6  0.018846      X X     X
 4    77.0  70.9   4.0  0.017353  X       X X X
 4    74.8  68.1   5.4  0.018179        X X X X
 4    72.7  65.4   6.7  0.018919  X X X     X
 5    77.0  68.9   6.0  0.017953  X X X X X
```

Data for Exercise 49

Obs	Springwood fibers	Summerwood fibers	% Springwood	% Springwood light absorption	% Summerwood light absorption	Specific gravity
1	573	1059	46.5	53.8	84.1	.534
2	651	1356	52.7	54.5	88.7	.535
3	606	1273	49.4	52.1	92.0	.570
4	630	1151	48.9	50.3	87.9	.528
5	547	1135	53.1	51.9	91.5	.548
6	557	1236	54.9	55.2	91.4	.555
7	489	1231	56.2	45.5	82.4	.481
8	685	1564	56.6	44.3	91.3	.516
9	536	1182	59.2	46.4	85.4	.475
10	685	1564	63.1	56.4	91.4	.486
11	664	1588	50.6	48.1	86.7	.554
12	703	1335	51.9	48.4	81.2	.519
13	653	1395	62.5	51.9	89.2	.492
14	586	1114	50.5	56.5	88.9	.517
15	534	1143	52.1	57.0	88.9	.502
16	523	1320	50.5	61.2	91.9	.508
17	580	1249	54.6	60.8	95.4	.520
18	448	1028	52.2	53.4	91.8	.506
19	476	1057	42.9	53.2	92.9	.595
20	528	1057	42.4	56.6	90.0	.568

50. The accompanying Minitab output resulted from applying both the backward elimination method and the forward selection method to the wood specific gravity data given in Exercise 49. Explain for each method what occurred at every iteration of the algorithm.

```
Response is spgrav on 5 predictors, with N = 20
Step            1         2         3         4
Constant     0.4421    0.4384    0.4381    0.5179

sprngfib    0.00011   0.00011   0.00012
T-Value        1.17      1.95      1.98

sumrfib     0.00001
T-Value        0.12

%sprwood   -0.00531  -0.00526  -0.00498  -0.00438
T-Value       -5.70     -6.56     -5.96     -5.20

spltabs     -0.0018   -0.0019
T-Value       -1.63     -1.76

sumltabs     0.0044    0.0044    0.0031    0.0027
T-Value        3.01      3.31      2.63      2.12

S            0.0180    0.0174    0.0185    0.0200
R-Sq          77.05     77.03     72.27     65.50
```

```
Response is spgrav on 5 predictors, with N = 20
Step            1         2
Constant     0.7585    0.5179

%sprwood   -0.00444  -0.00438
T-Value       -4.82     -5.20

sumltabs               0.0027
T-Value                  2.12

S            0.0218    0.0200
R-Sq          56.36     65.50
```

51. The article "The Analysis and Selection of Variables in Linear Regression" (*Biometrics*, 1976: 1–49) considered a data set of 32 observations on the following variables: y = fuel efficiency, x_1 = engine type (straight or V), x_2 = number of cylinders, x_3 = transmission type (manual or automatic), x_4 = number of transmission speeds, x_5 = engine size, x_6 = horsepower, x_7 = number of carburetor barrels, x_8 = final drive ratio, x_9 = weight, and x_{10} = quarter-mile time. Use the summary information (top of page 571) on the best model of each given size to select a model, and explain the rationale for your choice.

Number of predictors	Variables included	R^2	Adjusted R^2
1	9	.756	.748
2	2, 9	.833	.821
3	3, 9, 10	.852	.836
4	3, 6, 9, 10	.860	.839
5	3, 5, 6, 9, 10	.866	.840
6	3, 5, 6, 8, 9, 10	.869	.837
7	3, 4, 5, 6, 8, 9, 10	.870	.832
8	3, 4, 5, 6, 7, 8, 9, 10	.871	.826
9	1, 3, 4, 5, 6, 7, 8, 9, 10	.871	.818
10	All independent variables	.871	.809

52. Refer to the wood specific gravity data presented in Exercise 49. The following R^2 values resulted from regressing each predictor on the other four predictors (in the first regression, the dependent variable was x_1 and the predictors were $x_2 - x_5$, etc.): .628, .711, .341, .403, and .403. Does multicollinearity appear to be a substantial problem? Explain.

53. The article "Response Surface Methodology for Protein Extraction Optimization of Red Pepper Seed" (*Food Sci. and Tech.*, 2010: 226–231) gave data on the response variable y = protein yield (%) and the independent variables x_1 = temperature (°C), x_2 = pH, x_3 = extraction time (min), and x_4 = solvent/meal ratio.
    a. Fitting the model with the four x_i's as predictors yielded the following output:

```
Predictor     Coef   SE Coef      T       P
Constant    -4.586    2.542   -1.80   0.084
x1         0.01317  0.02707    0.49   0.631
x2          1.6350   0.2707    6.04   0.000
x3         0.02883  0.01353    2.13   0.044
x4         0.05400  0.02707    1.99   0.058


Source        DF       SS      MS      F      P
Regression     4  19.8882  4.9721  11.31  0.000
Res. Error    24  10.5513  0.4396
Total         28  30.4395
```

Calculate and interpret the values of R^2 and adjusted R^2. Does the model appear to be useful?

    b. Fitting the complete second-order model gave the following results:

```
Predictor       Coef      SE Coef        T       P
Constant     -119.49        18.53    -6.45   0.000
x1           -0.1047       0.2839    -0.37   0.718
x2            28.678        3.625     7.91   0.000
x3            0.4074       0.1303     3.13   0.007
x4            0.2711       0.2606     1.04   0.316
x1sqd      -0.000752     0.002110    -0.36   0.727
x2sqd        -1.6452       0.2110    -7.80   0.000
x3sqd      0.0002121    0.0005275     0.40   0.694
x4sqd      -0.015152     0.002110    -7.18   0.000
x1x2         0.02150      0.02687     0.80   0.437
x1x3        0.000550     0.001344     0.41   0.688
x1x4       -0.000800     0.002687    -0.30   0.770
x2x3        -0.05900      0.01344    -4.39   0.001
x2x4         0.03900      0.02687     1.45   0.169
x3x4        0.002725     0.001344     2.03   0.062

S = 0.268703  R-Sq = 96.7%  R-Sq(adj) = 93.4%

Analysis of Variance

Source        DF       SS      MS       F      P
Regression    14  29.4287  2.1020  29.11  0.000
Res. Error    14   1.0108  0.0722
Total         28  30.4395
```

Does at least one of the second-order predictors appear to be useful? Carry out an appropriate test of hypotheses.

    c. From the output in part (b), we conjecture that none of the predictors involving x_1 are providing useful information. When these predictors were eliminated, the value of SSResid for the reduced regression model is 1.1887. Does this support the conjecture?

    d. Here is output from Minitab's best subsets option, with just the single best subset of each size identified. Which model(s) would you consider using (subject to checking model adequacy)?

```
                                        1 2 3 4 x x x x x x
                                        s s s s 1 1 1 2 2 3
                  Mallows               x x x x q q q q x x x x x x
Vars  R-Sq  R-Sq(adj)    Cp      S      1 2 3 4 d d d d 2 3 4 3 4 4
  1   52.7    50.9     174.4  0.73030   X
  2   67.9    65.4     112.5  0.61349                 X              X
  3   77.5    75.0      73.1  0.52124   X             X              X
  4   83.4    80.7      50.8  0.45835   X     X       X   X
  5   90.9    88.9      21.4  0.34731   X             X   X        X X
  6   94.6    93.1       7.9  0.27422   X X X         X   X      X
  7   95.8    94.4       4.7  0.24683   X X           X   X        X X X
  8   96.2    94.6       5.1  0.24137   X X           X   X   X    X X X
  9   96.4    94.7       6.1  0.23962   X X X         X   X   X    X X X
 10   96.6    94.6       7.5  0.24132   X X X X X       X X        X X X
 11   96.6    94.4       9.4  0.24716   X X X X   X X X X          X X X
 12   96.6    94.1      11.2  0.25328   X X X X   X X X X X        X X X
 13   96.7    93.8      13.1  0.26041   X X X X X X X X X X        X X X
 14   96.7    93.4      15.0  0.26870   X X X X X X X X X X X X X X
```

54. It seems reasonable that the size of a cancerous tumor should be related to the likelihood that the cancer will spread (metastasize) to another site. The article "Molecular Detection of p16 Promoter Methylation in the Serum of Patients with Esophageal Squamous Cell Carcinoma" (*Cancer Res.*, 2001: 3135–3138) investigated the spread of esophageal cancer to the lymph nodes. With x = size of a tumor (cm) and $y = 1$ if the cancer does spread, consider the logistic regression model with $a = -2$ and $b = .5$ (values suggested by data in the article).
   a. Tabulate values of x, $\pi(x)$, the odds $\pi(x)/[1 - \pi(x)]$, and the log odds for $x = 0, 1, 2, \ldots, 10$.
   b. Explain what happens to the odds when x is increased by 1. Your explanation should involve the .5 that appears in the formula for $\pi(x)$.
   c. For what value of x are the odds 1? 5? 10?

55. Kyphosis refers to severe forward flexion of the spine following corrective spinal surgery. A study carried out to determine risk factors for kyphosis reported the accompanying ages (months) for 40 subjects at the time of the operation; the first 18 subjects did have kyphosis and the remaining 22 did not.

Kyphosis:
12	15	42	52	59	73
82	91	96	105	114	120
121	128	130	139	139	157

No kyphosis:
1	1	2	8	11	18
22	31	37	61	72	81
97	112	118	127	131	140
151	159	177	206		

Use the Minitab logistic regression output below to decide whether age appears to have a significant impact on the presence of kyphosis.

56. The following data resulted from a study commissioned by a large management consulting company to investigate the relationship between amount of job experience (months) for a junior consultant

Logistic Regression Table for Exercise 55

Predictor	Coef	StDev	Z	P	Odds Ratio	95% CI Lower	Upper
Constant	-0.5727	0.6024	-0.95	0.342			
age	0.004296	0.005849	0.73	0.463	1.00	0.99	1.02

Logistic Regression Table for Exercise 56

Predictor	Coef	StDev	Z	P	Odds Ratio	95% CI Lower	Upper
Constant	-3.211	1.235	-2.60	0.009			
age	0.17772	0.06573	2.70	0.007	1.19	1.05	1.36

and the likelihood of the consultant being able to perform a certain complex task.

Success: 8 13 14 18 20 21 21 22 25
          26 28 29 30 32

Failure: 4 5 6 6 7 9 10 11 11
          13 15 18 19 20 23 27

Interpret the Minitab logistic regression output (p. 572), and sketch a graph of the estimated probability of task performance as a function of experience.

57. Pillar stability is a most important factor to ensure safe conditions in underground mines. The authors of "Developing Coal Pillar Stability Chart Using Logistic Regression" (*Intl. J. of Rock Mechanics & Mining Sci.*, 2013: 55–60) used a logistic regression model to predict pillar stability. The article reported the following data on x_1 = pillar height to width ratio, x_2 = pillar strength to stress ratio, and pillar stability for 29 coal pillars.

ID	x1	x2	Stable?	ID	x1	x2	Stable?
1	1.80	2.40	Y	10	3.59	5.55	Y
2	1.65	2.54	Y	11	8.33	2.58	Y
3	2.70	0.84	Y	12	2.86	2.00	Y
4	3.67	1.68	Y	13	2.58	3.68	Y
5	1.41	2.41	Y	14	2.90	1.13	Y
6	1.76	1.93	Y	15	3.89	2.49	Y
7	2.10	1.77	Y	16	0.80	1.37	N
8	2.10	1.50	Y	17	0.60	1.27	N
9	4.57	2.43	Y	18	1.30	0.87	N

ID	x1	x2	Stable?	ID	x1	x2	Stable?
19	0.83	0.97	N	25	0.94	1.30	N
20	0.57	0.94	N	26	1.58	0.83	N
21	1.44	1.00	N	27	1.67	1.05	N
22	2.08	0.78	N	28	3.00	1.19	N
23	1.50	1.03	N	29	2.21	0.86	N
24	1.38	0.82	N				

The corresponding logistic regression output from R is given here:

Coefficients:

	Estimate	Std. Error	z value	Pr(>|z|)
(Intercept)	-13.146	5.184	-2.536	0.0112
x1	2.774	1.477	1.878	0.0604
x2	5.668	2.642	2.145	0.0319

a. Use the output to determine whether the two predictor variables appear to have a significant impact on pillar stability. Use $\alpha = .1$.

b. Provide interpretations for $e^{2.774}$ and $e^{5.668}$.

c. Determine an estimate (as the authors did) for the probability of pillar stability for each of the 29 pillars using the parameter estimates given in the output. Then label each pillar as "stable" if the estimated probability is at least .75 and "unstable" otherwise. How many of the pillars that were actually stable were correctly designated as "stable"? How many unstable pillars were correctly designated as "unstable"?

Supplementary Exercises

58. Suppose data was collected on y = bulk density (kg/m³) and x = moisture content (%) for a sample of six seeds of a particular type resulting in the accompanying scatterplot.

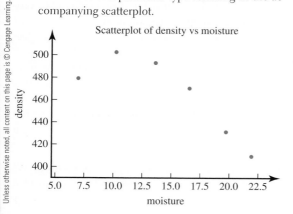

Scatterplot of density vs moisture

Here is the Mintab output from a request to fit a simple linear regression model of y on x:

The regression equation is
density = 545 - 5.46 moisture

Predictor	Coef	SE Coef	T	p
Constant	545.23	28.19	19.34	0.000
moisture	-5.463	1.786	-3.06	0.038

Noticing the relatively small *P*-value for the moisture predictor, a fellow student concludes that, based on the model utility test, there is a useful linear relationship between the two variables. Comment on the validity of this conclusion. How useful is this Minitab output (keeping in mind the scatterplot of the data)?

59. The accompanying data was read from a scatterplot in the article "Urban Emissions Measured with Aircraft" (*J. of the Air and Waste Mgmt. Assoc.*, 1998: 16–25). The response variable is ΔNO_y and the explanatory variable is ΔCO.

ΔCO:	50	60	95	108	135
ΔNO_y:	2.3	4.5	4.0	3.7	8.2
ΔCO:	210	214	315	720	
ΔNO_y:	5.4	7.2	13.8	32.1	

a. Fit an appropriate model to the data and judge the utility of the model.
b. Predict the value of ΔNO_y that would result from making one more observation when ΔCO is 400, and do so in a way that conveys information about precision and reliability. Does it appear that ΔNO_y can be accurately predicted? Explain.
c. The largest value of ΔCO is much greater than the other values. Does this observation appear to have had a substantial impact on the fitted equation?

60. Astringency is the quality in a wine that makes a wine drinker's mouth feel slightly rough, dry, and puckery. The paper "Analysis of Tannins in Red Wine Using Multiple Methods: Correlation with Perceived Astringency" (*Amer. J. Enol. Vitic.*, 2006: 481–485) reported on an investigation to assess the relationship between perceived astringency and tannin concentration using various analytic methods. Here is data provided by the authors on x = tannin concentration by protein precipitation and y = perceived astringency as determined by a panel of tasters.

x:	0.718	0.808	0.924	1.000
y:	0.428	0.480	0.493	0.978
x:	0.667	0.529	0.514	0.559
y:	0.318	0.298	−0.224	0.198
x:	0.766	0.470	0.726	0.762
y:	0.326	−0.336	0.765	0.190
x:	0.666	0.562	0.378	0.779
y:	0.066	−0.221	−0.898	0.836
x:	0.674	0.858	0.406	0.927
y:	0.126	0.305	−0.577	0.779
x:	0.311	0.319	0.518	0.687
y:	−0.707	−0.610	−0.648	−0.145
x:	0.907	0.638	0.234	0.781
y:	1.007	−0.090	−1.132	0.538
x:	0.326	0.433	0.319	0.238
y:	−1.098	−0.581	−0.862	−0.551

Relevant summary qualities are as follows:

$$\sum x_i = 19.404, \quad \sum y_i = -.549, \quad \sum x_i^2 = 13.248032$$
$$\sum y_i^2 = 11.835795, \quad \sum x_i y_i = 3.497811$$
$$S_{xx} = 13.248032 - (19.404)^2/32 = 1.48193150,$$
$$S_{yy} = 11.82637622$$
$$S_{xy} = 3.497811 - (19.404)(-.549)/32 = 3.83071088$$

a. Fit the simple linear regression model to this data. Then determine the proportion of observed variation in astringency that can be attributed to the model relationship between astringency and tannin concentration.
b. Calculate and interpret a confidence interval for the slope of the true regression line.
c. Estimate true average astringency when tannin concentration is .6, and do so in a way that conveys information about reliability and precision.
d. Predict astringency for a single wine sample whose tannin concentration is .6, and do so in a way that conveys information about reliability and precision.

61. In a discussion of the article "Tensile Behavior of Slurry Infiltrated Mat Concrete (SIMCON)" (*ACI Materials J.*, 1998: 77–79), the discussant presented data on y = toughness (psi) and x = aspect ratio. He stated that "a (simple linear) regression analysis clearly shows that the aspect ratio is not a reliable variable that can be used to predict toughness." The following observations were read from a graph in the article:

x:	500	500	500	500	500	715	715	715	715	715
y:	33	34	35	38	40	35	36	37	39	44

a. Why is the relationship between these two variables clearly not deterministic?
b. Fit the simple linear regression model, and state whether you agree with the discussant's assessment.
c. Even if the y values had been much closer together, so that the model could be judged useful, would there be any way to check model adequacy to decide whether a quadratic regression model would be more appropriate? Explain your reasoning.

62. The accompanying data on y = energy output (W) and x = temperature difference (K) was provided by the authors of the article "Comparison of Energy

and Exergy Efficiency for Solar Box and Parabolic Cookers" (*J. of Energy Engr.*, 2007: 53–62).

x:	23.20	23.50	23.52	24.30	25.10	26.20
y:	3.78	4.12	4.24	5.35	5.87	6.02
x:	27.40	28.10	29.30	30.60	31.50	32.01
y:	6.12	6.41	6.62	6.43	6.13	5.92
x:	32.63	33.23	33.62	34.18	35.43	35.62
y:	5.64	5.45	5.21	4.98	4.65	4.50
x:	36.16	36.23	36.89	37.90	39.10	4166
y:	4.34	4.03	3.92	3.65	3.02	2.89

The article's authors fit a cubic regression model to the data. Here is Minitab output from such a fit.

```
The regression equation is
y = -134 + 12.7 x - 0.377 x**2 + 0.00359 x**3

Predictor       Coef    SE Coef        T      P
Constant     -133.787      8.048  -16.62  0.000
x             12.7423      0.7750   16.44  0.000
x**2         -0.37652     0.02444  -15.41  0.000
x**3        0.0035861   0.0002529   14.18  0.000

s = 0.168354 R-Sq = 98.0% R-Sq(adj) = 97.7%


Analysis of Variance
Source       DF       SS      MS      F      P
Regression    3  27.9744  9.3248 329.00  0.000
Res. Error   20   0.5669  0.0283
Total        23  28.5413
```

a. What proportion of observed variation in energy output can be attributed to the model relationship?

b. Fitting a quadratic model to the data results in $R^2 = .780$. Calculate adjusted R^2 for this model and compare to adjusted R^2 for the cubic model.

c. Does the cubic predictor appear to provide useful information about y over and above that provided by the linear and quadratic predictors? State and test the appropriate hypotheses.

d. When $x = 30$, $s_{\hat{y}} = .0611$. Calculate a 95% CI for true average energy output in this case, and also a 95% PI for a single energy output to be observed when temperature difference is 30.

63. Secondary settling tanks play an important role in the performance of suspended-growth activated-sludge processes. The article "Sludge Volume Index Settleability Measures" (*Water Environ. Research*, 1998: 87–93) included a scatterplot of y = final settled height fraction versus x = initial solids concentration (g/L), from which the following data was read:

x:	.5	.9	1.1	1.7	2.0	2.2	2.7	3.0	3.3	4.2
y:	.06	.08	.10	.13	.15	.16	.18	.17	.15	.27
x:	4.5	5.3	5.8	5.9	6.2	6.8	7.2	9.1	9.4	10.4
y:	.30	.25	.31	.32	.48	.43	.32	.40	.61	.57

Summary quantities include $n = 20$,

$$\sum x_i = 92.2 \qquad \sum x_i^2 = 591.46$$
$$\sum y_i = 5.44 \qquad \sum y_i^2 = 1.9674$$
$$\sum x_i y_i = 33.577$$

a. The article included the statement "the linear correlation coefficient, $r^2 = .89$." Is this entire statement correct? If not, why, and what part is correct?

b. Carry out a test of appropriate hypotheses to see whether there is in fact a linear relationship between the two variables.

c. The standardized residuals from fitting the simple linear regression model are (in increasing order of x values) $-.04, -.05, .14, .13, .22, .21,$ $.11, -.37, -1.04, .36, .63, -1.08, -.43, -.34,$ $2.40, .88, -1.62, -2.04, 1.90,$ and $.05$. Does a plot of the standardized residuals versus x show a disturbing pattern? Explain.

64. The use of microorganisms to dissolve metals from ores has offered an ecologically friendly and less expensive alternative to traditional methods. The dissolution of metals by this method can be done in a two-stage bioleaching process: (1) microorganisms are grown in culture to produce metabolites (e.g. organic acids) and (2) ore is added to the culture medium to initiate leaching. The article "Two-Stage Fungal Leaching of Vanadium from Uranium Ore Residue of the Leaching Stage using Statistical Experimental Design" (*Annals of Nuclear Energy*, 2013: 48–52) reported on a two-stage bioleaching process of vanadium by using the fungus *Aspergillus niger*. In one study, the authors examined the impact of the variables x_1 = pH, x_2 = sucrose concentration (g/L), and x_3 = spore population (10^6 cells/ml) on y = oxalic acid

production (mg/L). The accompanying SAS output resulted from a request to fit the model with predictors x_1, x_2, and x_3 only.

Source	DF	Sum of Squares	Mean Square	F Value	Pr > F
Model	3	5861301	1953767	7.53	0.0052
Error	11	2855951	259632		
C. Total	14	8717252			

Fitting the complete second-order model resulted in SSResid = 541,632. Carry out a test at significance level .01 to decide whether at least one of the second-order predictors provides useful information about oxalic acid production.

65. The article cited in Exercise 64 also examined the effect of $x_1 = $ pH, $x_2 = $ sucrose concentration (g/L), and $x_3 = $ spore population (10^6 cells/ml) on $y = $ gluconic acid production (mg/L). The accompanying SAS output resulted from a request to fit the model with predictors x_1, x_2, and x_3 only.

Source	DF	Sum of Squares	Mean Square	F Value	Pr > F
Model	3	74027925	24675975	178.18	<.0001
Error	11	1523351	138486		
C. Total	14	75551276			

Fitting the complete second-order model resulted in SSResid = 805,534. Carry out a test at significance level .01 to decide whether at least one of the second-order predictors provides useful information about oxalic acid production.

66. The accompanying data was taken from the article "Applying Stepwise Multiple Regression Analysis to the Reaction of Formaldehyde with Cotton Cellulose" (*Textile Research J.*, 1984: 157–165). The

dependent variable is durable press rating, a quantitative measure of wrinkle resistance, and the four independent variables are formaldehyde concentration, catalyst ratio, curing temperature, and curing time, respectively.

a. Fitting the model with the four independent variables as predictors resulted in the following Minitab output. Does the fitted model appear to be useful?

The regression equation is
durpr = -0.912 + 0.161 formconc
+ 0.220 catratio + 0.0112 temp
+ 0.102 time

Predictor	Coef	StDev	T	p
Constant	-0.9122	0.8755	-1.04	0.307
formconc	0.16073	0.06617	2.43	0.023
catratio	0.21978	0.03406	6.45	0.000
temp	0.011226	0.004973	2.26	0.033
time	0.10197	0.05874	1.74	0.095

S = 0.8365 R-Sq = 69.2% R-Sq(adj) = 64.3%

Analysis of Variance

Source	DF	SS	MS	F	P
Regression	4	39.3769	9.8442	14.07	0.000
Error	25	17.4951	0.6998		
Total	29	56.8720			

b. Estimate, in a way that conveys information about precision and reliability, the average change in durability press rating associated with a 1-degree increase in curing temperature when concentration, catalyst ratio, and curing time all remain fixed.

c. Given that catalyst ratio, curing temperature, and curing time all remain in the model, do you think that formaldehyde concentration provides useful information about durable press rating?

Data for Exercise 66

Obs	Formaldehyde concentration	Catalyst ratio	Curing temperature	Curing time	Durable press rating
1	8	4	100	1	1.4
2	2	4	180	7	2.2
3	7	4	180	1	4.6
4	10	7	120	5	4.9
5	7	4	180	5	4.6

(Continued)

Obs	Formaldehyde concentration	Catalyst ratio	Curing temperature	Curing time	Durable press rating
6	7	7	180	1	4.7
7	7	13	140	1	4.6
8	5	4	160	7	4.5
9	4	7	140	3	4.8
10	5	1	100	7	1.4
11	8	10	140	3	4.7
12	2	4	100	3	1.6
13	4	10	180	3	4.5
14	6	7	120	7	4.7
15	10	13	180	3	4.8
16	4	10	160	5	4.6
17	4	13	100	7	4.3
18	10	10	120	7	4.9
19	5	4	100	1	1.7
20	8	13	140	1	4.6
21	10	1	180	1	2.6
22	2	13	140	1	3.1
23	6	13	180	7	4.7
24	7	1	120	7	2.5
25	5	13	140	1	4.5
26	8	1	160	7	2.1
27	4	1	180	7	1.8
28	6	1	160	1	1.5
29	4	1	100	1	1.3
30	7	10	100	7	4.6

d. Now consider models based not only on these four independent variables but also on second-order predictors (four x_i^2 predictors and six $x_i x_j$ predictors). Use a statistical computer package to identify a good model based on this candidate pool of predictors.

67. A study was carried out to investigate the relationship between brightness of finished paper (y) and the variables percentage of H_2O_2 by weight, percentage of NaOH by weight, percentage of silicate by weight, and process temperature ("Advantages of CEHDP Bleaching for High Brightness Kraft Pulp Production," *TAPPI*, 1964: 170A–173A). Each independent variable was allowed to assume five different values, and these values were coded for regression analysis as follows:

Variable	Coded value:	−2	−1	0	1	2
H_2O_2		.1	.2	.3	.4	.5
NaOH		.1	.2	.3	.4	.5
Silicate		.5	1.5	2.5	3.5	4.5
Temperature		130	145	160	175	190

The data follow:

Obs	H_2O_2	NaOH	Sili-cate	Temp-erature	Bright-ness
1	−1	−1	−1	−1	83.9
2	1	−1	−1	−1	84.9
3	−1	1	−1	−1	83.4
4	1	1	−1	−1	84.2
5	−1	−1	1	−1	83.8
6	1	−1	1	−1	84.7
7	−1	1	1	−1	84.0
8	1	1	1	−1	84.8
9	−1	−1	−1	1	84.5
10	1	−1	−1	1	86.0
11	−1	1	−1	1	82.6
12	1	1	−1	1	85.1
13	−1	−1	1	1	84.5
14	1	−1	1	1	86.0
15	−1	1	1	1	84.0
16	1	1	1	1	85.4
17	−2	0	0	0	82.9
18	2	0	0	0	85.5
19	0	−2	0	0	85.2
20	0	2	0	0	84.5
21	0	0	−2	0	84.7
22	0	0	2	0	85.0
23	0	0	0	−2	84.9
24	0	0	0	2	84.0
25	0	0	0	0	84.5
26	0	0	0	0	84.7
27	0	0	0	0	84.6
28	0	0	0	0	84.9
29	0	0	0	0	84.9
30	0	0	0	0	84.5
31	0	0	0	0	84.6

a. When the complete second-order coded model was fit, the estimate of the constant term was 84.67; the estimated coefficients of the linear predictors were .650, −.258, .133, and .108, respectively; the estimated quadratic coefficients were −.135, .028, .028, and −.072, respectively; and the estimated coefficients of the interaction predictors were .038, −.075, .213, .200, −.188, and .050, respectively. Calculate a point prediction of brightness when H_2O_2 is .4%, NaOH is .4%, silicate is 3.5%, and temperature is 175. What are the values of the residuals for the observations made with these values of the independent variables?

b. Express the estimated regression in uncoded form.

c. SSTo = 17.2567 and R^2 for the model of part (a) is .885. When a model that includes only the four independent variables as predictors is fit, $R^2 = .721$. Carry out a test at level .05 to decide whether at least one of the second-order predictors provides useful information about brightness.

68. Three sets of journal bearing tests were run on a Mil-L-8937-type film at each combination of three loads (psi) and three speeds (rpm). The wear life (hr) was recorded for each run, resulting in the following data ("Accelerated Testing of Solid Film Lubricants," *Lubrication Engr.*, 1972: 365–372):

Speed	Load (1000s)	Life	Speed	Load (1000s)	Life
20	3	300.2	60	6	65.9
20	3	310.8	60	10	10.7
20	3	333.0	60	10	34.1
20	6	99.6	60	10	39.1
20	6	136.2	100	3	26.5
20	6	142.4	100	3	22.3
20	10	20.2	100	3	34.8
20	10	28.2	100	6	32.8
20	10	102.7	100	6	25.6
60	3	67.3	100	6	32.7
60	3	77.9	100	10	2.3
60	3	93.9	100	10	4.4
60	6	43.0	100	10	5.8
60	6	44.5			

a. With w = wear life, s = speed, and l = load (in 1000s), fit the model with dependent variable w and predictors s and l, and assess the utility of the fitted model.

b. The cited article contains the comment that a lognormal distribution is appropriate for wear life, since $\ln(w)$ is known to follow a normal law. The suggested model is $w = [\delta/(s^\beta l^\gamma)]\varepsilon$, where ε denotes a random deviation and δ, β, and γ are parameters. Estimate the model parameters, and obtain a prediction interval for wear life when speed is 60 rpm and load is 6000 psi. (*Hint:* Transform the model equation so it has the appearance of the general additive multiple regression model equation.)

69. Normal hatchery processes in aquaculture inevitably produce stress in fish, which may negatively impact growth, reproduction, flesh quality, and susceptibility to disease. Such stress manifests itself in elevated and sustained corticosteroid levels. The article "Evaluation of Simple Instruments for the Measurement of Blood Glucose and Lactate, and Plasma Protein as Stress Indicators in Fish" (*J. of the World Aquaculture Society*, 1999: 276–284) described an experiment in which fish were subjected to a stress protocol and then removed and tested at various times after the protocol had been applied. The accompanying data on $x =$ time (min) and $y =$ blood glucose level (mmol/L) was read from a plot:

x: 2   2   5   7   12 13 17 18 23 24 26 28
y: 4.0 3.6 3.7 4.0 3.8 4.0 5.1 3.9 4.4 4.3 4.3 4.4

x: 29 30 34 36 40 41 44 56 56 57 60 60
y: 5.8 4.3 5.5 5.6 5.1 5.7 6.1 5.1 5.9 6.8 4.9 5.7

Use the methods developed in this chapter to analyze the data, and write a brief report summarizing your conclusions (assume that the investigators are particularly interested in glucose level 30 min after stress).

70. The article "Evaluating the BOD POD for Assessing Body Fat in Collegiate Football Players" (*Medicine and Science in Sports and Exercise*, 1999: 1350–1356) reports on a new air displacement device for measuring body fat. The customary procedure utilizes the hydrostatic weighing device, which measures the percentage of body fat by means of water displacement. Here is representative data read from a graph in the paper.

Obs	BOD	HW	Obs	BOD	HW
1	2.5	8.0	11	12.2	15.3
2	4.0	6.2	12	12.6	14.8
3	4.1	9.2	13	14.2	14.3
4	6.2	6.4	14	14.4	16.3
5	7.1	8.6	15	15.1	17.9
6	7.0	12.2	16	15.2	19.5
7	8.3	7.2	17	16.3	17.5
8	9.2	12.0	18	17.1	14.3
9	9.3	14.9	19	17.9	18.3
10	12.0	12.1	20	17.9	16.2

a. Use various techniques to decide whether it is plausible that the two techniques measure on average the same amount of fat.

b. Use the data to develop a way of predicting an HW measurement from a BOD POD measurement, and investigate the effectiveness of such predictions.

71. Curing concrete is known to be vulnerable to shock vibrations, which may cause cracking or hidden damage to the material. As part of a study of vibration phenomena, the paper "Shock Vibration Test of Concrete" (*ACI Materials J.*, 2002: 361–370) reported the accompanying data on peak particle velocity (mm/sec) and ratio of ultrasonic pulse velocity after impact to that before impact in concrete prisms:

Obs	ppv	Ratio	Obs	ppv	Ratio
1	160	.996	16	708	.990
2	164	.996	17	806	.984
3	178	.999	18	884	.986
4	252	.997	19	526	.991
5	293	.993	20	490	.993
6	289	.997	21	598	.993
7	415	.999	22	505	.993
8	478	.997	23	525	.990
9	391	.992	24	675	.991
10	486	.985	25	1211	.981
11	604	.995	26	1036	.986
12	528	.995	27	1000	.984
13	749	.994	28	1151	.982
14	772	.994	29	1144	.962
15	532	.987	30	1068	.986

Transverse cracks appeared in the last 12 prisms, whereas there was no observed cracking in the first 18 prisms.

a. Construct a comparative boxplot of ppv for the cracked and uncracked prisms, and comment. Then estimate the difference between true average ppv for cracked and uncracked prisms in a way that conveys information about precision and reliability.

b. The investigators fit the simple linear regression model to the entire data set consisting of 30 observations, with ppv as the independent variable and ratio as the dependent variable. Use a statistical software package to fit several different regression models, and draw appropriate inferences.

72. Have you ever wondered whether soccer players suffer adverse effects from hitting "headers"? The authors of the article "No Evidence of Impaired Neurocognitive Performance in Collegiate Soccer Players" (*The Amer. J. of Sports Medicine*, 2002: 157–162) investigated this issue from several perspectives.

 a. The paper reported that 45 of the 91 soccer players in their sample had suffered at least one concussion, 28 of 96 nonsoccer athletes had suffered at least one concussion, and only 8 of 53 student controls had suffered at least one concussion. Analyze this data and draw appropriate conclusions.

 b. For the soccer players, the sample correlation coefficient calculated from the values of x = soccer exposure (total number of competitive seasons played prior to enrollment in the study) and y = score on an immediate memory recall test was $r = -.220$. Interpret this result.

 c. Here is summary information on score on a controlled oral word association test for the soccer and nonsoccer athletes:

 $n_1 = 26$ $\bar{x}_1 = 37.50$ $s_1 = 9.13$
 $n_2 = 56$ $\bar{x}_2 = 39.63$ $s_2 = 10.19$

 Analyze this data and draw appropriate conclusions.

 d. Considering the number of prior nonsoccer concussions, the values of mean $\pm$ sd for the three groups were $.30 \pm .67$, $.49 \pm .87$, and $.19 \pm .48$. Analyze this data and draw appropriate conclusions.

Bibliography

Please see the bibliography for Chapter 3.

Appendix Tables

Table I The standard normal distribution (cumulative z curve areas)

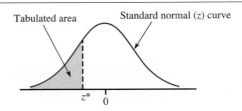

Tabulated area   Standard normal (z) curve

z^*	.00	.01	.02	.03	.04	.05	.06	.07	.08	.09
−3.8	.0001	.0001	.0001	.0001	.0001	.0001	.0001	.0001	.0001	.0000
−3.7	.0001	.0001	.0001	.0001	.0001	.0001	.0001	.0001	.0001	.0001
−3.6	.0002	.0002	.0001	.0001	.0001	.0001	.0001	.0001	.0001	.0001
−3.5	.0002	.0002	.0002	.0002	.0002	.0002	.0002	.0002	.0002	.0002
−3.4	.0003	.0003	.0003	.0003	.0003	.0003	.0003	.0003	.0003	.0002
−3.3	.0005	.0005	.0005	.0004	.0004	.0004	.0004	.0004	.0004	.0003
−3.2	.0007	.0007	.0006	.0006	.0006	.0006	.0006	.0005	.0005	.0005
−3.1	.0010	.0009	.0009	.0009	.0008	.0008	.0008	.0008	.0007	.0007
−3.0	.0013	.0013	.0013	.0012	.0012	.0011	.0011	.0011	.0010	.0010
−2.9	.0019	.0018	.0018	.0017	.0016	.0016	.0015	.0015	.0014	.0014
−2.8	.0026	.0025	.0024	.0023	.0023	.0022	.0021	.0021	.0020	.0019
−2.7	.0035	.0034	.0033	.0032	.0031	.0030	.0029	.0028	.0027	.0026
−2.6	.0047	.0045	.0044	.0043	.0041	.0040	.0039	.0038	.0037	.0036
−2.5	.0062	.0060	.0059	.0057	.0055	.0054	.0052	.0051	.0049	.0048
−2.4	.0082	.0080	.0078	.0075	.0073	.0071	.0069	.0068	.0066	.0064
−2.3	.0107	.0104	.0102	.0099	.0096	.0094	.0091	.0089	.0087	.0084
−2.2	.0139	.0136	.0132	.0129	.0125	.0122	.0119	.0116	.0113	.0110
−2.1	.0179	.0174	.0170	.0166	.0162	.0158	.0154	.0150	.0146	.0143
−2.0	.0228	.0222	.0217	.0212	.0207	.0202	.0197	.0192	.0188	.0183
−1.9	.0287	.0281	.0274	.0268	.0262	.0256	.0250	.0244	.0239	.0233
−1.8	.0359	.0351	.0344	.0336	.0329	.0322	.0314	.0307	.0301	.0294
−1.7	.0446	.0436	.0427	.0418	.0409	.0401	.0392	.0384	.0375	.0367
−1.6	.0548	.0537	.0526	.0516	.0505	.0495	.0485	.0475	.0465	.0455
−1.5	.0668	.0655	.0643	.0630	.0618	.0606	.0594	.0582	.0571	.0559
−1.4	.0808	.0793	.0778	.0764	.0749	.0735	.0721	.0708	.0694	.0681
−1.3	.0968	.0951	.0934	.0918	.0901	.0885	.0869	.0853	.0838	.0823
−1.2	.1151	.1131	.1112	.1093	.1075	.1056	.1038	.1020	.1003	.0985
−1.1	.1357	.1335	.1314	.1292	.1271	.1251	.1230	.1210	.1190	.1170
−1.0	.1587	.1562	.1539	.1515	.1492	.1469	.1446	.1423	.1401	.1379
−0.9	.1841	.1814	.1788	.1762	.1736	.1711	.1685	.1660	.1635	.1611
−0.8	.2119	.2090	.2061	.2033	.2005	.1977	.1949	.1922	.1894	.1867
−0.7	.2420	.2389	.2358	.2327	.2296	.2266	.2236	.2206	.2177	.2148
−0.6	.2743	.2709	.2676	.2643	.2611	.2578	.2546	.2514	.2483	.2451
−0.5	.3085	.3050	.3015	.2981	.2946	.2912	.2877	.2843	.2810	.2776
−0.4	.3446	.3409	.3372	.3336	.3300	.3264	.3228	.3192	.3156	.3121
−0.3	.3821	.3783	.3745	.3707	.3669	.3632	.3594	.3557	.3520	.3483
−0.2	.4207	.4168	.4129	.4090	.4052	.4013	.3974	.3936	.3897	.3859
−0.1	.4602	.4562	.4522	.4483	.4443	.4404	.4364	.4325	.4286	.4247
−0.0	.5000	.4960	.4920	.4880	.4840	.4801	.4761	.4721	.4681	.4641

Table I   The standard normal distribution *(continued)*

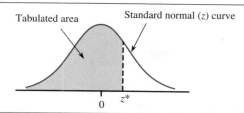

Tabulated area   Standard normal (z) curve

0   z*

z^*	.00	.01	.02	.03	.04	.05	.06	.07	.08	.09
0.0	.5000	.5040	.5080	.5120	.5160	.5199	.5239	.5279	.5319	.5359
0.1	.5398	.5438	.5478	.5517	.5557	.5596	.5636	.5675	.5714	.5753
0.2	.5793	.5832	.5871	.5910	.5948	.5987	.6026	.6064	.6103	.6141
0.3	.6179	.6217	.6255	.6293	.6331	.6368	.6406	.6443	.6480	.6517
0.4	.6554	.6591	.6628	.6664	.6700	.6736	.6772	.6808	.6844	.6879
0.5	.6915	.6950	.6985	.7019	.7054	.7088	.7123	.7157	.7190	.7224
0.6	.7257	.7291	.7324	.7357	.7389	.7422	.7454	.7486	.7517	.7549
0.7	.7580	.7611	.7642	.7673	.7704	.7734	.7764	.7794	.7823	.7852
0.8	.7881	.7910	.7939	.7967	.7995	.8023	.8051	.8078	.8106	.8133
0.9	.8159	.8186	.8212	.8238	.8264	.8289	.8315	.8340	.8365	.8389
1.0	.8413	.8438	.8461	.8485	.8508	.8531	.8554	.8577	.8599	.8621
1.1	.8643	.8665	.8686	.8708	.8729	.8749	.8770	.8790	.8810	.8830
1.2	.8849	.8869	.8888	.8907	.8925	.8944	.8962	.8980	.8997	.9015
1.3	.9032	.9049	.9066	.9082	.9099	.9115	.9131	.9147	.9162	.9177
1.4	.9192	.9207	.9222	.9236	.9251	.9265	.9279	.9292	.9306	.9319
1.5	.9332	.9345	.9357	.9370	.9382	.9394	.9406	.9418	.9429	.9441
1.6	.9452	.9463	.9474	.9484	.9495	.9505	.9515	.9525	.9535	.9545
1.7	.9554	.9564	.9573	.9582	.9591	.9599	.9608	.9616	.9625	.9633
1.8	.9641	.9649	.9656	.9664	.9671	.9678	.9686	.9693	.9699	.9706
1.9	.9713	.9719	.9726	.9732	.9738	.9744	.9750	.9756	.9761	.9767
2.0	.9772	.9778	.9783	.9788	.9793	.9798	.9803	.9808	.9812	.9817
2.1	.9821	.9826	.9830	.9834	.9838	.9842	.9846	.9850	.9854	.9857
2.2	.9861	.9864	.9868	.9871	.9875	.9878	.9881	.9884	.9887	.9890
2.3	.9893	.9896	.9898	.9901	.9904	.9906	.9909	.9911	.9913	.9916
2.4	.9918	.9920	.9922	.9925	.9927	.9929	.9931	.9932	.9934	.9936
2.5	.9938	.9940	.9941	.9943	.9945	.9946	.9948	.9949	.9951	.9952
2.6	.9953	.9955	.9956	.9957	.9959	.9960	.9961	.9962	.9963	.9964
2.7	.9965	.9966	.9967	.9968	.9969	.9970	.9971	.9972	.9973	.9974
2.8	.9974	.9975	.9976	.9977	.9977	.9978	.9979	.9979	.9980	.9981
2.9	.9981	.9982	.9982	.9983	.9984	.9984	.9985	.9985	.9986	.9986
3.0	.9987	.9987	.9987	.9988	.9988	.9989	.9989	.9989	.9990	.9990
3.1	.9990	.9991	.9991	.9991	.9992	.9992	.9992	.9992	.9993	.9993
3.2	.9993	.9993	.9994	.9994	.9994	.9994	.9994	.9995	.9995	.9995
3.3	.9995	.9995	.9995	.9996	.9996	.9996	.9996	.9996	.9996	.9997
3.4	.9997	.9997	.9997	.9997	.9997	.9997	.9997	.9997	.9997	.9998
3.5	.9998	.9998	.9998	.9998	.9998	.9998	.9998	.9998	.9998	.9998
3.6	.9998	.9998	.9999	.9999	.9999	.9999	.9999	.9999	.9999	.9999
3.7	.9999	.9999	.9999	.9999	.9999	.9999	.9999	.9999	.9999	.9999
3.8	.9999	.9999	.9999	.9999	.9999	.9999	.9999	.9999	.9999	1.0000

Table II The binomial distribution

$n = 5$

| | | | | | | π | | | | | | | |
x	0.05	0.1	0.2	0.25	0.3	0.4	0.5	0.6	0.7	0.75	0.8	0.9	0.95
0	.774	.590	.328	.237	.168	.078	.031	.010	.002	.001	.000	.000	.000
1	.203	.329	.409	.396	.360	.259	.157	.077	.029	.015	.007	.000	.000
2	.022	.072	.205	.263	.309	.346	.312	.230	.132	.088	.051	.009	.001
3	.001	.009	.051	.088	.132	.230	.312	.346	.309	.263	.205	.072	.022
4	.000	.000	.007	.015	.029	.077	.157	.259	.360	.396	.409	.329	.203
5	.000	.000	.000	.001	.002	.010	.031	.078	.168	.237	.328	.590	.774

$n = 10$

| | | | | | | π | | | | | | | |
x	0.05	0.1	0.2	0.25	0.3	0.4	0.5	0.6	0.7	0.75	0.8	0.9	0.95
0	.599	.349	.107	.056	.028	.006	.001	.000	.000	.000	.000	.000	.000
1	.315	.387	.268	.188	.121	.040	.010	.002	.000	.000	.000	.000	.000
2	.075	.194	.302	.282	.233	.121	.044	.011	.001	.000	.000	.000	.000
3	.010	.057	.201	.250	.267	.215	.117	.042	.009	.003	.001	.000	.000
4	.001	.011	.088	.146	.200	.251	.205	.111	.037	.016	.006	.000	.000
5	.000	.001	.026	.058	.103	.201	.246	.201	.103	.058	.026	.001	.000
6	.000	.000	.006	.016	.037	.111	.205	.251	.200	.146	.088	.011	.001
7	.000	.000	.001	.003	.009	.042	.117	.215	.267	.250	.201	.057	.010
8	.000	.000	.000	.000	.001	.011	.044	.121	.233	.282	.302	.194	.075
9	.000	.000	.000	.000	.000	.002	.010	.040	.121	.188	.268	.387	.315
10	.000	.000	.000	.000	.000	.000	.001	.006	.028	.056	.107	.349	.599

Table II The binomial distribution *(continued)*

$n = 15$

						π							
x	0.05	0.1	0.2	0.25	0.3	0.4	0.5	0.6	0.7	0.75	0.8	0.9	0.95
0	.463	.206	.035	.013	.005	.000	.000	.000	.000	.000	.000	.000	.000
1	.366	.343	.132	.067	.030	.005	.000	.000	.000	.000	.000	.000	.000
2	.135	.267	.231	.156	.092	.022	.004	.000	.000	.000	.000	.000	.000
3	.031	.128	.250	.225	.170	.064	.014	.002	.000	.000	.000	.000	.000
4	.004	.043	.188	.225	.218	.126	.041	.007	.001	.000	.000	.000	.000
5	.001	.011	.103	.166	.207	.196	.092	.025	.003	.001	.000	.000	.000
6	.000	.002	.043	.091	.147	.207	.153	.061	.011	.003	.001	.000	.000
7	.000	.000	.014	.040	.081	.177	.196	.118	.035	.013	.003	.000	.000
8	.000	.000	.003	.013	.035	.118	.196	.177	.081	.040	.014	.000	.000
9	.000	.000	.001	.003	.011	.061	.153	.207	.147	.091	.043	.002	.000
10	.000	.000	.000	.001	.003	.025	.092	.196	.207	.166	.103	.011	.001
11	.000	.000	.000	.000	.001	.007	.041	.126	.218	.225	.188	.043	.004
12	.000	.000	.000	.000	.000	.002	.014	.064	.170	.225	.250	.128	.031
13	.000	.000	.000	.000	.000	.000	.004	.022	.092	.156	.231	.267	.135
14	.000	.000	.000	.000	.000	.000	.000	.005	.030	.067	.132	.343	.366
15	.000	.000	.000	.000	.000	.000	.000	.000	.005	.013	.035	.206	.463

$n = 20$

						π							
x	0.05	0.1	0.2	0.25	0.3	0.4	0.5	0.6	0.7	0.75	0.8	0.9	0.95
0	.358	.122	.012	.003	.001	.000	.000	.000	.000	.000	.000	.000	.000
1	.377	.270	.058	.021	.007	.000	.000	.000	.000	.000	.000	.000	.000
2	.189	.285	.137	.067	.028	.003	.000	.000	.000	.000	.000	.000	.000
3	.060	.190	.205	.134	.072	.012	.001	.000	.000	.000	.000	.000	.000
4	.013	.090	.218	.190	.130	.035	.005	.000	.000	.000	.000	.000	.000
5	.002	.032	.175	.202	.179	.075	.015	.001	.000	.000	.000	.000	.000
6	.000	.009	.109	.169	.192	.124	.037	.005	.000	.000	.000	.000	.000
7	.000	.002	.055	.112	.164	.166	.074	.015	.001	.000	.000	.000	.000
8	.000	.000	.022	.061	.114	.180	.120	.035	.004	.001	.000	.000	.000
9	.000	.000	.007	.027	.065	.160	.160	.071	.012	.003	.000	.000	.000
10	.000	.000	.002	.010	.031	.117	.176	.117	.031	.010	.002	.000	.000
11	.000	.000	.000	.003	.012	.071	.160	.160	.065	.027	.007	.000	.000
12	.000	.000	.000	.001	.004	.035	.120	.180	.114	.061	.022	.000	.000
13	.000	.000	.000	.000	.001	.015	.074	.166	.164	.112	.055	.002	.000
14	.000	.000	.000	.000	.000	.005	.037	.124	.192	.169	.109	.009	.000
15	.000	.000	.000	.000	.000	.001	.015	.075	.179	.202	.175	.032	.002
16	.000	.000	.000	.000	.000	.000	.005	.035	.130	.190	.218	.090	.013
17	.000	.000	.000	.000	.000	.000	.001	.012	.072	.134	.205	.190	.060
18	.000	.000	.000	.000	.000	.000	.000	.003	.028	.067	.137	.285	.189
19	.000	.000	.000	.000	.000	.000	.000	.000	.007	.021	.058	.270	.377
20	.000	.000	.000	.000	.000	.000	.000	.000	.001	.003	.012	.122	.358

Table II The binomial distribution *(continued)*

$n = 25$

x	π												
	0.05	0.1	0.2	0.25	0.3	0.4	0.5	0.6	0.7	0.75	0.8	0.9	0.95
0	.277	.072	.004	.001	.000	.000	.000	.000	.000	.000	.000	.000	.000
1	.365	.199	.023	.006	.002	.000	.000	.000	.000	.000	.000	.000	.000
2	.231	.266	.071	.025	.007	.000	.000	.000	.000	.000	.000	.000	.000
3	.093	.227	.136	.064	.024	.002	.000	.000	.000	.000	.000	.000	.000
4	.027	.138	.187	.118	.057	.007	.000	.000	.000	.000	.000	.000	.000
5	.006	.065	.196	.164	.103	.020	.002	.000	.000	.000	.000	.000	.000
6	.001	.024	.163	.183	.148	.045	.005	.000	.000	.000	.000	.000	.000
7	.000	.007	.111	.166	.171	.080	.015	.001	.000	.000	.000	.000	.000
8	.000	.002	.062	.124	.165	.120	.032	.003	.000	.000	.000	.000	.000
9	.000	.000	.030	.078	.134	.151	.061	.009	.000	.000	.000	.000	.000
10	.000	.000	.011	.042	.091	.161	.097	.021	.002	.000	.000	.000	.000
11	.000	.000	.004	.019	.054	.146	.133	.044	.004	.001	.000	.000	.000
12	.000	.000	.002	.007	.027	.114	.155	.076	.011	.002	.000	.000	.000
13	.000	.000	.000	.002	.011	.076	.155	.114	.027	.007	.002	.000	.000
14	.000	.000	.000	.001	.004	.044	.133	.146	.054	.019	.004	.000	.000
15	.000	.000	.000	.000	.002	.021	.097	.161	.091	.042	.011	.000	.000
16	.000	.000	.000	.000	.000	.009	.061	.151	.134	.078	.030	.000	.000
17	.000	.000	.000	.000	.000	.003	.032	.120	.165	.124	.062	.002	.000
18	.000	.000	.000	.000	.000	.001	.015	.080	.171	.166	.111	.007	.000
19	.000	.000	.000	.000	.000	.000	.005	.045	.148	.183	.163	.024	.001
20	.000	.000	.000	.000	.000	.000	.002	.020	.103	.164	.196	.065	.006
21	.000	.000	.000	.000	.000	.000	.000	.007	.057	.118	.187	.138	.027
22	.000	.000	.000	.000	.000	.000	.000	.002	.024	.064	.136	.227	.093
23	.000	.000	.000	.000	.000	.000	.000	.000	.007	.025	.071	.266	.231
24	.000	.000	.000	.000	.000	.000	.000	.000	.002	.006	.023	.199	.365
25	.000	.000	.000	.000	.000	.000	.000	.000	.000	.001	.004	.072	.277

Table III   The Poisson distribution

					λ					
x	.1	.2	.3	.4	.5	.6	.7	.8	.9	1.0
0	.905	.819	.741	.670	.607	.549	.497	.449	.407	.368
1	.090	.164	.222	.268	.303	.329	.348	.359	.366	.368
2	.005	.016	.033	.054	.076	.099	.122	.144	.165	.184
3		.001	.003	.007	.013	.020	.028	.038	.049	.061
4				.001	.002	.003	.005	.008	.011	.015
5							.001	.001	.002	.003
6										.001

						λ					
x	2.0	3.0	4.0	5.0	6.0	7.0	8.0	9.0	10.0	15.0	20.0
0	.135	.050	.018	.007	.002	.001	.000	.000	.000	.000	.000
1	.271	.149	.073	.034	.015	.006	.003	.001	.000	.000	.000
2	.271	.224	.147	.084	.045	.022	.011	.005	.002	.000	.000
3	.180	.224	.195	.140	.089	.052	.029	.015	.008	.000	.000
4	.090	.168	.195	.175	.134	.091	.057	.034	.019	.001	.000
5	.036	.101	.156	.175	.161	.128	.092	.061	.038	.002	.000
6	.012	.050	.104	.146	.161	.149	.122	.091	.063	.005	.000
7	.003	.022	.060	.104	.138	.149	.140	.117	.090	.010	.001
8	.001	.008	.030	.065	.103	.130	.140	.132	.113	.019	.001
9		.003	.013	.036	.069	.101	.124	.132	.125	.032	.003
10		.001	.005	.018	.041	.071	.099	.119	.125	.049	.006
11			.002	.008	.023	.045	.072	.097	.114	.066	.011
12			.001	.003	.011	.026	.048	.073	.095	.083	.018
13				.001	.005	.014	.030	.050	.073	.096	.027
14					.002	.007	.017	.032	.052	.102	.039
15					.001	.003	.009	.019	.035	.102	.052
16						.001	.005	.011	.022	.096	.065
17						.001	.002	.006	.013	.085	.076
18							.001	.003	.007	.071	.084
19								.001	.004	.056	.089
20								.001	.002	.042	.089
21									.001	.030	.085
22										.020	.077
23										.013	.067
24										.008	.056
25										.005	.045
26										.003	.034
27										.002	.025
28										.001	.018
29											.013
30											.008

Table IV *t* critical values for confidence and prediction intervals

		80%	90%	95%	98%	99%	99.8%	99.9%
Central area = confidence/prediction level for two-sided interval:		80%	90%	95%	98%	99%	99.8%	99.9%
Cumulative area = confidence/prediction level for one-sided interval:		90%	95%	97.5%	99%	99.5%	99.9%	99.95%
	1	3.078	6.314	12.706	31.821	63.657	318.31	636.62
	2	1.886	2.920	4.303	6.965	9.925	22.326	31.598
	3	1.638	2.353	3.182	4.541	5.841	10.213	12.924
	4	1.533	2.132	2.776	3.747	4.604	7.173	8.610
	5	1.476	2.015	2.571	3.365	4.032	5.893	6.869
	6	1.440	1.943	2.447	3.143	3.707	5.208	5.959
	7	1.415	1.895	2.365	2.998	3.499	4.785	5.408
	8	1.397	1.860	2.306	2.896	3.355	4.501	5.041
	9	1.383	1.833	2.262	2.821	3.250	4.297	4.781
	10	1.372	1.812	2.228	2.764	3.169	4.144	4.587
	11	1.363	1.796	2.201	2.718	3.106	4.025	4.437
	12	1.356	1.782	2.179	2.681	3.055	3.930	4.318
	13	1.350	1.771	2.160	2.650	3.012	3.852	4.221
	14	1.345	1.761	2.145	2.624	2.977	3.787	4.140
	15	1.341	1.753	2.131	2.602	2.947	3.733	4.073
	16	1.337	1.746	2.120	2.583	2.921	3.686	4.015
Degrees of	17	1.333	1.740	2.110	2.567	2.898	3.646	3.965
freedom	18	1.330	1.734	2.101	2.552	2.878	3.610	3.922
	19	1.328	1.729	2.093	2.539	2.861	3.579	3.883
	20	1.325	1.725	2.086	2.528	2.845	3.552	3.850
	21	1.323	1.721	2.080	2.518	2.831	3.527	3.819
	22	1.321	1.717	2.074	2.508	2.819	3.505	3.792
	23	1.319	1.714	2.069	2.500	2.807	3.485	3.767
	24	1.318	1.711	2.064	2.492	2.797	3.467	3.745
	25	1.316	1.708	2.060	2.485	2.787	3.450	3.725
	26	1.315	1.706	2.056	2.479	2.779	3.435	3.707
	27	1.314	1.703	2.052	2.473	2.771	3.421	3.690
	28	1.313	1.701	2.048	2.467	2.763	3.408	3.674
	29	1.311	1.699	2.045	2.462	2.756	3.396	3.659
	30	1.310	1.697	2.042	2.457	2.750	3.385	3.646
	40	1.303	1.684	2.021	2.423	2.704	3.307	3.551
	60	1.296	1.671	2.000	2.390	2.660	3.232	3.460
	120	1.289	1.658	1.980	2.358	2.617	3.160	3.373
	∞	1.282	1.645	1.960	2.326	2.576	3.090	3.291

Table V Tolerance critical values for normal population distributions

	Two-sided intervals						One-sided intervals					
Confidence level	95%			99%			95%			99%		
% of population captured	≥90%	≥95%	≥99%	≥90%	≥95%	≥99%	≥90%	≥95%	≥99%	≥90%	≥95%	≥99%
2	32.019	37.674	48.430	160.193	188.491	242.300	20.581	26.260	37.094	103.029	131.426	185.617
3	8.380	9.916	12.861	18.930	22.401	29.055	6.156	7.656	10.553	13.995	17.370	23.896
4	5.369	6.370	8.299	9.398	11.150	14.527	4.162	5.144	7.042	7.380	9.083	12.387
5	4.275	5.079	6.634	6.612	7.855	10.260	3.407	4.203	5.741	5.362	6.578	8.939
6	3.712	4.414	5.775	5.337	6.345	8.301	3.006	3.708	5.062	4.411	5.406	7.335
7	3.369	4.007	5.248	4.613	5.488	7.187	2.756	3.400	4.642	3.859	4.728	6.412
8	3.136	3.732	4.891	4.147	4.936	6.468	2.582	3.187	4.354	3.497	4.285	5.812
9	2.967	3.532	4.631	3.822	4.550	5.966	2.454	3.031	4.143	3.241	3.972	5.389
10	2.839	3.379	4.433	3.582	4.265	5.594	2.355	2.911	3.981	3.048	3.738	5.074
11	2.737	3.259	4.277	3.397	4.045	5.308	2.275	2.815	3.852	2.898	3.556	4.829
12	2.655	3.162	4.150	3.250	3.870	5.079	2.210	2.736	3.747	2.777	3.410	4.633
13	2.587	3.081	4.044	3.130	3.727	4.893	2.155	2.671	3.659	2.677	3.290	4.472
14	2.529	3.012	3.955	3.029	3.608	4.737	2.109	2.615	3.585	2.593	3.189	4.337
15	2.480	2.954	3.878	2.945	3.507	4.605	2.068	2.566	3.520	2.522	3.102	4.222
16	2.437	2.903	3.812	2.872	3.421	4.492	2.033	2.524	3.464	2.460	3.028	4.123
17	2.400	2.858	3.754	2.808	3.345	4.393	2.002	2.486	3.414	2.405	2.963	4.037
18	2.366	2.819	3.702	2.753	3.279	4.307	1.974	2.453	3.370	2.357	2.905	3.960
19	2.337	2.784	3.656	2.703	3.221	4.230	1.949	2.423	3.331	2.314	2.854	3.892
20	2.310	2.752	3.615	2.659	3.168	4.161	1.926	2.396	3.295	2.276	2.808	3.832
25	2.208	2.631	3.457	2.494	2.972	3.904	1.838	2.292	3.158	2.129	2.633	3.601
30	2.140	2.549	3.350	2.385	2.841	3.733	1.777	2.220	3.064	2.030	2.516	3.447
35	2.090	2.490	3.272	2.306	2.748	3.611	1.732	2.167	2.995	1.957	2.430	3.334
40	2.052	2.445	3.213	2.247	2.677	3.518	1.697	2.126	2.941	1.902	2.364	3.249
45	2.021	2.408	3.165	2.200	2.621	3.444	1.669	2.092	2.898	1.857	2.312	3.180
50	1.996	2.379	3.126	2.162	2.576	3.385	1.646	2.065	2.863	1.821	2.269	3.125
60	1.958	2.333	3.066	2.103	2.506	3.293	1.609	2.022	2.807	1.764	2.202	3.038
70	1.929	2.299	3.021	2.060	2.454	3.225	1.581	1.990	2.765	1.722	2.153	2.974
80	1.907	2.272	2.986	2.026	2.414	3.173	1.559	1.965	2.733	1.688	2.114	2.924
90	1.889	2.251	2.958	1.999	2.382	3.130	1.542	1.944	2.706	1.661	2.082	2.883
100	1.874	2.233	2.934	1.977	2.355	3.096	1.527	1.927	2.684	1.639	2.056	2.850
150	1.825	2.175	2.859	1.905	2.270	2.983	1.478	1.870	2.611	1.566	1.971	2.741
200	1.798	2.143	2.816	1.865	2.222	2.921	1.450	1.837	2.570	1.524	1.923	2.679
250	1.780	2.121	2.788	1.839	2.191	2.880	1.431	1.815	2.542	1.496	1.891	2.638
300	1.767	2.106	2.767	1.820	2.169	2.850	1.417	1.800	2.522	1.476	1.868	2.608
∞	1.645	1.960	2.576	1.645	1.960	2.576	1.282	1.645	2.326	1.282	1.645	2.326

Sample size n

Table VI   Tail areas for t curves

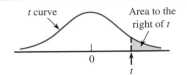

t \ df	1	2	3	4	5	6	7	8	9	10	11	12
0.0	.500	.500	.500	.500	.500	.500	.500	.500	.500	.500	.500	.500
0.1	.468	.465	.463	.463	.462	.462	.462	.461	.461	.461	.461	.461
0.2	.437	.430	.427	.426	.425	.424	.424	.423	.423	.423	.423	.422
0.3	.407	.396	.392	.390	.388	.387	.386	.386	.386	.385	.385	.385
0.4	.379	.364	.358	.355	.353	.352	.351	.350	.349	.349	.348	.348
0.5	.352	.333	.326	.322	.319	.317	.316	.315	.315	.314	.313	.313
0.6	.328	.305	.295	.290	.287	.285	.284	.283	.282	.281	.280	.280
0.7	.306	.278	.267	.261	.258	.255	.253	.252	.251	.250	.249	.249
0.8	.285	.254	.241	.234	.230	.227	.225	.223	.222	.221	.220	.220
0.9	.267	.232	.217	.210	.205	.201	.199	.197	.196	.195	.194	.193
1.0	.250	.211	.196	.187	.182	.178	.175	.173	.172	.170	.169	.169
1.1	.235	.193	.176	.167	.162	.157	.154	.152	.150	.149	.147	.146
1.2	.221	.177	.158	.148	.142	.138	.135	.132	.130	.129	.128	.127
1.3	.209	.162	.142	.132	.125	.121	.117	.115	.113	.111	.110	.109
1.4	.197	.148	.128	.117	.110	.106	.102	.100	.098	.096	.095	.093
1.5	.187	.136	.115	.104	.097	.092	.089	.086	.084	.082	.081	.080
1.6	.178	.125	.104	.092	.085	.080	.077	.074	.072	.070	.069	.068
1.7	.169	.116	.094	.082	.075	.070	.066	.064	.062	.060	.059	.057
1.8	.161	.107	.085	.073	.066	.061	.057	.055	.053	.051	.050	.049
1.9	.154	.099	.077	.065	.058	.053	.050	.047	.045	.043	.042	.041
2.0	.148	.092	.070	.058	.051	.046	.043	.040	.038	.037	.035	.034
2.1	.141	.085	.063	.052	.045	.040	.037	.034	.033	.031	.030	.029
2.2	.136	.079	.058	.046	.040	.035	.032	.029	.028	.026	.025	.024
2.3	.131	.074	.052	.041	.035	.031	.027	.025	.023	.022	.021	.020
2.4	.126	.069	.048	.037	.031	.027	.024	.022	.020	.019	.018	.017
2.5	.121	.065	.044	.033	.027	.023	.020	.018	.017	.016	.015	.014
2.6	.117	.061	.040	.030	.024	.020	.018	.016	.014	.013	.012	.012
2.7	.113	.057	.037	.027	.021	.018	.015	.014	.012	.011	.010	.010
2.8	.109	.054	.034	.024	.019	.016	.013	.012	.010	.009	.009	.008
2.9	.106	.051	.031	.022	.017	.014	.011	.010	.009	.008	.007	.007
3.0	.102	.048	.029	.020	.015	.012	.010	.009	.007	.007	.006	.006
3.1	.099	.045	.027	.018	.013	.011	.009	.007	.006	.006	.005	.005
3.2	.096	.043	.025	.016	.012	.009	.008	.006	.005	.005	.004	.004
3.3	.094	.040	.023	.015	.011	.008	.007	.005	.005	.004	.004	.003
3.4	.091	.038	.021	.014	.010	.007	.006	.005	.004	.003	.003	.003
3.5	.089	.036	.020	.012	.009	.006	.005	.004	.003	.003	.002	.002
3.6	.086	.035	.018	.011	.008	.006	.004	.004	.003	.002	.002	.002
3.7	.084	.033	.017	.010	.007	.005	.004	.003	.002	.002	.002	.002
3.8	.082	.031	.016	.010	.006	.004	.003	.003	.002	.002	.001	.001
3.9	.080	.030	.015	.009	.006	.004	.003	.002	.002	.001	.001	.001
4.0	.078	.029	.014	.008	.005	.004	.003	.002	.002	.001	.001	.001

Table VI   Tail areas for *t* curves *(continued)*

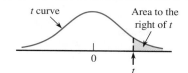

t \ df	13	14	15	16	17	18	19	20	21	22	23	24
0.0	.500	.500	.500	.500	.500	.500	.500	.500	.500	.500	.500	.500
0.1	.461	.461	.461	.461	.461	.461	.461	.461	.461	.461	.461	.461
0.2	.422	.422	.422	.422	.422	.422	.422	.422	.422	.422	.422	.422
0.3	.384	.384	.384	.384	.384	.384	.384	.384	.384	.383	.383	.383
0.4	.348	.347	.347	.347	.347	.347	.347	.347	.347	.347	.346	.346
0.5	.313	.312	.312	.312	.312	.312	.311	.311	.311	.311	.311	.311
0.6	.279	.279	.279	.278	.278	.278	.278	.278	.278	.277	.277	.277
0.7	.248	.247	.247	.247	.247	.246	.246	.246	.246	.246	.245	.245
0.8	.219	.218	.218	.218	.217	.217	.217	.217	.216	.216	.216	.216
0.9	.192	.191	.191	.191	.190	.190	.190	.189	.189	.189	.189	.189
1.0	.168	.167	.167	.166	.166	.165	.165	.165	.164	.164	.164	.164
1.1	.146	.144	.144	.144	.143	.143	.143	.142	.142	.142	.141	.141
1.2	.126	.124	.124	.124	.123	.123	.122	.122	.122	.121	.121	.121
1.3	.108	.107	.107	.106	.105	.105	.105	.104	.104	.104	.103	.103
1.4	.092	.091	.091	.090	.090	.089	.089	.089	.088	.088	.087	.087
1.5	.079	.077	.077	.077	.076	.075	.075	.075	.074	.074	.074	.073
1.6	.067	.065	.065	.065	.064	.064	.063	.063	.062	.062	.062	.061
1.7	.056	.055	.055	.054	.054	.053	.053	.052	.052	.052	.051	.051
1.8	.048	.046	.046	.045	.045	.044	.044	.043	.043	.043	.042	.042
1.9	.040	.038	.038	.038	.037	.037	.036	.036	.036	.035	.035	.035
2.0	.033	.032	.032	.031	.031	.030	.030	.030	.029	.029	.029	.028
2.1	.028	.027	.027	.026	.025	.025	.025	.024	.024	.024	.023	.023
2.2	.023	.022	.022	.021	.021	.021	.020	.020	.020	.019	.019	.019
2.3	.019	.018	.018	.018	.017	.017	.016	.016	.016	.016	.015	.015
2.4	.016	.015	.015	.014	.014	.014	.013	.013	.013	.013	.012	.012
2.5	.013	.012	.012	.012	.011	.011	.011	.011	.010	.010	.010	.010
2.6	.011	.010	.010	.010	.009	.009	.009	.009	.008	.008	.008	.008
2.7	.009	.008	.008	.008	.008	.007	.007	.007	.007	.007	.006	.006
2.8	.008	.007	.007	.006	.006	.006	.006	.006	.005	.005	.005	.005
2.9	.006	.005	.005	.005	.005	.005	.005	.004	.004	.004	.004	.004
3.0	.005	.004	.004	.004	.004	.004	.004	.004	.003	.003	.003	.003
3.1	.004	.004	.004	.003	.003	.003	.003	.003	.003	.003	.003	.002
3.2	.003	.003	.003	.003	.003	.002	.002	.002	.002	.002	.002	.002
3.3	.003	.002	.002	.002	.002	.002	.002	.002	.002	.002	.002	.001
3.4	.002	.002	.002	.002	.002	.002	.002	.001	.001	.001	.001	.001
3.5	.002	.002	.002	.001	.001	.001	.001	.001	.001	.001	.001	.001
3.6	.002	.001	.001	.001	.001	.001	.001	.001	.001	.001	.001	.001
3.7	.001	.001	.001	.001	.001	.001	.001	.001	.001	.001	.001	.001
3.8	.001	.001	.001	.001	.001	.001	.001	.001	.001	.000	.000	.000
3.9	.001	.001	.001	.001	.001	.001	.000	.000	.000	.000	.000	.000
4.0	.001	.001	.001	.001	.000	.000	.000	.000	.000	.000	.000	.000

Table VI   Tail areas for *t* curves (continued)

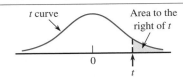

t \ df	25	26	27	28	29	30	35	40	60	120	∞(= z)
0.0	.500	.500	.500	.500	.500	.500	.500	.500	.500	.500	.500
0.1	.461	.461	.461	.461	.461	.461	.460	.460	.460	.460	.460
0.2	.422	.422	.421	.421	.421	.421	.421	.421	.421	.421	.421
0.3	.383	.383	.383	.383	.383	.383	.383	.383	.383	.382	.382
0.4	.346	.346	.346	.346	.346	.346	.346	.346	.345	.345	.345
0.5	.311	.311	.311	.310	.310	.310	.310	.310	.309	.309	.309
0.6	.277	.277	.277	.277	.277	.277	.276	.276	.275	.275	.274
0.7	.245	.245	.245	.245	.245	.245	.244	.244	.243	.243	.242
0.8	.216	.215	.215	.215	.215	.215	.215	.214	.213	.213	.212
0.9	.188	.188	.188	.188	.188	.188	.187	.187	.186	.185	.184
1.0	.163	.163	.163	.163	.163	.163	.162	.162	.161	.160	.159
1.1	.141	.141	.141	.140	.140	.140	.139	.139	.138	.137	.136
1.2	.121	.120	.120	.120	.120	.120	.119	.119	.117	.116	.115
1.3	.103	.103	.102	.102	.102	.102	.101	.101	.099	.098	.097
1.4	.087	.087	.086	.086	.086	.086	.085	.085	.083	.082	.081
1.5	.073	.073	.073	.072	.072	.072	.071	.071	.069	.068	.067
1.6	.061	.061	.061	.060	.060	.060	.059	.059	.057	.056	.055
1.7	.051	.051	.050	.050	.050	.050	.049	.048	.047	.046	.045
1.8	.042	.042	.042	.041	.041	.041	.040	.040	.038	.037	.036
1.9	.035	.034	.034	.034	.034	.034	.033	.032	.031	.030	.029
2.0	.028	.028	.028	.028	.027	.027	.027	.026	.025	.024	.023
2.1	.023	.023	.023	.022	.022	.022	.022	.021	.020	.019	.018
2.2	.019	.018	.018	.018	.018	.018	.017	.017	.016	.015	.014
2.3	.015	.015	.015	.015	.014	.014	.014	.013	.012	.012	.011
2.4	.012	.012	.012	.012	.012	.011	.011	.011	.010	.009	.008
2.5	.010	.010	.009	.009	.009	.009	.009	.008	.008	.007	.006
2.6	.008	.008	.007	.007	.007	.007	.007	.007	.006	.005	.005
2.7	.006	.006	.006	.006	.006	.006	.005	.005	.004	.004	.003
2.8	.005	.005	.005	.005	.005	.004	.004	.004	.003	.003	.003
2.9	.004	.004	.004	.004	.004	.003	.003	.003	.003	.002	.002
3.0	.003	.003	.003	.003	.003	.003	.002	.002	.002	.002	.001
3.1	.002	.002	.002	.002	.002	.002	.002	.002	.001	.001	.001
3.2	.002	.002	.002	.002	.002	.002	.001	.001	.001	.001	.001
3.3	.001	.001	.001	.001	.001	.001	.001	.001	.001	.001	.000
3.4	.001	.001	.001	.001	.001	.001	.001	.001	.001	.000	.000
3.5	.001	.001	.001	.001	.001	.001	.001	.001	.000	.000	.000
3.6	.001	.001	.001	.001	.001	.001	.000	.000	.000	.000	.000
3.7	.001	.001	.000	.000	.000	.000	.000	.000	.000	.000	.000
3.8	.000	.000	.000	.000	.000	.000	.000	.000	.000	.000	.000
3.9	.000	.000	.000	.000	.000	.000	.000	.000	.000	.000	.000
4.0	.000	.000	.000	.000	.000	.000	.000	.000	.000	.000	.000

Table VII   Chi-squared critical values

Right-tail area	df = 1	df = 2	df = 3	df = 4	df = 5
>.100	<2.70	<4.60	<6.25	<7.77	<9.23
0.100	2.70	4.60	6.25	7.77	9.23
0.095	2.78	4.70	6.36	7.90	9.37
0.090	2.87	4.81	6.49	8.04	9.52
0.085	2.96	4.93	6.62	8.18	9.67
0.080	3.06	5.05	6.75	8.33	9.83
0.075	3.17	5.18	6.90	8.49	10.00
0.070	3.28	5.31	7.06	8.66	10.19
0.065	3.40	5.46	7.22	8.84	10.38
0.060	3.53	5.62	7.40	9.04	10.59
0.055	3.68	5.80	7.60	9.25	10.82
0.050	3.84	5.99	7.81	9.48	11.07
0.045	4.01	6.20	8.04	9.74	11.34
0.040	4.21	6.43	8.31	10.02	11.64
0.035	4.44	6.70	8.60	10.34	11.98
0.030	4.70	7.01	8.94	10.71	12.37
0.025	5.02	7.37	9.34	11.14	12.83
0.020	5.41	7.82	9.83	11.66	13.38
0.015	5.91	8.39	10.46	12.33	14.09
0.010	6.63	9.21	11.34	13.27	15.08
0.005	7.87	10.59	12.83	14.86	16.74
0.001	10.82	13.81	16.26	18.46	20.51
<0.001	>10.82	>13.81	>16.26	>18.46	>20.51

Right-tail area	df = 6	df = 7	df = 8	df = 9	df = 10
>.100	<10.64	<12.01	<13.36	<14.68	<15.98
0.100	10.64	12.01	13.36	14.68	15.98
0.095	10.79	12.17	13.52	14.85	16.16
0.090	10.94	12.33	13.69	15.03	16.35
0.085	11.11	12.50	13.87	15.22	16.54
0.080	11.28	12.69	14.06	15.42	16.75
0.075	11.46	12.88	14.26	15.63	16.97
0.070	11.65	13.08	14.48	15.85	17.20
0.065	11.86	13.30	14.71	16.09	17.44
0.060	12.08	13.53	14.95	16.34	17.71
0.055	12.33	13.79	15.22	16.62	17.99
0.050	12.59	14.06	15.50	16.91	18.30
0.045	12.87	14.36	15.82	17.24	18.64
0.040	13.19	14.70	16.17	17.60	19.02
0.035	13.55	15.07	16.56	18.01	19.44
0.030	13.96	15.50	17.01	18.47	19.92
0.025	14.44	16.01	17.53	19.02	20.48
0.020	15.03	16.62	18.16	19.67	21.16
0.015	15.77	17.39	18.97	20.51	22.02
0.010	16.81	18.47	20.09	21.66	23.20
0.005	18.54	20.27	21.95	23.58	25.18
0.001	22.45	24.32	26.12	27.87	29.58
<0.001	>22.45	>24.32	>26.12	>27.87	>29.58

Table VII   Chi-squared critical values *(continued)*

Right-tail area	df = 11	df = 12	df = 13	df = 14	df = 15
>.100	<17.27	<18.54	<19.81	<21.06	<22.30
0.100	17.27	18.54	19.81	21.06	22.30
0.095	17.45	18.74	20.00	21.26	22.51
0.090	17.65	18.93	20.21	21.47	22.73
0.085	17.85	19.14	20.42	21.69	22.95
0.080	18.06	19.36	20.65	21.93	23.19
0.075	18.29	19.60	20.89	22.17	23.45
0.070	18.53	19.84	21.15	22.44	23.72
0.065	18.78	20.11	21.42	22.71	24.00
0.060	19.06	20.39	21.71	23.01	24.31
0.055	19.35	20.69	22.02	23.33	24.63
0.050	19.67	21.02	22.36	23.68	24.99
0.045	20.02	21.38	22.73	24.06	25.38
0.040	20.41	21.78	23.14	24.48	25.81
0.035	20.84	22.23	23.60	24.95	26.29
0.030	21.34	22.74	24.12	25.49	26.84
0.025	21.92	23.33	24.73	26.11	27.48
0.020	22.61	24.05	25.47	26.87	28.25
0.015	23.50	24.96	26.40	27.82	29.23
0.010	24.72	26.21	27.68	29.14	30.57
0.005	26.75	28.29	29.81	31.31	32.80
0.001	31.26	32.90	34.52	36.12	37.69
<0.001	>31.26	>32.90	>34.52	>36.12	>37.69

Right-tail area	df = 16	df = 17	df = 18	df = 19	df = 20
>.100	<23.54	<24.77	<25.98	<27.20	<28.41
0.100	23.54	24.76	25.98	27.20	28.41
0.095	23.75	24.98	26.21	27.43	28.64
0.090	23.97	25.21	26.44	27.66	28.88
0.085	24.21	25.45	26.68	27.91	29.14
0.080	24.45	25.70	26.94	28.18	29.40
0.075	24.71	25.97	27.21	28.45	29.69
0.070	24.99	26.25	27.50	28.75	29.99
0.065	25.28	26.55	27.81	29.06	30.30
0.060	25.59	26.87	28.13	29.39	30.64
0.055	25.93	27.21	28.48	29.75	31.01
0.050	26.29	27.58	28.86	30.14	31.41
0.045	26.69	27.99	29.28	30.56	31.84
0.040	27.13	28.44	29.74	31.03	32.32
0.035	27.62	28.94	30.25	31.56	32.85
0.030	28.19	29.52	30.84	32.15	33.46
0.025	28.84	30.19	31.52	32.85	34.16
0.020	29.63	30.99	32.34	33.68	35.01
0.015	30.62	32.01	33.38	34.74	36.09
0.010	32.00	33.40	34.80	36.19	37.56
0.005	34.26	35.71	37.15	38.58	39.99
0.001	39.25	40.78	42.31	43.81	45.31
<0.001	>39.25	>40.78	>42.31	>43.81	>45.31

Table VIII   *F* critical values

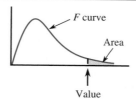

F curve

Area

Value

		Numerator df									
	Area	1	2	3	4	5	6	7	8	9	10
1	.10	39.86	49.50	53.59	55.83	57.24	58.20	58.91	59.44	59.86	60.19
	.05	161.4	199.5	215.7	224.6	230.2	234.0	236.8	238.9	240.5	241.9
	.01	4052	5000	5403	5625	5764	5859	5928	5981	6022	6056
2	.10	8.53	9.00	9.16	9.24	9.29	9.33	9.35	9.37	9.38	9.39
	.05	18.51	19.00	19.16	19.25	19.30	19.33	19.35	19.37	19.38	19.40
	.01	98.50	99.00	99.17	99.25	99.30	99.33	99.36	99.37	99.39	99.40
	.001	998.5	999.0	999.2	999.2	999.3	999.3	999.4	999.4	999.4	999.4
3	.10	5.54	5.46	5.39	5.34	5.31	5.28	5.27	5.25	5.24	5.23
	.05	10.13	9.55	9.28	9.12	9.01	8.94	8.89	8.85	8.81	8.79
	.01	34.12	30.82	29.46	28.71	28.24	27.91	27.67	27.49	27.35	27.23
	.001	167.0	148.5	141.1	137.1	134.6	132.8	131.6	130.6	129.9	129.2
4	.10	4.54	4.32	4.19	4.11	4.05	4.01	3.98	3.95	3.94	3.92
	.05	7.71	6.94	6.59	6.39	6.26	6.16	6.09	6.04	6.00	5.96
	.01	21.20	18.00	16.69	15.98	15.52	15.21	14.98	14.80	14.66	14.55
	.001	74.14	61.25	56.18	53.44	51.71	50.53	49.66	49.00	48.47	48.05
5	.10	4.06	3.78	3.62	3.52	3.45	3.40	3.37	3.34	3.32	3.30
	.05	6.61	5.79	5.41	5.19	5.05	4.95	4.88	4.82	4.77	4.74
	.01	16.26	13.27	12.06	11.39	10.97	10.67	10.46	10.29	10.16	10.05
	.001	47.18	37.12	33.20	31.09	29.75	28.83	28.16	27.65	27.24	26.92
6	.10	3.78	3.46	3.29	3.18	3.11	3.05	3.01	2.98	2.96	2.94
	.05	5.99	5.14	4.76	4.53	4.39	4.28	4.21	4.15	4.10	4.06
	.01	13.75	10.92	9.78	9.15	8.75	8.47	8.26	8.10	7.98	7.87
	.001	35.51	27.00	23.70	21.92	20.80	20.03	19.46	19.03	18.69	18.41
7	.10	3.59	3.26	3.07	2.96	2.88	2.83	2.78	2.75	2.72	2.70
	.05	5.59	4.74	4.35	4.12	3.97	3.87	3.79	3.73	3.68	3.64
	.01	12.25	9.55	8.45	7.85	7.46	7.19	6.99	6.84	6.72	6.62
	.001	29.25	21.69	18.77	17.20	16.21	15.52	15.02	14.63	14.33	14.08
8	.10	3.46	3.11	2.92	2.81	2.73	2.67	2.62	2.59	2.56	2.54
	.05	5.32	4.46	4.07	3.84	3.69	3.58	3.50	3.44	3.39	3.35
	.01	11.26	8.65	7.59	7.01	6.63	6.37	6.18	6.03	5.91	5.81
	.001	25.41	18.49	15.83	14.39	13.48	12.86	12.40	12.05	11.77	11.54
9	.10	3.36	3.01	2.81	2.69	2.61	2.55	2.51	2.47	2.44	2.42
	.05	5.12	4.26	3.86	3.63	3.48	3.37	3.29	3.23	3.18	3.14
	.01	10.56	8.02	6.99	6.42	6.06	5.80	5.61	5.47	5.35	5.26
	.001	22.86	16.39	13.90	12.56	11.71	11.13	10.70	10.37	10.11	9.89

Denominator df

Table VIII *F critical values (continued)*

		Numerator df									
Denominator df	Area	1	2	3	4	5	6	7	8	9	10
10	.10	3.29	2.92	2.73	2.61	2.52	2.46	2.41	2.38	2.35	2.32
	.05	4.96	4.10	3.71	3.48	3.33	3.22	3.14	3.07	3.02	2.98
	.01	10.04	7.56	6.55	5.99	5.64	5.39	5.20	5.06	4.94	4.85
	.001	21.04	14.91	12.55	11.28	10.48	9.93	9.52	9.20	8.96	8.75
11	.10	3.23	2.86	2.66	2.54	2.45	2.39	2.34	2.30	2.27	2.25
	.05	4.84	3.98	3.59	3.36	3.20	3.09	3.01	2.95	2.90	2.85
	.01	9.65	7.21	6.22	5.67	5.32	5.07	4.89	4.74	4.63	4.54
	.001	19.69	13.81	11.56	10.35	9.58	9.05	8.66	8.35	8.12	7.92
12	.10	3.18	2.81	2.61	2.48	2.39	2.33	2.28	2.24	2.21	2.19
	.05	4.75	3.89	3.49	3.26	3.11	3.00	2.91	2.85	2.80	2.75
	.01	9.33	6.93	5.95	5.41	5.06	4.82	4.64	4.50	4.39	4.30
	.001	18.64	12.97	10.80	9.63	8.89	8.38	8.00	7.71	7.48	7.29
13	.10	3.14	2.76	2.56	2.43	2.35	2.28	2.23	2.20	2.16	2.14
	.05	4.67	3.81	3.41	3.18	3.03	2.92	2.83	2.77	2.71	2.67
	.01	9.07	6.70	5.74	5.21	4.86	4.62	4.44	4.30	4.19	4.10
	.001	17.82	12.31	10.21	9.07	8.35	7.86	7.49	7.21	6.98	6.80
14	.10	3.10	2.73	2.52	2.39	2.31	2.24	2.19	2.15	2.12	2.10
	.05	4.60	3.74	3.34	3.11	2.96	2.85	2.76	2.70	2.65	2.60
	.01	8.86	6.51	5.56	5.04	4.69	4.46	4.28	4.14	4.03	3.94
	.001	17.14	11.78	9.73	8.62	7.92	7.44	7.08	6.80	6.58	6.40
15	.10	3.07	2.70	2.49	2.36	2.27	2.21	2.16	2.12	2.09	2.06
	.05	4.54	3.68	3.29	3.06	2.90	2.79	2.71	2.64	2.59	2.54
	.01	8.68	6.36	5.42	4.89	4.67	4.32	4.14	4.00	3.89	3.80
	.001	16.59	11.34	9.34	8.25	7.57	7.09	6.74	6.47	6.26	6.08
16	.10	3.05	2.67	2.46	2.33	2.24	2.18	2.13	2.09	2.06	2.03
	.05	4.49	3.63	3.24	3.01	2.85	2.74	2.66	2.59	2.54	2.49
	.01	8.53	6.23	5.29	4.77	4.44	4.20	4.03	3.89	3.78	3.69
	.001	16.12	10.97	9.01	7.94	7.27	6.80	6.46	6.19	5.98	5.81
17	.10	3.03	2.64	2.44	2.31	2.22	2.15	2.10	2.06	2.03	2.00
	.05	4.45	3.59	3.20	2.96	2.81	2.70	2.61	2.55	2.49	2.45
	.01	8.40	6.11	5.18	4.67	4.34	4.10	3.93	3.79	3.68	3.59
	.001	15.72	10.66	8.73	7.68	7.02	6.56	6.22	5.96	5.75	5.58
18	.10	3.01	2.62	2.42	2.29	2.20	2.13	2.08	2.04	2.00	1.98
	.05	4.41	3.55	3.16	2.93	2.77	2.66	2.58	2.51	2.46	2.41
	.01	8.29	6.01	5.09	4.58	4.25	4.01	3.84	3.71	3.60	3.51
	.001	15.38	10.39	8.49	7.46	6.81	6.35	6.02	5.76	5.56	5.39
19	.10	2.99	2.61	2.40	2.27	2.18	2.11	2.06	2.02	1.98	1.96
	.05	4.38	3.52	3.13	2.90	2.74	2.63	2.54	2.48	2.42	2.38
	.01	8.18	5.93	5.01	4.50	4.17	3.94	3.77	3.63	3.52	3.43
	.001	15.08	10.16	8.28	7.27	6.62	6.18	5.85	5.59	5.39	5.22

Table VIII *F* critical values *(continued)*

		Numerator df									
Denominator df	Area	1	2	3	4	5	6	7	8	9	10
20	.10	2.97	2.59	2.38	2.25	2.16	2.09	2.04	2.00	1.96	1.94
	.05	4.35	3.49	3.10	2.87	2.71	2.60	2.51	2.45	2.39	2.35
	.01	8.10	5.85	4.94	4.43	4.10	3.87	3.70	3.56	3.46	3.37
	.001	14.82	9.95	8.10	7.10	6.46	6.02	5.69	5.44	5.24	5.08
21	.10	2.96	2.57	2.36	2.23	2.14	2.08	2.02	1.98	1.95	1.92
	.05	4.32	3.47	3.07	2.84	2.68	2.57	2.49	2.42	2.37	2.32
	.01	8.02	5.78	4.87	4.37	4.04	3.81	3.64	3.51	3.40	3.31
	.001	14.59	9.77	7.94	6.95	6.32	5.88	5.56	5.31	5.11	4.95
22	.10	2.95	2.56	2.35	2.22	2.13	2.06	2.01	1.97	1.93	1.90
	.05	4.30	3.44	3.05	2.82	2.66	2.55	2.46	2.40	2.34	2.30
	.01	7.95	5.72	4.82	4.31	3.99	3.76	3.59	3.45	3.35	3.26
	.001	14.38	9.61	7.80	6.81	6.19	5.76	5.44	5.19	4.99	4.83
23	.10	2.94	2.55	2.34	2.21	2.11	2.05	1.99	1.95	1.92	1.89
	.05	4.28	3.42	3.03	2.80	2.64	2.53	2.44	2.37	2.32	2.27
	.01	7.88	5.66	4.76	4.26	3.94	3.71	3.54	3.41	3.30	3.21
	.001	14.20	9.47	7.67	6.70	6.08	5.65	5.33	5.09	4.89	4.73
24	.10	2.93	2.54	2.33	2.19	2.10	2.04	1.98	1.94	1.91	1.88
	.05	4.26	3.40	3.01	2.78	2.62	2.51	2.42	2.36	2.30	2.25
	.01	7.82	5.61	4.72	4.22	3.90	3.67	3.50	3.36	3.26	3.17
	.001	14.03	9.34	7.55	6.59	5.98	5.55	5.23	4.99	4.80	4.64
25	.10	2.92	2.53	2.32	2.18	2.09	2.02	1.97	1.93	1.89	1.87
	.05	4.24	3.39	2.99	2.76	2.60	2.49	2.40	2.34	2.28	2.24
	.01	7.77	5.57	4.68	4.18	3.85	3.63	3.46	3.32	3.22	3.13
	.001	13.88	9.22	7.45	6.49	5.89	5.46	5.15	4.91	4.71	4.56
26	.10	2.91	2.52	2.31	2.17	2.08	2.01	1.96	1.92	1.88	1.86
	.05	4.23	3.37	2.98	2.74	2.59	2.47	2.39	2.32	2.27	2.22
	.01	7.72	5.53	4.64	4.14	3.82	3.59	3.42	3.29	3.18	3.09
	.001	13.74	9.12	7.36	6.41	5.80	5.38	5.07	4.83	4.64	4.48
27	.10	2.90	2.51	2.30	2.17	2.07	2.00	1.95	1.91	1.87	1.85
	.05	4.21	3.35	2.96	2.73	2.57	2.46	2.37	2.31	2.25	2.20
	.01	7.68	5.49	4.60	4.11	3.78	3.56	3.39	3.26	3.15	3.06
	.001	13.61	9.02	7.27	6.33	5.73	5.31	5.00	4.76	4.57	4.41
28	.10	2.89	2.50	2.29	2.16	2.06	2.00	1.94	1.90	1.87	1.84
	.05	4.20	3.34	2.95	2.71	2.56	2.45	2.36	2.29	2.24	2.19
	.01	7.64	5.45	4.57	4.07	3.75	3.53	3.36	3.23	3.12	3.03
	.001	13.50	8.93	7.19	6.25	5.66	5.24	4.93	4.69	4.50	4.35
29	.10	2.89	2.50	2.28	2.15	2.06	1.99	1.93	1.89	1.86	1.83
	.05	4.18	3.33	2.93	2.70	2.55	2.43	2.35	2.28	2.22	2.18
	.01	7.60	5.42	4.54	4.04	3.73	3.50	3.33	3.20	3.09	3.00
	.001	13.39	8.85	7.12	6.19	5.59	5.18	4.87	4.64	4.45	4.29

Table VIII F critical values *(continued)*

	Area	Numerator df 1	2	3	4	5	6	7	8	9	10
30	.10	2.88	2.49	2.28	2.14	2.05	1.98	1.93	1.88	1.85	1.82
	.05	4.17	3.32	2.92	2.69	2.53	2.42	2.33	2.27	2.21	2.16
	.01	7.56	5.39	4.51	4.02	3.70	3.47	330	3.17	3.07	2.98
	.001	13.29	8.77	7.05	6.12	5.53	5.12	4.82	4.58	4.39	4.24
40	.10	2.84	2.44	2.23	2.09	2.00	1.93	1.87	1.83	1.79	1.76
	.05	4.08	3.23	2.84	2.61	2.45	2.34	2.25	2.18	2.12	2.08
	.01	7.31	5.18	4.31	3.83	3.51	3.29	3.12	2.99	2.89	2.80
	.001	12.61	8.25	6.59	5.70	5.13	4.73	4.44	4.21	4.02	3.87
60	.10	2.79	2.39	2.18	2.04	1.95	1.87	1.82	1.77	1.74	1.71
	.05	4.00	3.15	2.76	2.53	2.37	2.25	2.17	2.10	2.04	1.99
	.01	7.08	4.98	4.13	3.65	3.34	3.12	2.95	2.82	2.72	2.63
	.001	11.97	7.77	6.17	5.31	4.76	4.37	4.09	3.86	3.69	3.54
90	.10	2.76	2.36	2.15	2.01	1.91	1.84	1.78	1.74	1.70	1.67
	.05	3.95	3.10	2.71	2.47	2.32	2.20	2.11	2.04	1.99	1.94
	.01	6.93	4.85	4.01	3.53	3.23	3.01	2.84	2.72	2.61	2.52
	.001	11.57	7.47	5.91	5.06	4.53	4.15	3.87	3.65	3.48	3.34
120	.10	2.75	2.35	2.13	1.99	1.90	1.82	1.77	1.72	1.68	1.65
	.05	3.92	3.07	2.68	2.45	2.29	2.18	2.09	2.02	1.96	1.91
	.01	6.85	4.79	3.95	3.48	3.17	2.96	2.79	2.66	2.56	2.47
	.001	11.38	7.32	5.78	4.95	4.42	4.04	3.77	3.55	3.38	3.24
240	.10	2.73	2.32	2.10	1.97	1.87	1.80	1.74	1.70	1.65	1.63
	.05	3.88	3.03	2.64	2.41	2.25	2.14	2.04	1.98	1.92	1.87
	.01	6.74	4.69	3.86	3.40	3.09	2.88	2.71	2.59	2.48	2.40
	.001	11.10	7.11	5.60	4.78	4.25	3.89	3.62	3.41	3.24	3.09
∞	.10	2.71	2.30	2.08	1.94	1.85	1.77	1.72	1.67	1.63	1.06
	.05	3.84	3.00	2.60	2.37	2.21	2.10	2.01	1.94	1.88	1.83
	.01	6.63	4.61	3.78	3.32	3.02	2.80	2.64	2.51	2.41	2.32
	.001	10.83	6.91	5.42	4.62	4.10	3.74	3.47	3.27	3.10	2.96

Denominator df

Table IX(a) Studentized range critical values ($\alpha = .05$)

Error df	k=2	3	4	5	6	7	8	9	10	11	12	13	14	15	16	17	18	19	20
1	18.0	27.0	32.8	37.1	40.4	43.1	45.4	47.4	49.1	50.6	52.0	53.2	54.3	55.4	56.3	57.2	58.0	58.8	59.6
2	6.08	8.33	9.80	10.9	11.7	12.4	13.0	13.5	14.0	14.4	14.7	15.1	15.4	15.7	15.9	16.1	16.4	16.6	16.8
3	4.50	5.91	6.82	7.50	8.04	8.48	8.85	9.18	9.46	9.72	9.95	10.2	10.3	10.5	10.7	10.8	11.0	11.1	11.2
4	3.93	5.04	5.76	6.29	6.71	7.05	7.35	7.60	7.83	8.03	8.21	8.37	8.52	8.66	8.79	8.91	9.03	9.13	9.23
5	3.64	4.60	5.22	5.67	6.03	6.33	6.58	6.80	6.99	7.17	7.32	7.47	7.60	7.72	7.83	7.93	8.03	8.12	8.21
6	3.46	4.34	4.90	5.30	5.63	5.90	6.12	6.32	6.49	6.65	6.79	6.92	7.03	7.14	7.24	7.34	7.43	7.51	7.59
7	3.34	4.16	4.68	5.06	5.36	5.61	5.82	6.00	6.16	6.30	6.43	6.55	6.66	6.76	6.85	6.94	7.02	7.10	7.17
8	3.26	4.04	4.53	4.89	5.17	5.40	5.60	5.77	5.92	6.05	6.18	6.29	6.39	6.48	6.57	6.65	6.73	6.80	6.87
9	3.20	3.95	4.41	4.76	5.02	5.24	5.43	5.59	5.74	5.87	5.98	6.09	6.19	6.28	6.36	6.44	6.51	6.58	6.64
10	3.15	3.88	4.33	4.65	4.91	5.12	5.30	5.46	5.60	5.72	5.83	5.93	6.03	6.11	6.19	6.27	6.34	6.40	6.47
11	3.11	3.82	4.26	4.57	4.82	5.03	5.20	5.35	5.49	5.61	5.71	5.81	5.90	5.98	6.06	6.13	6.20	6.27	6.33
12	3.08	3.77	4.20	4.51	4.75	4.95	5.12	5.27	5.39	5.51	5.61	5.71	5.80	5.88	5.95	6.02	6.09	6.15	6.21
13	3.06	3.73	4.15	4.45	4.69	4.88	5.05	5.19	5.32	5.43	5.53	5.63	5.71	5.79	5.86	5.93	5.99	6.05	6.11
14	3.03	3.70	4.11	4.41	4.64	4.83	4.99	5.13	5.25	5.36	5.46	5.55	5.64	5.71	5.79	5.85	5.91	5.97	6.03
15	3.01	3.67	4.08	4.37	4.59	4.78	4.94	5.08	5.20	5.31	5.40	5.49	5.57	5.65	5.72	5.78	5.85	5.90	5.96
16	3.00	3.65	4.05	4.33	4.56	4.74	4.90	5.03	5.15	5.26	5.35	5.44	5.52	5.59	5.66	5.73	5.79	5.84	5.90
17	2.98	3.63	4.02	4.30	4.52	4.70	4.86	4.99	5.11	5.21	5.31	5.39	5.47	5.54	5.61	5.67	5.73	5.79	5.84
18	2.97	3.61	4.00	4.28	4.49	4.67	4.82	4.96	5.07	5.17	5.27	5.35	5.43	5.50	5.57	5.63	5.69	5.74	5.79
19	2.96	3.59	3.98	4.25	4.47	4.65	4.79	4.92	5.04	5.14	5.23	5.31	5.39	5.46	5.53	5.59	5.65	5.70	5.75
20	2.95	3.58	3.96	4.23	4.45	4.62	4.77	4.90	5.01	5.11	5.20	5.28	5.36	5.43	5.49	5.55	5.61	5.66	5.71
24	2.92	3.53	3.90	4.17	4.37	4.54	4.68	4.81	4.92	5.01	5.10	5.18	5.25	5.32	5.38	5.44	5.49	5.55	5.59
30	2.89	3.49	3.85	4.10	4.30	4.46	4.60	4.72	4.82	4.92	5.00	5.08	5.15	5.21	5.27	5.33	5.38	5.43	5.47
40	2.86	3.44	3.79	4.04	4.23	4.39	4.52	4.63	4.73	4.82	4.90	4.98	5.04	5.11	5.16	5.22	5.27	5.31	5.36
60	2.83	3.40	3.74	3.98	4.16	4.31	4.44	4.55	4.65	4.73	4.81	4.88	4.94	5.00	5.06	5.11	5.15	5.20	5.24
120	2.80	3.36	3.68	3.92	4.10	4.24	4.36	4.47	4.56	4.64	4.71	4.78	4.84	4.90	4.95	5.00	5.04	5.09	5.13
∞	2.77	3.31	3.63	3.86	4.03	4.17	4.29	4.39	4.47	4.55	4.62	4.68	4.74	4.80	4.85	4.89	4.93	4.97	5.01

Table IX(b) Studentized range critical values ($\alpha = .01$)

Error df	2	3	4	5	6	7	8	9	10	11	12	13	14	15	16	17	18	19	20
1	90.0	135	164	186	202	216	227	237	246	253	260	266	272	277	282	286	290	294	298
2	14.0	19.0	22.3	24.7	26.6	28.2	29.5	30.7	31.7	32.6	33.4	34.1	34.8	35.4	36.0	36.5	37.0	37.5	37.9
3	8.26	10.6	12.2	13.3	14.2	15.0	15.6	16.2	16.7	17.1	17.5	17.9	18.2	18.5	18.8	19.1	19.3	19.5	19.8
4	6.51	8.12	9.17	9.96	10.6	11.1	11.5	11.9	12.3	12.6	12.8	13.1	13.3	13.5	13.7	13.9	14.1	14.2	14.4
5	5.70	6.97	7.80	8.42	8.91	9.32	9.67	9.97	10.2	10.5	10.7	10.9	11.1	11.2	11.4	11.6	11.7	11.8	11.9
6	5.24	6.33	7.03	7.56	7.97	8.32	8.61	8.87	9.10	9.30	9.49	9.65	9.81	9.95	10.1	10.2	10.3	10.4	10.5
7	4.95	5.92	6.54	7.01	7.37	7.68	7.94	8.17	8.37	8.55	8.71	8.86	9.00	9.12	9.24	9.35	9.46	9.55	9.65
8	4.74	5.63	6.20	6.63	6.96	7.24	7.47	7.68	7.87	8.03	8.18	8.31	8.44	8.55	8.66	8.76	8.85	8.94	9.03
9	4.60	5.43	5.96	6.35	6.66	6.91	7.13	7.32	7.49	7.65	7.78	7.91	8.03	8.13	8.23	8.32	8.41	8.49	8.57
10	4.48	5.27	5.77	6.14	6.43	6.67	6.87	7.05	7.21	7.36	7.48	7.60	7.71	7.81	7.91	7.99	8.07	8.15	8.22
11	4.39	5.14	5.62	5.97	6.25	6.48	6.67	6.84	6.99	7.13	7.25	7.36	7.46	7.56	7.65	7.73	7.81	7.88	7.95
12	4.32	5.04	5.50	5.84	6.10	6.32	6.51	6.67	6.81	6.94	7.06	7.17	7.26	7.36	7.44	7.52	7.59	7.66	7.73
13	4.26	4.96	5.40	5.73	5.98	6.19	6.37	6.53	6.67	6.79	6.90	7.01	7.10	7.19	7.27	7.34	7.42	7.48	7.55
14	4.21	4.89	5.32	5.63	5.88	6.08	6.26	6.41	6.54	6.66	6.77	6.87	6.96	7.05	7.12	7.20	7.27	7.33	7.39
15	4.17	4.83	5.25	5.56	5.80	5.99	6.16	6.31	6.44	6.55	6.66	6.76	6.84	6.93	7.00	7.07	7.14	7.20	7.26
16	4.13	4.78	5.19	5.49	5.72	5.92	6.08	6.22	6.35	6.46	6.56	6.66	6.74	6.82	6.90	6.97	7.03	7.09	7.15
17	4.10	4.74	5.14	5.43	5.66	5.85	6.01	6.15	6.27	6.38	6.48	6.57	6.66	6.73	6.80	6.87	6.94	7.00	7.05
18	4.07	4.70	5.09	5.38	5.60	5.79	5.94	6.08	6.20	6.31	6.41	6.50	6.58	6.65	6.72	6.79	6.85	6.91	6.96
19	4.05	4.67	5.05	5.33	5.55	5.73	5.89	6.02	6.14	6.25	6.34	6.43	6.51	6.58	6.65	6.72	6.78	6.84	6.89
20	4.02	4.64	5.02	5.29	5.51	5.69	5.84	5.97	6.09	6.19	6.29	6.37	6.45	6.52	6.59	6.65	6.71	6.76	6.82
24	3.96	4.54	4.91	5.17	5.37	5.54	5.69	5.81	5.92	6.02	6.11	6.19	6.26	6.33	6.39	6.45	6.51	6.56	6.61
30	3.89	4.45	4.80	5.05	5.24	5.40	5.54	5.65	5.76	5.85	5.93	6.01	6.08	6.14	6.20	6.26	6.31	6.36	6.41
40	3.82	4.37	4.70	4.93	5.11	5.27	5.39	5.50	5.60	5.69	5.77	5.84	5.90	5.96	6.02	6.07	6.12	6.17	6.21
60	3.76	4.28	4.60	4.82	4.99	5.13	5.25	5.36	5.45	5.53	5.60	5.67	5.73	5.79	5.84	5.89	5.93	5.98	6.02
120	3.70	4.20	4.50	4.71	4.87	5.01	5.12	5.21	5.30	5.38	5.44	5.51	5.56	5.61	5.66	5.71	5.75	5.79	5.83
∞	3.64	4.12	4.40	4.60	4.76	4.88	4.99	5.08	5.16	5.23	5.29	5.35	5.40	5.45	5.49	5.54	5.57	5.61	5.65

k

Source: From E. S. Pearson and H. O. Hartley, *Biometrika Tables for Statisticians*, 1: 176–77. Reproduced by permission of the Biometrika Trustees.

Table X   Critical values for Dunnett's Method[a]

<table>
<tr><td colspan="10" align="center">Two-sided comparisons</td></tr>
<tr><td></td><td colspan="9" align="center">$k - 1 =$ number of treatment means (excluding control)</td></tr>
<tr><td>$n - k$</td><td>1</td><td>2</td><td>3</td><td>4</td><td>5</td><td>6</td><td>7</td><td>8</td><td>9</td></tr>
<tr><td>5</td><td>2.57</td><td>3.03</td><td>3.29</td><td>3.48</td><td>3.62</td><td>3.73</td><td>3.82</td><td>3.90</td><td>3.97</td></tr>
<tr><td>6</td><td>2.45</td><td>2.86</td><td>3.10</td><td>3.26</td><td>3.39</td><td>3.49</td><td>3.57</td><td>3.64</td><td>3.71</td></tr>
<tr><td>7</td><td>2.36</td><td>2.75</td><td>2.97</td><td>3.12</td><td>3.24</td><td>3.33</td><td>3.41</td><td>3.47</td><td>3.53</td></tr>
<tr><td>8</td><td>2.31</td><td>2.67</td><td>2.88</td><td>3.02</td><td>3.13</td><td>3.22</td><td>3.29</td><td>3.35</td><td>3.41</td></tr>
<tr><td>9</td><td>2.26</td><td>2.61</td><td>2.81</td><td>2.95</td><td>3.05</td><td>3.14</td><td>3.20</td><td>3.26</td><td>3.32</td></tr>
<tr><td>10</td><td>2.23</td><td>2.57</td><td>2.76</td><td>2.89</td><td>2.99</td><td>3.07</td><td>3.14</td><td>3.19</td><td>3.24</td></tr>
<tr><td>11</td><td>2.20</td><td>2.53</td><td>2.72</td><td>2.84</td><td>2.94</td><td>3.02</td><td>3.08</td><td>3.14</td><td>3.19</td></tr>
<tr><td>12</td><td>2.18</td><td>2.50</td><td>2.68</td><td>2.81</td><td>2.90</td><td>2.98</td><td>3.04</td><td>3.09</td><td>3.14</td></tr>
<tr><td>13</td><td>2.16</td><td>2.48</td><td>2.65</td><td>2.78</td><td>2.87</td><td>2.94</td><td>3.00</td><td>3.06</td><td>3.10</td></tr>
<tr><td>14</td><td>2.14</td><td>2.46</td><td>2.63</td><td>2.75</td><td>2.84</td><td>2.91</td><td>2.97</td><td>3.02</td><td>3.07</td></tr>
<tr><td>15</td><td>2.13</td><td>2.44</td><td>2.61</td><td>2.73</td><td>2.82</td><td>2.89</td><td>2.95</td><td>3.00</td><td>3.04</td></tr>
<tr><td>16</td><td>2.12</td><td>2.42</td><td>2.59</td><td>2.71</td><td>2.80</td><td>2.87</td><td>2.92</td><td>2.97</td><td>3.02</td></tr>
<tr><td>17</td><td>2.11</td><td>2.41</td><td>2.58</td><td>2.69</td><td>2.78</td><td>2.85</td><td>2.90</td><td>2.95</td><td>3.00</td></tr>
<tr><td>18</td><td>2.10</td><td>2.40</td><td>2.56</td><td>2.68</td><td>2.76</td><td>2.83</td><td>2.89</td><td>2.94</td><td>2.98</td></tr>
<tr><td>19</td><td>2.09</td><td>2.39</td><td>2.55</td><td>2.66</td><td>2.75</td><td>2.81</td><td>2.87</td><td>2.92</td><td>2.96</td></tr>
<tr><td>20</td><td>2.09</td><td>2.38</td><td>2.54</td><td>2.65</td><td>2.73</td><td>2.80</td><td>2.86</td><td>2.90</td><td>2.95</td></tr>
<tr><td>24</td><td>2.06</td><td>2.35</td><td>2.51</td><td>2.61</td><td>2.70</td><td>2.76</td><td>2.81</td><td>2.86</td><td>2.90</td></tr>
<tr><td>30</td><td>2.04</td><td>2.32</td><td>2.47</td><td>2.58</td><td>2.66</td><td>2.72</td><td>2.77</td><td>2.82</td><td>2.86</td></tr>
<tr><td>40</td><td>2.02</td><td>2.29</td><td>2.44</td><td>2.54</td><td>2.62</td><td>2.68</td><td>2.73</td><td>2.77</td><td>2.81</td></tr>
<tr><td>60</td><td>2.00</td><td>2.27</td><td>2.41</td><td>2.51</td><td>2.58</td><td>2.64</td><td>2.69</td><td>2.73</td><td>2.77</td></tr>
<tr><td>120</td><td>1.98</td><td>2.24</td><td>2.38</td><td>2.47</td><td>2.55</td><td>2.60</td><td>2.65</td><td>2.69</td><td>2.73</td></tr>
<tr><td>∞</td><td>1.96</td><td>2.21</td><td>2.35</td><td>2.44</td><td>2.51</td><td>2.57</td><td>2.61</td><td>2.65</td><td>2.69</td></tr>
</table>

[a] Reproduced with permission from C. W. Dunnett, "New Tables for Multiple Comparison with a Control," *Biometrics*, Vol. 20, No. 3, 1964, and from C. W. Dunnett, "A Multiple Comparison Procedure for Comparing Several Treatments with a Control," *Journal of the American Statistical Association*, Vol. 50, 1955.

Table XI   Control chart constants[a]

Sample size (n)	Process variation				Process average				Process standard deviation		
	D_3	D_4	B_3	B_4	A_2	A_3	A_6	A_7	d_2	c_4	d_3
2	0	3.267	0	3.267	1.880	2.659	1.880	1.880	1.128	0.7979	0.853
3	0	2.574	0	2.568	1.023	1.954	1.187	1.067	1.693	0.8862	0.888
4	0	2.282	0	2.266	0.729	1.628	0.796	0.796	2.059	0.9213	0.880
5	0	2.114	0	2.089	0.577	1.427	0.691	0.660	2.326	0.9400	0.864
6	0	2.004	0.030	1.970	0.483	1.287	0.549	0.580	2.534	0.9515	0.848
7	0.076	1.924	0.118	1.882	0.419	1.182	0.509	0.521	2.704	0.9594	0.833
8	0.136	1.864	0.185	1.815	0.373	1.099	0.434	0.477	2.847	0.9650	0.820
9	0.184	1.816	0.239	1.761	0.337	1.032	0.412	0.444	2.970	0.9693	0.808
10	0.223	1.777	0.284	1.716	0.308	0.975	0.365	0.419	3.078	0.9727	0.797
11	0.256	1.744	0.321	1.679	0.285	0.927	0.350	0.399	3.173	0.9754	0.787
12	0.283	1.717	0.354	1.646	0.266	0.886	0.317	0.382	3.258	0.9776	0.778
13	0.307	1.693	0.382	1.618	0.249	0.850	0.306	0.368	3.336	0.9794	0.770
14	0.328	1.672	0.406	1.594	0.235	0.817	0.282	0.356	3.407	0.9810	0.763
15	0.347	1.653	0.428	1.572	0.223	0.789	0.274	0.346	3.472	0.9823	0.756
16	0.363	1.637	0.448	1.552	0.212	0.763	0.257	0.337	3.532	0.9835	0.750
17	0.378	1.622	0.466	1.534	0.203	0.739	0.250	0.329	3.588	0.9845	0.744
18	0.391	1.608	0.482	1.518	0.194	0.718	0.237	0.322	3.640	0.9854	0.739
19	0.403	1.597	0.497	1.503	0.187	0.698	0.231	0.315	3.689	0.9862	0.734
20	0.415	1.585	0.510	1.490	0.180	0.680	0.218	0.308	3.735	0.9869	0.729
21	0.425	1.575	0.523	1.477	0.173	0.663	0.215	0.303	3.778	0.9876	0.724
22	0.434	1.566	0.534	1.466	0.167	0.647	0.204	0.298	3.819	0.9882	0.720
23	0.443	1.557	0.545	1.455	0.162	0.633	0.202	0.292	3.858	0.9887	0.716
24	0.451	1.548	0.555	1.445	0.157	0.619	0.192	0.288	3.895	0.9892	0.712
25	0.459	1.541	0.565	1.435	0.153	0.606	0.191	0.284	3.931	0.9896	0.708

[a] Values in this table were generated using MathCAD version 3.1 software.

Table XII Approximate critical values for the Ryan-Joiner test of normality

		α		
		.10	.05	.01
	4	.8951	.8734	.8318
	5	.9033	.8804	.8319
	6	.9114	.8893	.8409
	7	.9186	.8978	.8517
	8	.9248	.9054	.8622
	9	.9301	.9121	.8718
	10	.9347	.9179	.8804
	11	.9387	.9230	.8880
	12	.9422	.9275	.8947
	13	.9454	.9315	.9008
	14	.9481	.9351	.9061
n	15	.9506	.9383	.9109
	16	.9529	.9411	.9153
	17	.9549	.9437	.9192
	18	.9567	.9461	.9228
	19	.9584	.9483	.9260
	20	.9600	.9503	.9290
	25	.9662	.9582	.9407
	30	.9707	.9639	.9490
	40	.9767	.9715	.9597
	50	.9807	.9764	.9664
	60	.9835	.9799	.9709
	75	.9865	.9835	.9756

Source: Minitab Reference Manual.

Answers to Odd-Numbered Exercises

Chapter I

1. a.

```
 5 | 9
 6 | 3 8 8 5 3
 7 | 2 3 0 6 0 9 8 4 7 8 7    stem: ones
 8 | 1 2 7                    leaf: tenths
 9 | 0 7 7
10 | 7
11 | 6 3 8
```

A value close to 8.0 is representative. There appears to be a substantial amount of dispersion in the data.

b. There is clearly asymmetry, a skewness toward larger values (positive skewness).

c. No   **d.** .148, or roughly 15%

3.

```
3L | 1
3H | 5 6 6 7 8                      stem: tenths
4L | 0 0 0 1 1 2 2 2 2 2 3 4    leaf: hundredths
4H | 5 6 6 7 8 8 8
5L | 1 4 4
5H | 5 8
6L | 2
6H | 6 6 7 8
7L |
7H | 5
```

A specific gravity of roughly .45 is typical. The data spreads out quite a bit about this typical value. There is asymmetry in the distribution of values. The observation .75 appears at first glance to be a "mild" outlier, but this is simply a consequence of using repeated stems.

5. a. Two-digit stems. One-digit stems would give a display with too few rows to be informative, and three-digit stems would result in far too many rows.

b.

```
64 | 33 35 64 70
65 | 06 26 27 83        stem: thousands
66 | 05 14 94                and hundreds
67 | 00 13 45 70 70 90 98   leaf: tens and
68 | 50 70 73 90                ones
69 | 00 04 27 36
70 | 05 11 22 40 50 51
71 | 05 13 31 65 68 69
72 | 09 80
```

c.

```
64 | 3 3 6 7
65 | 0 2 2 8          stems: thousands
66 | 0 1 9                and hundreds
67 | 0 1 4 7 7 9 9    leaf: tens
68 | 5 7 7 9
69 | 0 0 2 3
70 | 0 1 2 4 5 5
71 | 0 1 3 6 6 6
72 | 0 8
```

The second display is essentially as informative as the first. With 200 observations, the first display would be very cumbersome.

7. a.

# Nonconforming	Frequency	Rel. freq.
0	7	.117
1	12	.200
2	13	.217
3	14	.233
4	6	.100
5	3	.050
6	3	.050
7	1	.017
8	1	.017
	60	1.001

b. .917, .867, 1 − .867 = .133

c. The histogram has a substantial positive skew. It is centered somewhere between 2 and 3 and spreads out quite a bit about its center.

9. a. .99 (99%), .71 (71%)

b. .64 (64%), .44 (44%)

c. Strictly speaking, the histogram is not unimodal, but is close to being so with a moderate positive skew. A much larger sample size would likely give a smoother picture.

11. a.

```
0 | 3 3 9 5 5 9 4 5 1 5 2 3
1 | 2 2 0 0 3 2 1 8 6 8 4   stem: thousands
2 | 1 4 1 2 3 4 4 7 7 1     leaf: hundreds
3 | 0 3 3 3 8 1 1
4 | 3 7
5 | 3 7 2 8 7
```

A typical value is one in the low 2000s; there is much variability in the data, no gaps, and the display is close to being unimodal with a positive skew.

b.

Class	Frequency	Rel. Freq.
0–<1000	12	.255
1000–<2000	11	.234
2000–<3000	10	.213
3000–<4000	7	.149
4000–<5000	2	.043
5000–<6000	5	.106
	47	1.000

.489, .149; see the description in (a)

13. a. 589/1570 = .3752

b. 1 − (589 + 190 + 176 + 157 + 115)/1570 = .2185

c. (115 + 89 + 57 + 55 + 33 = 31)/1570 = .2420

d. The shape of this histogram is positively skewed.

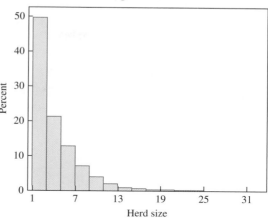

Histogram of Herd size

15. a. Yes, .518.

b. .152.

c. .408.

d. The distribution is heavily positively skewed. Though angles can range from 0° to 90°, approximately 85% of all angles are less than 30°.

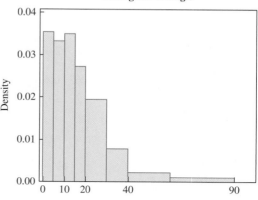

Histogram of Angle

17.

Class	Freq.	Rel. Freq.
4000–<4200	1	.01
4200–<4400	2	.02
4400–<4600	9	.09
4600–<4800	13	.13
4800–<5000	18	.18
5000–<5200	22	.22
5200–<5400	20	.20
5400–<5600	7	.07
5600–<5800	7	.07
5800–<6000	1	.01
	100	1.00

The histogram is quite symmetric and indeed approximately bell-shaped. A representative strength value is something in the neighborhood of 5000; the data spreads out rather substantially about this representative value.

19. a. $2 \times \frac{1}{2} = 1$ **b.** .5   **c.** 5   **d.** 5.8

21. a. The density curve is a triangle over the interval $[0, 10]$. Total area under curve $= \frac{1}{2}$ (base) (height) $= \frac{1}{2}(10)(.2) = 1$.

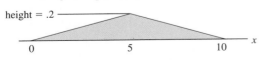

height = .2

x

0   5   10

b. Proportion $(x \le 3) = .18$, Proportion $(x \ge 7) = .18$, Proportion $(x \ge 4) = .68$, Proportion $(4 < x < 7) = .50$

c. 7.7639

23. b. .449, .699, .148   **c.** 115,129.25; 251.26

25. a. .75   **b.** .5 (50%)   **c.** .367

27. a. .70, .45   **b.** .10   **c.** .65   **d.** .45

29. x: 0   1   2
        $p(x)$: .3 .6 .1

31. a. .9625   **b.** .2912   **c.** .7881
     d. .3037   **e.** .0456   **f.** ≈ 0 **g.** ≈ 0

33. a. ≈ 1.04 or larger   **b.** $\approx -.675$ or smaller
     c. larger than 2.05 or smaller than -2.05

35. a. .8413   **b.** .9876   **c.** .0668

37. a. .9664   **b.** .2451   **c.** 45.62 km/h

39. a. .7967   **b.** .0004
     c. Those larger than .399

41. a. .8633   **b.** .8643, .8159

43. a. Proportion $(x > 120) =$ Proportion $(x \ge 120) = .9834$
     b. .0905
     c. 125.90

45. a.

0.02

0.01

0.00

100   125   150   175   200   225

b. .4602   **c.** .3636   **d.** 140.18

47. a. .0456   **b.** .8474   **c.** 6.592

49. a. .5517
     b. Proportion $(x > 200) =$ Proportion $(x \ge 200) = .1587$

51. a. .3099   **b.** $.4035 - .0678 = .3357$
     c. 90th percentile $= 1.3657$, 10th percentile $= .2501$

53. a. .3799   **b.** .1557   **c.** .5330

55. a. Proportion $(x \le 2) = .677$ (from Table II)
     b. Proportion $(x \ge 5) = .043$
     c. Proportion $(x \ge 11) = .000$

57. a. Proportion $(x \le 10) = .01$ (Table III, $\lambda = 20$)
     b. Proportion $(x > 20) = .428$
     c. Proportion $(10 \le x \le 20) = .556$, Proportion $(10 < x < 20) = .461$

59. Using Table III ($\lambda = 20$), Proportion $(x \ge 15) = .894$. Proportion $(x \le 25) = .902$.

61. a. The histogram is reasonably symmetric and bell-shaped. A representative value is about 90.
     b. Proportion $(x \ge 85) = .9231$. Proportion $(x < 95) = .9053$.
     c. $.0355 + .0414 + .1006 + .1775 + .2544/2 = .4822$

63. a. .4445   **b.** 2.107

65. a. .2946, .0708, .0222   **b.** .0348
     c. 254.3 separates the fastest 10% of all times from the slowest 90%.   **d.** The distribution is quite positively skewed.

67. a. .82   **b.** .18   **c.** .65, .27

69. b. .8647, .1353, .4712

71. b. .491, .269   **c.** 5.12
     d. 15.85 separates the largest 10% from the others.

73. b. x: 1   3   4   6   12
        $p(x)$: .30 .10 .05 .15 .40
     c. .30

75. .4423

Chapter 2

1. a. $\bar{x} = 640.5$, $\tilde{x} = 582.5$. The average sale price for a home in this sample was \$640,500. Half the sales were for less than \$582,500.
     b. Mean becomes 610.5, median is unchanged.
     c. $\bar{x}_{tr(20)} = 591.2$ (\$591,200).
     d. $\bar{x}_{tr(15)} = (591.2 + 596.3)/2 = 593.75$ (\$593,750).

3. a. 2 | 0 4 5 6 6 7 7 8
          3 | 0 1 2 3 3 4 4 6 6 6 6 7
          4 | 4 6 7 8   stem: ones
          5 | 3   leaf: tenths
          6 |
          7 |
          8 |
          9 |
         10 | 1

Due to the strong positive skew, the sample mean will be greater than the sample median.
b. $\bar{x} = 3.654$, $\tilde{x} = 3.35$
c. By any amount. By no more than 6.7.

5. Due to the unusually large observation 59.31, the sample mean will be greater than the sample median. Since the mean can be inflated when an unusually large observation exists, the median (31.28) appears to be a more representative value.

7. $\tilde{x} = 68.0$, 20% $\bar{x}_{tr} = 66.2$, 30% $\bar{x}_{tr} = 67.5$

9. a. 4/3 because of skewness
b. 1.414, so $\mu < \tilde{\mu}$ because of negative skewness.
c. .615, .707

11. $\mu = 1.614$, $\tilde{\mu} = 1.64$, .032 (a bit more than 3% of all weeks)

13. 1.8

15. a. $\bar{x} = 1939.367$, and the deviations are 66.733, 125.833, 179.533, -252.767, 27.533, and -146.867.
b. 27747.695, 166.576

17. a. Group 1 has mean = 9.86, SD = 2.67. Group 2 has mean = 8.93, SD = 2.37.
b. Group 1 has range = 7, Group 2 has range = 8.
c.

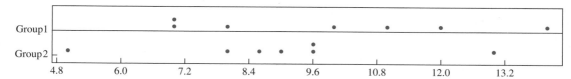

d. The standard deviation measures spread by incorporating the deviation of each observation from the sample mean. Many observations of Group 2 are clustered near its sample mean of 8.93, whereas the observations of Group 1 are farther away from its sample mean of 9.86. So, although Group 2 data exhibits a larger range, it also yields the smaller standard deviation.

19. The sample mean of 17.67 can be considered a representative value for this data. The standard deviation is 6.41. In general, the size of a typical deviation from the sample mean is about 6.41. Some observations may deviate from 17.67 by a little more than this, some by less.

21. 76,683 and 76,910

23. a. .785
b. .688

25. a. 1.72
b. .3, 0

27. .423

29. $\sigma^2 = n\pi(1 - \pi) = (25)(.20)(.80) = 4$ $\sigma = 2$
$P(x > \mu + 2\sigma) = P(x > 5 + 2(2)) = P(x > 9) = .017$

31. .135

33. a. Lower quartile = 122, upper quartile = 135, IQR = 13
b. The proximity of the upper quartile to the median suggests a negative skew. The variation seems quite large and there do not appear to be any outliers.
c. Observations less than 102.5 and greater than 154.5 would be outliers, and observations less than 83 and greater than 174 would be extreme outliers.
d. Decrease the maximum by any amount and the IQR remains unchanged.

35. $\mu = 3.51$, $\sigma = .146$

37. min = 16; lower quartile = 87; median = 140; upper quartile = 210; max = 403. A mild high outlier is above 394.5 N and an extreme high outlier is above 579 N. The value 403 N is a mild outlier. The distribution has positive skew.

39. The most noticeable feature of the comparative boxplots is that machine 2's sample values have considerably more variation than do machine 1's sample values. However, a typical value, as measured by the median, seems to be about the same for the two machines. The only outlier that exists is from machine 1.

41. The endotoxin concentration in urban homes generally exceeds that in farm homes. The range of endotoxin concentrations for urban homes exceeds that for farm homes. For the urban homes data, there is one mild outlier (1) and one extreme outlier (104). For the farm homes data there is one mild outlier (64).

43. a. $IQR = (q_u - q_l) = (133.44 - 97.43) = 36.01$
b. $IQR = (13.34 - 9.74) = 3.6$

45. The general pattern is reasonably straight and a departure from linearity is not clear-cut. One should not rule out normality of the tension distribution.

47. The plot shows some nontrivial departures from linearity, especially in the lower tail of the distribution. This indicates a normal distribution might not be a good fit to the population distribution of clubhead velocities for female golfers.

49. The corresponding probability plot appears sufficiently straight to lead us to agree with the argument that the distribution of fracture toughness in concrete specimens could well be modeled by a Weibull distribution

51. Clearly, the variable IDT is not normally distributed, since its normal quantile plot is nonlinear. IDT is likely to be lognormally distributed since the normal quantile plot of ln(ITD) is quite linear.

53. a. Clearly, the variable, hourly median power, is not normally distributed, as the normal quantile plot is curvilinear.
 b. By taking the natural logarithm of the variable and constructing a normal quantile plot, the plot looks quite linear indicating that it is plausible that these observations were sampled from a lognormal distribution.

55. The corresponding histogram shows the noise distribution is bimodal (but close to unimodal) with a positive skew and no outliers. The mean noise level is 64.89 dB and the median noise level is 64.7 dB. The IQR of the noise measurements is about $70.4 - 57.8 = 12.6$ dB.

57. b. $\bar{x}_{16} = 12.53125$, $s_{16} = .532$

59. a. The initial Se concentrations in the treatment and control groups are not that different. The median initial Se concentrations for the treatment and control groups are 10.3 mg/L and 10.5 mg/L, respectively, each with IQR of about 1.25 mg/L. So, the two groups of cows are comparable at the beginning of the study.
 b. The final Se concentrations of the two groups are extremely different. The median final Se concentration for the control group is 9.3 mg/L, the median Se concentration in the treatment group is now 103.9 mg/L, nearly a 10-fold increase.

61. a.

Percentage within	Chebyshev's Rule	Empirical Rule
1σ	No statement	About 68%
2σ	At least 75%	About 95%
3σ	At least 89%	About 99.7%

Chebyshev's inequality is more conservative than is the empirical rule.

 b.

Percentage within	Chebyshev's Rule	Exponential
1σ	No statement	86.47%
2σ	At least 75%	95.02%
3σ	At least 89%	98.17%

 c. Chebyshev's inequality may not accurately estimate any particular distribution as it must accommodate all distributions.

63. a. $\bar{x}_{tr(6.7)} = 10.67$, $\bar{x}_{tr(13.3)} = 10.58$
 b. $\bar{x}_{tr(10)} = \left(\dfrac{(10.67 + 10.58)}{2} \right) = 10.625$
 c. Interpolate between $\bar{x}_{tr(6.25)}$ and $\bar{x}_{tr(12.5)}$ to obtain $\bar{x}_{tr(10)}$

65. The mean and the midrange are sensitive to outliers. The median, the trimmed mean, and the midhinge are not sensitive to outliers.

67. a. Aortic root diameters for males have mean 3.64 cm, median 3.70 cm, standard deviation 0.269 cm, and IQR 0.40. The corresponding values for females are $\bar{x} = 3.28$ cm, $\tilde{x} = 3.15$ cm, $s = 0.478$ cm, and IQR $= 0.50$ cm. Aortic root diameters are typically smaller for females than for males, and females show more variability. The distribution for males is negatively skewed, while the distribution for females is positively skewed.
 b. For females ($n = 10$), the 10% trimmed mean is the average of the middle 8 observations: $\bar{x}_{tr(10)} = 3.24$ cm. For males ($n = 13$), the $1/13$ trimmed mean is $40.2/11 = 3.6545$, and the $2/13$ trimmed mean is $32.8/9 = 3.6444$. Interpolating, the 10% trimmed mean is $\bar{x}_{tr(10)} = 0.7(3.6545) + 0.3(3.6444) = 3.65$ cm.

69. .0228, .1587

Chapter 3

1. The scatterplot exhibits a negative linear association between the variables.

3. The scatterplot exhibits a positive linear association between the variables. One unusual observation (with # beds $= 68$) deviates from the linear pattern.

5. b. Yes
 c. There appears to be an appropriate quadratic relationship (points fall closest to a parabola).

7. The scatterplot exhibits a negative linear association between the variables.

9. a. Positive **b.** Negative **c.** Positive
 d. Little or none **e.** Negative
 f. Little or none

11. $r = .4806$, a weak to moderate linear correlation exists

13. If, for example, 18 is the minimum age of eligibility, then for most people $y \approx x - 18$.

15. $-.9$

17. a. .733 **b.** .9985

19. a. $\hat{y} = -305.88 + 9.96x$. The coefficient of determination is .124, which is quite low. The linear regression model accounts for only 12.4% of the variability of colony density.
b. $\hat{y} = 34.37 + .78x$. The coefficient of determination is .024, which is much lower than before. The linear regression model accounts for only 2.4% of the variability of colony density. The elimination of the observation has a drastic impact on the regression model.
21. a. The scatterplot reveals a roughly positive linear relationship.
b. $\hat{y} = -31.80 + .987x$. A one-MPa increase in cube strength is associated with a .987 MPa increase in the predicted axial strength for these asphalt samples.
c. $r^2 = .630$. That is, 63.0% of the observed variation in axial strength of asphalt samples of this type can be attributed to its linear relationship with cube strength.
d. $s_e = 6.625$.
23. a. $\hat{y} = 11.013 - .448x$. A one percent increase in fiber weight is associated with a .448 MPa increase in the predicted compressive strength.
b. .694
c. $\hat{y} = 8.101$ MPa
d. The observed range for x was 0 to 10%. 25% is well outside this range and the extrapolated prediction could be unreliable. For $x = 25$, $\hat{y} = -.187$ MPa, a nonsensical value.
25. a. No, if the values of C_c were perfectly linearly related to the e_0 values, then one line would exactly satisfy all points in the scatterplot.
b. $\hat{y} = -.144 + .337x$
c. .874
d. $\hat{y} = .227$ when $x = 1.10$. Predicting y when $x = .80$ would not be advisable as this is an example of extrapolation.
27. Data set #1: scatterplot yields a rough linear relationship. Data set #2: scatterplot reveals a quadratic relationship, so a linear relationship does not hold. Data set #3: scatterplot shows a clear outlier. Without this observation, a linear relationship holds very well. Data set #4: scatterplot (containing a clear outlier) shows a linear relationship does not hold.
29. a. It is not appropriate to fit a straight line to this data as there is clear curvature to the scatterplot.
b. A scatterplot of $(x, 1/y)$ yields rough linearity. The least squares line is $1/\hat{y} = .105 - 21.02x$ with corresponding $r^2 = .868$.

31. b. The $\ln(x)$ versus y transformation seems to do the best job, though it yields a somewhat low $r^2 = .497$.
c. $\hat{y} = .0197 - .0013*\ln(5000) = .0086$
33. a. No, there is a quadratic relationship between strength and thickness, so a quadratic model should be fit.
b. $\hat{y} = 14.521 + .0432x - .00006x^2$. At $x = 500$, $\hat{y} = 21.121$. The residual plot shows no unusual pattern and $R^2 = .780$. The quadratic fit seems adequate.
35. a. $\hat{y} = 4.479$. Residual $= 4.454 - 4.479 = -.0025$
b. $1 - .03836/5.1109 = .9925$.
37. a. 92.34% of the observed variability in hydrocarbon deposition can be attributed to the given multiple regression model involving x_1 and x_2.
b. $\hat{y} = 37.476$
c. Yes, it is legitimate to interpret b_2 in this way.
39. a. For $\hat{y} = a + b_1x_1 + b_2x_2 + b_3x_3$, $R^2 = .0165$. The second model gives $R^2 = .9866$. Clearly, the second model yields a superior fit to the data.
b. $\hat{y} = .3569$, residual $= -.1549$.
c. $\hat{y} = .1801$, residual $= -.0219$.
d. The larger residual magnitude based on $\hat{y} = a + b_1x_1 + b_2x_2 + b_3x_3$ is reasonable given the corresponding low coefficient of determination.
41. a. $a = 89.111$, $b_1 = -.050$, $b_2 = 6.564$, $b_3 = -27.418$, $R^2 = .9175$
b. $a = 55.703$, $b_1 = .018$, $b_2 = 8.719$, $b_3 = -11.313$, $b_4 = -.005$, $b_5 = -.033$, $b_6 = .105$, $R^2 = .9237$
c. $a = 81.233$, $b_1 = .123$, $b_2 = -6.837$, $b_3 = -42.035$, $b_4 = -.005$, $b_5 = -.033$, $b_6 = .105$, $b_7 = -.0001$, $b_8 = 1.945$, $b_9 = 10.241$, $R^2 = .9679$
43. a. .030 **b.** .120 **c.** .105 **d.** 2.80
e. 4.90
45. 9375, .302, no
47. b. 35, 5, -6 **c.** .632
49. a. $\hat{y} = 1.6932 + .0805x$
b. $\hat{y} = -20.0514 + 12.1149x$
c. .975 for both regressions
51. a. 109.07 **b.** $R^2 = .893$
c. $\hat{y} < 0$, which is ridiculous.
53. a. No
b. $\ln(y) = -7.2557 + 8328.4/x$, $\hat{y} = 74.6$, $r^2 = .953$
55. $\ln(y) = -3.7372 - .12395\ln(x)$, $r^2 = 46.9$, $y = .00829$ when $x = 5000$

Chapter 4

1. Operational definitions are used to define measurement procedures. Benchmarks are existing objects or procedures used to compare two or more products or processes.

3. Example: Temperature at 2:00 p.m. in a fixed, unshaded area on top of City Hall

5. ISP ppm is an operational definition.

7. Here is one possibility. Divide the one-square-mile area of forest into 100 smaller regions each of equal size (each area would be $(1/10)^{th}$ of a mile by $(1/10)^{th}$ of a mile). Call each region a cluster. Randomly sample n of these clusters. Within each sampled cluster study all of the trees that are growing.

9. **a.** Both methods are capable of generating random samples from the block of trees. By Researcher A's suggestion the chance a tree is selected is:
$\left(\dfrac{5}{40}\right)\left(\dfrac{6}{25}\right) = 0.03$. Each tree has this same chance of selection, making this sampling scheme a random sampling scheme. By Researcher B's suggestion the chance of selection $\left(\dfrac{30}{1000}\right) = 0.03$ is for each tree.

   b. Stratified random sampling.

11. Use $=$RANDBETWEEN $(1,1000)$ which uses sampling with replacement.

13. **a.** 10,000   **b.** 70   **c.** .0786   **d.** .0098

15. **a.** Since w_i can be shown to be equal to $\dfrac{N_i\sigma_i}{\sum\limits_{i=1}^{k} N_i\sigma_i}$,
   this yields Neyman allocation.

   b. Since w_i can be shown to be equal to
   $\dfrac{N_i}{N_1 + N_2 + \cdots + N_k} = \dfrac{N_i}{N}$, this yields proportional allocation.

17. 2.576, the required sample size increases

19. Biases tend to be eliminated when several measurements are averaged; but more importantly, the variation between repeated measurements gives a measure of experimental error.

21. **a.** Variation in fuel efficiency between 100-mile segments can be quantified if one measures fuel efficiency every segment. If one measures efficiency at the end of the 500-mile course, there is no measure of experimental error.

   b. The researcher should consider specifying those variables that may affect fuel efficiency.

Examples include: type and condition of the vehicle, tire pressure, driving speed and style, environmental conditions, etc...

   c. To draw conclusions about the effectiveness of the new fuel additive, the researcher may want to assess the effectiveness under different experimental conditions by introducing experimental factors and blocking variables. For example, the researcher may wish to determine the effect that "vehicle type" has on the response.

23. Two basic experimental design principles are violated; replication and randomization.

25. $\bar{x} = .3024$, accuracy $= (\bar{x} - x) = (.3024 - .300) = .0024$. Precision $= s = .0024083$.

27. **a.**

Measurement, m	Relative Error
.301	.333%
.303	1%
.299	−.333%
.305	1.67%
.304	1.33%

   b. The maximum absolute error you would expect in a measured reading of 70 degrees Fahrenheit from this thermometer is: $(70)(.04) = 2.8$ degree Fahrenheit

29. **b.** The Youden plot for this data shows many points near the 45 degree line, indicating that several of the laboratories are following slightly different versions of the test procedure. Lab 19 clearly made unusual measurements.

31. Suppose you take a random sample of size n with replacement. Then according to Rule 1 in Section 4.2, the complement of this random sample is also a random sample. Notice that the complement will contain no duplicates. Finally, using Rule 1 again, the complement of the complement will be a random sample and is equivalent to the original random sample but with duplicates discarded (i.e., a random sample without replacement).

33. **a.** The background samples of air would be used as a benchmark of the ambient levels of Cr(VI) in the air. Then the background samples can be compared to the plant samples in order to estimate the increase in Cr(VI) pollutant at chromite ore plants.

   b. ASTM Standard Test Method D5281-92 is the operational definition for how measurements are to be made. Using this method, the authors hope to reduce measurement variation so that any changes in Cr(VI) concentrations can be attributed to the chromite ore plants and not to variation in the measurement system.

c. The location at which an air sample is taken can be considered an experimental factor (i.e., independent variable). The six sampling periods illustrate the experimental principle of replication. Distinguishing between wet and dry days constitutes blocking.

Chapter 5

1. a. There are 10 possible such samples of size 3: {a, b, c}, {a, b, d}, {a, b, e}, {a, c, d}, {a, c, e}, {a, d, e}, {b, c, d}, {b, c, e}, {b, d, e}, {c, d, e}
   b. A = {{a, b, c}, {a, c, d}, {a, c, e}}
   c. A' = {{a, b, d}, {a, b, e}, {a, d, e}, {b, c, d}, {b, c, e}, {b, d, e}, {c, d, e}}
3. a. A and B is the event "either 4 or 5 defectives in the sample."
   b. A or B is the event "there is at least one defective in the sample."
   c. A' is the event "there are at most 3 defectives in the sample."
5.

7. The event A and B is the shaded area where A and B overlap in a Venn diagram. Its complement consists of all events that are either not in A or not in B (or not in both). That is, the complement can be expressed as A' or B'.
9. a. 1159 *distinct* joints were identified by the inspectors together.
   b. A and B' contains $724 - 316 = 408$ solder joints.
11. $P(A_1$ or A_2 or A_3 or . . . or $A_k)$
    $\leq P(A_1) + P(A_2) + P(A_3) + \cdots + P(A_k)$
    $= .01 + .01 + \cdots + .01 = 10(.01) = .10$
13. a. $P(A|E') = P(A$ and $E')/P(E') = P(A)/P(E') = .20/(1 - .10) = .20/.90 = 20/90$.
    $P(B|E') = P(B$ and $E')/P(E') = P(B)/P(E') = .25/(1 - .10) = .25/.90 = 25/90$.
    $P(C|E') = P(C$ and $E')/P(E') = P(C)/P(E') = .15/(1 - .10) = .15/.90 = 15/90$.
    $P(D|E') = P(D$ and $E')/P(E') = P(D)/P(E') = .30/(1 - .10) = .30/.90 = 30/90$.
    b. $P(A|B, D, E$ not chosen$) = P(A)/(1 - (.25 + .30 + .10)) = .20/.35 = 20/35$.
    $P(C|B, D, E$ not chosen$) = P(C)/(1 - (.25 + .30 + .10)) = .15/.35 = 15/35$.

15. a. $(.80)(.60) = .48$
    b. $.95 + (.05)(.80)(.60) = .974$
    c. $P(F|I) = P(F$ and $I)/P(I) = .95/.974 = .9754$
17. The probabilities of independent events A and B must satisfy the equation $P(A$ and $B) = P(A) \cdot P(B)$. If A and B were also mutually exclusive, then $P(A$ and $B)$ would equal 0, which would mean that $P(A) \cdot P(B) = P(A$ and $B) = 0$. But, $P(A) \cdot P(B) = .5 \cdot .6 = .3$. So, A and B cannot be mutually exclusive.
19. a. $(.42)(.42) = .1764$ b. .01, .0016, .1936
    c. $.1764 + .01 + .0016 + .1936 = .3816$
    d. $1 - (.3816) = .6184$
21. $.81 + .99 - .8019 = .9981$
23. a. .9042   b. .7660
25. Using the addition law for exclusive events, $P(B) = P(A$ and $B) + P(A'$ and $B)$, which can be rearranged as $P(A'$ and $B) = P(B) - P(A$ and $B)$. Using the fact that A and B are independent, $P(A$ and $B) = P(A) \cdot P(B)$, so, $P(A'$ and $B) = P(B) - P(A$ and $B) = P(B) - P(A) \cdot P(B) = [1 - P(A)] \cdot P(B) = P(A') \cdot P(B)$, which shows that A' and B are independent.
27. a. Discrete   b. Continuous   c. Discrete
    d. Discrete   e. Continuous   f. Continuous
    g. Discrete
29. a. 2.3   b. .81   c. 88.5 lb
31. a. $k = 1/15$ b. .40
    c. $11/3 = 3.667$ d. 1.2472
33. a. Mean = 2.85; standard deviation = 1.6797
    b. .05702   c. .77883
35. a. Binomial; mean = 50
    b. Normal approximation (with continuity correction) to binomial gives .0287.
37. a. $1 - .736 = .264$
    b. 1, because there will be no defectives in *any* sample
    c. .086 (for 5%); .624 (for 20%); .989 (for 50%)
39. a. Binomial with $n = 25$, $\pi = 1/5$
    b. Mean = 5; standard deviation = 2
    c. Closest integer score S that satisfies $P(x \geq S) = .01$ is $S = 11$.
41. a. Median = 346.57 hours
    b. Median is smaller than mean.
    c. Median = $-\ln(.50)/\lambda = .693/\lambda = .693\mu$
43. a. $P(x = 5) = .40$; $P(x = 6) = .35$; $P(x = 7) = .25$
    b. $P(y = 10) = .40$; $P(y = 15) = .40$; $P(y = 20) = .20$
    c. No, because $P(x = 5$ and $y = 10) = .20 \neq (.40)(.40) = P(x = 5)P(y = 10)$

45. a. 0, because x cannot take values between .2 and .3
  b. .36498
  c. .3544 (without continuity correction); .5098 (with continuity correction)
47. a. Mean of sampling distribution should be closer to 4.
  b. Mean of sampling distribution based on $n = 100$ will be closer to 4.
  c. Variance of sampling distribution based on $n = 2$ will be larger.
49. a. Mean = .80; standard deviation = .08
  b. Mean = .20; standard deviation = .08
  c. Mean = .80; standard deviation = .04
51. a. .6826   **b.** .9544
53. a. .0228   **b.** .0228; same as in (a)
  c. 8.8225 hours   **d.** $0.15 per package
55. a. .9803; .4803   **b.** 31.91, or $n = 32$
57. a. Sampling distribution is approximately normal with $\mu_p = .02$ and $\sigma_p = .014$
  b. .7611
59. a. $\mu_x = \exp(\mu + \sigma^2/2) = .099308$ **b.** .2643
61. a. 2/3   **b.** 7/9
63. As long as $P(A)$ and $P(B)$ are both positive, A and B cannot be independent.
65. $P(A \text{ or } B) = 1 - P(A')P(B')$ if A and B are independent.
67. a. $b = 2$ **b.** $\mu = 4/3$
  c. $\sigma^2 = 32/243$, so $\sigma = .36289$
69. a. .1396   **b.** .8604   **c.** .0099
71. a. The shape of the histogram should be symmetric and bell shaped
  b. The shape of the histogram should be positively skewed.
  c. For the uniform distribution, a sample size of 10 is sufficiently large to produce a reasonable normal sampling distribution of $\bar{x}$. However, for the exponential distribution, a sample size of 10 is not yet sufficiently large to produce a normal sampling distribution of $\bar{x}$.
73. a. 0   **b.** .0038   **c.** 6
75. For flights coming into DC: $P(1\,|\,\text{late}) = .4918$, $P(2\,|\,\text{late}) = .2459$, $P(3\,|\,\text{late}) = .2623$
  For flights coming into LA: $P(1\,|\,\text{late}) = .3125$, $P(2\,|\,\text{late}) = .375$, $P(3\,|\,\text{late}) = .3125$
77. $P(A) = .45$, $P(B) = .32$

Chapter 6

1. Tolerance $= \pm(.05)(560) = 28$ ohm, so LSL $= 532$ and USL $= 588$.
3. a. The envelope puts an upper specification limit of 4.00 inches on the width of a folded letter.

b. Possible penalties: refold letter (rework), bend letter to fit envelope (lower quality), reprint and fold new letter (scrap and rework).
5. a. Attributes data   **b.** Variables data
  c. Attributes data   **d.** Attributes data
  e. Attributes data   **f.** Variables data
  g. Attributes data   **h.** Variables data
  i. Variables data
7. Some unacceptable parts whose *true* lengths are .02 inch or less below the LSL will give measured lengths above the LSL (and will then be incorrectly classified as acceptable). Conversely, some acceptable parts whose true lengths are less than .02 inch below the USL will have measured lengths above the USL (which incorrectly classifies them as unacceptable).
9. Method 2 would be a better rational subgrouping scheme.
11. a. $P(z > 3) = .0013$ **b.** $P(z > 3.09) = .001$
13. Chart #1: Test #3 is found [Six points in a row are steadily increasing, starting with point #3.]
  Chart #2: Even though there are no tests found, Test #7 (which requires that 15 points in a row be in zone C) seems likely to occur.
  Chart #3: Test #2 is found [Nine points in a row on one side of the centerline, starting with point #2.]
  Chart #4: Both Tests #5 and 6 are found starting with point #1.
15. Centerline $\bar{R} = \left(\dfrac{85.2}{30} \right) = 2.84$

$$UCL_R = D_4\bar{R} = (2.282)(2.84) = 6.48$$

$$LCL_R = D_3\bar{R} = (0)(2.84) = 0$$

17. a. On the s chart no rules for statistical control are broken. So, we would conclude that the process variation is in statistical control.
  b. The control limits of Exercise 16(b) are based on a different formula compared to that used in 17(b). However, the control limits in both exercises are similar in values.
19. a. Centerline $= 1.2642$, UCL $= 2.4905$, LCL $= .0379$
  b. Centerline $= 96.503$, UCL $= 98.1300$, LCL $= 94.8760$
21. a. If each x_i value is transformed into $y_i = b(x_i - a)$, where a and b are constants and $b > 0$, then for any set of n values, $\bar{y} = b(\bar{x} - a)$ and $\bar{R}_y = b\bar{R}_x$. From these two relationships,

simple algebra will show that, for example, $\bar{x}$ > UCL (of the x-data) if and only if $\bar{y}$ > UCL (of the y-data). That is, the $\bar{x}$ charts based on untransformed data and transformed data will give the same signals. In the same manner, it can be shown that the R charts for both transformed and untransformed data give the same signals.

b. R chart: centerline = 4.200, UCL = 10.8108, LCL = 0
$\bar{x}$ chart: centerline = .1833, UCL = 4.4799, LCL = −4.1133
There are no "out-of-control" conditions in either chart.

c. R chart: centerline = .0042, UCL = .01081, LCL = 0
$\bar{x}$ chart: centerline = .25418, UCL = .25848, LCL = .2499
There are no "out-of-control" conditions in either chart.

23. a. .04932   **b.** $P(x$ > USL) = .0228, $P(x$ < LSL) = .1075

25. If a process is not in statistical control, then we do not have stable output and so there is no use in comparing this output to the specifications.

27. Since the C_p > 1, we know that the process has the *potential* of meeting both specifications. However, since the C_{pk} < 1, the process is not actually capable of meeting both specifications.

29. .0035

31. C_p = .785, C_{pu} = .733, C_{pl} = .837, C_{pk} = .733. The process does not have good capability.

33. a. To compute capability indexes on the transformed process data, the control chart statistics from the transformed data should be used. So, $\bar{\bar{x}}$ = .1833, $\bar{R}$ = 4.200. Also, the process specifications need to be transformed. So, USL = 10 and LSL = −10. Using these values the C_p indexes can be computed for the transformed data.

b. C_p = 1.34, C_{pu} = 1.32, C_{pl} = 1.37, C_{pk} = 1.32. The process is capable.

35. C_p = 10.789, C_{pk} = 1.842. The process is capable.

37. $9/(n + 9)$

39. a. $\bar{p} = \left(\dfrac{49}{1500}\right)$ = .032667, LCL = 0, UCL = .1081

b. There are no signs of any 'out of control' conditions and we conclude that the process is in statistical control.

41. a. .01043

b.
$$UCL = .01043 + 3\sqrt{\frac{(.01043)(.98957)}{n_i}}$$
$$LCL = .01043 - 3\sqrt{\frac{(.01043)(.98957)}{n_i}}$$

c. On day 21, the proportion of keyboards failing inspection is .0189. This value is above the upper control limit and the production process of that day should be investigated.

43. A c-chart is required in this case.
Computations are: $\bar{c} = \left(\dfrac{1,179}{25}\right)$ = 47.16
$UCL = 47.16 + 3\sqrt{47.16}$ = 67.76
$LCL = 47.16 - 3\sqrt{47.16}$ = 26.56
When analyzing the control chart we do not see any "out of control" conditions. However, the last 6 days of production have produced below average numbers of flaws and these days may need to be investigated.

45 a. $\bar{u} = \left(\dfrac{91}{15.8}\right)$ = 5.759. The general control chart formulas are:
$$UCL = 5.759 + 3\sqrt{\frac{5.759}{n_i}}$$
$$LCL = 5.759 - 3\sqrt{\frac{5.759}{n_i}}$$

b. Panels #7 and 8 seem to have significantly larger flaw rates than the process average. Test #5 (two out of three points in a row outside of 2 standard deviations) is observed.

47. a. $R(400,000) = e^{-(400,000/600,000)^4}$ = .82075

b. $R(800,000) = e^{-(800,000/600,000)^4}$ = .0424

c. $R(600,000) = e^{-(600,000/600,000)^4} = 1/e$ = .3679

d. $Z(t) = \dfrac{(\alpha/\beta^\alpha)t^{\alpha-1}exp\{-(t/\beta)^\alpha\}}{exp\{-(t/\beta)^\alpha\}} = \dfrac{\alpha}{\beta^\alpha}t^{\alpha-1}$. The failure function is increasing.

49. b. The normal failure laws have an **increasing** rate.

51. b. $R(t) = [R_1(t)]^3 = (1 - [1 - e^{-.025t}] \cdot [1 - e^{-.025t}])^3$

53. The shape of the distribution of fill volumes that pass inspection is truncated on the left, since bottles with fill volumes below the lower specification have been inspected out, resulting in the left part of the distribution being "cut-off." The histogram illustrates a normal distribution that has been truncated at the LSL.

55. As the drill wears out it may not be able to drill the hole diameters properly. On a control chart this problem will likely manifest itself as a slow trend down in the hole diameters that are being drilled. That is, the hole diameters may get smaller and smaller. Test #3 on the conditions for an "out of control" process may occur.

57. **b.** When analyzing the control chart we do not see any "out of control" conditions. We conclude that the milling process is in statistical control.

59. **a.** $C_p = 1.33$, $C_{pu} = 1$, $C_{pl} = 1.67$, $C_{pk} = 1$.

61. It can be shown that $P(x > t_1 + t_2 | x > t_1)$
$$= \frac{P(x > t_1 + t_2)}{P(x > t_1)} = e^{-t_2 \lambda} = P(x > t_2)$$

63. **a.** Since this system is connected in series, the overall reliability is $R(t) = R_1(t) \cdot R_2(t)$. Note that $R_1(t) \le 1$ and $R_2(t) \le 1$, and so $R(t) \le R_1(t)$ and $R(t) \le R_2(t)$. Thus, the overall reliability never exceeds the reliability of any of its individual components. That is, $R(t) \le \min\{R_1(t), R_2(t)\}$.

b. In the case where the components are not necessarily independent, then $R(t) = P(T > t) = P(\text{both components last longer than } t) = P(T_1 > t$ and $T_2 > t)$. Since $\{T_1 > t$ and $T_2 > t\}$ is the intersection of the two events $\{T_1 > t\}$ and $\{T_2 > t\}$, it's probability cannot exceed $P(T_1 > t)$ or $P(T_2 > t)$. That is, $R(t) \le \min[P(T_1 > t), P(T_2 > t)] = \min[R_1(t), R_2(t)]$.

Chapter 7

1. Yes, because the length x can also be thought of as a sample average based on a sample size of $n = 1$.

3. **a.** .4714
   b. .8414 ($n = 50$); .9544 ($n = 100$) approximately 1 ($n = 1000$)
   c. The probability that the sample mean lies within ± 1 unit of μ increases as the sample size n increases.

5. **a.** $\sqrt{n} = 2(1.645)$, so $n \ge 11$
   b. 80%: $\sqrt{n} = 2(1.282)$, so $n \ge 7$
   95%: $\sqrt{n} = 2(1.960)$, so $n \ge 16$
   99%: $\sqrt{n} = 2(2.576)$, so $n \ge 27$
   c. Increasing the probability that $\bar{x}$ lies within 1 unit of μ requires corresponding increases in the sample size n.

7. **a.** 99.8% **b.** 99.5% **c.** 85% **d.** 68%

9. **a.** Increased interval width
   b. Decreased interval width
   c. Increased interval width

11. **a.** Narrower **b.** No **c.** No **d.** No

13. **a.** (12.69, 14.97). We are 99% confident the average backpack weight of 6th graders is between 12.69 and 14.97 pounds
   b. (13.26, 16.25).
   c. The average backpack weight as a percentage of body weight of 6th graders seems well above the recommendation as 10% is well outside the interval (13.26, 16.25).

15. **a.** (1398.90, 1455.10). We are 95% confident that the true average FEV_1 level for the given population is between 1398.90 and 1455.10 ml.
   b. 158

17. $s \pm (z \text{ critical value}) \left(\dfrac{s}{\sqrt{2n}} \right) = (3.332, 4.128)$

19. 390.74 min

21. 4.062 kip

23. **a.** (.50, .56). We are 99% confident the proportion of all adult Americans who have watched streamed programming is between 50 and 56%.
   b. 664

25. **a.** .042
   b. If we were to sample repeatedly, the calculation method in (a) is such that π will exceed the calculated lower confidence bound for 95% of all possible random samples of $n = 143$ individuals who received ceramic hips.

27. **a.** $p_1 - p_2$
   $$\pm (z \text{ critical value}) \sqrt{\frac{p_1(1 - p_1)}{n_1} + \frac{p_2(1 - p_2)}{n_2}}$$
   b. (−.118, .136), no
   c. (−.118, .135)

29. **a.** $(A, B) = \ln(p_1/p_2)$
   $$\pm (z \text{ critical value}) \sqrt{\frac{n_1 - u}{n_1 u} + \frac{n_2 - v}{n_2 v}}, \, (e^A, e^B)$$
   b. (.970, 1.349), yes

31. 271

33. (.012, .056), using a 95% confidence level

35. 4.3, no

37. **a.** 2.228 **b.** 2.086 **c.** 2.845
   d. 2.680 **e.** 2.485 **f.** 2.571

39. **a.** 1.812 **b.** 1.753 **c.** 2.602 **d.** 3.747
   e. 2.1716 (from Minitab) **f.** Roughly 2.43

41. **a.** Yes, a normal quantile plot shows a somewhat linear relationship.
   b. (106.4, 109.1). Based on this interval, 107 is a plausible value but 110 is not plausible for the true average work of adhesion.

43. a. We are 95% confident that the true average mileage is between 46,145.4 and 86,296.8.
b. We are 95% confident that the mileage for a single vehicle is between 0 and 148,995.4. This interval is much wider than the interval from part (a).

45. a. Using a normal probability plot, we ascertain that it is plausible that this sample was taken from a normal population distribution.
c. 38.78
d. 42.29, a higher upper bound than that found in part (c).

47. a. A 95% prediction interval for the amount of warpage of a single piece of laminate is .0635 $\pm$ .0137
b. (.0464, .0806)

49. (3.43, 4.13). Thus, with 95% confidence, we can say that the true average firmness for zero-day apples exceeds that of 20-day apples by between 3.43 and 4.13 N.

51. a. The most notable feature of these boxplots is the larger amount of variation present in the mid-range data as compared to the high-range data. Otherwise, both boxplots look reasonably symmetric and there are no outliers present.
b. A 95% confidence interval for (μ mid range $-$ μ high range) is (-7.84, 9.54). Since plausible values for ($\mu_1 - \mu_2$) are both positive and negative (i.e., the interval spans zero), we would conclude that there is not sufficient evidence to suggest that μ_1 and μ_2 differ.

53. Assuming sample "1" corresponds to the lab method, the CI says we're 95% confident that the true mean arsenic concentration measurement using the lab method is between 6.498 μg/L and 11.102 μg/L higher than using the field method.

55. a. A 95% confidence interval for μ_d is (-14.83, 26.50). Since this interval contains negative and positive values, there is not sufficient evidence to suggest that μ_d is different from zero.
b. A 95% prediction interval for the difference d is (-48.85, 60.51).

57. a. (2.03, 6.10). We are 95% confident that the true mean difference between dominant and nondominant arm translation for pitchers is between 2.03 and 6.10.
b. (-0.54, 1.01). We are 95% confident that the true mean difference between dominant and nondominant arm translation for position players is between $-.54$ and 1.01.

c. Let μ_1 and μ_2 represent the true mean differences in side-to-side AP translation for pitchers and position players, respectively. To generate a confidence interval for $\mu_1 - \mu_2$, we use the *differences* utilized in parts (a) and (b). A 95% confidence interval for $\mu_1 - \mu_2$ is (1.69, 5.98). Since both endpoints are positive, we concur with the authors' assessment that this difference is greater, on average, in pitchers than in position players.

59. a. (-3.85, 11.35)
b. (7.02, 10.06)

61. a. 95% bootstrap interval is (431.82, 445.65) based on 200 bootstrap replications. (Note that all bootstrap intervals will differ slightly from one another.)
b. t interval: (430.51, 446.08); bootstrap interval: (431.82, 445.65)

63. a. MLE for π is x/n.
b. x/n is an unbiased estimator of π.
c. MLE for $(1 - \pi)^5$ is $(1 - x/n)^5$.

65. a. MLE is $\bar{x} + 11.645(s^*)$, where s^* equals
$$s\sqrt{\frac{n-1}{n}},$$
where s is the sample standard deviation of the data.
b. 403.3

67. a. $\hat{\theta} = \min(x_1, x_2, \ldots, x_n)$; $\hat{\lambda} = 1/(\bar{x} - \hat{\theta})$
b. $\hat{\theta} = .64$; $\hat{\lambda} = 1/(5.58 - .64) = .202$

69. $\lambda = 2$ is too large; the resulting kernel density will not show much detail in the data.

71. a. The kernel density graph will have a very choppy appearance.
b. Larger values of λ will result in smoother kernel density curves.

73. λ will have to be raised.

75. a. $\mu > 134.78$
b. Tensile strengths should be normally distributed.
c. A histogram of the data appears approximately bell-shaped, so the normality assumption is a good one for this data.
d. $\mu > 127.81$

77. (-299.3, 1517.9)

79. (1024.0, 1336.0), yes

81. a. A normal probability plot shows it is reasonable to assume the sample was taken from a normal population distribution.
b. Letting d = peak ER velocity–peack IR velocity, a 95% confidence interval for μ_d is (34.1, 130.9). Since both endpoints are positive, we conclude that IR and ER differ significantly, with ER being the higher of the two.

83. a. .5   b. .25   c. $(.5)^n$ d. $(.5)^n$
   e. $1 - 2(.5)^n$, $100[1 - 2(.5)^n]$
   f. (28.7, 42.0), 99.8%
   g. (28.62, 40.28), narrower than (f)
85. a. 1/2   b. 1/3   c. $1/(n + 1)$, $1/(n + 1)$
   d. $1 - 2/(n + 1)$, $100[1 - 2/(n + 1)]\%$,
   (28.7, 42.0), 81.8%
87. No, (69.80, 88.80), 99.97%
89. a. (38.46, 38.84)
   b. (Answers will vary): For a simulation programmed in R using 1000 bootstrapped means,
   a 95% bootstrap interval for the population
   mean was (38.51, 38.81). This interval agrees
   closely with the interval from part (a).
91. a. (.296, .324)
   b. Since the interval value dips below 30%, we
   cannot conclude that the 2002 percentage is
   more than 1.5 times the 1998 percentage.

Chapter 8

1. a. Yes   b. No   c. Yes   d. No
   e. Yes   f. Yes   g. Yes   h. Yes
   i. Yes   j. Yes
3. $H_0: \mu = 40$ versus $H_a: \mu \neq 40$
5. $H_0: \mu = 120$ versus $H_a: \mu < 120$. Type I: Conclude that the new system does reduce average
   distance when in fact it does not. Type II: Conclude that the new system does not reduce average
   distance when in fact it does.
7. With μ_1 for regular and μ_2 for special, $H_0: \mu_1 - \mu_2 = 0$ versus $H_a: \mu_1 - \mu_2 > 0$. Type I: Conclude that
   the special outperforms the regular laminate when
   this is not the case. Type II: Conclude that the
   regular laminate is at least as good as the special
   laminate when in fact the special does yield an
   improvement.
9. a. Reject H_0.   b. Reject H_0.
   c. Don't reject H_0.   d. Reject H_0.
   e. Do not reject H_0.
11. a. 1.83, .0336   b. 4.22, approximately 0
   c. 1.33, .0918
13. a. $H_0: \mu = .85$ versus $H_a: \mu \neq .85$
   b. Don't reject H_0, because P-value $> \alpha$. Same
   conclusion and reason.
15. $H_0: \mu = 55$ versus $H_a: \mu \neq 55$, $z = -5.25$,
   P-value ≈ 0, reject H_0
17. a. Using software, $\bar{x} = 0.75$, $\tilde{x} = 0.64$, $s = .3025$,
   $IQR = .505$. These summary statistics, as well
   as a boxplot (not shown) indicate substantial
   positive skewness, but no outliers.

b. No, it is not plausible from the results in part a
   that the variable ALD is normal. However, since
   $n = 49$, normality is not required for the use of
   z inference procedures.
c. $H_0: \mu = 1.0$ versus $H_a: \mu < 1.0$. $z = -5.79$; at
   any reasonable significance level, we reject the
   null hypothesis. Yes, the data provides strong evidence that the true average ALD is less than 1.0.
19. a. P-value $= P(t > 3.2) = .003$, reject H_0.
   b. P-value $= P(t > 1.8) = .055$, do not reject H_0.
   c. P-value $= P(t > -.2) = .578$, do not reject H_0.
21. a. P-value $= 2P(t > 1.6) = 2(.068) = .136$, do not
   reject H_0.
   b. P-value $= 2P(t < -1.6) = 2(.068) = .136$, do
   not reject H_0.
   c. P-value $= 2P(t < -2.6) = 2(.008) = .016$, do
   not reject H_0.
   d. P-value $= 2P(t < -3.9) \approx 2(0) \approx 0$, reject H_0.
23. $H_0: \mu = 30$ versus $H_a: \mu < 30$. $t = 0.84$, P-value $=$
   .209, do not reject H_0.
25. $H_0: \mu = 181$ versus $H_a: \mu > 181$. $t = 1.91$,
   P-value $= .041$, reject H_0.
27. a. 17   b. 21   c. 18   d. 26
29. $H_0: (\mu_1 - \mu_2) = 0$ versus $H_a: (\mu_1 - \mu_2) < 0$.
   $t = -2.46 \approx -2.5$, df $= 15$, P-value $= .012$.
   Do not reject H_0.
31. a. Normal quantile plots show sufficient linearity
   for each data set. Therefore, it is plausible that
   both samples have been selected from normal
   population distributions.
   b. The comparative boxplot does not suggest a
   difference between average extensibility for the
   two types of fabrics.
   c. $H_0: (\mu_H - \mu_P) = 0$ versus $H_a: (\mu_H - \mu_P) \neq 0$.
   $t = -.38$, df $= 10$, P-value $= .71$. Do not reject
   H_0.
33. $H_0: (\mu_1 - \mu_2) = 0$ versus $H_a: (\mu_1 - \mu_2) > 0$. When
   assuming unequal variances, $t = 3.6362$, the
   corresponding df is 37.5, and the P-value for our
   upper-tailed test would be $[(.0008)/2] = .0004$.
   (Note: P-value $= P(t > 3.6362) = .0004$) Reject
   H_0. We could have committed a Type I error.
35. $H_0: (\mu_H - \mu_{NH}) = 0$ versus $H_a: (\mu_H - \mu_{NH}) > 0$.
   $t = 2.09$, df $= 17$, P-value $= .026$. Do not reject
   H_0.
37. a. Use $t = \dfrac{(\bar{x}_1 - \bar{x}_2) - \Delta}{s_p\sqrt{\dfrac{1}{n_1} + \dfrac{1}{n_2}}}$ with corresponding

   df $= (n_1 + n_2 - 2)$. s_p is defined in Exercise 54
   in Chapter 7.

b. $t = 3.6362$, df $= (n_1 + n_2 - 2) = (20 + 20 - 2) = 38$, P-value $= P(t > 3.6362) = (.0008/2) = .0004$. Reject H_0.

c. $t = 2.01074$, df $= (n_1 + n_2 - 2) = (9 + 18 - 2) = 25$, P-value $= P(t > 2.01074) = .0276$. Do not reject H_0.

39. a. For MSD, $H_0: \mu_d = 0$ versus $H_a: \mu_d \neq 0$. $t = .85$, df $= 20$, P-value $= .408$. Do not reject H_0.

b. For RULA, $H_0: \mu_d = 0$ versus $H_a: \mu_d \neq 0$. $t = -4.47$, df $= 20$, P-value $< .001$. Reject H_0.

c. Measurements were taken before and after intervention. The intervention in the form of a short oral presentation would most likely not lead to instant reductions in musculoskeletal disorders (MSD). However, such an intervention could cause an immediate change in one's posture, and therefore have a major impact on one's RULA score.

41. $H_0: \mu_d = 0$ versus $H_a: \mu_d > 0$. $t = 2.68$, df $= 12$, P-value $= .01$. Reject H_0.

43. First, compute the percent change, (measured–stated)/stated, for each meal. Let μ denote the true average percentage change for all supermarket convenience meals. $H_0: \mu = 0$ versus $H_a: \mu \neq 0$. $t = 3.90$, df $= 9$, P-value $= .004$. Reject H_0.

45. $H_0: \pi_1 = .40 \quad \pi_2 = .30 \quad \pi_3 = .20 \quad \pi_4 = .10$ versus H_a: The Statistics Department's expectations are not correct.
$\chi^2 = 1.57$, df $= 3$, P-value $> .10$. Do not reject H_0.

47. $H_0: \pi_1 = .177, \pi_2 = .032, \pi_3 = .734, \pi_4 = .057$. The alternative hypothesis is that at least one of these proportions is incorrect. $\chi^2 = 19.6$, P-value $< .001$. Reject H_0.

49. $H_0: \pi_1 = 3/9, \pi_2 = 4/9, \pi_3 = 2/9$. The alternative hypothesis is that at least one of these proportions is incorrect. $\chi^2 = .6875$, P-value $> .10$. Do not reject H_0.

51. This is a χ^2 test of the homogeneity of several proportions. The hypotheses to test are:
H_0: the groups are homogeneous with respect to side effects versus
H_a: the groups are not homogeneous with respect to side effects
Counseling Group: 24 had at least one side effect, 31 had none. No-Counseling Group: 8 had at least one side effect, 44 had none. $\chi^2 = 10.177$, P-value $= .0014$. Reject H_0.

53. This is a χ^2 test of the homogeneity of several proportions. The hypotheses to test are:
H_0: the genders are homogeneous with respect to neck pain versus
H_a: the genders are not homogeneous with respect to neck pain
$\chi^2 = 142.1$, P-value $< .001$. Reject H_0.

55. a. The four expected counts are 1.56, 2.44, 32.44, 50.56. Two of the cells have expected counts less than 5.

b. Fisher's exact test P-value $= .02083$. We conclude at $\alpha = 5\%$ that surgery method affects the provision of ondanestron.

57. a. i. Since $r^* > .9347$, P-value $> .10$
ii. Since $.8804 < r^* < .9180$, $.01 < $ P-value $< .05$
iii. Since $r^* > .9662$, P-value $> .10$
iv. Since $r^* < .9408$, P-value $< .01$

b. i. Fail to reject H_0, since P-value $> .05$
ii. Reject H_0, since P-value $< .05$
iii. Fail to reject H_0, since P-value $> .05$
iv. Reject H_0, since P-value $< .05$

59. The Ryan-Joiner test P-value is larger than .10, so we conclude that this data could reasonably have come from a normal population. We can safely use a one-sample t-test to test hypotheses about the value of the true average compressive strength.

61. a. The Ryan-Joiner test P-value is less than .01, so it is implausible that this data came from a normal population. In particular, the observation 65 is a clear outlier.

b. The Ryan-Joiner test P-value is larger than .10, so we conclude that the data without the outlier could reasonably have come from a normal population.

63. $\hat{\lambda} = .0742$, combine the last six intervals to obtain $X^2 = 1.34$ based on 2 df, so a truncated exponential distribution is plausible.

65. a. .2514, .0918, .0038, .0004
b. .9515, .8413, .3669, .1587

67. Two-sided test of $H_0: \mu = 100$ versus $H_a: \mu \neq 100$ at $\alpha = .01$ with $n = 15$ is proposed. σ is thought to be between .8 and 1. The first two printouts show that power of detecting shifts of $.5\sigma$ or $.8\sigma$ will be very low. Last printout shows that power can be increased to 90% by increasing sample size to 42 (for a $.5\sigma$ shift) and 19 (for a $.8\sigma$ shift).

69. P-value $= .056$, so at significance level .05, H_0 cannot be rejected.

71. a. The corresponding probability plot suggests the data is consistent with a normally distributed population. So, we are comfortable proceeding with the t procedure.

b. $H_0: \mu = 0.6$ versus $H_a: \mu < 0.6$, $t = -2.14$, P-value $= .0495$, reject H_0 when $\alpha = 5\%$, do not reject H_0 when $\alpha = 1\%$.

c. In this context, a Type I error would be to conclude that less than 10% of the tube's contents remain after squeezing, on average, when in fact 10% (or more) actually remains. When we rejected H_0 at the 5% level, we may have committed a Type I error. A Type II error occurs if we fail to recognize that less than 10% of a tube's contents remains, on average, when that's actually true (i.e., we fail to reject the false null hypothesis of $\mu = 0.6$ oz). When we failed to reject H_0 at the 1% level, we may have committed a Type II error.

73. a. Since the mean, $\bar{x} = 215$, is so much lower than the midrange (about 585), one would suspect the distribution is positively skewed. However, it is not necessary to assume normality if the sample size is "large enough," due to the central limit theorem. Since $n = 47$, we can proceed with a test of hypothesis about the true mean consumption.

b. $H_0: \mu = 200$ versus $H_a: \mu > 200$, $z = .44$, P-value $= .33$, do not reject H_0.

75. $H_0: \mu = 1.75$ versus $H_a: \mu \neq 1.75$, $t = 1.70$, P-value $= .102$, do not reject H_0.

77. $H_0: \pi = 1/3$ versus $H_a: \pi < 1/3$, $z = -1.35$, P-value $= .0885$, do not reject H_0.

79. $H_0: (\mu_1 - \mu_2) = 0$ versus $H_a: (\mu_1 - \mu_2) \neq 0$, $z = 2.25$, P-value $= .0244$, reject H_0.

81. a. $H_0: (\mu_1 - \mu_2) = 0$ versus $H_a: (\mu_1 - \mu_2) \neq 0$. Using the unpooled t-test statistic we have $t = 2.84$ and df $= 18$. This results in a P-value $= 2[P(t > 2.8)] = 2(.006) = .012$. These values differ slightly from $t = 2.51$ and P-value $= .019$.

b. $H_0: (\mu_1 - \mu_2) = 25$ versus $H_a: (\mu_1 - \mu_2) > 25$, the unpooled t-test statistic value is $.556$, P-value $= .278$, do not reject H_0.

83. a. $H_0: \mu_{37,dry} - \mu_{22,dry} = 100$ v. $H_a: \mu_{37,dry} - \mu_{22,dry} > 100$. The relevant test statistic value is $t = 2.58$, df $= 9$, P-value $= .015$, reject H_0 when $\alpha = 5\%$.

b. $H_0: \mu_{22,wet} - \mu_{37,wet} = 100$ v. $H_a: \mu_{22,wet} - \mu_{37,wet} > 50$. The relevant test statistic value is $t = .46$, df $= 9$, P-value $= .328$, do not reject H_0.

85. $H_0: \mu_d = 0$ versus $H_a: \mu_d > 0$, $\bar{d} = .821$, $s_d = 2.52$, $t = 1.22$, P-value $= .126$, do not reject H_0.

87. $H_0: \mu_1 - \mu_2 = 10$ versus $H_a: \mu_1 - \mu_2 > 10$, $t = 2.49$, df $= 5$, P-value $= .027$, reject H_0 when $\alpha = 5\%$.

89. a. $H_0: \pi_1 = .27477$, $\pi_2 = .20834$, $\pi_3 = .15429$, $\pi_4 = .3626$. The alternative hypothesis is that at least one of these proportions is incorrect. $\chi^2 = 9.02$, $.01 < P$-value $< .05$. Reject H_0 when $\alpha = 5\%$. Thus, the above model is questionable.

b. $H_0: \pi_1 = .45883$, $\pi_2 = .18813$, $\pi_3 = .11032$, $\pi_4 = .24272$. The alternative hypothesis is that at least one of these proportions is incorrect. $\chi^2 = .157$, P-value $> .10$. Do not reject H_0. Thus, the proposed model appears to fit the data quite well.

Chapter 9

1. a. $H_0: \mu_A = \mu_B = \mu_C$; where $\mu_i =$ average strength of wood of Type i.

b. Use either Type A or B, but choose the less expensive of the two types.

c. Choose the least expensive of the three types.

3. The two ANOVA tests will give *identical* conclusions.

5. There is no way of knowing whether there is a statistically significant difference between the means. When there is no difference, the "pick the winner" strategy doesn't allow you to choose between the population based on other criteria (e.g., cost, time, etc.).

7. $F_{.05}(5, 8) \neq F_{.05}(8, 5)$ and $F_{.01}(5, 8) \neq F_{.01}(8, 5)$

9. $F = 4.12$ exceeds $F_{.05}(3, 30)$, so conclude that there is a difference among the means.

11. a.

Source	df	SS	MS	F
Treatments	5	3575.065	715.013	51.3
Error	150	2089.350	13.929	
Total	155	5664.415		

b. $H_0: \mu_1 = \cdots = \mu_6$ versus H_a: at least two of the μ_i's are different.

c. P-value is $P(F_{5,150} \geq 51.3) \approx 0$, reject H_0

13. a. $H_0: \mu_1 = \cdots = \mu_5$ versus H_a: at least two of the μ_i's are different.

b. $F = 4.14$, using software P-value $= .0061$, reject H_0

15. $H_0: \mu_1 = \mu_2 = \mu_3 = \mu_4$ versus H_a: at least two of the μ_i's are different. $F = 2.31$, P-value $> .10$, do not reject H_0

17. a. SST, SSTr, and SSE are each multiplied by a factor of $(2.54)^2$, but the F ratio does *not* change.

b. Changing the units of measurement will change the sum of squares column in the ANOVA table, but the degrees of freedom and F ratio will remain unchanged.

19. SSTr = .2164; SSE = .3870; F ratio = 4.194 is significant at 5% level. There *is* evidence of a difference between the means.

21. a. SSTr = 25.80; SSE = 115.48; SST = 141.28; F ratio = 2.35 is not significant at 5% level. There is no evidence of a difference between the means.

b. Favorable, because pegs have same strength regardless of positioning

23. a. The ANOVA table entries will be unchanged by the calibration error.

b. If all data points are shifted (up or down) by the same amount δ, this will not affect any of the entries in the ANOVA table; however, the means of each sample *will* shift by an amount equal to δ.

25. An effects plot does not show the within-samples variance, only the between-groups variation.

27. $q_{.05}(5, 15) = 4.37$, so $T = 36.09$.

$$\underline{437.5 \quad 462.0 \quad 469.3} \quad \underline{512.8 \quad 532.1}$$

29. $T = 36.09$ as in Problem 27.

$$\underline{427.5 \qquad 462.0} \qquad 469.3 \qquad 502.8 \qquad 532.1$$

31. $q_{.05}(6,150) \approx q_{.05}(6,120) = 4.10$, $T = 3.00$.

$$14.18 \quad \underline{17.94 \quad 18.00 \quad 18.00} \quad \underline{25.74 \quad 27.67}$$

33. $q_{.05}(3, 6) = 4.34$, $T = 7.92$. There are 2 distinct sets: Set 1 (42.67, 43.33), Set 2 (53.67).

$$\underline{42.67 \quad 43.33} \quad 53.67$$

35. a. F ratio for Brands = 95.57 is significant at α = 1%. There is a difference between the brands.

b. F ratio for Humidity = 278.20 is significant at α = 1%. Humidity levels do affect power consumption, so it was wise to use humidity as a blocking factor.

37. a. F ratio for Brand = 8.96 is significant at α = 5%. There is a difference among lathe brands.

b. F ratio for Operators = 10.78 is significant at α = 5%. There is a difference among operators.

39. a.

	Df	Sum Sq	Mean Sq	F-value	Pr(F)
DESIGN	3	519515	173171.67	35.46	<.0001
PERSON	20	100460	5023.00	1.03	0.445
Residuals	60	293009	4883.48		

b. Yes, P-value <.0001.

c. Corresponding F ratio = 1.03 with P-value = .445. The person-to-person differences in RPN are not confirmed by the data.

41. a. F ratio for Methods = 8.69 is significant at α = 5%. Curing methods do have differing effects on strength.

b. F ratio for Batches = 7.22 is significant at α = 5%. Different batches do have an effect on strength.

c. F ratio for Methods = 2.83, which is not significant at α = .05. Conclusion: Curing method does not have an effect on strength.

43. a. H_0: $\mu_1 = \mu_2 = \mu_3$ versus H_a: at least two of the population means differ

b.

Source	df	SS	MS	F
Factor	2	591.2	295.6	1.3
Error	21	4773.3	227.3	
Total	23	5364.5		

c. Corresponding P-value > .10. Do not reject H_0.

45. a. $F_{.05}(1, 10) = 4.96$ and $t_{.025}(10) = 2.228$. $(2.228)^2 \approx 4.96$; the equality is approximate because the F and t table entries are rounded.

b. $F_{.05}(1, df_2)$ approaches 3.8416, the square of 1.96.

47. MSTr = 140, so if F ratio > $F_{.05}(2, 12) = 3.89$, then MSE < 140/3.89 = 35.99. $q_{.05}(3, 12) = 3.77$, so to have $T > 10$ (i.e., largest mean − smallest mean), MSE must exceed $(5)(10/3.77)^2 = 35.18$. Therefore, if 35.18 < MSE < 35.99, then the two conditions will be satisfied. In terms of SSE, 422.16 < SSE < 431.88.

49. For condition 1 to be satisfied it can be shown that SSE < 385.60. For condition 2 to be satisfied it can be shown that SSE > 422.15. Therefore, no SSE value exists that can satisfy both conditions.

51. a.

Source	df	SS	MS	F	P-value
Drying method	4	14.962	3.741	36.70	0.000
Fabric type	8	9.696	1.212	11.89	0.000
Error	32	3.262	0.102		
Total	44	27.920			

b. The null hypothesis of interest is H_0: there are no differences in mean smoothness scores for the five drying methods. The F-ratio for "drying method" is $F = 36.7$, P-value < .001. reject H_0.

Chapter 10

1. Replication allows you to obtain an estimate of the experimental error.

3. a. Surface is a dome over the x–y plane

b. Maximum occurs at $x = 2$, $y = 5$.

c. Contours are circles centered at $(x, y) = (2, 5)$.

5. When the lines in the AB interaction plot are parallel, the effect of changing factor A (factor B) from one level to another will be the *same* for each level of factor B (factor A).

7.

Source	DF	SS	MS	F
Factor A	4	20	5	2.5
Factor B	4	64.8	16.2	8.1
Interaction	16	15.2	.95	.475
Error	50	100	2	
Total	74	200		

9. **a.** F ratio for Interaction is 1.545, which does not exceed $F_{.05}(2, 12) = 3.89$, so there is no evidence of interaction between the factors.
   b. F ratio for Formulation = 376.25, F ratio for Speed = 19.269. Both factors have an effect on yield.

11. **a.** F ratio for the Interaction = 1.10, which is not significant at $\alpha = .05$.
   b. F ratio for the Aggregate Content = 56.06, which is significant at $\alpha = .05$.
   c. F ratio for the Asphalt Grade = 14.12, which is significant at $\alpha = .05$.

13 **b.** At the .01 level, there is not a statistically significant interaction between adhesive and condition's effects on shear bond strength. Ignoring the interaction effect, condition (dry/moist) is not statistically significant, while adhesive (OBP/SBP) is highly statistically significant.
   c. Using Tukey's procedure for a one–way ANOVA using the four groups (OBP–D, OBP–M, SBP–D, SBP–M) yields the following result:

OBP–D	OBP–M	SBP–M	SBP–D
39.9	46.1	50.8	53.0

15. **a.** Software gives the following table:

Source	DF	SS	MS	F	P-value
A	2	210.67	105.33	0.53	0.60
B	2	132.17	66.09	0.33	0.72
C	2	2586.35	1293.18	6.45	0.01
AB	4	57.48	14.37	0.07	0.99
AC	4	636.84	159.21	0.79	0.54
BC	4	875.00	218.75	1.09	0.38
ABC	8	888.52	111.06	0.55	0.81
Error	27	5416.67	200.62		
Total	53	10803.70			

   b. There are no significant interaction effects.
   c. The only significant main effect is for C (quill gap) having F ratio = 6.45.

17. **a.**

Source	DF	SS	MS	F
A	2	14,144.44	7,072.22	61.06
B	2	5,511.27	2,755.64	23.79
C	2	244,696.39	2,348.20	1,056.27
AB	4	1,069.62	267.20	2.31
AC	4	62.67	15.67	.14
BC	4	311.67	82.92	.72
ABC	8	1,080.77	135.10	1.17
Error	27	3,127.50	115.83	
Total	53	270,024.33		

   b. All F ratios for interaction terms are smaller than corresponding tabled $F_{.05}$ values.
   c. All three main effects are significant.

19. **a.**

Source	DF	SS	MS	F
A	2	12.896	6.448	1.04
B	1	100.042	100.042	16.10
C	3	393.417	131.139	21.10
AB	2	1.646	.823	.13
AC	6	71.021	11.837	1.90
BC	3	1.542	.514	.08
ABC	6	9.771	1.628	.26
Error	72	477.50	6.215	
Total	95	1,037.833		

   b. Main effects for factors B and C are significant.
   c. None of the interaction terms are significant.

21. **a.** Software gives the following table:

Source	DF	SS	MS	F	P-value
A	2	124.60	62.30	4.85	0.04
B	2	20.61	10.30	0.80	0.48
C	2	356.95	178.47	13.89	0.00
AB	4	57.49	14.37	1.12	0.41
AC	4	61.39	15.35	1.19	0.38
BC	4	11.06	2.76	0.22	0.92
Error	8	102.78	12.85		
Total	26	734.88			

   b. The appropriate F ratios for the AB, AC, and BC interactions are 1.12, 1.19, and .22, respectively. These F ratios are all not statistically significant at $\alpha = .05$.
   c. The main effects for A (paste thickness) and for C (laser power) have corresponding F ratios 4.85 and 13.98 respectively. These F ratios are all statistically significant at $\alpha = .05$.

23. a.

Source	df	SS	MS	F	P-value
A	2	34436	17218	436.92	0.000
B	2	105793	52897	1342.3	0.000
C	2	516398	258199	6552.04	0.000
AB	4	6868	1717	43.57	0.000
AC	4	10922	2731	69.29	0.000
BC	4	10178	2545	64.57	0.000
ABC	8	6713	839	21.3	0.000
Error	27	1064	39		
Total	53	692372			

b. The appropriate F ratios for the AB, AC, BC, and ABC interactions are 43.57, 69.29, 64.57, and 21.3, respectively. These F ratios are all statistically significant at $\alpha = .01$.

c. The appropriate F ratios for the A, B, and C main effects are 436.92, 1342.3 and 6552.04, respectively. These F ratios are all statistically significant at $\alpha = .01$.

25.

AB	AC	AD	BD	CD	ABD	ACD	BCD	ABCD
1	1	1	1	1	−1	−1	−1	1
−1	−1	−1	1	1	1	1	−1	−1
−1	1	1	−1	1	1	−1	1	−1
1	−1	−1	−1	1	−1	1	1	1
1	−1	1	1	−1	−1	1	1	−1
−1	1	−1	1	−1	1	−1	1	1
−1	−1	1	−1	−1	1	1	−1	1
1	1	−1	−1	−1	−1	−1	−1	−1
1	1	−1	−1	−1	1	1	1	−1
−1	−1	1	−1	−1	−1	−1	1	1
−1	1	−1	1	−1	−1	1	−1	1
1	−1	1	1	−1	1	−1	−1	−1
1	−1	−1	−1	1	1	−1	−1	1
−1	1	1	−1	1	−1	1	−1	−1
−1	−1	−1	1	1	−1	−1	1	−1
1	1	1	1	1	1	1	1	1

27. a.

Source	df	SS	MS	F	P-value
A	1	1685.1	1685.1	102.38	0.000
B	1	21272.2	21272.2	1292.36	0.000
C	1	5076.6	5076.6	308.42	0.000
AB	1	36.6	36.6	2.22	0.174
AC	1	0.4	0.4	0.03	0.877
BC	1	109.2	109.2	6.63	0.033
ABC	1	23.5	23.5	1.43	0.266
Error	8	131.7	16.5		
Total	15	28335.3			

b. At $\alpha = .01$, all three main effects are important, since each of their P-values is less than .001. No significant interaction effects exist, when testing at $\alpha = .01$.

29. a. Let A = storage time, B = storage temp, C = packaging type.

Term	Effect
A	−0.03125
B	−0.24625
C	−0.21125
AB	−0.03125
AC	0.02375
BC	−0.21125
ABC	0.02375

b.

Source	DF	SS	MS	F	P-value
A	1	0.003906	0.003906	25.00	0.001
B	1	0.242556	0.242556	1552.36	0.000
C	1	0.178506	0.178506	1142.44	0.000
AB	1	0.003906	0.003906	25.00	0.001
AC	1	0.002256	0.002256	14.44	0.005
BC	1	0.178506	0.178506	1142.44	0.000
ABC	1	0.002256	0.002256	14.44	0.005
Error	8	0.001250	0.000156		
Total	15	0.613144			

All interaction and main effects are significant at $\alpha = .01$.

d. A (storage time), B (storage temp), and C (packaging type) should be set to their low values.

All contrasts and effects are shown here:

Effect Name	Contrast	Effect	Effect Name	Contrast	Effect	Effect Name	Contrast	Effect
A	−33.84	−4.23	AC	−4.76	−0.595	ABC	−1.20	−0.150
B	−6.94	−0.8675	AD	−2.56	−0.320	ABD	−1.408	−0.185
C	3.98	0.4975	BC	24.62	3.0775	ACD	1.20	0.150
D	150.94	18.8675	BD	15.42	1.9275	BCD	5.98	0.7475
AB	−13.8	−1.725	CD	3.26	0.4075	ABCD	2.04	0.255

31. a.

Term	Effect
A	$-.625$
B	9.625
C	-4.625
AB	$.175$
AC	1.125
BC	-1.825
ABC	2.525

b. Factors B and C appear to be significant.
c. B at its high level and C at its low level.
d. $\hat{y} = 14.313 + 4.813x_B - 2.313x_C$
e.

Term	Effect
A	$.002$
B	-2.693
C	$.067$
AB	$-.132$
AC	$.108$
BC	$.063$
ABC	$-.058$

Only factor B appears to be significant. Set factor B at its low level. Prediction equation is $\hat{y} = 4.371 - 1.347x_B$.

33. a. Let A = time, B = current, C = EC area, D = volume, E = arsenic.

Term	Effect	Term	Effect	Term	Effect
A	20.019	AB	-3.169	BD	2.906
B	26.119	AC	-1.181	BE	-1.456
C	2.131	AD	2.131	CD	1.069
D	-17.531	AE	-1.281	CE	-0.594
E	-2.519	BC	-1.256	DE	-1.331

b. The important effects appear to be the main effects A (time), B (current), and D (volume).
c. A (time) and B (current) should be set to their high values. D (volume) should to set to its low value.
d. Grand mean = 71.953, Coefficient for $A = (20.019/2) = 10.010$, Coefficient for $B = (26.119/2) = 13.060$, Coefficient for $D = (-17.531/2) = -8.766$. So, the prediction equation is:
$\hat{y} = 71.953 + 10.010x_A + 13.060x_B - 8.766x_D$

35.

A	B	C	D	E
-1	-1	-1	-1	1
1	-1	-1	1	-1
-1	1	-1	1	1
1	1	-1	-1	-1
-1	-1	1	1	-1
1	-1	1	-1	1
-1	1	1	-1	-1
1	1	1	1	1

37. By multiplying each of the 2^{5-1} effects through by the defining relation $I = ACE = BDE = ABCD$ you obtain the following alias structure:
$A = CE = BCD = ABDE$, $B = DE = ACD = ABCE$, $C = AE = ABD = BCDE$, $D = BE = ABC = ACDE$, $E = AC = BD = ABCDE$, $AB = CD = ADE = BCE$, $AD = BC = ABE = CDE$

39. a. $k = 5$ and $p = 1$
b. Let A = temp, B = pH, C = yeast, D = Tryptone, and E = Nitsch. $A = BCDE$, $B = ACDE$, $C = ABDE$, $D = ABCE$, $E = ABCD$, $AB = CDE$, $AC = BDE$, $AD = BCE$, $AE = BCD$, $BC = ADE$, $BD = CE$, $BE = ACD$, $CD = ABE$, $CE = ABD$, $DE = ABC$
c. The four–way interactions are confounded with the main effects. The two–way interactions are confounded with the three–way interactions. So, if all interactions consisting of three or more factors are negligible, none of the estimates of the remaining effects will be confounded with one another.

41. a. $k = 3$ and $p = 1$
b. Let A = anode height, B = board orientation, C = anode placement. The design generator in this design is $E = ABCD$. The alias structure is:
$A = -BC$, $B = -AC$, $C = -AB$
c.

Effect Name	Effect
A	-3.135
B	-1.135
C	-4.925

d. SSE $= (-4.925)^2 = 24.26$, SSTo $= (s^2)(3) = (3.4338)^2(3) = 35.37$, SSA $= (-3.135)^2 = 9.83$, SSB $= (-1.135)^2 = 1.29$. When testing at $\alpha = .05$, neither factor A nor factor B is important, since their corresponding P-values (.639 and .856) are so large.
e. Based on our analysis in part (d), we cannot conclude that factors A or B are significant. Also, we assumed factor C was not significant in order to test for the significance of factors A and B.

f. Since we have found no significant differences between the factors, the decision about how to *minimize* the variation in plating thickness would not be made using the statistical analysis from early parts of this problem. However, based solely on the sign of the effect for each factor, one might conclude that all three factors should be set at the high level ($+1$), in order to minimize the variation in plating thickness.

43. a. Let A = temp, B = pH, C = yeast, D = Tryptone, and E = Nitsch. Using the DOE command in Minitab, the effects estimates are:

Term	Effect	Coef	Term	Effect	Coef
Constant		50.288	AD	-1.400	-0.700
A	23.750	11.875	AE	-1.050	-0.525
B	6.850	3.425	BC	12.450	6.225
C	-0.675	-0.337	BD	14.100	7.050
D	-18.725	-9.363	BE	-2.700	-1.350
E	-5.725	-2.863	CD	-4.125	-2.062
AB	-8.075	-4.037	CE	8.225	4.112
AC	4.200	2.100	DE	-6.675	-3.337

Estimates of the three and four-way interaction terms would be determined by the estimates of the corresponding aliased term. See Exercise 39 for alias structure.

b. $A(=BCDE)$, $D (=ABCE)$

d. Settings that maximize percent protection are: A high and D low.

45. a.

Source	df	SS	MS	F
A	2	30,763.00	15,381.50	3.79
B	3	34,185.60	11,395.20	2.81
AB	6	43,581.20	7,263.53	1.79
Error	24	97,436.80	4,059.87	
Total	35	205,966.60		

b. The F ratio for the interaction effect is 1.79 which is not significant at the 5% level.

c. The F ratio for the A main effect is 3.79 which is significant at the 5% level.

d. The F ratio for the B main effect is 2.81 which is not significant at the 5% level.

47. a.

Source	df	SS	MS	F
A	2	2.0742	1.0371	162.38
B	2	0.080570	0.0403	6.31
C	2	0.2604	0.130195	20.38
AB	4	0.0143	0.0036	0.56
AC	4	0.145137	0.0363	5.68
BC	4	0.0194	0.0049	0.76
Error	8	0.0511	0.006387	
Total	26	2.4195		

b. The significant effects are the A and C main effects.

49. a.

Source	df	SS	MS	F
A	2	326.67	163.34	5.89
B	2	43.83	21.92	0.79
C	2	123.84	61.92	2.23
AB	4	48.51	12.13	0.44
AC	4	168.26	42.06	1.52
BC	4	23.49	5.87	0.21
Error	8	221.68	27.71	
Total	26	956.28	36.78	

b. The two–way interactions AB, AC, and BC have corresponding F ratios .44, 1.52, and .21 respectively. None of these values are significant at the 5% level.

c. Only main effect A is significant at the 5% level.

51. a. Let A = time, B = pressure, C = temp. Using the DOE command in Minitab, the effects estimates are:

Term	Effect	Coef	Term	Effect	Coef
Constant		163.88	AB	-63.25	-31.62
A	19.25	9.63	AC	3.75	1.88
B	99.25	49.62	BC	-47.25	-23.62
C	41.25	20.63	ABC	1.25	0.62

c. B, C, AB, BC

d. Settings that maximize lignin removal are: B high and C high.

53. Caution: test runs are not in Yates order; pooled SS for 2-factor interactions is 18.12 with 10 degrees of freedom, so MSE = 1.812; SSA = .856, SSB = 11.391, SSC = 1.380, SSD = 44.56, SSE = 14.25. Factors B, D, and E are significant at $\alpha = .01$.

Chapter 11

1. **a.** .095   **b.** .475
   c. .830, 1.305   **d.** .4207
3. **a.** $V = \gamma_0 \cdot \gamma_1^{1/T} \cdot \varepsilon$ **b.** 26.341
5. **a.** Yes, a linear model seems appropriate.
   b. $\hat{y} = .1012 + .4607x$
   c. .3085
   d. .0011
7. **a.** Yes, a linear model seems appropriate for each pair of variables.
   b. $\hat{y}_{0\%} = 123.501 - 8.711x$, $\hat{y}_{20\%} = 158.570 - 13.562x$, $\hat{y}_{40\%} = 167.282 - 17.113x$
   c. As timber damage increases, the linear relationship between pile length and critical rating becomes increasingly negative.
   d. $s_e = .45$ at 0%, $s_e = 3.01$ at 20%, $s_e = 4.7$ at 40%. As timber damage increases, the estimated value s_e also increases.
9. **a.** For a one unit increase in inverse foil thickness, one would expect a .260 unit increase in flux. 98% of the observed variation in flux can be attributed to the simple linear regression relationship between flux and inverse foil thickness.
   b. 5.712
   c. 11.302
11. $H_0: \alpha = 0$ versus $H_a: \alpha \neq 0$, $t = \left(\dfrac{a}{s_a}\right) = \left(\dfrac{-1.128}{2.368}\right) =$
    $-.48$, P-value $= .642$, do not reject H_0.
13. **a.** Method 1: Hypothesis Test, $H_0: \beta = 0$ versus $H_a: \beta \neq 0$, $t = 54.56$, P-value $< .0001$, reject H_0 and conclude that there is a useful linear relationship between these two variables. Method 2: A confidence interval for $\beta = b \pm (t$ critical value)$\cdot s_b$. A 95% confidence interval for β is: $.87825 \pm (2.179)(.01610) = (0.8432, 0.9133)$, using t critical value for df $= (n - 2) = (14 - 2) = 12$. The plausible values are all positive so we conclude there is a useful linear relationship between the two variables.
    b. The t ratio for testing model utility would be the same value regardless of which of the two variables was defined to be the independent variable. This can be easily seen by looking at the t test statistic for testing if the population correlation coefficient is equal to zero. In that equation the only values required are the sample size (n) and the sample correlation coefficient (r). Both r and n are not dependent on which variable was the independent variable.

15. As we saw in Exercise 13(b), the t ratio for testing the model utility is dependent only on the sample size and the sample correlation coefficient. Neither of these quantities is unit dependent. So, multiplying the dependent variable by a constant will have no effect on the t test statistic.
17. $H_0: \rho = 0$ versus $H_a: \rho > 0$, $t = 5.25$, P-value $< .0001$, reject H_0.
19. **a.** $b = 1.378$. There is, on average, a 1.378% increase in reported nausea for each unit increase in motion sickness dose.
    b. $t = 3.422$. Yes, there is a useful relationship between the two variables.
    c. It would be possible, but not advisable because $x = 5$ is outside the range of the x data.
    d. $b = 1.424$
21. **a.** The scatterplot appears to be quite linear.
    b. .931
    c. If increasing velocity by 900 cm/sec results in an average change in the response of .6, then our true population slope coefficient is $\beta = .6/900 = 6.667 \times 10^{-4}$. $H_0: \beta = 6.667 \times 10^{-4}$ versus $H_a: \beta < 6.667 \times 10^{-4}$, $t = -.6016$, P-value $> .10$, do not reject H_0.
    d. We are 95% confident that the true average change in mist associated with a 1 cm/sec increase in velocity is between 4.26×10^{-4} and 8.159×10^{-4}.
23. **a.** A 95% prediction interval is (3.2833, 3.6067).
    b. The interval when the temperature is 1200 degrees will be wider than when the temperature is 1500 degrees. This is because 1200 degrees is 200 degrees away from the mean temperature of 1400 degrees whereas 1500 degrees is only 100 degrees away from the mean temperature.
25. The mean x value is 40.3. Intervals with x values farther away from this mean are wider. Also, prediction intervals are wider than confidence intervals. And, 99% intervals are wider than 95% intervals. Therefore, (i) will be wider than (iii), (i) will be more narrow than (ii), (ii) will be wider than (iv), (iii) will be more narrow than (iv) and (v).
27. **a.** $t = 16.2$, P-value ≈ 0, conclude there is a useful linear relationship.
    b. (.879, .947)   **c.** (.780, 1.046)
29. **a.** 4.9 hr
    b. When number of deliveries is held fixed, the average change in travel time associated with a 1-mile increase in distance traveled is .060 hr.

When distance traveled is held fixed, the average change in travel time associated with one extra delivery is .900 hr.

c. .9861

31. b. For $x = 350, y = 106.5$. For $x = 485, y = 65.325$. The mean free flow percentage is higher when viscosity is 350.

c. The change in mean free flow percentage as viscosity increases from 450 to 460 is $81.2 - 86.5 = -5.3$. The change in mean free flow percentage as viscosity increases from 460 to 470 is $75.3 - 81.2 = -5.9$.

33. a. 77.3   **b.** 40.4

35. a. To test $H_0: \beta_1 = \beta_2 = 0$ versus H_a: *at least one of β_1 and β_2 is not zero*, F = MSRegr/MSResid = 1260.71 (from printout), P-value < .001, reject H_0.

b. A 95% confidence interval for β_2: $b_2 \pm$ (t-critical)s_{b_2} = .002775 ± (2.093)(.001121) = (0.00043, 0.00512).

c. For $x_1 = 11.5$ and $x_2 = 40, \hat{y} = 4.478$. A 95% confidence interval for true average deposition rate is 4.478 ± (2.093)(.02438) = (4.42697, 4.52903).

d. $s_e = .04485$, prediction interval is
$$4.478 \pm (2.093) \sqrt{(.04485)^2 + (.02438)^2}$$
= (4.371, 4.585).
This interval contains the interval from part (c) as expected.

37. a. $F = 87.6$, P-value = 0; there does appear to be a useful linear relationship between y and at least one of the predictors.

b. .935   **c.** (9.095, 11.087)

39. b. .9986

c. P-value = 0; judge the model useful.

d. $t = 48$, P-value = 0; the quadratic predictor does appear useful.

e. (20.00, 21.14), (19.44, 21.70)

41. $F = 3.44, .05 <$ P-value $< .10$, conclude that the second-order predictors do not provide useful information.

43. a. The variable "supplier" has three categories, so we need two indicator variables: $x_2 = 1$ for supplier 1 (0 otherwise), $x_3 = 1$ for supplier 2 (0 otherwise). Likewise for "lubrication" we have two indicator variables: $x_4 = 1$ for lubricant 1 (0 otherwise), $x_5 = 1$ for lubricant 2 (0 otherwise).

b. $H_0: \beta_1 = \beta_2 = \beta_3 = \beta_4 = \beta_5 = 0$ versus H_a: at least one $\beta_i \neq 0$. $F = 20.67$, P-value < .001, reject H_0.

c. (13.80, 19.09)

d. .741, a negligible drop in R^2, suggesting the lubrication regimen indicator variables are not important. A formal "full" versus "reduced" model test confirms this suggestion.

e. The corresponding "full" versus "reduced" model test uses the null hypothesis that the interaction terms are *not* statistically significant contributors to the model. $F = 3.19, .01 <$ P-value $< .05$, reject H_0 at the .05 level and conclude that the interaction terms, as a group, do contribute significantly to the regression model.

45. a. Since the plot of normal quantiles versus standardized residuals looks linear, we would conclude that the standardized residuals are normally distributed.

b. The plot of x versus the standardized residuals has no discernible pattern. So, we would conclude that our simple linear regression model assumptions are being met.

47. a. We would recommend the model with $k = 2$. This model has a substantially higher R^2 adjusted value over the model with $k = 1$. And, the models with $k = 3$ and $k = 4$ give little improvement.

b. No, a forward selection method would not have considered the $k = 2$ model described in the example. Forward selection would let x_4 enter the model first and would not delete it at the next stage.

49. The model with four variables including all but the summerwood fiber variable would seem best. R^2 is as large as any of the models, including the 5 variable model. R^2 adjusted is at its maximum and CP is at its minimum. As a second choice, one might consider the model with $k = 3$ which excludes the summerwood fiber and springwood % variables.

51. The model using the three variables x_3, x_9, x_{10} would seem best. It has an adjusted R^2 only slightly smaller than the largest adjusted R^2. As a second choice, the two predictor model is also quite good.

53. a. $R^2 = 1 - \dfrac{\text{SSE}}{\text{SST}} = 1 - \dfrac{10.5513}{30.4395} = .653$ or

65.3%, while adjusted $R^2 = 1 - \dfrac{\text{MSE}}{\text{MST}} =$

$1 - \dfrac{10.5513/24}{30.4395/28} = .596$ or 59.6%. Yes, the model
appears to be useful.

b. The corresponding "full" versus "reduced" model test uses the null hypothesis that the 10 second-order interaction terms are *not* statistically significant contributors to the model. $F = 13.21$, P-value $< .001$, reject H_0 at the .01 level and conclude that at least one of the second-order terms is a statistically significant predictor of protein yield.

c. We want to compare the "full" model with 14 predictors in (b) to a "reduced" model with 5 fewer predictors $(x_1, x_1^2, x_1x_2, x_1x_3, x_1x_4)$. $F = .62$, P-value $> .10$, fail to reject H_0; therefore, it indeed appears that the five predictors involving x_1 could all be removed.

d. The "best" models seem to be the 7-, 8-, 9-, and 10-variable models. All of these models have high adjusted R^2 values and low Mallows' CP values compared to the other models.

55. H_0: $\beta = 0$ versus H_a: $\beta \neq 0$, $z = .73$, P-value is .463, do not reject the null hypothesis. There is insufficient evidence to claim that age has a significant impact on the presence of kyphosis.

57. a. For $x_1 =$ pillar height to width ratio, H_0: $\beta_1 = 0$ versus H_a: $\beta_1 \neq 0$, $z = 1.878$, P-value $= .0604$, reject H_0. For $x_2 =$ pillar strength to stress ratio, H_0: $\beta_2 = 0$ versus H_a: $\beta_2 \neq 0$, $z = 2.145$, P-value $= .0319$, reject H_0. Each of the variables appears to have a significant impact on pillar stability.

b. The odds of pillar stability changes by the multiplicative factor $e^{2.774} = 16.02$ when x_1 increases by 1 and x_2 remains fixed. The odds of pillar stability changes by the multiplicative factor $e^{5.668} = 289.46$ when x_2 increases by 1 and x_1 remains fixed.

c. The table of observations with corresponding probabilities and labels is shown below. Based on this, only two observations had a label that did not match actual stability status. The pillar with ID #3 was labeled as "unstable" when in fact it was stable. The pillar with ID #28 was labeled as "stable" when in fact it was unstable.

ID	x_1	x_2	Stable?	Prob	Label
1	1.8	2.4	Y	0.996	stable
2	1.65	2.54	Y	0.997	stable
3	**2.7**	**0.84**	**Y**	**0.29**	**unstable**
4	3.67	1.68	Y	0.999	stable
5	1.41	2.41	Y	0.988	stable
6	1.76	1.93	Y	0.936	stable
7	2.1	1.77	Y	0.938	stable
8	2.1	1.5	Y	0.765	stable
9	4.57	2.43	Y	1	stable
10	3.59	5.55	Y	1	stable
11	8.33	2.58	Y	1	stable
12	2.86	2	Y	0.998	stable
13	2.58	3.68	Y	1	stable
14	2.9	1.13	Y	0.787	stable
15	3.89	2.49	Y	1	stable
16	0.8	1.37	N	0.041	unstable
17	0.6	1.27	N	0.014	unstable
18	1.3	0.87	N	0.01	unstable
19	0.83	0.97	N	0.005	unstable
20	0.57	0.94	N	0.002	unstable
21	1.44	1	N	0.03	unstable
22	2.08	0.78	N	0.05	unstable
23	1.5	1.03	N	0.041	unstable
24	1.38	0.82	N	0.009	unstable
25	0.94	1.3	N	0.04	unstable
26	1.58	0.83	N	0.017	unstable
27	1.67	1.05	N	0.072	unstable
28	**3**	**1.19**	**N**	**0.872**	**stable**
29	2.21	0.86	N	0.105	unstable

59. a. A simple linear regression model seems to fit the data well. The least squares regression equation is: $\hat{y} = -.220 + .0436x$. The model utility test obtained from Minitab produces a t test statistic equal to 12.72. The corresponding P-value is extremely small. So, we have sufficient evidence to claim that ΔCO is a good predictor of ΔNO$_y$.

b. $\hat{y} = -.220 + .04362(400) = 17.228$. A 95% prediction interval produced by Minitab is (11.953, 22.503). Since this interval is so wide, it does not appear that ΔNO$_y$ is accurately predicted.

c. The large ΔCO value has extremely high leverage. The least squares line that is obtained when excluding the value is $\hat{y} = 1.00 + .0346x$. The R^2 value with the value included is 96% and is reduced to 75% when the value is excluded. The value of s_e with the value included is 2.024 and with the value excluded is 1.96. So, the large ΔCO value does appear to affect our analysis in a substantial way.

61. a. Same x values yet different y values

b. $b = .01023$, $s_b = .009577$, $t = 1.1$, P-value $\approx .3$; model cannot be judged useful.

63. a. The statement is incorrect. r^2 is not the "linear correlation coefficient." r^2 is the coefficient of determination. The <u>linear</u> correlation coefficient is r and $r = \sqrt{.89} = .9434$.

b. H_0: $\beta = 0$ versus H_a: $\beta \neq 0$. The value of the t test statistic equals 12.06. The corresponding P-value is extremely small. So, we reject the null hypothesis and conclude that there is a linear relationship between the two variables.

c. As x increases so does the variation in the standardized residuals. This fact is inconsistent with our constant variance assumption of a least squares regression analysis.

65. a. The full model contains $k = 9$ predictors. The reduced model contains 3 predictors. H_0: $\beta_4 = \cdots = \beta_9 = 0$ versus H_0: At least one of the β's is not zero.

$$F = \left[\frac{(1523351 - 805534)/6}{(805534)/(15 - 10)} \right] = .743, \ P\text{-value} > $$

.10, do not reject H_0. There is not sufficient evidence to claim that the second-order predictors provide useful information beyond what is contained in the three first-order predictors.

67. a. $\hat{y} = 84.67 + .650 - .258 + .133 + .108 - .135 + .028 + .028 - .072 + .038 - .075 + .213 + .200 - .188 + .050 = 85.39$

The value of the residual for the one observation made under the specified conditions is: $(y - \hat{y}) = (85.4 - 85.39) = .01$

b. Let z_1, z_2, z_3, z_4 denote the uncoded variables. Then, $z_1 = .1x_1 + .3$, $z_2 = .1x_2 + .3$, $z_3 = x_3 + 2.5$, $z_4 = 15x_4 + 160$. Equivalently, $x_1 = 10z_1 - 3$, $x_2 = 10z_2 - 3$, $x_3 = z_3 - 2.5$, $x_4 = (z_4 - 160)/15$. Substitution yields the following least squares regression coefficients:

Term	Coefficient
Constant	76.437
z_1	-7.35
z_2	9.61
z_3	$-.915$
z_4	.09632
z_1^2	-13.452
z_2^2	2.798
z_3^2	.02798
z_4^2	$-.0003201$
$z_1 z_2$	3.750
$z_1 z_3$	$-.7500$
$z_1 z_4$	.14167
$z_2 z_3$	2.000
$z_2 z_4$	$-.1250$
$z_3 z_4$	.00333

c. The full model contains $k = 14$ variables. The reduced model contains 4 variables. H_0: $\beta_5 = \cdots = \beta_{14} = 0$ versus H_0: At least one of the β's is not zero. SSResid(full) = 1.9845, SSResid(reduced) = 4.8146.

The value of the test statistic is:

$$F = \left[\frac{(4.8146 - 1.9845)/10}{(1.9845)/(31 - 15)} \right] = 2.28,$$

$.05 < P$-value $< .10$, do not reject the null hypothesis. There is not sufficient evidence at the 5% level to claim that the second-order predictors provide useful information beyond what is contained in the four first-order predictors.

69. A plot of y versus x suggests that simple linear regression model *may* be appropriate, but a graph of the residuals versus fitted values questions the validity of a simple linear regression model. Fitting higher order models (such as second and third-order) may be more appropriate.

The second-order model has $R^2 = 65.3\%$ and adjusted $R^2 = 62\%$, whereas the third-order model has $R^2 = 70.7\%$ and adjusted $R^2 = 66.3\%$. Comparing adjusted R^2 values the third-order model seems to perform slightly better. From the second-order model, we predict y (at $x = 30$) to be $3.45 + .0618(30) - .000377(30^2) = 4.9647$. For the third-order model, our estimate for y is $3.94 - .045(30) = .0041(30^2) - .000048(30^3) = 4.984$. Both models appear to give roughly the same estimate.

71. a. The boxplot shows that the shapes of the ppv for the cracked and uncracked prisms appear to be fairly symmetric. The boxplot further suggests that the ppv for the cracked prisms tend to be greater than the ppv for the uncracked prisms. Let μ_1 = the true mean ppv for uncracked prisms and μ_2 = the true mean ppv for cracked prisms. A 95% confidence interval for $(\mu_2 - \mu_2)$, using the critical t value = 2.093 based on 19 df, is:

$$(482.7 - 827.4) \pm 2.093 \sqrt{\frac{233.7^2}{18} + \frac{295.3^2}{12}} \Rightarrow$$

$-344.7 \pm 2.093(101.494)$, or $(-557.127, -132.273)$.

b. Using Minitab, we can use the best subsets option using the PPV, PPV^2, the indicator variable Crack? (0 if there is no crack present and 1 if there is a crack), and the interaction term PPV*Crack?. The best subsets regression suggests that the single quadratic term PPV^2 is the single most useful predictor. The quadratic regression model, which has the R^2 value of 61.2%, has the equation $\hat{y} = .996719 - .00000001(PPV)^2$. The next most useful single predictor is the PPV term. This simple linear regression model, which has the R^2 value of 57.7%, has the equation $\hat{y} = 1.00161 - .000018(PPV)$. Models involving more than 1 term don't appear to explain the ratio variable any more significantly, since the R^2 values of such models are not much different than the model that simply uses PPV^2 or PPV.

Index